Lecture Notes in Networks and Systems

Volume 501

The series "Lecture Notes in Networks and Systems" publishes the latest developments in Networks and Systems—quickly, informally and with high quality. Original research reported in proceedings and post-proceedings represents the core of LNNS.

Volumes published in LNNS embrace all aspects and subfields of, as well as new challenges in, Networks and Systems.

The series contains proceedings and edited volumes in systems and networks, spanning the areas of Cyber-Physical Systems, Autonomous Systems, Sensor Networks, Control Systems, Energy Systems, Automotive Systems, Biological Systems, Vehicular Networking and Connected Vehicles, Aerospace Systems, Automation, Manufacturing, Smart Grids, Nonlinear Systems, Power Systems, Robotics, Social Systems, Economic Systems and other. Of particular value to both the contributors and the readership are the short publication timeframe and the world-wide distribution and exposure which enable both a wide and rapid dissemination of research output.

The series covers the theory, applications, and perspectives on the state of the art and future developments relevant to systems and networks, decision making, control, complex processes and related areas, as embedded in the fields of interdisciplinary and applied sciences, engineering, computer science, physics, economics, social, and life sciences, as well as the paradigms and methodologies behind them.

Indexed by SCOPUS, INSPEC, WTI Frankfurt eG, zbMATH, SCImago.

All books published in the series are submitted for consideration in Web of Science.

For proposals from Asia please contact Aninda Bose (aninda.bose@springer.com).

More information about this series at https://link.springer.com/bookseries/15179

Radek Silhavy
Editor

Software Engineering Perspectives in Systems

Proceedings of 11th Computer Science On-line Conference 2022, Vol. 1

Springer

Editor
Radek Silhavy
Faculty of Applied Informatics
Tomas Bata University in Zlin
Zlin, Czech Republic

ISSN 2367-3370 ISSN 2367-3389 (electronic)
Lecture Notes in Networks and Systems
ISBN 978-3-031-09069-1 ISBN 978-3-031-09070-7 (eBook)
https://doi.org/10.1007/978-3-031-09070-7

This Springer imprint is published by the registered company Springer Nature Switzerland AG
The registered company address is: Gewerbestrasse 11, 6330 Cham, Switzerland

Preface

Software Engineering Perspectives in Systems section is presented in this volume. This proceeding is a Vol. 1 of the Computer Science On-line Conference 2022. Papers in this part discuss software engineering topics related to software analysis, design, and application of intelligent algorithms, machine, and statistical learning in software engineering research.

The volume constitutes the refereed proceedings of the software engineering and algorithms section of the 11th Computer Science On-line Conference 2022 (CSOC 2022), held online in April 2022.

CSOC 2022 conference intends to provide an international forum to discuss the latest high-quality research results in all areas related to computer science.

CSOC 2022 is held online and modern communication technology, which are broadly used, improves the traditional concept of scientific conferences. It brings equal opportunity to participate for all researchers around the world.

I believe that you find the following proceedings exciting and valuable for your research work.

April 2022 Radek Silhavy

Organization

Program Committee

Program Committee Chairs

Petr Silhavy	Tomas Bata University in Zlin, Faculty of Applied Informatics
Radek Silhavy	Tomas Bata University in Zlin, Faculty of Applied Informatics
Zdenka Prokopova	Tomas Bata University in Zlin, Faculty of Applied Informatics
Roman Senkerik	Tomas Bata University in Zlin, Faculty of Applied Informatics
Roman Prokop	Tomas Bata University in Zlin, Faculty of Applied Informatics
Viacheslav Zelentsov	Doctor of Engineering Sciences, Chief Researcher of St.Petersburg Institute for Informatics and Automation of Russian Academy of Sciences (SPIIRAS)
Roman Tsarev	Department of Information Technology, International Academy of Science and Technologies, Moscow, Russia
Stefano Cirillo	Department of Computer Science, University of Salerno, Fisciano (SA), Italy

Program Committee Members

Technical Program Committee Members

Ivo Bukovsky	Czech Republic
Maciej Majewski	Poland
Miroslaw Ochodek	Poland
Bronislav Chramcov	Czech Republic
Eric Afful Dazie	Ghana
Michal Bliznak	Czech Republic
Donald Davendra	Czech Republic
Radim Farana	Czech Republic
Martin Kotyrba	Czech Republic
Erik Kral	Czech Republic
David Malanik	Czech Republic
Michal Pluhacek	Czech Republic
Zdenka Prokopova	Czech Republic
Martin Sysel	Czech Republic
Roman Senkerik	Czech Republic
Petr Silhavy	Czech Republic
Radek Silhavy	Czech Republic
Jiri Vojtesek	Czech Republic
Eva Volna	Czech Republic
Janez Brest	Slovenia
Ales Zamuda	Slovenia
Roman Prokop	Czech Republic
Boguslaw Cyganek	Poland
Krzysztof Okarma	Poland
Monika Bakosova	Slovak Republic
Pavel Vaclavek	Czech Republic
Olga Brovkina	Czech Republic
Elarbi Badidi	United Arab Emirates

Organizing Committee Chair

Radek Silhavy — Tomas Bata University in Zlin, Faculty of Applied Informatics, email: radek@silhavy.cz

Conference Organizer (Production)

Silhavy s.r.o
Web site: https://www.openpublish.eu
Email: csoc@openpublish.eu

Conference Web site, Call for Papers

https://www.openpublish.eu

Contents

Framework for Disruptive Technologies Based E-Learning System Architecture for Education 4.0

Vijay Kumar Pandey[1](✉) and Vikash Kumar Singh[2]

[1] Department of Computer Science, Indira Gandhi National Tribal University, Amarkantaka, Madhya Pradesh, India
vijay.research4@gmail.com

[2] Dean Faculty of Department of Computer Science, Indira Gandhi National Tribal University, Amarkantaka, Madhya Pradesh, India

Abstract. With the advent of various technological advancements, the existing education system is consistently evolving to emerge as an innovative way of dispatching knowledge delivery services. The standards of Education 4.0 have further been recognized as a guideline towards improving the e-learning system integrated with disruptive technologies. A review of existing approaches to e-learning shows that there have been various dedicated attempts in this process; however, there is still an enormous scope of implementation and improvement. Therefore, the proposed study introduces an innovative model which implements disruptive technologies, i.e., big data approach and machine learning, using an experimental model with a complete adherence towards Education 4.0 standards. The study model can forecast the ultimate performance of a student using MOOCs before writing up the actual examination process. The outcome shows deep-neural network model offers better performance compared to support vector machines.

Keywords: Wireless image transmission · Wireless networks · Fading · Error-correction · Channel coding

1 Introduction

At present, the educational system has been significantly revolutionized with the advent of various technological advancements [1]. Along with the classroom and regular coaching, students adopt online coaching, which significantly adds knowledge and skills [2]. The inclusion of e-learning has contributed significantly to touching the education system as wand Education 4.0 is the latest standard for improving the teaching-learning experience with disruptive technologies in the current era of technological modernization. This standard is one of the mechanisms which imparts learning methods in compliance with the fourth industrial revolution [4-7]. This revolution system focuses on imparting artificial intelligence [8], smart technologies [9], as well as robotics [10] in order to incorporate technical advancement in the education system. It is necessary to target all

R. Silhavy (Ed.): CSOC 2022, LNNS 501, pp. 1–12, 2022.
https://doi.org/10.1007/978-3-031-09070-7_1

the educational universities to offer quality education to their students and adopt a cyber-physical system that is effective for threads knowledge delivery system in this process [3, 11]. In order to adopt this Education 4.0, the students must consider the adoption of this technological advancement in their subjects, which contributes towards modernizing the learning process as a whole. The prime target of adoption of Education 4.0 are: i) to promote distance learning with a higher quality of service, ii) to render more flexibility in personalized learning, iii) to offer more options for adoption of technical tools, iv) to induce learning with a case study, v) to provide enriched real-time experience, vi) to incorporate data analysis, and vii) to offer an effective examination pattern for assessing the grasping capability of students [12].

It should be noted that there are two forms of e-learning system based on, i.e., i) course-level and ii) program level [13]. The course-level-based e-learning mechanism is further classified into the classroom, synchronized distribution, web-enhanced, online, flexible mode, and hybrid mode [13]. The program-level-based e-learning is categorized as a classroom, multi-format, blended, and online [13]. It is found that the existing e-learning mechanism is witnessed with various challenges, e.g., transmission delivery, personalization, enabling technologies, cooperative learning options, and assessment techniques [13]. There is no doubt that various approaches are being designed and developed towards enriching the e-learning approaches [14-18]. The majority of the existing approaches infers that the adoption of data analytics as machine learning is gaining popularity for enriching the process management of e-learning systems as the core contributors of disruptive technologies. However, with various disruptive technologies, e-learning is still encountering various practical issues regarding its application. The prime challenge is to realize an effective solution towards delivering specific educational content to students based on their real-time needs. Another significant challenge is to carry out personalization.

Therefore, the proposed system introduces a novel model of e-learning which uses disruptive technology of big data and machine learning. The sole target is to obtain an applicable model in the adherence to Education 4.0 norms. The model assists in evaluating the students' performance in apriori, focusing on the inclusion of disruptive technologies in the e-learning system to improvise the teaching-learning experience for the upcoming education system. The paper's organization is as follows: Discussion of existing approaches towards revising e-learning system is carried out in Sect. 2 while its associated issues are discussed in Sect. 3. The proposed system is discussed in Sect. 4, while research methodology is discussed in Sect. 5. Result obtained is discussed in Sect. 6, while the paper's conclusion is summarized in Sect. 7.

2 Review of Literature

At present, there are various approaches where teaching and learning experience has been revolutionized to a large extent with the inclusion of various forms of disruptive technologies. At present, big data is found to be the most dominant scheme to improve the e-learning process. The work carried out by Otoo-Arthur [19], Udupi et al. [20], Zhang et al. [21], Zhou et al. [22], Yanhui [23], and Vyas et al. [24], Popchev et al. [25], and Dwivedi et al. [26] exhibit the contribution of big data in e-learning scheme where the idea is to extract knowledge from the educational data and make it applicable to various forms of applications in e-learning. Apart from the big data approach, e-learning schemes have also been investigated concerning Internet-of-Things, where the idea is basically over obtaining and processing the massive data from e-learning platforms. The work carried out by Babu [27], Wang et al. [28], Zahedi et al. [29], Yamada et al. [30], Matsuo et al. [31], and Njeru et al. [32] has integrated education system with IoT testbed to automate the process towards Education 4.0. Another dominant disruptive technique is machine learning towards pushing the education system in the direction of the Education 4.0 scheme. The work carried out by Rajkumar et al. [33], Lai et al. [34], Nafea et al. [35], Xie et al. [36], and Yang et al. [37] has used different variants of machine learning to improve further the educational outcomes to render them more as value-added data. Existing systems also wit- ness the dominant usage of artificial intelligence as a disruptive technology. The contribution of the researchers, e.g., Wu et al. [38], Colace et al. [39], Gaeta et al. [40], Chiu et al. [41], Barlybayev et al. [42], and Holmes et al. [43], have used artificial intelligence towards promoting the adoption of Education 4.0. Hence, there are various categories of studies in an existing system where different types of disruptive approaches have been used.

3 Research Problem

The research problems identified are multi-fold viz. i) although different variants of disruptive techniques are applied, but they are focused on addressing specific feature enhancement in e-learning and not subjected to a wholesome model of e-learning, ii) challenge associated with the delivery of the educational service is never addressed using disruptive technologies, without which Education 4.0 knowledge-delivery service is not fulfilled, iii) another significant issues is lack of focus on personalization of the modules on the basis of comfortable and applicable environment of users (students and instructors), iv) Classification being the consistent problem while using disruptive technologies even when there are various forms of machine learning approaches, v) Existing studies doesn't emphasize over the collaborative nature of complete environment without which Education 4.0 accomplishment cannot be claimed of, vi) assessment has also received much less attention using existing disruptive approaches ignoring global perception and more inclination to local assessment schemes for performance of knowledge-delivery system.

4 Proposed System

The prime aim of the proposed system is to apply the concept of Education 4.0 in improving e-learning systems using the disruptive technique. Artificial intelligence- based tools can be used to improve the personalized learning experience with more addictiveness in this direction. The proposed system develops a novel learning environment that has uniform tools for obtaining real-time information about the education system used for further updating policies. Figure 1 highlights the proposed model where disruptive technologies are used for e-learning.

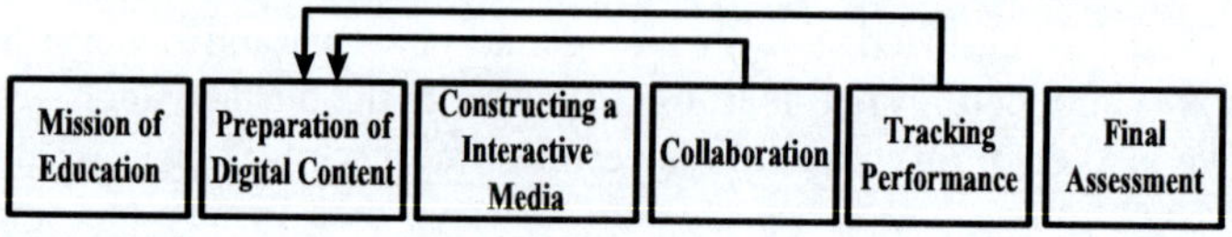

Fig. 1. Proposed model

The first module of the proposed model is about configuring the mission of education, which uses biometrics, activity diagram, and strategic learning mechanism for carrying out the orientation phase. The second module of the proposed system is about preparing digital content, whusese Natural Language Processing and ontologtor to construct the interactive subject materials. The third module of the proposed system is about constructing interactive media, which uses an adaptive assessment environment with multiple working groups and instructors for passively assessing the student performance using MOOC [44]. The fourth module of the proposed system is about collaboration with the aid of intelligent sensors as well as virtual reality. Various adoption of multi-dimensional environment can be used in making the collaboration more interactive and seamless. The fifth module of the proposed system is about tracking performance, where various bots can be used for self-monitoring along with time management and usage of wearable devices to track more adaptive information about student's performance. The sixth module of the proposed system is about the response system, which will use learning analytics and intelligent sensors, and wearable devices. The seventh module of the proposed system is about the final assessment that uses biometrand semantic order to extract information about student's performance dent during the assessment. The next section discusses the research methodology used in the study.

5 Research Methodology

The core idea of the proposed system is to introduce disruptive technologies over the online education delivery system towards the path of Education 4.0. The proposed system introduces an experimental model using disruptive technologies considering a case study where the emphasis is offered to adopting smart technology and machine learning. The sole target is to revolutionize the teaching and learning experience between the teachers and students globally. The idea of the proposed system is to develop a predictive tool capable of identifying the demands of the student's education skills to improve. This predictive model is designed and assessed using a cyclical model which where the machine learning is used to obtain the information about precise references for the education system. The implemented model is showcased in Fig. 2.

The proposed implementation targets minimizing the occurrences of failures during the classroom assessment (physical or virtual) using disruptive technologies of big data from the application viewpoint. It is already known now that the most dominant disruptive technology in big data is machine learning schemes. The proposed system uses two variants of machine learning, i.e., deep neural network and support vector machine, for this purpose. The development of the model is carried out considering technical subjects and non-technical subjects. Imparting knowledge to two discrete subjects will

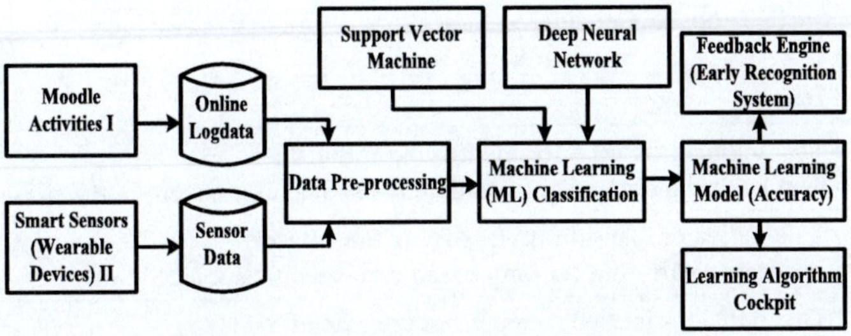

Fig. 2. Proposed model of Education 4.0

require formulating a strategy for usage learning, effective personalization, and a precise-productive engagement framework. Hence, the proposed system hypothesizes that the information collected by the various forms of sensing and wearable equipment can be supported by the proposed model with a multi-modal information environment.

The proposed study has been experimentally carried out considering the technical and non-technical subjects and the implementation of the standard framework of Education 4.0. The proposed system considers clusters of data of three discrete courses, as exhibited in Table 1.

Table 1. Clusters of multi-modal subject (data)

Subject	# Learning	# assessment	# Live	# Threat
Sub-1	114	110	90	20
Sub-2	109	96	77	27
Sub-3	59	52	34	19

Table 1 highlights the quantity of the dataset considered for learning operation using big data and a predictive identification system (live and threat). The tag #Learning represents the size of the dataset used for learning, #assessment represents the size of the dataset used for evaluation, #Live represents the size of the dataset collected for streaming e-learning classes, and #threat represents several subjects that are at risk for low performance by the students during the examination. The proposed system carry out a dual form of experiment analysis:

- To carry out predictive analysis for the students about to be promoted to the next study semester.
- To carry out online identification of the students dynamically who are weak in their study performance.

Table 2 represents various indicators used in a proposed model where an online MOOC-based tool is used to observe the students' performance and the generated outcome results in more information in building the dataset. In the notation of indicators

Table 2. Considered Indicators for Predictive Analysis

Indicators	Details
Class	This indicator is used for identifying if the student has been detained or promoted based on assessment in the prior semester
Subject	This indicator is used for classifying the dataset in order to carry out training based on a specific subject
Semester	This indicator is used to represent the student if they are part of the training or test dataset
Score (%)	This indicator represents the overall score obtained by the student in the final assessment
$A\alpha_1$	This indicator represents the total number of access made in the first month of observation
$M\alpha_1$	This indicator represents the total number of questions that are viewed in the first month of observation
$S\alpha_1$	This indicator represents the total number of the score achieved in online mock assessment in the first month of observation
$A\alpha_2$	This indicator represents the total number of access to resources in the second month of obser- vation
--	--
$S\alpha_3$	This indicator represents the total score achieved in the online mock assessment in the third month of observation

A, M, S represents accessing e-learning tools, e.g., MOOC, attending the mock test, and scores obtained in online analysis, respectively. The number *1, 2, 3* represents first, second, and the third month of observation respectively. In order to perform the classification of the proposed environmental information about subjects, the study makes use of multiple combinations of various indicators as well as processing elements in order to introduce the best suitable training data. The proposed system makes use of standard scores in order to standardize the dataset. In order to carry out prediction, the proposed system extracts information associated with the counts of the clicks for all the low-performing students. This can be obtained from the log files, which are collected over some time. For a practical analysis, the proposed system also aggregates the indicators $A\alpha$, $M\alpha$, and $S\alpha$ for the fourth and fifth months of observation. Finally, the proposed system arrives at the following best configuration of indicators which are as follows:

- Best Indicator-1: $A\alpha_3$ and $A\alpha_4$.
- Best Indicator-2: $A\alpha_3$, $A\alpha_4$, and $A\alpha_5$.
- Best Indicator-3: i) $A\alpha_3$ and $A\alpha_4$, ii) $M\alpha_3$ and $M\alpha_4$
- Best Indicator-4: i) $A\alpha_3$, $A\alpha_4$, and $A\alpha_5$, ii) $M\alpha_3$, $M\alpha_4$, and $M\alpha_5$
- Best Indicator-5: i) $A\alpha_3$ and $A\alpha_4$, ii) $S\alpha_3$ and $S\alpha_4$
- Best Indicator-6: i) $A\alpha_3$, $A\alpha_4$, and $A\alpha_5$, ii) $S\alpha_3$, $S\alpha_4$ and $S\alpha_5$

A closer look into the proposed predictive model formulation shows that the problem associated with learning can be modelled in the classification form of a binary attribute. There are possibilities of only two attributes which is *promoted* and *detained* as a consequence of the assessment. This logic can be used for tagging each student likewise using tagβ={*Promoted, Detained*}. The study considers that it possesses the system's accessibility to the status information of all the students. The target of the proposed machine learning approach is to look for a binary classifier so that a more significant deal of threat can be minimized. The term *threat* will mean the possibility of normal students to be performing below average in the primary assessment. The objective function constructed for this purpose is as follow:

$$fx =_a \gamma x_1 \cdot \mu(x_2) \tag{1}$$

In the above expression (1), the function $f(\mathrm{x})$ represents threat degree, where the argument x corresponds to a detained parameter of dual classification attribute β. The variable γ will represent the outlier function meant for offering a penalty iif. This ensures that the outcome of the prediction goes as accurate as possible. This function takes the input argument of the threat factor associated with the proportion of promotion represented as the x_1 parameter. The second variable, μ, is all about computing the probability of argument x_2 representing promotion and detained. The limit represents the feature space associated with the targeted students. The proposed system considers a deep neural network and support vector machine for this purpose. The proposed system makes use of the radial basis function.

6 Result Analysis

The proposed system constructs a synthetic dataset for analyzing this experiment concerning training operation being carried out on it. The analysis considers cross- validation over the training dataset to evaluate the kernel's trade-off attributes. The proposed system considers the gaussian distribution of the radial basis function as 10^{-5}-10^{-0}. The learning operation is carried out over the data that has been accessed by the students as well as the total number of the scores obtained over the mock test in MOOC along with the overall grades obtained by the student to represent promotion or detention over a specific subject. The proposed system uses MATLAB for applying a neural network where the threshold of hyper-parameter is initialized to 0.01 while the rate of the learning model is configured at 0.001. The neural network model uses the 9 hidden layers. The Moodle server restores all the weight factors of the neural network that has been trained in order to obtain the predictive values. The assessment is carried out considering three performance parameters PP1, PP2, and PP3, to represent the accuracy, true positive rate, and true negative rate, respectively. The analysis is carried out for Support Vector Machine and Deep Neural Network discretely. Tables 3, 4, 5 highlights outcome for Support Vector Machine.

The numerical outcome for a proposed system concerning Deep Neural Network is shown in Table 6. Accuracy parameter PP_1 is computed using a fraction of true negative and true positive divided by cumulative instances. The performance parameter of PP_2 represents true positive rate which is a direct representation of sensitivity, while the

Table 3. Performance of linear kernel system

Dataset	Indicators	PP_1	PP_2	PP_3
Sub-1	Access 10, Access 11	79%	90%	42%
Sub-2	$A\alpha_3$, $A\alpha_4$, $A\alpha_5$ $M\alpha_3$, $M\alpha_4$, $M\alpha_3$	72%	99%	8%
Sub-3	$A\alpha_3$, $A\alpha_4$ $S\alpha_3$, $S\alpha_4$	58%	54%	59%

Table 4. Performance of polynomial kernel system

Dataset	Indicators	PP_1	PP_2	PP_3
Sub-1	Access 10, Access 11	78%	-	-
Sub-2	$A\alpha_3$, $A\alpha_4$, $A\alpha_5$ $M\alpha_3$, $M\alpha_4$, $M\alpha_3$	74%	98%	19%
Sub-3	$A\alpha_3$, $A\alpha_4$ $S\alpha_3$, $S\alpha_4$	70%	96%	8%

Table 5. Performance of radial basis function kernel system

Dataset	Indicators	PP_1	PP_2	PP_3
Sub-1	Access 10, Access 11	81%	91%	51%
Sub-2	$A\alpha_3$, $A\alpha_4$, $A\alpha_5$ $M\alpha_3$, $M\alpha_4$, $M\alpha_3$	72%	99%	8%
Sub-3	$A\alpha_3$, $A\alpha_4$ $S\alpha_3$, $S\alpha_4$	73%	98%	8%

Table 6. Performance of deep neural network

Dataset	Indicators	PP_1	PP_2	PP_3
Sub-1	Access 10, Access 11	76%	-	-
Sub-2	$A\alpha_3$, $A\alpha_4$, $A\alpha_5$ $M\alpha_3$, $M\alpha_4$, $M\alpha_3$	89%	82%	94%
Sub-3	$A\alpha_3$, $A\alpha_4$ $S\alpha_3$, $S\alpha_4$	99%	99%	99%

performance parameters of PP_3 represent true negative rate which directly represents specificity. A closer look at the numerical performance of Support Vector Machine (Table 3-5) and Deep Neural Network (Table 6) shows that the deep neural network performs much better than the support vector machine. The results showcase that different subjects have different patterns of prediction. For example, the study shows that an effective classification performance carried out over the technical subject (i.e., sub-1) depends upon the indicators of the initial two months of examination while the non-technical subject (sub-2 and sub-3) depends upon initial two months or three months of data. The proposed study considers the dual examples of classification approaches using machine learning to ensure better performance. The system performance is found satisfactory when both the classifiers and any other variants of classifiers are used. However, the discrete evaluation of the study shows the deep neural network well manages that problem associated with the non-linearity decision making compared to the support vector machine. Apart from this, the outcome also shows that the performance of linear kernel systems in support vector machines provides nearly equivalent performance when it comes to non-technical subjects. From the perspective of non-technical subjects (sub-2 and sub-3), radial basis function performance is much better than others. Hence, the adoption of disruptive technologies of big data and machine learning offers significant.

7 Conclusion

The proposed study introduces disruptive technologies where big data and machine learning have been used to develop a novel model of e-learning systems. The idea of the proposed system is to carry out the extraction of apriori information associated with the performance of the student using MOOCs before writing the original examination. The idea is also to identify the performance of students who are identified to have a possibility of higher risk of lower performance. The modelling is carried out using an experimental analysis where wearable devices using smart sensors can capture data. The evaluation is carried out over technical and non-technical subjects where two forms of machine learning are used. The study outcome shows better accuracy for the deep learning-based approach compared to the support vector machine.

References

1. Almeida, F., Simoes, J.: The role of serious games, gamification and Industry 4.0 tools in the Education 4.0 paradigm. Contemp. Educ. Technol. **10**(2), 120–136 (2019)
2. Bielawski, L., Metcalf, D.S.: Blended learning: Integrating knowledge, performance, support, and online learning. Human Resource Development (2003)
3. Ciolacu, M., MugurSvasta, P., Berg, W., Popp, H.: Education 4.0 for tall thin engineer in a data driven society. In: 2017 IEEE 23rd International Symposium for Design and Technology in Electronic Packaging (SIITME), pp. 432–437. IEEE (2017)
4. Peña-Cabrera, M., Lomas, V., Lefranc, G.: Fourth industrial revolution and its impact on society. 2019 IEEE CHILEAN Conference on Electrical, Electronics Engineering, Information and Communication Technologies (CHILECON), Valparaiso, Chile, pp. 1–6 (2019). https://doi.org/10.1109/CHILECON47746.2019.8988083
5. Griffiths, F., Ooi, M.: The fourth industrial revolution - Industry 4.0 and IoT [Trends in Future I&M]. IEEE Instrum. Meas. Mag. **21**(6), 29–43 (2018). https://doi.org/10.1109/MIM.2018.8573590
6. Gooneratne, C., et al.: Drilling in the fourth industrial revolution — vision and Challenges. In: IEEE Eng. Manage. Rev. https://doi.org/10.1109/EMR.2020.2999420

7. Wagner, C., von Trotha, C., Palm, F., Epple, U.: Fundamentals for the next generation of automation solutions of the fourth industrial revolution. In: 2017 11th Asian Control Conference (ASCC), Gold Coast, QLD, pp. 2657–2662 (2017). https://doi.org/10.1109/ASCC.2017.8287596
8. Coşkun, S., Kayıkcı, Y., Gençay, E.: Adapting engineering education to industry 4.0 vision. Technologies **7**(1), 10 (2019)
9. Roblek, V., Mesko, M., Dimovski, V., Peterlin, J.: Smart technologies as social innovation and complex social issues of the Z generation. Kybernetes (2019)
10. Pachidis, T., Vrochidou, E., Kaburlasos, V.G., Kostova, S., Bonković, M., Papić, V.: Social robotics in education: State-of-the-art and directions. In: International Conference on Robotics in Alpe-Adria Danube Region, pp. 689–700. Springer, Cham (2018). https://doi.org/10.1007/978-3-030-00232-9_72
11. Kim, H., Shin, H., Kim, H.-S., Kim, W.-T.: VR-CPES: A novel cyber-physical education systems for interactive VR services based on a mobile platform. Mobile Information Systems 2018 (2018)
12. Mourtzis, D., Vlachou, E., Dimitrakopoulos, G., Zogopoulos, V.: Cyber-physical systems and education 4.0–the teaching factory 4.0 concept. Procedia Manufacturing **23**, 129–134 (2018)
13. Moubayed, A., Injadat, M., Nassif, A.B., Lutfiyya, H., Shami, A.: E-Learning: challenges and research opportunities using machine learning & data analytics. IEEE Access **6**, 39117–39138 (2018). https://doi.org/10.1109/ACCESS.2018.2851790
14. Rosenberg, M.J.: Beyond e-learning: Approaches and Technologies to Enhance Organizational Knowledge, Learning, and Performance. John Wiley & Sons (2005)
15. Nora, A., Snyder, B.P.: Technology and higher education: The impact of e-learning approaches on student academic achievement, perceptions and persistence. J. College Student Retention: Research, Theory & Practice **10**(1), 3- 19 (2008)
16. Ellis, R.A., Ginns, P., Piggott, L.: E-learning in higher education: some key aspects and their relationship to approaches to study. Higher Educ. Res. Dev. **28**(3), 303–318 (2009)
17. Garrison, D.R.: E-learning in the 21st Century: A Framework for Research and Practice. Taylor & Francis (2011)
18. Costa, A., Costa, A., Olsson, I.A.S.: Students' acceptance of e-learning approaches in laboratory animal science training. Laboratory animals **54**(5), 487-497 (2020)
19. Otoo-Arthur, D., van Zyl, T.L.: A Scalable Heterogeneous Big Data Framework for e- Learning Systems. In: 2020 International Conference on Artificial Intelligence, Big Data, Computing and Data Communication Systems (icABCD), Durban, South Africa, pp. 1–15 (2020). https://doi.org/10.1109/icABCD49160.2020.9183863
20. Udupi, P.K., Malali, P., Noronha, H.: Big data integration for transition from e-learning to smart learning framework. In: 2016 3rd MEC International Conference on Big Data and Smart City (ICBDSC), Muscat, pp. 1–4 (2016). https://doi.org/10.1109/ICBDSC.2016.7460379
21. Zhang, W., Qin, S.: A brief analysis of the key technologies and applications of educational data mining on online learning platform. In: 2018 IEEE 3rd International Conference on Big Data Analysis (ICBDA), Shanghai, pp. 83–86 (2018). https://doi.org/10.1109/ICBDA.2018.8367655
22. Zhou, D., Li, H., Liu, S., Song, B., Hu, T.:A map-based visual analysis method for patterns discovery of mobile learning in education with big data. In: 2017 IEEE International Conference on Big Data (Big Data), Boston, MA, pp. 3482-3491 (2017). https://doi.org/10.1109/BigData.2017.8258337
23. Yanhui, W.: Big data and foreign language e-learning on production-oriented approach. In: 2018 IEEE 3rd International Conference on Big Data Analysis (ICBDA), Shanghai, pp. 170–174 (2018). https://doi.org/10.1109/ICBDA.2018.8367671

24. Vyas, M.S., Gulwani, R.: Predictive analytics for E learning system. In: 2017 International Conference on Inventive Systems and Control (ICISC), Coimbatore, pp. 1–4 (2017). https://doi.org/10.1109/ICISC.2017.8068590
25. Popchev, I., Orozova, D., Stoyanov, S.:I oT and Big Data Analytics in E-Learning. In: 2019 Big Data, Knowledge and Control Systems Engineering (BdKCSE), Sofia, Bulgaria, pp. 1–5 (2019). https://doi.org/10.1109/BdKCSE48644.2019.9010666
26. Dwivedi, S., Roshni, V.S.K.: Recommender system for big data in education. In: 2017 5th National Conference on E-Learning & E-Learning Technologies (ELELTECH), Hyderabad, pp. 1–4 (2017). https://doi.org/10.1109/Eleltech.2017.8074993
27. Babu, N.S.C.: Keynote 1: Internet of Things(IoT) and augmented reality for e-learning. In: 2017 5th National Conference on E-Learning & E-Learning Technologies (ELELTECH), Hyderabad, pp. 1–10 (2017). https://doi.org/10.1109/ELELTECH.2017.8074987
28. Wang, Y., Allakany, A., Kulshrestha, S., Shi, W., Bose, R., Okamura, K.: Automatically Generate E-Learning Quizzes from IoT Security Ontology. In: 2019 8th International Congress on Advanced Applied Informatics (IIAI-AAI), Toyama, Japan, pp. 166-171 (2019). https://doi.org/10.1109/IIAI-AAI.2019.00042.
29. Zahedi, M.H., Dehghan, Z.: Effective E-learning utilizing Internet of Things. In: 2019 13th Iranian and 7th National Conference on e-Learning and e-Teaching (ICeLeT), Tehran, Iran, pp. 1–6 (2019). https://doi.org/10.1109/ICELET46946.2019.9091671
30. Yamada, M., Oda, T., Liu, Y., Matsuo, K., Ikeda, M., Barolli, L.: Performance Evaluation of an IoT-based e-Learning Testbed Considering OLSR Protocol in a NLoS Environment. In: 2016 19th International Conference on Network-Based Information Systems (NBiS), Ostrava, pp. 451–457 (2016). https://doi.org/10.1109/NBiS.2016.22
31. Matsuo, K., Yamada, M., Bylykbashi, K., Cuka, M., Liu, Y., Barolli, L.: Implementation of an IoT-based E-Learning testbed: performance evaluation using mean-shift clustering approach considering four types of brainwaves. In: 2018 32nd International Conference on Advanced Information Networking and Applications Workshops (WAINA), Krakow, pp. 203–209 (2018). https://doi.org/10.1109/WAINA.2018.00088
32. Njeru, A.M., Omar, M.S., Yi, S.: IoTs for capturing and mastering massive data online learning courses. In: 2017 IEEE/ACIS 16th International Conference on Computer and Information Science (ICIS), Wuhan, pp. 91–94 (2017). https://doi.org/10.1109/ICIS.2017.7959975
33. Rajkumar, R., Ganapathy, V.: Bio-inspiring learning style chatbot inventory using brain computing interface to increase the efficiency of e-learning. IEEE Access **8**, 67377–67395 (2020). https://doi.org/10.1109/ACCESS.2020.2984591
34. Lai, S., Sun, B., Wu, F., Xiao, R.: Automatic personality identification using students' online learning behavior. IEEE Trans. Learn. Technol. **13**(1), 26–37 (2020). https://doi.org/10.1109/TLT.2019.2924223
35. Nafea, S.M., Siewe, F., He, Y.: On recommendation of learning objects using felder-silverman learning style model. IEEE Access **7**, 163034–163048 (2019). https://doi.org/10.1109/ACCESS.2019.2935417
36. Xie, T., Zheng, Q., Zhang, W., Qu, H.: Modeling and predicting the active video- viewing time in a large-scale e-learning system. IEEE Access **5**, 11490–11504 (2017). https://doi.org/10.1109/ACCESS.2017.2717858
37. Yang, Z., Yang, J., Rice, K., Hung, J.-L., Du, X.: Using convolutional neural network to recognize learning images for early warning of at-risk students. In: IEEE Transactions on Learning Technologies, **13**(3), 617–630 (2020). https://doi.org/10.1109/TLT.2020.2988253
38. Wu, E.H., Lin, C., Ou, Y., Liu, C., Wang, W., Chao, C.: Advantages and constraints of a hybrid model K-12 e-learning assistant chatbot. IEEE Access **8**, 77788–77801 (2020). https://doi.org/10.1109/ACCESS.2020.2988252
39. Colace, F., De Santo, M.: Ontology for e-learning: a bayesian approach. IEEE Trans. Educ. **53**(2), 223–233 (2010). https://doi.org/10.1109/TE.2009.2012537

40. Gaeta, M., Orciuoli, F., Paolozzi, S., Salerno, S.: Ontology extraction for knowledge reuse: the e-learning perspective. IEEE Trans. Syst., Man, Cybernetics - Part A: Syst. Humans **41**(4), 798–809 (2011). https://doi.org/10.1109/TSMCA.2011.2132713
41. Chiu, P., Chang, J., Lee, M., Chen, C., Lee, D.: Enabling intelligent environment by the design of emotionally aware virtual assistant: a case of smart campus. IEEE Access **8**, 62032–62041 (2020). https://doi.org/10.1109/ACCESS.2020.2984383
42. Barlybayev, A., Kaderkeyeva, Z., Bekmanova, G., Sharipbay, A., Omarbekova, A., Altynbek, S.: Intelligent system for evaluating the level of formation of professional compe- tencies of students. IEEE Access **8**, 58829–58835 (2020). https://doi.org/10.1109/ACCESS.2020.2979277
43. Holmes, M., Latham, A., Crockett, K., O'Shea, J.D.: Near real-time comprehension classification with artificial neural networks: decoding e-learner non-verbal behavior. IEEE Trans. Learn. Technol. **11**(1), 5–12 (2018). https://doi.org/10.1109/TLT.2017.2754497
44. Hone, K.S., El Said, G.R.: Exploring the factors affecting MOOC retention: A survey study. Computers & Education **98**, 157–168 (2016)

Methodology for Placing Components of a Video Surveillance System for Smart City Based on a Composite Cost Optimization Model

V. Lakhno[1], V. Malyukov[1], O. Kryvoruchko[2], V. Chubaievskyi[2], A. Desiatko[2](✉), M. Misiura[1], and V. Pashorin[2]

[1] Department of Computer Systems and Networks, National University of Life and Environmental Sciences of Ukraine, Kyiv, Ukraine
mdmm@nubip.edu.ua

[2] Department of Software Engineering and Cybersecurity, Kyiv National University of Trade and Economics, Kyiv, Ukraine
ev_kryvoruchko@ukr.net, {chubaievskyi_vi,pashorin}@knute.edu.ua, desyatko@gmail.com

Abstract. The article proposes a methodology for optimizing the costs of building a video surveillance system (VSS) for Smart City. The technique assumes two-level optimization based on various methods of mathematical programming. This approach makes it possible to increase the design efficiency of a complex geographically distributed VSS Smart City. The available financial and technical capabilities are taken into account, as well as the list of negative factors that impede the deployment of such VSS or make it more expensive. The global level of optimization is proposed to be performed based on the application of the apparatus of game theory. The model for evaluating strategies for financing VSS for Smart City is based on the toolkit of bilinear differential quality games with several terminal surfaces, developed for the case of fuzzy information.

Keywords: Smart City · Video surveillance system · Cost optimization · Game theory

1 Introduction

Modern cities are complex multi-level structures that consist of many subsystems. One of the most important subsystems is the video surveillance subsystem. As a rule, such subsystems are implemented within the Safe City segment.

Video surveillance systems (hereinafter referred to as VSS), used within the Smart City concept [1, 2], implement the functions of collecting, transmitting and processing information, see Fig. 1.

Traditionally, such VSS systems form rather large volumes of transmitted data due to the video image format. Expansion of VSS capabilities, as a rule, is achieved through the use of personal computers with the connection to them of incoming video signals coming from video surveillance hardware systems.

R. Silhavy (Ed.): CSOC 2022, LNNS 501, pp. 13–23, 2022.
https://doi.org/10.1007/978-3-031-09070-7_2

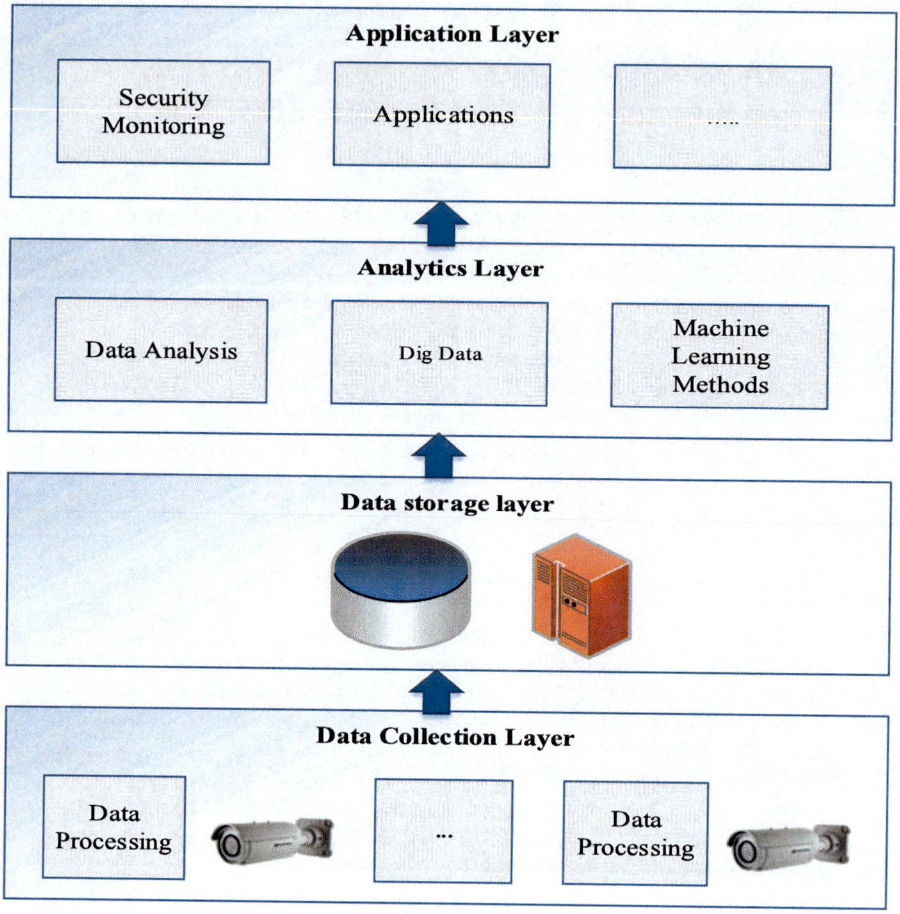

Fig. 1. Scheme of integrating a video surveillance system into the Smart City concept

The use of analog cameras for video surveillance tasks in Smart City is steadily declining due to the emergence of more functional IP cameras. The IP cameras use a local area network to transmit the signal. Moreover, the share of wireless networks (WN) for solving this problem is steadily increasing. Wireless CCTV cameras use different methods of data transmission. The most common option is the use of Wi-Fi technology [3, 4].

2 Review and Analysis of the Literature

The works [5, 6] show that the efficiency of the Smart City security subsystems is significantly influenced by new information technologies. Such technologies certainly include constant video surveillance and video registration systems. Such systems help to improve the safety of residents and visitors to the city. However, these works do not touch upon the issues of VSS optimization according to technical and economic criteria.

In [7–9], the main factors affecting the efficiency of deployment of VSS of complex objects are analyzed and can be useful for the economic justification of technical solutions for VSS. However, these works do not take into account the features of Smart City.

Taking into account the above, the task of improving the methodology for designing VSS for Smart City based on the composition of various mathematical methods remains relevant. Which implies the combination of, for example, linear programming methods and game theory models for a large geographically distributed VSS for Smart City.

3 The Purpose of the Study

The purpose of this study is to develop a methodology for solving the problem of optimizing the costs of VSS systems according to the criterion of minimum costs.

4 Methods and Models

The composition of various mathematical methods involves the combination of linear programming methods and game theory models for a large geographically distributed VSS, for example, Smart City. Moreover, in this case, the application of a sequential multistep procedure for optimizing the VSS according to the criterion "price-quality" will occur at the global level.

By composition, we mean the totality of the constituent parts of the model.

The choice of the placement of VSS cameras for individual Smart City objects is carried out based on the simplex method. And the final version for optimizing the costs of VSS is formed based on the theory of bilinear quality games.

Let us formulate the mathematical formulation of the problem.

t is necessary to minimize the costs of building a VSS based on maximizing the use of wireless networks (hereinafter WN) Wi-Fi for transmitting images to the monitors of the Smart City situation center and other decision-makers, subject to the maximum fulfillment of all the basic requirements for VSS, specified above.

The formalized formulation of the problem of minimizing the costs of air pumping systems can be presented as follows:

$$f(x) = \min(C, Z), \tag{1}$$

where $f(x)-$ is the objective function of the procedure for optimizing the costs of VSS;

$$(C, Z) = \sum_{i=1}^{n} (c, z)_i; \tag{2}$$

the dot product of vectors.

$$(c, z)_i = \sum_{j=1}^{m} c_{ij} z_{ij}, \ i = \overline{1, n}, \ j = \overline{1, m}; \tag{3}$$

here $(C, Z)-$ is the function that sums up the costs of the VSS elements.

Under the following restrictions:

$$A \cdot z \le b \tag{4}$$

$$x > 0, \tag{5}$$

where $A-$ is the number of channels through which the camera z can transmit the image to decision-makers; $b-$ the number of channels required to transmit the image to decision-makers.

In the simplest case, problem (1) can be successfully solved using traditional optimization methods.

For example, Fig. 1 shows a general view of a console application written in the algorithmic language C # for solving this problem using the simplex method.

Accordingly, in Fig. 1 in the window of the console application in the "basis" column (first column) we indicate the basic variables (in our case, these are the costs for VSS cameras and the organization of video transmission channels over the WN or using a traditional cable network). In the last row of the "basis" column, we write down the objective function $f(x)$. In the second column, we write down the free terms of the constraints b_i and the value $f(x)$ (at the first stage it is equal to 0, ie, no profit from the implementation of VSS based on Wi-Fi WN).

In the columns z_i for variables that are not basic, we indicate the coefficients for non-basic variables following the constraints on this problem. The columns of the basic variables will contain only 0 or 1. These elements will be located at the intersection of the column and the corresponding row of the variable that we assume to be the basic one.

Figure 2 shows an example of a calculation for the VSS variant with five surveillance cameras. As can be seen from the obtained result of the calculation, for such an organization of the VSS, 10 channels of video information transmission to decision-makers are sufficient.

However, the restrictions imposed on the classic simplex method do not allow using it for more complex situations of designing VSS for Smart City. The need to recalculate all the coefficients and free terms of the complete system of equations makes this method unacceptable for cases when the designed video surveillance system will have a large dimension. This will lead to an increase in the volume of calculations and slow down the search for the optimal solution.

Therefore, following the previously formulated proposal in this section of the work, it is proposed to use a composite model that combines the advantages of the simplex method and solutions based on game theory [10, 11].

Indeed, from the point of view of game theory, we are dealing with a situation when the side organizing the video surveillance system for Smart City is "opposed" by unfavorable factors. It is these factors that are considered in the game as the opposite side, therefore, to achieve the final goal, it is possible to consider the solution of this antagonistic game as a stage in solving the general problem of optimizing the costs of VSS for Smart City.

Then we will assume that player 1 is the side deploying VSS for Smart City ($PL1$), player 2 is external factors that can minimize or make the negative effect of the introduction of VSS based on Wi-Fi technology (denote $PL2$).

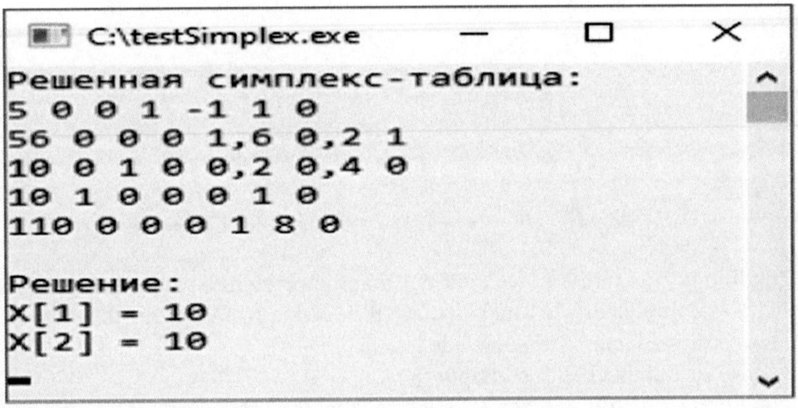

Fig. 2. An example of the implementation of a model for optimizing the costs of building a VSS

The dynamics of the interaction of the parties or players is given by the following system of bilinear differential equations with dependent motions.

The deployment time of the VSS is T.

$$\begin{cases} dh(t)/dt = -h(t) + B_1 \times h^+(t) + \\ +[(A_1 + R_1) - E] \times U(t) \times B_1 \times h^+(t) - \\ -[(A_2 + R_2) - E] \times V(t) \times B_2 \times f^{+,\xi}(t); \\ df^{\xi}(t)/dt = -f^{\xi}(t) + B_2 \times f^{+,\xi}(t) + \\ +[(A_2 + R_2) - E] \times B_2 \times f^{+,\xi}(t) - \\ [(A_1 + R_1) - E] \times U(t) \times B_1 \times h^+(t); \end{cases} \quad (6)$$

here $t \in [0, T]$, $T \in R_+$;

$h(t) \in R^n, f^{\xi}(t) \in R^n$, $U(t), V(t)$– square diagonal matrices of order n with positive entries $u_i(t), v_i(t) \in [0, 1]$;

B_1, B_2 - transformation matrices $PL1$ and $PL2$, which are square matrices of order n with positive entries g_1^{ij}, g_2^{ij}, respectively. These matrices determine the positive and negative effects of the implementation of the VSS based on WN for Smart City.

A_1, R_1– diagonal matrices with negative elements, corresponding to the negative influence of external factors $PL2$ in relation to investments $PL1$ in VSS for Smart City;

A_2, R_2– diagonal matrices with positive elements, corresponding to the positive impact of VSS based on WN and shares of return on investment $PL1$ with negative factors $PL2$;

E– unit matrix;

$$x^+ = \begin{cases} x, x \geq 0; \\ 0, x \prec 0, x \in R; \end{cases}$$

Interaction between players ends when the following conditions are met:

$$(h(t), f^{\xi}(t)) \in S_0, \text{ with certainty } \geq p_0, \quad (0 \leq p_0 \leq 1) \quad (7)$$

$$(h(t), f^{\xi}(t)) \in F_0, \text{ with certainty } \geq p_0 \tag{8}$$

where

$$S_0 = \bigcup_{i=1}^{n} \{(h, f^{\xi}) : (h, f^{\xi}) \in R^{2n}, h \succ 0, f_i^{\xi} = 0\},$$

$$F_0 = \bigcup_{i=1}^{n} (h, f^{\xi}) : (h, f^{\xi}) \in R^{2n}, f^{\xi} \succ 0, h_i = 0\},$$

If condition (7) is fulfilled, we consider that the procedure for investing in VSS for Smart City is completed. This means that all negative factors from the side $PL2$ are covered by positive factors from the side $PL1$.

The result is obtained with certainty $\geq p_0$.

If condition (8) is fulfilled, we also assume that the procedure for investing in VSS for Smart City is completed. This also means that $PL1$ did not have enough financial resources to continue the investment procedure and he was unable to overcome all technological and technical problems, as well as external unfavorable factors associated with the deployment of VSS at Smart City facilities. This subtask has been solved with certainty $\geq p_0$.

If both conditions (7) and (8) are not met, we assume that the procedure for investing in the deployment of VSS for Smart City continues.

The above process is considered within the framework of the scheme of a positional differential quality game with fuzzy information [5]. The information is unclear, since the party deploying the VSS for Smart City does not know in advance all the negative factors affecting its functionality.

The problem of the first player (further, $PL1$ bring the dynamical system with the help of its control strategies to the terminal surface S_0, no matter how the second player acts (further $PL2$). The problem $PL2$ is symmetric.

This means that two tasks are generated, from the point of view of the first player and from the point of view of the second player.

Within the framework of the article, we will consider the problem from the point of view of the first player. The solution of the problem from the point of view of the second player will not be considered due to symmetry.

The solution consists of constructing the sets of preferences of the players W_1 and W_2. Preference sets are sets of initial states of players and their strategies. These initial states allow objects to bring the system to one or another surface.

Setting the procedure for the interaction of players using a system of differential equations (SDE) generates at each moment of time t a set of fuzzy sets $\{X_t, m_t(.)\}$, that reflect the process of transition from the initial states of the players to the subsequent ones when the players apply control actions. The SDE solution (6) is implemented in software.

5 Computational Experiment

The software implementation of the model presented in the work is made in the Python programming language.

The results of modeling the sets of preferences of player 1 are presented in Fig. 3 for the multidimensional case of 5 variables (as negative factors on the part of player 2, the following were considered - a large extent of territories where Smart City objects are located; a high level of industrial noise; high air pollution, which prevents shooting, etc.

Also, the results obtained are partially shown in Table 1.

Table 1. Fragment of the table with the results of modeling the preference area *PL*1

№	U – multiple player strategies *PL*1	V – multiple strategies *PL*2	W_1 – player preference area 1 (*PL*1)	T – VSS deployment time for Smart City (months)
1	*E*	0	$5 \cdot H1(0) + 4,2 \cdot H2(0) >$ $> 6,5 \cdot Delta(P0)$	5,2
2	*E*	0	$3 \cdot H1(0) + 2,35 \cdot H2(0) >$ $> 5,35 \cdot Delta(P0)$	4,1
3	*E*	0	$H1(0) + 8,22 \cdot H2(0) >$ $> 1,7 \cdot Delta(P0)$	3,7
4	*E*	0	$1,25 \cdot H1(0) + 2,1 \cdot H2(0) >$ $> 5,65 \cdot Delta(P0)$	2,2
5	*E*	0	$8,99 \cdot H1(0) + 3,21 \cdot H2(0) >$ $> 5,71 \cdot Delta(P0)$	9,2
6	*E*	0	$3 \cdot H1(0) + 2,3 \cdot H2(0) >$ $> 0,71 \cdot Delta(P0)$	15,4
7	*E*	0	$11,1 \cdot H1(0) + 8,22 \cdot H2(0) >$ $> 5,1 \cdot Delta(P0)$	22,3
8	*E*	0	$2,1 \cdot H1(0) + 9,22 \cdot H2(0) >$ $> Delta(P0)$	2,9
9	*E*	0	$14,1 \cdot H1(0) + 8,23 \cdot H2(0) >$ $> 7,8 \cdot Delta(P0)$	8,9
10	*E*	0	$3,85, \cdot H1(0) + 4,22 \cdot H2(0) >$ $> 2,67 \cdot Delta(P0)$	13,2
11	*E*	0	$6,1 \cdot H1(0) + 7,32 \cdot H2(0) >$ $> 8,1 \cdot Delta(P0)$	24,1

(*continued*)

Table 1. (*continued*)

№	U – multiple player strategies *PL*1	V – multiple strategies *PL*2	W_1 – player preference area 1 (*PL*1)	T – VSS deployment time for Smart City (months)
12	*E*	0	$15,1 \cdot H1(0) + 1,3 \cdot H2(0) >$ $> 9,8 \cdot Delta(P0)$	12,2
13	*E*	0	$4,1 \cdot H1(0) + 2,31 \cdot H2(0) >$ $> Delta(P0)$	3,1
14	*E*	0	$0,31 \cdot H1(0) + H2(0) >$ $> 8,18 \cdot Delta(P0)$	33,1
15	*E*	0	$7,31 \cdot H1(0) + 2,02 \cdot H2(0) >$ $> 11,24 \cdot Delta(P0)$	15,2

Here *T* is the time during which the first player will bring the state of the system to his terminal surface given the data with the corresponding number in the table.

The obtained results of experimental calculations show the effectiveness of the proposed toolkit for solving the problem of assessing the prospects for deploying VSS for Smart City based on wireless networks. The problem is solved by taking into account multifactorial in a fuzzy formulation.

6 Discussion of the Results of Experimental Studies

Figure 3 shows a collection of points in three-dimensional space, which reflects the following situation. The dependence of financial resources on a variable is shown, which characterizes the degree of reliability of the procedure for financing the VSS based on the WN with a given level of reliability. The confidence level is described by a one-dimensional variable.

For each point corresponding to the financial resource, the value of the variable characterizing the reliability of the procedure for financing the VSS is given. There can be several of these values. Some of these values correspond to the set that guarantees the continuation of the VSS financing procedure. Then, choosing from these values the minimum (for each component), we obtain for each variant of the deployment of the VSS a set that will belong to the set of preferences of the first player, provided that all values of the player's variable are less than the previously given minimum values of the variables "reliability".

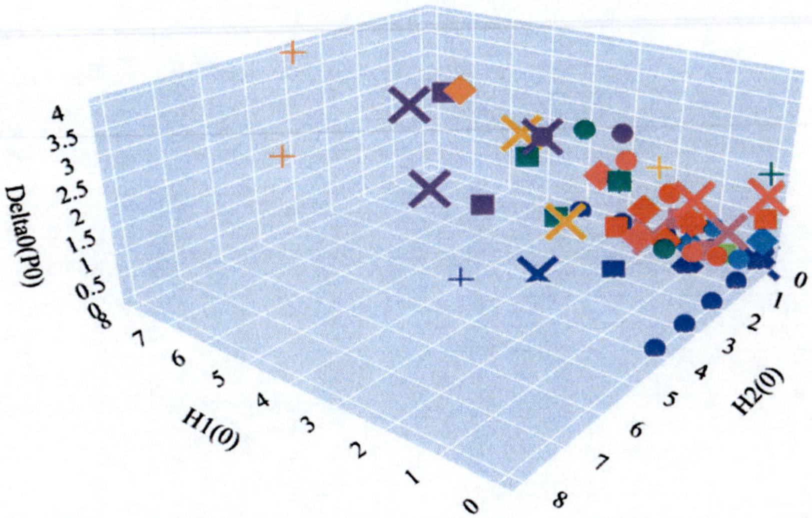

Axes: H1 (0) - the equivalent of resources $PL1$; H2 (0) - the equivalent of resources $PL2$; Delta0 (P0) - the number of resources $PL1$ spent on bringing the dynamic system to its terminal surface

Fig. 3. Dependence of the preference set W_1 for the first player for 5 variables for the second player

Thus, as a result of the research, theoretical and practical issues of designing a video surveillance system for Smart City based on the use of Wi-Fi technology were considered. The proposed model is part of the design algorithm for VSS for Smart City. A block diagram of the algorithm is shown in Fig. 4.

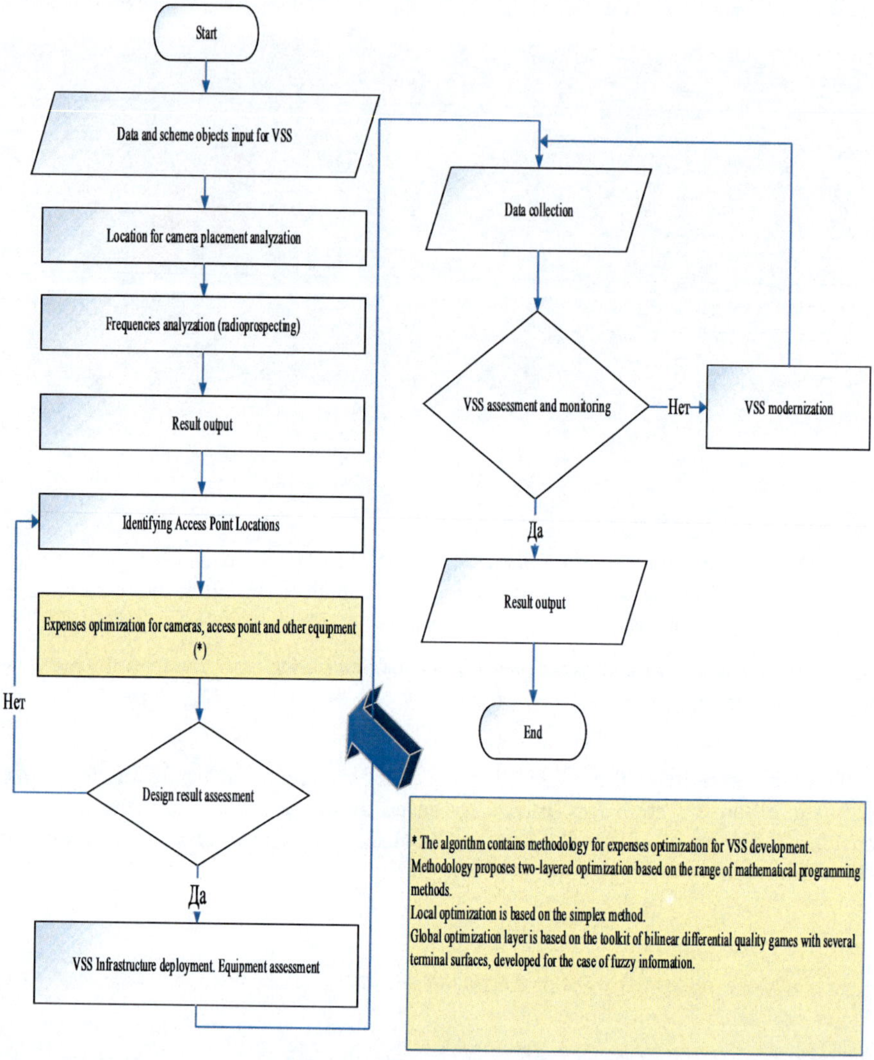

Fig. 4. Block diagram of the design algorithm of the VSS system for Smart City with a new model based on the toolkit of bilinear quality differential games with several terminal surfaces

7 Conclusion

A methodology for optimizing the costs of building a video surveillance system (VSS) for Smart City is proposed. The technique assumes two-level optimization based on various methods of mathematical programming. This, accordingly, makes it possible to increase the design efficiency of geographically complex VSS Smart City, taking into account the available financial and technical capabilities, and the list of negative factors that impede the deployment of such VSS or increase its cost.

It is shown that local optimization can be performed based on well-proven methods, for example, the simplex method. The global level of optimization is proposed to be performed based on the application of the apparatus of game theory. A new mathematical model has been developed for the choice of rational strategies by the party deploying the VSS, taking into account the multifactorial nature of this problem in a fuzzy statement. The model for evaluating strategies for financing VSS for Smart City is based on the toolkit of bilinear differential quality games with several terminal surfaces, developed for the case of fuzzy information. The difference between this model and the existing ones is the presented for the first time a new class of bilinear differential games in a fuzzy setting.

Computer modeling of various options for deploying VSS for Smart City and, accordingly, financing of this process has been carried out.

References

1. Kashef, M., Visvizi, A., Troisi, O.: Smart city as a smart service system: Human-computer interaction and smart city surveillance systems. Computers in Human Behavior **124** 106923 (2021)
2. Chui, K.T., Vasant, P., Liu, R.W.: Smart city is a safe city: information and communication technology–enhanced urban space monitoring and surveillance systems: the promise and limitations. In: Smart Cities: Issues and Challenges, pp. 111–124. Elsevier (2019)
3. Ye, Y., Ci, S., Katsaggelos, A.K., Liu, Y., Qian, Y.: Wireless video surveillance: a survey. IEEE Access **1**, 646–660 (2013)
4. Garcia-Sanchez, A.J., Garcia-Sanchez, F., Garcia-Haro, J.: Wireless sensor network deployment for integrating video-surveillance and data-monitoring in precision agriculture over distributed crops. Comput. Electron. Agric. **75**(2), 288–303 (2011)
5. Ferraz, F.S., Ferraz, C.A.G.: Smart city security issues: depicting information security issues in the role of an urban environment. In: 2014 IEEE/ACM 7th International Conference on Utility and Cloud Computing, pp. 842–847. IEEE (2014)
6. Zhang, K., Ni, J., Yang, K., Liang, X., Ren, J., Shen, X.S.: Security and privacy in smart city applications: challenges and solutions. IEEE Commun. Mag. **55**(1), 122–129 (2017)
7. Cheung, S.C., Venkatesh, M.V., Paruchuri, J.K., Zhao, J., Nguyen, T.: Protecting and managing privacy information in video surveillance systems. In: Protecting Privacy in Video Surveillance, pp. 11–33. Springer, London (2009) https://doi.org/10.1007/978-1-84882-301-3_3
8. Karimaa, A.: Video surveillance in the cloud: dependability analysis. In: Proceedings of the Fourth International Conference on Dependability, DEPEND, pp. 92–95 (2011)
9. Montero, D., et al.: Multi-stage dynamic batching and on-demand i-vector clustering for cost-effective video surveillance. In: VISIGRAPP (5: VISAPP), pp. 436–443 (2021)
10. Valeriy, L., et al.: Computer support system for choosing the optimal managing strategy by the mutual investment procedure in smart city. Advances in Intelligent Systems and Computing, 1194 AISC, pp. 278–287 (2021)
11. Lakhno, V., Malyukov, V., Kryvoruchko, O., Desiatko, A., Shestak, Y.: Smart city technology investment solution support system accounting multi-factories. In: Silhavy, R., Silhavy, P., Prokopova, Z. (eds.) Software Engineering Perspectives in Intelligent Systems. CoMeSySo 2020. Advances in Intelligent Systems and Computing, vol. 1294, pp. 1–11. Springer, Cham (2020). https://doi.org/10.1007/978-3-030-63322-6_1

The Information Technology for Customary Cadastre in Penglipuran Village Using Dynamic BIM

Ketut Tomy Suhari[1,2](✉), H. Z. Abidin[1], A. Y. Saptari[1], P. H. Gunawan[3], B. E. Leksono[1], and R. Abdulharis[1]

[1] Department of Geodesy and Geomatics Engineering, Faculty of Earth Sciences and Technology, Institut Teknologi Bandung, Bandung, Indonesia
{35120304,hzabidin,aysaptari,bleksono,rabdulharis}@gd.itb.ac.id

[2] Department of Geodesy Engineering, Faculty of Civil Engineering and Planning, Institut Teknologi Nasional Malang, Malang, Indonesia
ksuhari@lecturer.itn.ac.id

[3] School of Computing, Telkom University, Bandung, Indonesia
phgunawan@telkomuniversity.ac.id

Abstract. Tri Hita Karana is known for harmonizing the relationship between human God and nature in Hindu Bali tradition, Indonesia. This traditional philosophy is implemented in some aspects of life. For instance, the Balinese community designs their house in customary spatial using this philosophy. This tradition faces various threats of extinction due to modernization which continues to develop over time. Such modern buildings that come from outside investors can destroy the existing traditional order. In this study, researchers propose a technology, namely Dynamic BIM, to record the spatial structure of conventional Balinese buildings that adopt the Tri Hita Karana concept. So with this technology, it can accommodate changes in function and use of space before and after. Moreover, it can provide historical and cultural information on ownership. In addition, this technology can make efforts to conserve customary space.

Keywords: Tri Hita Karana · Balinese · Cadastre · Dynamic BIM · Technology

1 Introduction

Bali is known as a favorite destination for vacation, and it has full of beautiful nature and a unique tradition. Balinese people have a philosophy of life known as Tri Hita Karana (THK), i.e., the concept of harmonization between humans, God, and nature. Indeed, this philosophy has been implemented in several aspects of Balinese life and culture. For instance, Balinese customary spatial

R. Silhavy (Ed.): CSOC 2022, LNNS 501, pp. 24–36, 2022.
https://doi.org/10.1007/978-3-031-09070-7_3

is designed using the concept of THK. Traditionally, Balinese houses are consist of Pura or Temple for the manifestation of God, a living room for human interaction, and a garden for preserving nature.

Currently, the land system (land registration) performs digital data management in a 2D model. This model has some weaknesses in several aspects of document legalization. So it is necessary to transition into a 3D model to provide better visuals, legal certainty and keep the essence of existing buildings. In addition, this model can help some related parties to assess, assure mastery, and easy to find the implementation. Various researchers offer various concepts of 3D cadastre and their application in multiple fields, which are integrated into various existing models and technologies [10,13,24,27,32]. According to [5], the use of Building Information Modeling (BIM) is a feasible approach in managing land and property information in the administration of high-rise buildings. BIM is a digital form of a building that contains building knowledge and manages assets in it. BIM consists of three components, namely technology, processes, and digital information, that are used to improve the operation of assets in the [19] building.

The BIM platform uses a proprietary format in model storage and refers to a collaborative and multi-disciplinary process for managing buildings in different phases of the building's lifespan [9]. BIM products can be made different according to the needs of the actors who use them, such as architects, civil engineers, surveyors, and others. The BIM process can integrate different products with a common data environment to provide a complete building model. One of the organizations that promote BIM is known as BuildingSMART, where this organization develops Industry Foundation Classes (IFC) standards as an open data model and vendor to facilitate interoperability across various BIM platforms [16].

This research is reinforced by Suhari (2020) [25], explaining that BIM can be used to model, inventory, and manage 3D cadastral information data that is integrated with customary law and local wisdom in Bali. However, setting a static BIM is still problematic and cannot accommodate changes that will occur continuously due to various existing factors. Various researchers discuss the static to dynamic changes by providing data and driven models. In the paper, [7], structural monitoring of train bridges with time series sensors using parametric methods and R-Tree model is elaborated, moreover in [6], the database system is modified to be able to perform queries on the 3D land system by utilizing the Boundary Representation Model (Brep). In addition, BIM can be developed into workflow activities on 3D land systems and dynamic systems with GIS (see [17,20] for more detail).

Tourist destinations that are proven to be a source of income for local communities do not automatically realize the preservation of local culture but are highly dependent on the participation of residents (stakeholders). Sustainable Development Goals (SDGs) are efforts to create prosperity for the world community, as well as efforts to protect the environment and create sustainable development for future generations [3,14,23]. The concept of sustainable development proposed by the United Nations Environment Program and the World

Tourism Organization (2005) states that there are 3 (three) dimensions or pillars in the conceptual system of sustainable development ([2,18], namely: economic sustainability, social-cultural sustainability, and ecological sustainability.

The involvement of the Penglipuran tourism village (Bali-Indonesia) residents is very high, so to maintain cleanliness, they work voluntarily to clean the environment. The local government of Bangli Regency, Kubu Village, is also building various infrastructures such as road construction and community economic needs. The proceeds from the sale of rural tourism tickets are mostly used for the needs of the Penglipuran tourism village community in the form of necessities [15]. To maintain the tourist village, Penglipuran residents maintain a clean environment and preserve the traditional structure of their house building. This village has its uniqueness and is attractive to tourists because it can maintain its culture and customs by presenting old buildings and accepting structural modernization [12,22,31]. According to Primadi [22], the influence of modernization has brought concerns about the continued development of tourism in the village. The existence of the traditional houses of the residents has been threatened by the action of modernization and the improvement of the economic level of the local community, as well as the shift in the cultural values of the traditional society towards the modern one.

This study aims to present the concept of dynamic BIM, which can be a link to existing problems and solutions by modifying the idea of land administration and spatial unit [1,11]. Thus it can create the concept of customary cadastre. It will present this technology can take inventory of digital data in the form of 3D and accommodate changes in the use and use of space based on applicable customary law. Therefore, this concept can apply customary cadastre and be used to reference conservation actions in traditional areas.

2 Materials and Methods

2.1 Tri Hita Karana Concept in Balinese Traditional Cadastre

The concept of Tri Hita Karana (THK) is a harmonious relationship between humans and God, fellow humans and humans, and nature. The Balinese people apply this concept to spatial planning and cardinal directions that the ancestors of the Balinese people have inherited. The spatial layout of the Balinese THK is limited to the space on the earth that is stepped on and the sky above it. So this traditional concept then developed from the spatial orientation of "sky-earth" in the early Bali period, "mountain-sea" during the Bali Aga period, and "sunrise and sunset" to the time of Bali Arya/Majapahit [29]. The Tri Mandala pattern is a reference pattern for building layouts in traditional Balinese architecture. The three aspects of the Tri Mandala are Utama Mandala, Madya Mandala, and Nista Mandala. According to Parwata (2011) [21,26] explains that the elaboration of the teachings of Tri Hita Karana and its relation to the concept of Tri Mandala is the relationship between humans and God, which is carried out in the dimensions of the *Utama Mandala* space, Human relations are carried out in the *Madya Mandala* space. Meanwhile, the *Nista Mandala* dimension is the relationship between humans and their environment.

In Hindu belief, the concept of horizontal orientation is based on two axes, namely the axis at sunrise and at sunset Kangin-Kauh (East-West) and the mountain-sea axis called Kaja-Kelod (north-south). The ritual axis or sun (sunrise-sunset) has a meaning from the naming of Kangin, which is the direction of the sun's rising and is considered the most important area. The rising of the sun means the start of a day, meaning the beginning of life with new energy. At the same time, Kauh is the direction of sunset which means the end of all daily activities, replaced with a night. This condition means that the direction of Kauh is a dirty area or a dirty place. Kaja is the direction of the Mountain, which is considered the highest direction and is considered the Main area. In contrast, Kelod is the direction of the sea, which is regarded as a direction that represents bad or dirty things and has a Nista value. Figure 1 shows an illustration of the cardinal orientations of the two axes and their relationship to the Tri Mandala.

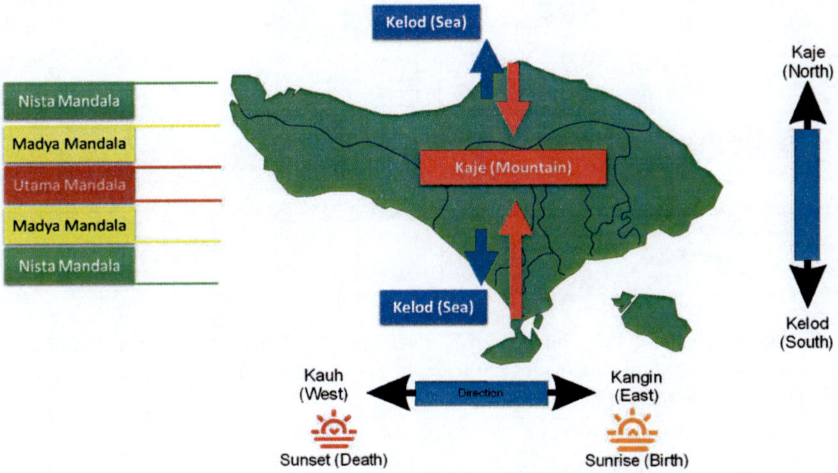

Fig. 1. The cardinal orientation of the two axes and the relationship on the Tri Mandala

Traditional Balinese architecture has a basis in development, namely determining functions and zones by using the Tri Angga concept as the vertical axis and the Tri Mandala concept as the horizontal axis so that the parts of the building have a high traditional value [28]. Vertically (the Tri Angga concept), facilities in Bali illustrate the human body, namely Utama Angga is identified with the head located at the top, the Madya Value is the body part is located in the middle. The legs are found below, which is called Nista. Horizontally (the Tri Mandala concept) forms zoning with a hierarchy of values corresponding to the natural axis's value, such as the upstream or inner part of Utama value, the middle part is of Madya value, and the downstream or outer part will be Nista. Figure 2(a) illustrates the use of the Tri Angga, Tri Bhuana, and Tri Mandala concepts.

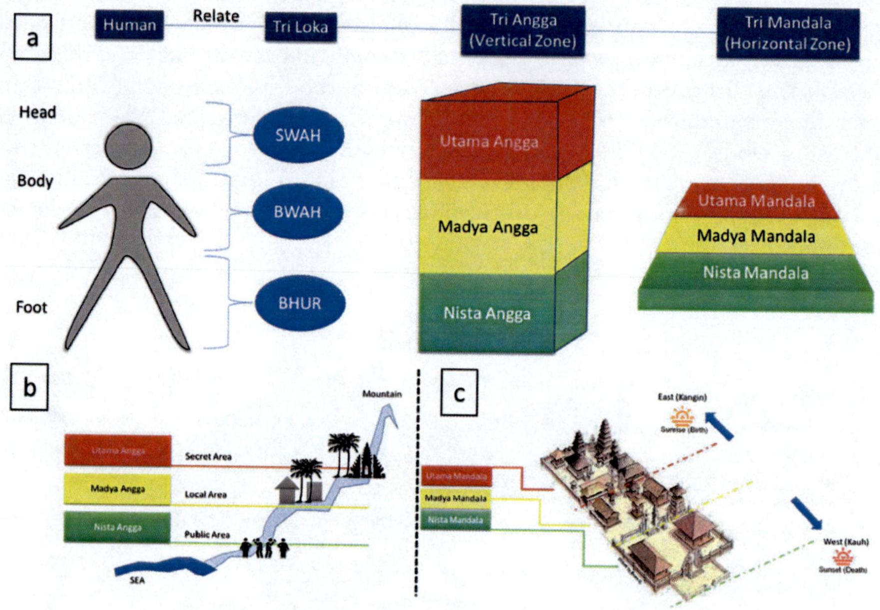

Fig. 2. (a) The use of the concept of Tri Angga, Tri Bhuana and Tri Mandala; (b) illustration of sea-mountain orientation with Tri Angga; (c) Illustration of kaje-kelod using Tri Mandala

Figure 2(b) shows an illustration of the regional scale. A mountain is a sacred place, so that it is in the Utama Mandala or a limited place is only allowed for prayer and spiritual purposes. A temple is a sacred place so that in the construction and establishment of the temple, it must follow several requirements with the structure of the temple building following the Tri Mandala concept. While in Fig. 2(c) is the pattern of Balinese community buildings, which are divided into three zones following the THK concept.

The regional regulation (Perda) of the province of Bali number 16 of 2009 concerning the spatial plan of the province of Bali states that the height of buildings that utilize air space above the earth's surface is limited to a maximum of fifteen (15) meters or the same as buildings with 4–5 floors. Non-commercial buildings can exceed 15 m, such as temples and public facilities. This is an effort made by the government to limit the population not to exceed the capacity given

Bali's relatively narrow and limited area. Apart from the government, this is the aspiration of the Balinese indigenous people who are very concerned about spatial planning and culture. Figure 3 illustrates the restrictions on building height according to Perda no 16/2009.

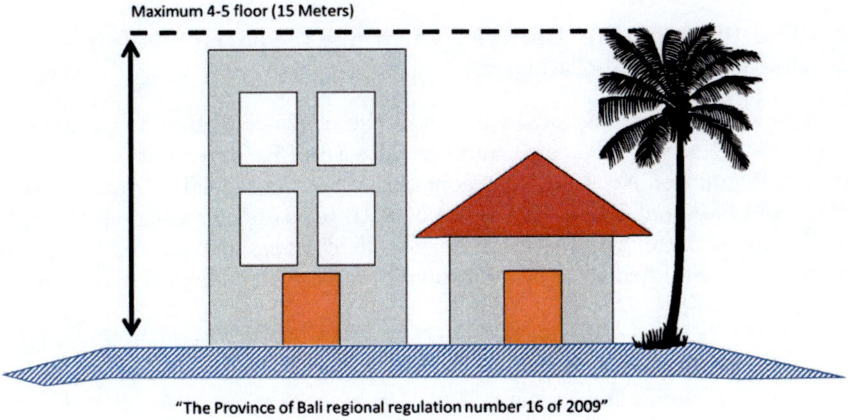

Fig. 3. Illustration of restriction on building height.

Penglipuran Village is a traditional village located in the Kubu Village area, Bangli District, Bangli Regency, Bali Province. The location of this village reaches 112 ha with an altitude of about 500–600 m above sea level. Penglipuran Village is one of the traditional villages well-known as a tourist destination in Bali because of its people who still preserve and practice local wisdom and traditional Balinese culture in their daily lives. The people of Penglipuran Village follow a customary rule known as Awig-Awig. This Awig-Awig embodies local knowledge, which is a strong foundation in good natural resource and environmental management and is customary law in the form of regulations or laws drawn up and stipulated by members of the village community, Banjar, and Subak regarding the rules of life. Society in the fields of religion, culture, and socio-economics in Bali.

In the spatial concept of the Penglipuran Traditional Village, the environmental arrangement is oriented towards the north. This is influenced by the old cultural heritage that places the north as the highest and sacred place. In this case, the symbolic orientation is Mount Batur. This mountain is located north of the village of Penglipuran, which they believe has magical and religious powers. So that the pattern of placement of village buildings always crosses from north to south with the north as a sacred part.

Penglipuran Village follows the development procedure according to the architectural rules that exist in Awig Awig. In addition to the outline of the Penglipuran Village landscape, the Tri Hita Karana concept is also applied in the arrangement of the Penglipuran Village community's house complex. The idea used is still in the division of Utama, Madya, and Nista zones. These zones also affect community activities, such as the Madya zone, which is the zone where most human activities are carried out, the Utama zone where people worship, and the Nista zone where people carry out activities such as putting livestock. An illustration of the zoning in Penglipuran Village can be seen in [8].

2.2 The Relationship Between Customary Spatial Planning and Customary Cadastre

The regulations governing local wisdom in the province of Bali are local regulations (Perda) No. 5 of 2005 concerning Architectural Requirements for Buildings, Regional Regulation No. 16 of 2009 concerning Spatial Planning for the Province of Bali, and Regional Regulation No. 4 of 2019 regarding Traditional Villages in Bali. These regulations represent all development, management, and living activities that are regulated to create harmony and can implement Tri Hita Karana in life (see Fig. 4).

Then, if it is related to 3D Cadastre and BIM and IFC modeling technology, local wisdom regulations that regulate all development and life activities require using the Tri Hita Karana concept. Where in it there are traditional and modern architecture with the orientation of Tri Mandala, Tri Angga, Sangga Dewata, and Tri Loka [26]. This confirms that in addition to spatial boundaries, there are also rules in life that provide information about mandala zones such as sections passed by guests or the public (Nista), fellow family members or relatives (Madya), and designated places (Utama).

In addition, the above information provides a new semantic inland technology where 2D Cadastre mapping does not provide detailed information regarding the use of the land. So it is necessary to make a transition to 3D Cadastre, which adds information related to 3R to local wisdom on a parcel and the building that is in it. For example, housing or apartments that a local wisdom policy can influence.

Fig. 4. The relationship between regulations, local wisdom and cadastral.

3 Results and Discussions

3.1 The Concept of Dynamic BIM and Sustainable Development Goals

Sustainable development has a positive impact on the Penglipuran Traditional Village. The creation of a prosperous life by reducing poverty and income inequality or unemployment, increasing food security, and improving physical and spiritual health due to the increasing number of tourists who want to visit Penglipuran village. Applying the Tri Hita Karana concept in the Penglipuran traditional village to organize its customary space can create a harmonious life and social welfare. For example, it can overcome various potential conflicts, disputes, and land cases in the future by following the applicable regulations and the existing land management system in the area and, on the other hand, implementing the government's strategic program "Complete Systematic Land Registration" (or known as PTSL) in traditional villages and with the Bali regional regulation No. 4. In 2019 concerning Traditional Villages in Bali, certificates with the subject of traditional villages have the right to have ownership of land rights. This can create an arrangement of life that is just and dignified with legal certainty that it can control, own, use, and utilize the land.

Over time, Penglipuran village has become a unique traditional village because it applies a spatial arrangement with the Tri Hita Karana concept in each of its land divisions and accepts structural and infrastructure modernization because the Penglipuran traditional village becomes a tourist village [4]. However, there is a threat that when fully taking modernization, Penglipuran village will no longer have popularity because it does not maintain its uniqueness. In addition, another impact is that rural communities can experience poverty economically, and residents prefer to work outside their area. From these problems, a liaison is needed to accommodate changes in the function and use of space before and after as well as providing historical and cultural information on ownership, so that it can make efforts to conserve customary space (see Fig. 5)

The image 6 explains the modification of the [11] theory, related to the relationship between land administration and sustainable development. The position in this research is on the application of customary cadastre, which is useful for the conservation of customary space and is carried out in a multipurpose cadastre with the help of a dynamic BIM model with spatial-temporal dynamics information. This is in the provision of land information infrastructure, which is modified with the theory of spatial system administrative unit [1] which supports and is in modern cadastral infrastructure. This customary cadastre is needed to provide legal certainty to indigenous peoples, know their rights, limitations, and responsibilities, and provide historical information on ownership and their traditional buildings and culture.

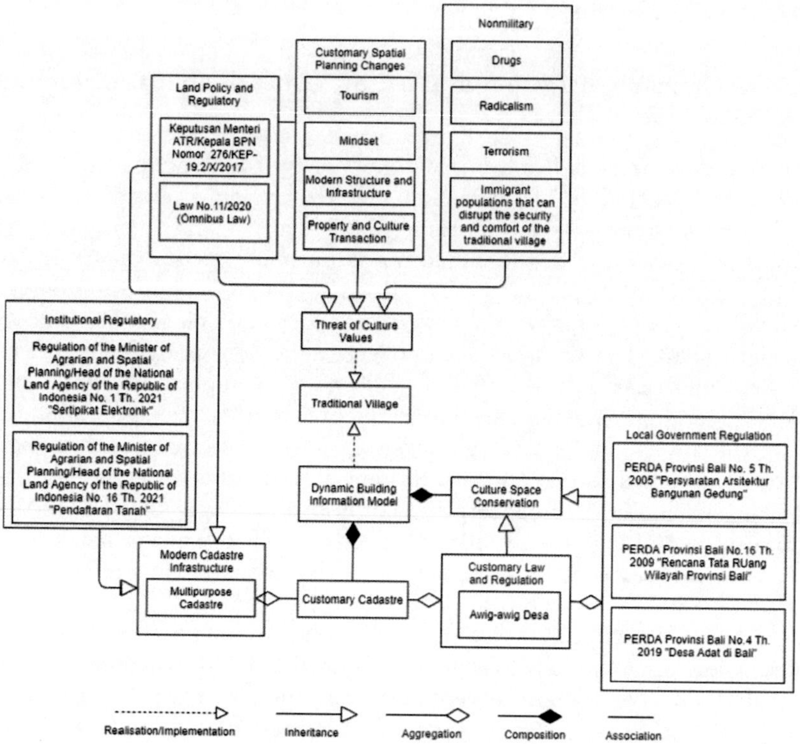

Fig. 5. Threats to the value of the traditional village

3.2 Implementing Dynamic BIM Concept for Customary Cadastre (Restrictions)

Implementing the customary cadastre in Penglipuran village can be done by using the Building Information Modeling approach. BIM can model, manage, and digital inventory data. Future research will implement Dynamic BIM, which adopts the R-tree model to facilitate effective data collection. In addition, it can be integrated with GIS to enable decision-making in giving sanctions to someone who builds a modern building exceeding 15 m. In addition, it can provide a rule that it is not allowed to pass or make on a family temple in the ownership of one plot of land. Figure 7 describes the concept of the process of determining the height limit of space and confirms the zone area in one ownership which contains three zones, namely Utama, Madya, and Nista mandala, with a BIM model that is integrated with R-Tree and GIS. The height of the space from the temple is not limited. The GIS function will provide a height limit of the information so that traditional villages such as Penglipuran village can maintain culture and accept modernization with certain limitations. Thus, Dynamic BIM can accommodate changes in the use and use of space based on applicable customary law and conservation in customary space.

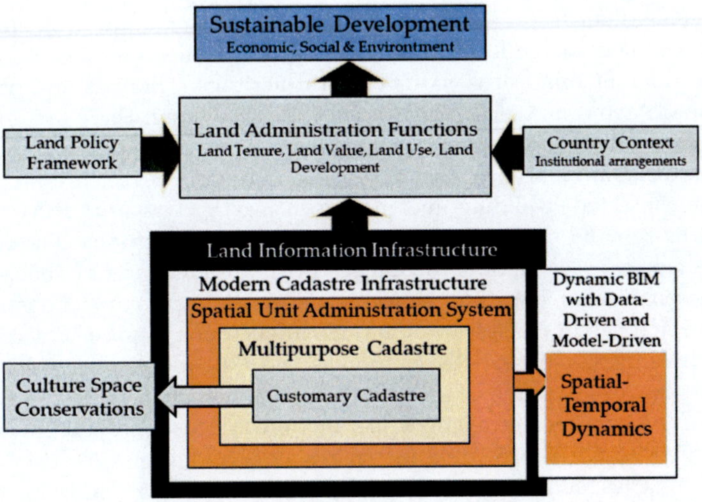

Fig. 6. The concept of dynamic BIM in customary cadastre towards sustainable development ([1,11,30] Integrated and Modified by Author)

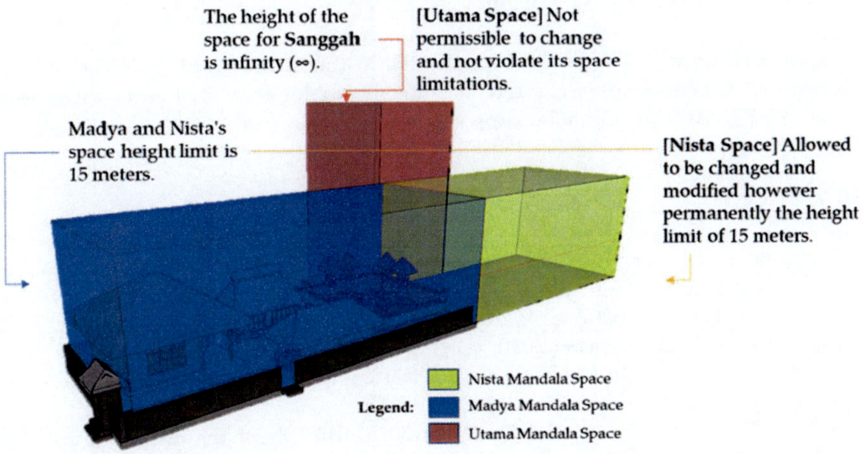

Fig. 7. The concept of setting boundaries related to the construction of structures and infrastructure.

4 Conclusion

Penglipuran traditional village is unique and attractive to tourism because it can maintain its culture and customs by presenting old buildings and accepting modernization in structure (guest-houses, restaurants, souvenir shops) and infrastructure (internet/wifi, modern facilities, and a combination of modern development). However, with this beauty, various potential threats are felt in traditional villages, such as economic, social, and environmental problems (SDGs) there are changes in modernization. In addition, other threats can be in the form of a different mindset towards cultural heritage buildings and dependence on tourism to avoid financial poverty. On the other hand, there are government policies related to investors and mass certification. Strengthened by local regulations, traditional villages in Bali have legal certainty over their land rights and cannot be contested. Building and implementing the Dynamic BIM model can become a liaison and bridge the existing problems and solutions. The technology it will present can accommodate changes in the use and use of space-based on applicable customary law. The concept can apply customary cadastre and be used as a reference in conservation actions in customary space. Future research will prioritize using existing R-tree models and GIS to be integrated into BIM so that it can be dynamic. The definition of dynamic here is (1) being able to provide information on changes in function and use of space in a customary administration, (2) providing information on land administration related to rights, boundaries, and responsibilities, (3) providing changes in geometric shape after and before using the BRep model, (4) retrieve data quickly and efficiently by adopting the existing R-tree model, (5) make decisions regarding restrictions on the construction of structures and infrastructure under the applicable Customary Law by utilizing GIS technology.

Acknowledgements. The authors would like to thank the grant from the Ministry of Education, Culture, Research, and Technology, Republic of Indonesia with Decree No. 8/E1/KPT/2021 for financial support.

References

1. Abdulharis, R., Sarah, K., Hendriatiningsih, S., Djunarsjah, E., Hernandi, A.: Spatial unit administration: an administrational framework of land, water and air space unit towards sustainable development of archipelago state? In: The XXIV FIG International Congress 2010, pp. 1–20 (2010)
2. Agustina, M.D.P.: Optimalisasi desa wisata untuk meningkatkan kesejahteraan masyarakat (2021)
3. Andari, R.N., Ella, S.: Model Desa Cerdas Untuk Membangun Indonesia Maju. Syiah Kuala University Press (2021)
4. Andayani, A.A.I., Martono, E., Muhamad, M.: Pemberdayaan masyarakat melalui pengembangan desa wisata dan implikasinya terhadap ketahanan sosial budaya wilayah (studi di desa wisata penglipuran bali). Jurnal Ketahanan Nasional **23**(1), 1–16 (2017)
5. Atazadeh, B., Kalantari, M., Rajabifard, A.: Comparing three types of BIM-based models for managing 3d ownership interests in multi-level buildings (2016)
6. Atazadeh, B., Rajabifard, A., Zhang, Y., Barzegar, M.: Querying 3d cadastral information from BIM models. ISPRS Int. J. Geo Inf. **8**(8), 329 (2019)
7. Delgado, J.M.D., Butler, L., Brilakis, I., Elshafie, M., Middleton, C.: Structural performance monitoring using a dynamic data-driven BIM environment. J. Comput. Civ. Eng. **32**, 04018009 (2018)
8. Dwijayasastra, N.: Kajian arsitektur hijau desa adat penglipuran bali (2013)

9. Eastman, C.M., Eastman, C., Teicholz, P., Sacks, R., Liston, K.: BIM Handbook: A Guide to Building Information Modeling for Owners, Managers, Designers, Engineers and Contractors. Wiley (2011)
10. El-Mekawy, M., Paasch, J.M., Paulsson, J.: Integration of 3d cadastre, 3d property formation and BIM in Sweden. In: 4th International Workshop on 3D Cadastres, Dubai, United Arab Emirates, 9–11 November 2014, pp. 17–34 (2014)
11. Enemark, S., Williamson, I., Wallace, J.: Building modern land administration systems in developed economies. J. Spat. Sci. **50**(2), 51–68 (2005)
12. Hendriyati, L.: Upaya masyarakat di desa wisata penglipuran dalam menjalankan sapta pesona. J. Tourism Econ. **3**(1), 49–57 (2020)
13. Ho, S., Rajabifard, A., Stoter, J., Kalantari, M.: Legal barriers to 3d cadastre implementation: what is the issue? Land Use Policy **35**, 379–387 (2013)
14. Iskandar, A.H.: SDGs DESA: Percepatan Pencapaian Tujuan Pembangunan Nasional Berkelanjutan. Yayasan Pustaka Obor Indonesia (2020)
15. Mahagangga, I.G.A.O., Suryawan, I.B., Nugroho, S., Sudana, I.P.: Pemetaan jalur paket wisata pedesaan di desa wisata penglipuran, kecamatan bangli, kabupaten bangli. Jurnal Udayana Mengabdi **15**(2), 1–6 (2016)
16. Liebich, T.: IFC4-the new buildingSMART Standard. AEC3 Ltd., High Wycombe, UK (2013)
17. Ma, Z., Ren, Y., Xiang, X., Turk, Z.: Data-driven decision-making for equipment maintenance. Autom. Constr. **112**, 103103 (2020)
18. Maylinda, E.: Pembangunan Pariwisata Berkelanjutan Berbasis Kearifan Lokal Di Desa Adat Osing Kabupaten Banyuwangi Provinsi Jawa Timur. Ph.D. thesis, IPDN Jatinangor (2021)
19. Mordue, S., Swaddle, P., Philp, D.: Building Information Modeling for Dummies. Wiley (2015)
20. Olfat, H., Atazadeh, B., Shojaei, D., Rajabifard, A.: The feasibility of a BIM-driven approach to support building subdivision workflows-case study of Victoria, Australia. ISPRS Int. J. Geo Inf. **8**(11), 499 (2019)
21. Parwata, I.W.: Rumah tinggal tradisional bali dari aspek budaya dan antropometri. Mudra Jurnal Seni Budaya **26**(1), 95–106 (2011)
22. Primadi, I.G., Adhimastra, I.N.: Pengaruh moderenisasi terhadap rumah tinggal tradisional di desa adat penglipuran kabupaten bangli. Jurnal Anala **5**(1), 12–28 (2017)
23. Sadikin, A.: Analisis Hukum Internasional Terkait Deforestasi Dan Hak-Hak Masyarakat Adat Hutan Amazon Di Brazil. Ph.D. thesis, Universitas Hasanuddin (2021)
24. Stoter, J., Ploeger, H., van Oosterom, P.: 3d cadastre in the Netherlands: developments and international applicability. Comput. Environ. Urban Syst. **40**, 56–67 (2013)
25. Suhari, K.T.: Pengembangan Building Information Model untuk Kadaster Tiga Dimensi Berdasarkan Kearifan Lokal dengan Konsep Tri Hita Karana di Bali. Master's thesis, Geodesy and Geomatics Engineering, Faculty of Earth Sciences and Technology, Institut Teknologi Bandung (2020)
26. Suhari, K.T., Saptari, A.Y., Laksono, B.E., Saputra, N.R.M., Gunawan, P.H.: Implementation of 3d cadastre with indigenous knowledge concept of "tri hita karana" in Bali Island. Indonesia. Int. Arch. Photogram. Remote Sens. Spat. Inf. Sci. **XLII–4/W16**, 615–619 (2019)
27. Sun, J., Mi, S., Olsson, P.O., Paulsson, J., Harrie, L.: Utilizing BIM and GIS for representation and visualization of 3d cadastre. ISPRS Int. J. Geo-inf. **8**(11), 503 (2019)

28. Susanta, I.N.: Makna dan konsep arsitektur tradisional bali dan aplikasinya dalam arsitektur bali masa kini. SPACE **4**(2), 199–212 (2017)
29. Wahana, N.P.P.D., Sari, S.M., Rakhmawati, A.: Wujud ajaran tri hita karana pada interior pura agung jagad karana surabaya. Intra **3**(2), 520–530 (2015)
30. Williamson, I.P., Enemark, S., Wallace, J., Rajabifard, A.: Understanding land administration systems. Coordinates **4**(11), 29–32 (2008)
31. Yasmine, B., Putri, B., Chandra, N., Zuhdi, R.: Konsep sustainability tourism dengan tema smart cultural tourism pada desa penglipuran, bali. MATRAPOLIS: Jurnal Perencanaan Wilayah dan Kota **2**(1), 7–15 (2021)
32. Ying, S., Guo, R., Li, L., He, B.: Application of 3d GIS to 3d cadastre in urban environment (2012)

Building an Agent Patient for an Agent-Based Model of the Functioning of a Medical Center During a Pandemic Outbreak

Vladimir A. Sudakov[1], Yurii P. Titov[1(✉)], and Elena A. Davydkina[2]

[1] Plekhanov Russian University of Economics, Stremyanny lane, 36, Moscow 117997, Russian Federation
sudakov@ws-dss.com, kalengul@mail.ru

[2] Moscow Aviation Institute, Volokolamskoe sh, 4, Moscow 125080, Russian Federation

Abstract. Pandemics, large-scale outbreaks of diseases with serious economic, social and political consequences. Decisions made in a pandemic era have a significant impact on the spread of the disease. It is very difficult to determine the consequences of this or that decision, since the situation is changing rapidly. In addition, there is a constant shortage of various resources, human and medical. The article considers the agent "Patient" for use in the agent-based model of the functioning of a medical center. Not only its states are proposed, but also rules of transitions and decision making. To make decisions, fuzzy sets are used that describe the rules and approaches of the patient's reaction to the appearance of various symptoms in him. This approach makes it easy to modify the model when symptoms change. For the introduction of additional conditions associated with political and social decisions, a composite event is introduced, which determines the entire set of patient decisions. So for the SARS-Covid-19 pandemic, conditions such as "Do nothing", "Go to the pharmacy", "Contact the clinic" and "Call an ambulance" are added to "Test", "Vaccine". Within the framework of the proposed model, a separation of the patient's response and various assumptions appears. Based on the results of the model's work, it is proposed to determine the required number of medical personnel of various qualifications, to take into account the need for retraining, to determine the number of necessary wards and hospitals, as well as expendable resources and personal protective equipment.

Keywords: Agent-based simulation model · Decision making · Fuzzy sets · Medical center model

1 Introduction

Throughout world history, the world's population has experienced many infectious epidemics and pandemics. At the end of the 20th century, due to the development of public institutions, industry and science, it seemed that the development of infectious pandemics was impossible in developed countries. But in 2019, the world community faced a new infectious disease SARS-Covid-19. The disease quickly spread throughout the

R. Silhavy (Ed.): CSOC 2022, LNNS 501, pp. 37–47, 2022.
https://doi.org/10.1007/978-3-031-09070-7_4

world, despite measures taken at the earliest stages. Although SARS-Covid-19 is still a serious problem for global society, it is worth considering the vulnerability of even modern society to outbreaks of new infectious diseases. It is proposed to develop a model describing the processes of combating the disease at the global economic level. [1–4] Within the framework of the work, a model of the functioning of a medical institution is considered.

The medical center is based on the regular work of medical personnel in the diagnosis and treatment (mainly procedural and medication) of various diseases in the surrounding population. According to the results of diagnostics, inpatient, outpatient treatment of the patient is also possible, most often with a move to another institution. But during a pandemic, this interaction pattern is subject to changes that should be taken into account when creating a simulation model [5, 6].

- Changes in the rules for diagnosing a disease. These are regular and population-wide tests carried out in separate premises. Establishing a preliminary diagnosis based on the results of one test and the use of non-drug treatment. [7]
- Increased risk of contamination of medical personnel. This feature determines the needs of medical personnel in protective equipment. Pandemics are characterized by some panic and reinsure of medical personnel in relation to personal protective equipment. [8, 2]
- Alternative treatment options. Various hypotheses related to the rules of treatment and patient care are being developed and refuted quite quickly. At the same time, there is a need for continuous training of medical personnel. As a result, many alternative treatments for the disease are possible, with varying efficacy.
- Creation of separate centers. In almost all pandemics, humanity has tried to isolate sick people and medical personnel in contact with infected people. For this, organizational measures were taken to create separate medical centers for pandemic patients.
- Increasing the degree of protection of medical personnel and stricter requirements for patients. Since accidental contact with the patient is possible, the degree of protection also increases for medical personnel who are not directly related to the treatment of the pandemic. There are additional requirements for ordinary citizens. So, for example, with SARS-Covid-19, an increase in body temperature led to a person's isolation.
- Concomitant diseases and the period of rehabilitation of patients. Many patients were able to recover from the pandemic, but, due to weak, post-hospital immunity, they fell ill with concomitant diseases.
- Panic in society and, including among medical personnel. The increase in the number of patients also increases the likelihood of accidentally contracting the disease while moving around the city. This creates additional panic and distrust of the people around you. In the situation with SARS-Covid-19, the situation worsened when an asymptomatic manifestation of the disease was discovered. As a result, many people hid the disease, or did not even know about it.

The developed simulation model should also take into account various mutations of the virus. Due to the large number of sick people during a pandemic, mutations of the virus are quickly detected. If in a normal disease such mutations would lead to isolated cases, then in a pandemic these mutations lead to a sharp increase in the number of cases.

These mutations can cause other symptoms and require additional testing, as well as the possibility of triage of patients to different centers. These features require taking into account the need for rapid modification when developing a model.

The proposed model of a medical center is necessary to build a set of models for the rational use of resources in order to combat global outbreaks of pandemics. The considered medical center is a kind of universal model that does not take into account the movement of medical personnel and patients in the center, small time delays. When developing a model, it is necessary to take into account the possibility of integrating it into other simulation models to assess the global parameters of the pandemic counteraction system. [9] At the same time, most of modern research is aimed at building agent-based models of human interaction and the spread of infection [10–12].

2 Methods

It is proposed to build a simulation model on the basis of an agent-based approach, since it will best describe the interaction of various people during a pandemic and is simply expanding. Within the framework of the simulation model of the medical center, it is proposed to use 2 classes of agents - people (patients) and medical personnel.

Patients form a Marsk input stream for a medical center, but for the convenience of integration with other models, a separate agent should be used that has states: healthy and the presence of symptoms of a certain disease. The transition between these states in the medical center model is carried out in the form of a probabilistic transition to ensure the flow of human visitors. To simulate this flow, it is necessary to simulate the appearance of symptoms in each person at every moment of time. This approach requires the process of advancing the model time with some fixed, small step. But, since the input flow to the medical center is Markov, it is possible to generate the time of onset of symptoms in each individual person using the corresponding distribution, for example, exponential. The parameter of such a distribution can be the intensity of the number of visits to the medical center.

The onset of symptoms does not mean that a person will decide to go to a medical center. Some symptoms will be treated with medication or expectantly, without going to a medical center. Other symptoms require going to a medical center for additional tests, and some call a special service at home. Since such approaches interact in different ways with medical personnel, then at the stage of the onset of symptoms, a person needs to make a decision about further actions.

If a person has symptoms that require contacting a medical center, then after a certain time (depending on the severity of symptoms), which is generated randomly (preferably using the normal distribution law), the person enters the medical center for further research. However, the number of studies may differ depending on the symptoms present. So, for example, in Russia, an appeal is made to a therapist, who can immediately write out a sick leave and a prescription for medicines, or he can send it to a profile specialist who conducts additional research. Based on the results, the patient is prescribed either drug treatment, most often with home isolation, or outpatient (inpatient) treatment. The departure of the medical team most often resolves the issues of visiting the clinic and can, both prescribe medication, and transport to the hospital.

In stationary conditions, the acquisition of additional diseases is possible. Although the likelihood of this phenomenon is significantly reduced with constant processing of premises and compliance with sanitary standards, it is not zero, since there is contact with other sick people. In addition, inpatient conditions are characterized by constant contact with medical personnel, doctors of various profiles. Various personal protective equipment (PPE) are used to protect medical personnel. PPE is divided into categories or classes (FFP1, FFP2 and FFP3), depending on the infectious diseases of the patients [13–16].

To formalize the process of modeling the behavior of the agent "Patient", let us single out the following events (Fig. 1). These events determine the patient's transition from one state to another. The "Symptoms appeared" event is a composite event, since it is recommended to define a class of possible symptoms for each disease. In this case, the type of disease can be modeled first, and then the time of occurrence of the event in accordance with the received type. An alternative option is to generate the time of occurrence of all symptoms and select only one symptom from the composite event.

After the onset of symptoms, a decision-making process takes place about the subsequent actions of the patient. The decision-making procedure is proposed to be built on the basis of fuzzy judgments. [17–21] This approach will take into account expert assessments that determine one of the possible management decisions for various symptoms. For example, at different body temperatures measured by the patient, decisions about action may be unclear. In this case, it is necessary to set membership functions for actions: do nothing, medication, go to a medical center, call an ambulance for each of the symptoms. When using fuzzy functions, qualitative indicators can serve as a carrier set (Fig. 2). In this case, you can set actions only when one symptom occurs. If the disease is manifested by several symptoms, then it is necessary to obtain a generalized membership function of the patient's actions. Generalization is carried out according to the formulas:

$$\mu_{n+1,j} = \mathrm{Min}(1; \sum_{i=1}^{n} \mu_{i,j}) \tag{1}$$

where $\mu_{n+1,j}$ – is the value of the j-th point of the generalized membership function; n is the number of symptom membership functions; $\mu_{i,j}$ - is the value of the j-th point of the i-th membership function.

After generalizing the membership functions, it is necessary to determine the probabilistic distribution of the patient's actions and generate a random event "patient's decision". The probability distribution can be obtained from the normalization of the values of the membership function for all actions.

$$P_j = \mu_{n+1,j} / \sum_{js} \mu_{n+1,js} \tag{2}$$

where, $\mu_{n+1,j}$ – the value of the j-th solution of the generalized membership function; P_j - the probability to choose the j-th solution.

All subsequent events, although they require the direct consent of the patient, are determined by the medical staff. The final diagram of transition states and events is shown in Fig. 1.

In the event of a pandemic disease, more often than not, work with this disease is not integrated into the overall picture. First, certain symptoms dramatically change the patient's behavior. With SARS-Covid-19, an increase in temperature and loss of taste and smell immediately led to treatments. Secondly, personnel and infrastructure were clearly divided according to diseases. So, for SARS-Covid-19, separate blocks, buildings and even entire hospitals are allocated. Medical professionals are dedicated to fighting the disease. These measures are designed to reduce the spread of infection.

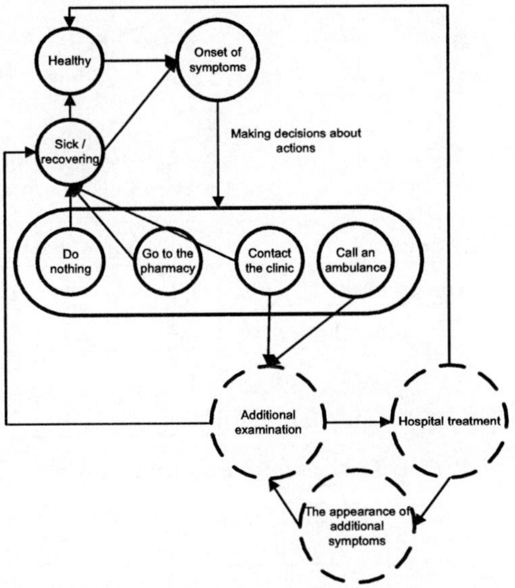

Fig. 1. Graph-diagram of the relationship of events for the agent "Patient"

Thirdly, the likelihood that there will be enough medical centers and medical personnel to fully cover the population during a pandemic is extremely small. As a result, alternative ways of interaction between patients and medical personnel are being developed. These are both telephone and video consultations. The ability to make a decision without interacting with a specialist. SARS-Covid-19, tests were developed that make it possible to make a decision about isolating a patient with certain symptoms, with the assumption that the disease will be defeated by the patient's immunity. In such cases, the self-isolated patient monitored his health and, in case of deterioration, called the medical service. At the same time, the state controls the implementation of the measures taken.

As a result, in the case of a pandemic, the graph-diagram of the relationship between states and events for the agent "Patient" will look differently (Fig. 3). An additional solution "test" appears. The rules for referring to personnel are divided. It should also be noted that the staff themselves, i.e. agents vary.

Vaccination allows you to reevaluate decision making for different symptoms, as well as reduce the risk of certain symptoms of the disease.

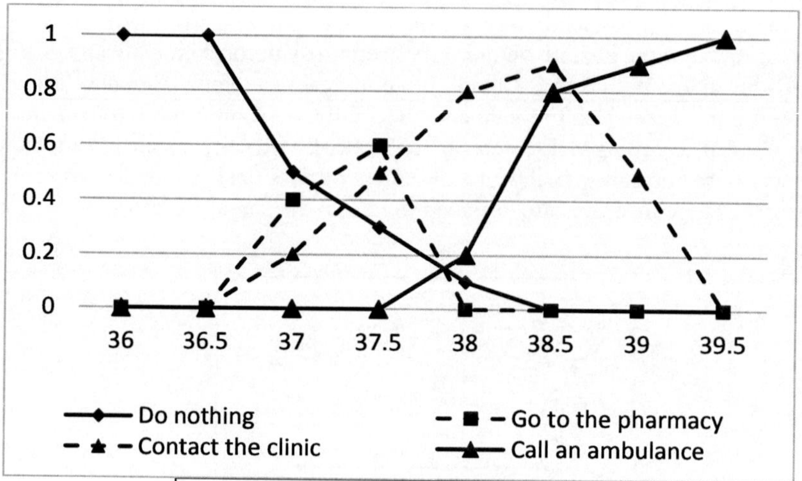

Action	Have a cough	No cough
"Do nothing"	0,2	1
"Go to the pharmacy"	0,7	0
"Contact the clinic"	0,5	0
"Call an ambulance"	0,1	0

Fig. 2. Fuzzy set of actions of patients with various symptoms.

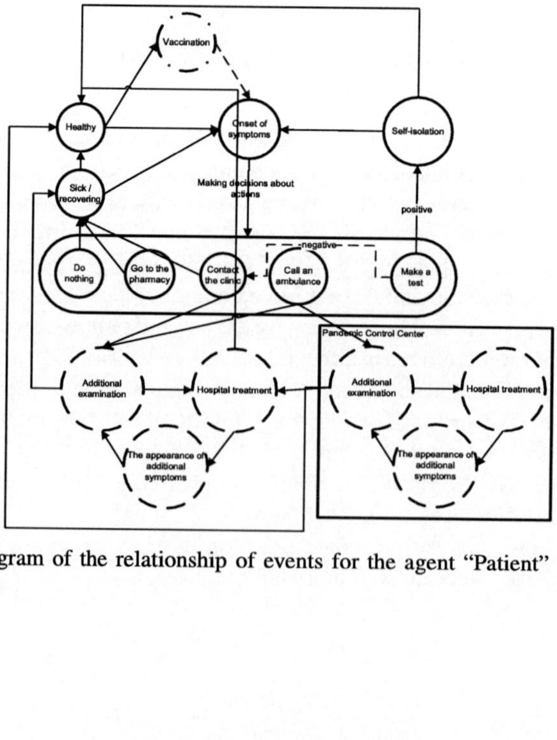

Fig. 3. Graph-diagram of the relationship of events for the agent "Patient" in the presence of pandemic diseases

3 Results

To implement the simulation model, it is proposed to apply a discrete event-driven approach. [22] The modeling process is carried out by advancing the model time and changing the state of the system. To advance the model time, it is necessary to implement the "List of future events" and the basic procedures for working with it: adding a new event and extracting the first (critical event from the list). The various events in Fig. 3 are represented as classes that inherit from the base event class. The list of future events contains events of various classes. But they are all ordered by the time of the accomplishment. Each event has a unique handler that changes the state of the model. The algorithm of the model is shown in Fig. 4.

In event handlers, new events are scheduled for the list of future events and the state of the agent changes. The agent is represented as an object to which there is a pointer in the event. The mapping of objects and basic data structures in the system is shown in Fig. 5. New events are planned in accordance with the arcs of the graph shown in Fig. 3. The event "symptoms appear" is scheduled when processing the events "Healthy", "Self-isolation" and "Sick/recovery "According to the flow of diseases (Markov flow). Similarly, the transition in the "Healthy" state from the "Sick/Recovery" state is determined by the Markov flow. At the same time, it is possible to obtain the corresponding time parameters from other models of the spread of infections. The transition from the "Self-isolation" state to the "Healthy" state is carried out by directive, depending on the regulations, for example, 2 weeks. The decision on what to do with the onset of symptoms and vaccination is made based on the parameters of the specific agent and its rules. And the planning of the time to change the state is carried out based on the geographical availability of the required establishments. The transition to the states "Additional examination", "Inpatient treatment" and "The appearance of additional symptoms" is determined by the work of other agents and the medical center, therefore, the transitions are carried out by receiving appropriate messages from other agents.

The system: distribution laws for transitions from one state to another, geographic location, the agent's confidence in the need for vaccination and the frequency of tests for the disease. Also, the agent has fuzzy sets for making decisions. These sets are specified in the form of arrays of points consisting of pairs: value of the membership function, parameter. This information is used to determine the action, and the corresponding event, that the agent will take when symptoms appear. The agent is easily typed by means of certain typical sets of parameters.

To implement the model, software has been developed in the C # language. It should be noted that the implementation of this model in other programming languages is not easy. When simulating in the system, a sufficiently large number of agents, about 100,000, of various types were used. During the modeling, all agents were divided into 10 types according to their preferred behavior, and the agent was assigned to a certain type according to the probability distribution. Such a large number of agents allows us to consider the work of a medical center in conditions close to reality. Nevertheless, the memory consumption was not catastrophic due to the fact that the agents had no more than 3 events in the list of future events. The adequacy of the model was checked on the basis of tracing printing and varying parameters.

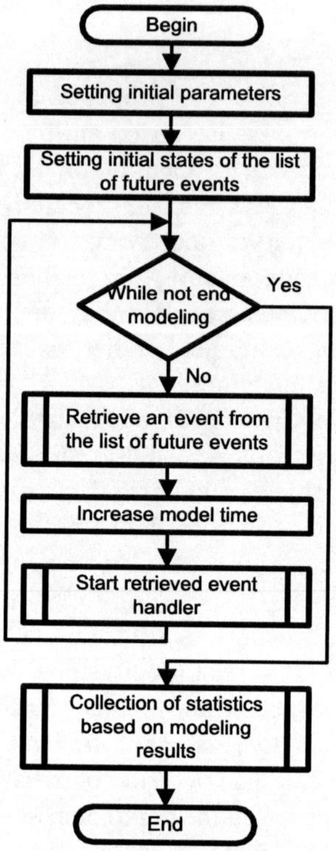

Fig. 4. Algorithm of the agent-based model based on the event-based approach

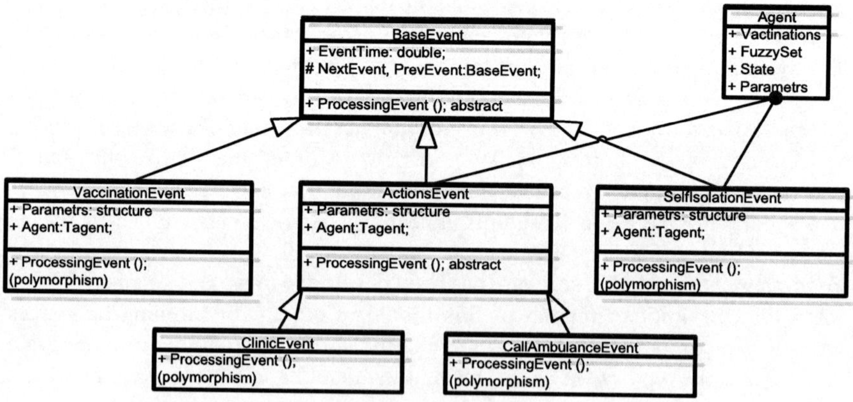

Fig. 5. Description of the class diagram of the developed software

At the moment, the program has defined fuzzy sets for 12 symptoms characteristic of SARS-Covid-19. The software highlights 4 more diseases with similar symptoms: adenovirus, rhinosincial infection, influenza, adeno flu.

The agent object "patient" has parameters that determine the behavior of the agent in.

4 Discussion

As part of the work, a fairly simple agent-based patient model has been implemented. Due to the use of fuzzy sets, it was possible to get rid of a large number of agent events, which makes it easy to check the adequacy of the resulting simulation model. The proposed model is easily modified for various administrative rules and virus mutations. But this approach imposes restrictions on the accuracy of the specified parameters, making the model unstable with respect to them. Nevertheless, the presence of many studies on this topic makes it possible to collect fairly accurate values, both for membership functions and for system parameters.

Further development of the proposed model consists in creating a complex of models that interact with each other and allow describing the processes taking place in the medical center. It is possible to expand it by intelligent agents capable of analyzing the environment themselves. To do this, you need to use the fixed step method and create an event at which the agents are polled for making or changing decisions. Such an event will be generated at regular intervals, and agents need to write a handler for this event.

At this stage, in addition to developing a model of a medical center, the possibility of integrating this model with models of the spread of infections is being considered in order to accurately determine the moment of onset of symptoms and obtain more accurate results.

5 Conclusion

The paper presents a description of the "Patient" agent for a simulation model of a medical center operating during a pandemic. This model takes into account many features and allows flexible configuration of the agent. The main setting of the behavior of the "Patient" agent is to set the rules for making decisions about their actions when various symptoms appear. For the convenience of setting rules, fuzzy sets are used to determine the behavior of an agent not only for single symptoms, but also for a complex of different symptoms. The disease model, i.e. the appearance of symptoms in the "Patient" agent is planned to be taken from higher-level models or statistical data if the model of the medical center works autonomously. It also becomes possible to quickly modify the model by changing the fuzzy sets that describe the rules of the agent's behavior. This feature makes it easy to readjust the model for different mutations of the disease.

The emergence of vaccinations against a pandemic is also easily reflected in this system. It is necessary to change the likelihood of some symptoms and the rules for making decisions when they appear. So, for example, with SARS-Covid-19 after vaccination, contacting the center for testing for a pandemic practically does not occur when the temperature rises.

Based on the decision-making results, one event is implemented, which are part of a composite event, which includes such actions of the patient as "Do nothing", "Go to the pharmacy", "Contact the clinic", "Call an ambulance" and "Make a test." At the same time, for some decisions and determination of further conditions, interaction with the "Medical Worker" agent is required, who is responsible for further decision-making.

It should also be noted that the health center model aims to address resource management issues in order to tackle global pandemic outbreaks. The proposed models are necessary to determine the volume of contacts of medical workers, the volume of workers involved in the fight against a pandemic, as well as to analyze the consequences of applying different levels of security and organizational decisions.

Acknowledgments. The reported study was funded by RFBR and CNPq, FASIE, DBT, DST, MOST, NSFC, SAMRC according to the research project № 20-51-80002.

References

1. Sureka, B., Nag, V.L., Garg, M.K., et al.: Rational use of PPE and preventing PPE related skin damage. J. Fam. Med. Primary Care, (January 2021). https://doi.org/10.4103/jfmpc.jfmpc_1772_20
2. Sasidharan, S.: PPE -A hindrance to therapeutic alliance! article in Turkish. J Emerg. Med. March (2021). https://doi.org/10.5152/TJAR.2021.883
3. Personal protective equipments (PPEs) for COVID-19: a product lifecycle perspective. Int. J. Prod. Res. May (2021). https://doi.org/10.1080/00207543.2021.1915511
4. Sciacca, D.: A dynamic supply chain network for PPE during the Covid-19 pandemic. J. Appl. Numer. Optim. (2021). https://doi.org/10.23952/jano.3.2021.2.09
5. Aly, S., Talutis, S.D., et al.: The Boston Medical Center coronavirus disease 2019 (COVID-19) Procedure Team: optimizing the surgeon's role in pandemic care at a safety-net hospital. Surgery **168**(3), p. 404–407 (2020). https://doi.org/10.1016/j.surg.2020.05.030
6. Rubin, R.: COVID-19's crushing effects on medical practices, some of which might not survive. JAMA **324**, 321–323 (2020)
7. Benaderette, S.: Tests Covid-19: priorisation toute ! Option/Bio **31**(621–622), 1–3 (2020). https://doi.org/10.1016/S0992-5945(20)30184-7
8. Kotian, R.P., Debnath, M., Zosangliani, D'souza, B., Faujdar, D.: A web survey to assess the use efficacy of personnel protective materials among allied health care workers during COVID-19 pandemic at North-East India medRxiv 2020.06.08.20125708. https://doi.org/10.1101/2020.06.08.20125708
9. Sudakov, V.A., Sivakova, T.V.: Applying an agent-based approach to modeling COVID-19. In: Silhavy, R., Silhavy, P., Prokopova, Z. (eds.) CoMeSySo 2021. LNNS, vol. 232, pp. 913–919. Springer, Cham (2021). https://doi.org/10.1007/978-3-030-90318-3_72
10. Abdulla, M.: Alsharhan survey of agent-based simulations for modelling COVID-19 pandemic advances in science. Technol. Eng. Syst. J. **6**(2), 439–447 (2021). https://doi.org/10.25046/aj060250
11. Wolfram, C.: An agent-based model of COVID-19. Complex Syst. **29**(1), pp. 87–105 (2020). https://doi.org/10.25088/ComplexSystems.29.1.87
12. Li, T., Liu, Y., Li, M., Qian, X., Dai, S.Y.: Mask or no mask for COVID-19: a public health and market study. PLoS ONE **15**(8), e0237691 (2020). https://doi.org/10.1371/journal.pone.0237691
13. Zhang, M., et al.: Knowledge, attitude, and practice regarding COVID-19 among healthcare workers in Henan. China. J Hosp Infect. **105**(2), 183–187 June 1, 2020

14. Roy, D., Tripathy, S., Kar, S.K., Sharma, N., Verma, S.K., Kaushal, V.: Study of knowledge, attitude, anxiety & perceived mental healthcare need in Indian population during COVID-19 pandemic. Asian J Psychiatr. **1**(51), 102083 June 2020
15. Zhong, B.L., et al.: Knowledge, attitudes, and practices towards COVID-19 among Chinese residents during the rapid rise period of the COVID-19 outbreak: a quick online cross-sectional survey. Int J. Biol. Sci. **16**(10), 1745–1752 (2020)
16. Zhao, J., et al.: Antibody responses to SARSCoV-2 in patients of novel coronavirus disease 2019. Clin. Infect. Dis. (2020)
17. Tavares, F.F., Betti, G.: The pandemic of poverty, vulnerability, and COVID-19: evidence from a fuzzy multidimensional analysis of deprivations in Brazil. https://doi.org/10.1016/j.worlddev.2020.105307
18. Mandla, M., Sharma, N.: Fuzzy modelling of clinical and epidemiological factors for COVID-19. https://doi.org/10.21203/rs.3.rs-29370/v1
19. Clemente-Suárez, V.J., et al.: Performance of fuzzy multi-criteria decision analysis of emergency system in COVID-19 pandemic. An Extensive Narrative Review. https://doi.org/10.3390/ijerph18105208
20. Van Khanh, T., Kim Ngan, V., Van Tam, V.: Knowledge and attitude toward COVID-19 among healthcare workers at District 2 Hospital, Ho Chi Minh City. Asian Pac. J. Trop. Med. [cited 2020 Jun 8]; 13(March):6–11 (2020). http://www.apjtm.org
21. Osipov, V., Posadskii, A., Sivakova, T.: Methods of rational decision making in multi-agent systems for evaluating the effectiveness of innovations. AIP Conf. Proc. **2181**, 020031 (2019). https://doi.org/10.1063/1.5135691
22. Khakhulin, G.F.: Fundamentals of designing simulation models: a Tutorial. - 2nd edition. Moscow NPK "Potok", pp. 228 (2002)

The Necessity of Information Extraction from Big Data Systems for the Purpose of Business Process Optimization

Kamelia Shoilekova(✉) and Boyana Ivanova

Angel Kanchev University of Ruse, 8 Studentska Street, 7000 Ruse, Bulgaria
{kshoylekova,bivanova}@uni-ruse.bg

Abstract. The paper analyzes the role of big data as a competitive asset for business decision making. It presents a compiled classification of the basic Big Data sources. The focus is laid on the main characteristics of big data and most frequently applied data processing technologies that would lead to business processes optimization.

Keywords: Big Data · Business process · 5Vs model

Our everyday dealings are marked with the usage of more and more electronic devices while business processes are being further digitalized which is a prerequisite for accumulation of big data sets both from the field of economy, and from the personal and social life of people.

The European Commission announced prognosis information that the total amount of data on world's scale will grow up six times within 2018–2025 [1].

1 Big Data Arrays as Part of the World Around Us

Over the last few years, IT specialists and business analyzers focused their attention on big data processing, side by side with the cloud services and object virtualization, because big data proved its role as a significant asset to firms' strategies in the modern competitive environment [12]. Big data systems help firms or business organizations develop insights, which may affirm or destruct them.

Big data analysis allow a business organization make better decisions, detect and predict changes and become aware of future opportunities. Big data reveals the limitations of the existing strategies for knowledge extraction confronting us with different unknown situations connected to the availability of vast data amounts [10, 11]. Despite of the serious efforts made in the field of big data processing, we must agree that much of work has still to be done for overcoming the challenges related to data heterogeneity, stability, velocity, preciseness, veracity, confidentiality, and instructive value. Information retrieval is the capability to extract useful knowledge from big data/streams that was not possible up to now, because of the huge data volumes and varieties or the high stream velocity.

R. Silhavy (Ed.): CSOC 2022, LNNS 501, pp. 48–54, 2022.
https://doi.org/10.1007/978-3-031-09070-7_5

In this context, big data processing has a key role for economy digital transformation. Artificial intelligence technologies depend on the way of data management. Digital services as part of our everyday life continually change it and rely on data processing progress (Fig. 1).

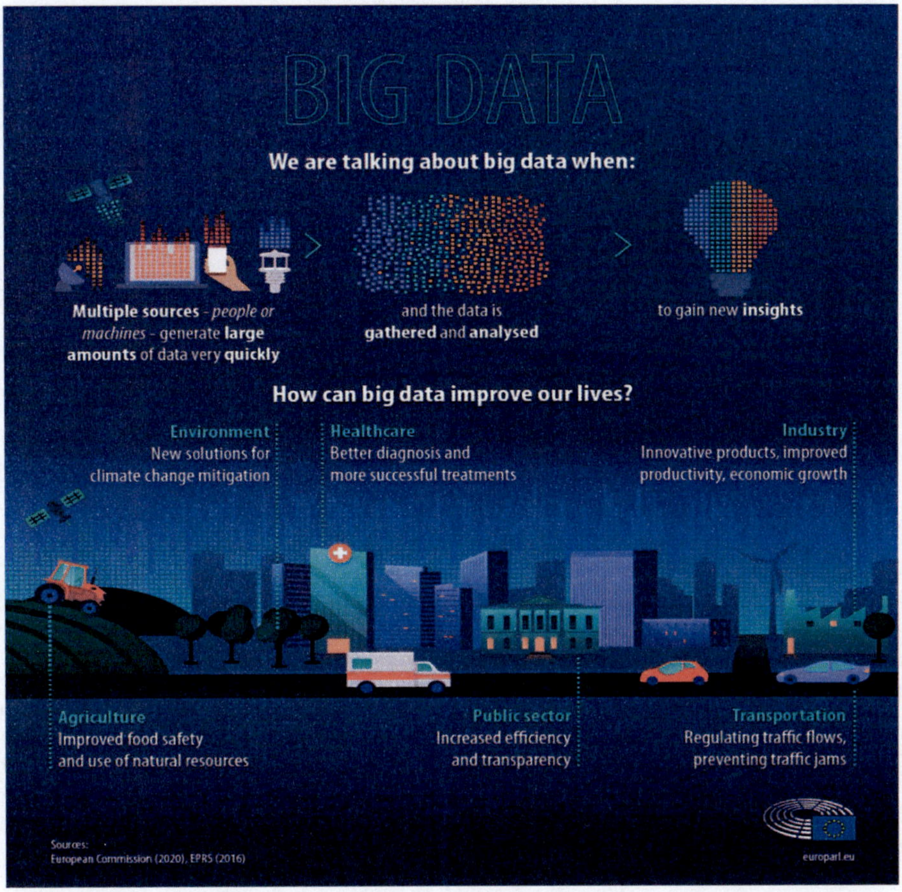

Fig. 1. Big Data [1]

The huge amount of information representing big data is unlike the traditional perceptions. Big data volumes cannot be stored and processed by means of conventional relational databases and SQL applications, but they can be described with the following basic characteristics: volume, velocity, value, and veracity of the data. These characteristics are known as 5Vs model (Fig. 2). Data volume that is usually huge and continually growing defines data quantity. Variety refers to data type that is regarded most frequently as structured, semi-structured and non-structured, etc. Velocity indicates the data quantity that is generated by data sources, handled, and processed per time unit. Value is connected with the process of valuable information retrieval from large groups of social data and it is usually named "big analytical data". Value is the most essential characteristic of each application based on big data volume as it allows generating beneficial business information. Veracity refers to data truthfulness, accuracy, and reliability.

Fig. 2. 5V's model [2]

Based on all the above stated, it becomes clear that new algorithms have to be developed for identifying frequently encountered forms, schemes, patterns as well as for detecting regularities, text analysis, etc., aiming to obtain meaningful knowledge from big data. In fact, knowledge extraction from available data is the fundament of the modern business intelligence. Since the existing algorithms for information retrieval do not work, or, if work, they do this inefficiently, it follows that projects and applications connected to information retrieval from big data is a serious challenge for organizations, which strive to optimize their business processes to the maximum [9].

In other words, big data is defined as a huge data amount demanding new technologies and architectures that enable the meaningful data (knowledge) extraction through processes of data capturing and analysis. Big data is important because the more information is gathered and piled, the more precise and detailed the outcome is, and hence, a better possibility for business processes optimization can be ensured.

2 Classification of Basic Big Data Sources

The analysis of several scientific references [3–5] leads to the conclusion that big data volumes can be extracted from three basic Big Data sources:

- Social networks (information generated by the population) – the information is based mainly on human experience. In the past, it was described in books and works of art, photographs, audio and video records. Nowadays, the information is almost entirely digitalized and it is stored everywhere, i.e. starting from personal computers to social networks and cloud technologies. The data is non-structured and difficult for processing and managing.

- Traditional business systems and websites (data generated by information systems) – the contained information is in result of economic activities, such as user registration, production of goods, queries, financial transactions, etc. The data is structured in a definite format (mainly tables), the data relations are defined, as well as the metadata for its content. It is stored in relational databases.
- Internet of things (data generated by machines/sensor devices) – machines and sensor devices generate the information by measurements and records of events and situations in the physical world. The data is well structured and suitable for computer processing but its size and velocity are beyond the capabilities of the conventional methods for data processing.

Big Data classification is important because it determines the strategy of access to data sources, the way of gathering information from them, defining cognitive tasks for the business processes in various spheres of the public life. At the same time, new data sources form a new information environment for management of financial flows, investments, labour force and exchange of material resources, goods and services.

3 Big Data and Knowledge Retrieval

Knowledge retrieval from big data is essential for large business intelligence (BI). If big data characteristics of volume, velocity, and variety (structured, semi-structured and non-structured) are not taken into account, a wrong perception of the situation may be formed that shall lead to wrong conclusions due to partial data consideration and/or lack of a comprehensive approach. Information retrieval from big data provides us with the possibility to discover hidden (latent) knowledge in it. The knowledge gained is helpful for making proper decisions leading to business processes optimization.

Real-time data analysis is necessary when we have data flows from various sources. Since many organizations produce real-time data, there are all prerequisites for keeping it never at rest. The dynamical data growth requires a platform (software framework) for its processing (Fig. 3). The platform enables data processing and ensures relevant stages for data extraction. Furthermore, the data extraction platform helps to find profitable business knowledge in big data systems.

Figure 4 represents current data growth by 2020. It can be seen that the data is traditionally divided into structured, semi-structured, and non-structured type. Data quantity grows exponentially within the interval of 2006–2020 and it is measured in Exabyte (EB). Besides the big data rise within that period, a similar growth is observed concerning the operations for information retrieval. Therefore, the information retrieval "adds value" to the initiatives, i.e. it truly contributes to the profits of businesses and companies when being applied to business intelligence. There are three models applicable as ETL (Extract, Transform, Load) procedure.

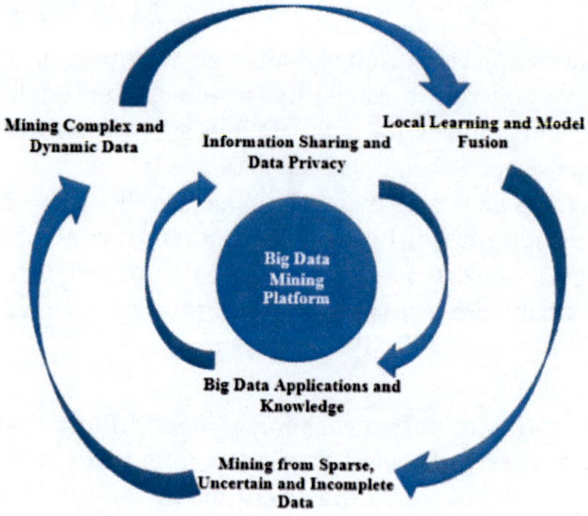

Fig. 3. Big Data processing framework

Fig. 4. Data growth by 2020

Characterization of the most frequently applied technologies for big data processing:

- Massive parallel processing (MPP). The architecture of relational database for MPP distributes data to a number of independent servers or nodes in a way transparent for users. Analytical MPP systems are frequently applied for Big Data operations, usually named “share-nothing databases”, since the nodes, which represent relevant clusters, work independently, communicate via a network, but do not share resources on disc or memory [6].
- NoSQL (Not only Structured Query Language) is an approach in database designing and managing. NoSQL incorporates a wide range of technologies and architectures intended to solve the problems related to Big Data scalability and representation, which cannot be successfully realized by relational databases. NoSQL is helpful for access to and analysis of a vast quantity of non-structured data, or data, which is stored at a distance on several virtual servers [7].
- Hadoop is a scalable, with no single point of failure, distributed system for data storage and processing, convenient for cloud operations. It consists of two basic components:

- Hadoop Distributed File System (HDFS) is multiple-ensured, highly reliable, clustered file system, self-healing from faults and destruction, providing a very high aggregate bandwidth, fitted to handle huge files.
- Map-Reduce is a programming model and an associative application for processing and generating big data volumes by means of a parallel, distributive algorithm for a cluster [8].

4 Conclusion

Information retrieval from big data can provide comprehensive knowledge, beneficial for business analyses and intelligence that can be used for making well-informed decisions and for optimizing business processes in organizations. Entrepreneurs in the real world are constantly in need of platforms for decision making that would lead to optimizing the organization activities. Most often the companies find problem solutions by implementing distributed programming platforms, such as Hadoop, and using LaaS for cloud computing and services.

Big Data characteristics including volume, velocity, and variety have an impact on the IT systems and infrastructures. Future IT infrastructures will be determined mainly by the new business models, which will be introduced for producing statistics from Big Data sources.

References

1. https://www.europarl.europa.eu/news/bg/headlines/society/20210211STO97614/gholemi-masivi-ot-danni-polzi-i-predizvikatelstva-infoghrafiki
2. Ibtissame, K., Yassine, R., Habiba, C.: Real time processing technologies in big data: Comparative study. In: IEEE International Conference on Power, Control, Signals and Instrumentation Engineering (ICPCSI), Chennai, India, (2017) https://doi.org/10.1109/ICPCSI.2017.8392202
3. Devlin, B., Rogers, S., Myers, J.: Big Data comes of age. Enterprise management associates (EMA) и 9sight Consulting Research, November (2012). http://9sight.com/pdfs/Big_Data_Comes_of_Age.pdf
4. Big Data and the SDGs. United Nations Global Pulse. (2016) https://www.slideshare.net/unglobalpulse/bigdata-and-the-sdgs
5. Vale, S.: Classification of types of Big Data (2013). https://statswiki.unece.org/display/bigdata/Classification+of+Types+of+Big+Data
6. Fernández, F.: Parallel and distributed computational intelligence, ISBN 3–642–10674–9 (2010)
7. Leavitt, N.: Will NoSQL databases Live up to their promise? IEEE Computer, february. (2010) http://leavcom.com/pdf/NoSQL.pdf
8. Dean, J., Ghemawat, S.: MapReduce: simplified data processing on large clusters. OSDI'04: Sixth Symposium on Operating System Design and Implementation, San Francisco (2004). https://www.usenix.org/legacy/publications/library/proceedings/osdi04/tech/full_papers/dean/dean_html
9. Atanasova, G., Hristova, P.: The significance of software psychology knowledge In computer science related courses, INTED2020 Proceedings, pp. 3225–3229 (2020)

10. Krastev, G., Voinohovska, V.: Smart mobile application for public transport schedules – data organization and program realization. In: International Congress on Human-Computer Interaction. Optimization and Robotic Applications (HORA), 1–4 (2020). https://doi.org/10.1109/HORA49412.2020.9152884
11. Krastev, G., Voinohovska, V.: Smart mobile application for public transport Schedules - Logical Model, TEM J., ISSN 2217-8309 (2020) https://doi.org/10.18421/TEM92-15
12. Petrov, V.l., Rusev, R., Rusev, I.: anaesthesiology in the big data era - are we ready for that?, Anaesthesiol. Intensive Care. ISBN: 1310–4284 (2021)

Usability of Dashboards in the Measurement Process - An Experience of Use in a Brazilian Software Development Company

Adriano Bessa Albuquerque, Gleidson Sobreira Leite(✉), Daniel Morais, and Leonardo Pontes

Universidade de Fortaleza, Fortaleza, CE, Brazil
adrianoba@unifor.br, gleidson.sleite@gmail.com

Abstract. Information technology companies have to know what they need to change if they want to continually improve. For this, it is essential to measure process, services and products in order to understand and control them. However, the difficulties of these measurements are not only related to data collection and analysis, but also related to the presentation of the results. In this scenario, the use of BI dashboards has become an increasing trend over the years mainly due to its advantages, such as greater agility in collecting and organizing data, among others. However, even with the benefits provided by the use of BI tools, it is also important to consider the presentation of the obtained results, as well as the navigation among them, that is, the usability of the dashboards. Considering the aspects of the usability in BI dashboards, this paper presents an experience performed in a Brazilian software company which addressed the usability evaluation of the developed dashboards in its measurement process.

Keywords: Business intelligence · Dashboard usability · Information technology · Software measurement

1 Introduction

According to [1], it is important to measure to understand and control processes, products and projects. Measures provides information about objects and events, making them understandable and controllable, and measurement is the process of assigning numbers or symbols to selected attributes of an entity, creating a mapping from the real world to a formal mathematical world [2].

[3] adds that measuring software is expensive and involves a considerable effort in terms of human resources and, in many cases, measurements are not always useful for knowledge, control or decision making. Some steps are crucial for a measurement program to be able to be implemented and established in a company, such as: the collection, analysis and presentation of results.

According to [4], the level of automation can also positively influence the success of the measurement, since with more automation there is less interference in the development team's daily activities, less cost, less risk of not collecting the necessary data, and more correct these data can be.

R. Silhavy (Ed.): CSOC 2022, LNNS 501, pp. 55–67, 2022.
https://doi.org/10.1007/978-3-031-09070-7_6

However, the presentation of results, that is, the care with the usability of dashboards has also been an important aspect to be addressed, since the usability level of the developed boards can interfere in how users perceive the results as reliable and valid. This concern is essential, especially with the advancement of interaction possibilities that BI (Business Intelligence) tools make available, as it is the case of Power BI, Pentaho, among others.

Motivated by this scenario and seeking to contribute to the body of knowledge related to studies regarding usability in dashboards, this study presents the execution of a real case scenario experience aiming the application of an evaluation of the usability of dashboards built in a Brazilian software company, in a scenario in which it built indicators to assist in the measurement and decision-making of the company.

To enable the evaluation (which was carried out by a company manager), criteria were selected from the literature review and pointed out in this study.

It is also expected that this study can be adopted to help both practitioners and researchers in order to identify possible applications trends and improvements.

This paper is organized as follow: the methodology and procedures performed in the work (Sect. 2); background and related work (Sect. 3); the experience and discussions (Sect. 4), followed by the final considerations (Sect. 5).

2 Research Methodology

To accomplish the objective of this work, five specific actions were carried out. They were:

1) Bibliographic research in academic studies regarding measurement and metrics, usability in dashboards and results presentation.
2) Definition of the scope and delimitation of the experience, as well as its planning
3) Selection of the company and negotiations with its senior management to apply the experience
4) Selection of criteria to be adopted in the evaluation of the usability of dashboards by the company manager
5) Conduction of the experience and presentation of the results, challenges and limitations.

3 Background and Related Work

As mentioned by [4], the main steps of a measurement process are indicator specification (operational definition), collection, analysis, and presentation of results, where one of the requirements of a measurement process is that it generates reliable and valid results.

Reliability indicates the consistency of measurement results performed when using the same measurement method on the same object and can be expressed in terms of the size of the standard deviation of repeated measurements. In the case of validity, this indicates whether the measurement was able to generate the intended data. It can also be considered that the level of usability of developed dashboards interferes with how much users perceive the results as reliable and valid [4].

For the measurements to be really effective, [5] defined a set of guidelines, namely: (i) use different strategies for different cultures; (ii) do not take measurements in the dark; (iii) be judicious in the measurement process; (iv) be simple and straightforward; (v) change measurements and their models as the situation becomes more understood; (vi) using some metric is better than not using any; (vii) measure twice, act once; (viii) keep metrics close to developers; (ix) capture as much as possible without inconveniencing the developers and (x) start the measurement process with whoever is in need of help.

The first step in any measurement process is to discover information needs, and then define what to measure and how to measure it. For this, it is very important to prepare a measurement plan, consistently defining the data that will be collected, the way to collect them and the mechanism for analyzing the results [6].

Some approaches have already been defined to support the preparation of measurement plans, highlighting the GQM - Goal Question Metrics [7] and the PSM - Practical Software Measurement [8].

The main characteristic of a metric is the type of its scale, as the scale defines the set of analyzes that can be performed using the collected data.

The four main types of scale are [4]:

- Nominal scale: is related to the classification or categorization of elements. The categories are exhaustive and mutually exclusive and do not imply that one is better or greater than another;
- Ordinal scale: the elements can be compared in terms of order, it can be deduced that if A > B and B > C then A > C. However, this scale does not provide information about the magnitude of the differences between the elements;
- Interval scale: indicates the exact difference between measurement points. If A equals 5 defects per KLOC and B equals 3.5 defects per KLOC, it can be said that A has 1.5 more defects per KLOC than B;
- Rational scale: when an absolute or non-arbitrary zero can be found on an interval scale, this becomes a rational scale. It is the highest measurement level scale.

In addition to defining the indicators that should be used, the way in which data collection is organized is also essential for the measurement to be successful, and it is very important to identify those responsible for each metric, the moment of measurement and define the procedures for storage and presentation of results.

Support tools, the way to capture and store data and monitoring adherence to established procedures for data collection and storage are also fundamental for the quality of the measurement process and the reliability of results.

In the case of data analysis, its objective is to make some pattern or relationship more visible, so that it can be used to make judgments about the attributes being measured [9]. The analysis makes it possible to improve the knowledge of reality and helps to propose the necessary improvements. However, the analysis depends on the presentation of the results, that is, it depends on the level of usability of the dashboards.

Data analysis can be classified into quantitative and qualitative. In quantitative analysis, numbers and statistical techniques are used, and it is quite common to represent measurements over time, establish limits for control and analyze values that are outside the expected range [10].

It is noteworthy that in any type of analysis, it is very important to consider the context in which the data were captured, as it allows for a greater meaning to the analysis and makes it possible to carry out comparisons between results of different measurements. No count or measurement makes sense without its context.

Regarding the presentation of results, its great importance can be understood, as it is the moment when the results of the measurement process will support decisions, whether at the strategic, tactical or operational level.

As in software development, aspects related to usability must be addressed both at the project level, with the use of prototypes, and in validation, preferably using checklists with heuristics that the company considers relevant.

Currently, when using BI tools to support measurement processes, companies have more and more alternatives to present their data with a level of usability where the results, positive or negative, are more readily visible and become more understandable. BI solutions are applied in different sectors, from hospitals to large industries and in the banking sector [11].

In other words, the importance of both data for companies and BI (Business Intelligence) is clear, being relevant for companies in any sector, whether in industry, commerce or services [11].

Business Intelligence (BI) is a term that has gained popularity in recent years, with other related concepts: Data Warehouse, Data Mart, Data Deluge, Data Quality, ETL (Extract, Transform, Load), and Data Science. BI encompasses processes, tools and technologies essential for transforming data into information and knowledge [12].

[12] stated that there are several nuances when approaching BI, which can be seen as a technical component or as a corporate approach. The authors state that the focus of BI is to support managers in their day-to-day decisions, required by the business. [12] complements that BI encompasses the process of collecting a large amount of data, analyzing it, and producing reports that summarize the results.

Therefore, in the BI Presentation Area, where the production of reports is inserted, dashboards are important elements, as it is from them that users start their analysis, using more aggregated information, being able to perform various actions, including, for example, drill down.

Therefore, a dashboard is a visual presentation of the most important information needed to achieve one or more goals; consolidated and organized in a single screen so that information can be monitored quickly [13].

With the increase in the number of BI tools, dashboards allow you to present results in an increasingly sophisticated way, providing different resources [14].

Regarding the usability of dashboards, [14] presented the elements that make up a dashboard. They are: (i) Visual elements; (ii) Functional elements and (iii) Interactive elements of analysis. The authors also highlighted the impact that the design of a dashboard can have in situations that require attention and on the performance of an activity.

[15] defined a model to build BI dashboards considering dashboard relevance, dashboard efficiency and dashboard balancing as being components of the User Experience with a dashboard (Fig. 1). For example, one of the questions related to relevance is

whether the dashboard generates positive cognitive effects, while in relation to efficiency it is considered, among others, whether the dashboard facilitates interaction with the data. As for balancing, there is an issue that investigates whether the dashboard has a reasonable amount of parameters.

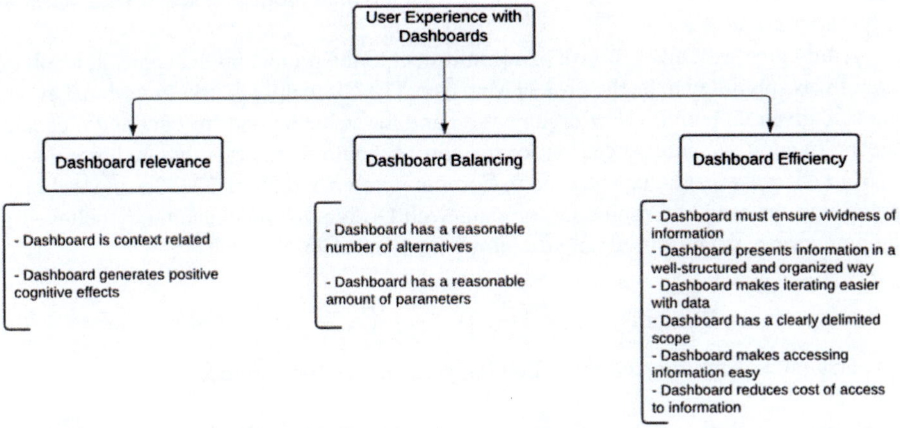

Fig. 1. A quality model and construct for designing comprehensive BI dashboards [15]

[16] investigated the visual behavior of dashboard users focusing on the human characteristic of having limitations with attention. For this, they carried out an exploratory study considering eye movement, which according to the authors is essential during the design of dashboards. [17] defined a framework to measure the usability and efficiency of dashboards.

4 The Experience

4.1 Scenario

In the begining of 2019, a software company from Ceará (in Brazil) started a Management project by Indicators, with the objective of building dashboards with the possibility of supporting strategic and tactical decision-making, via quantitative data.

The company has been operating since 1993 in the development of systems for the public sector, specifically covering City Halls, Chambers, Local Authorities and Special Funds. Headquartered in Fortaleza, Ceará, Brazil, is the leader in some markets in the North and Northeast of the country, being present in three more states: Maranhão, Pará and Rio Grande do Norte.

The company currently has more than 130 employees, and its organizational structure includes the following sectors: Development, Infrastructure, Service, Sales, Contract, Finance, Training, Marketing and Human Resources.

To initiate and conduct the Management Project by Indicators, a consultancy was hired and followed a process of implementation of Management by Indicators, adapted from previous experience in the IT Area of a Brazilian public company. The consultancy used knowledge from the Academy, via a professor and graduate students from a University of Ceará.

First, there was a meeting with the sponsor to understand what their motivation and objectives with that project was, as well as the opportunity for the consultancy to clarify how the work would be carried out, what the final product was and what possible difficulties could arise.

After agreement and promise of commitment to the project by the sponsor, a collaborator was appointed to be the Project Manager. The company's Marketing Manager was chosen, given its high level of organization and the value he had for data and measurements. In addition, another reason for the choice would be the fact that he has a degree in Marketing, a postgraduate degree in Communication and Digital Media and technical courses in the area of Graphic Design and Web Design, as the consultancy believed in the importance of presenting and disseminating the results.

4.2 Experience Planning

To carry out the experiment, the following actions were performed:

- Meetings with the sectors of the selected company, and identification of the main indicators to be considered for measurements
- Development of a low-fidelity dashboard prototype to consider the level of usability of the dashboards
- Validation of prototypes, construction of dashboards and presentation to the sponsor (selected company)
- Definition of criteria to be adopted in the analysis of the usability of dashboards obtained from a literature review
- Evaluation of the usability of the produced dashboards, and presentation of results.

4.3 Experience Execution

After planning, the consultancy, together with the Project Manager, started meetings with each sector of the company that would have its results monitored by Management by Indicators. The meetings were held with the following sectors: Development, Infrastructure, Service, Sales, Contracts and Marketing.

The first meetings with the sectors aimed to identify their main indicators. For this, the GQM approach [7] was adopted. After identifying the indicators, another meeting was scheduled to establish the operational definition of each indicator, that is, a form was defined for each indicator identified.

One of the most controversial points at the meeting to identify the indicators was to analyze the feasibility of working with such an indicator, considering the possibility of collecting them automatically and the level of reliability of the collected measure.

As can be seen in one of the Development Sector sheets, each sheet contains (Fig. 2): The name of the indicator, its description, the established goal, the sector that will manage the indicator results, the frequency of collection, the calculation formula and the collection procedure.

Indicator: Rate of Service Orders resolved on Time			
Description:	The indicator measures the percentage of Work Orders that were resolved within 2 days	**Objective:**	80%
Sector :	**Frequency:**	**Calculation Formula:**	
Development	Monthly	(Amount of Service Orders resolved within the deadline / Amount of Service Orders opened in the month)*100	
Collection Procedure:	> Number of Service Orders resolved within the deadline: Git > Number of Service Orders opened in the month: Git		

Fig. 2. Example of indicator form adopted by the company

The consultancy opted for a leaner operational description to make it simpler, making it easier to develop the culture of Management by Indicators in the company.

After the construction of the form, a meeting was held to validate what had been specified by the consultant and the Project Manager. If the specification was successfully validated, the consultancy started to develop a low-fidelity prototype of the dashboards, considering how the results should be presented in the presentation support tool.

The decision to develop prototypes was motivated by the fact that the consultancy considers the level of usability of the dashboards to be fundamental, since both strategic and tactical decisions would be influenced by the results of the indicators. In addition, the consultancy believed that good usability could positively influence how much the users, who were managers, would trust and find the results valid.

After validating the prototypes, the construction of the dashboards in the PowerBI tool began. At the beginning of the project, a comparative evaluation was carried out between the PowerBI and Pentaho tool, the first one having been chosen, as, despite having a cost to acquire it, it was easy to build dashboards.

Every time after building the dashboards of a sector, the manager was called to a meeting to validate that the usability was good, considering the graphs and summarized data. Finally, the consultancy asked the sector manager to assess the completeness and correctness of what was being presented by the tool and scheduled another meeting where the requested changes would be presented. In this first moment, usability evaluations were carried out in an ad-hoc manner.

After carrying out all these steps for all sectors involved with Management by Indicators, including the Development Sector, the Infrastructure Sector and the Customer Service Sector, the dashboards built were presented to the sponsor.

Afterwards, the consultancy decided to prepare presentations for the top management, with the analyzes being carried out by the consultancy itself, in order to facilitate the understanding of the results. The motivation for the preparation of presentations came after verifying that the top management was not analyzing what was being made available in the tool. The consultancy considered that carrying out the analyzes itself would be a way of teaching how to analyze data, in addition to helping to create a culture of Management by Indicators.

Another type of meeting was established, in which the sector manager, the sponsor, the consultancy and the project manager are part of, with the aim of improving the contextualization of the results obtained, since in this meeting a qualitative analysis of what was obtained. The motivation for this meeting was the consultancy's belief that any analysis must be made considering a context, which in this case would be given by the sector manager.

Finally, to carry out the usability evaluation/analysis of the built dashboards, a set of criteria (see Table 1) obtained from a literature review were identified and selected, so that the selected Marketing Manager could evaluate all dashboards available in the company's PowerBI.

Table 1. Selected criteria adopted in the evaluation of the dashboards usability

Criteria	References	Criteria	References
1. Results are displayed quickly	[18]	11. Graphics are self explanatory	[23–27]
2. Data is presented in a single dimension	[18, 19]	12. The dashboard aesthetics are pleasing	[26]
3. There is a help mechanism within the dashboards	[18]	13. The colors red and yellow were used to highlight data that you want to draw attention to	[24]
4. Dashboard behaviors are homogeneous in terms of performance	[18, 23, 24]	14. There was no need to double-click in any situation	[24]
5. The aesthetics of dashboards are homogeneous	[18, 20, 24]	15. The colors were not repeated in the same graph	[19]
6. The dashboards were adapted to the user's characteristics	[19]	16. Blank spaces were used as vent areas	[19]
7. The presentation of data on dashboards is at most three levels deep	[19]	17. Only horizontal texts were used	[19]
8. Users can customize the presentation of data in dashboards	[19, 20]	18. Different colors were used for different situations	[27]
9. The dashboard aesthetics can be considered "clean"	[19, 21–23, 25, 26]	19. The data was presented on a single page, avoiding scrolling pages	[27]
10. The most important information has been visually highlighted	[23]		

The second column of Table 1 presents the bibliographic reference from which each question in the questionnaire was taken. All dashboard usability directives were collected from the literature by the consultancy.

In the evaluation, the Marketing Manager evaluated the presence or not of each characteristic in the built dashboards, using the following Likert scale: It does not have this characteristic; It has little of this feature; It reasonably possesses this feature; It has this feature a lot; and It fully possesses this feature.

Figure 3 illustrate one of the dashboards of the Development Sector (Information that was not possible to be disclosed was marked in yellow).

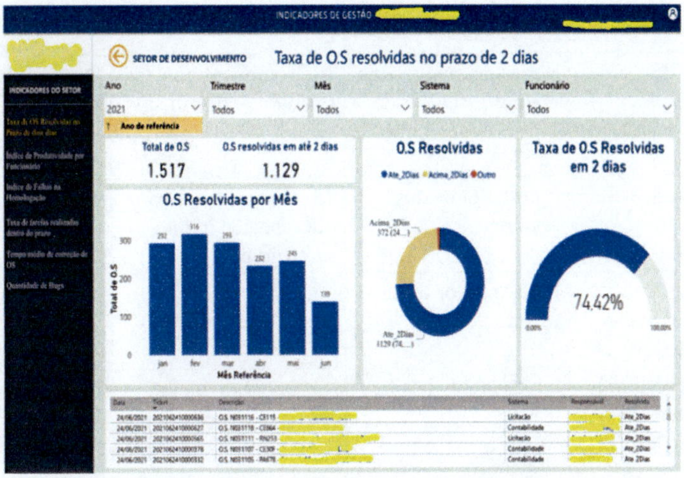

Fig. 3. Example of a dashboard produced during the experiment

4.4 Results and Discussions

The evaluation result (presented in Table 2) showed that no characteristic was considered as "Totally possesses this characteristic", that is, at the time of dashboard prototypes and validation by sector managers, none of these best practices was considered essential. Perhaps aspects of usability of dashboards are not considered by companies as they are in software development.

According to the results presented in Table 2, 57% of the characteristics were considered as "Has a lot of this characteristic", 26% were considered "Reasonably possesses this characteristic", 5% were considered as "Has little this characteristic" and finally, only 10% were considered as "Does not have this feature".

The results showed that at times the user is required to double-click to continue browsing and that care was not taken not to repeat colors in the same graphic, that is, the same color was used with different shades, which could negatively interfere with the perception of the graph and its understanding.

However, it was found that care was taken in relation to performance, aesthetics and consistency between the different dashboards, trying to ensure standardization.

Table 2. Usability evaluation results

Criteria	Result	Criteria	Result
1. Results are displayed quickly	It has a lot of this feature	11. Graphics are self explanatory	It has a lot of this feature
2. Data is presented in a single dimension	It has a lot of this feature	12. The dashboard aesthetics are pleasing	It has a lot of this feature
3. There is a help mechanism within the dashboards	It has little this feature	13. The colors red and yellow were used to highlight data that you want to draw attention to	It has a lot of this feature
4. Dashboard behaviors are homogeneous in terms of performance	It reasonably has this feature	14. There was no need to double-click in any situation	Does not have this feature
5. The aesthetics of dashboards are homogeneous	It has a lot of this feature	15. The colors were not repeated in the same graph	Does not have this feature
6. The dashboards were adapted to the user's characteristics	It has a lot of this feature	16. Blank spaces were used as vent areas	It has a lot of this feature
7. The presentation of data on dashboards is at most three levels deep	It reasonably has this feature	17. Only horizontal texts were used	It has a lot of this feature
8. Users can customize the presentation of data in dashboards	It reasonably has this feature	18. Different colors were used for different situations	It has a lot of this feature
9. The dashboard aesthetics can be considered "clean"	It reasonably has this feature	19. The data was presented on a single page, avoiding scrolling pages	It reasonably has this feature
10. The most important information has been visually highlighted	It has a lot of this feature		

Regarding aesthetics, what contributed to the well-treated aspects of this dimension was the fact that the design and construction had the interference of the Marketing Sector, which has professionals specialized in Advertising, that is, professionals who know and value factors linked to harmony, beauty, proportion, among others.

In the case of performance, the involvement of analysts and programmers from the Development Sector was essential, as well as the involvement of the Infrastructure Sector manager.

5 Conclusion

In order to enable greater understanding and control of processes, products and projects in technology companies, the importance of applying measurements in order to obtain more reliable information for decision-making and improvements is essential.

In this context, not only activities such as identification of indicators, data collection, analysis are essential, but also how the results are presented, that is, the concern with the usability of the tools (e.g. dashboards) adopted to present these results.

Motivated by this scenario and seeking to contribute to the body of knowledge related to studies regarding usability in dashboards, this study presented the execution of a real case scenario experience aiming the application of an evaluation of the usability of dashboards built in a Brazilian software company.

During the execution of the experience, 19 usability criteria where identified and selected from the literature, and adopted in a usability evaluation of constructed dashboards.

Among some lessons learned from the study and the experience carried out, we can mention: (i) the usability of dashboards must be properly treated so that it positively influences the users' confidence in the results; (ii) already consolidated software engineering practices should be used in the specification, design and construction of BI dashboards; (iii) low-fidelity prototyping and validations, both of the prototype and of the built dashboards, help to better understand the users' needs, as well as to define the usability project properly; (iv) the presence of an infrastructure professional helps a lot when identifying indicators and preparing indicator sheets, to assess whether automated data collection is possible; (v) after the implementation of the project, it is necessary that periodic checks are carried out in relation to the correctness of the data.

As a limitations of this work, we can mention the fact that the experiment was carried out in only one company, as well as the evaluation was carried out only by one expert.

In addition to the limitations pointed out, as future works, it is recommended that the scope of this study be expanded to a greater number and variety of companies, the evaluations should be performed by a group of specialists of the selected companies, and to conduct a survey with the company's managers so that it is possible to know how useful the selected criteria is in usability analysis/evaluation, as well as to identify improvements.

References

1. Fenton, N.E., Pfleeger, S.L.: Software metrics: a rigorous and practical approach, September 2020. Third Edition, CRC Press, Florida, EUA
2. Christensen, M.J., Thayer, R.H.: The project manager's guide to software engineering best practices. IEEE Comput. Soc. Press, New York (2001)

3. Rocha, A.R.C., Souza, G.S., Barcellos, M.P.: Software measurement and statistical process control. Ministry of Science, Technology and Innovation SEPIN PBQP Software, Brasília – Brazil, ISSN 1679–187 (2012)
4. Albuquerque, A.B.: Evaluation and improvement of organizational process assets in software development environments. D.Sc. Thesis, COPPE, UFRJ, Rio de Janeiro, Brazil (2008)
5. Pfleeger, S.L.: Use realistic, effective software measurement. In: Constructing superior software, software quality institute series. The Software Quality Institute. November 4. New Riders Pub; 1st edition (1999)
6. Solinger, R., Berghout, E.: The goal/question/metric method: a practical guide for quality improvement of software development. January 16. McGraw-Hill Education, Europe, Middle East & Africa (1999)
7. Basili, V., et. al.: Goal question metric paradigm. In: Marciniak, J.J. (ed), Encyclopedia of software engineering, vol. 1, pp. 528–532, Wiley (1994)
8. PSMSC. Practical software & systems measurement: objective information for decision makers. October 2004. www.psmsc.com. Accessed 20 Oct 2021
9. Probst, G., Raub, S., Rombardt, K.: Managing knowledge – Building blocks for success, 1ed. December 21. Wiley, United States (1999)
10. Keeni, G.: The evolution of quality processes at Tata Consultancy Services. In IEEE Software, vol. 17, no. 4, pp. 79–88, July-Aug. (2000). https://doi.org/10.1109/52.854073
11. Moro, S., Cortez, P., Rita, P.: Business intelligence in banking: a literature analysis from 2002 to 2013 using text mining and latent dirichlet allocation. Expert Systems with Applications, Issue 3, pp. 1314–1324. February 15, (2015). https://doi.org/10.1016/j.eswa.2014.09.024
12. Guarda, T., et al.: Pervasive business intelligence: a key success factor for business. In: Rocha, Á., Reis, L.P. (eds.) Developments and advances in intelligent systems and applications. SCI, vol. 718, pp. 81–89. Springer, Cham (2018). https://doi.org/10.1007/978-3-319-58965-7_6
13. Hockenhull, M., Cohn, M.: Speculative data work & dashboards: designing alternative data visions. In: Proceedings ACM Human Computer Interaction, Vol. 4, pp. 1–31. December, (2020) https://doi.org/10.1145/3434173
14. Nadia, M., Maedchea, A., Schiederb, C.: The effect of interactive analytical dashboard features on situation awareness and task performance, Decision Support System, Epub, May 19, (2020) https://doi.org/10.1016/j.dss.2020.113322
15. Burnay, C., Bouraga, S., Faulkner, S., Jureta, I.: User-experience in business intelligence - a quality construct and model to design supportive BI dashboards. In: Dalpiaz, F., Zdravkovic, J., Loucopoulos, P. (eds.) RCIS 2020. LNBIP, vol. 385, pp. 174–190. Springer, Cham (2020). https://doi.org/10.1007/978-3-030-50316-1_11
16. Toeirini, P., Moritz, L.: Designing user-adaptive information dashboards: considering limited attention and working memory, In: Proceedings of the 27th European Conference on Information Systems (ECIS), Stockholm & Uppsala, Sweden, June 8–14, (2019)
17. Oliveira, M., Cardoso, E., Santana, M.: Measuring dashboards usability and efficiency: framework developed using business intelligence design principles and human-machine interaction, In: Proceedings of 15th Conference of the Portuguese Association of Information Systems. Lisbon, Portugal. (2020) https://doi.org/10.18803/capsi.v15.338-362
18. Magdalena, R., Ruldeviyani, S.D., Bernardo, C.: Methods to enhance the utilization of business intelligence dashboard by integration of evaluation and user testing. In: 2019 3rd International Conference on Informatics and Computational Sciences (ICICoS), pp. 1–6. October 29–30, Semarang, Indonesia (2019). https://doi.org/10.1109/ICICoS48119.2019.8982481

19. Orlando, T., Sunindyo, W.: Designing dashboard visualization for heterogeneous stakeholders (case study: ITB central library), In: 2017 International Conference on Data and Software Engineering (ICoDSE), pp. 1–6. November 1–2. Palembang, Indonesia (2017) https://doi.org/10.1109/ICODSE.2017.8285872
20. Ma, R. et al.. LADV: Deep learning assisted authoring of dashboard visualizations from images and sketches. In: IEEE Transactions on Visualization and Computer Graphics, vol. 27, no. 9, pp. 3717–3732, September, 1, (2021). https://doi.org/10.1109/TVCG.2020.2980227
21. Bafna, A., Parkhe, A., Lyer, A., Halbe, A.: A novel approach to data visualization by supporting Ad-hoc query and predictive analysis: (An Intelligent Data Analyzer and visualizer). In: 2019 International Conference on Intelligent Computing and Control Systems (ICCS), May 15–17, Madurai, India. pp. 113–119 (2019). https://doi.org/10.1109/ICCS45141.2019.9065380
22. Sluijter, J., Otten, M.: Business intelligence (BI) for personalized student dashboards. In: Proceedings of the Seventh International Learning Analytics & Knowledge Conference, March 13–17, Vancouver British Columbia Canada. pp. 562–563 (2017). https://doi.org/10.1145/3027385.3029458
23. Manwaring, E., Carter, N., Maynard, K.: Redesigning educational dashboards for shifting user contexts. In: Proceedings of the 35th ACM International Conference on the Design of Communication, August 11–13. Halifax Nova Scotia Canada (2017) https://doi.org/10.1145/3121113.3121210
24. Dwoman, G.: Arbitration of a help system, Interactions, Volume 14, Issue 1, pp. 39–42. Association for Computing Machinery, New York, United States. (2007) https://doi.org/10.1145/1189976.1189999
25. Jooste, C.: A collage of change: utilizing appreciative inquiry in a user experience design study for big data interfaces - A reality. In: Conference: International Conference On Interfaces and Human Computer Interaction, July 16. Porto, Portugal (2019) https://doi.org/10.33965/ihci2019_201906L014
26. François, M., Osiurak, F., Fort, A., Crave, P., Navarro J.: Usability and acceptance of truck dashboards designed by drivers: two participatory design approaches compared to a user-centered design. Int. J. Ind. Ergonomics **81** (2021) https://doi.org/10.1016/j.ergon.2020.103073
27. Mazor, I., Heart, T., Even, A.: Simulating the impact of an online digital dashboard in emergency departments on patients length of stay. J. Decis. Syst. **25**(sup1), 343–353 (June 2016). https://doi.org/10.1080/12460125.2016.1187422

Simulation of the Hybrid Optical Amplification Connections for Performance Analysis

Martin Bonkalo(✉) and Rastislav Róka

Slovak University of Technology, Bratislava, Slovakia
martin.bonkalo@gmail.com, rastislav.roka@stuba.sk

Abstract. In this contribution, an attention is focused on hybrid connections of optical amplifiers that have a great potential to be implemented in future long-haul optical transmission networks. Specifically, we evaluate possibilities for hybrid amplification connections and analyze appropriate source wavelengths and pumping power levels for the optical signal amplification. Based on simulation results, appropriate features of pumping sources for resulting optical signal amplification are analyzed and a suitable arrangement of hybrid connections is designed.

Keywords: The hybrid amplification connection · The EDFA simulation model · The ROA model · The optical transmission path · Numerical simulations

1 Introduction

Every second, a large amount of data is transferred in the world, and every user needs the fastest and highest quality of data transport. The only transmission medium that is capable to satisfy the increasing requirements of end-users is the optical fiber incorporated into fully optical transmission networks. Increased network capacity and flexibility can be reached as well using advances in optical fibers, amplification, and regenerations, coherent transceivers, digital and optical signal processing, modulation formats, forward error correction [1–4]. Optical amplifiers are a key element of long-haul systems [5] and have contributed to the success of optical data transmission together with low loss optical fibers. Very important is their capability to accommodate dense wavelength division multiplexing (DWDM) techniques for transmitting multiple channels in a single optical fiber over long distances [6]. From an amplification perspective, increasing the capacity per fiber can be supported by several different approaches. An effort on hybrid amplification approaches, i.e., a combination of optical amplifications can help to avoid unacceptable short span lengths or reaches with new more spectrally efficient modulation formats. So, this research has led to efficient and very powerful solutions for long-distance and high-capacity networks, however, many practical issues must be solved together with a need for cost and energy efficient pump sources [7]. Since a practical implementation is quite expensive, an evaluation and then simulation experiments is a better alternative to assess the suitability and applicability of solutions for any hybrid amplification connections.

R. Silhavy (Ed.): CSOC 2022, LNNS 501, pp. 68–78, 2022.
https://doi.org/10.1007/978-3-031-09070-7_7

Nowadays, it is more and more important to avoid expensive practical testing and demonstrations due to requirements on fully optical transmission networks. Therefore, simulation platforms are increasing their role. However, it is important to search for a simulation platform that will be able to accurately describe the behavior and limitations of optical amplifiers in real optical transmission systems under various working conditions. For future long-haul optical transmission networks and/or for utilization spectrally efficient, multilevel modulation formats, optical signals can be pre-amplified before the opto-electronic conversion by a combination of the erbium doped fiber amplifier (EDFA) and/or Raman optical amplifier (ROA). So, possibilities for their numerical simulations can be shortly introduced by following related works. In [8], an observation of the EFDA performance and a determination of its characteristics at 1550 nm wavelength range is experimentally studied using the OptiSystem software. A block scheme is not fully presented, moreover, many parameters of the optical fiber are also not introduced. Besides, a way for acquiring presented EDFA characteristics is not presented. In [9], optical amplifications are implemented into future next-generation passive optical networks and the performance of EDFA boosters and hybrid optical amplifiers is analyzed. The analyzed architecture of hybrid passive optical networks is not presented in a sufficient way, many necessary parameters are mismatched or missing. For the EDFA simulation, the OptiSystem software is also used. However, no verification of pre-prepared blocks is realized and presented.

Based on previous related works and our experiences with various simulation software programs, a simulation platform based on the Matlab R2020B Simulink environment is selected for designing and creating an optical transmission path [10]. In the first phase, our work was focused on the optical transmission medium, primarily on linear and non-linear effects present in the single-mode optical fiber. Then, their environmental negative influences on transmitted optical signals are analyzed. Its summary presents the extended base of simulation tools for analyzing signal processing techniques in the optical transmission path [11]. In the next phase, a work is focusing on significant components including optical amplifiers in the optical transmission path with an emphasis placed on their specific features influencing transmitted optical signals. In [12], these innovations in the simulation of the optical transmission path are presented and, specifically, the dynamic performance analysis of EDFA properties for system applications is focused. The EDFA model in the steady state can be consequently applied for a purpose of advanced simulations performed in the complete optical transmission path. Its possible utilization in the analysis based on simulation is presented in [13] that deals with the integration of polarization multiplexed 16-ary quadrature amplitude modulation format in optical metropolitan networks enhanced with the signal amplification.

This contribution continues in mentioned works and presents innovations in the enhanced simulation tool suitable for performance analysis of hybrid amplification connections in the optical transmission path.

This contribution is organized as follows: In Section II, features of optical amplifier models, as well as possible hybrid amplification connections for performance analysis in the optical transmission path are presented. In Section III, the optical transmission path incorporated both EDFA and ROA simulation models for analyzing hybrid amplification connections is presented. In Section IV, simulation results for hybrid amplification connections are concretely analyzed and summarized.

2 Features of Optical Amplifier Models and Possibilities for hybrid Amplification Connections

Advances in optical amplifier research can have a very beneficial impact on data transmission cost and energy efficiency in long-haul optical transmission networks utilizing DWDM techniques with fixed and/or variable wavelength grid. Therefore, it is necessary to realize experimental efforts in an appropriate simulation platform for the optical transmission path that can accurately describe the behavior of optical amplifiers at the signal transmission under varying working conditions. First, the EDFA simulation model is presented. EDFA amplifiers are commonly used in commercial optical transmission systems due to their high optical amplification compared to costs against optical amplifiers utilizing other mechanisms in hybrid amplification solutions [10]. A simulation of the erbium-doped fiber amplification for performance analysis in Matlab is presented in detail in [12]. A complete list of parameters for the simulation of output EDFA characteristics and their extracted values [10, 12] is displayed in Table 1.

Input channels on various wavelengths have different values of the EDFA signal gain caused by the spectral dependency of absorption and emission cross-sections and by the overlap factor. Therefore, this presumption must be emphasized at the model's modification for multichannel amplification using the DWDM technique.

Second, the ROA simulation model is introduced. A Raman optical amplifier can be characterized as a pumping laser that emits light waves into an optical fiber. The operation of the pumping laser depends on the achieved gain, especially on the pumping wavelength. As a transmission medium, no special fiber is needed unlike the EDFA amplifier [14]. In Table 2, parameters for the simulation of output ROA characteristics are presented.

The Raman gain parameter plays a key role in the ROA amplification efficiency and can be expressed by the equation

$$g_{R_{eff}} = \frac{g_R}{A_{eff}} \tag{1}$$

where $g_{R_{eff}}$ is the Raman gain efficiency, g_R is the Raman gain and A_{eff} is the effective fiber area. It´s the most important factor of the Raman amplifier booster because it says that the Raman amplification is more effective with increased fiber length. The longer optical fiber will be used, the signal amplification will be higher.

For effective utilization of the optical transmission medium, hybrid amplification approaches are one of the future challenges. Therefore, we are focusing on analyzing a combination of EDFA and ROA amplifiers in the optical transmission path. So, there are two possible hybrid amplification connections for analysis. In the first one, both EDFA and ROA amplifiers are connected at the end of the optical transmission path realized with the Single-mode optical fiber block (Fig. 1).

Table 1. Parameters for the simulation of output EDFA characteristics

Parameter	Unit	Extracted value
r	μm	1.4
$\Gamma_{p(980)}$	-	0.64
$\Gamma_{p(1480)}$	-	0.43
Γ_s	-	0.4
ρ	ions/m^3	6.3×10^{24}
τ_e	ms	10
σ_{1-2}	m^2	2.4×10^{-25}
σ_{2-1}	m^2	3.8×10^{-25}
$\sigma_{1-3(980)}$	m^2	2.7×10^{-25}
$\sigma_{1-3(1480)}$	m^2	1.5×10^{-25}
λ_s	nm	1550
λ_{p1}	nm	980
λ_{p2}	nm	1480

Table 2. Parameters for the simulation of output ROA characteristics

Parameter	Unit	Extracted value
Ps	mW	4
$Aeff$	μm^2	80
Pp	mW	350
a_s	dB/km	0.19
a_p	dB/km	0.24
g_R	m/W	7×10^{-14}
G_{R_eff}	1/W.km	0.72
λ	nm	1550
v_s	THz	193.5
v_p	THz	206.7
λ_{p1}	nm	980
λ_{p2}	nm	1480

In the second one, the EDFA amplifier is placed at the beginning of the optical transmission path followed by the Single-mode optical fiber block and the ROA amplifier (Fig. 2).

In the next sections, simulation models are concretely introduced and hybrid amplification connections are analyzed.

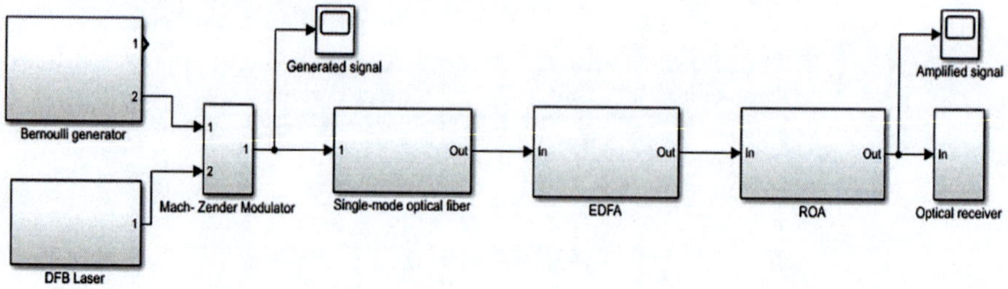

Fig. 1. The SMF + EDFA + ROA hybrid connection

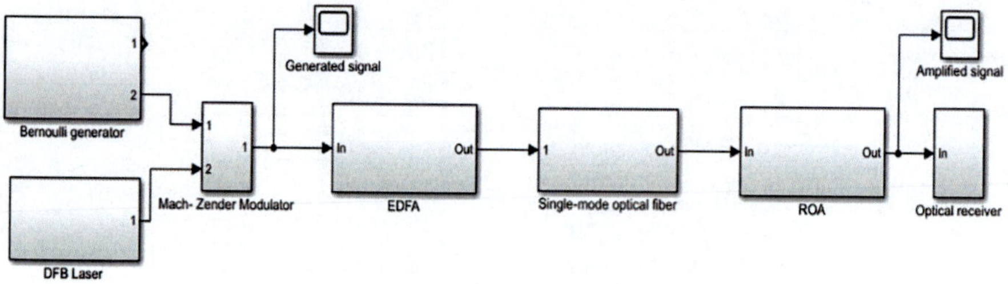

Fig. 2. The EDFA + SMF + ROA hybrid connection

3 The OA Models in the Optical Transmission Path

The presented EDFA and ROA simulation models are realized in the Matlab R2020 B Simulink environment because of its simplicity, features, and functionalities. The EDFA model included in the optical transmission path is created on principles for the steady-state modeling with a single input channel [12]. The ROA model is built in the same way, the ROA block also includes amplification and noise blocks.

Simulations are performed in the Matlab R2020 B Simulink environment on the simulation platform based on previous works [11, 15–17]. Figure 3, there is presented a complete block scheme for the optical transmission path including main optical components. This optical transmission path will be used for all realized simulations.

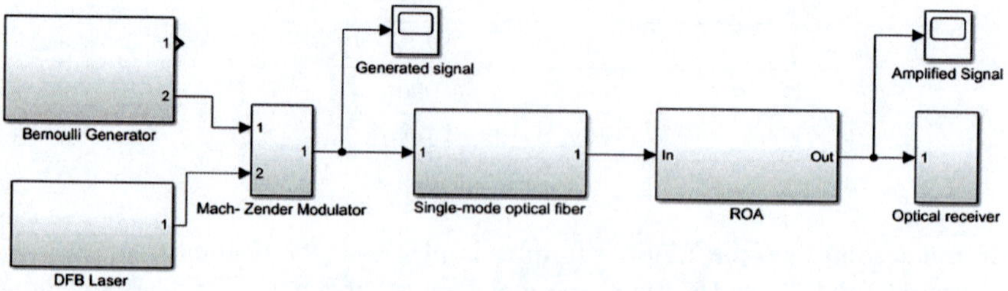

Fig. 3. The ROA amplification block in the optical fiber transmission path

The simulated optical transmission path consists of the following functional blocks representing distinguished optical components:

- Bernoulli generator
- DFB Laser
- Mach-Zender Modulator
- Single-mode optical fiber
- EDFA / ROA optical amplifiers
- Optical receiver
- Measurement units

The Bernoulli Generator block stands for a random data stream that is used to control arms of the Mach-Zehnder modulators (MZM). This data stream is shaped as a Gaussian wave to maintain a shape of real generated data signals.

The DFB Laser block represents the continuous-wave light stream produced by a distributed feedback laser that is modulated information data stream using the MZM. The polarization beam splitter divides a signal from the DFB laser into two polarization planes that are entering MZM V and MZM H branches. The SMF block represents the transmission medium of the single-mode optical fiber including relevant environmental negative effects [18, 19]. Simulation parameters (according to the ITU-T G.652.D) for the single-mode fiber transmission channel are presented in Table 3. Bear in mind that some parameters can vary depending on various wavelengths of optical signals.

Table 3. Simulation parameters for the optical single-mode fiber channel

Parameter	Unit	Extracted value
L	km	10
λ	nm	1550
α	dB/km	0.21
CD	ps/km	10
PMD	ps/(nm.$\sqrt{}$km)	0.1
Brillouin gain gB	-	4×10^{-11}
Raman gain gR	-	7×10^{-14}
Kerr effect $\gamma.L_{eff}$	-	2.23427×10^{-11}
Four Wave Mixing	-	1.1695×10^{-8}

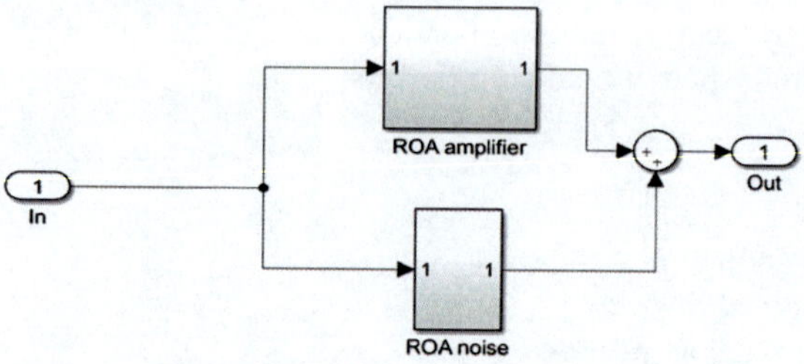

Fig. 4. The ROA block scheme

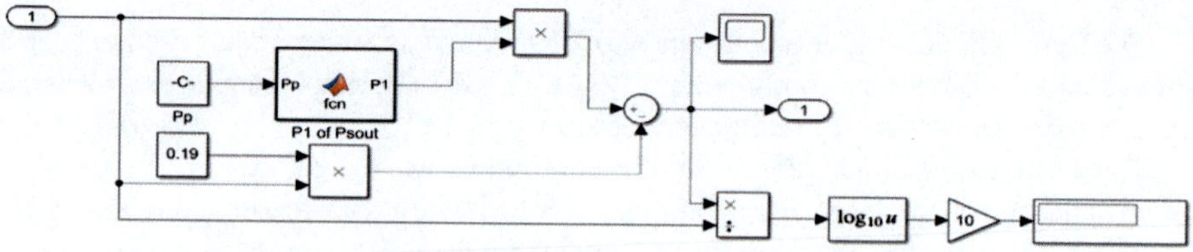

Fig. 5. The detailed scheme of the ROA amplifier block

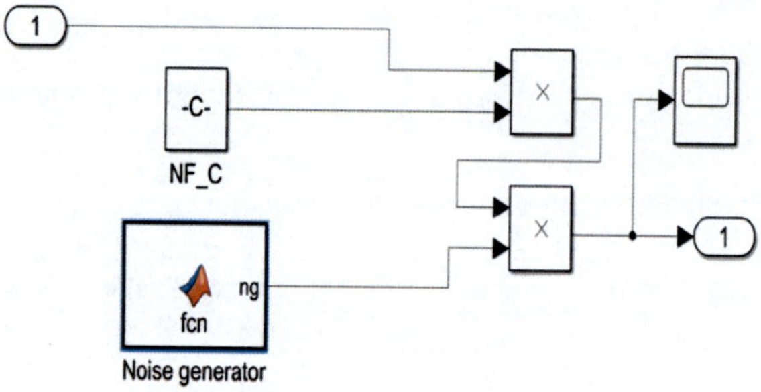

Fig. 6. The detailed scheme of the ROA noise block

The Optical receiver block is represented by a preamplifier and the PIN diode with corresponding noise sources. The optical signal can be pre-amplified before the opto-electronic conversion by the ROA block. Measurement units present Eye diagram and Constellation diagrams blocks used for evaluating transmitted signals and for estimating BER values.

The ROA block (Fig. 4) consists of the ROA amplifier block (Fig. 5) and the ROA noise block (Fig. 6) that involves noise components. Most of the parameters for the simulation of output ROA characteristics are defined in Table 2. However, parameters can differ according to applied optical fibers. Also, some parameters can vary depending on various wavelengths of optical signals and on parameters of Raman amplification, such as a pumping wavelength.

4 Results for Analyzing Hybrid Optical Amplification Connections

All simulations are realized in Matlab R2020B. Two steps of simulations are executed and described more in the following paragraphs. The first step is focused on selecting a more suitable hybrid amplification connection for practical utilization from a viewpoint of cumulative signal amplification. Consequently, the second step is dedicated to analyzing and selecting appropriate pumping power for signal amplification.

For the simulation, we used the optical transmission path model with EDFA and ROA simulation blocks, where the working signal wavelength is $\lambda_{s1} = 1550$ nm, the generated input signal $P_s = 4$ mW and the fiber length $L = 100$ km.

After initial simulations, a selection of more suitable hybrid amplification connections can be realized. Based on simulation results, it can be confirmed that the EDFA + SMF + ROA connection is more powerful compared to the SMF + EDFA + ROA connection. For the most efficient way of utilization for hybrid amplification connections, appropriate parameters of the pumping source must be analyzed and selected. The first series of simulations are realized at the signal 1550 nm wavelength range. Results are displayed in Fig. 7.

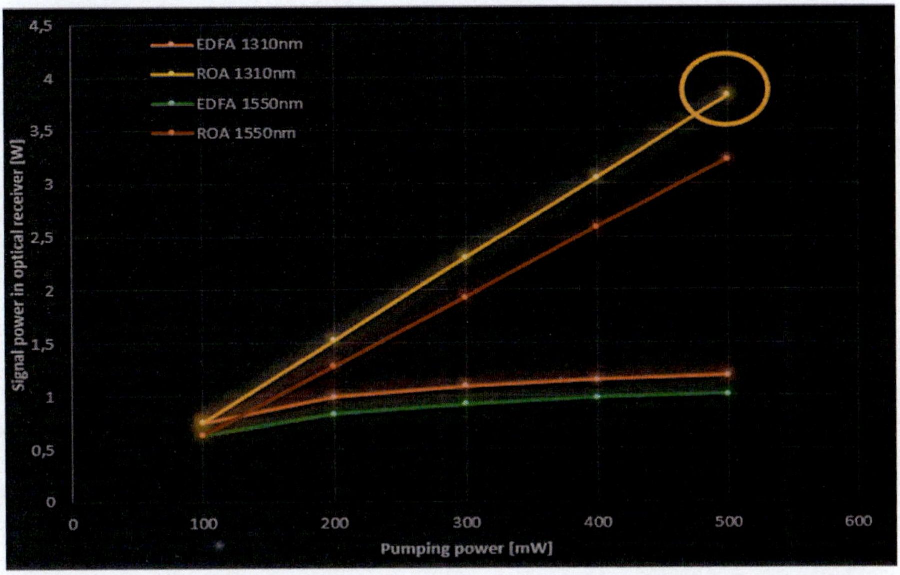

Fig. 7. Dependencies of signal power levels in the connection of SMF + ROA + EDFA

After changing the signal wavelength λ_{s2} in the DFB Laser block to the 1310 nm value, the second series of simulations can be realized (Fig. 8). As can be seen in the following graph highlighted with yellow and orange color, a signal power level in the optical receiver will be higher.

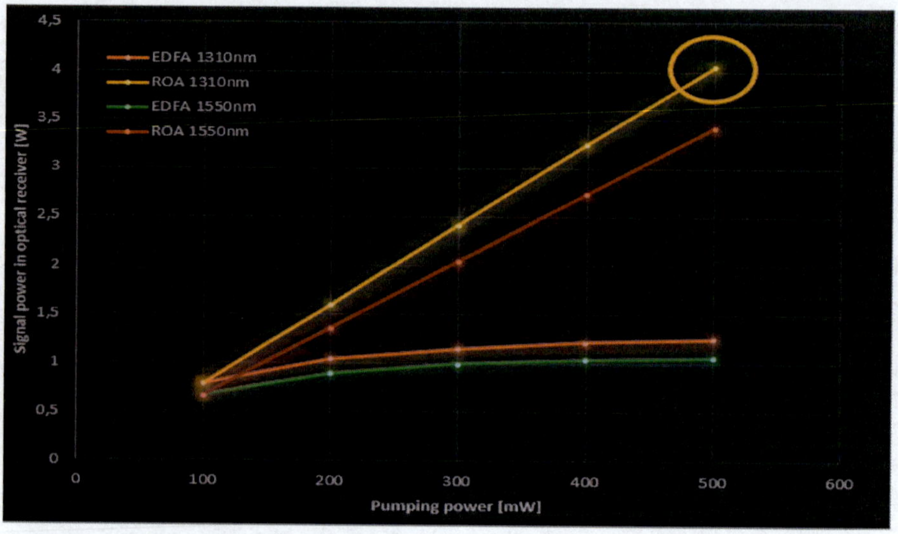

Fig. 8. Dependencies of signal power levels in the connection of EDFA + SMF + ROA

5 Conclusion

This contribution presents innovations in the simulation of the optical transmission path focused on the hybrid optical amplification connections, based on performance analysis of EDFA and ROA amplifiers. The created EDFA and ROA simulation models are expandable by a simple configuration of functions in the Matlab Simulink program environment. Both EDFA and ROA models can execute simulations using different parameters, such as pumping wavelengths and pumping power levels.

Hybrid optical amplification connections are the future of amplifying in the optical transmission path. Based on simulation results for performance analysis, we can conclude that the hybrid EDFA + SMF + ROA connection is more beneficial when the EDFA amplifier is connected as a booster and the ROA amplifier as a preamplifier. After changing the signal wavelength, this hybrid connection is still more efficient in terms of the resulting signal amplification.

Presented simulations of the hybrid optical amplification connections can be used for analyzing and selecting appropriate parameters of the pumping source for the Erbium Doped Fiber and Raman amplifiers, concretely a pumping wavelength and a pumping power level, utilized at the signal transmission with advanced modulation formats in optical long-haul and enhanced metropolitan networks.

Acknowledgment. This paper is a part of the "Utilization of Web-based Training and Learning Systems at the Development of New Educational Programs in the Area of Optical Wireless Technologies" KEGA agency project - 034STU-4/2021.

References

1. Agrell, E., et al.: Roadmap of optical communications. J. Opt. **18**(6), 063002 (2016). https://doi.org/10.1088/2040-8978/18/6/063002
2. Freeman, R.L.: Fiber-Optic Systems for Telecommunications. Wiley-Interscience, New York (2002).ISBN 978-0471414773
3. Kartalopoulos, S.V.: DWDM - Networks, Devices and Technology. IEEE Press & Wiley-Interscience, New Jersey (2003). ISBN 978-0471269052
4. Mukherjee, B.: Optical WDM Networks. Springer Science+Business Media, New York, USA (2006). ISBN 978–0387290553
5. Desurvire, E.: Erbium-Doped Fiber Amplifiers: Principles and Applications. Wiley, New York (2002). ISBN 978-0-471-26434-7
6. Zyskind, J., Srivastava, A.: Optically Amplified WDM Networks. Academic Press, New York (2010). ISBN 9780123749659
7. Krummrich, P.M.: Amplification and Regeneration. In: Roadmap of optical communications. IOP Publishing, UK. https://doi.org/10.1088/2040-8978/18/6/063002
8. Ajibodu, F.A., Adetona, Z.A., Ojo, B.A.: Performance analysis of erbium doped fiber amplifier's and FWM characteristics in a passive optical network. Int. J. Optoelectronic Eng. **6**(1), 1–7 (2016). https://doi.org/10.5923/j.ijoe.20160601.01
9. Aryandhika, R.B., Isnawati, A.F., Zulherman, D.: Performance analysis of erbium-doped fiber amplifier (EDFA) and hybrid optical amplifier in NG-PON2 based on TWDM-PON system. IOP Conference Series Material Science and Eng. **982**, 012027 (2020). https://doi.org/10.1088/1757-899X/982/1/012027
10. Binh, L.: Optical fiber communication systems with MATLAB and Simulink Models. CRC Press, Taylor and Francis Group, Boca Raton (2015). https://doi.org/10.1201/b17781
11. Róka, R.: Enhanced simulation tools for analyzing signal processing techniques in the optical transmission path. In: Handbook of research on 5G networks and advancements in computing, electronics, and electrical engineering. IGI Global, Hershey (2021). https://doi.org/10.4018/978-1-7998-6992-4.ch014
12. Róka, R.: Simulation of the optical erbium-doped fiber amplification for performance analysis. Journal of Electrical Engineering – EČ, **72**(3), 168–175 (2021). https://doi.org/10.2478/jee-2021-0023
13. Róka, R.,Šalík, P.: An effective integration of the PM 16-QAM modulation in enhanced metropolitan networks with the EDFA amplification. Journal of Electrical Engineering – EČ, **71**(5), 317–325 (2020). https://doi.org/10.2478/jee-2020-0043
14. FiberLabs, Inc.: Erbium-Doped Fiber Amplifier (EDFA). /www.fiberlabs.com/glossary/erbium-doped-fiber-amplifier
15. Róka, R., Čertík, F.: Simulation and analysis of the signal transmission in the optical transmission medium. In: SIMULTECH, Colmar, France, pp. 219–226 (2015). https://doi.org/10.5220/0005569602190226
16. Šalík, P., Róka, R., Tomáš, G.: Simulation platform of the optical transmission system in Matlab Simulink. Procedia Computer Science **134**, 196–203 (2018). https://doi.org/10.1016/j.procs.2018.07.162
17. Čučka, M., Šalík, P., Róka, R., Münster, P., Filka, M.: Simulation models of pulse generator for OTDR in Matlab and VPIphotonics. In: TSP, Athens, Greece, pp. 179–183 (2018). https://doi.org/10.1109/TSP.2018.8441274

18. Róka, R.: The environment of fixed transmission media and their negative influences in the simulation. Int. J. Mathematics Computers in Simulation **9**, 190–205 (2015)
19. Róka, R., Šalík, P.: Impact of environmental influences on multilevel modulation formats at the signal transmission in the optical transmission medium. Int. Journal of Communication Networks and Information Security - IJCNIS, **9**(1), 76–87 (2017). https://doi.org/10.54039/ijcnis.v9i1.2134

Determination of Phobic Disorders Using Virtual Reality Technologies

Artem Obukhov(✉), Denis Dedov, Nikita Chebotov, Alexandra Nazarova, and Daniil Teselkin

Tambov State Technical University, 392000 Tambov, Russian Federation
obuhov.art@gmail.com

Abstract. The article considers the possibility of using virtual reality technologies to diagnose phobic disorders by modeling stressful situations. This makes it possible to determine a phobia in a virtual environment in conditions that are safe for humans, and to predict in advance the occurrence of a panic state that is dangerous both for the person himself and for his environment. The study considers possible approaches to fixing the state of a phobic disorder, as well as the developed software for virtual scenes to determine acrophobia and claustrophobia. The presented virtual scenes simulate the main scenarios for the emergence of these phobias, confronting a person with their manifestations in virtual space. As a metric in this work, we used an estimate of the heart rate. As a result, it was found that virtual reality technologies make it possible to induce signs of a phobia in a person suffering from a corresponding disease, which can be recorded by an expert. Experiments have also shown that the heart rate metric is not effective enough. In further studies, it is necessary to expand the list of metrics for assessing the human condition for a comprehensive, automated and more accurate diagnosis.

Keywords: Virtual reality · Phobic disorders · Detection of phobias · Simulation of stressful situations

1 Introduction

In the past few decades, there has been an increase in the number of mental disorders, in particular anxiety-phobic disorders, which is partially explained by the improvement in the quality of medical diagnostics. However, the living conditions of society in modern realities are characterized by an increase in the number of stress factors, an increase in information flow, a deterioration in the general state of the environment and imperfection of a person's psychophysical adaptation to them.

Accordingly, this leaves its mark on those areas of professional activity in which the influence of stress factors is great. An unexpectedly manifested phobia in the workplace does not allow a person to work constructively in a group, perform labor functions, and adequately respond to the environment. Subjective ideas about a particular frightening situation cause corresponding behavioral reactions to it. Timely identification and informing a person about the presence of a phobia that is incompatible with a specific

R. Silhavy (Ed.): CSOC 2022, LNNS 501, pp. 79–89, 2022.
https://doi.org/10.1007/978-3-031-09070-7_8

professional activity makes it possible to timely adjust the choice of specialty, start treatment of anxiety-phobic disorder, and improve the quality of life.

Recognition of this group of neurotic disorders causes certain difficulties, because often the boundary between normal psycho-physical reactions to stress and pathology is blurred. Solving this problem requires a multidisciplinary approach to diagnosis.

The use of virtual reality (VR) systems in the study and therapy of patients with neurotic disorders has been going on for several decades [1–4]. VR technologies make it possible to reproduce a complex effect on the main human modalities (auditory, visual, kinesthetic), as well as are highly valid: plunging into digital reality, the user perceives virtual stimuli as somewhat weakened real ones and reproduces the corresponding behavioral responses to it. Extensive possibilities for modeling objects and processes in virtual reality allow the user to interact with a variety of stimuli that in the real world would be unattainable, dangerous or stressful.

1.1 Overview of Phobias that Are Dangerous for Human Labor Activity

The presence of certain non-hypertrophied fears is present in each individual (for example, fear of unusual phenomena, sudden changes in the environment, disease, hunger, etc.) and works as a protective mechanism for the instinct of self-preservation. In the face of real danger in the human body, all life support systems are activated, allowing to overcome certain difficulties. However, in the presence of a phobia - an irrational, uncontrollable fear - a person is both unable to ensure his own survival in a moment of danger, and during his life tries to avoid a repetition of collision with the object of his fear, which is often unjustified and leads to wrong decisions.

It is most common for people who study or work in specialties that require increased responsibility from them to fall into panic states. At the same time, the occurrence of this condition for the designated category of persons threatens with serious consequences for the health, safety and life of both the person himself and his environment.

Consider the main phobias that pose a danger to human labor activity:

- Pyrophobia. Relevant for firefighters. Fear of fire is a natural defense mechanism of a person and turns into pathology in cases where thoughts about fire and burns become intrusive. Often, this phobia is comorbid with other obsessive fears (for example, claustrophobia, amaxophobia, aerophobia), i.e. those stressful situations in which the occurrence of fire is likely.
- Claustrophobia. Matters in the mining industry. The claustrophobia has a growing sense of anxiety when in confined spaces (room, train carriage, shower stall). If possible, a person leaves open the doors in the premises and tries to stay as close to the exit as possible. The situation is aggravated if there are no windows in the room and the person is unable to leave it.
- Acrophobia. Relevant for installers, builders, pilots. The fear of heights is a natural human fear that contributes to its survival. However, pathological forms of acrophobia interfere with a person's life. So, starting from the "highly specialized" fear of flying on an airplane (aerophobia), the problem of acrophobia can reach extreme stages when it is difficult for a person to go down the stairs or stand on a stool.

- Hemophobia. Plays a large role in the medical field. This phobia is co-morbid with other phobias associated with invasive intervention (trypanophobia - fear of injections, tomophobia - fear of surgical intervention, traumophobia). In the presence of fear of blood, a person's reaction to it (or something that may be indirectly related to blood) is usually hypertrophied, for example, when a hematophobe receives a wound, a degree of shock is disproportionate to the actual loss of blood.

Thus, an actual problem is to identify such disorders not in real conditions of work or educational activity, but in artificially simulated conditions with the supervision of medical specialists and experts.

1.2 Overview of Existing Approaches to the Diagnosis of Phobias in Virtual Reality

Identifying various phobias is not a trivial task, since not every situation can be reliably and, at the same time, safe to simulate. One of the ways to solve this problem is to use virtual reality technologies [5, 6]. Consider some research in this area.

- The most common phobia test in VR is the acrophobia test [7, 8]. The scenario consists in a brief introduction of the user into VR (for example, ascending to a height in an elevator, pressing buttons), then the user is at a high point and can view the scene from a bird's eye view. Then it all depends on the user - he can get off the support surface and "fall" or stop just looking at the scene. The individual reactions of the user, which are recorded by medical workers, and his subjective assessment in the aggregate will characterize the degree of acrophobia.
- Glossophobia or fear of public speaking [9] can be overcome in the same way as most other phobias - by direct regular interaction with the object of phobias, increasing the frequency and intensity of these interactions. In the Limelight VR application, there are several possible scenarios (scene, meeting) where the user can exercise with the virtual audience.
- For the treatment of arachnophobia, various applications for virtual reality are used, in which the user can gradually, in accordance with his sensations, let the virtual spider come closer to him and thereby fight his phobia [10–13].
- Also, to work with people suffering from various anxiety disorders, they use the virtual reality platform for psychology and mental health Psious. For example, to work with aerophobia, this platform presents a simulator for practicing the scenario of a person's flight on an airplane. First, a person goes through several scenes where he is not on board the plane (in his house, on a taxi ride, at the airport) - these scenes are necessary for working out anticipatory anxiety. Next, a person is immersed on board, the plane takes off and falls into the turbulence zone, after which it lands.
- There are also simulators for astraphobia (fear of thunder and lightning), nyctophobia (fear of the dark) and all kinds of zoophobia.

It should be noted that in the examples considered, one of the key questions is an objective assessment - whether a person experiences a certain phobia in the process of passing through a virtual scene or his state is associated with other external or internal factors. Therefore, at the stage of identifying the presence of phobias, it is important to determine the tools, technologies and approaches for assessing the presence of a phobia in a person.

1.3 Overview of Approaches to Detecting a Phobic Condition in Humans

In the modern world, phobias [14] are diagnosed by a psychologist or psychiatrist through observation and questionnaires [15]. This requires an integrated approach, a lot of time. Also, one cannot be limited to the conclusions of one doctor, because an incorrect diagnosis can lead to serious consequences for the patient. Often, a person may not know about the presence of a phobia or not give it due importance, not seek medical help.

With the development of psychotherapy, humanity has identified most of the symptoms of phobias. The presence of a combination of signs such as increased heart rate, anxiety, shortness of breath, hyperhidrosis (increased sweating), nausea may well indicate that a person has a phobia. Accordingly, confronting the subject with various stressful situations in virtual reality and comparing the above parameters with the norm, we will be able to diagnose phobia with greater accuracy.

Various devices and technologies can be used to capture phobic signs.

1. Pulse. The simplest approach is to measure the heartbeat, since it is enough to use any heart rate monitor that provides the required accuracy. Even fitness trackers can be used as such a device, since they are quite convenient for constantly monitoring the heart rate and detecting sudden surges [16].
2. Temperature. Until recently, the determination of the subject's temperature was a rather laborious process, since mercury and electronic thermometers took a long time to determine the temperature, being rather fragile and inconvenient devices. At the same time, specialized devices were too cumbersome and unsuitable for accurate measurement of human temperature [17]. However, with the onset of the pandemic, a need arose for the ability to quickly determine a person's temperature and infrared non-contact thermometers were developed. Although they have a rather large error, they are able to record an increase in body temperature by more than half a degree, which makes it possible to identify the fact of manifestation of a phobic disorder.
3. Electroencephalogram. The electroencephalogram (EEG) allows to determine the nature of brain activity, namely, to record the state of anxiety [18, 19]. EEG, in contrast to positron emission tomography and functional magnetic resonance imaging, is capable of real-time (for example, every millisecond) providing the brain's response to stimuli. It is equally important that the process of registration of electroencephalography, even with sensitivity to motion that creates interference, gives a person more mobility and less discomfort. This does not create additional distractions when immersed in a stressful situation.
4. Hyperhidrosis. The fixation of hyperhidrosis is possible so far only by test methods, for example, a sensor glued with a plaster to the human skin [20]. But so far, reliable data on increased sweating and nausea can only be obtained with the help of a doctor after observation during the test and after talking with the subject, respectively.

Among the considered methods, the easiest to implement is the determination of the user's temperature during immersion in virtual reality. However, the addition of temperature and EEG measurements in the future will improve the accuracy of assessing the presence of a phobic disorder.

2 Methods

The aim of the study is to develop software that will make it possible to diagnose with high accuracy the presence of phobic disorders in a user using virtual reality technologies. Let's consider the main components of the subject area and form an optimization problem.

To identify a phobic disorder, we use the accuracy criterion A for recognizing a phobia. This criterion depends on the hardware and software used to solve the problem of determining the phobia.

Hardware H is divided into components for imaging (VR headset, monitor), sensors (EEG, thermometer, heart rate monitor, and others), and controls (VR controllers, keyboard, mouse, and so on).

S software consists of many visual effects E, virtual scene objects O, scripts of program code, scripts D. Virtual objects shape the scene, visual effects add realism to it, and scripts "animate" the scene by allowing the user to interact with its elements.

S software consists of many visual effects E, virtual scene objects O, scripts of program code, scripts D. Virtual objects shape the scene, visual effects add realism to it, and scripts "animate" the scene by allowing the user to interact with its elements.

$$A(H, S, F) \to \max \tag{1}$$

Let us consider possible approaches to the choice of function A. The calculation of this criterion is carried out in accordance with the percentage of correctly diagnosed manifestations of phobias to the total number of those tested:

$$A = \frac{\sum_{i=1}^{N} m_i}{N} \tag{2}$$

where N is the total number of diagnosed users;

m_i- assessment of the correctness of the diagnosis, determined by the formula:

$$m_i = \begin{cases} 1, \text{if } \forall i (P_i \geq P_i^* + \Delta P_i), \\ 0, \text{otherwise.} \end{cases} \tag{3}$$

where P_i is the value of the i-th metric for assessing the state of a person after passing through the virtual scene;

P_i^* is the value of the i-th metric for assessing the state of a person before the passage of the virtual scene (normal value in a calm state);

ΔP_i is a permissible change in the i-th metric for assessing a person's condition, the excess of which determines the presence of a phobia.

Thus, the presence of a phobic disorder in a patient is determined by his physiological parameters (temperature, pulse, etc.). If, while a patient is in a virtual environment, his physiological parameters change markedly and go beyond the normal range, then this confirms the presence of a phobia in a person.

In addition to the choice of software and hardware parameters affecting the accuracy of diagnostics, there are a number of conditions for a virtual scene that it must satisfy in order to gain advantages over both traditional exposure therapy and existing solutions for VR psychotherapy:

1. The virtual environments provided in the software must contain the necessary set of virtual objects O and visual effects E, which determine the level of detail and graphical execution for the maximum possible immersion of the patient in the virtual environment. A high level of detail in the environment is determined by filling the environment with three-dimensional objects, animations and sounds, as well as human interaction with the virtual environment and receiving a response from the environment. Primitive graphics and scanty content of the virtual scene will not be able to convince the user of the credibility of what he saw. Therefore, when developing virtual scenes, you need to pay attention not only to the object causing a certain phobia, but also to the rest of the environment on the stage.
2. The complexity of the scenes being developed should be moderate. A moderate complexity of a scene means such a number of different three-dimensional objects and their detail, images and their resolutions, sounds and animations, so that the scenes are as believable as possible for the user and not too demanding on computer resources. For example, the higher the detail of a three-dimensional object, the more resources the computer's video card will need to display these objects on the scene, and there may be several such objects on the scene.
3. The software should provide interaction of the patient with objects in the virtual environment. In addition to the graphic component, for the credibility of the scene, it is necessary that the user can interact with various objects in the virtual environment. For example, pick up a ball, press the elevator button. It is necessary that the objects of the environment can interact with each other. For example, throwing a ball to the floor, it will bounce upward, as it happens in reality.
4. The software must ensure the interaction of the therapist with the virtual environment. This is required so that the virtual environment does not follow the same scenario. For example, a scene of a flight on an airplane for aerophobia therapy. The therapist can increase the number of flights or shorten the flight time if the patient is very scared, or add turbulence if this is what causes the fear during the flight.

Next, we will carry out the development of the necessary software and the choice of hardware for the diagnosis of two phobias - acrophobia and claustrobia.

3 Results

3.1 Software and Hardware Structure for Diagnosing Phobias in Virtual Reality

We divide the components necessary for solving the research problem into two components in accordance with formula (1): software S, consisting of the necessary models of virtual objects and visual effects and scripts for the correct functioning of the VR scene; hardware *H*, which includes equipment for displaying a virtual scene, as well as specialized components to enhance the immersion in the scene. The general scheme is shown in Fig. 1.

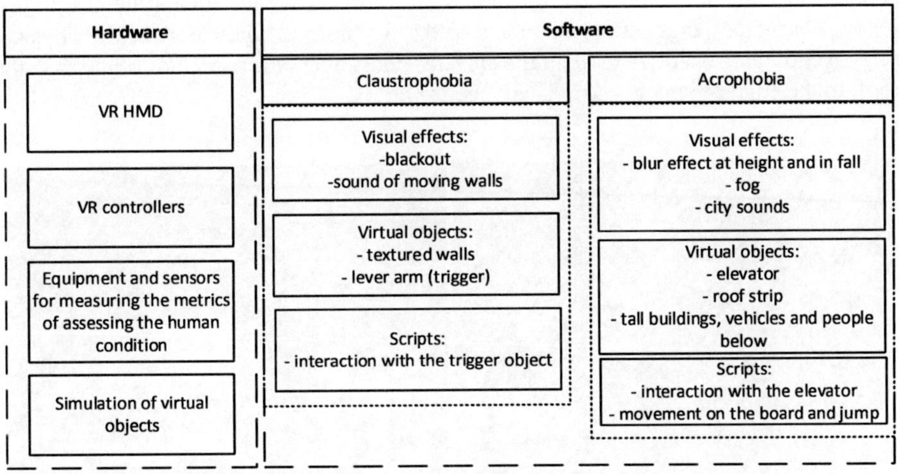

Fig. 1. The structure of software and hardware for identifying phobias

The hardware consists of a headset and VR controllers used to both play the VR scene and interact with its objects. This category also includes various third-party equipment that simulates the necessary objects for each phobia. For example, a shaking plank in a scene with acrophobia, movable panels to simulate moving scenes, various kinds of liquids, warm or cold objects of different densities and textures. Hardware also includes devices for collecting metrics for analyzing the presence of a phobia: heart rate monitor, thermometer, encephalograph, and others.

The software part will vary for each phobia. The software components can be grouped into three blocks. The first block includes visual effects, for example, to test a person for acrophobia, the scene will include blurring of vision at heights and in the fall, fog and sounds of the city. The second block contains virtual objects, for example, for claustrophobia, virtual objects are represented by a lever, interaction with which starts the script and moves the walls. The third block includes files of program code for the implementation of various kinds of scenarios. So, for the script to test the subject for acrophobia, scripts for interacting with the elevator, imitating the activity of the city, and falling are necessary. In case of claustrophobia, scripts move the walls.

3.2 Results of Detecting Claustrophobia and Acrophobia Using Virtual Reality Technologies

Next, it is necessary to develop virtual scenes and conduct experimental research. The tests were carried out on 10 respondents, 1 of whom has acrophobia, 1 is claustrophobic, and 8 have no pronounced phobias. A handheld heart rate monitor was used to measure quality metrics. And the medical professional acted as an expert.

Consider the scenes implemented for a virtual reality VR HMD to detect acrophobia and claustrophobia.

The scene for testing a person for acrophobia includes a scenario of falling from a height. The user starts the scene in a closed elevator and spends some time in it to test for the absence of anticipatory fears and to fix the "normal" indicators of the physical and psychological state. After that, the elevator doors open, the user finds himself on the roof, to the edge of which a strip is attached (Fig. 2).

Fig. 2. Scene for testing acrophobia (beginning).

A view of the city opens from a tall building and, stepping on the plank, a person can fully look around. In the process, the user can either step off the support and be in free fall with subsequent "death", or remain in place and leave the scene (Fig. 3).

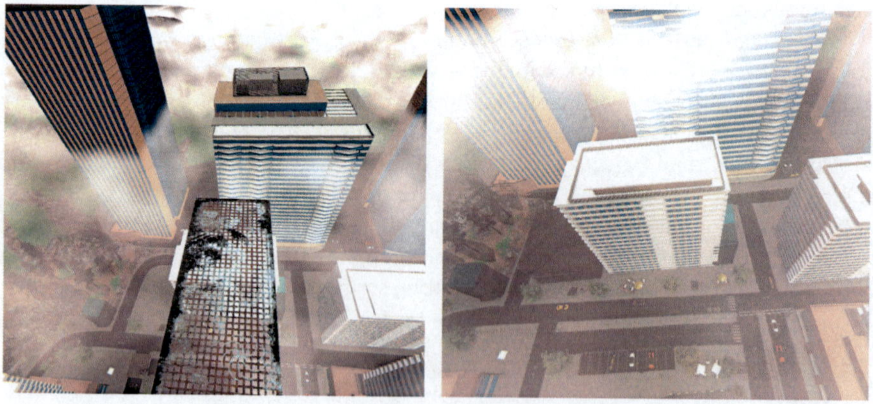

Fig. 3. Scene for testing acrophobia (completion).

Next, we will consider the results of the tests carried out on the first stage. As a metric of the presence of a phobia in a person, the readings of his pulse were used, and his behavior in a stressful situation was monitored by a medical professional. Respondents who do not have a fear of heights did not show significant changes in heart rate and pronounced behavioral reactions. And when testing a claustrophobic respondent, an increase in pulse rate and tachycardia was detected, an increase in sweating was noted (on the inner surface of the arms), and he was not able to make a jump from the bar.

When testing a person for the presence of claustrophobia, the subject, putting on a VR HMD, finds himself in a dimly lit enclosed space with a roar in the center. To run the claustrophobic test script, the user must interact with the lever. After that, the walls will begin to move towards the center (Fig. 4). Returning the lever to its original position stops the movement of the walls and, accordingly, the test.

Fig. 4. Claustrophobic testing scene.

Similarly, 9 respondents not prone to claustrophobia showed no dramatic changes in heart rate. And just like in the previous testing, the claustrophobic respondent had an increased heart rate, sweating, and corresponding behavioral reactions in the form of disorientation and mild panic were recorded.

4 Discussion

When summing up, it was found that a noticeable change in the pulse rate in people with phobic disorders persists after repeated passage of the virtual scene. On the other hand, in some cases, a significant change in heart rate was also observed in respondents without phobias once when immersed in virtual reality for the first time, however, already during the second pass, a decrease in the influence of stress factors on a person was observed (i.e. immersive the impact on a person is reduced). However, measuring only the heart rate is often not an adequate measure for assessing the presence of fear. Similar conclusions were obtained in earlier studies [21]. Pulse measurement is an extremely subjective metric and often demonstrates the general degree of anxiety of a person who is in VR, and does not indicate the intensity of fear at certain points in time. Also, the use of non-professional means to record changes in a person's physical condition (fitness trackers) does not allow obtaining sufficiently accurate results.

Also, visual control of a medical specialist was possible to establish the presence / absence of a phobic disorder. This means that the immersive effect of VR really causes the user to have corresponding negative associations and increased anxiety. This provides an opportunity for further research in the chosen area.

A promising direction in the development of this work is the use of EEG for assessing brain activity, as well as measuring body temperature, sweating, etc. Thus, only an integrated approach will allow reaching a sufficiently accurate level of diagnosis of phobic disorders.

5 Conclusion

In accordance with the structure of this article, research has been carried out in the field of diagnostics of phobic disorders using virtual reality technologies.

To simulate certain possible phobias, the corresponding software was developed in the Unity development environment with the necessary hardware and software, which made it possible to achieve a sufficient degree of user immersion in the simulated environment.

When diagnosing phobic disorders, subjective assessments of the presence / absence of phobias were taken by a medical professional, and a heart rate monitor was used to measure qualitative metrics. On the basis of his indicators, graphs of the dependence of heart rate on time were built. As a result of practical tests on a group of 10 respondents, it was found that heart rate does not correlate with the degree of anxiety. For users with acrophobia and claustrophobia, the graphs of heart rate changes were not too different from those of the rest of the respondents. Also, the graphs obtained did not allow us to accurately identify the time of exposure of a phobia to a person, reflecting only a gradual increase in heart rate values. Therefore, in further research, it is necessary to expand the list of metrics for assessing the human condition for complex automated and more accurate diagnostics.

This work was supported by the laboratory of medical VR simulation systems for training, diagnostics and rehabilitation.

References

1. Jerdan, S., Grindle, M., van Woerden, H., Boulos, M.: Head-mounted virtual reality and mental health: critical review of current research. JMIR Serious Games **6**(3), e9226 (2018)
2. Jadhav, L., Parekh, K., Gupta, V., Chandak, S.: Using virtual reality for therapeutic treatment of phobia. In: 2020 International Conference on Convergence to Digital World-Quo Vadis (ICCDW). pp. 1–5. IEEE, Mumbai (2020)
3. Rizzo, A., et al.: Virtual reality applications for the assessment and treatment of PTSD. In: Handbook of Military Psychology, pp. 453–471. Springer, Cham (2017). https://doi.org/10.1007/978-3-319-66192-6_27

4. Inozu, M., Celikcan, U., Akin, B., Cicek, N.: The use of virtual reality (VR) exposure for reducing contamination fear and disgust: can VR be an effective alternative exposure technique to in vivo? J. Obsessive-Compulsive Related Disorders **25**, 100518 (2020)
5. Horváthová D., Siládi V., Lacková E.: Phobia treatment with the help of virtual reality. In: 2015 IEEE 13th International Scientific Conference on Informatics, pp. 114–119. IEEE, Poprad (2015)
6. Wechsler, T., Kümpers, F., Mühlberger, A.: Inferiority or even superiority of virtual reality exposure therapy in phobias? —a systematic review and quantitative meta-analysis on randomized controlled trials specifically comparing the efficacy of virtual reality exposure to gold standard in vivo exposure in agoraphobia, specific phobia, and social phobia. Front. Psychol. **10**, 1758 (2019)
7. Abdullah, M., Shaikh, Z.: An effective virtual reality based Remedy for acrophobia. Int. J. Adv. Comput. Sci. Appl. **9**(6), 162–167 (2018)
8. Donker, T., et al.: Effectiveness of self-guided app-based virtual reality cognitive behavior therapy for acrophobia: a randomized clinical trial. JAMA Psychiat. **76**(7), 682–690 (2019)
9. Stupar-Rutenfrans, S., Ketelaars, L., van Gisbergen, M.: Beat the fear of public speaking: Mobile 360 video virtual reality exposure training in home environment reduces public speaking anxiety. Cyberpsychol. Behav. Soc. Netw. **20**(10), 624–633 (2017)
10. Miloff, A., Lindner, P., Hamilton, W., Reuterskiöld, L., Andersson, G., Carlbring, P.: Single-session gamified virtual reality exposure therapy for spider phobia vs. traditional exposure therapy: study protocol for a randomized controlled non-inferiority trial. Trials **17**(1), 1–8 (2016)
11. Rosenbaum, D., et al.: Neuronal correlates of spider phobia in a combined fNIRS-EEG study. Sci. Rep. **10**(1), 1–14 (2020)
12. Shiban, Y., Fruth, M., Pauli, P., Kinateder, M., Reichenberger, J., Mühlberger, A.: Treatment effect on biases in size estimation in spider phobia. Biol. Psychol. **121**, 146–152 (2016)
13. Notzon, S., et al.: Psychophysiological effects of an iTBS modulated virtual reality challenge including participants with spider phobia. Biol. Psychol. **112**, 66–76 (2015)
14. Eaton, W., Bienvenu, O., Miloyan, B.: Specific phobias. The Lancet. Psychiatry **5**(8), 678–686 (2018)
15. Appukuttan, D.: Strategies to manage patients with dental anxiety and dental phobia: literature review. Clin. Cosmet. Investig. Dent. **8**, 35–50 (2016)
16. Turki, A., Jani, M., Ding, K., Zhang, R., Behbehani, K.: An investigation of heartrate sensing accuracy by wrist-worn fitness tracking devices. In: 2019 41st Annual International Conference of the IEEE Engineering in Medicine and Biology Society (EMBC), pp. 3337–3340 (2019)
17. Pušnik, I., Miklavec, A.: Dilemmas in measurement of human body temperature. Instrum Sci. Technol. **37**(5), 516–530 (2009)
18. Wojcik, G., et al.: New protocol for quantitative analysis of brain cortex electroencephalographic activity in patients with psychiatric disorders. Front. Neuroinform. **12**, 27 (2018)
19. Ribas, V., et al.: Pattern of anxiety, insecurity, fear, panic and/or phobia observed by quantitative electroencephalography (QEEG). Dementia & Neuropsychologia **12**, 264–271 (2018)
20. Currano, L.J., Sage, F.C., Hagedon, M., Hamilton, L., Patrone, J., Gerasopoulos, K.: Wearable sensor system for detection of lactate in sweat. Sci. Rep. **8**(1), 1–11 (2018)
21. Felnhofer, A., Kothgassner, O., Hetterle, T., Beutl, L., Hlavacs, H., Kryspin-Exner, I.: Afraid to be there? Evaluating the relation between presence, self-reported anxiety, and heart rate in a virtual public speaking task. Cyberpsychol. Behav. Soc. Netw. **17**(5), 310–316 (2014)

A Detection of a Sequence of Jump-Like Pulses Against Random Distortions

Yury Korchagin[1], Konstatnin Titov[1], Olga Zavalishina[1], Alexander Faulgaber[2](✉), and Dmitry Shepelev[2]

[1] Voronezh State University, Universitetskaya sq. 1, 394018 Voronezh, Russia
[2] National Research University "MPEI", Krasnokazarmennaya str. 14, 111250 Moscow, Russia
a.n.faulgaber@mail.ru

Abstract. In this paper, a wide class of effects defined as a sequence of jump-like pulses with a harmonic filling and discontinuous points of the first order at the moments of appearance and disappearance is considered. This kind of pulses is often characterized by the unknown amplitudes and initial phases observed against the Gaussian random distortions. To detect such pulses, the algorithm synthesized based on the maximum likelihood method is proposed. The structure and the main statistical properties of the introduced detection algorithm are studied. The technique for the calculation of the detection characteristics is presented. The dependence of the type I error probability upon the threshold as well as the dependence of the type II error probability upon the signal-to-noise ratio are demonstrated. A statistical simulation for the obtained algorithms is carried out. It is established that the proposed detector is operable while the theoretical formulas describing its characteristics are in good agreement with the corresponding experimental data in a wide range of values for the parameters of the analyzed pulse sequence. #CSOC1120.

Keywords: Pulse sequence · Jump-like signal · Signal detection · Maximum likelihood method · Parametric prior uncertainty · Type I and type II errors · Statistical simulation

1 Introduction

One of the important problems of the automatic control theory is synthesis and analysis of the algorithms for processing jump-like (stepwise) changes of the diagnosed processes and systems in the presence of random distortions [1]. The jump-like disturbances can be of different physical nature and thus they can be considered as either information processes to be detected, or as interferences in the control channel. In the latter case, the detection and measurement of the disturbance parameters are useful for their timely prevention. In a number of practical applications, the abrupt changing of the object parameters can be the result of an addition of a certain number of elementary random or deterministic effects and can be interpreted as a pulse sequence. The examples of such disturbances are a flash of optical noise [2], a radar signal reflected by a composite

R. Silhavy (Ed.): CSOC 2022, LNNS 501, pp. 90–104, 2022.
https://doi.org/10.1007/978-3-031-09070-7_9

object [3, 4] and the signals propagating in the hydro-acoustic medium [4], etc. To successfully detect a pulse sequence against random distortions, numerous practical applications of communication and navigation theory are required [5, 6]. In this case, the pulse sequence can be considered non-deterministic, that is, the one described by a deterministic function with the random or the unknown parameters. For example, in radar systems, the initial phase of the signal is a random parameter due to the reflection of the signal from the object. The signal amplitude and duration can also be random due to the addition of copies of the signal reflected from different parts of an extended object. In radio communication and data transmission systems operating in a complex interference environment, the signal parameters undergo significant changes when passing through the propagation channel and become random parameters for each of the pulses.

Currently, in [7], the wavelet-based algorithms are proposed for an online detection of a train of the unknown pulse signals distorted by a random noise, while the algorithms for detecting the sequence of narrowband radio pulses with the unknown amplitudes and initial phases are studied in detail in [5, 6]. In the papers [8–11], the algorithms for detecting the sequence of UWB pulses are tested. In [12, 13], both synthesis and analysis of the algorithms for detecting the sequence of IR UWB signals without a harmonic filling is carried out. The present paper deals with a large class of effects defined as a sequence of jump-like pulses with a harmonic filling, having the discontinuous points of the first order at the moment of appearance and disappearance.

2 The Problem Statement

The sequence of jump-like pulses with a harmonic filling as a deterministic function is represented as

$$s(t) = \sum_{k=0}^{\nu-1} s_k(t - kT_0, a_k, \phi_k), \tag{1}$$

where ν is the number of pulses in the sequence, T_0 is the pulse-repetition period. Each pulse in the sequence is a single harmonic oscillation [14] determined by the following expression

$$s_k(t, a_k, \phi_k) = \begin{cases} a_k \cos(\omega t - \phi_k)\,, & 0 \le t \le \tau\,, \\ 0\,, & t < 0\,,\ t > \tau\,, \end{cases} \quad k = \overline{0, \nu - 1}, \tag{2}$$

where a_k and ϕ_k are the amplitude and initial phase of the k-th pulse in the sequence; ω and τ are the frequency and the duration that are equal for all the pulses, and the condition $\tau \le T_0$ is satisfied, i.e., the pulse duration does not exceed the pulse-repetition period. It is presupposed that the duration τ and the frequency ω are known a priori.

In this paper, the detection problem is considered for the case of the sequence (1) with the unknown amplitudes and initial phases observed against the white Gaussian noise. Similarly, to [5, 6], the two most frequent cases can be identified. In the first case, both the amplitudes and the phases of the signal (1) are unknown on the receiving side but they are equal for all the pulses in the sequence: $a_k = a$, $\phi_k = \phi$, $k = \overline{0, \nu - 1}$. Such sequence is called a slowly fluctuating [8, 9] or a coherent [15] one. Meanwhile, the signal (1) depends on the two unknown parameters $s(t) = s(t, a, \phi)$, and its detection algorithms are studied in [16]. In the second case, when a rapidly fluctuating or an incoherent sequence is observed, the unknown amplitudes a_k and phases ϕ_k changes from one pulse to another independently of each other and the detected signal (1) includes 2ν of the unknown parameters: $s(t) = s\left(t, \vec{a}, \vec{\phi}\right)$ where $\vec{a} = \| a_0, a_1, \ldots, a_{\nu-1} \|$, $\vec{\phi} = \| \phi_0, \phi_1, \ldots, \phi_{\nu-1} \|$ are the vectors of the amplitudes and phases of the pulse sequence. This paper is devoted to the detection of the rapidly fluctuating (non-coherent) sequence of pulses.

It is assumed that the sequence (1) is observed against the random noise process $n(t)$ within the time interval $t \in [0, T]$ in the following form:

$$\xi(t) = \gamma_0 S\left(t, \vec{a}_0, \vec{\phi}_0\right) + N(t), \tag{3}$$

In (3), $S\left(t, \vec{a}_0, \vec{\phi}_0\right)$ and $N(t)$ are the realizations of the deterministic function (1) and the random process $n(t)$, respectively; $\vec{a}_0 = \| a_{0k} \|$, $\vec{\phi}_0 = \| \phi_{0k} \|$, $k = \overline{0, \nu - 1}$ are the row vectors of the real values for the amplitudes and phases of the pulses in the sequence (1); a_{0k}, ϕ_{0k} are the real values of the amplitudes and initial phases of each of the pulses; γ_0 is the discrete state parameter taking the value $\gamma_0 = 0$ in the signal absence and $\gamma_0 = 1$ in the signal presence; $n(t)$ is approximated by the Gaussian white noise with the one-sided spectral density N_0. After the reception of the realization $\xi(t)$, the receiver' task is to make a decision on the signal (1) presence or absence in it. It is obvious that such task can be reduced to estimating the state parameter γ_0 based on the observed data $\xi(t)$.

The detection of the pulse sequence against the Gaussian white noise, similarly to papers [5, 17, 18], can be considered as a detection task under the parallel signal processing. Meanwhile, the channels are assigned over the time period T_0 (1) and are tuned to the intervals $t \in [T_0 k, T_0(k+1)]$, $k = \overline{0, \nu - 1}$. In this case, the observation interval increases with the number of pulses in the sequence, if the pulse duration and the repetition period are fixed. The processes $\xi_k(t)$ and $n_k(t)$ are now defined as the realizations of the sections $\xi(t)$ and $n(t)$ within the time intervals $t \in [T_0 k, T_0(k+1)]$, $k = \overline{0, \nu - 1}$, wherein the time reference point in each time channel coincides with the interval start $t' = t - kT_0$. Then in each channel, the sequence (1) observed against the random distortions can be presented in the form

$$\xi_k(t) = \gamma_0 s_k(t, a_{0k}, \phi_{0k}) + n_k(t), \tag{4}$$

where $t \in [0, T_0]$. The cross-correlation functions of the noise in channels with the numbers m and k are equal to $K_{mk}(t_1, t_2) = N_0 \delta_{mk} \delta(t_1 - t_2)/2$, where δ_{mk} is Kronecker symbol, and $\delta(t)$ is Dirac delta function.

3 The Maximum Likelihood Detection

In order to synthesize the algorithm of the information impact detection (that is the estimate of parameter γ_0), one uses the maximum likelihood (ML) method [8, 9, 19]. According to this method, while using the observed data (4), it is necessary to generate the logarithm of the functional of likelihood ratio (FLR) $L(\gamma, \vec{a}, \vec{\gamma})$ depending on the current values γ, $\vec{a}$, $\vec{\phi}$ of the unknown state parameter γ_0 and signal parameters $\vec{a}_0$, $\vec{\phi}_0$. The ML estimate $\hat{\gamma}$ of the parameter γ_0 is determined as the value of the variable γ, under which the FLR logarithm takes its maximum value. According to [15, 18], while the realization (3) is observed, the FLR logarithm can be represented as follows

$$L\left(\gamma, \vec{a}, \vec{\phi}\right) = \frac{2\gamma}{N_0} \sum_{k=0}^{\nu-1} \int_0^{\tau} \xi_k(t)\, a_k \cos(\omega t - \phi_k)\, \mathrm{d}t - \frac{\gamma}{N_0} \sum_{k=0}^{\nu-1} \int_0^{\tau} a_k^2 \cos^2(\omega t - \phi_k)\, \mathrm{d}t. \tag{5}$$

As it is stated in the papers [16, 17, 19], the ML algorithm for the deterministic function detection problem (the estimate of the parameter γ_0) can be represented in the following way

$$\hat{\gamma} = \begin{cases} 1, & L > h, \\ 0, & L < h, \end{cases} \tag{6}$$

where

$$L = \sup_{\vec{a},\, \vec{\phi}} L\left(\vec{a}, \vec{\phi}\right) = L\left(\hat{\vec{a}}, \hat{\vec{\phi}}\right), \tag{7}$$

$L\left(\vec{a}, \vec{\phi}\right) = L\left(\gamma = 1, \vec{a}, \vec{\phi}\right)$, $\left(\hat{\vec{a}}, \hat{\vec{\phi}}\right) = \arg\sup_{\vec{a},\, \vec{\phi}} L\left(\vec{a}, \vec{\phi}\right)$ is the ML estimates of the unknown signal (1) amplitude and initial phase. The threshold h in (6) is chosen according to the preset optimality criteria [5, 6, 18].

By substituting (1) in (5), one can rewrite the FLR logarithm in the form

$$L\left(\vec{a}, \vec{\phi}\right) = \sum_{k=0}^{\nu-1} a_k (X_k \cos\phi_k + Y_k \sin\phi_k) - \frac{1}{2} \sum_{k=0}^{\nu-1} a_k^2 (Q_1 + P_{c1} \cos 2\phi_k + P_{s1} \sin 2\phi_k), \tag{8}$$

where

$$X_k = \frac{2}{N_0} \int_0^{\tau} \xi_k(t) \cos\omega t\, \mathrm{d}t,\; Y_k = \frac{2}{N_0} \int_0^{\tau} \xi_k(t) \sin\omega t\, \mathrm{d}t,\; Q_1 = \tau/N_0,$$

$$P_{c1} = \tau \sin(4\pi\kappa)/4N_0\pi\kappa,\; P_{s1} = \tau\,[1 - \cos(4\pi\kappa)]/4N_0\pi\kappa,\; \kappa = \omega\tau/2\pi.$$

Here the value κ is introduced, being equal to the number of the harmonic oscillation periods (1) that fit into the duration of each pulse of the sequence τ.

To obtain the decision determining statistics (7), it is necessary to maximize the FLR logarithm (8) by the variables a_k and ϕ_k. Setting in (8)

$$\begin{cases} u_k = a_k \cos\phi_k \, , \\ v_k = a_k \sin\phi_k \, , \end{cases} \begin{cases} a_k = \sqrt{u_k^2 + v_k^2} \, , \\ \phi_k = atan(v_k / u_k) \, , \end{cases}$$

one gets

$$L(\vec{a}, \vec{\phi}) = \sum_{k=0}^{\nu-1} (X_k u_k + Y_k v_k) - \frac{1}{2} \sum_{k=0}^{\nu-1} \left[(Q_1 + P_{c1})\, u_k^2 + (Q_1 - P_{c1})\, v_k^2 + 2P_{s1} u_k v_k \right]. \tag{9}$$

The next step is to introduce the row vectors as

$$\vec{u} = \| u_k \|, \ \vec{v} = \| v_k \|, \ \vec{X} = \| X_k \|, \ \vec{Y} = \| Y_k \| \tag{10}$$

along with the block vectors $\vec{w} = \| \vec{u} | \vec{v} \|$, $\vec{Z} = \left\| \vec{X} \middle| \vec{Y} \right\|$ where $(\vec{u} | \vec{v})^T = \begin{pmatrix} \vec{u}^T \\ \vec{v}^T \end{pmatrix}$ and T denotes the transpose of matrix.

The value Q_1 in (9) can be can be transformed into the matrix $\mathbf{Q_1} = \| Q_1 \delta_{ik} \|$. In the sequel, the matrices are denoted in bold. In the case of equal duration and shape of each pulse in the sequence, the matrix $\mathbf{Q_1}$ displays equal elements on the diagonal line, i.e. $\mathbf{Q_1} = Q_1 \mathbf{I}$ where $\mathbf{I}$ is the identity matrix of the dimension $\nu \times \nu$. Now, one can rewrite the first term in (9) as

$$\sum_{k=0}^{\nu-1} (X_k u_k + Y_k v_k) = \vec{X}\, \vec{u}^T + \vec{Y}\, \vec{v}^T = \vec{Z}\, \vec{w}^T. \tag{11}$$

Turning to the last term in (9), it should be noted that each element of this sum can be represented as

$$\sum_{k=0}^{\nu-1} Q_1 u^2 = \vec{u} \mathbf{Q_1} \vec{u}^T. \tag{12}$$

Using (11), (12) the functional (9) can be rewritten in the following way.

$$L(\vec{u}, \vec{v}) = \vec{X}\, \vec{u}^T + \vec{Y}\, \vec{v}^T - \vec{u}\, (\mathbf{Q_1} + \mathbf{P_{c1}})\, \vec{u}^T \big/ 2 - \vec{v}\, (\mathbf{Q_1} - \mathbf{P_{c1}})\, \vec{v}^T \big/ 2 - \vec{u}\, \mathbf{P_{s1}} \vec{v}^T,$$

After that the block (cellular) matrix of size $2\nu \times 2\nu$ is introduced:

$$\mathbf{R} = \begin{pmatrix} \mathbf{Q_1} + \mathbf{P_{c1}} & \mathbf{P_{s1}} \\ \mathbf{P_{s1}} & \mathbf{Q_1} - \mathbf{P_{c1}} \end{pmatrix} = \begin{pmatrix} \mathbf{A} & \mathbf{B} \\ \mathbf{C} & \mathbf{D} \end{pmatrix}, \tag{13}$$

and the second sum in (9) is represented as

$$\sum_{k=0}^{\nu-1}\left[(Q_1+P_{c1})\,u_k^2+(Q_1+P_{c1})\,v_k^2+2P_{s1}u_kv_k\right]$$
$$=\vec{u}\,(\mathbf{Q_1}+\mathbf{P_{c1}})\,\vec{u}^T+\vec{v}\,(\mathbf{Q_1}-\mathbf{P_{c1}})\,\vec{v}^T+ \tag{14}$$
$$+2\vec{u}\,\mathbf{P_{s1}}\vec{v}^T=(\,\vec{u}\,|\,\vec{v})\begin{pmatrix}\mathbf{A}\;\mathbf{B}\\ \mathbf{C}\;\mathbf{D}\end{pmatrix}\begin{pmatrix}\vec{u}^T\\ \vec{v}^T\end{pmatrix}=(\,\vec{u}\,|\,\vec{v})\begin{pmatrix}\mathbf{A}\vec{u}^T\;\mathbf{B}\vec{v}^T\\ \mathbf{C}\vec{u}^T\;\mathbf{D}\vec{v}^T\end{pmatrix}=$$
$$=\vec{u}\left(\mathbf{A}\vec{u}^T+\mathbf{B}\vec{v}^T\right)+\vec{v}\left(\mathbf{C}\vec{u}^T+\mathbf{D}\vec{v}^T\right)=\vec{u}\mathbf{A}\vec{u}^T+\vec{u}\mathbf{B}\vec{v}^T+\vec{v}\mathbf{C}\vec{u}^T+\vec{v}\mathbf{D}\vec{v}^T.$$

Thus, from (9), (10), (11), (14), it follows that

$$L(\vec{u},\vec{v})=L(\vec{w})=\vec{Z}\,\vec{w}^T+\vec{Y}\,\vec{v}^T-\vec{w}\,\mathbf{R}\,\vec{w}^T\big/2. \tag{15}$$

As a result of FLR logarithm maximization (15) by the components of the vector $\vec{w}$ one gets

$$L=\sup_{\vec{w}}L(\vec{w})=\sup_{\vec{a},\,\vec{\phi}}L\left(\vec{a},\vec{\phi}\right)=\vec{Z}\,\mathbf{R}^{-1}\vec{Z}^T\big/2, \tag{16}$$

where $\mathbf{R}^{-1}$ is the inverse matrix for $\mathbf{R}$.

To calculate the inverse matrix $\mathbf{R}^{-1}$ for the block matrix of the type (13), the Frobenius formula [20] is used:

$$\mathbf{R}^{-1}=\begin{pmatrix}\mathbf{A}^{-1}+\mathbf{A}^{-1}\mathbf{B}\mathbf{H}^{-1}\mathbf{C}\mathbf{A}^{-1} & -\mathbf{A}^{-1}\mathbf{B}\mathbf{H}^{-1}\\ -\mathbf{H}^{-1}\mathbf{C}\mathbf{A}^{-1} & \mathbf{H}^{-1}\end{pmatrix}, \tag{17}$$

where $\mathbf{H}=\mathbf{D}-\mathbf{C}\mathbf{A}^{-1}\mathbf{B}$.

The block matrices **A**, **B**, **C**, **D** are the quasi-diagonal (block-diagonal) ones, and in the case of the equal duration and shape of each pulse in the sequence (1), they possess the same elements on the diagonal lines. It is important that, according to the definition in (13), $\mathbf{A}=(Q_1+P_{c1})\,\mathbf{I}$, $\mathbf{B}=\mathbf{C}=P_{s1}\,\mathbf{I}$, $\mathbf{D}=(Q_1-P_{c1})\,\mathbf{I}$, and the elements of the block matrix (17) can be calculated as.

$$\mathbf{A}^{-1}=\mathbf{I}/(Q_1+P_{c1}),\;\mathbf{H}=\mathbf{D}-\mathbf{C}\mathbf{A}^{-1}\mathbf{B}=\left(Q_1^2-P_{c1}^2-P_{s1}^2\right)\mathbf{I}\big/(Q_1+P_{c1}),$$

$$\mathbf{H}^{-1}=(Q_1+P_{c1})\,\mathbf{I}\big/\left(Q_1^2-P_{c1}^2-P_{s1}^2\right)=\boldsymbol{\varepsilon},$$

$$\mathbf{A}^{-1}+\mathbf{A}^{-1}\mathbf{B}\mathbf{H}^{-1}\mathbf{C}\mathbf{A}^{-1}=(Q_1-P_{c1})\,\mathbf{I}\big/\left(Q_1^2-P_{c1}^2-P_{s1}^2\right)=\boldsymbol{\gamma},$$

$$-\mathbf{A}^{-1}\mathbf{B}\mathbf{H}^{-1}=-\mathbf{H}^{-1}\mathbf{C}\mathbf{A}^{-1}=-P_{s1}\mathbf{I}\big/\left(Q_1^2-P_{c1}^2-P_{s1}^2\right)=\boldsymbol{\mu},$$

From here

$$\mathbf{R}^{-1}=\begin{pmatrix}\boldsymbol{\gamma}\;\boldsymbol{\mu}\\ \boldsymbol{\mu}\;\boldsymbol{\varepsilon}\end{pmatrix}. \tag{18}$$

The inverse matrix $\mathbf{R}^{-1}$ contains four diagonal blocks, just like the matrix $\mathbf{R}$ itself. Whereas $\vec{Z}$ is the block vector $\vec{Z} = \left\| \vec{X} \,\middle|\, \vec{Y} \right\|$, using (18), one can rewrite the expression (16) in the form

$$L = \left(\vec{X} \boldsymbol{\gamma} \vec{X}^T + \vec{X} \boldsymbol{\mu} \vec{Y}^T + \vec{Y} \boldsymbol{\mu} \vec{X}^T + \vec{Y} \boldsymbol{\varepsilon} \vec{Y}^T \right) \Big/ 2. \tag{19}$$

As it can be seen, all the terms in (19) are represented as $\vec{F} \boldsymbol{\alpha} \vec{G}^T$ where $\vec{F}$ ($\vec{G}^T$) is one of the vectors $\vec{X}$ or $\vec{Y}$, and $\boldsymbol{\alpha}$ is the diagonal matrix with the elements $m_{kj} = m_k \delta_{kj}$.

It is not difficult to show that $\vec{X} \boldsymbol{\alpha} \vec{Y}^T = \sum\limits_k m_k x_k y_k$. Then the expression (19) can be represented in the form

$$L = \sum_{k=0}^{\nu-1} \frac{(Q_1 - P_{c1}) X_k^2 + (Q_1 + P_{c1}) Y_k^2 - 2 P_{s1} X_k Y_k}{2\left(Q_1^2 - P_{c1}^2 - P_{s1}^2\right)}. \tag{20}$$

Now, accounting for (20), the ML detection algorithm (6) transforms into the following:

$$L \underset{"0"}{\overset{"1"}{\gtrless}} h, \qquad L = \sum_{k=1}^{\nu} L_k, \tag{21}$$

where

$$L_k = \frac{(Q_1 - P_{c1}) X_k^2 + (Q_1 + P_{c1}) Y_k^2 - 2 P_{s1} X_k Y_k}{2\left(Q_1^2 - P_{c1}^2 - P_{s1}^2\right)}. \tag{22}$$

4 The Detection Efficiency

The detection efficiency can be described in terms of the type I and type II error probabilities α and β [5, 18], respectively. The error probabilities in the ML detection algorithm (21) are determined as $\alpha = P\{ L > h \,|\, \gamma_0 = 0\}$, $\beta = P\{ L < h \,|\, \gamma_0 = 1 \}$. Therefore, to determine the detection efficiency, the study of the properties of the random variables (RVs) X_k and Y_k is required. By substituting $\xi_k(t)$ (4) in (8), one can represent the RVs X_k and Y_k as the sums of the deterministic and random components:

$$X_k = \gamma_0 S_{xk} + N_{xk}, \quad Y_k = \gamma_0 S_{yk} + N_{yk},$$

where

$$S_{xk} = \langle X_k \rangle = \frac{2a_{0k}}{N_0} \int_0^{\tau} \cos(\omega t - \phi_0) \cos \omega t \, \mathrm{d}t = a_{0k} \left[(Q_1 + P_{c1}) \cos \phi_0 + P_{s1} \sin \phi_0 \right],$$

$$S_{yk} = \langle Y_k \rangle = \frac{2a_{0k}}{N_0} \int_0^{\tau} \cos(\omega t - \phi_0) \sin \omega t \, \mathrm{d}t = a_{0k} \left[(Q_1 - P_{c1}) \sin \phi_0 + P_{s1} \cos \phi_0 \right],$$

$$N_{xk} = \frac{2}{N_0} \int_0^\tau n_k(t) \cos \omega t \, dt, \; N_{yk} = \frac{2}{N_0} \int_0^\tau n_k(t) \sin \omega t \, dt.$$

The RVs N_{xk} and N_{yk} are the linear transformations of the Gaussian white noise and, therefore, are also the Gaussian ones, with the zero expectations, the dispersions $\sigma_{xk}^2 = \langle N_{xk}^2 \rangle = Q_1 + P_{c1}$, $\sigma_{yk}^2 = \left\langle N_{yk}^2 \right\rangle = Q_1 - P_{c1}$ and the cross correlations $K_{km} = \langle N_{xk} N_{ym} \rangle = P_{s1} \delta_{km}$.

Similarly to [21], one introduces the variables.

$$J_k = -P_{s1} X_k + (Q_1 + P_{c1}) Y_k, \; Z_k = g_1 X_k, \; g_1^2 = Q_1^2 - P_{c1}^2 - P_{s1}^2. \tag{23}$$

Then, by applying (23), the expression (22) is transformed into the following:

$$L_k = \left(Z_k^2 + J_k^2 \right) \Big/ 2 g_1^2 (Q_1 + P_{c1}). \tag{24}$$

To calculate the type I and type II error probabilities, one needs to find the probability density of the variable L_k (24) which is expressed in terms of the variables Z_k and J_k. Therefore, it is required to calculate the joint probability density of the RVs Z_k and J_k.

As it can be seen, the RVs Z_k and J_k are the linear combinations of the Gaussian RVs X_k and Y_k. Therefore, they are also Gaussian ones and can be completely described through their first two moments.

$$m_{Z_k} = \langle Z_k \rangle = \gamma_0 S_{xk} g_1, \; m_{J_k} = \langle J_k \rangle = -\gamma_0 P_{sk} S_{xk} + \gamma_0 (Q_1 + P_{c1}) S_{yk}.$$

$$\sigma_{Z_k}^2 = \sigma_{jk}^2 = \left\langle (Z_k - m_{Z_k})^2 \right\rangle = \left\langle (J_k - m_{J_k})^2 \right\rangle = g_1^2 (Q_1 + P_{c1}), \tag{25}$$

$$\left\langle (Z_k - m_{Z_k}) \left(J_k - m_{J_k} \right) \right\rangle = 0.$$

According to (25), the joint Gaussian probability density for the RVs Z_k and J_k is determined as.

$$W_k(z, j) = \exp\left[-\left(z - m_{Z_k} \right)^2 \Big/ 2\sigma_{Z_k}^2 - \left(j - m_{J_k} \right)^2 \Big/ 2\sigma_{J_k}^2 \right] \Big/ 2\pi \sigma_{Z_k} \sigma_{J_k}.$$

The next step is to introduce the new variables Λ_k, θ_k that are related to the variables Z_k, J_k by the relations: $Z_k = \Psi_1(\Lambda_k, \theta_k) = \Lambda_k \cos \theta_k$, $J_k = \Psi_2(\Lambda_k, \theta_k) = \Lambda_k \sin \theta_k$. Then, for L_k (24) one gets

$$L_k = \Lambda_k^2 \Big/ 2 g_1^2 (Q_1 + P_{c1}). \tag{26}$$

For the joint probability density of the RVs Λ_k and θ_k, there can be written the following expression:

$$W_{\Lambda_k \theta_k}(\Lambda_k, \theta_k) = W_{Z_k J_k}(\Psi_1(\Lambda_k, \theta_k), \Psi_2(\Lambda_k, \theta_k)) \, |D_k|. \tag{27}$$

In (27), the Jacobian transformation is $|D_k| = \begin{vmatrix} \partial\Psi_1/\partial\Lambda_k & \partial\Psi_1/\partial\theta_k \\ \partial\Psi_2/\partial\Lambda_k & \partial\Psi_2/\partial\theta_k \end{vmatrix}$, while $\Lambda_k \geq 0$ and $\theta_k \in [-\pi, \pi]$.

Due to using the functions $\Psi_1(\Lambda_k, \theta_k)$, $\Psi_2(\Lambda_k, \theta_k)$ in (27), one can obtain

$$W_{\Lambda_k\theta_k}(\Lambda_k, \theta_k) = \frac{\Lambda_k}{2\pi g_1^2(Q_1 + P_{c1})} \exp\left[-\frac{\left(\Lambda_k \cos\theta_k - m_{Z_k}\right)^2 + \left(\Lambda_k \sin\theta_k - m_{j_k}\right)^2}{2g_1^2(Q_1 + P_{c1})}\right]. \tag{28}$$

$\Lambda_k \geq 0$, $\theta_k \in [-\pi, \pi]$. The probability density (28) determines the RVs Λ_k and θ_k for the both hypotheses (that is, in both absence and presence of the signal (1)).

In order to calculate the type I error probability, the probability density $W(L_k|\gamma_0 = 0)$ of the RV L_k (26) should be found under $\gamma_0 = 0$, i.e., when the signal (1) is absent in the observed realization (3). For this purpose one assumes that $\gamma_0 = 0$ in (28). Further, after integrating the expression (28) by the variable θ_k, one gets.

$$W(L_k|\gamma_0 = 0) = \int\limits_{-\pi}^{\pi} W_{\Lambda_k\theta_k}(\Lambda_k, \theta_k)\, d\theta_k = \frac{\Lambda_k}{g_1^2(Q_1 + P_{c1})} \exp\left[-\frac{\Lambda_k^2}{2g_1^2(Q_1 + P_{c1})}\right], \Lambda_k \geq 0.$$

Passing from the probability density of the RV Λ_k to the probability density of the RV L_k (26), as it is carried out in [21], one obtains.

$$W(L_k|\gamma_0 = 0) = \exp(-L_k),\ L_k \geq 0. \tag{29}$$

By applying (29), the probability density for the sum (21) can be found. Indeed, the value L (20) under $\gamma_0 = 0$ is the sum of the exponentially distributed random values. The exponential distribution (29) is the special case of the gamma distribution $\Gamma(p, \theta)$ [22], described by the probability density.

$$w(x) = x^{p-1} \exp(-x/\theta) \Big/ \theta^p \Gamma(p),\ x \geq 0, \tag{30}$$

where $\Gamma(p)$ is the gamma-function [23]. As it can be seen from (30), under $p = 1$ and $\theta = 1$, this probability density is equal to $w(x) = \exp(-x)$. Under the fixed θ, the gamma distribution is infinitely divisible [22]. Therefore, the sum L (21) of the finite number of terms L_k, $k = \overline{0, \nu - 1}$, obeying the gamma distribution $\Gamma(1, 1)$, is also described by the gamma distribution that takes the form: $\Gamma\left(\sum\limits_{k=0}^{\nu-1} 1, 1\right) = \Gamma(\nu, 1)$. Thus, the probability density of the RV L (21) in the signal (1) absence can be represented as follows.

$$W(L|\gamma_0 = 0) = L^{\nu-1} \exp(-L) \Big/ \Gamma(\nu),\ L \geq 0.$$

With this probability density in mind, the type I error probability for the detector (21) is calculated by the formula

$$\alpha = P\{L > h|\gamma_0 = 0\} = 1 - \int\limits_0^h L^{\nu-1} \frac{\exp(-L)}{\Gamma(\nu)}\, dL. \tag{31}$$

To find the type II error probability, one needs calculating the probability density $W(L\,|\gamma_0 = 1)$ of the RV L_k (26) under $\gamma_0 = 1$, i.e., when the signal (1) is present in the observed realization (3).

By substituting $\gamma_0 = 1$ in (28) and integrating the obtained expression by the variable θ_k, for the probability density of the RV Λ_k it can be written.

$$W(\Lambda_k) = \frac{\Lambda_k}{g_1^2(Q_1 + P_{c1})} \exp\left[-\frac{\Lambda_k^2 + m_{Z_k}^2 + m_{J_k}^2}{2g_1^2(Q_1 + P_{c1})}\right] I_0\left[\frac{\Lambda_k\sqrt{m_{Z_k}^2 + m_{J_k}^2}}{g_1^2(Q_1 + P_{c1})}\right],\ \Lambda_k \geq 0.$$

where $I_0(x)$ is the zero-order modified Bessel function [23].

Passing from this probability density of the RV Λ_k to the probability density of the RV L_k (26) results in.

$$W(L_k\,|\gamma_0 = 1) = \exp\left(-z_k^2 V_k\big/2\right) \exp(-L_k)\, I_0\left(z_k\sqrt{2L_k V_k}\right),\ L_k \geq 0,$$

Here the notations are the following:

$$V_k = \left[g_1^2 A_{ck}^2 + ((Q_1 + P_{c1})A_{sk} - P_{s1}A_{ck})^2\right]\Big/ g_1^2 Q_1(Q_1 + P_{c1}),$$

$$A_{ck} = (Q_1 + P_{c1})\cos\phi_{0k} + (Q_1 + P_{c1})\cos\phi_{0k},\ A_{ck} = (Q_1 + P_{c1})\cos\phi_{0k} + P_{s1}\sin\phi_{0k},$$

and $z_k^2 = Q_1 a_{0k}^2$ is the signal-to-noise ratio (SNR) at the output of the ML receiver for each of the pulses in the sequence.

One introduces the characteristic function of the RV L_k as follows [22].

$$\theta_k(u) = \langle \exp(juL_k)\,|\,\gamma_0 = 1\rangle = \int\limits_0^\infty \exp\left(jux - z_k^2 V_k\big/2 - x\right) I_0\left(z_k\sqrt{2xV_k}\right) \mathrm{d}x.$$

As it follows from [22], the characteristic function of the sum (21), while the RVs L_k are statistically independent, is equal to

$$\theta_L(u) = \prod_{k=0}^{\nu-1} \theta_k(u) = \prod_{k=0}^{\nu-1} \int\limits_0^\infty \exp\left(jux - z_k^2 V_k\big/2 - x\right) I_0\left(z_k\sqrt{2xV_k}\right) \mathrm{d}x. \quad (32)$$

Then the probability density of the RV L can be found as the inverse Fourier transform of its characteristic function (32):

$$W(L|\gamma_0 = 1) = \frac{1}{2\pi}\int_{-\infty}^{\infty} \exp(-juL) \prod_{k=0}^{\nu-1} \int\limits_0^\infty \exp\left(jux - z_k^2 V_k/2 - x\right) I_0\left(z_k\sqrt{2xV_k}\right) \mathrm{d}x\mathrm{d}u. \quad (33)$$

By applying (33) the type II error probability for the detector (21) can be found

$$\beta = \int_0^h W(L\,|\,\gamma_0 = 1)\,\mathrm{d}L =$$

$$= \frac{1}{2\pi}\int_0^h \int_{-\infty}^{\infty} \exp(-juL) \prod_{k=0}^{\nu-1} \int_0^{\infty} \exp\left(jux - z_k^2 V_k \big/ 2 - x\right) I_0\left(z_k\sqrt{2xV_k}\right) \mathrm{d}x\mathrm{d}u\mathrm{d}L. \tag{34}$$

It should be noted that the type II error probability using the expression (34) can be calculated, if only the numerical method is applied. To simplify the calculations, it is useful to transform the expression (34) into the form

$$\beta = \int_{-\infty}^{\infty} \chi(u)\,\mathrm{d}u = \int_{-\infty}^{0} \chi(u)\,\mathrm{d}u + \int_0^{\infty} \chi(u)\,\mathrm{d}u, \tag{35}$$

where

$$\chi(u) = \frac{1}{2\pi}\int_0^h \exp(-juL) \prod_{k=0}^{\nu-1} \int_0^{\infty} \exp\left(jux - z_k^2 V_k \big/ 2 - x\right) I_0\left(z_k\sqrt{2xV_k}\right) \mathrm{d}x\mathrm{d}L.$$

After changing the variables $t = u/(u-1)$ in the first integral and $t = u/(u+1)$ in the last one in (35), for the type II error probability one gets

$$\beta = \frac{1}{\pi}\int_0^1 \int_0^h \left[\prod_{k=0}^{\nu-1} \int_0^{\infty} \left(j\frac{t}{1-t}x - \frac{z_k^2}{2}V_k - x\right) I_0\left(z_k\sqrt{2xV_k}\right)dx\right] \cos\left(L\frac{t}{1-t}\right) \frac{dLdt}{(1-t)^2}. \tag{36}$$

If the real values of the amplitudes and the initial phases of the pulses are equal, i.e. $a_{0k} = a_0$, $\phi_{0k} = \phi_0$, $k = \overline{0, \nu - 1}$, then the formulas (34) and (36) can be significantly simplified as follows.

$$\beta = \int_0^h W(L|\gamma_0 = 1)\,\mathrm{d}L = \frac{\exp(-z^2 V/2)}{2\pi}\int_0^h \int_{-\infty}^{\infty} \exp(-juL)\left[\int_0^{\infty} \exp(jux - x) I_0\left(z\sqrt{2xV}\right)\mathrm{d}x\right]^{\nu} \mathrm{d}u\mathrm{d}L.$$

and.

$$\beta = \frac{\exp(-z^2 V/2)}{\pi}\int_0^1 \int_0^h \int_0^{\infty} \exp\left(j\frac{t\nu}{1-t}x - \nu x\right) \frac{I_0^{\nu}\left(z\sqrt{2xV}\right)}{(1-t)^2} \cos\left(L\frac{t}{1-t}\right) \mathrm{d}x\mathrm{d}L\mathrm{d}t.$$

If one substitutes $\nu = 1$ in the expressions for the type I and type II error probability in (31) and (34), there can be obtained the same forms as for the single pulse studied in [19, 21].

To analyze the cumulative ML algorithm for detecting the non-coherent jump-like pulse sequence (1), the statistical simulation of the detection algorithm (21) is carried out. During the simulation, the decision determining statistics (FLR logarithm) is developed at the output of the receiver for the pulse sequence (20). It can be represented in the following form:

$$L = \left[1 \Big/ 2 \Big((4\pi\kappa)^2 - 2(1 - \cos(4\pi\kappa)) \Big) \right] \times$$

$$\times \sum_{k=0}^{\nu-1} \left[(4\pi\kappa - \sin(4\pi\kappa)) x_k^2 + (4\pi\kappa + \sin(4\pi\kappa)) y_k^2 - 2(1 - \cos(4\pi\kappa)) x_k y_k \right],$$

where $x_k = X_k \sqrt{N_0/\tau}$, $y_k = Y_k \sqrt{N_0/\tau}$ are the independent Gaussian random variables with the mathematical expectations.

$$s_{xk} = \gamma_0 z_k \left[\left(1 + \frac{\sin(4\pi\kappa)}{4\pi\kappa} \right) \cos\phi_0 + \frac{1 - \cos(4\pi\kappa)}{4\pi\kappa} \sin\phi_0 \right],$$

$$s_{yk} = \gamma_0 z_k \left[\left(1 - \frac{\sin(4\pi\kappa)}{4\pi\kappa} \right) \sin\phi_0 + \frac{1 - \cos(4\pi\kappa)}{4\pi\kappa} \cos\phi_0 \right]$$

and the dispersions

$$\sigma_{xk}^2 = 1 + \sin(4\pi\kappa)/4\pi\kappa, \; \sigma_{yk}^2 = 1 - \sin(4\pi\kappa)/4\pi\kappa.$$

The decision determining statistics (33) has been developed repeatedly for the different values of γ_0 and compared with the threshold h selected by Neumann-Pearson criterion [18] based on the solution of the equation $\alpha(h) = 1 - \int_0^h L^{\nu-1} \exp(-x)\, dx / \Gamma(\nu)$. If the threshold increases under $\gamma_0 = 0$, then the type I error is fixed. If the value of the FLR logarithm does not exceed the threshold under $\gamma_0 = 1$, the type II error is recorded. During the simulation, it is assumed that the pulse amplitudes in the preset sequence (1) form the equalities: $a_{0k} = a_0, k = \overline{0, \nu - 1}$; while the initial phases of all the pulses are zeroes, that is: $\phi_{0k} = 0$. Thus, the equality $z_k = z_1$ is true for each of the pulses, and SNR for the whole sequence is equal to $z^2 = \nu z_1^2$.

In Fig. 1, the dependences of the type I error probability (31) on the threshold under the different values of the pulse numbers in the sequence are presented. The dot-and-dash curve corresponds to the single pulse: $\nu = 1$; the dashed curve corresponds to the two pulses in the sequence: $\nu = 2$; the solid one – to the five pulses, i.e. $\nu = 5$; and the dotted one – to the ten pulses, i.e. $\nu = 10$.

In Fig. 2, the dependences of the type II error probability (34) on SNR under the fixed level of the type I error probability $\alpha = 10^{-3}$ are drawn. It also shows the results of the simulation by the markers. The dot-and-dash curve and the squares correspond to the single pulse when $\nu = 1$; the dashed curve and the circles correspond to the two pulses in the sequence, i.e. $\nu = 2$; the solid curve and the triangles correspond to the case when $\nu = 5$; the dotted curve and the rhombuses correspond to the case when $\nu = 5$. The curves in Figs. 1, 2 are calculated at $\kappa = 0.3$.

Figure 1 and Fig. 2 demonstrate that, with the increase in the number of pulses in the sequence, the type I error probability is increasing while the type II error probability is decreasing. The type of the dependence in Fig. 1 is confirmed by the results obtained in the paper [1] for the pulses considered under the assumption that $\kappa >> 1$. It follows from the fact that it is more difficult to detect the signal when the total energy is distributed between ν pulses, compared with the case when the energy is concentrated in a single pulse. On the contrary, the detection probability increases with the number of pulses in a sequence under the specified average energy of the pulse.

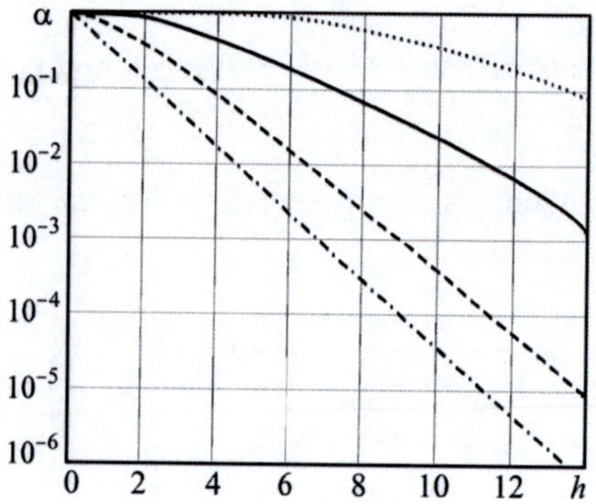

Fig. 1. The dependence of the type I error probability on the threshold h

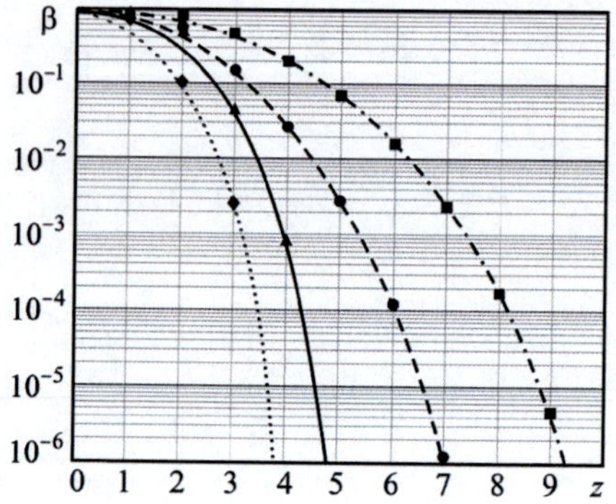

Fig. 2. The dependence of the type II error probability on SNR

Figure 2 demonstrates that the fixed type I error probability (at the level of the type II error probability $\beta = 10^{-6}$, that is typical, for example, for radio communication systems) is reached for a single pulse at SNR $z = 9.3$, for 2 pulses at SNR $z = 7.1$, for 5 pulses at SNR $z = 4.8$, for 10 pulses at SNR $z = 3.7$. Therefore, data transfer by the sequence of 10 pulses instead of the single pulse allows improving the detection quality by 2.5 times.

5 Conclusion

The paper focuses on the maximum likelihood algorithm for detecting a rapidly fluctuating sequence of jump-like pulses, observed against the background of a random process. The structure and statistical characteristics of the detection algorithm are obtained. The expressions for the type I and type II error probabilities in the case of rapidly fluctuating sequences of jump-like pulses coincide with the similar expressions used in the case of a single pulse, with the difference being that SNR, in terms of the power, increases in proportion to the number of pulses in the sequence. Thus, the choice of a sequence of jump-like pulses instead of a single signal can significantly increase the efficiency of an information transmission. The expressions for detection characteristics are closed and accurate, but for plotting graphical dependencies, the methods of numerical integration have been used, and that results in improving the accuracy of the values obtained. To confirm the correctness of the expressions and to refine the values of the type II error probability at the fixed SNR values, a statistical computer simulation of a sequence detection algorithm has been carried out.

It should be noted that due to the specifics of many practical applications, frequency, duration and shape of pulses of the received sequence may not be known. Thus, the synthesized maximum likelihood detection algorithm should be considered quasi-optimal. This will undoubtedly lead to the losses in its performance. However, a quantitative study of the losses in detection efficiency due to a priori ignorance of the pulse shapes, as well as their frequencies and durations, is a separate independent problem requiring further research. The results of detecting a sequence of jump-like pulses, that are presented in this paper, can help to improve the efficiency of the existing communication and data transmission systems. They can also be of great use in designing new telecommunication, radar, subsurface and remote sensing systems, etc.

Acknowledgements. The work was supported by grant of the President of the Russian Federation for state support of young Russian scientists – candidates of sciences (project MK-767.2021.1.6) as well as the Ministry of Education and Science of the Russian Federation (research project No. FSWF-2020–0022).

References

1. Basseville, M., Nikiforov, I.V.: Detection of Abrupt Changes: Theory and Application. Prentice-Hall, Hoboken (1993)
2. O'Neill, E.L.: Introduction to Statistical Optics. Dover Publications, Mineola (2004)
3. Bakut, P.A., Bolshakov, I.A., Tartakovsky, G.P., et. al.: Problems of the Statistical Theory of Radar. Defense Technical Information Center, Virginia (1964)
4. Van Trees, H.L.: Detection, Estimation, and Modulation Theory, Part III: Radar-Sonar Signal Processing and Gaussian Signals in Noise. Wiley, New York (2001)
5. Helstrom, K.: Statistical Theory of Signal Detection, 2nd edn. Elsevier, Amsterdam (2013)
6. Finkelstein, M.I.: Fundamentals of Radar Equipment. Radio i Svyaz', Moscow (1983).(in Russian)
7. Ohsumi, A., Ijima, H., Kuroishi, T.: Online detection of pulse sequence in random noise using a wavelet. IEEE Trans. Signal Process. **47**(10), 2526–2531 (1999)
8. Trifonov, A.P., Bespalova, M.B.: Efficiency of superbroadband detection of a target with unknown distance. Izvestiya Vysshikh Uchebnykh Zavedenij. Radioelektronika **44**(5), 20–26 (2001)
9. Trifonov, A.P., Bespalova, M.B.: Superbroadband target detection in treatment by broken pulses. Izvestiya Vysshikh Uchebnykh Zavedenij. Radioelektronika. **46**(6), 3–10 (2003)
10. Trifonov, P.A.: Detection sequence of ultrawideband signals with unknown time of arrival at presence of narrow-band interference. Physics of Wave Processes and Radio Systems **10**(1), 27–32 (2007). (in Russian)
11. Ali Tayaranian Hosseini, S.M., Amindavar, H., Ritcey, J.A.: Robust detection in ultra-wideband impulse radar using DPSS-MMSE estimator. Eurasip Journal on Advances in Signal Processing 2016 (1), 1–15 (2016)
12. Zhou, Q., Zou, Z., Chen, Q., Tenhunen H., Zheng, L.: Low complexity burst packet detection for wireless-powered UWB RFID systems. In: 2015 IEEE International Conference on Ubiquitous Wireless Broadband, ICUWB, pp. 1–5. IEEE, Canada (2015)
13. Zhan, H., Ayadi, J., Farserotu, J., Boudec, J.L.: Impulse radio ultra-wideband ranging based on maximum likelihood estimation. IEEE Trans. Wireless Commun. **8**(14), 5852–5861 (2009)
14. Trifonov, A.P., Rudnev, P.E.: Detection of an ultra-wideband quasi radio signal against background random noise. Radiophys. Quantum Electron. **52**(10), 671–681 (2009)

15. Radzievsky, V.G., Trifonov, P.A.: Processing of Ultra-Wideband Signals Against the Background of Noise. Radiotekhnika, Moscow (2009).(in Russian)
16. Korchagin, Y.E., Titov, K.D.: Detection of coherent ultra-wideband quasi-radiosignal of a waveform shape. Radiotekhnika **4**, 26–32 (2019). (in Russian)
17. Korchagin, Y.E., Titov, K.D.: Detection of an ultra-wideband quasi radio signal with unknown duration against the background of white noise. Radiophys. Quantum Electron. **61**(13), 853–866 (2019)
18. Akimov, P.S., Bakut, P.A., Bogdanovich, V.A., et al.: Signal detection theory. Radio i Svyaz', Moscow (1984, in Russian)
19. Trifonov, A.P., Korchagin, Y.E., Titov, K.D.: Efficiency of the detection of a specific wideband signal under a priori parametric uncertainty. Tech. Phys. **63**(9), 1199–1204 (2018)
20. Gantmacher, F.R.: Matrix theory, Vol. 2. American Mathematical Society, Providence (2000)
21. Korchagin, Y.E., Titov, K.D.: Synthesis and Analysis of the Algorithms for Processing Ultra-Wideband Quasi-Radio Signals. VSU Publishing house, Voronezh (2019).(in Russian)
22. Korolyuk, V.S., Portenko, N.I., Skorokhod, A.V., Turbin, A.F.: Handbook on Probability Theory and Mathematical Statistics. Nauka, Moscow (1985).(in Russian)
23. Abramowitz, M., Stegun, I.A.: Handbook of mathematical functions with formulas, graphs and mathematical tables. National Bureau of Standards. Applied Mathematics Series **55**, USA (1964)

Assessment of Factors Affecting the Volume of Rural Production: Information by Regions of Russia

Iuliia S. Pinkovetskaia[1](✉), Magomedsaid A. Yakhyaev[2], Natalia V. Berezina[3], and Ildar G. Nuretdinov[4]

[1] Ulyanovsk State University, Ulyanovsk 432000, Russia
pinkovetskaia@gmail.com
[2] Institute of Social Sciences, Moscow 105005, Russia
[3] Chuvash State University, Cheboksary 428015, Russia
[4] Ulyanovsk State Agrarian University Named After P. A. Stolypin, Ulyanovsk 432000, Russia

Abstract. The study was based on the development of regression models that characterize the activities of rural enterprises in the regions of Russia. Official statistical information on 65 regions of Russia for 2017 and 2018 was used. The conducted research made it possible to identify factors (fixed production assets cost, employees salary expenses and ratio of crop production to livestock production) that affect the volume of production in the rural sector in the regions of Russia and suggest using three-factor regression models of high quality to describe this influence. It is proved that the economy of the country's regions has not reached saturation with rural products and there are significant reserves for further development of this sector. The developed regression models are effective management tools that allow assessing the level of use of financial and labor resources. The acquired new knowledge and tools for assessing the activities of agriculture in the Russian regions are of scientific and practical importance. They can be used in research of the rural sector of the economy, monitoring of production volumes, in determining the needs for resources necessary for the development of agriculture, substantiating plans and programs for its development.

Keywords: Regression models · Agriculture · Production factors · Salary expenses · Russia

1 Introduction

To date, agriculture in Russia has received significant development. Currently, it is among the top four countries that have the largest areas of arable land. According to some estimates, about 9% of the world's farmland is located in Russia [1]. The solution of managerial tasks in the Russian economy requires an understanding of the factors that affect the volume of rural production. In Russia, there is currently an urgent need for accelerated development of agriculture. The efficiency of rural production, as one of the directions of increasing the productivity and competitiveness of this branch of the

R. Silhavy (Ed.): CSOC 2022, LNNS 501, pp. 105–116, 2022.
https://doi.org/10.1007/978-3-031-09070-7_10

economy, is directly related to the use of resources, with the degree of their involvement in the production process. To a large extent, efficiency depends on the quantitative and qualitative ratio of resources among themselves, on their balance. Determining the cost structure that ensures an increase in output per unit of resource becomes an urgent task of the management system. Therefore, in recent years, one of the most pressing problems is to determine the growth reserves of this sector of the economy in each region of Russia. The justification of these reserves, as well as the resources necessary for the effective functioning of agriculture, can be based on such economic and mathematical models as regression models.

The purpose of our research was to develop economic and mathematical models to assess the impact of indicators characterizing the use of resources on rural production in the regions of Russia. Our study responds to the calls for taking into account the regional characteristics of rural production, formulated in publications [2, 3].

Our article makes a certain contribution to the knowledge about the regional peculiarities of the development of agriculture in Russia. The theoretical contribution is related to the methodology proposed by the authors, which makes it possible to assess the dependence of agricultural production volumes on factors such as fixed production assets cost, employees salary expenses and ratio of crop production to livestock production based on the development of economic and mathematical models representing regression models. Based on empirical data in the course of the study, new knowledge was obtained about the impact of each factor on the volume of rural production. In addition, regions were identified in which high and low values of resource efficiency were noted.

2 Review of Literature

Scientific research conducted in the twenty-first century has shown the possibility of applying regression models in the economic analysis of the activities of enterprises and entrepreneurs that operate in the agricultural sector. Regression models are economic and mathematical models of production processes and quantitatively express a stable natural relationship between factors describing capital and labor costs and an indicator characterizing the volume of agricultural production [4–6]. Most scientific publications considered data for a number of years (the so-called time series). For example, in the article [7], according to the agricultural sector of India, the efficiency of farms was evaluated using the Cobb-Douglas regression model. When constructing this function, data for the period from 2005 to 2014 were analyzed. Modeling of agricultural production in China was described in the article [8]. While data were analyzed in 12 prefectures during the period from 2009 to 2019. In contrast to the above articles, the study [9] developed models using spatial data for 25 provinces of Cambodia. Four models were built corresponding to the information for each of the four years from 2012 to 2015.

Capital and labor costs were used as factors influencing agricultural production volumes in most previously performed scientific studies [10–13]. As a factor of the regression model in the article [14], the ratio of crop production volumes to livestock production volumes was used. This approach gives opportunity to complex mutual connections are taken into account. Namely, the consumption of feed by farm animals and the creation of organic fertilizers by them.

A certain place in scientific publications on the problem of the development of regression models of the agricultural sector is occupied by studies in Russia. Table 1 shows an analysis of the agricultural sector's production volumes using examples from Russian studies.

Table 1. Characteristics of Russian studies

Authors	Factor of capital	Factor of labor	Research object
1	2	3	4
Shestakov and Yakovlev [15]	capital expenditures	labor costs	agricultural production volumes for 2005–2018 in Russia as a whole
Tolmachev [16]	indices of the physical volume of fixed assets	index of total working time expenditures	indices of the physical volume of agricultural products in Russia for 1996–2008
Potapov [17]	costs of mechanical engineering products, fuel and energy resources, chemical products	-	gross agricultural output in Russia in 2011–2015
Naumov [18]	fixed capital	number of employees	production volumes in agriculture in the Chelyabinsk region for 2005–2015
Kutenkov [19]	cost of fixed assets	number of people employed in agriculture per 100 hectares of acreage	three groups of regions of Russia, data for 2017
Zyukin and Zhilin [20]	production funds	value of labor costs	volume of production of the agricultural complement of the Kursk region for 2000–2011
Zhilyaskova [21]	fixed assets	number of employees	agricultural production volumes in the Rostov region for the period 2004–2006
Germanova and Rudaya [22]	fixed assets	number of employed workers	Production volumes in agriculture of the Krasnodar Territory for 2000–2014

Source: Compiled by the authors.

The data in Table 1 show that in most cases the objects of research are agricultural sectors in specific regions (five cases). The other three publications discuss regression models for Russia as a whole. The initial data in seven studies were time series, only one publication used spatial data for one year. In most studies (seven cases), the number of employees was used as labor costs. In two publications, the values of working time costs were considered. Data on fixed assets of agricultural enterprises were used as capital factors in six publications. In two cases, the costs of production assets were considered, and in one case, the costs of purchasing products from the machine-tool, fuel and chemical sectors. It should be noted that previous Russian scientific publications did not pay sufficient attention to the comparative analysis of regional features of agricultural production in Russia.

3 Materials and Methods

Rural complexes located in the regions of Russia were considered as the object of the study. The use of initial data for several years (time series) is complicated by the fact that inflationary processes that have taken place should be taken into account. In addition, it is necessary to proceed from the assumption that the conditions of functioning of the object under consideration for a certain period of time will be identical or, at least, undergo few changes, which in practice is not always fulfilled. Time series are often limited in length, especially since due to crisis phenomena in the economy, the dynamics of changes in indicators experiences significant fluctuations. When using the values of fixed production assets cost as a factor describing capital expenditures, the main problem is the reliability of information regarding the share of fixed production assets cost actually used in production processes of the economic system under consideration. The assumption of the full use of fixed production assets cost does not always correspond to their actual utilization. The situation is similar with the second factor. The number of people directly employed in production processes does not always coincide with the actual labor costs, since workers are often not employed all day. This leads to erroneous indicators when assessing labor factors.

The number of factors, in accordance with the recommendation of Granberg [23], should be small, since in this case the necessary calculations and interpretation of the results are simplified. Taking into account the analysis, we consider fixed production assets cost, employees salary expenses and ratio of crop production to livestock production as factors of regression models. Correlation analysis has shown that these factors have the greatest impact on the volume of rural production in the regions. At the same time, there is no mutual connection (collinearity) between them. Employees salary expenses employed in agriculture is a complex indicator that takes into account not only labor costs for production, but also the characteristics of a particular region (price level, employment and other socio-economic aspects). In addition, the use of employees salary expenses as a factor ensures the same dimensionality of indicators of regression models, which, as shown in the article [24], ensures high quality of construction of the corresponding models.

In our study, spatial data were used to characterize the factors under consideration and the resulting indicators for rural sectors in the regions of Russia. It should be noted that spatial data allows you to get away from the problems that are characteristic of time series. The advantages of using spatial data in the evaluation of regression models are described in detail in the work [25].

Our study included the following stages:

1. Collection and processing of initial statistical data. Formation of arrays of information based on data characterizing the activities of enterprises and entrepreneurs in the rural sector of 65 regions of Russia. These arrays describe the values of production volumes, fixed production assets cost, employees salary expenses and ratio of crop production to livestock production.
2. Linearization of the data obtained at the first stage, which characterize the independent factors and resulting variables for rural production in the regions.
3. Development of regression models using the least squares method.
4. Evaluation of the quality of functions using correlation and determination coefficients, Fisher-Snedekor and Student tests, as well as the corresponding significance levels.
5. Checking the developed functions for the presence of autocorrelation, heteroscedasticity and multicollinearity, as well as determining whether the distributions of residuals for each of the regressions are functions of the normal distribution.
6. Consideration of theoretical and practical results arising from the analysis of the developed regression models and the possibilities of their use.

The study used data from the Federal State Statistics Service on rural activity in the regions of Russia for 2017 and 2018 [26]. The work is based on information on 65 regions of Russia in which rural production has received the greatest development. In our study, three hypotheses were tested:

- the first hypothesis is that regression models can be used to model the production volumes of the rural sector in the regions;
- the second hypothesis is that regression models demonstrate the presence of stable dependence of rural production volumes on factors such as fixed production assets cost, employees salary expenses and ratio of crop production to livestock production;
- the third hypothesis is that factor fixed production assets cost in both regression models affects on turnover greater than the factors of wage of employees and the ratio of crop production and livestock production.

In the course of the study, two regression models were developed, reflecting the dependence of rural production volumes on fixed production assets cost, employees salary expenses and ratio of crop production to livestock production by regions of Russia. The functions constructed by the authors have a specification similar to the well-known Cobb-Douglas functions. The parameters of regression models were determined using the regression analysis methodology [27]. The first function describes the activity of sets of all enterprises and entrepreneurs that are located in each of the 65 regions under consideration for 2017, and the second function - according to data for 2018.

4 Results of the Calculation Experiment

In the course of the computational experiment, two regression models were developed that reflect the dependence of rural production volumes on fixed production assets cost, employees salary expenses, ratio of crop production to livestock production in Russian regions. The formulas and tables given in the article are developed by the author.

The first regression model reflects the dependence of the production volume rural industry in 2017:

$$y_1(x_1, x_2, x_3) = 10.373 \times x_1^{0.412} \times x_2^{0.245} \times x_3^{0.229}, \tag{1}$$

y_1- the turnover of all enterprises and entrepreneurs in the rural sector located in a certain region of Russia per year, billion rubles;
x_1- the fixed production assets cost of all regional enterprises and entrepreneurs in the rural sector per year, billion rubles;
x_2- rural industry employees salary expenses in the region per year, billion rubles;
x_3- ratio of crop production to livestock production in the region per year.

The second regression model reflects the dependence of the production volume rural industry in 2018:

$$y_2(x_4, x_5, x_6) = 9.432 \times x_3^{0.409} \times x_4^{0.262} \times x_6^{0.202}, \tag{2}$$

y_2- the turnover of all enterprises and entrepreneurs in the rural sector located in a certain region of Russia per year, billion rubles;
x_4- the fixed production assets cost of all regional enterprises and entrepreneurs in the rural sector per year, billion rubles;
x_5- rural industry employees salary expenses in the region per year, billion rubles;
x_6- ratio of crop production to livestock production in the region per year.

Table 2 shows the analysis of the function's quality. It presents the calculated values of the correlation and determination coefficients, Fisher-Snedecor and Student's tests (column 2), as well as the significance of the Fisher-Snedecor test and p-values for Student's test (column 3).

The correlation coefficients more 0.9 and close to 1 in both functions. Regression models are known to be of high quality when the coefficient of determination more than 0.8; for both functions this requirement is met. The coefficient of determination characterizes the proportion of dispersion, which is caused by the influence of the considered factors. The difference between 1 and this coefficient describes the influence of factors that are not included in the regression equation. Thus, the effect of variables not included in the functions under consideration is less than 15 percent. The calculated statistic values (119 and 132) are higher than the table value of the Fisher-Snedecor test, which is 3.98 at a significance level of 0.05. For three functions, all calculated Student test values for the coefficient and the exponents are in the range from 2.85 to 15.87; in absolute value they exceed the table amount, which is 1.99 at a significance level of 0.05.

Table 2. Values of calculated statistics

Characteristics	Calculated values		Significance level
	Function (1)	Function (2)	
1	2	3	4
Determination coefficient	0.854	0.867	-
Correlation coefficient	0.924	0.931	-
Standard error	0.338	0.323	-
Calculated value of the Fisher-Snedecor test	119.365	132.103	less than 0.01
Calculated value of the Student's test for y-intersection	14.465	15.875	less than 0.01
Calculated value of the Student's test for x_1 and x_4	5.922	6.671	less than 0.01
Calculated value of the Student's test for x_2 and x_5	3.126	3.600	less than 0.01
Calculated value of the Student's test for x_3 and x_6	3.092	2.857	less than 0.01

Source: Indicators are determined by the authors based on models (1)–(2)

Results presented in Table 2 allow us to conclude that there is a high quality correlation between the resulting values and the three factors of the functions (1) and (2). All levels of significance given in column 3 of Table 2 have values less than 0.01. Therefore, the coefficients of the developed functions and the degree values in these functions are statistically significant with the precision of 99 percent. Functions (1) and (2) were checked using the Durbin-Watson test, which showed the absence of autocorrelation; the Breusch–Pagan test indicated the absence of heteroscedasticity. Collinearity does not exist, which is proved by the criterion of the inflation dispersion factors (VIF).

In the process of the source data approximation using the least squares method, residues are obtained, showing deviations of the calculated values from the source data. Checking the distribution of these residues by regression models is carried out based on the histograms assessment, normal distribution functions and tests of normality for these functions. Residue histogram charts are characterized by the maximum heights of the constructed rectangles in the middle of the histogram and the minimum heights of the rectangles located on the right and left sides of the histogram (so-called "tails"). Histograms are symmetric in relation to the middle. They demonstrate that residues are concentrated around zero. The same conclusion can be made after conducting the analysis of the density functions of the normal distribution (3) and (4), which have average values close to zero.

The distribution function of residuals corresponding to the regression model (1) is shown below:

$$y_3(x_7) = \frac{4.875}{0.101 \cdot \sqrt{2\pi}} \cdot e^{\frac{-(x_7-0.004)^2}{2\times 0.101\times 0.101}}. \quad (3)$$

The test shows the high quality of the function (3) and confirms the normality of the distribution described by this function. These conclusions are made after a check using three tests:

- the calculated value of 0.97 according to the Shapiro-Wilk test is greater than the tabular one –0.93;
- the calculated value of 1.49 for the Pearson test is less than the tabular one –9.49;
- the calculated value of 0.05 by Kolmogorov-Smirnov test is less than the tabular one –0.15.

The distribution function of residuals corresponding to the regression model (2) is shown below:

$$y_4(x_8) = \frac{6.508}{0.097 \cdot \sqrt{2\pi}} \cdot e^{\frac{-(x_8-0.005)^2}{2\times 0.097\times 0.097}}. \quad (4)$$

The test shows the high quality of the function (4) and confirms the normal distribution, described by this function. These conclusions are made after a check using three tests:

- the calculated value of 0.96 according to the Shapiro-Wilk test is greater than the tabular one –0.93;
- the calculated value of 2.64 in the Pearson test is less than the tabular one –9.49;
- the estimated value of 0.09 by Kolmogorov-Smirnov test is less than the tabular one –0.15.

The data obtained allow us to make a general conclusion that the developed functions (1) and (2) fully meet the econometric requirements and, therefore, can be used to describe the dependencies of rural regression models in the regions on fixed production assets cost, employees salary expenses and ratio of crop production to livestock production. Consequently, the first hypothesis was confirmed.

The developed regression models (1)–(2) prove the influence of fixed production assets cost, employees salary expenses and ratio of crop production to livestock production on the volume of production of enterprises and entrepreneurs belonging to the rural sector of the economy of the regions of Russia. It should be noted that in both functions there are small differences in the values of both coefficients and degrees. Thus, it can be concluded that the developed functions show the existence of established stable dependencies of rural production volumes in the regions on the factors under consideration for the period from 2017 to 2018. Thus, the second hypothesis was confirmed.

The values of degrees for three factors in the functions are positive, therefore it can be stated that the stimulation of rural production in the regions can be provided by an increase in the employee's salary expenses, fixed production assets cost and ratio of crop production to livestock production. The regression models for all the considered values of the factors do not reach their maximum. This is confirmed by the fact that the values of the maximum return on three factors for all functions are positive on the considered ranges of changes in the values of the factors. Thus, it can be concluded that the economy of the Russian regions has not reached saturation with rural products. In all regions there are significant reserves for the further development in this sector of the economy, including on the basis of increasing the number of enterprises and the number of employees in them, as well as increasing ratio of crop production to livestock production.

The factor fixed production assets cost in both regression models affects turnover to a greater extent than the factors wage of employees and the ratio of crop production to livestock production. This follows from comparing the values of the degrees, in the first function 0.412 is greater than 0.245 and 0.229, and in the second function 0.409 is greater than 0.262 and 0.202, respectively. Therefore, the third hypothesis has been confirmed. A comparison of the returns on scale according to data for 2017 and 2018 shows that the values of this indicator (equal to the sum of the values of the degrees in the functions) are close to each other and amount to 0.886 (function 1) and 0.873 (function 2). This suggests that with a simultaneous increase in three factors, the growth in rural production over the years under review was almost the same. The return on scale in rural over the years under review was less than 1. This is due to the relatively small number of people employed in every rural enterprise, which leads to a combination of the functions performed. According to the authors of the study [28], this leads to a relatively low level of personnel training, a decrease in labor productivity and, as a result, low resource efficiency in such enterprises. To increase rural production in the Russian regions, it is advisable to ensure the simultaneous growth of three factors, that is, fixed production assets cost, employees salary expenses and ratio of crop production to livestock production. It should be noted that for regions with an excess of able-bodied population in rural areas, the main direction of rural production development is associated with attracting new workers, including the creation of family businesses. In regions where there are not enough potential workers, the main direction of increasing rural production is associated with fixed production assets cost. The cross-derivatives of the regression models for each of the three factors are positive for all values of the range of changing factors, so increasing one factor improves the conditions for using other factors. The second derivatives of all isoquants are positive. The level of convexity decreases with an increase in the volume of production, which indicates an increase in the elasticity of replacement factors: with the growth of rural production, the possibility of replacing one factor with other factors increases.

The use of regression models is possible when solving such an important task as ranking regions by resource efficiency, namely fixed production assets cost and employees salary expenses. In addition, regression models can be used to compare the actual volume of production of rural sector enterprises in the region and the amount of rural production in the same region, projected on the basis of the regression model. In our opinion, a relatively large positive value of this value (that is, the excess of the actual turnover over the estimated one) indicates a good use of available resources in the region. And accordingly, a large negative value of this value allows us to conclude that there are problems with the functioning of enterprises specialized in rural in the relevant region.

A comparative analysis of the actual values of production volumes and the data predicted on the basis of the regression model (1) showed a high level of resource efficiency in 2017 in the following regions: Volgograd region (9.7%), republic of Bashkortostan (10.4%), Orenburg region (12.0%), republic of Tatarstan (14.5%), Kurgan region (14.9%), Altai republic (19.7%), Trans-Baikal territory (20.7%), republic of Kalmykia (32.4%), republic of Tyva (33.9%). The deviations of the actual values from the predicted values are indicated in parentheses. The low level of use of the considered factors of production was in such regions as Khabarovsk territory (–21.0%), Primorsky territory (–14.1%), Vologda region (–12.4%), Kostroma region (–12.0%), Kaluga region (–11.4%), Kaliningrad region (–11.2%), Vladimir region (–11.0%), republic of Adygea (–10.2%), Kirov region (–10.0%).

A comparative analysis of the actual values of production volumes and the data predicted on the basis of the regression model (2) showed a high level of resource efficiency in 2018 in the following regions: Belgorod region (9.6%), Orenburg region (10.2%), republic of Bashkortostan (10.9%), republic of Tatarstan (12.6%), Kurgan region (13.9%), republic of Sakha (16.5%), Altai republic (17.9%), republic of Kalmykia (29.2%), republic of Tyva (37.9%). The low level of use of the considered factors of production was in such regions as Khabarovsk territory (–17.8%), Primorsky territory (–15.9%), Kostroma region (–12.6%), Vologda region (–12.2%), republic of Adygea (–11.2%), Kaluga region (–11.0%), Kaliningrad region (–10.9%), Kirov region (–10.4%), Vladimir region (–10.2%). The above lists showed that most of the regions in 2018 retained their characteristics shown in 2017.

5 Conclusions

The conducted research has a certain scientific and practical significance. The scientific significance of the study is as follows:

- methodological aspects of the evaluation of functions describing the volumes of rural production in the regions of Russia are considered. It is proposed to use fixed production assets cost as a capital factor and employees salary expenses as a labor factor. In addition, as a third factor, it seems appropriate to consider the ratio of crop production to livestock production. At the same time, spatial data (by regions) is used for one year;
- in the course of the research, two three-factor regression models similar to the Cobb-Douglas functions were developed. These functions describe the dependence of production volumes in the rural sectors of each of the regions on the factors under consideration. Using a number of tests, the high quality of both developed regression models and their good approximation of the empirical data were confirmed;
- regression models have proved that there are significant reserves for the further development of rural sectors of the economy in all discussing Russian regions;

- an increase in one of the factors of the regression model improves the conditions for using other factors. With the growth of rural production in the regions, the possibility of replacing one factor with another factors;
- using the regression models, the regions of the country with high and low levels of efficiency in the use of available resources were identified.

The developed regression models are effective management tools that allow assessing the level of use of financial and labor resources in agriculture in specific regions of Russia. The results of the work can be in demand in the current activities of state, municipal and public organizations related to the regulation and support of agriculture, including when adjusting their actions based on scientific data.

The practical significance of the work lies in the possibility of using the results obtained to justify resources and monitor the level of efficiency of agriculture. The results of the study can be used by state and regional authorities to monitor the effectiveness of fixed production assets cost and employees salary expenses, i.e. to assess how well these resources are used. In addition, the functions allow you to identify an imbalance in the values of factors for each of the regions. The functions can be used in the justification of programs to increase fixed production assets cost and employees salary expenses, the formation of plans and programs for the further development of agriculture.

There were limitations in the research process, since 65 regions of Russia were considered in which rural production has received significant development. At the same time, data on 17 regions of Russia in which the rural sector has not received significant development were not taken into account when constructing regression models. Further research may be related to the development of similar models in the years following the publication of the relevant official statistics.

References

1. Petrikov, A.V.: The necessity and main features of the new agrarian policy in Russia. APC Econ. Manage. **12**, 24–34 (2020)
2. Margono, H., Sharma, S.C.: Technical Efficiency and Productivity Analysis in Indonesian Provincial Economies. Discussion Papers, Paper 26, Southern Illinois University, Carbondale USA (2004)
3. Zhang, D., Xie, J., Ermanno, A.: An efficiency and productivity analysis of the agricultural sector in alabama. Int. J. Appl. Econ. **14**(2), 19–36 (2017)
4. Ahmetov, K., Madiev, G., Bekbossynova, A.: A Systematic assessement of the resource potential of agriculture based on correlation-regression analysis and modeling of production functions. Probl. AgriMarket **3**, 58–67 (2019)
5. Parlinska, M., Dareev, G.: The agricultural production in mathematical models. Problems of World Agriculture (Problemy Rolnictwa Światowego) Warsaw University of Life Sciences **11**(26), 1–5 (2011)
6. Petrick, M., Kloss, M.: Identifying factor productivity from micro-data: The case of EU agriculture, Discussion Paper, No. 171, Leibniz Institute of Agricultural Development in Transition Economies (IAMO), Halle (Saale) (2021). http://nbn-resolving.de/urn:nbn:de:gbv:3:2-90157. Accessed 10 Oct 2021
7. Ghoshal, P., Goswami, B.: Cobb-Douglas production function for measuring efficiency in Indian agriculture: A region-wise analysis. Econ. Aff. **62**(4), 573–579 (2017)

8. Binghun, W., Zhou, E.: Research of total factor productivity and agricultural management based on malmquist-DEA modeling. Hindawi, Mathematical Problems in Engineering, vol. 2021, Article ID 2828061 (2021)
9. Kea, S., Li, H., Pich, L.: Technical efficiency and its determinants of rice production in Cambodia. Economies **4**(4), 1–17 (2016)
10. Czyzewski, B., Majchrzak, A.: Economic size of farms and adjustments of the total factor productivity to the business cycle in polish agriculture. Agric. Econ. (Zemědělská ekonomika) **63**, 93–102 (2017)
11. Nowak, A., Kijek, T., Domacska, K.: Technical efficiency and its determinants in the European Union. Agric. Econ. **61**, 275–283 (2015)
12. Prager, D.L., Foltz, J.D., Barham, B.L.: Making time for agricultural and life science research: Technical change and productivity gains. Am. J. Agr. Econ. **97**(3), 743–761 (2015)
13. Rezitis, A.N., Kalantzi, M.A.: Investigating technical efficiency and its determinants by data envelopment analysis: An application in the greek food and beverages manufacturing industry. Agribusiness **32**(2), 254–271 (2016)
14. Carpentier, A., Gohin, A., Sckokai, P., Thomas, A.: Economic modelling of agricultural production: Past advances and new challenges. Rev. Agric. Environ. Stud. **96**(1), 131–165 (2015)
15. Shestakov, R.B., Yakovlev, N.A.: Analysis of production potential in agriculture based on production function modeling. Bull. Rural Dev. Soc. Policy **3**(27), 9–12 (2020)
16. Tolmachev, M.N.: Problems of building production functions in Russian agriculture. Acc. Stat. **4**(24), 88–94 (2011)
17. Potapov, A.P.: Estimation of agricultural production volumes depending on the structure of resource costs. Econ. Sci. **6**(187), 74–79 (2020)
18. Naumov, I.V.: Problems of forecasting gross output in the regional socio-economic system. J. Econ. Theor. **4**, 68–83 (2017)
19. Kutenkov, R.P.: Methodology and results of factor forecasting of the dynamics of gross output and labor productivity in agriculture of the regions of the Russian Federation using production functions. Ostrovsky Readings **1**, 99–103 (2020)
20. Zyukin, D.A., Zhilin, V.V.: Cobb-Douglas function when assessing the development of agriculture of Kursk area. Curr. Dir. Sci. Res. XXI Century Theor. Pract. **4–2**(9–2), 299–302 (2014)
21. Zhilyaskova, N.P.: Production function in agriculture. Econ. Bull. Rostov State Univ. **6**(4), 63–67 (2008)
22. Germanova, O.E., Rudaya, Y.N.: Dynamics of parameters and type of technological progress in agriculture. Reg. Econ. South of Russia **3**(17), 158–172 (2017)
23. Granberg, A.G.: Modeling of the Socialist Economy. Ekonomika, Moscow (1988)
24. Felipe, J., McCombie, J.: Problems with Regional Production Functions and Estimates of Agglomeration Economies: A Caveat Emptor for Regional Scientists, Levy Economics Institute of Bard College Working Paper № 725 (2021). http://www.levyinstitute.org/pubs/wp_725.pdf. Accessed 10 Oct 2021
25. Charoenrat, T., Harvie, C.: Technical efficiency of Thai manufacturing SMEs: A Stochastic Frontier Analysis. Australas. Acc. Bus. Finan. J. **7**(1), 97–122 (2013)
26. Federal State Statistics Service (2021). https://rosstat.gov.ru/. Accessed 15 Oct 2021
27. Pindyck, R., Rubinfeld, D.: Microeconomics. Pearson, New York (2013)
28. International Labour Conference, 104th Session. Report IV Small and medium-sized enterprises and decent and productive employment creation. International Labour Office (ILO), Geneva (2015)

Towards More Precise Descriptions of Design Patterns

Ed van Doorn[1(✉)] and Sylvia Stuurman[2]

[1] The Hague University of Applied Sciences, Johanna Westerdijkplein 75, 2521 The Hague, The Netherlands
ed.vandoorn@ou.nl

[2] Faculteit Bètawetenschappen (vakgroep Informatica), Open University of the Netherlands, Valkenburgerweg 177, Heerlen, The Netherlands
Sylvia.Stuurman@ou.nl

https://www.hhs.nl, https://www.ou.nl

Abstract. The description of design patterns is not sufficiently precise to detect design patterns in class diagrams and source code. A proposal for a more precise description is given using two new concepts: *repeating group* and *peninsula*, and other types of associations than Gamma et al. have prescribed. Design patterns were originally modeled in OMT. Nowadays UML is used as a modeling language. The concept of interface, which is present in UML but not in OMT, is necessary to detect design patterns in class diagrams. Principles to translate OMT class diagrams to UML are described. This results in a description of design patterns in XML format, which will be useful for detecting design patterns in UML class diagrams and source code.

Keywords: Design patterns · Software analyze · Maintenance

1 Introduction

Automatic detection of design patterns can be useful during refactoring or preparation of architectural changes because it helps to understand and maintain software [1,2]. In an educational environment, automatic detection of design patterns may be helpful during elaborating exercises in which specifically described design patterns have to be implemented.

For automatic detection tools of design patterns, precise description of patterns are necessary. The original description of design patterns by Gamma et al. [10] do not contain enough information to base precise description. They specify design patterns by a combination of natural language, diagrams, and examples in C^{++} source code, and show inherent ambiguities. Ambiguity will be discussed in Sect. 2.2.

Some design patterns e.g. Abstract Factory, Factory Method, and Template Method have variants. Precise descriptions have to be appropriate for variants as well but are not specified by Gamma et al. It is clear that it is impossible to

R. Silhavy (Ed.): CSOC 2022, LNNS 501, pp. 117–140, 2022.
https://doi.org/10.1007/978-3-031-09070-7_11

evaluate the usefulness of an automatic detection tool when it is not clear which precise description the tool is based on. However, when such tools are presented, an explicit description is almost always omitted [3,5–8].

A survey [9] analyses 31 tools and associated algorithms. Various aspects are compared, for instance, four detection approaches are classified into the structural, behavioral, and semantic analysis. However, nothing is mentioned about the descriptions that are used for automatic detection. One of the conclusions of the survey is:

'*However, the reported accuracy is not balanced (i.e. high precision and low recall or vice versa). One possible solution is to use a common formalized definition of GoF patterns.*'

An example of low precision description is the Bridge design pattern [10]. As shown by Fig. 1. *Implementor* can have more than one realization, but how many realizations or subclasses may have *Abstraction*? These multiplicities should be specified.

The diagram is just an example of the Bridge pattern. It does not fully specify the design pattern because different occurrences of the Bridge pattern are possible. The diagram depicts an aggregate between *Abstraction* and *Implementor*. In Sect. 2.1 will be explained that instead of an aggregation an association is more appropriate.

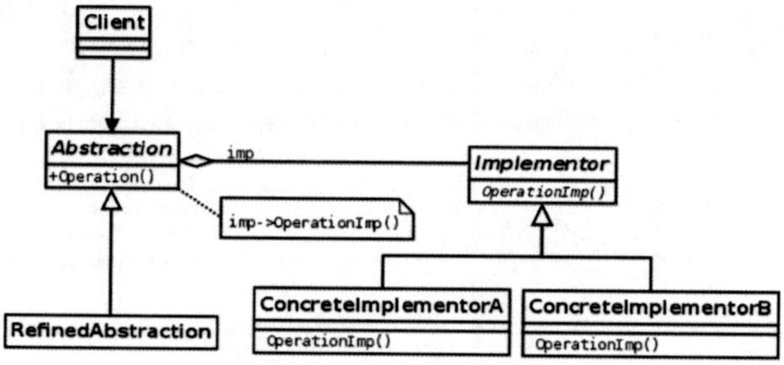

Fig. 1. Bridge pattern [10]

We have explained why a precise description of design patterns is both necessary and lacking. A description without ambiguities and vagueness is needed. The contribution of this research is a specification of the elements which are needed for a precise description of design patterns in a transparent way.

By transparent way, we mean a clear motivated selection of elements and the meaning of the elements, and also clear arguments for the necessity of the elements to specify design patterns.

Important motivation of our selections is our future work. We will create a detection tool for detecting design patterns in UML class diagrams and source code. So, our description contains structural elements, attributes, and operations. Features of programming languages that were not present in 1995, the year in which Gamma et al. published their book, such as functional features, will not be considered. This article focuses on object-orientation.

New elements are *repeating group* and *peninsula*. In Sect. 2.2 we introduce a rule to convert an abstract class in an OMT class diagram to an abstract class or an interface in a UML class diagram. This rule will be applied to all Gang of Four design patterns. In Sect. 3 we motivate our choice for XML as a specification language for design patterns and show some examples of these specifications. We also specify the structure of the specification language. In Sect. 4 the motivation of our choices and results are discussed. In Sect. 5 the conclusions are described.

2 Methods

To create precise definitions of design patterns based on the work of Gamma et al. [10], we describe the next steps. First, we show in Sect. 2.1 how we select the elements that are needed to specify structural and static aspects of design patterns. In Sect. 2.2 we discuss the vagueness caused by unclear use of abstract classes and interfaces and how we solve that by applying a rule for conversion of abstract classes in an OMT class diagram to an abstract class or interface in an UML class diagram. The rule is applied to all Gang of Four design patterns.

2.1 Specification of Elements

In this section, the elements that are needed for a more precise description of design patterns are listed.

As mentioned in Sect. 1, a survey [9] analyses 31 tools and associated algorithms. The tool with the most precise specification of design patterns is described by Yu et al. [12]. They define the concept of a subpattern that is used to define a structural feature model for each design pattern. For every design pattern, they also define a template of method signature to describe the sequence of method calls. However, their description has several shortcomings. They do not use the UML concept interface and multiplicities. Contrary to Gamma et al. none of their definitions has methods with parameters. The keyword 'static' which is necessary to define the Singleton pattern, is missing.

Some classes in design patterns have operations which names indicate a clear prescribed intention. An example is the Prototype pattern [10] that contains the operation 'Clone'. In Java, the method 'clone' has the same intended functionality and so is regularly used in an instance of this pattern. In a Python source, 'deepcopy' is used to duplicate an object. So, it should be possible to prescribe names of methods. If a detection algorithm uses a description with a prescribed name, false positive detection of this design pattern will often be prevented. This specification element is not used by Yu et al.

Therefore, a better specification of elements has to be made.

For now, we restrict ourselves to structural aspects of UML class diagrams and source code: relations between classes, attributes, and signature of operations. A signature of an operation is its name, input, and output. For now, the implementation of operations is left out because they refer to behavior and dynamic aspects of design patterns.

- Abstract class, concrete class, and interface It is evident that they are part of the structural aspect of design patterns and so should occur in the proposal of a general and more precise specification of design patterns.
- Relations
  - Association
    An association is a structural relationship between classes, which of course is needed to describe a design pattern. An aggregation is a special kind of association, representing a structural relationship between a whole and its parts [14]. A composition is a form of aggregation, with strong ownership and coincident lifetime as part of the whole [11].
    The last two concepts semantically differ from an association, but are these two concepts needed to describe design patterns? See i.e. Fig. 1. Is this still an example of the Bridge pattern when the aggregate between *Abstraction* and *Implementor* is replaced by an association? The necessity of the characteristic *whole - part* of an aggregate is not derivable from the description given by Gamma et al. Therefore, the more general concept association satisfies as well.
    Composition is for two reasons not necessary as a defining element of a more formal description of design patterns. First, Gamma et al. used OMT to model design patterns but composition is not an element of OMT. Second, the composition is responsible for creating and removing objects. Therefore, this type of association describes a structural and behavioral aspect. This article focuses only on structural aspects as stated in Sect. 1, hence this type of association is left out.
    These arguments apply to class diagrams. In source code, the 1-* association and aggregate are declared identical. A composition is implemented identically to an aggregate for the structural aspect, and by methods for the behavior aspect. So, for source code the differences between the three types of associations are not relevant for the structural aspects of design patterns.
    These arguments also apply to the other design patterns which contain an aggregate: Builder, Decorator, Flyweight, Command, Interpreter, Memento, Observer, State, and Strategy. The Composite design pattern, see Fig. 2, needs other arguments. The class *Composite* contains the methods *Add(Component)* and *Remove(Component)*. So, the class *Composite* is responsible for the duration of objects of *Component* within the design pattern but not of the *creation* of those objects. Therefore, the aggregate is not a composition.
    This article concludes that for the description of design patterns no aggregate or composition is needed, and an association is sufficient. Due to the differences in semantics, association, aggregate, and composition are useful for modeling a system.
  - Multiplicity
    Class diagrams that are used by Gamma et al. contain several categories of multiplicities. The one-one association exists in the class diagram of Abstract Factory pattern. See Fig. 3. This Figure also contains the one-or-

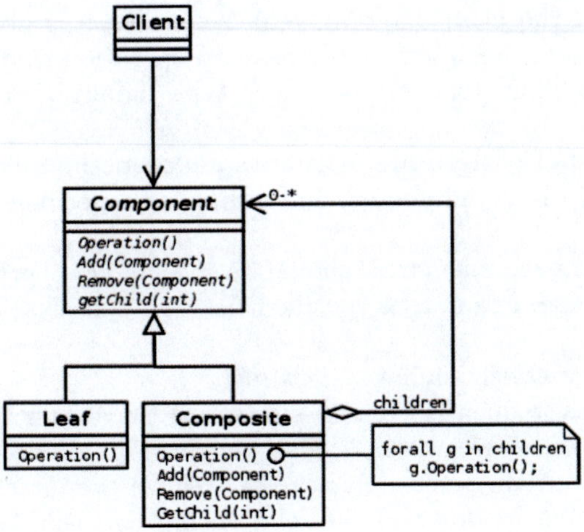

Fig. 2. Composite [10]

more multiplicity for the association between *client* and *AbstrProd*. The zero-or-more multiplicities occur in the class diagram of the Composite pattern. So, multiplicities are needed.

- Inheritance
  It is evident that inheritance is needed in the description language.
- Dependencies
  Dependencies did not occur in OMT [11]. In 1996 Rumbaugh introduced four kinds of dependencies [15]. Gamma et al. only use one of these dependency associations: the case of creating an object of another class. This dependency does also occurs in UML. The dependencies are used in the specification language.

– Recursive structures
  Recursive structures occur e.g. in the design patterns Composite and Decorator. Recursion occurs when classes and/or interfaces are interconnected through a circle of associations. The simplest case is a class that has a reflexive association. So, recursive structures can be realized by classes and/or interfaces and associations.
– Attributes and operations
  - Attributes
    Only the name and not the type of attributes occur in class diagrams, e.g. in Singleton, Decorator, Flyweight, Proxy, Memento, and Observer. The class Singleton in the Singleton pattern should have an attribute of the class Singleton. So types of attributes are necessary to describe design patterns.
  - Operations
    The name and/or type of parameters of operations sometimes occur in class diagrams, e.g. in Composite, Flyweight, Interpreter, and Visitor.

In none of the class diagrams, the return type is given. However, the design pattern Singleton should have an operation that returns an object of the class Singleton. So, the return type should be specified.

- Modifiers
  The Singleton pattern has attributes and a method for which the access modifiers (private, protected, and public) are specified in the code example [10].
  Likewise, the modifier static should be specifiable otherwise, the Singleton design pattern can not be specified. See the specification of this pattern in Sect. 3.

– Repeating groups with similar structures
Design patterns frequently contain structures that occur more than once in the design pattern. First, a simple example. See Fig. 1. The Bridge design pattern [10] may contain several classes that realize the interface *Implementor*. So, *ConcreteImplementor* and the relationship between *ConcreteIplementor* and *Implementor* may occur several times. They form a repeating group.
Second, a more complex example. See Fig. 3 that shows an example of an Abstract Factory [10]. On the left side, there are four ellipses with classes and relationships. These ellipses have similar structures: a superclass with some subclasses, e.g. *AbstractProd_A* and *A1, A2, A3*. In fact, there are three repeating groups. *A1, A2, A3*, with their relationship to *AbstractProd_A* forms an occurrence of the first repeating group. To be precise: Products with their relationships to their superclass *AbstractProduct* form a repeating group. Notice that the superclass *AbstractProduct* is not a part of the repeating group.
The ellipses consist of *AbstractProd*, his subclasses, and their interrelationships, and the relationship between *AbstractProd* and *Client* form the second repeating group. Likewise, a *ConcreteFactory* with his relationship to *AbstractFactory* forms the third repeating group. The dependencies between a *ConcreteFactory* and a Product e.g. *A1, A2, A3, B1* ... will be denoted the XML description in Sect. 3.
The repeating group is an example of the *open/closed principle* [16]. Repeating groups also occur in Builder, Factory Method, Prototype, Composite, Decorator, Interpreter, Chain of Responsibility, State, and Strategy. So, repeating groups that contain classes, interfaces and associations should be definable items in the description of a design pattern. The concept of repeating group contributes to the specification of variants of design patterns. Variants of design patterns apply to Bridge, Decorator, Factory Method, Facade, Chain of Responsibility, Mediator, Prototype, State, Strategy, and Visitor.

– Peninsula
The peninsula is a necessary concept to precisely describe the Facade pattern [10]. Figure 4 shows an example of a Facade pattern. The classes belong to four sets. The first set only consists of the class *Facade*. The second set consists of the classes *Out_1*, *Out_2*, and *Out_3*. The third set consists of the classes *In_1*, *In_2*, *In_3*. The fourth set consists of the classes *Extra_1* and *Extra_2*.

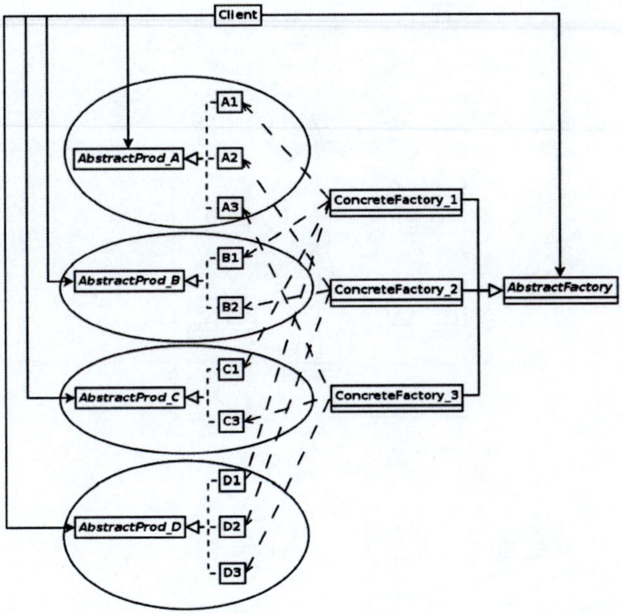

Fig. 3. Abstract Factory with repeating groups

The difference between the third and fourth set is simple. The classes in the third set have an association with *Facade* and the classes in the fourth set may not have an association with *Facade*.
The classes the third and fourth may not have associations to the classes of the second set and so form together with *Facade* a *peninsula*. Examples of forbidden associations are shown in Fig. 5. The specification of the Facade pattern is given and discussed in Sect. 3.
If a specification of a design pattern contains the specification element *peninsula*, a search algorithm for design patterns should report an error when it detects a forbidden association.

2.2 Abstract Classes and Interfaces

Gamma et al. use OMT, C^{++}, and natural language to specify design patterns. They use the concept of interface, but OMT nor C^{++} contain this concept. So, their specification of design patterns is not quite clear.

The concepts of interface and abstract class occur in the currently dominant modeling language UML, and object-oriented languages such as Java and C#. An interface, in this context, is a description of a set of public methods. For the conversions from OMT to UML, we have to choose between an interface and an abstract class.

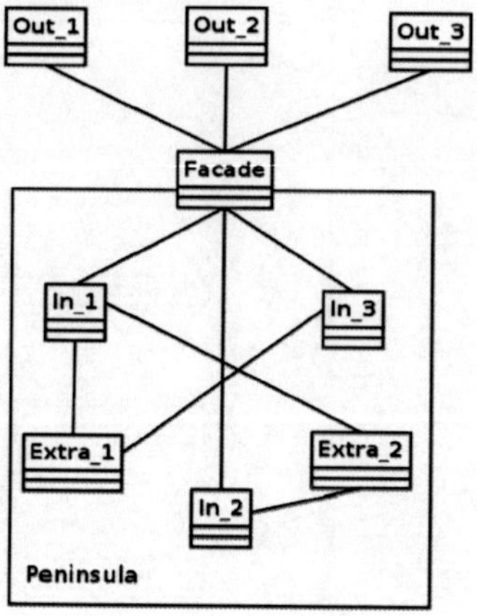

Fig. 4. Facade [10]

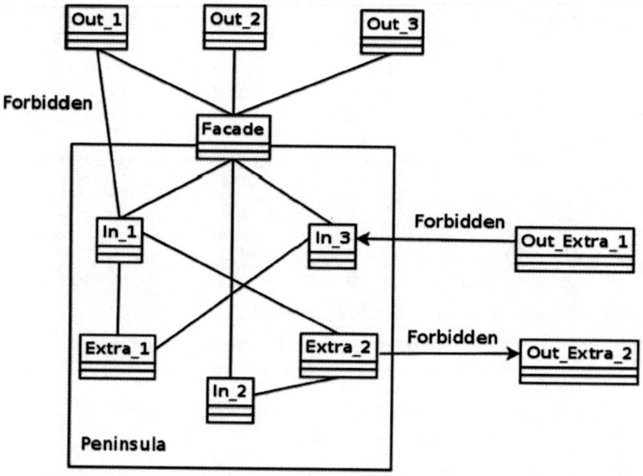

Fig. 5. Facade with forbidden associations

The textual descriptions of Gamma et al. provide clues for the question of whether an interface in the OMT class diagram may be replaced by an interface in a UML class diagram or whether an interface is in fact preferred. This article uses the description of the *intent* and sometimes the *notes on implementations* to check what the original description tells us about the question of an interface or an abstract class.

Furthermore, the next rule can be used for conversion. An abstract class in the OMT class diagram may be replaced by an interface in a UML if

- all operations are abstract and
- the specific class does not need an association with another class and
- there is no need to maintain a state that should be specified in this class.

Note that all requirements must be met to be applied, otherwise an abstract class in OMT should be converted to an abstract class in UML.

This rule is used to convert abstract classes in all Gang of design patterns.

Abstract Factory. The intent of Abstract factory is '*Provide an interface for creating families of related or dependent objects without specifying their concrete classes.*'

An abstract factory does not need to have an association and does not need to have concrete methods (Table 1).

Table 1. Abstract factory

Role	Abstract class	Interface	Concrete
Abstract factory	✓	✓	×

Builder. The intent of this pattern does not give a hint for a decision: '*Separate the construction of a complex object from its representation so that the same construction process can create different representations.*'

The Participant section states '*specifies an abstract interface for creating parts of a Product object.*'

The Director needs an association to the builder. The builder does not need an association, and does not need concrete methods (Table 2).

Table 2. Builder

Role	Abstract class	Interface	Concrete
Director	×	×	✓
Builder	✓	✓	×

Factory Method. The intent states '*Define an interface for creating an object, but let subclasses decide which class to instantiate. Factory Method lets a class defer instantiation to subclasses.*'

In the section on Participants is stated '*defines the interface of objects the factory method creates.*'

The notes on Implementation state that there are two major varieties of the Factory method pattern: '*(1) the case when the Creator class is an abstract class and does not provide an implementation for the factory method it declares, and (2) the case when the Creator is a concrete class and provides a default implementation for the factory method. It's also possible to have an abstract class that defines a default implementation, but this is less common.*'

That means that Creator can either be an abstract class or an interface. Both choices determine a specific variety of the pattern (Table 3).

Table 3. Factory method

Role	Abstract class	Interface	Concrete
Creator	✓	✓	×
Product	✓	✓	×

Prototype. The notes on the Participants state for Prototype '*declares an interface for cloning itself.*'

Therefore, Prototype may be an interface or an abstract class (Table 4).

Table 4. Prototype

Role	Abstract class	Interface	Concrete
Protoype	✓	✓	×

Singleton. The Singleton ensures that a class only has one instance. This class is, therefore, by definition, concrete (Table 5).

Table 5. Singleton

Role	Abstract class	Interface	Concrete
Singleton	×	×	✓

Adapter. The intent states '*Convert the interface of a class into another interface clients expect. Adapter lets classes work together that couldn't otherwise because of incompatible interfaces.*'

In the class variant, the Adapter implements the interface of the Adaptee and subclasses theTarget, or the other way around. In the object variant, the Adapter has an association to the Adaptee, and is a subclass of the Target.

In the diagrams, both Adapter and Adaptee are concrete classes. Target is an abstract class in the diagram. Nothing prevents Target from being an interface.

The Adapter can be a concrete class. It has at least one concrete method. Therefore, it may also be an abstract class (Table 6).

Table 6. Adapter

Role	Abstract class	Interface	Concrete
Adapter	×	×	✓
Adaptee	×	×	✓
Target	✓	✓	✓

Bridge. The intent states '*Decouple an abstraction from its implementation so that the two can vary independently.*'

The Abstraction has an association to an object of type Implementor, and so should be an abstract class. The Implementor 'defines the interface for implementation classes', and can thus be an interface as well (Table 7).

Table 7. Bridge

Role	Abstract class	Interface	Concrete
Abstraction	✓	×	×
Implementor	✓	✓	×

Composite. The intent is '*Compose objects into tree structures to represent part-whole hierarchies. Composite lets clients treat individual objects and compositions of objects uniformly.*'

The notes on participants state '*declare the interface for objects in the composition.*
implements default behavior for the interface common to all classes, as appropriate.'

So Component may be an abstract class, and, when there is no default behavior for all objects, an interface. Leaf and Composite are depicted as concrete classes (Table 8).

Table 8. Composite

Role	Abstract class	Interface	Concrete
Component	✓	✓	×

Decorator. The intent is '*Attach additional responsibilities to an object dynamically. Decorators provide a flexible alternative to subclassing for extending functionality.*'

The Component '*defines the interface for objects that can have responsibilities added to them dynamically.*' That means that the Component can both be an abstract class or interface (Table 9).

The Decorator has an association with Component, and can therefore only be an abstract class.

Table 9. Decorator

Role	Abstract class	Interface	Concrete
Component	✓	✓	×
Decorator	✓	×	×

Facade. The intent states '*Provide a unified interface to a set of interfaces in a subsystem. Facade defines a higher-level interface that makes the subsystem easier to use.*'

The Facade '*delegates client requests to appropriate subsystem objects*', and thus, obviously, needs associations to other classes. It can only be an abstract (or concrete) class (Table 10).

Table 10. Facade

Role	Abstract class	Interface	Concrete
Facade	✓	×	✓

Flyweight. The Flyweight '*declares an interface through which flyweights can receive and act on an extrinsic state.*' That means that Flyweight is preferably an interface (Table 11).

Table 11. Flyweight

Role	Abstract class	Interface	Concrete
Flyweight	✓	✓(preferably)	×

Proxy. The Subject '*defines the common interface for RealSubject and Proxy so that a Proxy can be used anywhere a RealSubject is expected*', and can, therefore, be both an abstract class or an interface.

The Proxy '*maintains a reference that lets the proxy access the real subject. Proxy may refer to a Subject if the RealSubject and Subject interfaces are the same*', and can therefore only be an abstract class. The Proxy may also be a concrete class (Table 12).

Table 12. Proxy

Role	Abstract class	Interface	Concrete
Subject	✓	✓	×
Proxy	✓	×	✓

Chain of Responsibility. The Handler has an association to another Handler, and can, therefore, only be an abstract class (Table 13).

Table 13. Chain of responsibility

Role	Abstract class	Interface	Concrete
Handler	✓	×	×

Command. Command '*declares an interface for executing an operation.*' Therefore, Command may be an interface or an abstract class (Table 14).

Table 14. Command

Role	Abstract class	Interface	Concrete
Command	✓	✓	×

Interpreter. An AbstractExpression '*declares an abstract Interpret operation that is common to all nodes in the abstract syntax tree*', and therefore, may be an abstract class or interface (Table 15).

Table 15. Interpreter

Role	Abstract class	Interface	Concrete
AbstractExpression	✓	✓	×

Iterator. An Iterator '*defines an interface for accessing and traversing elements*', and may therefore be an abstract class or an interface.

An Aggregate '*defines an interface for creating an Iterator object*', and may therefore also be both an abstract class or an interface (Table 16).

Table 16. Iterator

Role	Abstract class	Interface	Concrete
Iterator	✓	✓	×
Aggregate	✓	✓	×

Mediator. A Mediator '*defines an interface for communicating with Colleague objects*' and may thus be both an interface and an abstract class.

'*Each Colleague class knows its Mediator object.*' A Colleague thus needs an association to Mediator, and can therefore only be an abstract class (Table 17).

Table 17. Mediator

Role	Abstract class	Interface	Concrete
Mediator	✓	✓	✗
Colleague	✓	✗	✗

Memento. Both Originator and Memento are classes that do not have subclasses and only have concrete operations. So, they must be concrete (Table 18).

Table 18. Memento

Role	Abstract class	Interface	Concrete
Memento	✗	✗	✓
Originator	✗	✗	✓

Observer. The Subject '*knows its observers*' and should therefore be an abstract class.

The Observer '*defines an updating interface for objects that should be notified of changes in a subject*' and can, therefore, also be an interface (Table 19).

Table 19. Observer

Role	Abstract class	Interface	Concrete
Subject	✓	✗	✗
Observer	✓	✓	✗

State. The Context ' *maintains an instance of a ConcreteState subclass that defines the current state*', and should therefore be an abstract class.

The State '*defines an interface for encapsulating the behavior associated with a particular state of the Context.*', and can also be an interface (Table 20).

Table 20. State

Role	Abstract class	Interface	Concrete
Context	✓	✗	✗
State	✓	✓	✗

Strategy. The Context '*maintains a reference to a Strategy object*' and should, therefore, be an abstract class. It can also be a concrete class because the pattern does not depend on subclasses of Context.

Strategy '*declares an interface common to all supported algorithms*', and can be an interface as well (Table 21).

Table 21. Strategy

Role	Abstract class	Interface	Concrete
Context	✓	×	✓
Strategy	✓	✓	×

Template Method. AbstractClass '*implements a template method defining the skeleton of an algorithm*' and can, therefore, not be an interface (nor a concrete class) (Table 22).

Table 22. Template method

Role	Abstract class	Interface	Concrete
AbstractClass	✓	×	×

Visitor. Visitor '*declares a Visit operation for each class of ConcreteElement in the object structure.*' It does not need an association or an implementation of methods. It can, therefore, be an abstract class or interface.

Element '*defines an Accept operation that takes a visitor as an argument.*' Here, the same applies (Table 23).

Table 23. Visitor

Role	Abstract class	Interface	Concrete
Visitor	✓	✓	×
Element	✓	✓	×

Therefore, translations exist that result in only an abstract class, only an interface, or abstract class and interface. These three possibilities are used in the next section.

3 Results

In this Section, we motivate our choice for XML as a specification language for design patterns. We chose a meta-language to specify the XML-structure. Some typical examples of specifications of design patterns are given.

Specialized languages to express precise descriptions of patterns exist. The Balanced Pattern Specification Language (BPSL) is used to describe the structural and behavioral aspects of design patterns with first-order logic i.e. predicates and temporarily logic [13]. They apply multiplicities and methods with parameters. The concepts of abstract class, interface, aggregate, 'static' and the possibility to use a prescribed name are missing. It can be concluded that BPSL could be extended to overcome all these shortcomings. However, this would result in a large list of first-order logical predicates to describe a design pattern.

For future work, maintenance of the description of design patterns, and reading and parsing a file that contains the description of design patterns should be relatively easy. Therefore XML is chosen to describe design patterns.

For every Gang of Four design pattern, a template is defined. See http://92.109.194.125/DesignPatterns/templates. Compared to the list of first-order logical predicates these descriptions are concise and easy to maintain.

For programming languages that are frequently used, i.e. Java, Python, C^{++}, C, and PHP a SAX (Simple API for XML) parser can be easily implemented.

To formalize the specification of XML description of design patterns, there are several methods: XML schema, BNF, EBNF (extended Backus-Naur form) [17], PEG (Parsing Expression Grammar) [18], and syntax diagram.

All methods are well defined but have some disadvantages. An XML schema that contains recursive structures is not easy to create and read. BNF has some disadvantages which are overcome by EBNF. Although EBNF is well defined, in practice, there are some variants. PEG is not well understood as a language specification tool [19]. A Syntax diagram is very easy to read but takes a lot of space.

We have chosen EBNF because it is well defined, the description is free available and the specification of the XML-description of design patterns is concise. The specification of the XML structure is available at http://24.132.55.38/DesignPatterns/EBNF_structure.

As representative examples, descriptions of the Singleton and Facade, Visitor, and Template Method patterns are given below. They are representative because all XML-tags and attributes occur in at least one example.

Singleton[10]

C^{++}source:

```
class Singleton {
   public:
      static Singleton* Instance();
   protected:
      Singleton();
```

```
  private:
     static Singleton* _instance;
};
```

Formal XML description:

```
<template name="Singleton">
   <class name="Singleton">
      <attribute name="uniqInstance"
         type    ="Singleton"
         modifier="private"
         isStatic="true"/>
      <operation name="Singleton"
            modifier="private"/>
      <operation name="getInstance"
            modifier="public"
            isStatic="true"/>
   /class>
</template>
```

Facade [10] See Fig. 4
This example shows the application of the *peninsula* and the *repeating group*.

```
<template name="Facade">
   <class name="Facade"/>

   <peninsula>
      <repeating_group>
         <class name="In"/>

         <edge node1="Facade"
               node2="In"
               type ="ASSOCIATION_DIRECTED"/>
      <repeating_group/>
   </peninsula>

   <repeating_group>
      <class name="Out"/>

      <edge node1="Facade" node2="Out"
            type ="ASSOCIATION_DIRECTED"/>
   <repeating_group/>
</template>
```

It looks like the *Extra_* classes in Fig. 4 are missing, but there are not. Essential to the Facade patterns are the *Facade* class, the classes that are associated with the *Facade*, and the concept of the peninsula. Other classes may occur, provided that the concept of the peninsula is not violated.

Visitor [10] See Fig. 6
The formalized description of this design pattern illustrates the application of the *correspondsTo* attribute.

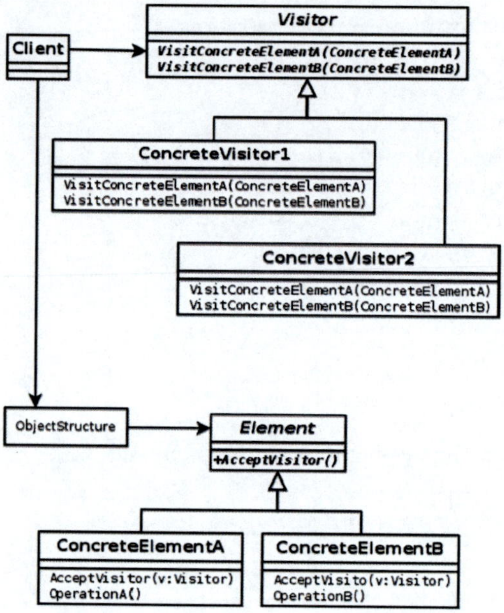

Fig. 6. Visitor [10]

```
<template name="Visitor">
  <class name="Client"/>

  <class name="Visitor"
         type="ABSTRACT_INTERFACE">
    <operation name="visit"
        correspondTo="Element">
        <parameter name=""
            type="ConcreteElement"/>
        <!--
          correspondTo="Element" and
          type="ConcreteElement"
          means that for every
          specialization of "Element",
```

```
        i.c. "ConcreteElement",
        there is an operation
        "visit" with parameter of
        that specialization of
        Element.
      -->
    </operation>
  </class>

  <class name="Element"
        type="ABSTRACT_INTERFACE">
    <operation name="acceptVisitor"
          isAbstract="true">
        <parameter name=""
          type="Visitor"/>
    </operation>
  </class>

  <class name="ObjectStructure"/>

  <repeating_group>
    <class name="ConcreteVisitor"/>

    <edge node1="ConcreteVisitor"
          node2="Visitor"
          type ="INHERITANCE"/>
  </repeating_group>

  <repeating_group>
    <class name="ConcreteElement">
      <operation name="operation"/>
    </class>

    <edge node1="ConcreteElement"
          node2="Element"
          type ="INHERITANCE"/>
  </repeating_group>

  <edge node1="Client"
        node2="Visitor"
        type ="ASSOCIATION_DIRECTED"/>
  <edge node1="Client"
        node2="ObjectStructure"
        type ="ASSOCIATION_DIRECTED"/>
  <edge node1="ObjectStructure"
```

```
        node2="Element"
        type ="ASSOCIATION_DIRECTED"/>
</template>
```

Template Method [10] See Fig. 7
This definition illustrates the application of the *calledBy* and the *occursMultiple* attributes.

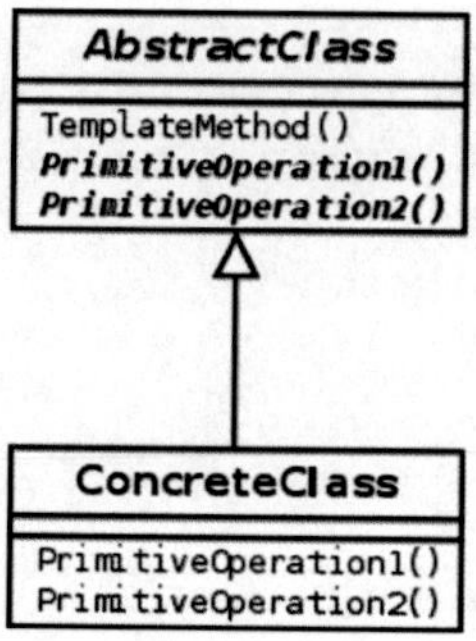

Fig. 7. Template method [10]

```
<template name="Template Method">
   <class name="AbstractClass"
          type="ABSTRACT">
      <operation name="templateMethod"/>
      <operation name="primitiveOperation"
          occursMultiple="true"
          isAbstract     ="true"
          calledBy       ="templateMethod"/>
   </class>

   <class name="ConcreteClass"/>

   <edge node1="ConcreteClass"
         node2="AbstractClass"
         type ="INHERITANCE"/>
</template>
```

Abstract Factory [10] See Fig. 3
The repeating groups are discussed in Sect. 1. Notice the recurrent occurrences of the repeating groups and the location of the description of the dependency between ConcreteFactory and Prod.

```
<template name="Abstract Factory">
  <class name="Client"/>
  <class name="AbstractFactory"
         type="ABSTRACT_INTERFACE"/>

  <repeating_group>
    <class name="AbstrProd"
           type="ABSTRACT_INTERFACE"/>

    <repeating_group>
      <class name="Prod"/>
      <edge node1="Prod"
            node2="AbstrProd"
            type ="INHERITANCE"/>
      <edge node1="ConcreteFactory"
            node2="Prod"
            type ="DEPENDENCY"/>
    </repeating_group>

    <edge node1="Client"
          node2="AbstrProd"
          type ="ASSOCIATION_DIRECTED"/>
  </repeating_group>

  <repeating_group>
    <class name="ConcreteFact"/>
    <edge node1="ConcreteFact"
          node2="AbstractFactory"
          type ="INHERITANCE"/>
  </repeating_group>

  <edge node1="Client"
        node2="AbstractFactory"
        type ="ASSOCIATION_DIRECTED"/>
</template>
```

In preparation for the software to detect design patterns in the future and to verify that the XML structure is usable, a SAX-parser was written in Java that is able to transform every GoF template to an intermediate Java structure.

4 Discussion

Our contribution is a specification of the elements which are needed for a precise description of design patterns in a transparent way.

Are the decisions to create the transparent precise specifications clearly motivated, necessary, and correctly made? We have shown that for our future work, automatically detecting design patterns in class diagrams and source code, the specifications of design patterns by Gamma et al. are not sufficiently clear and precise. The decisions are transparently i.e., clearly and fully described, made. The choice between abstract class and interface is elaborated.

The concept of *repeating group* is necessary to formally describe several design patterns like the Bridge pattern and the Abstract Factory. Gamma et al. describe these patterns by example. Decisions about their description are transparent.

Are the resulting descriptions in XML useful? Of course, the final answer can be given after finishing our future work, but all but two design patterns are formally and generally described in XML. The descriptions are more general because variants of a design pattern such as the Bridge pattern and the Abstract Factory are each given by one specification. The State and Strategy patterns are structurally identical. The definition of a design pattern should comprise more than the specification of structural and statical aspects.

Part of future work, building a pattern detection tool, may therefore produce for these patterns false negatives.

Are all decisions correct? To detect design patterns we will make no differences between the relationships aggregate, composite, and the 1-$*$ associations although they semantically differ. However, in source code, the differences are not noticeable.

Are all necessary elements to describe design patterns specified? This article has specified the elements based on class diagrams. All but two of the Gang of Four design patterns can be specified. It is not known whether other than Gang of Four design patterns need more structural elements to be defined. All UML-elements, except for notes, and all known structures that may occur in a class diagram can be described by the proposal. Therefore, this article expects that all design patterns that can be modeled in a UML class diagram can be specified.

5 Conclusions

The descriptions of design patterns, that can be used to detect them in class diagram and in source code, need an update for two reasons. The de facto standard description given by Gamma et al. is based on OMT, which is out of date, diagrams, and examples in source code. In Sect. 1 we indicated that for tools that can detect design patterns, it is not sufficiently specified what descriptions of design patterns they use. Even the best tool in the discussed survey has descriptions with shortcomings.

The contribution of this research, a specification of the elements which are needed for a precise description of design patterns, resulted in a more precise and formalized description of design patterns using the new concepts *repeating group, peninsula*, and the characteristics *correspondTo*, *calledBy*, and *occursMultiple* as shown in Sect. 3. Conversion of an abstract class in an OMT class diagram to an abstract class of an interface in a UML class diagram is possible by a given rule. Two descriptions of the Factory method show that variants of design patterns can be defined.

The formalized language and the description of the design patterns are made publicly available which will help us to realize future work.

References

1. Stuurman, S.: Design for Change. Ph.D. thesis, Open University, Department of Computer Science (2015)
2. Uchiyama, S., Kubo, A., Washizaki, H., Fukazawa, Y.: Detecting design patterns in object-oriented program source code by using metrics and machine learning. J. Softw. Eng. Appl. **7**(12), 983–998 (2014)
3. Tsantalis, N., Chatzigeorgiou, A., Stephanides, G., Halkidis, S.: Design pattern detection using similarity scoring. IEEE Trans. Software Eng. **32**(11), 896–909 (2006)
4. Prechelt, L., Kramer, C.: Functionality versus practicality: employing existing tools for recovering structural design patterns. J. Univers. Comput. Sci. **4**(12), 866–882 (1998)
5. Niere, J., Schäfer, W., Wadsack, J.P., Wendehals, L., Welsh, J.: Towards pattern-based design recovery. In: Proceedings of the 24th International Conference on Software Engineering, ICSE 2002, New York, NY, USA, pp. 338–348. ACM (2002)
6. Doorn, E.: Supporting design process by automatically detecting design patterns and giving some feedback. Master's thesis, Open University, Heerlen, The Netherlands (August 2016)
7. Doorn, E., Stuurman, S., van Eekelen, M.: Static detection of design patterns in class diagrams. In: Rahimi, E., Stikkolorum, D. (eds.) 8th Computer Science Education Research Conference, CSERC 2019, 18 November 2019–20 November 2019, pp. 79–88. ACM, United States (November 2019)
8. Xiong, R., Lo, D., Li, B.: Distinguishing similar design pattern instances through temporal behavior analysis. In: 2020 IEEE 27th International Conference on Software Analysis, Evolution and Reengineering (SANER), pp. 296–307 (2020)
9. Al-Obeidallah, M.G., Petridis, M., Kapetanakis, S.: A survey on design pattern detection approaches. Int. J. Softw. Eng. (IJSE) **7**(3), 41–59 (2016)
10. Gamma, E., Helm, R., Johnson, R., Vlissides, J.: Design Patterns: Elements of Reusable Object-Oriented Software. Addison-Wesley Longman Publishing Co., Inc., Boston, MA, USA (1995)
11. Rumbaugh, J., Blaha, M., Premerlani, W., Eddy, F., Lorensen, W.: Object-Oriented Modeling and Design. Prentice-Hall, Englewood Cliffs, NJ (1991)
12. Yu, D., Zhang, Y., Chen, Z.: A comprehensive approach to the recovery of design pattern instances based on sub-patterns and method signatures. J. Syst. Softw. **103**, 1–16 (2015)
13. Taibi, T., Ngo, D.C.L.: Formal specification of design patterns - a balanced approach. J. Object Technol. **2**(4), 127–140 (2003)
14. Booch, G., Rumbaugh, J., Jacobson, I.: The Unified Modeling Language User Guide. Addison-Wesley, Upper Saddle River, NJ (2005)
15. Rumbaugh, J.: OMT Insights: Perspective on Modeling from the Journal of Object-Oriented Programming. SIGS Reference Library (SIGS) (1996)
16. Meyer, B.: Object-Oriented Software Construction, 2 edn. Prentice Hall (1997)
17. ISO/IEC 14977:1996: Information technology - syntactic metalanguage - Extended BNF (1996)

18. Ford, B.: Parsing expression grammars: a recognition-based syntactic foundation. In: Jones, N.D., Leroy, X. (eds.) POPL, pp. 111–122. ACM (2004)
19. Redziejowski, R.: Some aspects of parsing expression grammar. Fund. Inform. **85**, 441–454 (2008)

Design and Implementation of Hyperledger-Based Certificate Management System

Jong Hyuk Park[1](✉) and In Whee Joe[2]

[1] MNI Lab, Hanyang University, Seoul 80305, CO, Korea
johnpy2930@gmail.com
[2] Department of Computer Science, Hanynag University, Seoul, Korea

Abstract. In this paper, we propose the design and implementation of a blockchain-based certificate management system as a Hyperledger Fabric platform to complement the data security and reliability of the centralized architecture. HyperCMS configured a network of one channel, one orderer, one organization for the certificate management system, and implemented the certificate enrollment and lookup chaincode. The original certificate file is compressed by the LZMA compression algorithm and stored on the blockchain ledger. As a result, HyperCMS implemented a blockchain-based certificate management system that includes data reliability and security.

Keywords: Hyperledger Fabric · Smart contract · Blockchain

1 Introduction

This paper proposes the design and implementation of the blockchain based certificate management system. the blockchain based certificate management system complements data security and reliability of the existing certificate management system. The existing certificate management system has different verification processes for different companies and institution. The typical verification process of the existing certificate management system is as follows. certificate information and document authenticity number are stored in database of companies and institutions. The certificate is verified by looking up the document authenticity number in the database of companies and institutions.

This data approach with a centralized architecture causes some problems. First, traditional databases using a centralized architecture are governed by a single center [3]. Second, Independent database operation are not efficient for sharing data between different organizations. Third, the existing databases are vulnerable to security related to data tampering and leakage. Because of these problems, companies and institutions require the certificate management system that ensures data reliability and security.

This paper proposes a new system design approach for Bitcoin-derived blockchain technology. Blockchain has been implemented in various ways, including public blockchain and private blockchain. Public blockchain include Bitcoin [1], Ethereum

R. Silhavy (Ed.): CSOC 2022, LNNS 501, pp. 141–148, 2022.
https://doi.org/10.1007/978-3-031-09070-7_12

[4], and all participants can crate blocks of blockchain if only a few conditions are met. private blockchain can only create blocks of the blockchain by authorized participants, and other participants cannot create blocks and cannot share the blockchain ledger [6]. For this reason, private blockchain can provide higher throughput of the system and more intimate services to companies and institutions than public blockchain.

Hyperledger Fabric is a private blockchain framework and uses a modular architecture [5]. This is a platform for developing blockchain solutions. HyperCMS is a blockchain-based certificate management system implemented using the Hyperledger Fabric platform. Hyperledger Fabric's chaincode, channel, and authentication complement the data reliability and security of existing certificate management systems [7]. HyperCMS provides certificate registration, modification, and retrieval functions, and all transactions occurring in each function are recorded in the blockchain ledger.

The following sections of this paper consist of HyperCMS scenario, architecture and performance analysis.

2 System Overview

2.1 Application Scenario

The entities in HyperCMS Scenario consists of certificate issuing company or institution, certificate acquirer and certificate management system (Fig. 1 and 2).

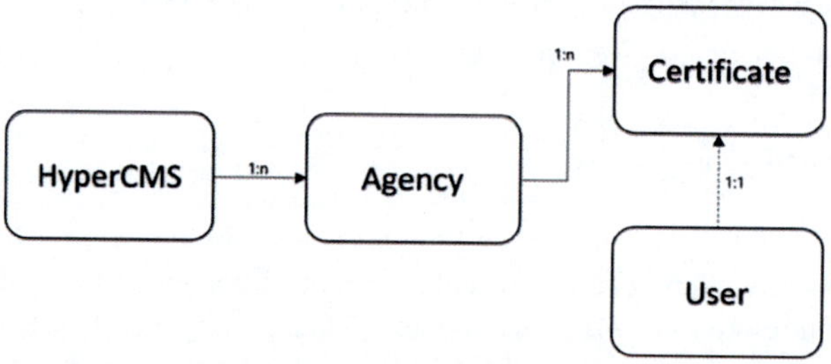

Fig. 1. Application structure

Certificate issuing Company or Institution. HyperCMS provides certificate registration and modification function to registered companies and institutions. First, companies and institutions must go through the HyperCMS sign up process.

Certificate Acquirer. HyperCMS provides certificate acquirer with certificate retrieval function for certificate forgery verification. The certificate acquirer must know the identification key of the certificate issued by the certificate issuing companies and institutions.

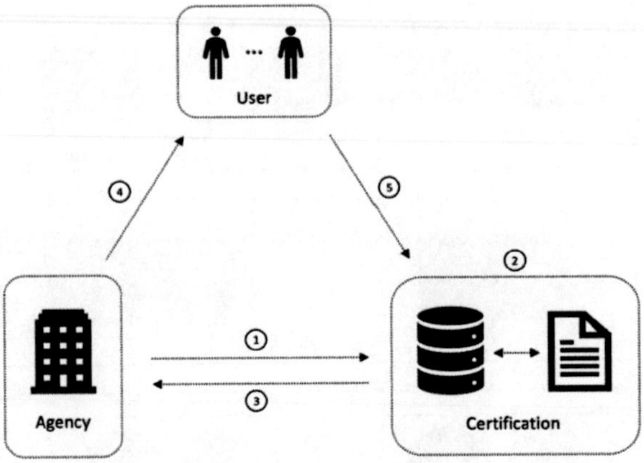

Fig. 2. Application scenario

Certificate Management System. HyperCMS provides certificate management functions such as certificate registration, modification and retrieval to certificate acquirer and the certificate issuing companies and institutions. The certificate registration function stores the certificate information and the compressed original certificate file in the blockchain ledger and issues an identification key. The certificate retrieval function finds certificate information and compressed original certificate file through the identification key in the blockchain ledger. The certificate modification function can change the certificate information and compressed original certificate file, and all transactions are stored in the blockchain ledger.

2.2 Application Architecture

All transactions that occur in the blockchain network are recorded in the blockchain ledger. the blockchain ledger is distributed and replicated to all participants in the blockchain network. This approach ensures data reliability and security in certificate management systems. HyperCMS is implemented as a private blockchain-based hyperledger fabric platform. Certificate management system logic is implemented in chaincode written in the Node.js language (Fig. 3).

Order. The orderer keeps the blockchain ledger up to date. The orderer sorts all transactions occurring in all channels in chronological order and creates blocks. The block is then sent to all peers in the blockchain network. After receiving the latest block, the peer completes the verification process and updates the blockchain ledger. Updating data in the blockchain ledger requires consensus between all peers in the blockchain network. Consensus algorithms of Hyperledger Fabric include Solo, Kafka, and PBFT.

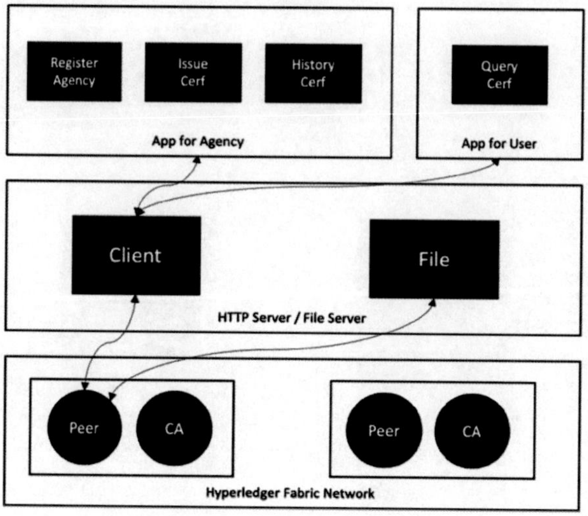

Fig. 3. Application architecture

Channel. All peers communicate on the channel. This channel has a blockchain ledger and can only be accessed by participants who participated in the channel. This provides data confidentiality that cannot be provided by public blockchains. The Channel is created through CSCC (Configuration System Chain Code), and blockchain ledgers used in the channel are created through Genesis Block. The Genesis Block contains settings for channel members, channel policies and the role of each peer. This setting allows channel participants to share blockchain ledgers.

Smart Contract. Hyperledger Fabric's smart contracts design blockchain-based program logic and are implemented in chaincode written in Java, Go, and Node.js general-purpose programming languages [2]. Chaincode stores or reads data in the blockchain ledger. And all transactions that occur in this process are stored in the blockchain ledger. HyperCMS's certificate registration, modification, and retrieval functions are implemented in chaincode written in Node.js.

Ledger. HyperCMS uses Hyperledger Fabric's ledger system. It consists of World State and Transaction Log. World state is a database that stores the ledger state at a specific point in time.The transaction log records all transactions that are protected by encryption and cannot be changed.

Application. HyperCMS implements applications that interact with the blockchain ledger through the Hyperledger Fabric Client SDK. The Hyperledger Fabric Client SDK consists of RESTful APIs and can invoke various functions such as channel creation, chaincode installation, and execution. HyperCMS implements a certificate registration-only application interface for companies and institutions, and a certificate retrieval-only application interface for certificate acquirers.

LZMA is a data compression algorithm developed by Igor Pavlov, and is used for the 7z format. HyperCMS stores the original certificate file on the blockchain ledger for the security and stability of the data. HyperCMS performance affects transaction size, with smaller transaction sizes resulting in faster performance. HyperCMS uses the highly compressed LZMA compression algorithm to make the original certificate file 40% smaller. As a result, HyperCMS uses LZMA open source to compress the original certificate file and store it in the blockchain ledger.

3 Implementation of Hyperets

3.1 Blocahin Network

HyperCMS is implemented with Hyperledger Fabric platform. HyperCMS's smart contracts are designed to perform certificate management system logic.

Smart Contract Design. Smart contracts are written in chaincode and interact with the blockchain to achieve program logic. it supports several programming languages. The smart contracts in HyperCMS are designed to perform the functions required for a certificate management system and the chaincode is written in Node.js. The certificate registration logic of HyperCMS records the certificate issuing company or institution infomation, issued certificate information, and compressed original certificate file along with the identifier key in the blockchain ledger. HyperCMS's certificate retrieval logic retrieves the certificate issuing company or institution infomation, issued certificate information, and compressed original certificate file from the blockchain ledger with the identification key.

Blockchain Network Configuration. HyperCMS is implemented as a Hyperledger Fabric platform and includes essential components such as Order, Channel, and Peer. HyperCMS is a network consisting of one orderer node, one channel, and one organization. The organizations include local CA and single peers, which represent the certificate issuing company or institution. The orderer node adopts the solo consensus algorithm as the basic consensus algorithm.

3.2 Client Application

HyperCMS client provides an interface to interact with the blockchain ledger through RESTFul API. The HyperCMS server is implemented as a web application framework through the Node.js language. The HyperCMS server registers the administrator through the Hyperledger Fabric CA SDK and invokes the chaincode of a specific channel through the Hyperledger Fabric SDK. This interacts with the blockchain ledger. Data is returned in JSON format (Fig. 4).

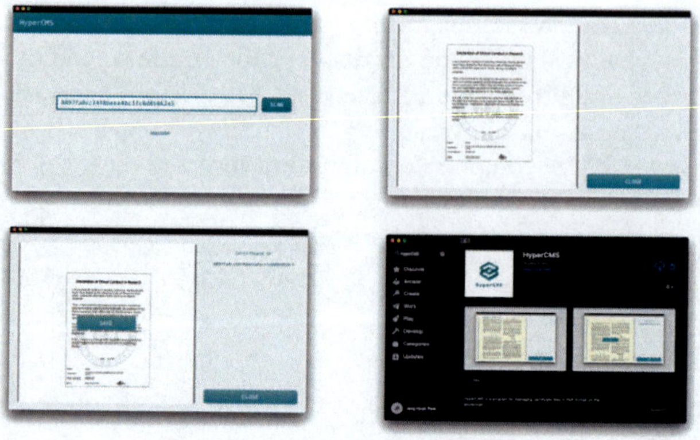

Fig. 4. HyperCMS client interface

4 Experimental Result and Analysis

4.1 Performance of Compression Algorithms

Four compression algorithms were compared to analyze performance of LZMA compression algorithm. Specific files were compressed with the Snappy, QLZ, ZLIB and LZMA compression algorithms and compared to the compressed file size. The file sizes compressed by the four compression algorithms are graphed. As a result, the highly compressed compression algorithm is the LZMA compression algorithm, which reduces the file size by 35.79% (Fig. 5).

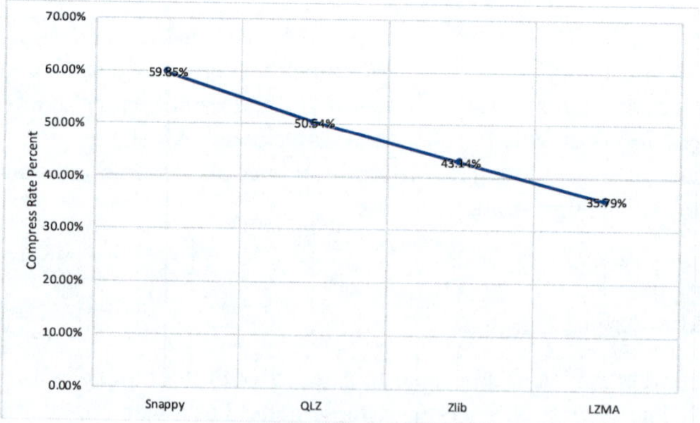

Fig. 5. Compression algorithms comparison

4.2 System Throughput of Create Operation

The graph shown in Fig. 6 shows response and request per second according to the number of users. The creation Transaction saves the compressed certificate file. It shows the responses and requests per second when the compressed certificate file of the creation transaction is 0k, 10k, or 30k. The load tool sequentially increases the number of users to 300 and increases the number of the creation transactions. It shows that the RPS continues to increase and maintains up to 200 RPS. It shows a constant difference in RPS based on the compressed certificate file size. RPS depends on system performance.

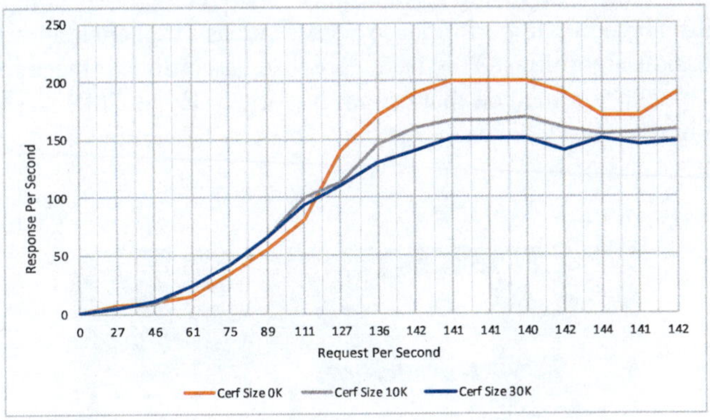

Fig. 6. Create operation request

4.3 System Throughput of Query Operation

The graph shown in Fig. 7 shows the query transaction response and request per second according to the number of users. It shows that RPS continues to increase and maintains up to 300 RPS. The creation transaction must be agreed upon by all peers in the blockchain network until the block is created in the blockchain ledger. Query transactions, on the other hand, get data without the agreement of all peers. For this reason, query transactions have higher RPS than creation transactions.

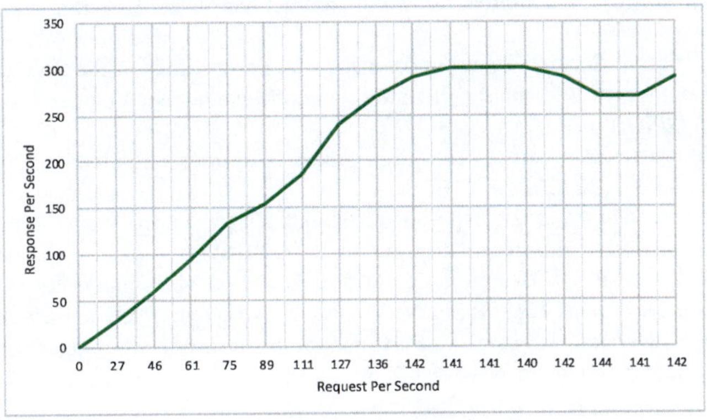

Fig. 7. Query operation request

5 Conclusion

HyperCMS adopted the Hyperledger Fabric platform to use the private blockchain method. Smart Contract, which performs the function of certificate management system, was implemented through Hyperledger Fabric SDK. HyperCMS representative smart contract functions are certificate registration and certificate issuance. HyperCMS is related to the block size of the private blockchain. The LZMA compression algorithm maximizes HyperCMS performance because it has a higher compression rate than other compression algorithms. This paper proposes a certificate management system that adopts a blockchain method that cannot be forged. HyperCMS was designed and implemented in a private blockchain method. This can be verified in the App Store on MacOS as shown in Fig. 8.

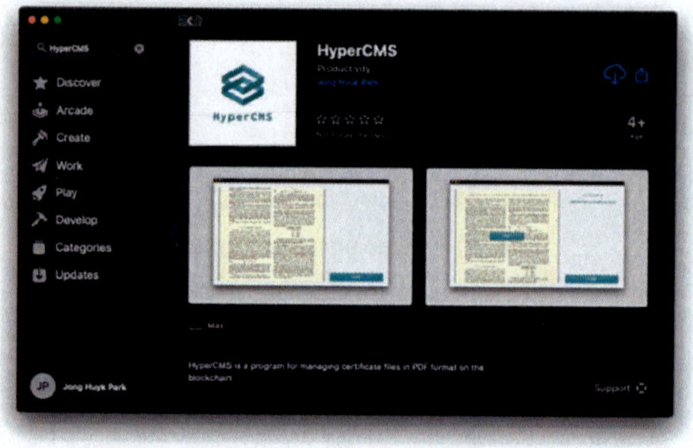

Fig. 8. HyperCMS for MacOS

References

1. Nakamoto, S.: Bitcoin: A peer-to-peer electronic cash system, White Paper (2008)
2. Clack, C.D., Bakshi, V.A., Braine, L.: Smart contract templates: Essential require-ments and design options (2016)
3. Imran, S., Hyder, I.: Security issues in databases. In: Proceedings of the Second International Conference on Future Information Technology and Management Engineering, Sanya, 2009, pp. 541–545 (2009)
4. Wood, G.: Ethereum: A secure decentralised genera-lised transaction ledger. Ethereum Project Yellow Paper **151**, 1–32 (2014)
5. Cachin, C.: Architecture of the hyperledger blockchain fabric. In: Workshop on Distributed Crypto-currencies and Consensus Ledgers, vol. 310 (2016)
6. Tsai, W., Bai, X., Yu, L.: Design issues in permission-ed blockchains for trusted computing. In: Nasir, Q., Qasse, I.A., Talib, M.A., Nassif, A.B. (eds.) 2017 IEEE Symposium on Service-Oriented System Engineering, Performance Analysis of Hyperledger Fabric Platforms, Security and Communication Net-works, vol. 2018, Article ID 3976093, p. 14(SOSE), San Francisco, CA, 2017, pp. 153–159 (2018)
7. Androulaki, E., Barger, A., Bortnikov, V.: Hyperledger fabric: A distributed operating system for permissioned blockchains. In: Proceedings of the Thirteenth EuroSys Conference (EuroSys 2018), Article 30, p. 15. ACM, New York (2018)

A Multi-criteria Model Application in the Prioritization of IT Problems in Public Sector Institutions – An Experience

Gleidson Sobreira Leite(✉), Adriano Bessa Albuquerque, Plácido Rogério Pinheiro, and Everardo Barros Filho

Universidade de Fortaleza, Fortaleza, CE, Brazil
gleidson.sleite@gmail.com, {adrianoba,placido}@unifor.br

Abstract. Technology is an essential component for public organizations to provide services efficiently to society. However, it also brings with it a great challenge for managers: dealing with problems and deciding on their order of priority to be solved so that they minimize, as much as possible, the impacts caused by them. Due to the high volume, the need for quick decision-making, and the existence of both simple and complex problems that have different impacts for each institution, there is an opportunity to apply models that help decision-making on the prioritization on which problems to deal first considering several criteria. In this scenario, this paper presents a multi-criteria decision model to assist decision makers on the prioritization of IT problems in order to make the decision more impartial, efficient, and less subjective. Exploring the main concepts of multi-criteria methodology and adopting the MACBETH approach and software, an experience was performed in the IT area of a Public Prosecutor Office in Brazil. After the ordering of real problems, indications of applicability and robustness of the constructed model were obtained.

Keywords: Decision support · Information Technology · Multicriteria approach · Problem Management

1 Introduction

Over the past few years, Information Technology (IT) has been growing and transforming, becoming increasingly essential for organizations, due to society's demands for more and better services, whether in the public or private sector, also increasing the need for new technological solutions that aim to meet such anxieties and needs of its customers/users.

In the public sector, Information Technology Management has been following changes in the business strategies of these organizations to meet the new requirements regarding transparency efficiency, security, availability, and efficiency, where IT-related products and services are increasingly occupying strategic positions [1].

The IT sectors have undergone constant processes of modification, updating and innovation inherent to their nature. As it could not be different, the employment and

R. Silhavy (Ed.): CSOC 2022, LNNS 501, pp. 149–167, 2022.
https://doi.org/10.1007/978-3-031-09070-7_13

treatment are given to IT sectors in administrative organizations (whether public or private) also evolved following such dynamics [1].

For [2], IT is essential in perpetuating public organization due to its services. This means that, in addition to improvements in the organization's internal environment, by increasing organizational efficiency (optimization of processes, structure, communication and the elimination of bureaucracy), the strategic use of IT and the administration of IT resources can improve the service pro-vided to the population and the services provided to citizens.

According to [3], it is possible to adopt an IT service management system structured similarly and regardless of the type or size of organizations, even if they are public, private, non-profit or supplier of IT services with only one person responsible for supporting the IT services. For each case, there is a requirement to provide IT services economically, with the best cost/benefit ratio. Also, at the same time, they are reliable, consistent from the point of view of their users and suitable for the organization, that is, aligned with the strategic objectives defined from the organization's business strategy.

There are also day-by-day management problems, where resources are often scarce, and there is a mistaken view of Information Technology Management by some departments. Numerous factors must be care-fully analyzed not to cause damage, which is often irreversible.

One of the main libraries that bring together good IT Service Management practices, ITIL (Information Technology Infrastructure Library), has Problem Management as one of its processes, which is responsible for managing the entire life cycle of the IT problem from its detection until its closure. As per ITIL, issues need to be prioritized considering criteria of severity, urgency, and impact [4].

However, in the day-to-day of IT, several problems occur while need to be timely resolved so as not to affect the organizations' business. For this, it is necessary to prioritize the most relevant problems to be treated accorded to pre-define criteria adopted in the organization. However, considering several and different criteria, the best order for prioritizing these problems is a complex process and, many times, decisions related to prioritizing demands regarding the solution of these problems end up being taken subjectively, which can lead to different prioritization for similar problems at different times.

Through the aid of multicriteria decision support methods, it is possible to assist decision making regarding the priority of similar problems in technology sectors to be treated.

Motivated by this scenario, in this paper, a multi-criteria analysis model to support decision making regarding prioritization of IT problems is presented and resulted from an experience performed in a selected Brazilian public institution to enable the construction and analysis of the model.

Exploring the main characteristics of multi-criteria decision methodology and adopting the MACBETH approach and software, the model aims to suggests the prioritization (ordering) of IT problems based on the defined criteria (and decision makers preferences), in order to make the ordering more impartial and efficient, and less subjective.

This paper is organized as follow: the methodology and procedures performed in the work is presented in Sect. 2; background and related work in Sect. 3; the proposed model and analysis in Sect. 4, followed by the final considerations in Sect. 5.

2 Research Methodology

To achieve the objective of this study and following the premises proposed by [5], the steps presented in Fig. 1 (and detailed in Sect. 4) where performed, where the model construction was based on the phases of the decision supporting process proposed by [6] (structuring, evaluation, and recommendation).

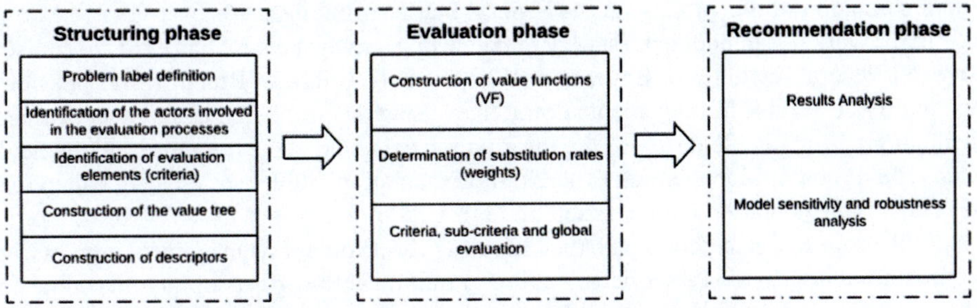

Fig. 1. Phases and stages of the multicriteria evaluation model

3 Background and Related Work

Information Technology Governance comprises the set of mechanisms to ensure that the use of IT adds value to organizations' business, whether public or private, with acceptable risks. They are organizational structures, processes, controls, and other components that aim to avoid or mitigate deficiencies that are still common in an organization's IT management, such as lack of personnel training, the inadequacy of the planning process, recurrence of unsuccessful projects and contracts that do not meet the needs of the business [7].

Over the past few years, due to the importance of Information Technology (IT), current organizations have been forced to adopt measures to manage their technological environment better. According to [8], with the reality that information technology has become essential for organizations, IT environments have become increasingly complex. The management of the various components that make up these environments must be the most efficient and adequate to meet the expectations of the business areas. They are people, processes, and tools of the most diverse types that need to be managed.

According to [3], IT Service Management is, in short, the management of the integration between people, processes and technologies, components of an IT service. These aim to enable delivery and support of IT service focused on customer needs and in line with the organization's business strategy, aiming at achieving cost and performance objectives by establishing service level agreements between the IT area and the other business areas of the organization.

According to [8], problem management aims to prevent failures, optimize the impact of those that could not be avoided and eliminate their recurrences. These activities have two aspects: a proactive one that seeks to prevent failures that may arise; and another reactive one that works to identify the root cause of the incidents and resolve them definitively in the shortest possible time.

From this scenario emerges the importance of dealing in the best possible way with daily problems with regard to information technology services.

The work developed by [9], for example, proposed elaborating an implementation guide for the problem management process, based on ITIL good practices, for the Federal Institutes of Education, Science and Technology, to boost and facilitate the implementation of this process in the Federal Public Administration.

In his work, [9] conducted a survey of data using survey questionnaires to specialists from 31 Federal Institutes of Education (IFE) from all regions of Brazil. It was possible to identify IT Service Management adoption level and the problem management process in the IFEs. From the field survey, the information collected in the literature and the field study, the Problem Management Process Implementation Guide was created, called the GPITIL Implementation Guide, based on ITIL good practices.

[10] proposed a methodology that supports the Problem Management process in companies that provide IT services, aiming to minimize the effects of any inaccuracies in the categorization of incidents, thus contributing to the reduction in investigation time and the reduction of the impact of recurring problems on the business. To apply the proposed methodology, [10] specified the Problem and Incident Analysis Tool – FAPI. This allows analyzing problems and incidents and generating a file that, through statistical analysis, allows investigating possible deviations in the categorization of these incidents.

In addition, to investigate the occurrence of deviations in the categorization of incidents, a file was generated by the FAPI tool with data on incidents and problems. This file was submitted to data mining, from which association rules were intended. These rules were validated and analyzed to verify the existence of categorization errors. Also, to assess the effectiveness of the proposed methodology and tool, a case study was carried out on an IT service provider company.

Through his work, [8] defined an approach to prioritize and select IT infrastructure problems using the multi-criteria methodology to optimize the impact of technology failures on organizations' businesses. In his work, [8] used questionnaires and interviews to raise organizational criteria relevant to the decision-making process of the organization's Problem Management with the specialists working in the process. Then, it was suggested the weighting of criteria regarding the degree of impact on the company's business through the multi-criteria tool and standardization so that the decision-making process could become more efficient and objective when defining the priority of action on IT infrastructure problems. The evaluation of the proposed method was carried out through the practical application of multi-criteria at the stage of the selection process and prioritization of IT infrastructure problems.

[11] carried out a study on how Knowledge Management can be used to support and improve the Problem Management process based on ITIL. The authors examined how an IT ser-vice provider identified, created, stored, shared, and used knowledge about software issues through a case study. From this study, a list was generated containing lessons learned, suggestions for improvement and information on how the Knowledge Management models can be combined with the Problem Management process to improve the quality of IT services.

However, to the best of our knowledge, despite the existence of several studies that present or adopt different applications (or approaches) reggarding IT problem managment in industrial or academic environments, there is also the opportunity to apply methodologies to help decision making regarding the prioritization of IT problems.

To this end, exploring the main concepts of multi-criteria methodology and adopting the MACBETH (Measuring Attractiveness by a Categorical Based Evaluation Technique) approach and software, this paper presents a model that suggests the prioritization of IT problems based on multiple criteria defined by specialists in order to make the decision of process ordering more impartial, efficient, and less subjective.

3.1 Multi-criteria Decision Support

Decision making is a critical factor in aligning the technology areas with the organizations' strategic objectives. Deciding which failures to act and in which order requires a series of analyses, from business impact to available resources. Decision making is based on multiple choices that lead to the final decision [12, 13].

Making decisions involves everyday activities in people's daily lives. Many of these decisions become more complex because they are necessary to solve problems, considering the most diverse points of view, with specific forms of evaluation, which can generate. As a result, a set of alternative solutions may consider multiple conflicting criteria [14, 15].

[16] emphasize that a decision needs to be taken, constantly faced with a problem that has more than one alternative to solve it. In this way, the decision theory is defined to foresee possible scenarios under procedures and analysis methods that ensure the decisions' coherence, effectiveness, and efficiency [17].

[16] and [18] mention that the multiple criteria decision analysis – MCDA helps decision making regarding choice, ordering or classification of actions, to analyze a more complex decision problems, and aim to select alternatives of action and points of view considering various criteria and existing structures grouping the value judgments of agents.

The decision-making model comprises the following components: criteria, weights, and classifications given for each alternative in each criterion. Assuming knowledge of the decision-making actors' preferences and the quality of the evaluation, it is possible to admit that action is as good, better, or worse than another, that is, to rank the alternatives. Multicriteria decision support (MDS) allows the prioritization of alternatives in a situation of conflicting criteria, seeking to satisfy constraints with conflicting goals, that is, a trade-off. It should be noted that MDS begins the search for alternative(s) to solve the problem by the non-dominated alternatives, that is, alternatives that outperform other alternatives in all criteria, showing the hierarchy of preference of alternatives [17, 19].

3.2 MACBETH - Measuring Attractiveness by a Categorical Based Evaluation Technique (Approach and Software)

According to [16], the MACBETH approach is a decision support methodology that allows evaluating options considering multiple criteria, where is based on weighting the criteria and validating the options in qualitative judgments on differences in attractiveness. That is, the technique proposes the comparison of two options and the definition of the difference in attractiveness between them, according to the following options: null, very weak, weak, moderate, strong, very strong or extreme [20–22].

Thus, it is possible to numerically represent the decision makers judgments about the global attractiveness of the actions, requiring only qualitative judgements about the difference of elements attractiveness, to generate value points and weights for each criterion [16, 23].

Following the concepts of the MACBETH methodology, [16] created a software, currently called M-MACBETH, which is a multi-criteria decision support software that allows the structuring of value trees, the construction of the criteria classification, the development of value functions, the weighting of criteria, and extensive sensitivity and robustness analyzes on the intrinsic and relative value of options [20–22].

According to [8], given two alternatives, the decision-maker must say which is the most attractive. In this case, it has a greater degree of confidence, which is the degree of this attractiveness in a semantic scale corresponding to an ordinal scale. In this aspect, through the application of the M-MACBETH tool, an analysis of cardinal coherence (transitivity) and semantics (relationships between differences) can be observed, suggesting, in case of inconsistency, possible solutions. The decision-maker can graphically adjust the value of the assigned grades within the allowed ranges. After this adjustment, with the introduction of expert knowledge, the construction of the cardinal scale of values is characterized. The difference in attractiveness is of outstanding importance in this methodology.

In this study, the MACBETH (Measuring Attractiveness by a Categorical Based Evaluation-Technique) approach and software was adopted for the following reasons: i) availability of a tool that is already widely used academically (M-MACBETH); ii) possibility of adapting the methodology to the problem under study (ordering); iii) easy adoption by non-experts; and iv) feasibility of adoption in the proposed scenario.

4 Proposed Model and Analysis

Applying the steps of the multi-criteria methodology, the criteria, sub-criteria, scales, and weights are obtained, resulting in the proposed model. Thus, the execution of the sorting activity (prioritization) of the proposed approach is carried out by inserting the criteria, sub-criteria, scales, and weights in the M-MACBETH tool, followed by the insertion of process data (scale values), obtaining thus the list of problems sorted based on the preferences of the decision-makers automatically.

To allow the construction of the model and execution of the experience, a Brazilian public institution (Public Prosecutor Office) was selected which, among its infrastructures, has an advanced technology sector that provides services to an entire Brazilian state.

According to art. 127 of the Federal Constitution of 1988, the Public Prosecutor Office is a permanent institution, essential to the jurisdictional function of the State, responsible for defending the legal order, the democratic regime, and unavailable social and individual interests [24].

Another characteristic present in the organization's IT sector was the lack of an approach to prioritizing problems. Most of the time, problems were subjectively prioritized, meaning that the prioritization or-der defined at that time was not the best to reduce business impacts. It is common for business area managers to judge their problems as a priority. However, when there is a defined formal prioritization approach that clearly shows the weight of the impacts caused by each problem, it facilitates understanding among managers.

Although the incident and problem management system used in the organization has a knowledge base, it was hardly kept up to date. It is known that an updated knowledge base allows the anticipation of problems and the creation of a solid and responsive structure to deal with them in an agile and efficient way.

The absence of performance metrics and indicators, both for the team and for the applications, was also a point that contributed to the low performance of the process and the non-fulfilment of goals established by the strategic planning.

After constructing the multi-criteria model based on the preferences of specialists from selected institution, the model was applied in a real case scenario to obtain evidence if the approach supports decision making regarding the prioritization of IT problems to be addressed. After application, sensitivity and robustness analysis was performed.

The elaboration of the model followed the premises sugested by [5], was based on the phases of the decision supporting method proposed by [6] and is presented in the following sub-sessions.

4.1 Structuring Phase

Initial phase of the decision-making process in which, according to [5, 6, 25], has great importance in the construction of the multicriteria decision support model and includes the activities of contextualization and identification of the problem (definition of the problem label), identification of the actors involved in the evaluation processes, identification of elements (criteria) of evaluation, construction of the tree of values and construction of descriptors.

Problem Label Definition

Considering that the objective of this research is to present a model that suggests the prioritization of IT Probems, the label defined for the model was "Prioritization of IT Problems".

Identification of the Actors Involved in the Evaluation Processes

For the model construction, some actors participated in the decision-making process, being classified, folowing [6] premisses, into interveners, who directly participate in the decision-making process contributing with suggestions for the construction of the model, and those who act indirectly (acted), being affected decisions and being able to put pressure on stakeholders.

Interveners can be classified into decision makers, who have decision-making power, representatives, who represent decision-makers through designation, and facilitators, who drive the decision or evaluation process. Table 1 presents the participating actors in the context of this work.

Table 1. Actors of the process

Actors	Description
Decision makers	CIO's and Managers
Representatives	Managers, Cordinators and Analysts
Facilitators	Authors of this paper
Indirect actors	Brazilian Public Institution sectors

Identification of Evaluation Elements (Criteria) and Construction of the Value Tree
The preparation of this activity was divided into three stages. The first stage consisted of a meeting with the main IT leaders who would act as decision-makers and experts (three specialists were selected, one with a master's degree and two with specialization degree). Using the Brainstorming technique, they were motivated to list the main criteria and sub-criteria they considered relevant when prioritizing the institution's IT problems. It is important to emphasize that some criteria mapped by the work of [8] were used as the basis for this research.

In the second, a questionnaire made available via the web was applied to all leaders who acted in any way in solving IT problems of the selected institution, where they were motivated to confirm or not the criteria and sub-criteria raised and to indicate new ones, if deemed necessary (eight experts participated, three with a master's degree and five with specialization degree).

After organizing the survey's previously carried out answers, the third stage was carried out through a meeting, with only the main decision-makers of the process (three specialists), to reach a consensus on the criteria and sub-criteria that would finally be used in the survey. The result of the meeting was a complete cognitive map, organized and without redundancies.

According to [5] and [6], a fundamental point of view is an isolated individual point of view, or a set of points of view that actors agree that their impacts and options are analyzed separately, independently of other points of view.

The fundamental viewpoints cover all aspects that must be considered in the decision context and provide a basis for modelling these aspects. For cases in which more than one point of view is needed to translate the actors' values, reflecting their judgment, elementary points of view can be adopted to compose a fundamental point of view that allows them to explain their value and operate on it.

After defining the fundamental points of view, the necessary decompositions were carried out, creating the elementary points of view, both being the basic structure of the multi-criteria evaluation model.

The fundamental and elementary points of view are presented on Fig. 2 and the Multicriteria evaluation model value tree in Fig. 3.

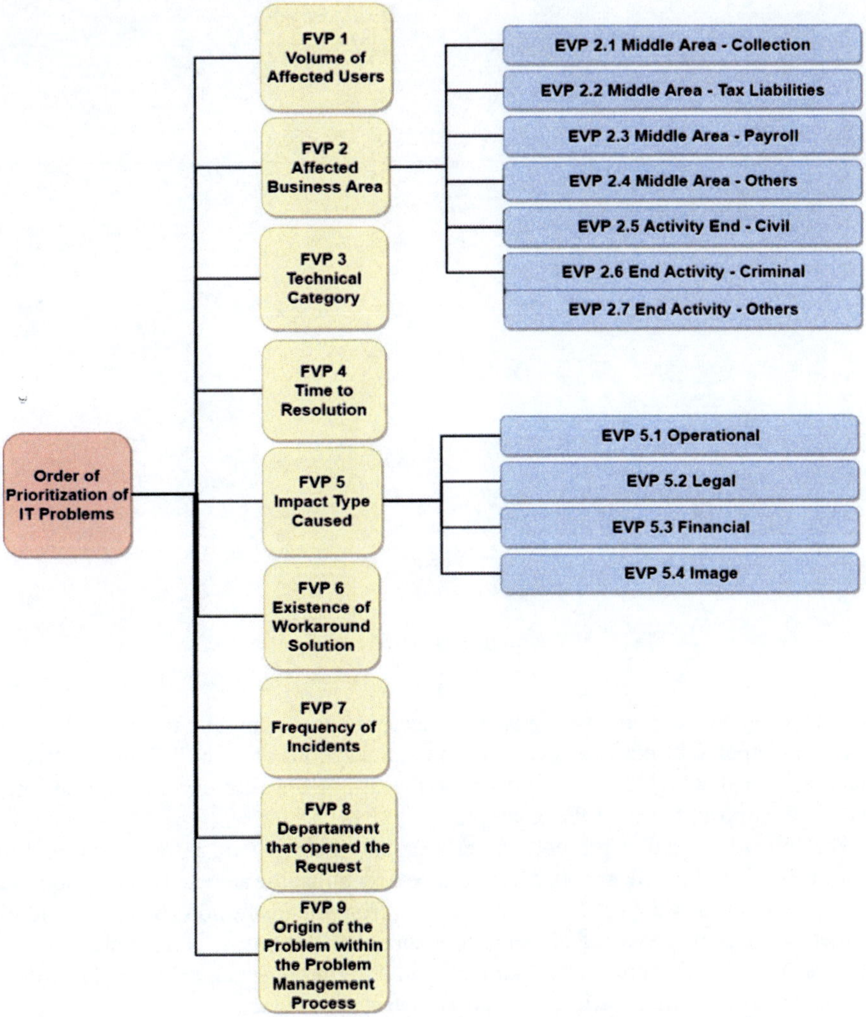

Fig. 2. Fundamental and elementary points of view

Construction of Descriptors

As mentioned by [7], after the definition of the fundamental (FVP) and elementary (EVP) points of view, it is necessary to construct a criterion, consisted of a descriptor, which

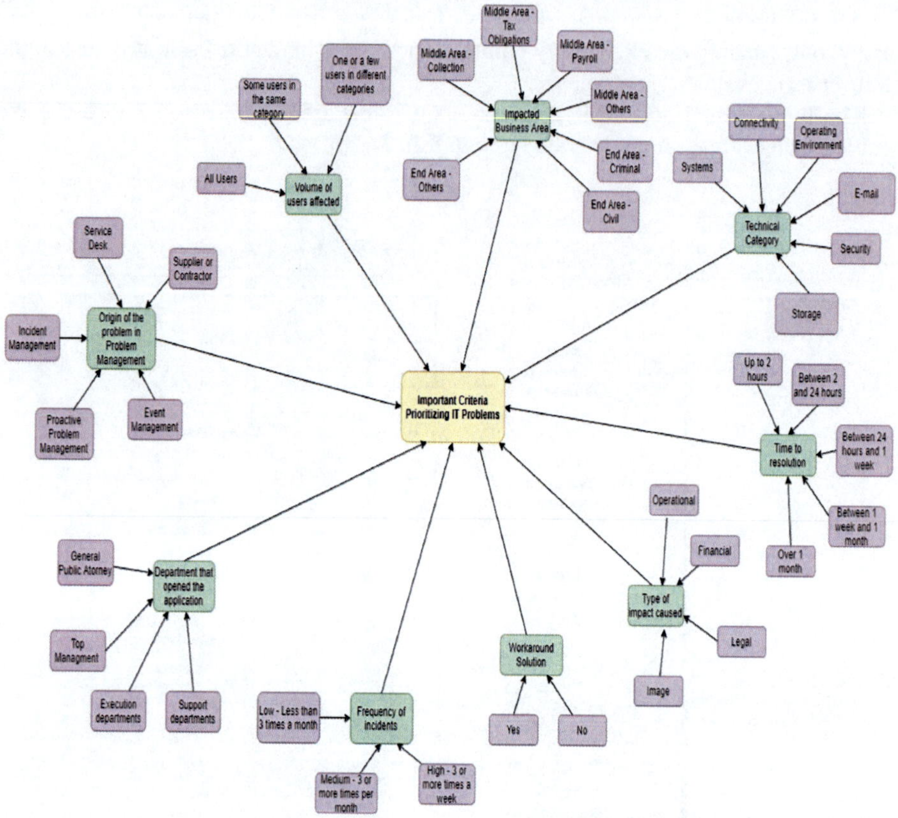

Fig. 3. Multicriteria evaluation model value tree

promote the understanding of what will be measured corresponding to a set of levels of impact that reports the performance of the potential actions of each FVP and EVP, and a value function, that highlights the information regarding the differences in attractiveness between the impact levels of the descriptors.

The following section presents the criteria and their general information, and the descriptors defined by the experts. It should be added that the selected criteria and their descriptors (values) were established by experts from the institution. As a result, to use the methodology in scenarios of other companies/institutions, it is necessary to carry out the step of defining criteria, sub-criteria and descriptors again, using the knowledge base obtained from the experience of this study as a reference.

- **PVF 1 – Volume of affected users**: This criterion refers to the volume of users affected by the issue. Options: All institution users; Some users of the same category (Organ, Department, User Profile); One or a few users from different categories.
- **PVF 2 – Affected Business Area**: This criterion refers to the business area of the organization affected by the occurrence of the problem. It was broken down into

seven sub-criteria. They are Middle Area – Collection; Middle Area – Tax Obligations; Middle Area – Payroll; Middle Area – Others; End Area – Civil; End Area – Criminal; End Area – Others.

- **PVF 3 – Technical Category**: This criterion corresponds to the category of the problem in its technical aspect. Options: Connectivity, Operating Environment (Includes all hardware, servers, computers and their operating systems), Systems, Security, Storage, E-mail.
- **PVF 4 – Time to resolution**: This criterion corresponds to the estimated time to resolve the issue in its entirety. Options: Up to 2 h; Between two and 24 h; Between 24 h and one week; Between one week and one month and Above one month.
- **PVF 5 – Impact Type Caused**: This criterion refers to the impacts that bring risks and losses to the institution's operation to fulfil its objectives before the State. It was broken down into four sub-criteria. They are Operational, Legal, Financial, Image.
- **PVF 6 – Existence of Workaround Solution**: This criterion refers to the existence of an alternative, even if temporary, to work around the failure.
- Options: Yes, No.
- **PVF 7 – Frequency of incidents**: This criterion corresponds to the number of times that incidents related to the problem occur. Options: 3 or more times a week (High), three or more times a month (Average), Less than three times a month (Low).
- **PVF 8 – Department that opened the request**: This criterion refers to the agency or sector of the employee that opened the request. The descriptors were defined from the Organizational Structure defined by Law n. 8,625/93 - National Organic Law of the Public Prosecution Service, which provides for general rules for the organization of the Public Prosecution Service of the States. Options: Attorney General's Office, Execution Bodies (e.g. Attorney General's Office, Attorney's Office and Decon), Superior Administration (Superior Council of the Public Ministry, College of Attorneys of Justice and General Internal Affairs Unit of the Public Ministry), Auxiliary Bodies (Support Centers, Tender Committee, Study and Functional Improvement Center and Administrative Support Bodies).
- **PVF 9 - Origin of the problem within the Problem Management Process**: This criterion refers to the activity of the Problem Management process in which the problem was detected. Options: Service Desk, Incident Management, Proactive Problem Management, Event Management and Vendor or Contractor.

4.2 Evaluation Phase

In the evaluation phase, a survey was carried out with the CIOs (Chief Information Officer) of the State Public Prosecutors in the country, where they were motivated to classify as to the degree of importance of each criterion and their respective sub-criteria and also indicate new ones if deemed necessary with the result of the survey, the data were treated and tabulated to allow a better analysis of the main experts and decision-makers of the problem management process. They met once again to discuss the degree of importance or attractiveness among the criteria.

To further consolidate and enrich the survey, the information collected covered questions about the profile of respondents. They were asked to answer questions such as position or function in the institution, length of experience in the function, education level and certifications in best practices.

The questionnaire was structured in three parts:

(i) Profile of the respondent.
(ii) Classification of criteria regarding the degree of importance and
(iii) Classification of sub-criteria according to the degree of importance.

The questionnaire was answered by 17 (seventeen) leaders, or their representatives distributed by the country's Public Ministry. Notably, 64.60% of the experts interviewed held Secretary, Deputy Director, and Manager positions. Regarding the interview specialist's length of experience, it is noteworthy that 64.7% of respondents had worked for more than nine years as an IT leader.

Regarding the level of education, 94.1% of the interviewed specialists had a postgraduate degree, and 5.9% had a master's degree. That is, all respondents had a university degree. And 52% of the experts inter-viewed had certifications in COBIT and ITIL.

Construction of Value Functions and Determination of the Weights
According to [26, 27], the adequate method for constructing value functions is semantic judgment, as it helps the decision-maker identify their preferences during the evaluation of potential actions from a certain point of view.

The replacement rates were defined together with the specialists by ordering and verifying the attractiveness among the criteria. To perform the semantic judgment, the differences in attractiveness levels proposed in the MACBETH approach were adopted as value functions: null, very weak, weak, moderate, strong, very strong or extreme.

The determination of the numerical values of the weights of the performance levels of the criteria was generated using the M-MACBETH software after the results of the evaluation judgments by selecting the levels of attractiveness among the criteria through a semantic matrix. As qualitative judgments are issued by the evaluator and entered into the tool, the software automatically checks their consistency and offers suggestions to eliminate it when it finds an inconsistency.

Figure 4 presents two examples of the results of the evaluation carried out for the criteria "Volume of users affected" and "Frequncy of incidents".

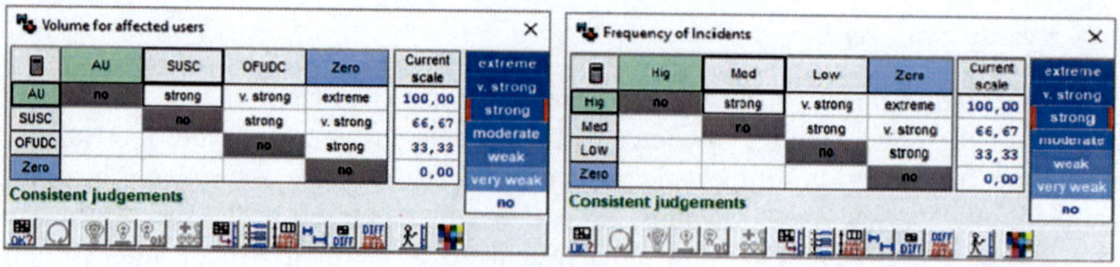

Fig. 4. Scales of performance levels (descriptors) of the criteria "Volume of Users affected" and "Frequency of Incidents" (Judgment matrix)

Criteria, Sub-criteria, and Global Evaluation

After determining the weights of the criteria (performance levels), the assessment phase of the attractiveness between the criteria began to identify their weights for use in the problem prioritization activity. The experts performed the same peer-review procedure, assisted by the facilitator responsible for entering the data into the M-MACBETH software. Initially, the priority of the criteria was identified, where, for each criterion, the specialists were asked about the priority level of importance that one criterion would have concerning the others (Fig. 5).

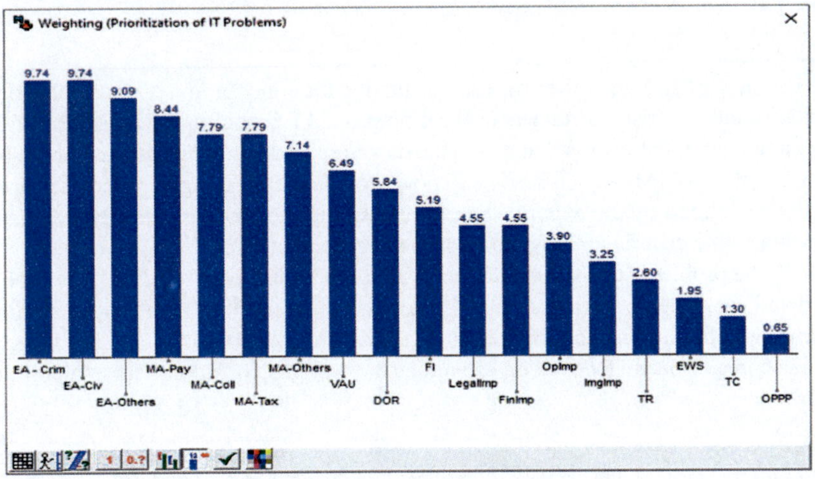

Fig. 5. Weighting of criteria

4.3 Recommendation Phase

In this phase, the practical application of the proposed model occurs. With that, the procedure for including the data of the alternatives begins (for the studied scenario would be information about the IT Problems), framing the options (descriptors) as presented in Sect. 4.1. The procedures performed in this phase involve the analysis of results, sensitivity and robustness.

Results Analysis

Figure 6 presents information about the data entered by the facilitator in the M-MACBETH tool according to information provided by decision makers. It should be added that only six IT Problems were selected for the purpose of illustrating the execution of the method.

Table of performances

Options	VAU	EA - Crim	EA-Civ	EA-Others	MA-Pay	MA-Coll	MA-Tax	MA-Others	TC	TR	LegalImp	OpImp	FinImp	ImgImp	EWS	FI	DOR	OPPP
Pro1	AU	Hig	Hig	Hig	Low	Low	Low	Med	OE	Bet2A24H	Hig	Hig	Low	Low	No	Low	GPA	SD
Pro2	SUSC	Zero	Zero	Zero	Zero	Hig	Zero	Med	Sys	Bet2A24H	Hig	Low	Hig	Hig	No	Med	SD	IM
Pro3	SUSC	Zero	Zero	Zero	Med	Med	Low	Hig	Sec	Bet24HA1W	Hig	Hig	Med	Low	No	Med	GPA	EM
Pro4	SUSC	Hig	Hig	Low	Zero	Zero	Zero	Low	Sys	Bet1WA1M	Hig	Hig	Low	Low	Yes	Hig	ED	SD
Pro5	SUSC	Zero	Zero	Zero	Hig	Zero	Zero	Med	Sto	Bet24HA1W	Hig	Hig	Hig	Low	Yes	Med	GPA	PPM
Pro6	OFUDC	Zero	Zero	Zero	Zero	Zero	Zero	Hig	Sys	Bet1WA1M	Low	Low	Low	Hig	No	Hig	SD	SD

Fig. 6. IT Problems information according to defined performance levels

It is noteworthy that, as the proposal resulted in the construction of a weighted matrix of criteria and descriptors related to the context of IT probelms prioritization, for the next applications, the idea is that it is necessary to include only the data as shown in the Fig. 6, and the tool (M-MACBETH) will now provide the ordering of processes based on the criteria defined by the experts. Thus streamlining the process prioritization activity according to the criteria already defined by the specialists.

With the problems classified and their data entered in M-MACBETH, the tool generated and suggested ordering. Figure 7 presents the result of the ordering according to the criteria defined by specialists, discriminating the overall scores and the individual criteria scores of each IT problem, as well as the weights considered for each criterion.

Table of scores

Options	Overall	VAU	EA - Crim	EA-Civ	EA-Others	MA-Pay	MA-Coll	MA-Tax	MA-Others	TC	TR	LegalImp	OpImp	FinImp	ImgImp	EWS	FI	DOR	OPPP
[all upper]	100.00	100.00	100.00	100.00	100.00	100.00	100.00	100.00	100.00	100.00	100.00	100.00	100.00	100.00	100.00	100.00	100.00	100.00	100.00
Pro1	71.23	100.00	100.00	100.00	100.00	33.33	33.33	33.33	66.67	83.33	80.00	100.00	100.00	33.33	33.33	50.00	33.33	100.00	100.00
Pro4	52.88	66.67	100.00	100.00	33.33	0.00	0.00	0.00	33.33	66.67	40.00	100.00	100.00	33.33	33.33	100.00	100.00	50.00	100.00
Pro3	50.20	66.67	0.00	0.00	0.00	66.67	66.67	33.33	100.00	50.00	60.00	100.00	100.00	66.67	33.33	50.00	66.67	100.00	40.00
Pro5	45.24	66.67	0.00	0.00	0.00	100.00	0.00	0.00	66.67	33.33	60.00	100.00	100.00	100.00	33.33	100.00	66.67	100.00	60.00
Pro2	39.89	66.67	0.00	0.00	0.00	0.00	100.00	0.00	66.67	66.67	80.00	100.00	33.33	100.00	100.00	50.00	66.67	25.00	80.00
Pro6	27.07	33.33	0.00	0.00	0.00	0.00	0.00	0.00	100.00	66.67	40.00	33.33	33.33	33.33	100.00	50.00	100.00	25.00	100.00
[all lower]	0.00	0.00	0.00	0.00	0.00	0.00	0.00	0.00	0.00	0.00	0.00	0.00	0.00	0.00	0.00	0.00	0.00	0.00	0.00
Weights :		0.0649	0.0974	0.0974	0.0909	0.0844	0.0779	0.0779	0.0714	0.0130	0.0260	0.0455	0.0390	0.0455	0.0325	0.0195	0.0519	0.0584	0.0065

Fig. 7. IT Problems ordering results and score distribution

According to the results, it was possible to verify that, based on the defined criteria, problems 1 and 4 had the highest priority for resolution and problem six the lowest. Also noteworthy is that the Affected Business Area criterion (AN-AF-Crim and AN-AF-Civ), with 0.0974 scale factors, is considered the most important and the Origin of the Problem in the Management Process Problems (Problem Origin) with a scale factor of 0.0065, the least important.

It was also possible to observe that no problem obtained a global score equal to 100, which would mean the worst scenario of all concerning the impacts caused on the organization. However, the highest overall score (71,23) was for Problem 1. This was the most serious and should be the first problem to be solved among the six analyzed problems and considering the defined criteria and weights.

Model Sensitivity and Robustness Analysis

According to [20] and [28], the most popular type of sensitivity analysis of weights consists of analyzing the changes that may occur in the global ordering of proposals when the relative weight of a given criterion (or sub-criterion) is modified, maintaining the proportion among the weights of the other. The authors also state that, in addition to the classical sensitivity analysis, it is very interesting to analyze the robustness of the results, causing several weights to vary simultaneously but respecting the or-der of the weights defined in the initial proposal.

According to [28] and [29], the sensitivity analysis is adopted to assess whether a model is robust to changes in the replacement rates (weights) of the criteria were to verify its robustness, small changes are made (e.g., 10% more or less) in the criteria weights, analyzing the variations that occurred in the global scores of the alternatives.

To perform the robustness analysis of the proposed model, for each criterion, 10% variation tests of the weights were performed, up and down, to see if these changes cause any significant change in the ordering of the problems that more received importance [29].

Figure 8 presents a graph of sensitivity analysis on the weight and overall ordering with the position of the problems sorted on the cut-offline of the original weight (6.49) of the Volume of affected users' criterion.

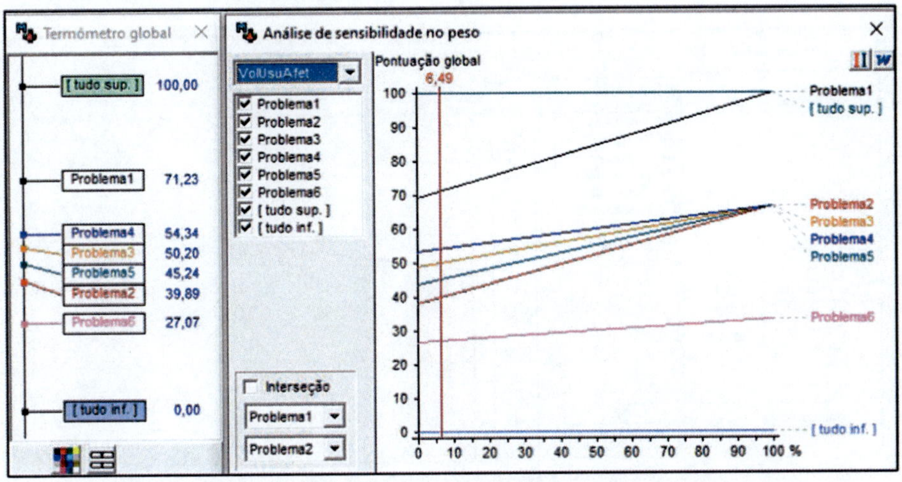

Fig. 8. Sensitivity analysis and general order in original weight 6.49

Figure 9 presents the graph and the general ordering after the 10% weight increase (7,14), in which it was found that there were no changes in the order of problems. Figure 10 shows the result after the 10% decrease (5.84), also without causing changes in the order of problems.

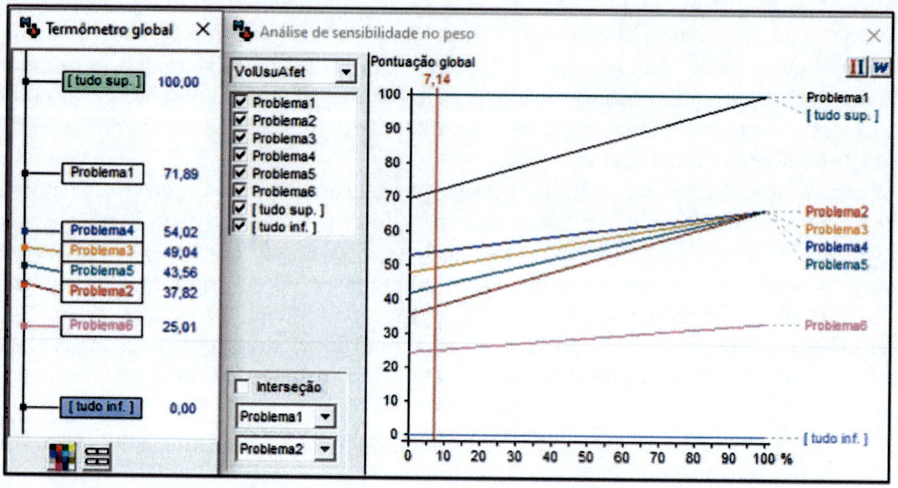

Fig. 9. Sensitivity analysis and general order in weight with 10% increment - 7.14

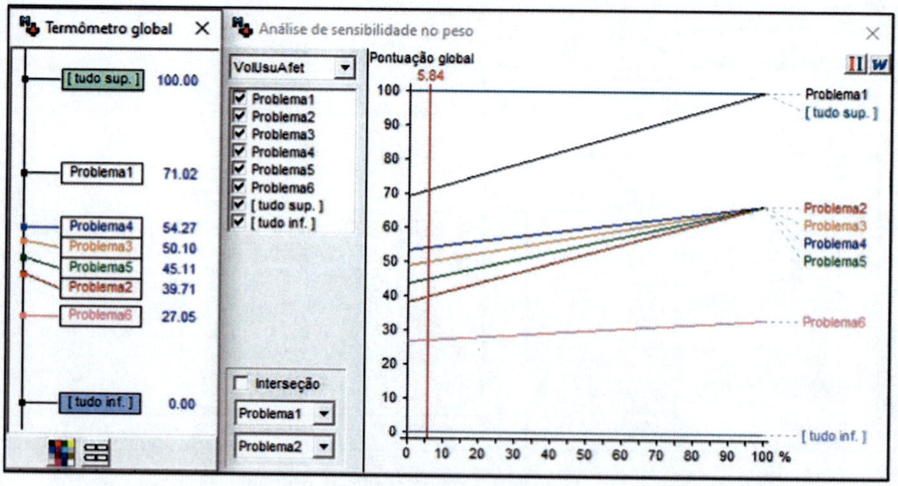

Fig. 10. Sensitivity analysis and general order in weight with 10% decrement - 5.84

The performed tests, also replicated to the other criteria, indicated evidence (subsidies) that the model is robust, since there were no significant changes in the ordering of the IT problems.

5 Conclusion

The problem management process is a fundamental part of optimizing the performance and availability of IT solutions for an organization that provides services to society. Acting on the main problems causing the most significant impact on business generation is of fundamental importance given the reality of the demand for assistance by the population.

In the day-to-day of IT sectors or companies, several problems occur while need to be timely resolved so as not to affect the organizations' business. For this, it is necessary to prioritize the most relevant problems to be treated accorded to pre-define criteria adopted in the organization.

Motivated by this scenario, in this paper, a multi-criteria analysis model to support decision making regarding prioritization of IT problems was presented, and an experience was performed in a selected Brazilian public institution to enable the construction and analysis of the model.

Specialists from IT sectors participated in the construction of the model in relation to the definition, structuring and evaluation of criteria, as well as in an application of the model in the analysis of its applicability resulting in the ordering of IT problems in a real scenario.

Through the aid of multicriteria decision support methods, results obtained showed indications of applicability and robustness of the method, and that it is possible to assist decision making regarding the priority of similar problems in technology sectors to be treated.

However, it should be added that, for the application of the model in other IT companies/sectors, it is necessary to perform the steps carried out in the experience aimed at developing the criteria and values of their weights, thus enabling the necessary adaptation to the real scenario of each IT companies/sector.

Future work recommends expanding the scope of this work by applying the model in a broader scope of IT companies, as well as in the identification of new criteria that can serve as a knowledge base for selection by specialists interested in adopting the model.

References

1. Canabarro, D.R., Cepik, K.: IT governance: Transforming public administration in Brazil. Federal University of Rio Grande do Sul, UFRGS/CEGOV. Porto Alegre, Brazil. ISBN: 978–85-386–0254–5 (2014)
2. Campos, M.V.S, Towers, I.V.: Management of IT services in public administration: Case study at the national land transport agency. Technologies in Projection Magazine, Brazil (2011)
3. Magalhães, I.L., Pinheiro, W.B.: IT service management in practice: An approach based on ITIL. Novatec, São Paulo, Brazil. ISBN: 978–85–7522–106–8 (2007)
4. ITIL: Welcome To The Itil Open Guide! (2021). https://www.itlibrary.org/
5. Bana, C.A., Ensslin, L., Correa, E.C., Vansnick, J.C.: Decision support systems in action: Integrated application in a multicriteria decision aid process. Eur. J. Oper. Res. **113**(2), 315–335 (1999). https://doi.org/10.1016/S0377-2217(98)00219-7
6. Ensslin, L., Montineller, N.G., Noronha, S.M.: Decision support: Methodology for structuring problems and multi-criteria evaluation of alternatives. Insular, Florianópolis, Brazil (2001)

7. Brazil: Federal Court of Accounts - Survey report, TC 008.127/2016–6. 2016, 109f. https://portal.tcu.gov.br/lumis/portal/file/fileDownload.jsp?fileId=FF80808163 E13A259230F7 (1 Jun 2020)
8. Silva, C.F.G.: A multi-criteria model applied to prioritization of information technology infrastructure problems. Master's Thesis University of Fortaleza. Management Ceará, Brazil (2017)
9. Sena, P., Viviane, D.: ITIL-based Problem Management for Federal Institutes of Education, Science and Technology: An Implementation Proposal. Master's Thesis Federal University of Pernambuco Recife, Brazil (2017)
10. Carvalho, F.N.: A methodology to support the problem management process in large companies providing ICT services. Master's Thesis Federal University of Pernambuco. Computer science Recife, Brazil (2011)
11. Jäntti, M., Miettinen, A., Pylkkänen, N., Kainulainen, T.: Improving the problem management process from knowledge management perspective. In: Münch, J., Abrahamsson, P. (eds.) PROFES 2007. LNCS, vol. 4589, pp. 389–401. Springer, Heidelberg (2007). https://doi.org/10.1007/978-3-540-73460-4_33
12. Roy, B.D.: The assessment of socioeconomic development. Technical Manual II: Methods and Techniques. Framing Instruments for Assessment Conclusions: Multicriteria Analysis (1993)
13. Laxxus, et al.: Multiple Criteria Decision Analysis: State of the Art Surveys. Springer Verlag, Boston, Dordrecht, London (2005)
14. Simão, M.F.: An approach structured on verbal analysis to support the allocation of tasks in projects of distributed software development. Doctorate Thesis, University of Forta-leza (UNIFOR), Ceará, Brazil (2017)
15. Morais, D.C.: Multicriteria group modeling for strategic planning of loss control in water supply. 148 f. Thesis (Doctorate) - Postgraduate Program in Production Engineering, Federal University of Pernambuco, Recife, Brazil (2006)
16. Bana, C.A., Sanchez, L.R., Vansnick, J.C., Corte, J.M.: Introducción a MACBETH. In: Leyva López, J.C. (Ed.) Análisis Multicriterio para la Toma de Decisiones: Méto-dos y Aplicaciones. México: Plaza y Valdés, pp. 233–241 (2011)
17. Gomes, L.F.A.M., Gomes, C.F.S.: Management decision making: a multicriteria approach, 4th edn. Publisher Atlas, São Paulo, Brazil (2012)
18. Cavalcanti, R.C.: Multi-criteria system to support the purchase of multi-family urban properties in the Recife real estate market based on the SMARTS method. Dissertation (Masters in Production Engineering) – Federal University of Pernambu-co (UFPE). Recife, Brazil (2007)
19. Bouyssou, D.: Building Criteria: A Prerequisite for MCDA in Readings in Multiple Crite-ria Decision Aid. Springer Berlin Heidelberg, Berlin, Heidelberg, pp. 58–80 (1990). https://doi.org/10.1007/978-3-642-75935-2_4
20. Bana, C.A., Corte, J.M., Vansnick, J.C.: MACBETH. Int. J. Inf. Technol. Decis. Mak. **11**(2), 359–387 (2012)
21. Bana, C.A., Vansnick, J.C.: The MACBETH approach: Basic ideas, software, and an application. In: Meskens, N., Roubens, M.R. (eds) Advances in Decision Analysis. Mathematical Modeling: Theory and applications. Kluwer Academic Publishers, Dordrecht, pp 131–157 (1999)
22. Bana, C.A., Corte, J.M, Vansnick, J.C.: M-MACBETH version 1.1 user's guide (2005)
23. Almeida, L.H.P.: McDSDS: A Multi-criteria Model for Planning Distributed Software Development Projects with Scrum. Masters Thesis (Master in Applied Computer Science) – University of Fortaleza (UNIFOR), Ceará, Brazil (2011)
24. BRAZIL: Constitution (1988). Constitution of the Federative Republic of Brazil. Brasília, DF: Federal Senate: Graphic Center (1988)

25. Figueira, J., Greco, S., Ehrogott, M.: Multiple criteria decision analysis: State of the art surveys, vol. 78. Springer New York, New York (2005)
26. Quirino, M.G.: An approach structured on verbal analysis to support the allocation of tasks in projects of distributed software development. Doctorate Thesis, University of Fortaleza (UNIFOR), Ceará, Brazil (2017)
27. Quirino, M.G.: Incorporation of Subordination Relationships in the Roberts Ordering Matrix in MCDA when the Asymmetry and Negative Transitivity Axioms are Violated. 211f. Doctorate Thesis (PhD in Production Engineering). Federal University of Santa Catarina, Florianópolis, Brazil (2002)
28. Goodwin, P., Wrigthat, G.: Decision Analysis for Managment Judgement. J. Wiley, Chichester (1991)
29. Bana, C.A., Correa, E.C., Corte, J.M., Vansnick, J.C.: Facilitating bid evaluation in public call for tenders: A social-technical approach. Omega **30**, 227–242 (2002)

Model Train Controlled by Microcomputer

Iva Kavankova(✉)

Tomas Bata University in Zlin, nam. T.G. Masaryka 5555,
760 01 Zlin, Czech Republic
kavankova@utb.cz

Abstract. The paper focuses on the controlling model trains by the camera. The paper deals with the analysis, design, and implementation of the embedded system for modeling track management. The main part is image processing in real-time with the following evaluation. Designed and realized product contains model railroad with sensor-controlled by an open-source platform called Arduino Mega 2560 and microcomputer Raspberry Pi 3 Model B. Programming language Wiring and scripting language Python are used for final system development.

Keywords: Trains · Microcomputer · Arduino · Raspberry Pi · Camera · PWM · Modeling · Image processing · IoT · User interface

1 Introduction

Nowadays trains are the most popular, the most secure and the fastest terrestrial transporting vehicle. Their popularity has also increased due to electrification and reduced environmental impact. However, this increase in rail traffic worldwide also leads to problems such as congestion and almost maximum capacity utilization, which has logically resulted in delays. Finding solutions in real traffic seems inappropriate, so models of trains and lines have begun to be developed. It is possible to model different situations and look for optimal solutions [1,2].

Thanks to modern technologies, this branch is developing quickly, expecting innovations and automation. Previously, the use of analog controlling [3,4], which over time replaced digital controlling [5,6]. That is much more expensive because DCC (Digital Command Control) central and decoders for locomotives and action elements are necessary. Analog control usually uses a 12 V power supply, which affects the speed of the train. However, this method is limited by the number of trains on the track. Digital control allows you to control multiple sets because they are equipped with decoder units. Despite this fact, the author uses analog controlling and 50 years old trains and railway tracks and presents the possibility of controlling any model and fully replacing the digital control version. For this purpose, the platforms Arduino [7–9] and Raspberry [10–12] disposes of the new types of microcomputers were used. Arduino controls the

R. Silhavy (Ed.): CSOC 2022, LNNS 501, pp. 168–176, 2022.
https://doi.org/10.1007/978-3-031-09070-7_14

railway track because it is easy to use and has a lot of input pins. Raspberry Pi was chosen for the more considerable computing power that was necessary for real-time image processing. Author uses for this experiment the knowledge described in the publication [13].

The paper aims to control trains with a microcomputer and monitor them by the camera above the railway track. The camera works as a motion sensor; trains are monitored by image evaluation. For more considerable comfort, the control will be possible via Wi-Fi (Wireless Fidelity), for example, using a mobile phone.

2 Controlling of Model Railway

2.1 Model Railway

The work dealt with designing and implementing the railway model, buildings, and machinery to reflect reality. The critical factor is the scale, which is a ratio between real objects and models. There are two ways how to control model trains, namely analog or digital control. [14]

Analog control uses constant voltage between 0–12 V, and the speed of trains is affected by voltage magnitude. However, there is possible to have a railway only with one locomotive (otherwise, it is necessary to split the track into independently powered parts). [15]

Digital control is much more complicated. Every locomotive and action element needs its own decoder, which decodes the commands for a specific device. The benefit of this solution is to control more trains at the same time. It is also necessary to have a DCC station, which modulates DCC and sends it to amplifiers, that powers the track. [16]

Communication is via packets that are put together from the coded block. The packet comprises the preamble, start bit, address, data, control, and stop bit. [17] The following example shows a packet for a locomotive for speed and direction (Table 1).

Table 1. Example of packet for locomotive, [5]

Preamble	Start bit	Address	Start bit	Data	Start bit	Control	Stop bit
11111111111	0	0AAAAAAA	0	01DUSSSSS	0	EEEEEEEE	1

2.2 Microcontrollers and Microcomputers

Arduino is an open-source platform with a graphical development environment where the most significant focus is on easy usage. There is possible to program in C++ language with Wiring framework. The program is divided into two specific blocks. The first block is function void setup(), which runs only at the beginning of the program run. The second part has a function void loop(), which runs in an infinity loop until reset or disconnection of voltage. [18] There are many types of Arduino; for this paper, Arduino Mega is used.

Raspberry Pi is a simple computer in size of a credit card which contains HDMI output (High-Definition Multi-media Interface) to connect on monitor, USB ports for connecting mouse and keyboard. After an internet connection, there is the possibility to connect via SSH (Secure Shell) protocol and work remotely. [19]

There are many generations of Raspberry Pi; this paper uses Raspberry Pi 3 Model B.

The main difference between Arduino and Raspberry Pi is that Arduino is the only microcontroller, whereas Raspberry Pi is a microcomputer (own operating system). Arduino is used for more manageable tasks, and Raspberry is better for much more complicated tasks. Another big difference is the price. [20]

3 Evaluation of the Current State

Aim of this thesis is to study and controlling model train on the track. After that implement the whole system and demonstrate on the model railway. Another part is to create user friendly interface to control the trains comfortably. There are used two platforms - Arduino and Raspberry Pi. For their communication was created protocol. Also there is added sensor for watching movement on the track. There is used camera, which is taking pictures of track and after that the situation is evaluated.

Communication between them might be via network, which is Raspberry Pi supports; however Arduino not, it needs additional Ethernet Shield. For this reason was used communication via USB interface, which both have. Communication will be realized in form of special sequence of strings for controlling the railway.

Arduino will be controlling the model railway, switching switches and power parts. Raspberry Pi will be controlling camera and following image processing. Format of the string in communication protocol will be string of character, which will describe position or state of parts or switches.

Due to high financial demands, it was decided, that trains will not be DCC controlled. Considering analog controlling, it will be necessary to split railway into independently powered parts.

4 Design and Implementation

This part is about design, hardware assembly, programming algorithms and testing. Final product is model track controlled by web interface via Wi-Fi connection.

4.1 Decomposition of Problems

Firstly, the whole project must be divided into logical parts, namely the software and hardware parts. The software part has a bigger priority because it is essential for the second part. Decompositions of problems are in the following (Table 2).

Table 2. Decomposition of problems

Hardware part	Software part
Railway, switches	Controlling railway
Trains	Image processing
Controlling	Evaluation
Camera	User interface
Microcomputers	Web server

The construction of the railway was designed into six independently powered parts. The number was determined because Arduino Mega has 12 PWM (Pulse Width Modulation) pins (for each position is necessary to have two pins).

4.2 Control Board

The Control board is based on the Arduino Mega platform. For power were used three pieces of double H-bridges L298N for the switches were used board, which contain eight pieces of relay JQC3F-05VDC-C (two of them are unused because there are only six switches). For easier manipulation, there are added LED (Light-Emitting Diode) diodes; if the green is on, the part of the railway is powered clockwise. In the case of yellow diodes, the railway is fed anti-clockwise.

4.3 Image Processing

Microcomputer Raspberry Pi was chosen for image processing because of the more considerable computing power and memory. For taking pictures was chosen module with the camera, which is not the original Picamera, because of the high price and inability to replace the lens (there might be problems with positions of the camera against the railway). The camera is taking pictures in resolution 640×480 pixels which is a sufficient resolution for recognizing moving trains are calibration marks.

The image must be divided into logical parts because of the motion analysis of trains. There were used five calibration marks that split the image into subparts. The better representation shows Fig. 1.

Because of this approximation, the small parts in the corners are lost; however, it does not represent any significant trouble as the train is detected before and after blind sections. This problem can be solved by adding more calibration marks. One Special mark is in the middle; it is used to detect the left or right side.

The position of trains is given by difference on two consecutive photos because of the different speeds of the locomotive as Fig. 2 illustrates; it is necessary to recognize them. It is solved by measuring the longer side of them (because locomotives are of varying sizes). First, it is an essential photo of an empty railway. This is done by calibration script, which is mandatory before every

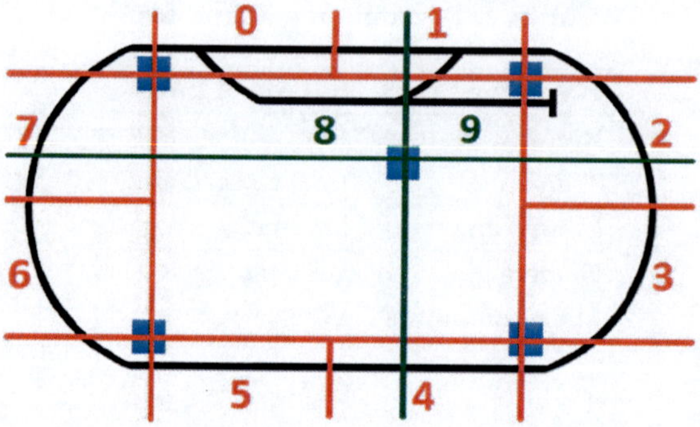

Fig. 1. Graphic representation of subparts

usage. After that, two consecutive images are taken, and a difference is made; therefore, it is possible to see their position.

Fig. 2. Difference of two pictures - movement of one locomotive (highlighted with morphologic filter)

After that, the approximate center of locomotives was calculated, and concrete coordinates were added. Subparts are also defined in coordinates to be compared, and the result is the subsection where the train is. Locomotives can not be in the same subpart if it happens; the system detects the potential risk of accident and stops the movement of the trains. There is no possibility to ride them again automatically; it is necessary to move one locomotive to a different subpart (because of the powered railway - they cant ride on the same subpart because each of them has a different speed). According to this, the initial condition of replacing locomotives must be indifferent subpart (otherwise, it is not possible to start the railway). It is the responsibility of users.

4.4 Communication Protocol

Own new protocol called ATRAB (Arduino To Raspberry And Back) was created for the communication. Both microcomputers are cooperating with the help of the existing library - Serial. After evaluating the situation, Raspberry sends Arduino a string of numbers in format XXXXXXYYY, where X represents each subpart, Y represents switches.
X can acquire:

- 0 - subpart is not powered
- 1 - subpart is powered clockwise
- 2 - subpart is powered anti-clockwise

Y can acquire:

- 0 - do nothing
- 1 - switch a switch to narrow direction
- 2 - switch a switch to opposite direction

For example, code 000111201 means that: first, second and third subparts are not powered; the fourth, fifth, and sixth subparts are powered; the first switch is switched to the opposite direction, the second switch is doing nothing, and the third switch is switched to narrow direction. This protocol was designed for more upgrades, like adding semaphores are other train stuff by adding other numbers.

After receiving this code, Arduino executes all tasks and sends to Raspberry Pi message back:

- 1 - command was successful
- 2 - command has not required format

4.5 User Interface

Control was realized as the web user interface, and for this purpose, a web server on Raspberry Pi was created, which allows this function. This part of the paper was not the major one, so it was only a simple interface. Emphasis was placed on simplicity and intuitiveness so that children could control the railway (Fig. 3).

First, it is necessary to use the CAL button to calibrate and show an actual photo of the railway (calibration photo). Button 1, 2, and 3 contain three demo programs with an infinite loop. The red button serves for immediate stop trains. Description of buttons:

- CAL - calibration button, takes a picture of the railway and shows it. After that, the user could decide whether the user is satisfied (if not, it is possible to move the camera). Other settings like exposition, ISO, resolution can be changed directly in the calibration script
- 1 - run the first program, where both locomotives wait for each other at the station and go on the loop alternately
- 2 - run the second program, where is shown blind subpart of the railway; one train goes into it, and drawback

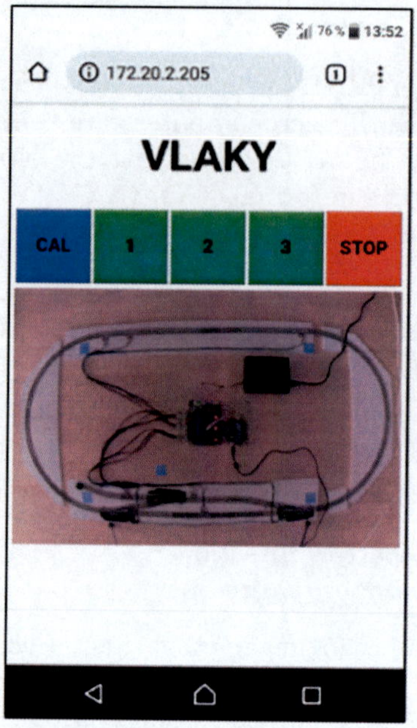

Fig. 3. Screenshot from user interface

- 3 - run the third program, similar to the first program, but the trains are riding the loop in different directions
- STOP - trains stop immediately

HTTP (Hypertext Transfer Protocol) and CSS (Cascading Style Sheets) are used for the user interface. Framework Flask was used for the backend.

5 Conclusion

The paper aims at analyzing, implementation, and demonstration of controlling trains by microcomputer. There were studied and compared options of controlling model railway. It included examining whether it is possible to use commercially available microcomputers for these purposes eventually set limits for their use.

Two platforms were used to design the train control by microcomputer, namely Arduino Mega and Raspberry Pi. Arduino controls the whole track, and Raspberry Pi makes image processing. Both platforms communicate with each other, and for this purpose, a new protocol called ATRAB was designed.

The user interface is realized like a web application where users can calibrate and execute prepared demo programs. There was demonstrated controlling model train by microcomputers and realized with communication between two different microcomputers. The camera above the track prevents accidents between trains.

There are a lot of possibilities for how to continue, for example, to use DCC controlling. There is the option to apply technology Bluetooth thanks to Raspberry Pi support. The popular function is also a stream of the railway (for example, on the streaming platform Youtube). Another improvement is to protect H-bridges against the high electric current - this is possible with sensors ACS714, which can measure the current. It is possible to recognize the trains using convolutional networks.

The benefit of this work is the analysis of the problem with controlling model trains by microcomputer, created function embedded system and successful communication between Arduino and Raspberry Pi platforms.

Acknowledgments. The work was funded with the support of the Internal Grant Agency of Tomas Bata University under project No. IGA/FAI/2022/004.

References

1. Van den Boom, T., de Schutter, B.: Modeling and control of railway networks. In: Proceedings of the 2004 American Control Conference, vol. 6, pp. 5728–5733. IEEE (2004). https://doi.org/10.23919/ACC.2004.1384769, [viewed 8 March 2022]. ISBN 0-7803-8335-4
2. Bart, K., van den Boom, T., de Schutter, B.: Distributed model predictive control for railway traffic management. In: Transportation Research Part C: Emerging Technologies, vol. 68, pp. 462–489 (2016), https://doi.org/10.1016/j.trc.2016.05.006, [viewed 8 March 2022], ISSN 0968090X
3. Yin, J, Su, S., Xun, J., Tang, T., Liu, R.: Data-driven approaches for modeling train control models: Comparison and case studies. ISA Trans. **98**, 349–363 (2020). https://doi.org/10.1016/j.isatra.2019.08.024, [viewed 8 March 2022], ISSN 00190578
4. Ursu, M.-P., Novac, O.-C., Oproescu, M., Buidoso, G., Hathazi, F.I.: Comparative Study of the Analog and Digital Operation for Miniature Railway Systems. In: 2019 15th International Conference on Engineering of Modern Electric Systems (EMES), vol. 2019, pp. 137–140. IEEE (2019). https://doi.org/10.1109/EMES.2019.8795182, [viewed 8 March 2022], ISBN 978-1-7281-0773-8
5. Ursu, M.-P.: Digital electronic modules for command and control of miniature railway systems. In: 2021 16th International Conference on Engineering of Modern Electric Systems (EMES), vol. 2021–6-10, pp. 1–4. IEEE (2021). https://doi.org/10.1109/EMES52337.2021.9484138, [viewed 8 March 2022], ISBN 978-1-6654-4995-3
6. Ursu, M.P., Condruz, D.A.: Digital intelligent booster for DCC miniature train networks. In: IOP Conference Series: Materials Science and Engineering, p. 227 (2017). https://doi.org/10.1088/1757-899X/227/1/012133, [viewed 8 March 2022], ISSN 1757–8981
7. Badamasi, Y.A.: The working principle of an Arduino. In: 2014 11th International Conference on Electronics, Computer and Computation (ICECCO), vol. 2014, pp. 1–4. IEEE (2014). https://doi.org/10.1109/ICECCO.2014.6997578, [viewed 8 March 2022], ISBN 978-1-4799-4106-3
8. Kondaveeti, H.K., Kumaravelu, N.K., Vanambathina, S.D., Mathe, S.E., Vappangi, S.: A systematic literature review on prototyping with Arduino: Applications, challenges, advantages, and limitations. Comput. Sci. Rev. 40 (2021). https://doi.org/10.1016/j.cosrev.2021.100364, [viewed 8 March 2022], ISSN 15740137

9. Lopez-belmonte, J., Marin-Marin, J.-A., Soler-Costa, R., Moreno-guerrero, A.-J.: arduino advances in web of science. a scientific mapping of literary production. IEEE Access **8**, 128674–128682 (2020). https://doi.org/10.1109/ACCESS.2020.3008572, [viewed 8 March 2022], ISSN 2169–3536
10. khalifa, F., Eman Badr, A., Hesham, N.: ELMAHDY. A survey on human detection surveillance systems for Raspberry Pi. In: Image and Vision Computing, vol. 85, pp. 1–13 (2019). https://doi.org/10.1016/j.imavis.2019.02.010, [viewed 8 March 2022], ISSN 02628856
11. Jolles, J.W.: Broad-scale applications of the raspberry pi: a review and guide for biologists. Methods Ecol. Evolu. **12**(9), 1562–1579 (2021). https://doi.org/10.1111/2041-210X.13652, [viewed 8 March 2022], ISSN 2041–210X
12. Petrov, N., Dobrilovic, D., Kavalić, M., Stanisavljev, S.: Examples of raspberry pi usage in internet of things. In: Proccedings of the ICAIIT 2016 University "St. Kliment Ohridski" Bitola, Macedonia, vol. 2016, pp. 112–119 (2016). https://doi.org/10.20544/AIIT2016.15, [viewed 08 March 2022], ISBN 9789989870750
13. Kavánková, I.: Model Train Controlled by microcomputer. Brno, 2019. Bachelor thesis. Brno University of Technology, Faculty of information technology. Supervisor Prof. Dr. Ing. Pavel Zemçík (in Czech)
14. Model making, March 2008. https://cs.wikipedia.org/wiki/Model%C3%A1%C5%99stv%C3%AD, [visited 28 March 2019] (in Czech)
15. Electrical connection of the track. http://masinky2003.wz.cz/zapojeni_stanice.html, [visited 8 April 2019]. (in Czech)
16. DCC (Digital Command Control), January 2014. http://www.vlacky.com/dcc/ [visited 28 March 2019]. (in Czech)
17. NMRA Standards and Recommended Practices, January 2014. https://nmra.org/index-nmra-standards-and-recommended-practices, [visited 11 April 2019]
18. What is Arduino? 2019. https://www.arduino.cc/en/Guide/Introduction, [visited 30 March 2019]
19. Okoi, M.D.: 10 Operating Systems You Can Run on Raspberry Pi in 2019, March 2019. https://www.fossmint.com/operating-systems-for-raspberry-pi, [visited 6 April 2019]
20. Justo, P.D.: Raspberry Pi or Arduino Uno? One Simple Rule to Choose the Right, December 2015, https://makezine.com/2015/12/04/admittedly-simplistic-guideraspberry-pi-vs-arduino/, [visited 07 April 2019]

Energy-Based Leader Election (E-LE) for Group Management of IoT

Juhyun Maeng and Inwhee Joe(✉)

Hanyang University, 222, Wangsimni-ro, Seongdong-gu, Seoul, Republic of Korea
{jhmaeng,iwjoe}@hanyang.ac.kr

Abstract. Internet of Things (IoT) shares and collects data on existing systems. IoT applications are changing our daily lives a lot. In this situation, IoT devices may exchange private data. Therefore, it is necessary to protect the collected data from unauthorized nodes. Groups can be formed to allow only authenticated nodes to access data so that they can use the data. In this paper, all group nodes share data with a secure communication link through the same key. This key is generated by the elected leader, and shared among all group nodes. In particular, the leader is replaced periodically. This is because, if the leader is malicious, the common key is shared with the external node, so that the collected data may be leaked to the unauthorized node. Therefore, the group is managed stably by periodically replacing the leader. We analyze the number of messages according to the leader election, and compare the number of messages according to the number of nodes as well as the number of conflicts. Performance evaluation shows that the proposed energy-based leader election (E-LE) method reduces the number of messages compared to the existing raft.

Keywords: Leader election · Group management · Data security · IoT

1 Introduction

Internet of Things (IoT) consists of devices connected through the Internet. IoT devices share and collect data in existing systems [1]. Various IoT applications are changing our daily lives. IoT applications are diverse, such as smart city, smart agriculture, smart medical, and so on [2]. In addition, there are applications such as vehicle monitoring, controlling, and healthcare. In this application, IoT devices may exchange private data. In this environment, it is necessary to prevent unauthorized nodes from accessing data [3]. Since IoT devices are limited in terms of resources, it's consumption must be reduced. To this end, data is protected from unauthorized nodes by forming groups between nodes having the same key. In addition, resource consumption of IoT devices is reduced by reducing the number of messages according to the election of the leader who manages the group.

R. Silhavy (Ed.): CSOC 2022, LNNS 501, pp. 177–184, 2022.
https://doi.org/10.1007/978-3-031-09070-7_15

2 Related Work

2.1 Group Management

Group management is implemented through group key management (GKM). This GKM can be classified into three categories. All group nodes share the same traffic encryption key (TEK) [4]. Group nodes with common keys share data through secure communication links. Therefore, data may be protected from unauthorized nodes. To this end, the leader to manage the group is elected. The leader generates a common key and shares the key among all group nodes. However, centralized architecture can be vulnerable to single point of failure (SPOF) [5]. So the leader must be replaced periodically.

2.2 Raft

Raft is one of the consensus algorithms. This algorithm elects the leader through voting between nodes. As shown in Fig. 1, in the raft, each node has three states of follower, candidate, and leader. The initial state of all nodes is follower. If election timeout of a particular node becomes 0, the state of the node is changed to candidate. Next, election period begins. Candidate node votes for itself and requests follower node to vote. The follower node votes for the candidate node. After all nodes vote, the election period ends. The node that received the most votes becomes the leader node [6].

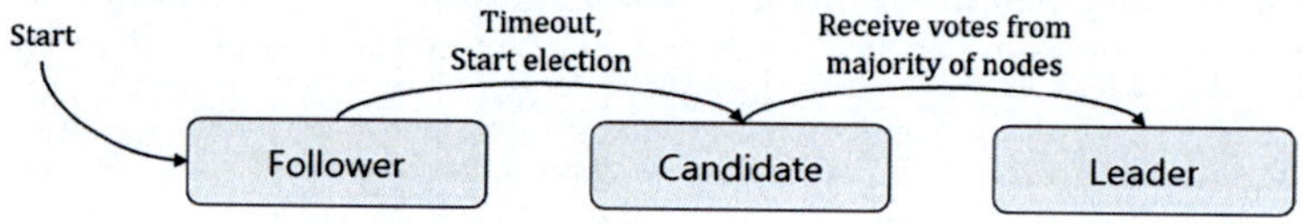

Fig. 1. State change of raft [7].

If multiple candidate nodes compete to be elected as a leader, multiple leaders may be elected. This is because there may be no candidate node that obtains more than half of the votes [8]. This paper proposes an energy-based leader election (E-LE).

3 Proposed Method

In this paper, we assume that, as shown in Fig. 2, the proposed system consists of a core system, a leader, where $leader_i, \forall_i \in \{0, 1, 2, ..., n_g - 1\}$ represents the leader of each group, and a group is expressed as $group_i, \forall_i \in \{0, 1, 2, ..., n_g - 1\}$. A group includes multiple nodes. Under this configuration, the procedure for leader election is as follows. The core system randomly elects a leader among all group nodes and informs the node of the result. Then, the leader generates a common key and transmits the key to all group nodes so that group communication between nodes may be performed. Where, the leader periodically measures the number of transmissions to the group (s_g), the number of messages (m_g), and the amount of energy of each node (e_n). Using the measured value, the result value calculated according to the following equation is notified to the core system.

$$(n_g - s_g) + (n_g - m_g) \tag{1}$$

If the result value is not 0, it may be expected that the leader has transmitted a common key to an unauthorized node. This is because more common keys were generated than n_g and keys were sent to external nodes. If the result value received by the core system is not 0, or there is no message transmitted from the leader within a certain period of time, the process of re-electing the leader begins. First, the core system requests information from the group to get energy amount information of each node. The group notifies the core system of information on the amount of energy of all nodes, where $e_i, \forall_i \in \{0, 1, 2, ..., n_g - 1\}$ represents the energy of each nodes. The core system excludes the existing leader from the group and re-elects the node with the most energy as the leader.

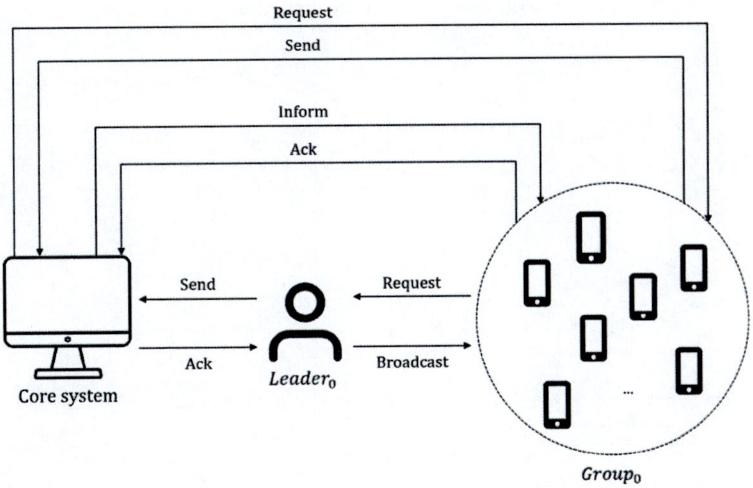

Fig. 2. General system model.

As shown in Fig. 3, the core system monitors the leader through state information and re-elects the optimal leader if a problem occurs. Also, as shown in Fig. 4, the leader periodically provides state information to the core system, and the leader manages the group.

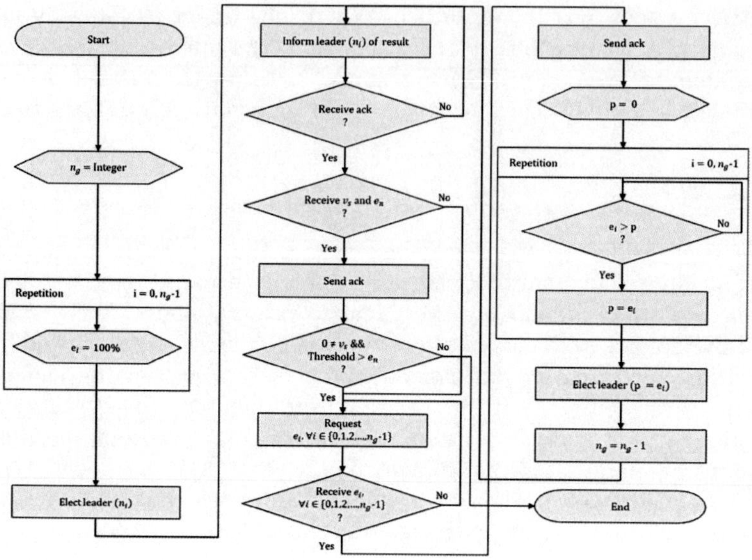

Fig. 3. Core system flow chart.

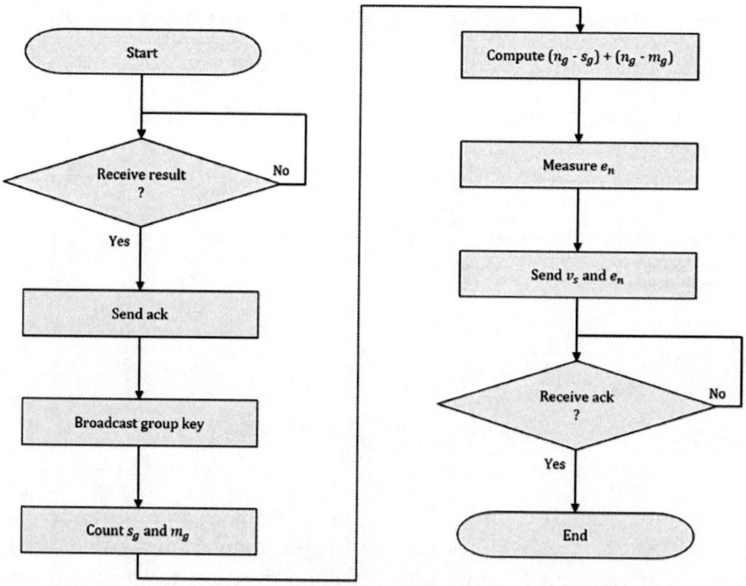

Fig. 4. Leader flow chart.

4 Performance Evaluation

In this section, we show the results of analyzing the proposed method. The values used in the experiment are shown in Table 1.

Table 1. Simulation parameter's values.

Symbol	Value	Description
e_n	Variable	The amount of energy in each node
n_g	Variable	The number of all group nodes
s_g	Variable	The number of transmissions to each group
m_g	Variable	The number of messages transmissioned to each group
c	Variable	The number of conflicts according to the leader election
v_s	1	The number of messages from abnormality detection results
r_e	1	The number of request messages for the amount of energy in each node
e_l	1	The number of notification messages for the leader election result
ack	2	The number of response messages from the elected leader

Where, e_n is the energy amount of each node, and n_g represents the number of nodes in each group. In addition, s_g is the number of transmissions by the leader to all group nodes, and m_g is the number of transmitted messages. c is the number of votes again because the number of votes by candidate nodes becomes the same when electing a leader in the raft. v_s is a value for monitoring a leader from generating more common keys than all group nodes and transmitting common keys to non-group nodes. r_e represents the number of messages requesting energy amount information from all group nodes to elect the leader again. In addition, e_l is the number of messages the core system informs the leader of the election result, and ack represents the number of confirmation messages between the core system and the leader. The number of messages exchanged between the core system, the existing leader, and the group to re-elect the leader of the proposed model is calculated by the following generalization equation.

$$v_s + r_e + n_g + e_l + 2(ack) \tag{2}$$

The number of messages exchanged between the candidate node and the follower node to re-elect the leader at raft is calculated by the following generalized equation.

$$n_g(n_g + 1) + n_g \tag{3}$$

If the number of votes received by several candidate nodes in raft is the same, the leader re-election process is repeated until a new leader is elected. Therefore, the number of messages exchanged between the candidate node and the follower node is calculated by the following generalized equation.

$$(c + 1)(n_g(n_g + 1) + n_g) \tag{4}$$

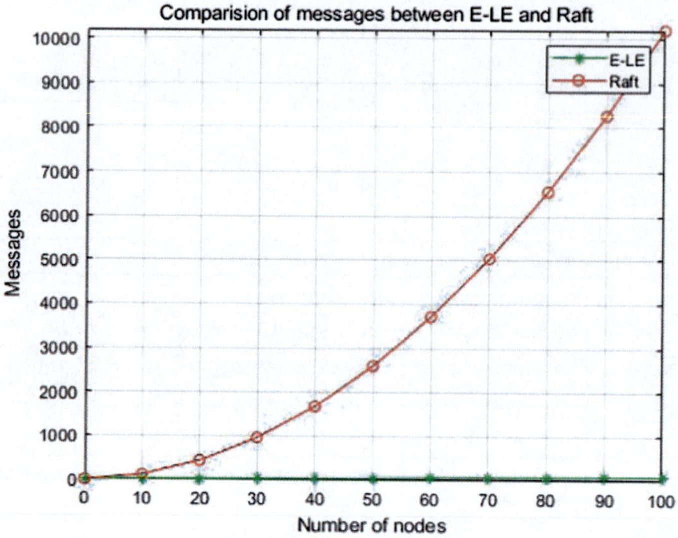

Fig. 5. Comparison of messages according to the number of nodes.

Figure 5 is the result of comparing the number of messages according to the increase in the number of nodes in the two methods for re-election of a leader. In the E-LE, messages for re-election of a leader are exchanged between the core system, the existing leader, and the group. The number of messages is calculated through the values in Table 1 and Eq. 2. In raft, messages are exchanged between the candidate node and the follower node for re-election of the leader. The number of messages is calculated according to the values in Table 1 and Eq. 3. E-LE reduces the number of messages compared to the raft. This is because even if the number of nodes increases, the number of messages exchanged between the core system and the existing leader is constant, and only the energy amount information of each node obtained by the core system increases according to the number of nodes. However, as the number of nodes increases, the follower node increases, and the number of voting request messages and voting messages between the candidate node and the follower node increases by the number of nodes.

Figure 6 is the result of comparing the number of messages according to the increase in the number of conflicts with E-LE to elect the leader again when the number of votes by several candidate nodes in the raft is the same. Where n_g is 100. In E-LE, the number of messages is calculated through the values in Table 1 and Eq. 2. In raft, the number of messages is calculated according to the values in Table 1 and Eq. 4. E-LE reduces the number of messages compared to raft. This is because E-LE does not repeat the leader's re-election process even if the number of conflicts increases, so that the number of messages exchanged between the core system, the existing leader, and the group does not increase as much

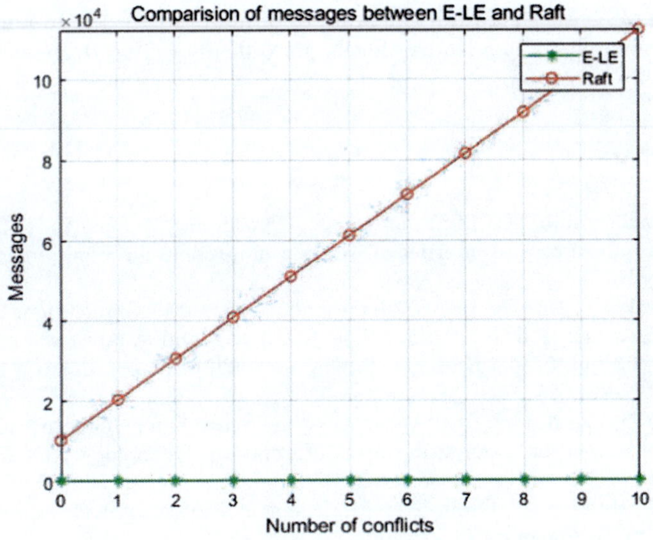

Fig. 6. Comparison of messages according to the number of conflicts.

as the number of conflicts. However, as the number of conflicts increases, the process of re-electing a leader until a new leader is elected is repeated, and the number of voting requests and voting messages between candidate and follower nodes increases by the number of conflicts.

5 Conclusion

In this paper, we knew the problem of increasing the message of the existing raft electing the leader through voting. This is because as the number of nodes and the number of conflicts increase, the number of voting request messages and voting messages increases. Therefore, we proposed energy-based leader election (E-LE). It can reduce the number of messages by electing a leader without voting and periodically change the leader to prevent system failure caused by single point of failure (SPOF). System stability can be improved by periodically changing the leader through E-LE. In addition, managing a group with an elected leader can protect the collected data from unauthorized nodes. As the number of nodes increases, the number of messages exchanged between nodes increases, so it is difficult to apply the existing raft to resource-constrained Internet of Things (IoT) devices. The number of messages in the proposed E-LE according to the number of nodes and the number of conflicts was analyzed. If the leader is re-elected, the number of messages is reduced because the messages exchanged between the core system and the existing leader are constant. Therefore, E-LE is suitable for application to resource-constrained IoT devices because it reduces the number of messages even if the number of nodes increases. In addition, since the group is managed by a leader that is periodically replaced, the collected data may be stably protected.

References

1. Muhammad Ayoub, K., Muhammad, A., Mazliham, M., Hafiz, R.: A review of middleware platforms in internet of things a non-functional requirements approach (2021)
2. Chakraborty, S., Das, P., Pal, S.: IoT foundations and its application. In: Pattnaik, P.K., Kumar, R., Pal, S., Panda, S.N. (eds.) IoT and Analytics for Agriculture. SBD, vol. 63, pp. 51–68. Springer, Singapore (2020). https://doi.org/10.1007/978-981-13-9177-4_3
3. Al-Taleb, N., Zagrouba, R.: Authentication Scheme for IoT. In: 2020 International Conference on Computing and Information Technology (ICCIT-1441), pp. 1-5 (2020)
4. Yacine, C., Hamida, S.: Group Key Management Protocols: A Novel Taxonomy. In: Information Technology - IT. 2 (2005)
5. Chen, C.-M., Deng, X., Gan, W., Chen, J., Islam, S.K.H.: A secure blockchain-based group key agreement protocol for IoT. J. Supercomput. **77**(8), 9046–9068 (2021). https://doi.org/10.1007/s11227-020-03561-y
6. Back, J., Choi, S., Shin, Y.: A study on the election of suitable leader vehicle in vehicle platooning using the raft algorithm. In: 2019 Eleventh International Conference on Ubiquitous and Future Networks (ICUFN), pp. 756-761 (2019)
7. The Raft Consensus Algorithm. https://www.andrew.cmu.edu/course/14-736-s20/applications/ln/riconwest2013.pdf
8. Fu, W., Wei, X., Tong, S.: An improved blockchain consensus algorithm based on raft. Arab. J. Sci. Eng. **46**(9), 8137–8149 (2021). https://doi.org/10.1007/s13369-021-05427-8

Development of Navigation Systems in Educational Buildings of Higher Educational Institutions

Artem Obukhov(✉), Denis Dedov, Artem Sveshnikov, and Nadezhda Vekhteva

Tambov State Technical University, 392000 Tambov, Russian Federation
obuhov.art@gmail.com

Abstract. The article discusses the problems of navigation associated with the process of rebuilding the structure of educational institutions, which violate the generally accepted standards for numbering rooms inside a building. The study of the influence of the problem of university navigation on the quality of student learning in the first days of training was carried out using surveys of several control groups. The current solutions to this problem in various universities, which are aimed at using special services, are considered. On the basis of the analysis carried out, two conceptually new methods are proposed, which to a certain extent improve some elements of existing methods or form completely new methods of solution. The first solution is a virtual interactive excursion system, the second solution is a navigation system using stands and a mobile application. A description of the developed systems is presented, their functionality and scope are considered.

Keywords: Adaptation of students · Navigation in the university · Virtual excursion · 3D models

1 Introduction

Upon admission to higher educational institutions, applicants face many stressful situations, adaptation can take several weeks or even months, which negatively affects the effectiveness of the educational process. Higher education institutions use many different solutions to simplify the navigation of applicants and students who find themselves in unfamiliar surroundings, but not all of the existing solutions are effective enough. Let's consider the main components of this issue and the problems that students face in the adaptation process.

In the course of its long existence, the university undergoes many changes in its structure: it receives new buildings - buildings - into its ownership. Buildings that are already part of it are undergoing changes, extensions appear, some premises are transferred from one department to another. These and other factors lead to the fact that the numbering system is constantly changing, its consistency and orderliness decreases.

The standard room naming system, which is used not only in universities, but also in any other building (for example, hotels or hospitals) has worked well. The first digit

R. Silhavy (Ed.): CSOC 2022, LNNS 501, pp. 185–199, 2022.
https://doi.org/10.1007/978-3-031-09070-7_16

coincides with the floor number, and the next digits are the ordinal numbers of the offices already on this floor. However, installing various room modifications, the system stops working. For example, office # 306 may not be located between # 305 and # 303, but on the opposite end of the building on a different floor. It is also quite difficult to determine the description of important premises that do not have serial numbers. Such premises include, for example, a dining room or a wardrobe.

This problem is of much greater importance for first-year students. It is very important for them to show themselves on the positive side in the first stages of training. Modern culture, parenting upbringing, stories of senior students - all this creates the impression that if a student cannot create a positive reputation for himself in the first days of training, then he will never be able to earn it. This determines the relevance of the research - software systems and software tools that simplify the process of student adaptation in the university environment. Within the framework of this study, we will focus on the issue of creating digital services for educational institutions of higher educational institutions.

2 Methods

At the first stage, we will analyze the subject area and determine the relevance of the problem of navigation in the structure of the university, as an important component of the adaptation process of students.

2.1 Analysis of the Relevance of the Problem of Navigation of Students

To determine the level of anxiety of students, a survey structure was formed for young people studying at the university. An important criterion is not the current age of students, but the age of their admission to an educational institution. This characteristic will be used in the future when considering the results of the survey.

The survey contains the following questions:

1) Did the respondent experience difficulties in finding the rooms he needed in the first days of his university/college studies. This question is key to the research problem. It is important to bear in mind that such a question is not an attempt to negatively assess the organization, since the structure changes in the course of the activity of any organization.
2) Has the respondent ever met the wrong numbering of the rooms? (for example, room # 305 is not located between # 304 and # 306, but completely at the other end of the floor).
3) Did the interviewee have difficulty finding key places (such as printing services, dining room, wardrobe, etc.).
4) Did the respondent feel any discomfort from a possible delay and the need to make excuses to the teacher because of navigation problems. This moment will help to understand how serious a problem for students is disorientation in area.
5) Does the interviewee think that punctuality, at least in the first days of university/college, is of great importance to the student's reputation. This question determines the relevance of the problem, taking into account the life views of students.

6) Age of the respondent at the time of admission to an educational institution. Young people between the ages of 14 and 18 within a short period of time (a year or two) can completely change their attitude to interacting with other people [1].
7) Gender. Before conducting the survey, a hypothesis was put forward that the view of the female representatives is very different from the male in matters of punctuality [2, 3];

There are several options for solving the problem of adaptation and navigation of students in an unfamiliar structure of premises in the university building. During the analysis, the following were selected from a variety of solution paths: installation of a large map-plan in the main hall of the building, marks on the parquet with guide lines that help to navigate while already in the current space; development of appropriate software solutions using various methods of visualization and interactive control.

Thus, in the course of the study, a survey was conducted in which students of several groups of students at the university, on average, from 16 to 22 years old, took part. During the week, 81 people took part in the survey, among whom 45 people entered the first course at the age of 17–18 years (55%), 33 people at the age of 15–16 years and (40%) and 3 people at the age of 19–20 years old (5%). Among them: 66 males (82%) and 15 females (18%).

The survey results are presented in Fig. 1.

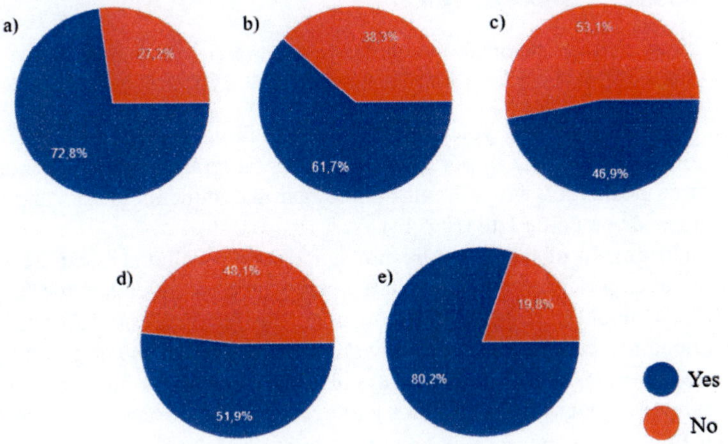

Fig. 1. Poll results.

Difficulties with finding the necessary audiences in the first days of training were experienced by 72.8% of the respondents (Fig. 1a). This is a fairly high figure, indicating that only a quarter of students can cope with the current navigation in a particular educational institution.

A smaller number of respondents can recall specific violations in the numbering of rooms - 61.7% to the question "Have you ever met an incorrect numbering of classrooms?" answered positively (Fig. 1b).

The last question regarding navigation is "Have you had any difficulty in finding functional places such as printing, dining, etc.?" Less than half of the respondents answered it positively - 46.9% (Fig. 1c). This suggests that there are significantly fewer problems with finding important places for all students.

Thus, it can be concluded that approximately two-thirds of students experience problems with navigation in an educational institution. At the same time, there are fewer problems with common areas, since people can consult with a large number of people to find out the right direction. Also places like this get more attention in terms of signposts.

The next category of questions concerns students' attitudes towards possible navigation problems. To the question "Did you feel any discomfort from a possible delay and the need to make excuses to the teacher because you cannot find the classroom you need?" 51.9% of the respondents answered positively (Fig. 1d).

At the same time, a relationship between male and female sex was noticed and a difference was found in the answer to this question. Among girls, 13 out of 15 respondents gave a positive answer to this question, which is 87%. At the same time, out of 66 young men, only 29 responded positively, which is 44% (Fig. 2). This indicator may, with some probability, mean that girls are more responsible for being late, since the sample for such an analysis is too small.

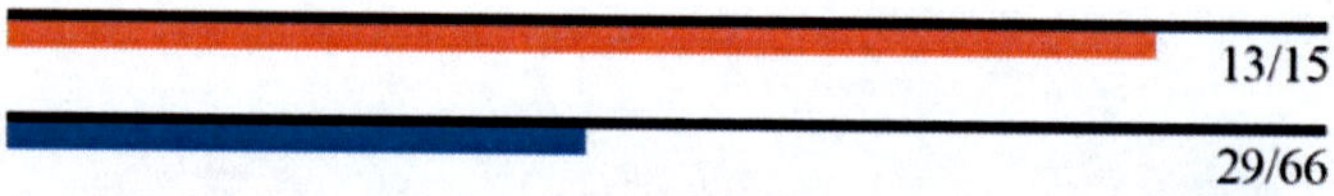

Fig. 2. The ratio of boys and girls when answering question No. 4.

To the last question, "Do you think that punctuality, at least in the first days of university/college, is of great importance to a student's reputation?" 80.2% of respondents answered positively, which confirms the fact that students understand that their reputation depends on being late (Fig. 1e).

Based on the results of surveys of freshmen, it can be concluded that there is a strong emotional stress experienced by a fairly large percentage of students in the first days of university life. One of the reasons for this stress is the desire to establish oneself in the team and among teachers, to form a positive reputation. One of the important points of a student's reputation is punctuality. Surveys also show that part of the emotional stress is the pressure of parents who too often try to exceed the importance of certain criteria of life at the university.

Due to the strong emotional stress, an urgent task is to reduce it. Unfamiliar floor plans are an important part of emotional pressure. It is possible to reduce this pressure due to the remote acquaintance of the freshman with the university in a comfortable environment for him [4–8], which will simplify the adaptation of the student in an unfamiliar university environment.

2.2 Existing Approaches to the Adaptation of Students to the University Structure

One of the important components of the adaptation process, as shown by the survey results, is the solution to the issue of students' navigation in the academic buildings of the university. Consider the existing approaches to solving this problem.

A popular solution is the implementation of virtual tours, which are offered by the official websites of major universities for key premises. The importance of these premises is determined by the university administration. Also on the Internet you can find aggregators of these services. These sites contain a large number of virtual tours of many universities. One such example is the "You Visit" site, which includes a "Colleges" section [9]. Many American universities can be found on this service. Virtual tours are designed in the same visual style. They are panoramic images available for 360-degree viewing and available for VR headsets. The tour is not only visualized, but also accompanied by a voice assistant (Fig. 3).

Fig. 3. "You Visit" online system interface.

However, despite the existence of such aggregators, many universities are trying to create their own similar systems on their own. [10–13].

These systems allow you to show a person how the university works from the inside. However, in essence, they are a set of photographs tied to a map or placed on a diagram. Universities, laying out such excursions, strive to show the level of technical or scientific development, objects of cultural value. These excursions are not intended to make navigation easier for the student. Due to the fragmentation of such photographs, the immersion effect is completely absent, which is why it is impossible to talk about the continuity of the information received, which results in the lack of usefulness of these services as navigation services.

Based on the analysis performed, several conceptually new solutions to the above-described problems can be proposed [14, 15].

A promising approach is to demonstrate various 3d models to the student. Designing the interior of a building in virtual space is a feasible task at the current level of development of 3D modeling technologies. It takes three to six man-months to create a detailed model with realistic lighting for a single building, while creating a low-poly schematic model is possible in one man-month. Thus, the creation of a virtual university model even for a small development team is possible in a fairly short time frame.

Within the research, university students can be divided into two parts. The first group of students are people who prefer to do everything in advance - they try to be punctual. While the second part of the students are people who prefer to solve emerging problems as they come. They may, for one reason or another, not familiarize themselves with any information about the structure of the premises in advance. For these two categories, two completely different approaches need to be prepared.

2.3 Statement of the Problem of Creating an Interactive 3D Excursion System

It was previously determined that a fairly large percentage of students experience a strong emotional stress during the first days of university. For the first category of students, for this reason, it is necessary to reduce the load. The unfamiliar layout of the premises is the most important part of the emotional pressure, which can be reduced in only one way - to make the freshman feel like he is in a place that he has already been to several times.

To solve this problem, we use a 3d building model with a full realistic visualization of the interior decoration of the premises, available for research in the form of an excursion with an imitation of "live" movement.

A virtual 3d model of the interior layout of a building must meet specific conditions:

1. The application must ensure the continuity of the content provided. The user must perceive the model as a whole. Thus, the continuity of the study of the model can be perceived as an analogue of movement through the building, imitating the passage of ways by a real person. This aspect is achieved by combining models of individual rooms and corridors into a single whole model, which is significantly different from the methods that exist at the moment.
2. The model presented in the appendix must have a refined detail and emphasis on any objects inside the building. Accents are an extremely important thing from a psychological point of view. Many students note that they memorized very well some places completely by accident for some unique details, for example, such a detail may be a plate different from the standard, which for some reason changed the color of the background or text. The 3d model is an excellent tool for highlighting different accents. Game designers and level designers have shaped most of these techniques into complete sets of methods. One example of such methods is directional lighting, which involves working with light: natural or artificial directional light that stands out from the overall picture allows you to subtly highlight certain important points that are worth paying attention to. A place that should be indicated may be the inconsistency of the rooms of the offices with each other, caused by the redevelopment or rebuilding of a certain section of the building. These problem areas are identified by interviewing students.
3. The application should include a complete and detailed route for the study of the model, which must be manually set for the user at the design stage. It is not enough to give a spatial model to the user for research - he simply will not understand what

to do with it and, ultimately, after looking at it for a couple of minutes, he will safely forget about it. A smart route should include all important objects for the user (in this case, for a freshman), as well as simulate his route in reality, predicting the list of his interests. Routes will be obtained through a survey of students.

Based on the above, we can conclude that the 3d model should not only exist as an object, but also have the functionality to simulate the user's movement through the territories, as well as include tools for interacting with the model. Computer games are one of the most successful ways to simulate the presence of a person in the world of three-dimensional models. Thus, we get one more task within the framework of the study:

4. Integration of the created 3d model into the computer game engine. It is with the help of this action that you can achieve maximum immersion and better orientation of the user inside the space, which will lead to obtaining the necessary experience of exploring a new territory.

The application should also be as accessible as possible to the end user. This defines the following condition:

5. Ensure the most comfortable and intuitive delivery of the application to the end user, regardless of his personal computer skills. The solution is to create a subsection for the university website, which will contain detailed instructions for setting up and installing the application, which would be understandable to everyone, regardless of their skills.

Thus, the task of developing this system is an acceptable and affordable solution to the problem of adaptation of students in the field of navigation, it will make it possible to conduct a virtual 3d excursion with a specific and well-designed route that takes into account all the features of the terrain identified by interviewing university students.

A virtual tour requires only a personal computer. The student may not be interested in learning about the structure of the premises, but will need specific locations and audiences at a specific point in time. For example, the schedule contains the number of the classroom in the building, where he has never been. You can get knowledge about its location from evacuation plans or by asking the university staff, but the use of information technology allows you to do this more comfortably and quickly.

2.4 Statement of the Problem of Creating an Information Router Application

The second important component of the problem of students' navigation at the university is not only familiarity with the structure of the buildings in the game mode, but also obtaining information about the location of a specific object. To do this, consider the concept of a second system aimed at solving this problem.

You can find the right place using information touch terminals or interactive kiosks, which have a floor plan, indicating or even finding the right place. Also, software products for mobile devices can help with navigation in the premises, with the help of which a route can be laid to the desired point.

Unfortunately, at the moment most of the existing popular navigators do not allow positioning the user indoors. The solution can be fixed Wi-Fi stations and high-frequency

Bluetooth emitters, as well as magnetic beacons. Material or technical opportunities for their purchase, installation and settings may not be available at the university. Under such conditions, the sensors built into the smartphone can be used. At the moment they are not perfectly accurate, and if small deviations in the space for street maps are insignificant for positioning, then for small sizes of corridors, passages and rooms this can be critical. Therefore, the search for solutions to realize the function of automatic positioning and tracking the movement of the device indoors is still ongoing.

Most often, in order to find a specific auditorium, a canteen or special premises in an educational institution, the student does not need to be guided along the route, the main thing is to know on which floor and in which part of the building to start searching. In this case, an information application is sufficient, which can search for the desired object according to the specified parameters inside the room and give information about its position. The parameters will be the location of the person and the place where you need to go.

The problem statement when creating an application for navigation is as follows:

1. The main problem in the development of such a software product is the choice of a visual and informative route display option. This can be a sound router, text route descriptions, 2D floor plans, and even 3D maps. Creating a floor plan with the positioning of objects on it - offices and auditoriums - will solve the previously described problem. But the creation of only floor projections will not be sufficient for the user to understand the features of the structure of the building, which has a non-standard layout of the premises. To solve this problem, you can recreate a simplified three-dimensional model of the building, inside which the route from one point to another will be displayed. Due to this, the user of this application will be able to more accurately project his path onto the real environment.

   Different types of models can be used to display the layout of a building.

   Wireframe - The model is built from vertices and edges between them, forming a mesh of the building. In this case, the entire building will be clearly visible, but many lines can create incorrect illusions of volume, that is, it will be difficult to understand which objects are closer to the beholder and which ones are farther from him. This model is only suitable for a small room.

   The second type is the polygon model. It consists of vertices, edges, and surfaces. Such a model is almost an exact repetition of an object in the real world. If you neglect the realistic display of the physical properties of the model, such as color, reflections, density or thickness of some elements of objects, you can assign a single-color material with some transparency to the polygonal model. In this case, the walls of the premises will be slightly translucent through, for example, like glass. Landmark objects are given a lower opacity and a different color to make them stand out from other model details. The room frame can also be highlighted with color so that the boundaries of rooms and corridors are visually better visible and create contrast.

   When creating a three-dimensional model, an important parameter is the number of polygonal surfaces, since the amount of memory occupied by the model depends on it. Their number can be reduced by removing unnecessary objects from the model, as well as simplifying the polygonal mesh. A solid wall can be built from one plane, rather than several hundred. Therefore, in order to create an optimal three-dimensional model of the educational institution building for the navigator, it is necessary to exclude objects that will not be used when laying the path. At the end of the simulation, it is necessary to optimize the polygonal mesh, which will not violate the integrity of the display.

It is also better to exclude polygons in the mesh structure whose vertices are not on the same plane; as well as extruded polygons. This is due to the fact that often the algorithm for rendering a model occurs by transforming its coordinate data into a raster image in a certain way. In order for the computer to correctly analyze sharp corners and vertices in space, it is necessary to take into account these specifics of specifying the polygonal mesh, as well as assign materials to the model objects that will allow you to easily see the route requested by the user.

2. Developing a model that can adequately convey information about the structure of the corridors, the location of the stairs, the dining room, the cloakroom, and other places is only one part of the router application. It is also important to implement convenient tools for viewing this model in order to view it from different angles. This is the advantage of 3D maps over 2D ones [16, 17].

   One possible solution to navigate in the space of the building model viewport is to use pushbutton controls. In this case, you should take into account the size of the device on which the application will be launched. These can be smartphones or tablet computers. Therefore, the size of the buttons should be small enough to accommodate all the necessary elements, and large enough for the user to interact with them.

3. The next task is related to the correct viewing of the three-dimensional model. It can be difficult to rotate the camera around the model or to rotate the building model itself. The user can accidentally turn it over in such a way that it becomes impossible to determine where and what is. Taking this into account, it is necessary to form restrictions for viewing, for example, to allow rotation only around one axis: the model stands on an illusory platform and can rotate in a clockwise and counterclockwise direction, like a vase on a potter's wheel.
4. Another additional function in the application, if it is not designed for one building, but for the structure of educational institutions, consisting of several buildings, will be the choice of this building itself. And as in the case of a map, it would be best not just to select a building by its address, but also to visualize its external structure. To do this, you can use photographs or create a realistic model that reflects the exterior of the building. Due to this, the student can select the desired building from the list in the application and compare it visually with what is in front of him [18].

Thus, the problems of developing two systems are considered, aimed at solving the problem of navigation and adaptation of students in the structure of the university. On the basis of the theoretical studies carried out, the software implementation of each system was carried out.

3 Results

3.1 Implementation of an Application for an Interactive 3d Excursion

For the development of an interactive 3d tour, the Source graphic engine was chosen, which meets the task of providing a comfortable delivery of the application to the end user. This is because the engine is old enough to provide maximum performance even on the slowest personal computer. Using the Source toolkit, you can create a fairly high-quality and photorealistic picture.

A working fragment of a 3d model of one of the buildings of a higher educational institution has been recreated with detailed detailing and realistic lighting. To fulfill the condition of content continuity, all key rooms were relocated exactly as they were located in the building of the educational institution (Fig. 4). The placement of emphasis on important places was also taken into account, which was one of the development tasks. Indoor accents include light from windows, backlighting with various lamps in dark rooms. An example of an emphasis is shown in Fig. 5 (using lighting) and Fig. 6 (using virtual pointers, signs and inscriptions).

Fig. 4. Room model in the editor.

The developed application allows you to take a tour of the building, identify key places, the location of departments and classrooms.

3.2 Development of a Router Application for Navigation in Educational Buildings

Unity software was used as a development environment for a router application for navigation inside the buildings of a higher educational institution, which allows you to develop cross-platform 2D and 3D applications. Its tools allow you to create an application interface suitable for both smartphones and computers. For the application to work, simplified models of the internal and external structures of the building were created [19, 20].

An example of a model of the internal structure of a building can be seen in Fig. 7. In it, doors and stairs were highlighted in dark blue as landmark objects. They are clearly visible against the background of other elements.

Fig. 5. Demonstration of emphasis on the model

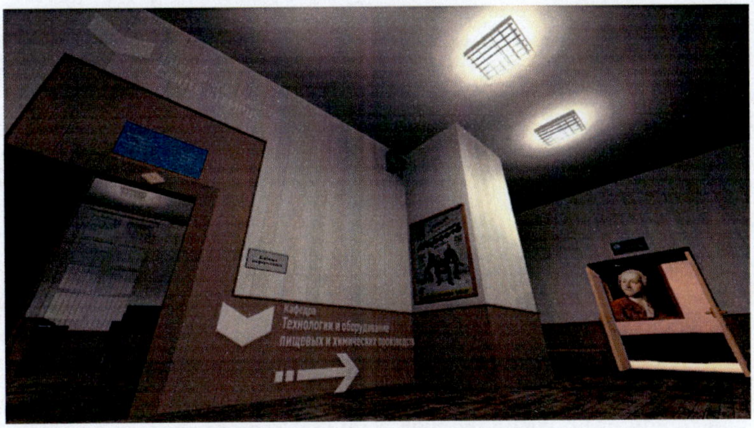

Fig. 6. Signs for important areas

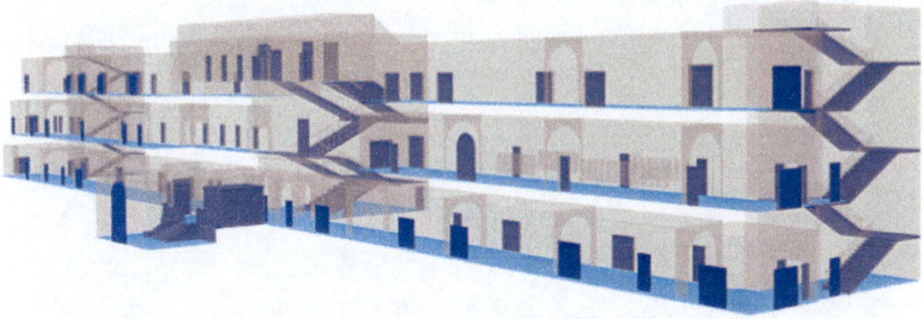

Fig. 7. Visualization of a simplified three-dimensional model of the internal structure of a building

An example of a router program interface that uses pushbutton control to navigate and set router parameters can be seen in Fig. 8. Some elements take up a lot of space. In this case, it is better to use gesture control.

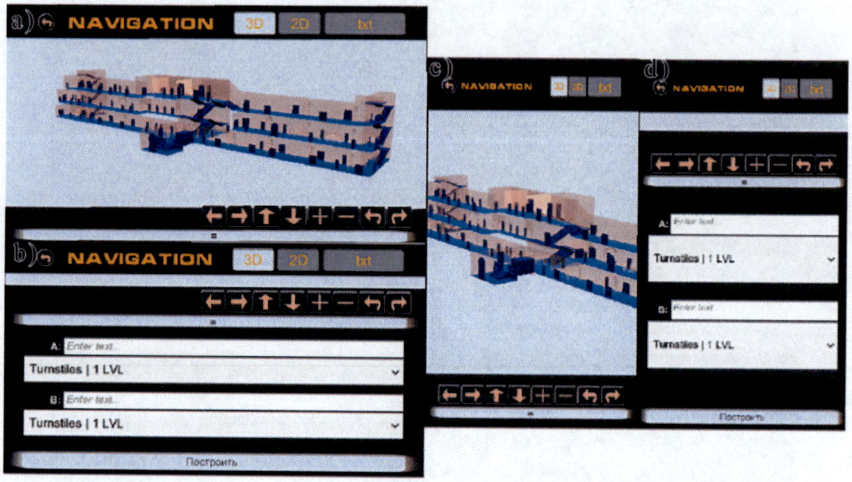

Fig. 8. An example of using button control in a router application: a-b) for landscape screen orientation; c-d) for horizontal screen orientation

Visualization of the main task, that is, the route from one point to another, must be performed inside the virtual rooms. For better visibility, the choice of material colors for the route should be done based on the color palette of materials on the building model - they should contrast with each other. If cold colors were used for the building (Fig. 8c), then it is better to use saturated colors of the warm spectrum to highlight the route. The starting and ending positions on the map should also be marked with objects such as spheres or cubes. You can highlight a door or a building element with color and light. In this case, you will most likely have to create additional copies of these elements, which will take up space in memory.

The created model of the appearance of the building is shown in Fig. 9. It was used to select the desired building for navigation.

Fig. 9. Visualization of a three-dimensional model of the exterior of the case.

As a result of testing, it was determined that some architectural elements are completely invisible on mobile platforms. In this regard, it was decided to exclude the model file itself from the project, and only visualization of the rotation of the outer body model was used to implement the selection of the hulls.

4 Discussion

During the survey, sufficient grounds were obtained for the need to find a solution to the problem of navigation in the structure of the university to improve the adaptation process of students.

The considered methods of solving the problem of adapting students to the complex structure of the premises of an unfamiliar university building, which exist and are used in our time, require the introduction of new technologies. The methods based on the use of software for a virtual excursion and for virtual positioning in the space of a building, presented in the work, are the result of the combination of 3D modeling technologies and game development.

In the course of testing the interactive excursion system, demonstrated to some students, certain results were obtained. The use of realistic visualization in virtual space can convey the necessary information to a person to a greater extent than a two-dimensional plan or photograph. A fuller sense of space is created. Also, realism is achieved through interactive interaction with three-dimensional elements. The closer the model is to the real world, the easier it is to remember and perceive.

The testing of the system by students with poor eyesight deserves special attention. After using the system, such students replied that if it existed when they entered the university, it would be easier for them to navigate in area. It was additionally noted that low vision makes it necessary to spend more time looking for the necessary premises, since in order for the student to determine which classroom is in front of him, he needs to come very close to each door, which increases the search time several times.

In addition, it was noted that the choice of the engine was rather successful, since the application performance on weak devices showed high rates. For example, the application was able to run on a laptop with an Intel N4100 mobile processor, at a frequency of 25–30 frames per second, which is a good indicator considering the power of this processor (2.4 GHz, 4 cores).

The router app system has also shown high relevance. Even employees and students who have been working and studying at the university for several years cannot know the location of absolutely all the classrooms in various buildings. Due to a fairly simple interface, the user can easily set the route and see its visualization, quickly find the desired place.

Viewing a three-dimensional map in the router application is made by push-button control and most of those who tried this method found it inconvenient. Users tried to mechanically move the building closer or further away by using gestures to move their fingers closer or away from each other at the moment they tap the screen. This control is often used in conventional navigators, therefore, it is necessary to change the button control existing in the second system to more familiar to most users, which will become a direction for further research.

Thus, both developed systems are relevant, successfully tested in real conditions and allow not only applicants to adapt to the new environment, but also people who are already familiar with the structure of the educational institution. Note that these systems need ongoing support and improvements as changes in room numbering or naming, or other building changes occur.

5 Conclusion

The paper considers the actual problem of navigation within the educational buildings of higher educational institutions. During the survey, it was determined that in the first days of training, students face big problems with navigation, which negatively affects the educational process. Therefore, for this problem, the adaptation of students to the structure of the university, it is necessary to implement a number of systems. Within the framework of this work, two systems are considered: a router application for navigating educational buildings and a virtual tour of the building. To a certain extent, this will help reduce the level of anxiety among students, which may be quite high anyway due to changes in activities, daily routine, new environment and other stress factors that appear upon admission.

The developed system of interactive excursions in virtual space allowed students to adapt to the unusual features of the building structure, to navigate in space in a real room, which will save time searching for the right place, reduce stress due to navigation in an unfamiliar environment.

A router application has been developed for indoor navigation. It made it possible to search for the desired audience or key point in educational buildings and build a route to it. The application is developed for mobile platforms, which allows you to quickly use it in the process of educational activities, and also does not require the presence or installation of additional devices.

The developed systems are aimed at solving the problem of students' navigation and adaptation to educational activities within the framework of higher educational institutions. This problem is complex and requires a comprehensive solution to ensure the high quality of the educational process. Further development of the systems is associated with the expansion of functionality, improvement of the interface, the addition of new digital services to help students in the process of integration into the university environment.

This work was supported by the laboratory of medical VR simulation systems for training, diagnostics and rehabilitation.

References

1. Ul'yanovskaya, S.I.: Morfogenez arhitekturnogo prostranstva obrazovatel'nyh uchrezhdenij. Arhitektura i sovremennye informacionnye tekhnologii. Morphogenesis of the Architectural Space of Educational Institutions **2**(47), 186–203 (2019)
2. Puhacheva, L.I.: Izuchenie psihologicheskih trudnostej pervokursnikov v usloviyah adaptacii k obucheniyu v vuze. Studying the psychological difficulties of freshmen in the context of adaptation to university studies. Special issue 1, ART 12507, 0.6 s. (2013)

3. Burov, E.B.: Adaptaciya studentov novogo nabora k obucheniyu v professional'noj obrazovatel'noj organizacii. Innovacionnoe razvitie professional'nogo obrazovaniya **2**(30), 119–124 (2021)
4. Zyryanov, S.B.: Psihologicheskie osobennosti studenta pervokursnika, faktory aktivizacii ego uspeshnoj ucheby i poznavatel'noj deyatel'nosti. Agrarnoe obrazovanie i nauka 2 (2013)
5. Vanyuhina, N.V., Alekseeva, E.S.: Programma formirovaniya otvetstvennosti v podrostkovom vozraste. ANI: pedagogika i psihologiya **5**(3, 16):240–244 (2016)
6. Pryadin, V.P., Smirnov, A.V., Smirnova, A.R.: Faktornyj analiz pokazatelej otvetstvennosti i zhiznestojkosti v razlichnyh polovozrastnyh gruppah. Teoretiko-metodologicheskie osnovy i sovremennye praktiki psihologicheskoj nauki **2**, 106–121 (2020)
7. Panina, E.V.: Issledovanie polovyh osobennostej samoregulyacii u podrostkov. Vestnik CHerepoveckogo gosudarstvennogo universiteta **1**(4), 132–136 (2011)
8. Smirnov, A.A., ZHivaev, N.N.: Adaptaciya studentov i obraz vuza: monografiya. YarSU 168 (2010)
9. Virtual College Tours, Virtual reality. https://www.youvisit.com/collegesearch/. Accessed 30 Nov 2021
10. Virtual'nye ekskursii po Universitetu – RTU MIREA. https://www.mirea.ru/mediapage/a-virtual-tour-of-the-university/. Accessed 4 Dec 2021
11. Virtual'nyj tur po universitetu. http://www.fa.ru/pages/vtour.aspx/. Accessed 4 Dec 2021
12. Virtual'nyj tur SPbGU. https://spbu.ru/virtual_tour/. Accessed 10 Dec 2021
13. 3D-ekskursiya po Universitetu | Universitet Profsoyuzov. Sankt-Peterburg - SPbGUP. https://www.gup.ru/uni/about/3dgallery/. Accessed 1 Dec 2021
14. Vaseeva, E.S., Smirnov, M.S.: Virtual'naya ekskursiya po obrazovatel'nomu uchrezhdeniyu: osobennosti proektirovaniya. Elektronnyj nauchnyj zhurnal «Nauka i perspektivy» no. 1 (2018)
15. Ustyuzhina, N.V.: Virtual'naya ekskursiya kak innovacionnaya forma obucheniya. Elektronnyj nauchnyj zhurnal «Nauka i perspektivy» no. 2 (2017)
16. Serova, M.: uchebnik-samouchitel' po trekhmernoj grafike v Blender 3d. Modelirovanie, dizajn, animaciya, speceffekty 272 (2021)
17. Zaharova, A.A.: SHklyar Postroenie mnogokomponentnyh vizual'nyh 3dmodelej s ispol'zovaniem raznorodnyh istochnikov informacii, na primere sozdaniya geologicheskih modelej. Bulletin of the Tomsk Polytechnic University **320**(5), 73–80 (2012)
18. Inzhenernaya grafika i trekhmernoe modelirovanie. Sbornik nauchnyh dokladov. Novosibirsk SGUGiT (2018)
19. Montilla, Y.: Maryury, Le´on-S´anchez, Camilo: 3D Modelling of a building oriented to indoor navigation system for users witch different mobility conditions. ISPRS Ann. Photogramm. Remote Sens. Spatial Inf. Sci. **VI-4**(2), 103–109 (2020)
20. Liu, L., Zlatanova, S.: Towards a 3D network model for indoor navigation, pp. 79–92. Urban and Regional Data Management, UDMS Annual (2011)

Investigation of the Mechanism of Errors Formation in Geometric Shape During Face Grinding

Hasan Huseynov[1], Tunzala Imanova[1], Durdana Rustamova[1], and Almaz Mehdiyeva[2](✉)

[1] Western Caspian University, Baku, Azerbaijan
durdana.rustamova@wcu.edu.az
[2] Azerbaijan State Oil and Industry University, Baku, Azerbaijan
almazmehdiyeva@yahoo.com

Abstract. The article discusses issues related to the theoretical description of the mechanism of the formation of errors in geometric shape during face grinding of high-precision slabs such as the gate valve of the once-through valve. It was found that the main reason for the formation of shape errors is the uneven abrasive effect on the treated surface due to a change in its configuration. The ways of increasing the accuracy of the flat grinding operation with the wheel end face are proposed.

Keywords: Surface grinding · Precision · Geometric shape · Gate · Abrasive action · Contact area · During face grinding · Micro- and macro deviations

1 Introduction

One of the most common methods for finishing flat surfaces of parts is flat grinding with the wheel end. For face grinding, high surface quality is achieved with high productivity and low cost. The mechanism for the formation of micro- and macro deviations of the profile during flat grinding with the end of the wheel has its own specific features. A wide range of stochastic phenomena occurring in the process of face grinding, in addition to the high efficiency of this process, also causes numerous disadvantages, the main ones of which are: high heat generation, unfavorable conditions for chips and heat removal, uneven abrasive effect on the work surface, etc. All these shortcomings, in aggregate, negatively affect the formation of geometric parameters of surfaces and physical and mechanical properties of the surface layer.

2 Statement of the Problem

Flat grinding with the wheel end is distinguished by a relatively larger contact area and, consequently, a greater length of technological input and output primitives. Violation of the stationarity of the grinded surface occurs along the length of processing within the

R. Silhavy (Ed.): CSOC 2022, LNNS 501, pp. 200–205, 2022.
https://doi.org/10.1007/978-3-031-09070-7_17

technological primitives of the input and output and changes in the configuration of the surface which should be grinded.

At face grinding, conditional strips with uneven abrasive effect on the treated surface have the form of concentric annular areas with a width equal to the lengthwise feed per one-wheel revolution [1–3].

3 Theoretical Part

Based on these considerations, we will determine the regularities of the distribution of the number of actually working grains over the working surface of the end of grinding wheel.

The area of the j-th conditional strip is determined by the formula

$$F_j = \prod S_{lw} D_{av_j}, \tag{1}$$

where S_{lw} is the lengthwise feed per revolution of the circle, mm/rev, D_{av_j} - the average diameter of the j-th conditional strip.

The average diameter of the current conditional strip we will express through the explicit parameters of the grinding process [4]

$$D_{av_j} = D_{cr} - (2j - 1)S_{lw}. \tag{2}$$

Then we get

$$F_j = \frac{\pi^2 V_u D_{cr}}{1000 V_k}\left[D_{cr} - (2j - 1)\frac{\pi^2 V_u D_{cr}}{1000 V_k}\right]. \tag{3}$$

The number of actually working grains on the working surface of the circle, taking into account the percentage of hitting already cut areas during repeated encounters, can be represented in the following form

$$i_\Phi = i\ F_1 + i\ F_2 q + i\ F_3 q^2 + \ldots + i\ F_j q^{j-1} \quad 0 < j \le b/S_{lw}, \tag{4}$$

where Fj is the area of the j-th conditional ring of the working end of the circle.

Let us input the value of F_j from formula (3) and D_{av} from formula (2) into formula (4)

$$i_a = i\pi S_{lw}\left[\left(D_{cr} + D_{cr}q + D_{cr}q^2 + \cdots + D_{cr}q^{j-1}\right) - \left(S_{lw} + 3S_{lw}q + 5S_{lw}q^2 + \cdots + (2j-1)S_{lw}q^{j-1}\right)\right] \quad 0 < j \le b/S_{lw} \tag{5}$$

Replacing the first bracket of formula (5) with the sum of the elements of a geometric progression, we obtain the formula for the number of actually working grains at the working end of the circle

$$i_a = i\pi S_{lw}\left[D_{cr}\frac{(1-q^j)}{1-q} - \sum_1^j (2j-1)S_{lw}q^{j-1}\right] \quad 0 < j \le b/S_{lw}. \tag{6}$$

The number of actually working grains at the working end of the grinding wheel, taking into account the formula for the number of active grains per 1 mm 2 of the wheel surface [6], can be presented in the following form

$$i_a = \frac{0{,}167\beta\sqrt{k}}{\alpha^{3/4}\sqrt{tg\gamma \overline{X}_j^2}\sqrt{1-\varepsilon}}\sqrt{\frac{\omega}{1000V_k}}\pi S_{lw}\left[\frac{D_{cr}(1-q^j)}{1-q} - \sum_1^j (2j-1)S_{cr}q^{j-1}\right] \tag{7}$$

$$0 < j \le b/S_{lw}.$$

The number of actually working abrasive grains within the contact area is determined by multiplying the number of actually working grains at the working end of the grinding wheel by the ratio of the arc length of the contact of the grains located along the average diameter, along the circumference, also along the average diameter of the grinding wheel end, i.e. [5, 6].

$$i_k = i_a \frac{L_{k_{av}}}{\pi D_{av}}. \tag{8}$$

For a technological input primitive

$$i_{k_{ex}} = \frac{0{,}167\beta\sqrt{k}}{\alpha^{3/4}\sqrt{tg\gamma \overline{X}_j^2}\sqrt{1-\varepsilon}}\sqrt{\frac{\omega}{1000V_k}}S_{lw}\frac{\alpha\sqrt{1+\frac{v_d^2}{(60V_{k_{av}})^2}}}{2}\left[\frac{D_{cr}(1-q^j)}{1-q} - \sum_1^j (2j-1)S_{lw}q^{j-1}\right]$$

$$0 < j \le b/S_{lw}. \tag{9}$$

For a technological prixmitive steady-state grinding

$$i_{k_{ycm}} = \frac{0{,}167\beta\sqrt{k}}{\alpha^{3/4}\sqrt{tg\gamma \overline{X}_j^2}\sqrt{1-\varepsilon}}\sqrt{\frac{\omega}{1000V_k}}S_{lw}\frac{\alpha\sqrt{1+\frac{v_d^2}{(60V_{k_{av}})^2}}}{2}\left[\frac{D_{av}(1-q^j)}{1-q} - \sum_{j=i}^{\frac{b}{S_{lw}}} (2j-1)S_{lw}q^{j-i}\right]$$

$$0 < j \le b/S_{lw} \tag{10}$$

For an output technological primitive

$$i_{k_{bblx}} = \frac{0{,}167\beta\sqrt{k}}{\alpha^{3/4}\sqrt{tg\gamma \overline{X}_j^2}\sqrt{1-\varepsilon}}\sqrt{\frac{\omega}{1000V_k}}S_{lw}\frac{\alpha\sqrt{1+\frac{v_d^2}{(60V_{k_{av}})^2}}}{2}\left[\frac{D_{cr}q^j(1-q^{\frac{b}{S}-j})}{1-q} - \sum_{j=i}^{\frac{b}{S_{lw}}} (2j-1)S_{lw}q^{j-i}\right]$$

$$0 < j \le b/S_{lw} \tag{11}$$

The analysis of the obtained formula shows that during flat grinding with the wheel end, the amount of actually working abrasive grains in the contact zone depends both on the dimensions and characteristics of the grinding wheel and on the elements of the cutting mode. The most significant effect on the number of active grains in the contact zone is exerted by: the width of the end of the cup grinding wheel, the ratio of the table speed to the wheel speed, i.e. the value of the so-called lengthwise feed per wheel revolution, the ratio of the diameter of the wheel to the width of the work piece, etc.

Having an analytical expression for the regularity of the change in the number of actually working abrasive grains within the technological primitives of the input, steady grinding and output, it is possible by changing the speed of the lengthwise feed, from the conditions for maintaining the stability of the number of active abrasive grains, to provide a uniform abrasive effect on the processed surface.

In the process of grinding, the cutting force causes the wrung out elements of the closed technological system machine-accessory-tool-part. Wrung out technological systems consist of elastic deformations of individual compliant elements-parts in the dynamic dimensional chain of the system, from gaps in joints and from contact deformations of mating surfaces. With repeated awards of the system, the latter are reduced to such a value that can be neglected. These displacements are considered mainly in the direction normal to the machined surface and are determined by the formula [6]

$$Y = \frac{P_y}{j} \tag{12}$$

where Py - is the normal component of the cutting force, N; j - rigidity of the technological system, N/mm.

It is expedient to consider the cutting force during grinding as the sum of the micro-cutting forces of individual single grains on the contact area. Let us assume that the magnitude of the micro-cutting force within the contact area of the grinding wheel with the machined surface is stable and equal to the arithmetic mean value ρ_{av} [6]

$$P_y = \rho_{av} i_k \tag{13}$$

where ρ_{av} is the average value of the micro-cutting force of a single grain, N; i_k is the number of actually working grains in the contact area of the grinding wheel with the machined surface.

Thus, the cutting force with the corresponding primitives can be expressed: for a technological input primitive

$$P_{yin} = \frac{\rho 0,167\beta\sqrt{k}}{\alpha^{3/4}\sqrt{tg\gamma X_j^2}\sqrt{1-\varepsilon}}\sqrt{\frac{\omega}{1000V_k}S_{lw}\frac{\alpha\sqrt{1+\frac{v_d^2}{(60V_{kav})^2}}}{2}}\left[\frac{D_{cr}(1-q^j)}{1-q} - \sum_{1}^{j}(2j-1)S_{lw}q^{j-1}\right]$$

$$0 < j \le b/S_{lw} \tag{14}$$

$0 < j \le b/S_{lw}$For technological primitive steady-state grinding.

$$P_{sted} = \frac{\rho 0,167\beta\sqrt{k}}{\alpha^{3/4}\sqrt{tg\gamma X_j^2}\sqrt{1-\varepsilon}}\sqrt{\frac{\omega}{1000V_k}S_{lw}\frac{\alpha\sqrt{1+\frac{v_d^2}{(60V_{kav})^2}}}{2}}\left[\frac{D_{cr}(1-q^j)}{1-q} - \sum_{j=i}^{\frac{b}{S}}(2j-1)S_{lw}q^{j-i}\right]$$

$$0 < j \le b/S_{lw}, \tag{15}$$

for an output technological primitive

$$P_{yout} = \frac{\rho 0,167\beta\sqrt{k}}{\alpha^{3/4}\sqrt{tg\gamma \overline{X}_j^2}\sqrt{1-\varepsilon}}\sqrt{\frac{\omega}{1000V_k}}S_{lw}\frac{\alpha\sqrt{1+\frac{V_d^2}{(60V_{kav})^2}}}{2}\left[\frac{D_{cr}q^j(1-q^{\frac{b}{S_{lw}}-j})}{1-q} - \sum_{j=i}^{\frac{b}{S_{lw}}}(2j-1)S_{lw}q^{j-i}\right]$$

$$0 < j \leq b/S_{lw} \tag{16}$$

A change in the cutting force along the length of the processing causes a change in the elastic wrung out of the technological system. The unevenness of the elastic wrung out causes the formation of errors in the geometric shape, which can be expressed by the following formula

$$\Delta_\Phi = \frac{P_{y\max} - P_{y\min}}{J_{av}} = \frac{\Delta P_y}{J_{av}} \tag{17}$$

4 Methodology

The research results [4, 5] show that when grinding parts such as the gate of a direct-flow valve, with a characteristic center hole in the middle of the sealing surface, the greatest value of the geometric shape errors is obtained in the axial section of the hole. Geometric errors in the whole area are elliptic form (Fig. 1).

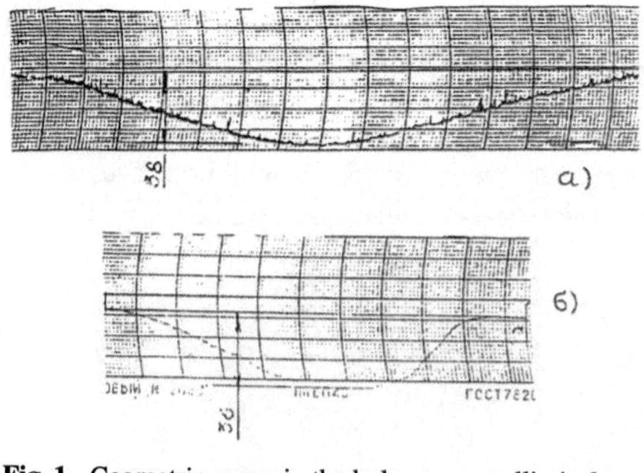

Fig. 1. Geometric errors in the hole area are elliptic form

5 Conclusion

The analysis of the obtained dependencies (14, 15, 16 and 17) for the characteristic areas of surfaces polished by the end of the wheel shows that the errors of the geometric shape are mainly obtained as a result of a change in the contact area of the grinding wheel with the processed surface, and from the uneven abrasive effect of the working end of the wheel on the length of contact. In this case, the most significant values of shape errors are obtained along the length of processing in the hole zone. Thus, knowing the magnitude and nature of the errors of the geometric shape in certain areas of the surface of the gate of a direct-flow gate valve polished by the end of the circle, one can judge its sealing properties as a whole. Having such dependencies, it is possible to purposefully control the grinding process to increase the accuracy of the geometric shape and thereby of the sealing properties of the processed surfaces.

The analysis of results of the conducted research shows that during flat grinding with the wheel end, the amount of actually working abrasive grains in the contact zone depends both on the size and characteristics of the grinding wheel and on the elements of the cutting mode. The most significant effect on the number of active grains in the contact zone is affected by: the width of the end of the cup grinding wheel, the ratio of the table speed to the wheel speed, that is, the value of the so-called lengthwise feed per wheel revolution, the ratio of the diameter of the wheel to the width of the work piece. Having an analytical expression for the regularity of the change in the number of actually working abrasive grains within the technological primitives of the input, steady grinding, output and hole, it is possible by changing the speed of the lengthwise feed, from the conditions for providing of the stability of the number of active abrasive grains, to provide a uniform abrasive effect on the processed surface. During the grinding of the parts such as a direct-flow gate valve, with a characteristic center hole in the middle of the sealing surface, the greatest value of geometric errors is obtained in the axial section of the hole. Geometric errors in the hole area are elliptical and reach up to 36–40 microns. Having an analytical expression for the regularity of the change in the value of the normal component of the cutting force depending on the number of active abrasive grains in the contact zone within the technological primitives of the input, steady grinding, output and hole, it can by changing the lengthwise feed rate, from the conditions for maintaining the stability of the number of active abrasive grains, to ensure a uniform abrasive effect on the work surface and thus the stability of the cutting force.

References

1. Hassan, H., Sahib, B., Krehel, R., Koc, M.: Increase of accuracy of geometric forms of surfaces ground by a grinding wheel of various grain sizes. Adv. Mech. Eng. **9**(8), 1–12 (2017)
2. Kai, A.J., Hansen, J.S., Martins, J.A.: Structural topology optimization for multiple load cases using a dynamic aggregation technique. Eng. Optim. **41**, 1103–1118 (2009). https://doi.org/10.1080/03052150902926827
3. Reznikov, A.N.: Abrasive and uncut diamond processing of materials. Machinery Construction **391**, 4 (1977). Loading automation of round grinding machine with N.P.C.p.42
4. Yakimov, A.V.: Optimization of the grinding process. M.: Mechanical Engineering (1975).173s
5. Husenov, H.A., Bagirov, S.A., Krehel, R., Kociško, M.: The increased production efficiency and optimization terms of stationarity by flat grinding with abrasive circle surface. Adv. Mater. Sci. Technol. (2014)

Unification of Contact Modeling Software for Analysis of an Infectious Diseases Spread

Esther Luna Colombini[1], Hanlie Smuts[2], and Vladimir Sudakov[3](✉)

[1] Institute of Computing (IC), University of Campinas, Av. Albert Einstein, 1251, Cidade Universitária, Campinas, SP 13083-852, Brazil

[2] University of Pretoria, Corner of Lynnwood Road and Roper Street, Hatfield 0083, Pretoria, South Africa

[3] Keldysh Institute of Applied Mathematics of Russian Academy of Sciences, Miusskaya sq., 4, 125047 Moscow, Russia

sudakov@ws-dss.com

Abstract. In the conditions of the COVID-19 pandemic the role of software allowing to predict the spread of the disease has significantly increased. Numerous simulation models have already been created and are still being developed. These models not only predict the number of newly infected people but also make it possible to estimate the consequences of various controlling factors: the introduction of restrictive measures from complete isolation and reduction of working days, the use of personal protective equipment, changes in the working hours and accessibility of large population facilities, stores, restaurants, optimization of transport. This paper develops a unified representation of basic metaclasses of objects necessary for modeling contacts of large populations and develops a software implementation of a unified simulation modeling system that uses these classes. As a result, a user of the system, by defining the characteristics of classes in a declarative style, can run different scenarios of situation development without using imperative programming. The system allows you to flexibly adjust the economic characteristics of the consequences of certain decisions. Using a multithreaded mode of event processing has accelerated calculations by 3 times. Connection of modules of multi-criteria analysis and meta-heuristic optimization for automation of rational management decisions search based on modeling results is planned in prospect. The result of this work may be useful as a self-sufficient software simulation tool, placed in open access on the portal ws-dss.com, and as a methodological basis for the reduction of existing models to a unified informational structure of their interaction.

Keywords: Software · Unification · Metaclass · Simulation · Infectious

1 Introduction

The relevance of creating models of the spread of infections in modern conditions is beyond doubt. There are two main classes of models: macromodels and micromodels. In macromodels, a person is usually not represented as an independent object, but only

R. Silhavy (Ed.): CSOC 2022, LNNS 501, pp. 206–214, 2022.
https://doi.org/10.1007/978-3-031-09070-7_18

the number of people in various states is calculated. The most typical macromodels are compartmental models [1]. Such models assume description of pandemics through a system of differential equations, which describe well the situation in typical conditions.

A significant advantage of models based on differential equations is the possibility of their analytical investigation. Nevertheless, all such models are characterized by another assumption - the characteristics and behavior of all individuals assigned to the same subgroup are assumed to be the same. In addition, it is problematic to assess the consequences of certain managerial decisions, for example, to change the operating mode of restaurants, using compartmental models.

Compartmental models contain a number of coefficients, if we knew how they would change, under our soft control actions, macromodels would be an excellent tool. But usually we can't know that.

Micromodels are primarily discrete simulation models that can be created in imperative or multi-agent programming paradigms, and easily allow us to account for our managerial decisions at the level of event processing rules and agent behavior.

Specialized languages and simulation tools were mainly created to solve business problems: production problems, logistic and transportation problems. A good but rather old toolkit is the GPSS system [2]. It allows you to describe many business-processes declaratively in terms of transactions, which are processed by certain blocks. There are similar solutions in modern systems, for example, simulation in BPMN notation [3]. This work aims to develop similar mostly declarative tools for another subject area - modeling of contacts of large masses of people and spread of infection among them.

2 Literature Review

The first contact simulation models were cellular automata [4], which are a set of cells, each of which takes a state from a finite set. Usually square cells combined into a rectangular lattice are used. Then the nodes of the lattice model individuals, each of which has a fixed position in space. Time for a cellular automaton moves discretely, at each step the infection can be transmitted from a sick individual to its susceptible neighbor with a certain probability. A cellular automaton for a large number of individuals would be quite inefficient in both speed and memory on conventional processors. The performance situation can be remedied by implementing cellular automata on GPUs. The main limitation of the cellular automaton is the modeling of the space around individuals as a lattice. In fact, for most diseases, it can be assumed that infection spreads through contact between individuals.

Contacts between individuals are not limited to immediate neighbors but are irregular in time and space. Network models [5], which present a graph scheme of possible contacts, allow us to consider them. The vertices of the graph here are objects with a set of individual properties, describing the state of separate individuals in any detail. Such models are also quite resource-intensive but are effectively implemented on GPUs [6]. Network models offer a general technology for modeling the environment but do not specify a way to organize the graph of contacts for the population of a particular territory.

The most detailed models today are agent-based models [7]. They do not use a centralized calculation of transmission probabilities but describe the behavior of each individual agent. Most agent-based models are discrete-event, that is, their modeling process is reduced to the sequential processing of a list of planned events. Each event entails some set of atomic actions that change the state of the system. Agents independently generate events related to their activities, which avoids a fixed discretization step and complex formulas of infection probability. Agent-based models are as realistic as possible that require substantial computational power. Their calibration and analysis, necessary for using the model as a forecasting tool, are based on conducting numerous numerical experiments. Agent-based models allow generating forecasts for any period of time, but their research and practical application are much more complicated than other techniques. Agent-based models are developed and tuned to solve specific problems, the peculiarities of which are reflected in the model itself and the values of its parameters.

The most natural way is to form connections between individuals and possible pathways of infection on the basis of their social affiliation. This method is used in population-based models [8].

The environment of population-based models is divided into locations - places where individuals can come into contact, e.g., schools, offices, living quarters, etc. Depending on an individual's age and social group, he or she is associated with these or those locations - daily "visits" them. Time in population models moves discretely, usually in increments of 12 or 24 h. At each step, the probability of infection of each individual is calculated by formulas that take into account the number of infected in the locations visited by the individual. Population models allow representing the structure of society of any complexity, but require careful adjustment of parameters based on empirical data.

AnyLogic is a quite powerful modeling tool, including for pandemics [9]. However, it does not contain ready-made special classes for creation of locations and requires significant resources for development and debugging of even a small city model. It is preferable to use specialized software tools to implement contact models. One specialized system: COVID-19 Outbreak Simulator is a population-based simulator that simulates the spread of SARS-CoV-2 virus in a dynamic and heterogeneous population, subject to preventative measures and post-outbreak action plans such as quarantine, testing, and removal of symptomatic cases [10]. However, like the other programs reviewed, they do not claim to be universal. The maximum that can be done is to vary the parameters. And to change the structure of considered object classes one has to modify the program code. Therefore, the actual task is to develop a unified software product, in which the user, by defining the characteristics of classes in declarative style, can run different scenarios of people's behavior without using imperative programming.

3 Method

The method for solving the problem of modeling contacts is to build variable models by automatically generating the necessary number of ready blocks from the description of different types of spatial, epidemic, trans-port interactions at the level of metaclass attributes. Let's describe some basic metaclasses.

All locations of people are tied to locations of a certain type. The metaclass `location_types` is used to describe location types. Its instances are specific location classes: Cinemas, Shops, Flats, Offices. In turn, when generating a specific model, specific objects of given classes are created: Movie theater 1, Movie theater 2,…., Movie theater N. The developer of the model is free to choose the degree of detail required. He can detail the enterprise to specific rooms, or consider the entire enterprise as a single location. The degree of such detail must be taken into account when calculating the probability of contamination. It is important that the unified software does not need to be modified when specific classes of locations are changed. Unified software contains only the definition of metaclasses, and specific classes are created from settings. Creating these settings does not require programming skills, but a deep understanding of the structure of modern urban agglomerations and modeling goals. Therefore, the description of classes is already the task of the subject area analyst. So if the goal of modeling is to find solutions to minimize the number of contacts in the process of moving around the enterprise, it is necessary to detail specific rooms. But if the problem of pandemic is solved at the level of a municipality, it may be excessive. Thus, the system is scalable to model objects of different sizes: specific buildings, cities, regions, countries, continents, or the entire planet.

The creation of specific locations from given types can be organized in two ways:

- by automatic random generation on a specified coordinate system with a specified distribution law;
- by importing real data from a geographic information system with linkage to location types.

The next important metaclass is `people_types`. Division of people into types allows to set estimations of probabilistic characteristics for behavior of people of the given type: how often they visit locations of certain types, how much time they spend there, on what days and hours they usually go there. The word "usually" refers to the definition of the parameters and the type of the distribution law of the corresponding characteristic when performing the simulation. These distribution laws are defined in another metaclass, `link_loc_people_types`.

The person itself is an instance of a class of a certain type of person. Where people types are spawned by the `people_types` metaclass. A person has the following attributes: gender, age, state of the person (according to the SEIR model), current location, there is no location detailing in the current implementation, although it is possible in the future. Human attributes can affect the distribution functions of human behavior characteristics described in classes belonging to `link_loc_people_types` metaclass, making them multidimensional.

To reflect the financial component of interactions, `link_loc_people_types` specifies the possible cash flows resulting from being in the corresponding location. For this purpose, the system has metaclasses of income and expenditure items that lead to expenditure and income accounts, which can be maintained by types of people and specific people. The movement on the account is possible both in plus and minus. The frequency of accruals is set in the `income.rate` attribute of the link_loc_people_types instance. Possible types of charges: per fact - for moving to a location, per day - once a day if the location was visited, per minute - per minute for being in a location, per hour - for every hour of being in a location.

From the unification point of view, vehicles are a special type of location, which does not have a permanent geographical position, but is able to move and all people move with it, if they stay in the location. Transport_types is a metaclass inherited from the metaclass (location_types). It can decrease people's income items and increase municipality's income items, if specified. Public transportation is bound to certain stops. The trans_station_link metaclass is used for this. The same location_types metaclass as for other locations is used to create stops. Stops can be separated into separate location types by vehicle type, or they can be unsegregated and use the same class. The defined sequence of stops defines the transport's march-route. The generation of routes, as well as the generation of locations can be organized in two ways: by pseudorandom number sensors or loaded from a real transport network.

The simulation is performed with a fixed time step, since it is expected to run the model on a substantial number of people, and a variable time step will not have a significant effect.

The framework software was implemented in C# language. It can function on all modern operating systems Ubuntu 20.04 and analogs, Windows 7/8/10, MacOS 10.15 and higher.

The program takes as input parameters a json-file in which all required class metaclasses should be described.

The work of the program is a repetitive cycle in which people move between locations and transmit infection with some probability if they are in the same location.

Before running the simulation, a table of routes is created, which contains information about the shortest path between any two locations. When creating the route table, the priority will be given to moving between stations (i.e. with the help of transport). The program supports work in several streams, which significantly reduces the calculation time.

When planning locations, taking into account the random laws of distribution of visiting time, conflicts may arise when a person has to leave a location where his time has not expired. Such situations can be created deliberately, for example, if a person needs to visit the canteen at lunchtime. In standard agent modeling such situations are solved in advance by describing the algorithm of human activities and a person moves between locations, following a given algorithm. In the proposed system, the algorithm is not implemented explicitly, and transitions are not always possible. As it happens in real life, we do not always manage to get to lunch. To implement this behavior, the mechanism of the stack is provided. It consists of locations that are supposed to be visited at the current moment, but in fact a person is only in one location at the top of the stack. Calculations include the following steps:

1. Processing people.

    a. Removing locations from the stack that a person must visit in a day, if the simulation time is longer than the defined time limit of being in a particular location.

b. If a transition to a new day has been made, a new list of locations the person must visit for the day is generated.
c. If the simulation time falls within the time limits of the planned location visit, then the location is added to the stack.
d. If the person is not currently at a location that is at the top of the stack, a move to that location is initiated.
e. If the person is infected and it's time to move to the next stage of the disease, this move is made.

2. Processing locations.

   a. If a new person appeared in the location, the attempt to transmit the virus between the people who arrived and those who were already in the location is performed.

3. Saving the results

   a. Recording the changes of the states.
   b. Recording the criteria for evaluating the simulation results.
   c. Recording the average number of people in the locations over a certain period of time.

Movement in space can be considered as movement between locations, and the movement between any locations is carried out either directly, or through stations.

To organize work in parallel, many people and locations are divided into groups according to the number of threads specified in the settings, the calculations with the groups are made in parallel.

4 Results

Figure 1 shows a fragment of the initial data to form the model. Access to the model is implemented through decision support web services, which are publicly available at [11]. These web services, through a RESTful API, allow you to generate a task to simulate a single model run. A set of runs is specified as multiple tasks. Tasks are executed asynchronously as separate operating system processes.

Figure 2 shows the output of some model runs. In those cycles of time, when the state of the model changes, information on the movement of people, transport, and changes in the state of infection are output.

The results are saved in the form of a table and output in the form of graphs. Examples of graphs are shown in Fig. 3.

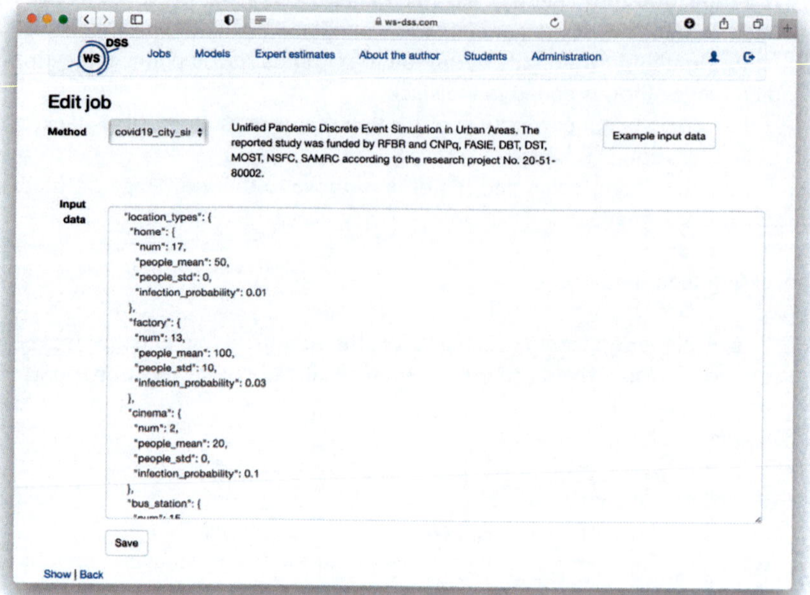

Fig. 1. Input data for model generation

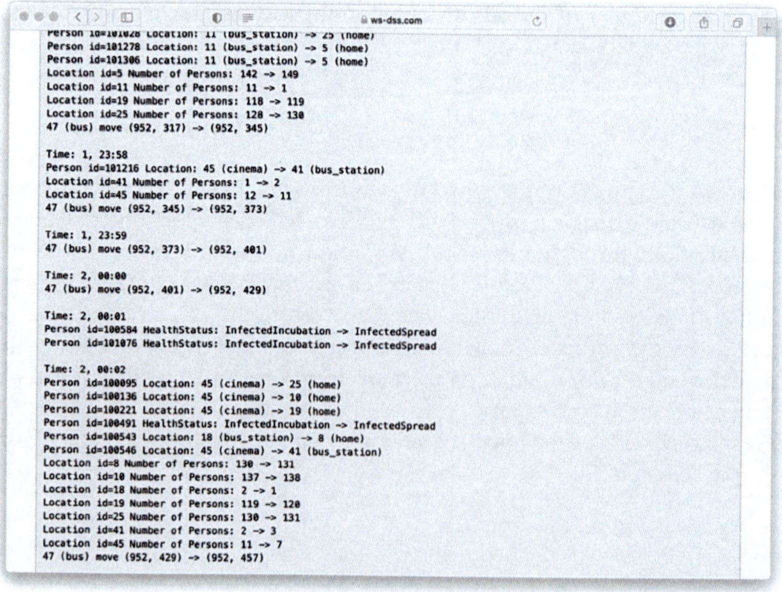

Fig. 2. Output of model runs

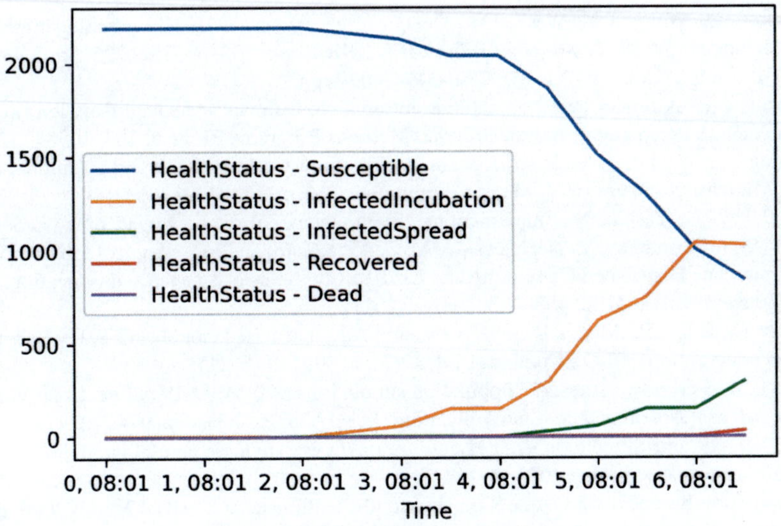

Fig. 3. Graphs with simulation results

5 Conclusion

Unification of modeling software allows to significantly increase the speed of model creation. The gain was evident in the modernization of the existing model. If development and testing of new classes for types of locations and types of vehicles in the source code of the specialized model required 2–3 days of work of the programmer and the analyst, then the declaration and adjustment of new classes in the unified model was carried out by one analyst in one hour.

The unified model allows us to create complex models of territorial formations of arbitrary scale based both on random number sensors and real data from geoinformation systems.

The use of parallel calculations in simulations, the use of web-services for running models and obtaining results allows organizing a full cycle of analysis of modeling results, modernization of model parameters and structures with integration of external systems, for operative support of decision-making on pandemic control.

Acknowledgements. The reported study was funded by RFBR and CNPq, FASIE, DBT, DST, MOST, NSFC, SAMRC according to the research project No. 20–51-80002.

References

1. Angulo, W., Ramírez, J., De Cecchis, D., et al.: A modified SEIR model to predict the behavior of the early stage in coronavirus and coronavirus-like outbreaks. Sci. Rep. **11**, 16331 (2021)
2. Ståhl, I., Born, R., Henriksen, J., Herper, H.: GPSS 50 years old, but still young. In: Proceedings of the 2011 Winter Simulation Conference (WSC), pp. 3947–3957. Phoenix, AZ, USA (2011)

3. Pufahl, L., Wong, T.Y., Weske, M.: Design of an extensible BPMN process simulator. In: Teniente, E., Weidlich, M. (eds.) BPM 2017. LNBIP, vol. 308, pp. 782–795. Springer, Cham (2018). https://doi.org/10.1007/978-3-319-74030-0_62
4. Schimit, P.: A model based on cellular automata to estimate the social isolation impact on COVID-19 spreading in Brazil. Comput. Methods Programs Biomed. **200**, 105832 (2021)
5. Tiraga, F., Kavi, C., Neudauer, N.: Comparing small world network and traditional models of Marburg virus disease. J. Math. Comput. Sci. **11**(6), 7773–7792 (2021)
6. Sudakov, V., Yashin, N.: Implementing the graph model of the spread of a pandemic on GPUS. In: Korenkov, V., Nechaevskiy, A., Zaikina, T. (eds.) Proceedings of 9th International Conference Distributed Computing and Grid Technologies in Science and Education, CEUR Workshop, vol. 3041, p. 488–493. Dubna (2021)
7. Kerr, C., Stuart, R., Mistry, D., et al.: Covasim: An agent-based model of COVID-19 dynamics and interventions. PLOS Comput. Biol. **17**(7), e1009149 (2021)
8. Peng, B., Pettit, R., Amos, C.: Population simulations of COVID-19 outbreaks provide tools for risk assessment and continuity planning. JAMIA Open **4**(3), ooaa074 (2021)
9. COVID-19 and Simulation – AnyLogic. https://www.anylogic.com/blog/covid-19-and-simulation. Accessed 08 Jan 2022
10. Population-based Forward-time Simulator for the Outbreak of COVID-19. https://github.com/ictr/covid19-outbreak-simulator. Accessed 08 Jan 2022
11. Web Services for Decision Support Systems. https://ws-dss.com. Accessed 08 Jan 2022

IoT Device Data Acquisition and Experimental Integration in Production Line Based on OPC UA Protocol

Peter Střelec, Tibor Horák(✉), Szabolcs Kováč, Pavol Tanuška, and Eduard Nemlaha

Institute of Applied Informatics, Automation and Mechatronics, Faculty of Materials Science and Technology in Trnava, Slovak University of Technology in Bratislava, Trnava, Slovakia
{peter.strelec,tibor.horak,szabolcs.kovac,pavol.tanuska, eduard.nemlaha}@stuba.sk

Abstract. Internet of Things (IoT) devices are increasingly being integrated into modern industrial production lines. The deployment of IoT devices complements existing data. Simultaneously, IoT device integration poses a risk that discourages the implementation of a change in the production line. The presented solution shows the integration of an IoT device into a production line. Data acquisition is proposed to validate the generated data. Demonstration of the integration of IoT devices is shown, based on the open platform communications united architecture (OPC UA) protocol, which represents an Industry 4.0 device. Taking into account the provided data, the integration of the OPC UA device and the possibility of integrating it into an existing Manufacturing Execution System (MES) system is experimentally verified. The experiment is built on the Node-Red flow base tool and presents a data acquisition solution and verification of the integration of the IoT device into the production line. The experiment also demonstrates how events can be used to control and verify the classification of image information from the IoT device. The experimental solution is completely independent of the production control process. However, it can run in parallel with the production control process implemented in MES.

Keywords: IoT · OPC UA · Data flow · Industry 4.0 · MES

1 Introduction

IoT devices and additional sensors integrated into the production line provide new additional data. Collecting data and delivering it to existing parts of the control for analysis is not sufficient [1]. These data can represent valuable information. Each added device must be monitored. Integration of IoT device sensors into higher layers of control is always a challenge. IoT device additions are desired due to the need to better understand the data, fill in the missing data, and support better data to analyze. IoT devices and sensors are using different types of communication protocols [2].

The data generated by IoT devices and complemented by sensors represent different data structures that need to be extracted, transformed, and loaded (ETL). Data analysis

R. Silhavy (Ed.): CSOC 2022, LNNS 501, pp. 215–223, 2022.
https://doi.org/10.1007/978-3-031-09070-7_19

is performed after data acquisition [3]. There are solutions like dashboard, or there are visualized data in the MES system. Data are classified as offline data, which are saved in a data warehouse. Data acquisition is typically used for analysis. The basic approach is batch processing, and the results are available with a delay [4].

The Manufacturing Execution System (MES) contains all the parts needed for analysis and data acquisition. However, interference with the control process of the MES system can be problematic. It represents complex changes that need to be applied in the control process. The data acquisition process provided by an MES system is an important part of the control process [5].

The control process is counting on received data and based on acquired data is executed next steps in the production process. Every change in the control process is risky and unwanted. It is recommended to verify the integration of the upcoming IoT device before the production deployment. Usually, a digital twin is used for such purposes. IoT devices outweigh the benefit. The cost of creating a digital twin of an IoT device is not cost-effective [6].

Whatever solution needs to be found to deploy the IoT device or the added sensors with the least risk. Data acquisition and data flow design can provide all the necessary steps to cover data acquisition and test integration of IoT devices into the production line [7].

2 Background

The flow-based programming approach (FBP), using the dataflow paradigm, represents a natural representation of the problem. This paradigm is used in process modeling, where a data flow is created using an oriented graph of data flows between executing operation [8]. Data flow-based programming typically envisions a network of asynchronous processes communicating through data streams of structured data blocks. FBP was created to automate the flow of data between systems.

FBP applications run fast and make optimal use of available computing resources. Data acquisition from various sources is the ideal use case to apply this technique [9].

The FBP method allows for processing data that may exist independently from multiple sources. It is possible to set the inputs, configure the protocol type, baud rate, load input registers, set the communication port, or set API calls as required, as shown in Fig. 1.

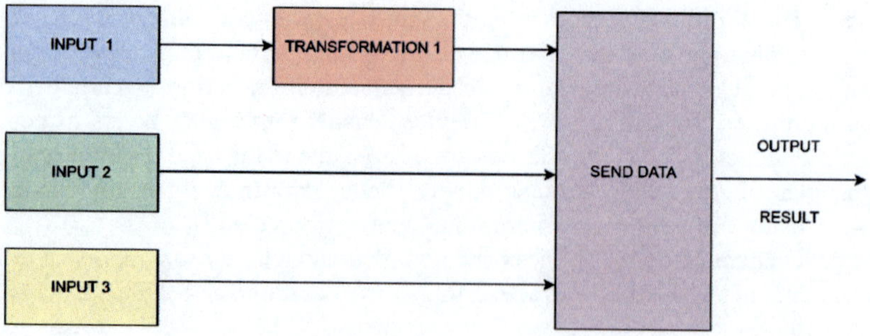

Fig. 1. The flow-based-programming basic concept for data acquisition

Data are fed into the input and passed on for processing. Input number one represents a situation where the received data need to be transformed into the required format. The transformation shown can be complemented by further logic, which is applied by functional programming. Inputs 2 and 3 represent data streams that do not need to be modified. The output is a stream of data that represent all the data in its original form, and at the same time, the values that were derived from the input data are added.

The data that are fed into the input can be transformed or an adjustment or calculation can be performed. The data flow output is the original data set plus the new value associated as a new pair of key values with it, which is the result of the operation that was performed. Each node can have multiple inputs or outputs. When multiple values are received, a defined node must make a decision.

The switch node determines the data flow according to defined conditions, and branching of the data flow occurs. A single output can be sent to multiple nodes to provide delivery for different nodes, systems, or consumers. With the data flow method, data can be acquired and processed in good performance, but it is not real time. The advantage is the simple implementation of data acquisition, which allows testing the integration of IoT devices or sensors into the production line.

2.1 Materials and Methods

Manufacturing, with the Internet of Things (IoT) elements in industry, is one of the most significant steps on the road to Industry 4.0 [10]. The manufacturing environment with IoT devices significantly increases the efficiency and quality of production and minimizes unnecessary costs. The combination of IoT device with production line can increase the quality of the control process. Data acquisition is one of prerequisite to successful integration of IoT device into the production process [11].

Data acquisition and IoT integration test was performed on a real model of production line. The production line consists of five zones. Each zone represents a certain part of the production process. Each part is controlled by an S7-300 PLC. This production line is controlled by Wonderware MES system and data acquisition is done by Historian, which is part of the MES. The production line is not Industry 4.0 compatible and control system is missing attributes of Industry 4.0 [12].

There is required change in recycling management, where frequently happen misclassifications of recycled containers and they are pushed into the wrong conveyor. Therefore, a IoT device was designed that can classify the container using computer vision and generate a control signal to ensure correct classification. The camera has an OPC UA interface available for integration [13]. The OPC UA standard has various options. It is possible to use generated events for control. The data that are generated can in turn provide additional data to facilitate the development of this device [14].

Ensuring connectivity and data acquisition from PLC controllers and IoT device was implemented by tool Node-Red. Data acquisition is implemented as a flow base program, represented by the flow diagram. Node-Red is built on the JavaScript library Node.js [15]. It's possible to run it in the cloud or even on less powerful hardware like a mini computer Raspberry Pi. This tool has extensive library and vendor support. Therefore, it is easy to install the necessary libraries to ensure that the required protocols are supported [16].

3 Experiment Design

Data acquisition has been implemented to enable the development of IoT devices. Simultaneously, it is necessary to keep the control part of the MES system and let the historian process collect the data. The data acquisition by the historian is used to validate new data collected.

Therefore, 3 PLC controllers related to the recycling part were selected. The IoT camera has been implemented only in the new system, because the historian in the first phase has not been configured to get data from IoT.

All devices are connected according to the block diagram shown in Fig. 2.

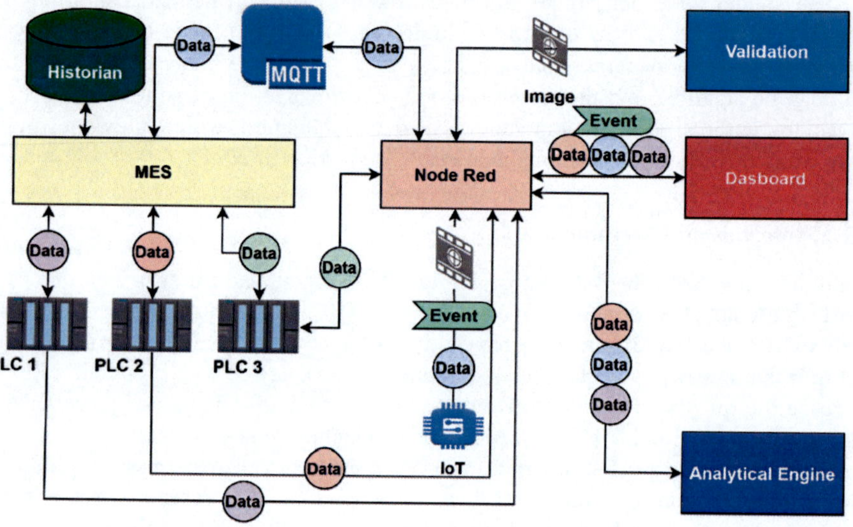

Fig. 2. The block scheme of connected devices.

Each PLC device was addressed and values were retrieved every 500 ms. according to a pattern from the MES system. The data are retrieved from the PLC program address space and transformed into the JSON format in Node-Red. When a key pair was created, its value and attributes were added with the time when the read was performed and the quality to represent any retrieval retries or possibly a read failure.

The IoT device was connected using an OPC UA client. The IoT device is an OPC UA server. This device was important in terms of improving the control process. It was ideal to be able to provide data to the controlling MES system. It is ensured using Node Red, where an MQTT client was implemented to deliver the data to the MQTT broker, from where the MES system consumed the data. Most of the production data are numeric, binary or text values. IoT generates image information and classifies objects represented there. The image data are binary, and there is a need to provide the transformation of the process. Messages used in Node-Red are based on the JSON structure. There-fore, image data are encoded to base64 and a part is joined to the JSON structure. Next, the

IoT devices provided data in the form of events. Internal variables of the IoT device are also available.

There is the possibility of the monitoring and debug internal behavior. Since it was necessary to provide debugging for this device, it was convenient to implement multiple levels of data collection. The camera captures an image, which is analyzed, and according to the computer vision evaluation, a final classification is created, with a percentage evaluation of the accuracy of the determination of the object and its color. The proposed data acquisition can be used to improve the function of the IoT device. The sensed image was sent to the Node red. Subsequently, it is reanalyzed by external tools. Possible misclassification was recorded, and then it was possible to use the samples identified in this way for subsequent learning of the computer vision model of the IoT device.

The main difference between retrieving data from PLC and IoT devices that supported the OPC UA interface is the ability to use events to perform data fetch operation from the address space of the OPC UA IoT device. If an event is generated by the IoT device, this event serves as an instruction to retrieve data. Moreover, the IoT device contains historical data and operation records that have been obtained. All these advantages were achieved using the OPC UA binary protocol. The OPC UA versions are fast and support all levels and parameters of previous OPC versions. The binary protocol runs on TCP port 4840 as shown in Fig. 3.

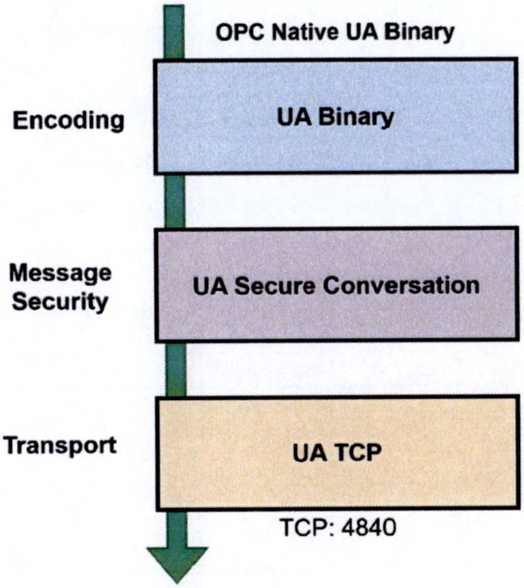

Fig. 3. OPC native UA binary

4 Experiment Implementation and Result

The first flow was implemented for the S7-300 PLC controllers. The data reading setup was implemented using node red and the data reading frequency was set to 500 ms. Data acquisition PLC data must define IP address, rack, slot, and port. PLC S7 Siemens S300 is the default port 102, Rack number, and slot is defined according to the real configuration. Values from the PLC data address space are read from all available variables immediately. The data are fed to the parsing function where the mapping is performed. Each PLC has its own function, due to different variables provided by each PLC. The second function generates insert statements which are written to the database. Data acquisition can be displayed on the dashboard for monitoring purposes. However, this functionality is only supplementary. The additional debug elements have a similar function to assist in creating a flow diagram. The designed PLC data acquisition flow is shown in Fig. 4.

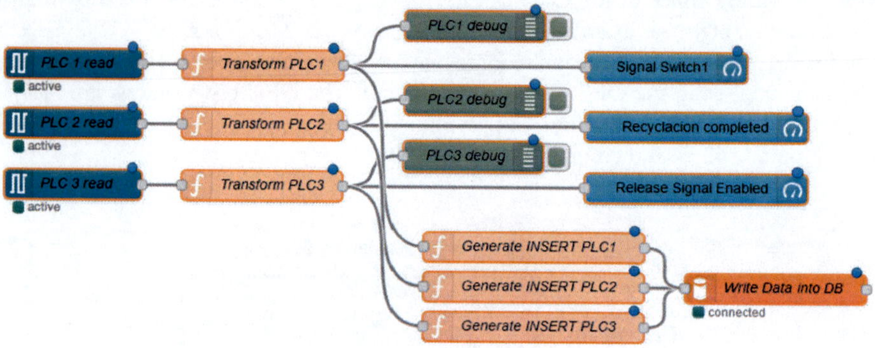

Fig. 4. Flow diagram for PLC data read.

The data stored in the database have the following mandatory attributes shown in Table 1.

Table. 1. Mandatory attributes for data acquisition.

Value	Description
ID	Unique record number
KEY	Key name
VALUE	Key value
VALUE_TYPE	Key value type stored in database
ANALOG_DIGITAL	Define the type of provided value
START_READ_TIMESTAMP	Timestamp of starting execution of read command
END_READ_TIMESTAMP	Timestamp of finish execution of read command
READ_TIMESTAMP	The timestamp provided by device
QUALITY	Quality of value
REPEAT	Number of repetitions to retrieve the value
ERROR	Reading error
DEBUG	Debug report on the reading process

The second process is different, as it deals with multiple data types. In addition to the classic data, events are processed, which are used for data acquisition, but also to verify the functionality of the IoT device. The data flow is shown in Fig. 5.

The two nodes named Events and OPC UA Client are connected to the OPC UA server, which represents the IoT device. If an IoT device needs to be explored, it is advisable to use a tool that enables connectivity to the OPC UA device. Suggested one is UaExpert® (https://www.unified-automation.com), which is designed as a generic test client supporting OPC UA features including DataAccess, Alarms, Events, Historical Access, and UA method calls. It is possible to test and map the OPC UA device decides which data and events need to be acquired, in case that not all the data provided by the device are necessary.

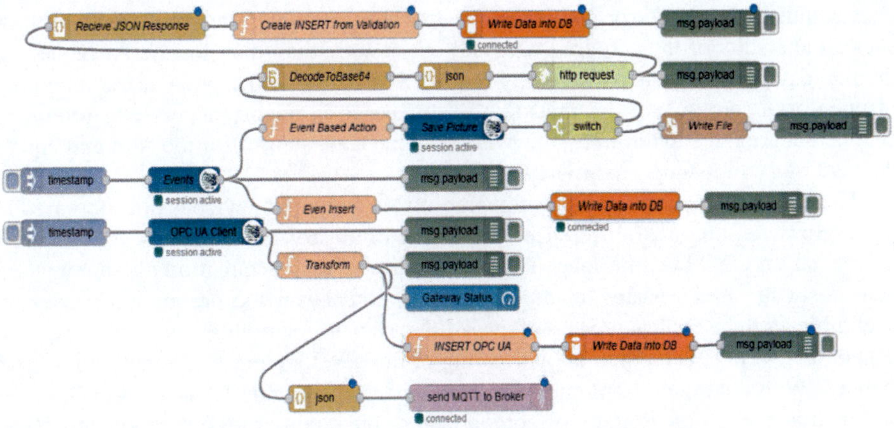

Fig. 5. OPC UA data acquisition for data and events.

OPC UA device and decide which data and events need to be acquired, in case that not all the data provided by the device are necessary. For testing, a combination of events, data, and captured image of the IoT device was selected. The data represent a combination of different types in terms of size, type by event generation, and content. Classical data such as image sequence, image name, and timestamps correspond to similar parameters as were captured on PLC devices.

The event has a control function. Data acquisition considers the event both as generated data and is also classified as the action trigger. In the test case, it triggers the acquisition of the captured image, its store on external storage, and the reanalysis of the image information from the external source. Data acquisition is implemented in such a way that data are recorded in the database but also sent to the MQTT broker. The existing MES system can read the data from the MQTT broker and further store or use subscribed data in the control process. The events that are captured from the OPC UA device are used to retrieve the image from the IoT device. The image represents a larger volume of data and is not stored in the database. Only the metainformation is stored. The image is stored on a network drive or in the cloud.

To validate the evaluation of image analysis, a flow is implemented, which decodes the information into base64 format, then a JSON message is generated, which is sent to the via HTTP element via RESTful API for reanalysis. The response is in the form of JSON where the object information is evaluated. The subject is determined by the percentage in which the evaluation of the object is accurate. The message also contains color determination. With this setup, it is possible to evaluate the IoT device and there is a possibility to create an additional learning set for AI computer vision and increase the model quality for image recognition. This validation process cannot be implemented in the MES system.

5 Conclusions

Data acquisitions are performed using the data flow method. The conceived data flow shows the results of the acquired data and the new training data for the AI model for the IoT device. The data collected by the MES system and the proposed solution are identical to a certain level of control. The difference in quality occurs when defining details that are not supported by the MES system. Time stamps for the start and end of the reading are not supported in MES.

The debug message from data acquisition from a specific device is not supported in MES. However, the main difference occurs when Industry 4.0 integrates. Here, an IoT device has an OPC UA interface. The possibilities of data acquisition and automating other tasks are much easier to implement. The impact on the production process is negligible until the control process of the MES system is not modified using the acquired data in this way. This method allows testing of the development and integration of IoT devices. With a suitable implementation, it is possible to modify data acquisition for verification of the functionality or comparison of the quality of the components. Here, it was used for the comparison of vision algorithms provided by the IoT device, against external recognition AI models.

Acknowledgments. The work was supported by Operational Program Integrated Infrastructure co-financed by the ERDF, grant number ITMS 313011W988: "Research in the SANET network and possibilities of its further use and development". This article was written thanks to the generous support under the Operational Program Integrated Infrastructure for the project: "Strategic research in the field of SMART monitoring, treatment and preventive protection against coronavirus (SARS-CoV-2)", Project no. 313011ASS8, co-financed by the European Regional Development Fund.

References

1. Aheleroff, S., et al.: IoT-enabled smart appliances under industry 4.0: a case study. Adv. Eng. Inform. **43**, 101043 (2020)
2. Mohammadi, M., Al-Fuqaha, A., Sorour, S., Guizani, M.: Deep learning for iot big data and streaming analytics: a survey. IEEE Commun. Surv. Tutor. **20**(4), 2923–2960 (2018)
3. Elmisery, A.M., Sertovic, M., Gupta, B.B.: Cognitive privacy middleware for deep learning mashup in environmental IoT. IEEE Access **6**, 8029–8041 (2018)

4. Urbina Coronado, P.D., Lynn, R., Louhichi, W., Parto, M., Wescoat, E., Kurfess, T.: Part data integration in the shop floor digital twin: mobile and cloud technologies to enable a manufacturing execution system. J. Manuf. Syst. **48**, 25–33 (2018)
5. Bratukhin, A., Sauter, T.: Functional analysis of manufacturing execution system distribution. IEEE Trans. Industr. Inf. **7**(4), 740–749 (2011)
6. Bécue, A., Maia, E., Feeken, L., Borchers, P., Praça, I.: A new concept of digital twin supporting optimization and resilience of factories of the future. Appl. Sci. **10**, 4482 (2020)
7. Evans, D., Eyers, D.M.: Efficient data tagging for managing privacy in the internet of things. IEEE Int. Conf. Green Comput. Commun. **2012**, 244–248 (2012)
8. Ortiz, G., Castillo, I., Garcia-de-Prado, A., Boubeta-Puig, J.: Evaluating a flow-based programming approach as an alternative for developing CEP applications in the IoT. IEEE Internet Things J. **9**, 11489–11499 (2022)
9. Lampa, S., Alvarsson, J., Spjuth, O.: Towards agile large-scale predictive modelling in drug discovery with flow-based programming design principles. J. Cheminform. **8** (2016)
10. Wollschlaeger, M., Sauter, T., Jasperneite, J.: The future of industrial communication: automation networks in the era of the Internet of Things and Industry 4.0. IEEE Ind. Electron. Mag. **11**(1), 17–27 (2017)
11. Rüßmann, M., et al.: Industry 4.0: the future of productivity and growth in manufacturing industries. Boston Consulting Group, pp. 54–89 (2015)
12. Tempest. Production Line Description: User Manual; Tempest: Ashland, OH, USA (2014)
13. Cavalieri, S., Salafia, M.G., Scroppo, M.S.: Integrating OPC UA with web technologies to enhance interoperability. Comput. Stand. Interfaces **2019**, 45–64 (2019)
14. Drahoš, P., Kučera, E., Haffner, O., Klimo, I.: Trends in industrial communication and OPC UA. In: 2018 Cybernetics & Informatics (K&I), pp. 1–5 (2018)
15. Lekić, M., Gardašević, G.: IoT sensor integration to Node-RED platform. In: 2018 17th International Symposium, pp. 1–5 (2018)
16. Rattanapoka, C., Chanthakit, S., Chimchai, A., Sookkeaw, A.: An MQTT-based IoT cloud platform with flow design by Node-RED. In: 2019 Research, Invention, and Innovation Congress, pp. 1–6 (2019)

Insights on Effectiveness Towards Research Approaches Deployed in Content Delivery Network

D. Priyanka[1(✉)], Channakrishnaraju[2], and B. K. Chethan[2]

[1] Sri Siddhartha Academy of Higher Education, Tumkur, India
prsh2019@gmail.com
[2] Sri Siddhartha Institute of Technology, Tumkur, India

Abstract. A dedicated content/service transmission right from server to end-user is facilitated by Content Delivery Network (CDN). Owing to the structure of the organized architecture of CDN, now it is possible to integrate them with cloud environment to deliver the contents in reduced time and higher reliability score. The present scenario of implementation has witnessed various schemes emphasizing content placement, replica management, caching, and multiple objectives towards formulating decisions to find the best technique to improve the data delivery system in CDN. However, they still lack essential features required for cloud-based CDN to cater to the exponentially rising demands of a massive user base worldwide. This manuscript explores various techniques, approaches, and models to put forward a clear picture of open-end research issues and the implication of existing CDN models. The paper offers a compact insight into the effectiveness of existing studies in CDN.

Keywords: Content Delivery Network · Replication · Content placement · Cloud computing

1 Introduction

Content Delivery Network (CDN) is sophisticated management of servers organized over different globe locations to ensure better traffic management [1]. The prime target is to minimize the delivery time of contents from the server to the user with the help of replication [2]. The contents are duplicated by the CDN from the servers and forwarded to the user from the datacenter located in its proximity. This system harnesses various attributes, e.g., availability, proximity, latency, and speed, to explore the better possible communication channel and path between the server and user. The prime point of deployment of CDN is within the data center to mitigate the issues connected with routing the contents and solving the user's request [3]. The implementation area of CDN is mainly witnessed over the educational institutions, e-commerce portals, entertainment media, and social networks where contents from one region are replicated to multiple geographical areas, which in turn reaches a massive number of users [4]. There are various practical benefits for using CDN viz. i) offering data privacy and integrity

R. Silhavy (Ed.): CSOC 2022, LNNS 501, pp. 224–243, 2022.
https://doi.org/10.1007/978-3-031-09070-7_20

when the content or services reaches directly from server to user [5], ii) improving the Quality-of-Experience (QoE) for user owing to capability to work through heaviest traffic condition [6], iii) effective management of redundancy to control the error formation in data transmission [7], iv) the concept of CDN also minimize the dependency toward extra usage of hardware [8], v) It also offers higher availability as well as higher scalability when CDN is deployed in large scale by integrating with cloud models [9], and vi) faster transmission of content to end-users. Various literature has been carried out to improve the performance of CDN systems using different variants of techniques and approaches [10]. However, there are certain increasing challenges in this area irrespective of the present state of solution. The *primary problem* is associated with CDN is to achieve maximized performance delivery of contents with minimal latency. Out of various responsible factors for these problems (e.g., bandwidth, networks, hardware, and software) [11]. The usage of optical fiber results in extra time is consumed for round trip and latency, resulting in assured latency. The *secondary problem* in CDN is associated with scaling up the operation to cater to the increasing base of users globally. Usage of intelligent failovers, routing schemes, load balancing assists in meeting the growing number of users. However, downtime still appears due to various extrinsic and intrinsic reasons. The *tertiary problem* in CDN is associated with reducing the risk, disaster recovery, and planning for sufficient redundancy. Irrespective of this fact, there is no exclusive discussion about the strength and weaknesses of the existing mechanism towards improving the performance of the CDN system.

Hence, this paper highlights various taxonomies of existing research standards being evolved to mitigate issues in CDN. The primary goa of this paper is to offer compact information about existing approaches towards CDN performance. The article discusses existing schemes of different variants and explores open-end research issues and briefing of study implications. The contribution of the paper is as follows:

- The paper contributes to discuss the CDN architecture along with its essential components highlighting the significant operation of CDN.
- The manuscript discusses about different methodologies being used for content management along with existing routing request.
- Discussion of existing models of CDN with respect to research and commercial application is carried out in this paper.
- Further, the paper depicts the updated trend of research problems in CDN along with highlights of identified research gap in existing studies.

The organization of the proposed study is as follows: Sect. 2 discusses insights on CDN architectures while the composition of CDN is briefed in Sect. 3. Management & Distribution of contents are discussed in Sect. 4, while Sect. 5 discusses Request-Routing techniques, and performance management is discussed in Sect. 6. Sect. 7 discusses existing CDN models in practice, followed by research trends in Sect. 8. Sect. 9 discusses open-end issues, while Sect. 10 summarizes the study findings as contributions and briefs about future work direction.

2 Insight of CDN Architecture

CDN is exclusively meant for content providers and customers with a demand for higher Quality-of-Service (QoS). From different ranges of study, the prime emphasis of researchers working on the specific goal of CDN is mainly towards performance, responsiveness, reliability, security, and scalability. Researchers use the standard architecture of CDN to reach the purposes mentioned above, as exhibited in Fig. 1.

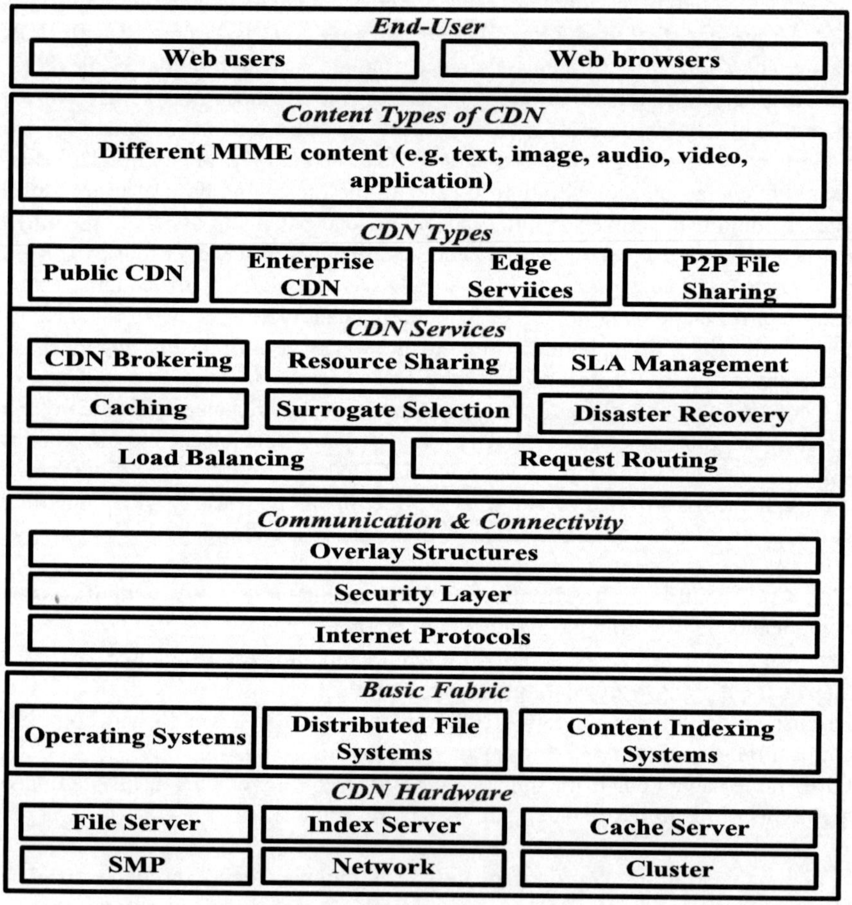

Fig. 1. Standard architecture of CDN

The topmost layer is reserved for *end-users* who use a specific URL of the content provider and access them on their respective browser. The next layer from the top is reserved for the *CDN layer* undertaking the various exclusive operation of CDN and further classified into content types, CDN types, and CDN services. Multiple forms of core services (e.g., selection of surrogate node, load balancing, caching, and request routing) are carried out by CDN. In contrast, user-specific services are carried out via brokering, resource sharing, and Service Level Agreements. The next layer is basically about *communication and connectivity* to facilitate various protocols required for data transmission. Finally, the bottom layer is the *fundamental fabric layer* which offers resources for infrastructure. This layer presents various computational resources, e.g., index servers, file servers, clusters, etc., are present in this layer. The existing solution towards research-based work in CDN is based on mainly four classes viz. i) composition of CDN, ii) management and distribution of contents, iii) request routing, and iv) performance measurement. This standard architecture has been discussed towards further inclusion of performance improvement as noticed in recent studies [12, 13]. An elaborate discussion of this is presented in the next section onwards.

3 Composition of CDN

The dynamic information associated with the network's condition and defined load over the cache server is incorporated within a CDN [14]. This redirects the request among other nodes and balances the traffic load over the stated surrogate nodes. Figure 2 highlights various essential elements of CDN, i.e., organization, relationships, content service types, interaction protocols, and servers.

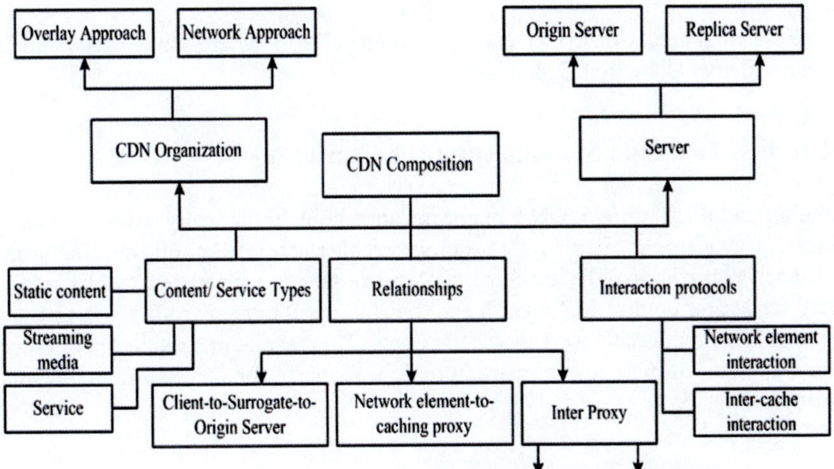

Fig. 2. Essential elements of the composition of CDN

Different types of elements of the design of CDN are as follows [15]:

- *Organization of CDN*: Existing system has witnessed two types of approaches for constructing CDN, i.e., network-based approach and overlay-based approach. The *network-based approach* identifies specific types of applications using switches and routers with specifically programmed codes. It also carries out the forwarding of the request message based on specified policies. The *overlay-based* approach handles the distribution of particular content found on the cache and application-based servers.
- *Relationships*: These elements are responsible for establishing connectivity among network elements, proxy caches, origin servers, surrogates, clients, etc. This component carries out the caching of the contents as well as replication. However, it is to be noted that the researchers mainly find the relationship between origin servers, surrogates, and clients.
- *Content service types*: All CDN-based applications are introduced with value-added services using their network resources as a discrete services distribution channel. The services offered by CDN could be in the form of e-commerce, file transfer, web storage, and directory.
- *Interaction Protocols*: These protocols are used for interacting among all the elements of CDN. There are two types of interaction protocol, i.e., i) inter-cache-based and ii) network element-based. The former protocol is classified into cache digest, hypertext caching, internet cache, and cache array routing protocols. The later protocol is further classified into SOCKS, web cache control, and network element control protocol.
- *Servers*: There are two types of servers in CDN, i.e., replica-based and origin-based servers. The origin-based server stores specific resource versions while the resource replica is stored in the replica server. The replica server can play the role of the cache server, web server, or media server.

The essential operation in CDN is highlighted in Fig. 3, where connectivity and setup of various elements are highlighted [16].

4 Studies Towards Management of Contents

Managing and distributing CDN content is carried out strategically using various processes, e.g., placement of surrogates, delivery, content selection, outsourcing content, and cache organization. The placement of surrogates is carried out to minimize the latency for access control and significant control over the bandwidth usage to transfer the contents from the replicated server to clients. This results in a reduction of communication cost and minimized infrastructure dependencies for facilitating better quality of services (Fig. 4).

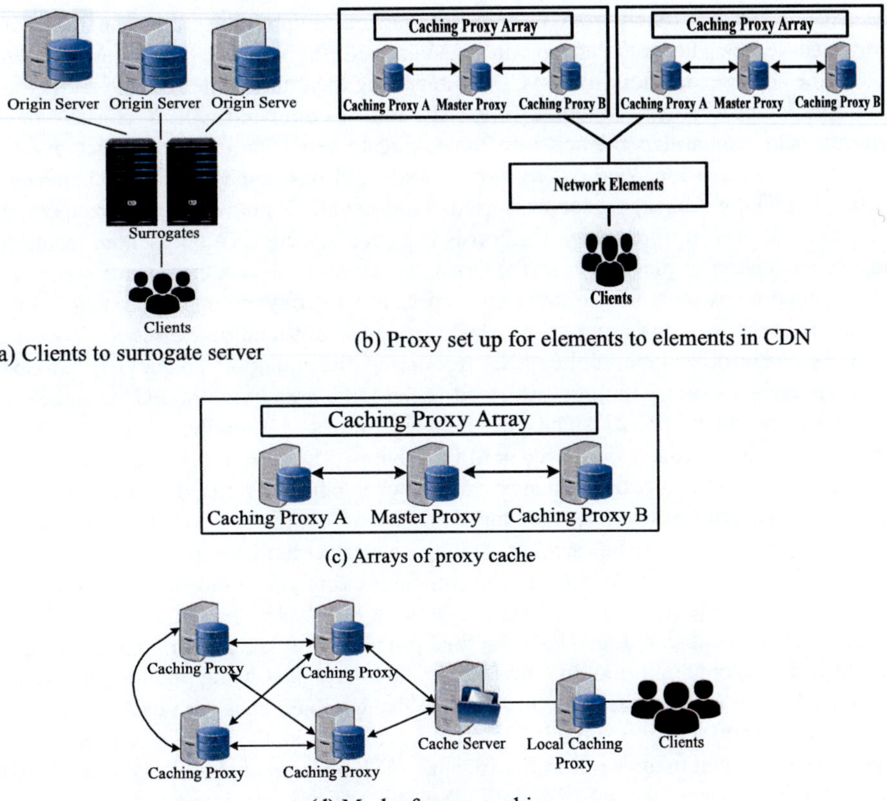

(a) Clients to surrogate server

(b) Proxy set up for elements to elements in CDN

(c) Arrays of proxy cache

(d) Mesh of proxy caching

Fig. 3. Essential operations in CDN [18]

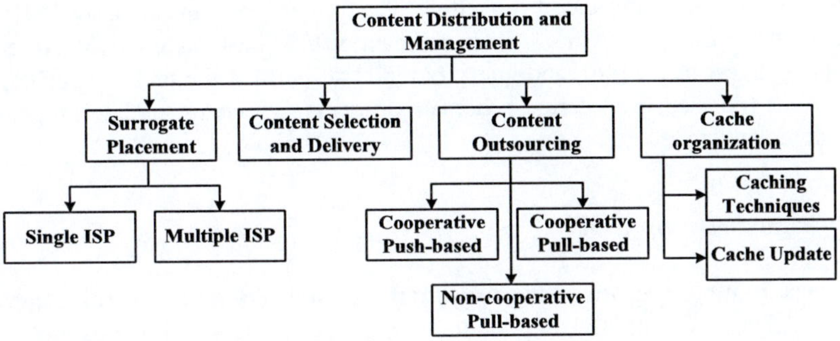

Fig. 4. Existing policy of management & distribution of contents in CDN [18]

At present, not much work is being carried out emphasizing the placement of a surrogate server. The work carried out by Eslami and Haghighat [19] has presented such a scheme considering cloud-based CDN integrating location problem of the facility and multiple knapsacks. The outcome shows the model compared with a heuristic-based approach to reduce dependencies on the surrogate server and reduced fluctuation in workload. The work carried out by Wong et al. [20] has emphasized QoS betterment deploying where a surrogate server is used. Sarkar et al. [21] have used an unsupervised clustering approach to optimize the surrogate server's usage. The presented technique makes a decision of placement of the surrogate server based on the cut-off score. The study outcome exhibited better outcomes compared to conventional approaches. Apart from the above existing approaches, the surrogate placement strategies are classified into placement of scalable replica [22], tree-based placement of replica [10], hot-spot-based placement [23], topology-informed based [24], greedy method [25], and center placement problem [26, 27]. The next delivery process and content selection can be carried out either partially site-based or full site-based. The partial-site-based approach of delivery and selection of content can be carried out using cluster-based [28], object-based [11, 29], popularity-based [30], and empirical-based [31]. In contrast, the full-site-based delivery and selection can be carried out using a complete replication process of total site content. The third process of outsourcing content is carried out either using a cooperative pull-based technique [32, 33], a non-cooperative pull-based technique [10, 34], or a suitable push-based technique [10]. The final process of organization of cache is carried out to facilitate content reliability, availability, and freshness using updating process of the cache [35–37]. The infrastructure of CDN is also integrated with caching policies. At present, various caching schemes have been evolved with the usage of learning-based approaches, as seen in the work of Kirilin et al. [38], Sadeghi et al. [39], Qiao et al. [40], Li et al. [41]. These schemes are mainly meant for offering an adaptive caching scheme harnessing the problem-solving capability of deep reinforcement learning approaches. Tao et al. [42] have also presented a unique content caching scheme especially focusing on Radio Access Network. Qin et al. [43] have worked on a model that adds novice users along with the cached contents in arbitrarily mode to enhance the rate of data delivery system over the wireless network. Apart from the studies mentioned above, after reviewing existing caching policies, it is noted that there are various other forms of caching, e.g., inter-caching and intra caching. The former type is query-based [44], while the latter scheme is classified into semi-hashing based [45], hashing based [46], directory-based [47], digest based [46], and query-based [11]. Further, the updated cache policy uses various reported invalidation schemes, on-demand updates, update propagation, and periodic updates.

5 Composition of CDN

The request routing mechanism can forward the client's request to a specific surrogate server to deliver contents. Existing studies show that the closest server cannot service client requests to act as the best surrogate node [48]. Therefore, using various metrics (a load of replica server, distance latency, network proximity, etc.), a system of routing requests in CDN ensures delivery of requests in its best form. Figure 5 highlights the mechanism of request routing. It should be noted that invoking the routing request is carried out only by obtaining the client's request, and it instructs for selecting a specific edge server as a response. Another mechanism called the request routing mechanism is carried out to update the selection process to the client. Request routing algorithms could be both adaptive [49] as well as non-adaptive [21, 50–54]. On the other hand, request routing mechanism could be further classified into six types viz. i) peering of CDN [55–60], ii) unicasting [60, 61], iii) URL rewriting [11], iv) HTTP Redirection [62], v) DNS-based request routing [63–65], and vi) load balancing [66].

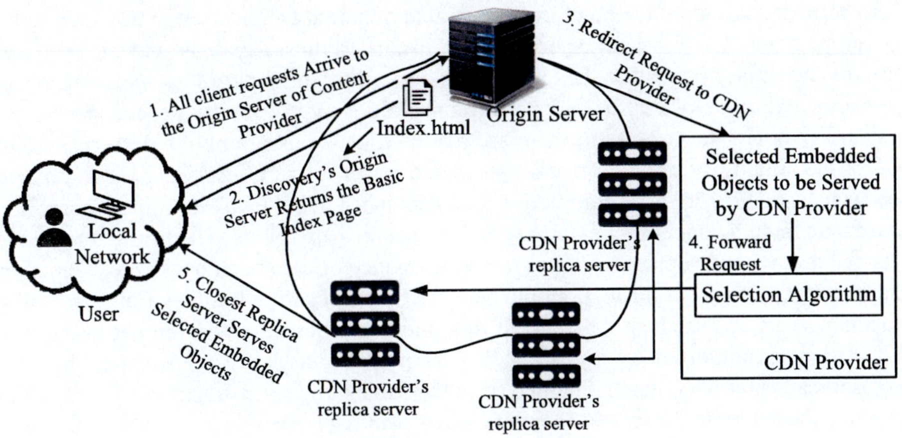

Fig. 5. Mechanism of request routing in CDN

Different schemes have been introduced in the area of request routing concerning CDN. The work carried out by Hu et al. [67] has presented a stochastic optimization model for dispatching the request associated with the video replication of the community. Study towards optimizing request routing has also been carried out by Fan et al. [68], emphasizing the request mapping along with routing response management in CDN over internet exchange point. Request routing operation is also dependable on resources, and this challenge is addressed in Chu et al. [69]. The study has optimized the routing request associated with utilities of the content provider to formulate a decentralized approach for addressing constraints connected to delay and bandwidth. The work presented by Kalghoum [70] where the author infers that usage of Software Defined Network (SDN) could significantly enhance the quality of the user while performing routing of the request. The study findings suggest that the reliability score improves significantly when the network address translation approach is integrated with SDN's domain name system. Another study has been carried out towards offering value-added points over-optimization in request routing in CDN, as seen in the work of Goudarzi et al. [71]. The study has focused on addressing energy efficiency issues in CDN and used the stochastic optimization principle for this purpose. Problems associated with scalability in request routing are discussed by Poirion and Ruosi [72] while working on virtual reality servers of streaming in CDN. The work carried out by Omotunde and Izang [73] has presented a selection mechanism of replication server using an adaptive request routing approach. The study has introduced a fuzzy logic mechanism for accepting and aborting the request execution in CDN. The work of Bilen et al. [74] has discussed a unique decomposition towards the request into multiple surrogate servers. The study has also used a queuing mechanism to control the CDN system's traffic state with the well-managed surrogate server.

6 Performance Measurement

The performance score is one of the essential parameters emphasized in the design and development of CDN-based experiments. It is carried out to evaluate the capability to provide necessary content to the clients with desired performance. This mechanism can perform assurance of end-to-end performance along with monitoring and prediction capability in CDN. At present, there are five metrics for assessing the performance of CDN, viz. reliability, utilization of the surrogate server, latency, reserved bandwidth, and cache hit ratio [75, 76]. The complete assessment mechanism involves both software and hardware factors concerning external and internal measurements. The former process is carried out by a third party to assess the performance of the specific website. The latter approach is carried out throughout the network by aggregating log files from streaming media servers and caches. Apart from this, the performance measurement process in CDN is also supported by acquiring various network statistics with respect to both dynamic and static feedback obtained from the surrogate server, monitoring traffic, and probing the network. System and network performance assessment use server load, frame rate, startup time, average bandwidth, packet loss, latency, and geographical proximity. Apart from this, there is also usage of various network simulators utilized to assess the performance of CDN [77–81]. However, there is a slide downside of the above-mentioned simulation-based approaches towards performance assessment. In case of non-inclusion of various essential factors (e.g., bottleneck condition, frequencies of session for each network elements), and involved nodes using internet-based protocol) the outcome of the simulation may not be considered as reliable. Hence, various performance metrics and research-based studies have explicitly focused on improving these performance metrics.

7 Existing CDN Model in Practice

At present, there are two types of CDN models in practice, i.e., research-based CDN model and commercial CDN model. The research-based CDN is used for educational purposes to carry out experiments and studies. The commercial-based CDN is used for offering professional services of content delivery of different types. Table 1 highlights some of the reputed models being used for both educational purposes and commercial purposes. A closer look into Table 1 data shows that still, there are a smaller number of models with more precise and effective solutions in commercial models of CDN. However, the capabilities and types of services keep improving after every usage.

Table 1. Existing CDN models in practice

	Name	Type of service	Capability	Solution
Research-based	Globule [81]	– Redirection of client requests to the replica server – Server monitoring	Collaborative CDN (open source)	-N/A-
	Coral [82]	– Facilitates replicated contents appropriately to the popularity of the client	Peer-to-peer CDN (Free) hosted by PlanetLab	-N/A-
	COMODIN [83]	– Offer playback service of media in collaborative form	Academic CDN on the global testbed	-N/A-
	CoDee N [84]	– HTTP request redirection – Caching of contents	Academic CDN testbed working on top of PlanetLab	-N/A-
Commercial Based	SyncCast [85]	– Offer a solution for content delivery via the internet	Used by the organization in online retail, music industry, media production, and motion picture	-N/A-
	Netli [86]	– Offers high-end internet services for businesses over the NetliOne platform	Presence of its clusters in 13 cities globally	-N/A-
	Mirror Image [87]	– Used for reporting services, web computing, streaming media, content delivery	Presence of its server (surrogate) in 22 countries	-N/A-
	Limelite Networks [88]	– Offers live video delivery (on-demand), games, etc.	Presence of its server (surrogate) in 72 countries	– Distributed delivery of content by Limelight ContentEdge using HTTP – Distributed music/video streaming – Customized CDN for distributed delivery

(*continued*)

Table 1. (*continued*)

Name	Type of service	Capability	Solution
Globix [89]	– Offers network services and internet infrastructure	Uses optical network over the mid-Atlantic region and IP backbone for transcontinental regions with > 1200 clients	-N/A-
EdgeStream [90]	– Video streaming application (disruptive)	Streaming video over consumer cable	Video-on-demand of EdgeStream
AppStream [91]	– Software distribution (on-demand) – Management tools for software license	Services 1000 reputed corporation, government, and educational institution	AppStream Software V5.0
Akamai [92]	Streaming, CDN services	85% market penetration with 20,000 servers in 71 countries	Managing dynamic & static contents by Edge Platform, Network Operation Control Center, Edge Control
Accellion [93]	Delivery of large files	Higher education, healthcare, manufacturing, media	Accellion Courier Secure File Transfer Appliances, Backup and recovery services

8 Existing Trend of Study

The adoption of CDN for dedicated transmission of heavier file systems has been investigated for more than a decade. The trend has majorly witnessed the migration of conventional CDN to cloud-based CDN systems. However, it should be noted that with a varied number of services offered by the cloud, e.g., multi-tenancy, the different customized definition of services, heavy traffic of concurrent users, and security, the anticipation towards a robust architecture of CDN is immensely demanded. Hence, it has been noticed that researchers have started focusing on a different set of problems to ensure a more effective and scalable operation of CDN. Figure 6 highlights the frequently adopted issues considered in research work published in the last five years. The prominent findings of current research trends are as follows:

- Lower Degree of Trends: Study towards network virtualization function, content outsourcing, request routing has witnessed lesser adoption in existing implementation studies.

- Higher Degree of Trends: Software-defined network, cache management, and learning approaches are increasingly adopted to improve performance in CDN.
- Evolving Trend: There is a strongly growing emphasis on quality of service toward performance analysis in CDN systems, mainly associated with energy factors associated with CDN elements.

Apart from this, it has been noticed that existing research studies have frequent adoption of Azure CDN [94], IBM CDN [95], Google CDN [96], and Amazon CloudFront [97]. Table 2 highlights the comparison of their essential characteristics. However, it is necessary to understand that an apriori analysis is essentially required to decide an appropriate CDN system in research work. Although, existing research work seems to deploy CDN for the application that encounters peak traffic conditions arising from multiple locations. All the current commercial CDN service providers named above perform redirection of request message to the cloud delivery server closer to the user. This is a good point as well as a constraint too. It is still feasible that the potential user location is quite far from the delivery server, leading to QoS and QoE issues. This is also the concern of existing trends of research work discussed in this paper, where various evolution of techniques is discussed.

9 Identified Research Gap

The open-end issues found in existing literature are as follows:

- *Scattered Flow of Problems*: The problems addressed in the existing CDN-based implementation scheme are highly scattered. The solution offered is symptomatic of the specific problem, and less correlation has been drawn with other essential issues. For example, the studies towards content placement are carried out without offering a better search optimization scheme in an extensive scale network. Problems addressed using the SDN system mainly emphasized its plane management and not much on managing request routing. The problem addressed towards caching management does not emphasize storage complexity towards cloud-based CDN. There is a need for a solution that can encapsulate a maximum number of connected issues altogether, not witnessed in existing studies.
- *Lesser Work towards Replica Management*: A decentralized scheme of replica management deals with the correlation of origin and replica server, whereas the inter-cache interaction and network-element interaction-based protocols do not offer many solutions. Emphasis towards selection and delivery of content is more compared to the replicated server. At the same time, studies towards different surrogate placement strategies consider replica management mainly focusing on its state of availability, location, popularity, and cluster, while less emphasis is on optimizing replica management. Often, the resource optimization scheme is missing in existing replication server management, which could reduce the sustainability of server capability over a more extended time.
- *Machine Learning in the Nascent Stage*: There is no doubt that the machine learning approach could offer the better problem-solving capability. However, the machine

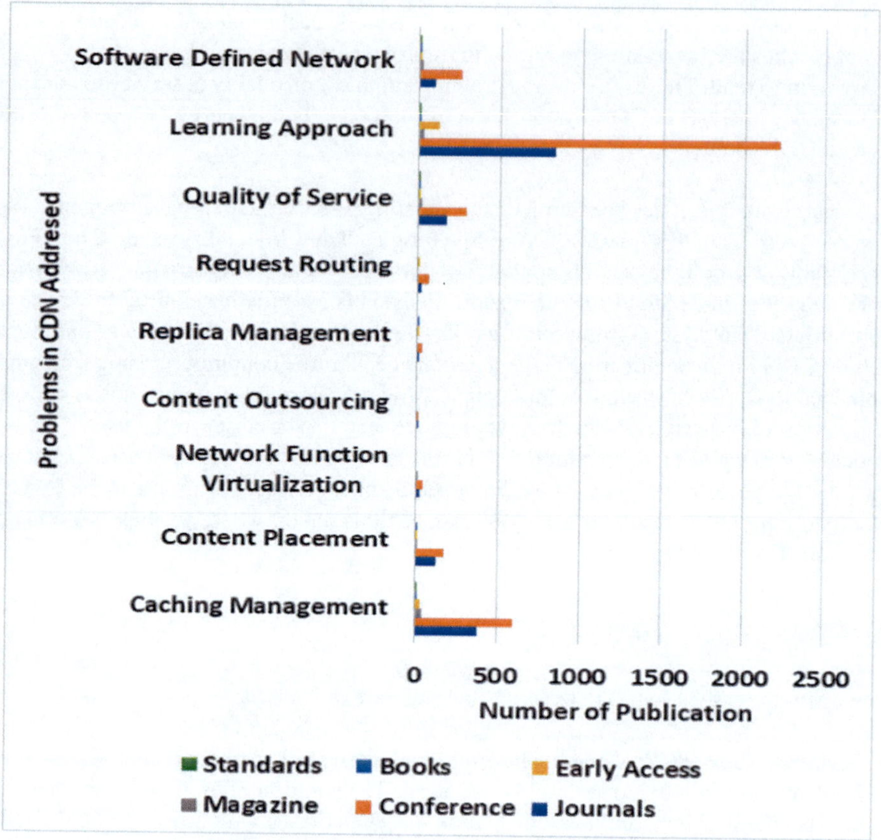

Fig. 6. Trend of problems in CDN (2016–2021)

learning scheme also demands trained data for providing better accuracy towards solving the problem. Hence, applying a machine learning strategy to solve the problem in CDN could require faster operation in large-scale decentralized networks. Apart from this, there is also a need to reduce resource dependency while applying such a scheme.

- *Lesser focus on QoS*: Adhering to QoS is one of the essential demands irrespective of any problem solution domain in the CDN system. However, existing approaches towards improving QoS are less found in the literature. There are no explicit algorithms in CDN to offer guaranteed QoS for large-scale deployment of cloud-based CDN systems. The performance parameters used in CDN and the test environment used for CDN are less emphasized to prove better QoS. Without claiming better QoS, the existing models cannot prove their reliability while working on large-scale deployment of CDN.

Table 2. Comparison of existing cloud services [98]

Characteristic	Azure CDN	IBM CDN	Google CDN	Amazon CDN
IPv6 support	Yes	Yes	No	Yes
Analytics	Yes	Yes	Yes	Yes
Geo Access Control	Yes	Yes	No	Yes
Large File Optimization	Yes	Yes	Yes	Yes
Field Level Encryption	No	No	No	Yes
Custom SSL	Yes	No	No	Yes
Certificate customization	Yes	No	No	Yes
Integration	Integrated with Azure Service	Akamai Integration	Integrated with Google Cloud	Integrated with AWS
Point of Presence	129	133	90	200

10 Conclusion

The working concept of conventional CDN is mainly targeted to deliver a massive-sized file to the end-user using a dedicated platform. Adoption of cloud-based CDN offers much supportive environment to tailor the functionality of CDN to suit up the large-scale deployment with higher reliability. However, there are still challenges connected with this deployment scheme, and researchers are consistently working towards it. The *contribution* of the proposed system is given in the form of research gap (viz. Scattered Flow of Problems, Lesser Work towards Replica Management, Machine Learning in Nascent Stage, Lesser focus on QoS) as well as some *notable study findings* of the proposed manuscript as follows:

- There is the adoption of a different form of technologies in CDN to improve its performance, viz. cloud computing, software-defined network, synthetic intelligence, record analytics, network virtualization function, machine learning, etc.,
- The modern version of CDN is also found to emphasize on cache recollection mechanism to process the contents faster and more appropriately.
- Looking into the existing picture of implementation, there is a potential need for better adherence to QoS for better SLA while working on cloud-based CDN.
- The majority of the computational decision-making problems associated with CDN can be effectively solved by machine learning approach; however, there is a need to select an appropriate algorithm to ensure a proper balance between computational performance and communication performance.

- Replica management has witnessed a lesser degree of literature in recent times and is a cost-effective solution to offer better service availability, which requires greater attention.

Hence, based on the above study findings, the future work of the study will be carried out towards evolving up with a novel computational model. The emphasis will address the identified research gap with a better balance between replica management system, optimization scheme, and QoS adherence policy in cloud-based CDN.

References

1. Zolfaghari, B., et al.: Content delivery networks: state of the art, trends, and future roadmap. ACM Comput. Surv. **53**(2), 1–34 (2020)
2. Jayakumar, S., Prakash, S., Akki, C.B.: An investigational study and analysis of cloud-based content delivery network: perspectives. Int. J. Adv. Comput. Sci. Appl. **9**(10), 307–314 (2018)
3. The Broadcaster's Guide to CDNs (Content Delivery Networks). https://www.dacast.com/blog/content-delivery-network-cdn/. Accessed 20 July 2021
4. Ansari, J.A.N., Khan, N.A.: Exploring the role of social media in collaborative learning the new domain of learning. Smart Learn. Environ. **7**(1), 1–16 (2020). https://doi.org/10.1186/s40561-020-00118-7
5. Orzechowski, L.: Senior Technical PgM, Akamai InfoSec, "Protection of Personal and Sensitive Data in Akamai CDN and Security Services", Akamai (2020)
6. Lu, W., Yin, B., Huang, G., Li, B.: Edge caching strategy design and reward contract optimization for uAV-enabled mobile edge networks. EURASIP J. Wirel. Commun. Netw. **2020**(1), 1 (2020). https://doi.org/10.1186/s13638-020-1655-2
7. Jin, H., Xu, D., Zhao, C., Liang, D.: Information-centric mobile caching network frameworks and caching optimization: a survey. EURASIP J. Wirel. Commun. Netw. **2017**(1), 1–32 (2017). https://doi.org/10.1186/s13638-017-0806-6
8. Tomarchio, O., Calcaterra, D., Modica, G.D.: Cloud resource orchestration in the multi-cloud landscape: a systematic review of existing frameworks. J. Cloud Comput. **9**(1), 1–24 (2020). https://doi.org/10.1186/s13677-020-00194-7
9. Kilanioti, I., et al.: Towards efficient and scalable data-intensive content delivery: state-of-the-art, issues and challenges. In: Kołodziej, J., González-Vélez, H. (eds.) High-Performance Modelling and Simulation for Big Data Applications. LNCS, vol. 11400, pp. 88–137. Springer, Cham (2019). https://doi.org/10.1007/978-3-030-16272-6_4
10. Salahuddin, M.A., Sahoo, J., Glitho, R., Elbiaze, H., Ajib, W.: A survey on content placement algorithms for cloud-based content delivery networks. IEEE Access **6**, 91–114 (2017)
11. Ghaznavi, M., Jalalpour, E., Salahuddin, M.A., Boutaba, R., Migault, D., Preda, S.: Content delivery network security: a survey. IEEE Commun. Surv. Tutor. **23**, 2166–2190 (2021). https://doi.org/10.1109/COMST.2021.3093492
12. Shen, Y., Ding, P., Zhang, X., Shi, X., Song, Y.: CDN accelerate research based on MEC for UHD video service. In: 2020 IEEE 6th International Conference on Computer and Communications (ICCC), pp. 1857–1862 (2020). https://doi.org/10.1109/ICCC51575.2020.9345095
13. Wen, Y., Chen, Y., Shao, M.-L., Guo, J.-L., Liu, J.: An efficient content distribution network architecture using heterogeneous channels. IEEE Access **8**, 210988–211006 (2020). https://doi.org/10.1109/ACCESS.2020.3037164

14. Al-Abbasi, A.O., Aggarwal, V., Ra, M.: Multi-tier caching analysis in CDN-based over-the-top video streaming systems. IEEE/ACM Trans. Netw. **27**(2), 835–847 (2019). https://doi.org/10.1109/TNET.2019.2900434
15. Zhao, M., Aditya, P., Chen, A., Lin, Y.: Peer-assisted content distribution in Akamai netsession. In: Proceedings of the 2013 Conference on Internet Measurement Conference. https://doi.org/10.1145/2504730.2504752
16. Jia, Q., Xie, R., Huang, T., Liu, J., Liu, Y.: The collaboration for content delivery and network infrastructures: a survey. IEEE Access **5**, 18088–18106 (2017)
17. Ding, C., Zhou, A., Huang, J., Liu, Y., Wang, S.: ECDU: an edge content delivery and update framework in Mobile edge computing. EURASIP J. Wirel. Commun. Netw. **2019**(1), 1–9 (2019). https://doi.org/10.1186/s13638-019-1590-2
18. Yin, H., Jiang, Z., Huang, H., Luo, Y., Zhang, X.: Adaptive content management for UGC video delivery in mobile Internet era. Mobile Inf. Syst. Article ID 3624860 (2016). https://doi.org/10.1155/2016/3624860
19. Eslami, G., Toroghi Haghighat, A.: A new surrogate placement algorithm for cloud-based content delivery networks. J. Supercomput. **73**(12), 5310–5331 (2017). https://doi.org/10.1007/s11227-017-2088-5
20. Wong, K.H.: Using surrogate servers for content delivery network infrastructure with guaranteed QoS. Ph.D. diss., UTAR (2012)
21. Sarkar, D., Rakesh, N., Mishra, K.K.: Population-based clustering to enhance the utilization of surrogate in Content Delivery Networks. Intell. Decis. Technol. **12**(4), 453–460 (2018)
22. Kilanioti, I., et al.: Towards efficient and scalable data-intensive content delivery: state-of-the-art, issues, and challenges. In: Kołodziej, J., González-Vélez, H. (eds.) High-Performance Modelling and Simulation for Big Data Applications, pp. 88–137. Springer, Cham (2019)
23. Wang, X., Mishra, B., Lushnikova, T., Narayana, J.L., Wang, G.: Amino acid composition determines peptide activity spectrum and hot-spot-based design of merecidin. Adv. Biosyst. **2**(5), 1700259 (2018)
24. Liu, H., et al.: Predicting the dissolution kinetics of silicate glasses by topology-informed machine learning. Npj Mater. Degrad. **3**(1), 1–12 (2019)
25. Chen, M., Wang, L., Chen, J., Wei, X., Lei, L.: A computing and content delivery network in the smart city: scenario, framework, and analysis. IEEE Network **33**(2), 89–95 (2019)
26. Xu, K., Li, X., Bose, S.K., Shen, G.: Joint replica server placement, content caching, and request load assignment in content delivery networks. IEEE Access **6**, 17968–17981 (2018)
27. Natalino, C., de Sousa, A., Wosinska, L., Furdek, M.: Content placement in 5G-enabled edge/core data center networks resilient to link cut attacks. Networks **75**(4), 392–404 (2020)
28. Wu, C., Yoshinaga, T., Ji, Y.: Cooperative content delivery in vehicular networks with integration of sub-6 GHz and mmWave. In: 2017 IEEE Globecom Workshops (GC Wkshps), pp. 1–6. IEEE (2017)
29. Baba, M.S., Herawan, T.: Object-based simulators for large scale distributed virtual environments. In: Abawajy, J.H., Othman, M., Ghazali, R., Deris, M.M., Mahdin, H., Herawan, T. (eds.) Proceedings of the International Conference on Data Engineering 2015 (DaEng-2015), pp. 11–19. Springer, Singapore (2019).
30. Sun, L., Ma, M., Wen, H., Pang, H., Wang, Z.: Beyond 1 million nodes: a crowdsourced video content delivery network. IEEE Multimedia **24**(3), 54–63 (2017)
31. Sarkar, D., Rakesh, N., Mishra, K.K.: Broadcast storm problem—a hidden consequence of content distribution in content delivery networks. In: Perez, G.M., Mishra, K.K., Tiwari, S., Trivedi, M.C. (eds.) Networking Communication and Data Knowledge Engineering. LNDECT, vol. 3, pp. 155–165. Springer, Singapore (2018). https://doi.org/10.1007/978-981-10-4585-1_13

32. Gupta, M., Garg, A.: A perusal of replication in the content delivery network. In: Lobiyal, D.K., Mansotra, V., Singh, U. (eds.) Next-Generation Networks, pp. 341–349. Springer, Singapore (2018)
33. Sarddar, D., Roy, S., Sen, P.: Edge multilevel edge server co-operation in content delivery network using hierarchical classification. Int. J. Grid Distrib. Comput. **10**(3), 41–52 (2017)
34. Bianco, A., Mashayekhi, R., Meo, M.: On the energy consumption computation in content delivery networks. Sustain. Comput. Inform. Syst. **16**, 56–65 (2017)
35. Wu, X., Li, Q., Li, X., Leung, V.C.M., Ching, P.C.: Joint long-term cache updating and short-term content delivery in cloud-based small cell networks. IEEE Trans. Commun. **68**(5), 3173–3186 (2020)
36. Zhou, L., Dan, W., Dong, Z., Li, X.: When collaboration hugs intelligence: content delivery over ultra-dense networks. IEEE Commun. Mag. **55**(12), 91–95 (2017)
37. Elayoubi, S.-E., Masucci, A.M., Roberts, J., Sayrac, B.: Optimal D2D content delivery for cellular network offloading. Mobile Netw. Appl. **22**(6), 1033–1044 (2017)
38. Kirilin, V., Sundarrajan, A., Gorinsky, S., Sitaraman, R.K.: RL-cache: learning-based cache admission for content delivery. IEEE J. Sel. Areas Commun. **38**(10), 2372–2385 (2020). https://doi.org/10.1109/JSAC.2020.3000415
39. Sadeghi, A., Wang, G., Giannakis, G.B.: Deep reinforcement learning for adaptive caching in hierarchical content delivery networks. IEEE Trans. Cogn. Commun. Networking **5**(4), 1024–1033 (2019). https://doi.org/10.1109/TCCN.2019.2936193
40. Qiao, G., Leng, S., Maharjan, S., Zhang, Y., Ansari, N.: Deep reinforcement learning for cooperative content caching in vehicular edge computing and networks. IEEE Internet Things J. **7**(1), 247–257 (2020). https://doi.org/10.1109/JIOT.2019.2945640
41. Li, L., et al.: Deep reinforcement learning approach for content caching in cache-enabled D2D networks. IEEE Internet Things J. **7**(1), 544–557 (2020). https://doi.org/10.1109/JIOT.2019.2951509
42. Tao, M., Gündüz, D., Xu, F., Roig, J.S.P.: Content caching and delivery in wireless radio access networks. IEEE Trans. Commun. **67**(7), 4724–4749 (2019). https://doi.org/10.1109/TCOMM.2019.2912384
43. Qin, Z., Gan, X., Fu, L., Di, X., Tian, J., Wang, X.: Content delivery in cache-enabled wireless evolving social networks. IEEE Trans. Wireless Commun. **17**(10), 6749–6761 (2018). https://doi.org/10.1109/TWC.2018.2863687
44. Hasselqvist, J., Helmertz, N., Kågebäck, M.: Query-based abstractive summarization using neural networks. arXiv preprint arxiv:arXiv:1712.06100 (2017)
45. Rezende, T.T.: Arquitetura híbrida terrestre/satélite para distribuição de conteúdos em áreas remotas com 5G (2020)
46. Vu, T.X., Chatzinotas, S., Ottersten, B.: Blockchain-based content delivery networks: content transparency meets user privacy. In: 2019 IEEE Wireless Communications and Networking Conference (WCNC), pp. 1–6. IEEE (2019)
47. Tseng, L., DeAntonis, J., Higuchi, T., Altintas, O.: Peer-assisted content delivery network by vehicular micro clouds. In: 2018 IEEE 7th International Conference on Cloud Networking (CloudNet), pp. 1–3. IEEE (2018)
48. Varshney, P., Gaur, S., and Pal, S.: A novel approach of load balancing in content delivery networks by optimizing the surrogate server. In: 2021 11th International Conference on Cloud Computing, Data Science & Engineering (Confluence), pp. 790–795. IEEE (2021)
49. Kannan, A., Vijayan, S., Narayanan, M., Reddiar, M.: Adaptive routing mechanism in SDN to limit congestion. In: Satapathy, S.C., Bhateja, V., Somanah, R., Yang, X.-S., Senkerik, R. (eds.) Information Systems Design and Intelligent Applications, pp. 245–253. Springer, Singapore (2019)

50. Ghasemi, H., Siavoshani, M.J., and Hadadan, S.: A novel communication cost aware load balancing in content delivery networks using honeybee algorithm. arXiv preprint arXiv:1902.04463 (2019)
51. Bajpai, V., Ahsan, S., Schönwälder, J., Ott, J.: Measuring YouTube content delivery over IPv6. ACM SIGCOMM Comput. Commun. Rev. **47**(5), 2–11 (2017)
52. Desai, R., Patil, B.P.: Adaptive routing based on predictive reinforcement learning. Int. J. Comput. Appl. **40**(2), 73–81 (2018)
53. Mazumdar, N., Nag, A., Nandi, S.: HDDS: hierarchical data dissemination strategy for energy optimization in dynamic wireless sensor network under harsh environments. Ad Hoc Netw. **111**, 102348 (2021)
54. Amadeo, M., Campolo, C., Molinaro, A.: A novel hybrid forwarding strategy for content delivery in wireless information-centric networks. Comput. Commun. **109**, 104–116 (2017)
55. Shojafar, M., Pooranian, Z., Naranjo, P.G.V., Baccarelli, E.: FLAPS: bandwidth and delay-efficient distributed data searching in Fog-supported P2P content delivery networks. J. Supercomput. **73**(12), 5239–5260 (2017)
56. Cui, X., Huang, X., Ma, Y., Meng, Q.: A load balancing routing mechanism based on SDWSN in smart city. Electronics **8**(3), 273 (2019)
57. Liu, X., Li, Z., Yang, P., Dong, Y.: Information-centric mobile ad hoc networks and content routing: a survey. Ad Hoc Netw. **58**, 255–268 (2017)
58. Stocker, V., Smaragdakis, G., Lehr, W., Bauer, S.: The growing complexity of content delivery networks: challenges and implications for the Internet ecosystem. Telecommun. Policy **41**(10), 1003–1016 (2017)
59. Xu, C., Quan, W., Vasilakos, A.V., Zhang, H., Muntean, G.-M.: Information-centric cost-efficient optimization for multimedia content delivery in mobile vehicular networks. Comput. Commun. **99**, 93–106 (2017)
60. Yang, H., Li, H., Wu, Q.: An integrating unicast and multicast solution by extending content delivery network to satellite. In: 2018 14th International Wireless Communications & Mobile Computing Conference (IWCMC), pp. 1414–1419. IEEE (2018)
61. Xu, Y., Li, X., Zhang, J.: Device-to-device content delivery in cellular networks: multicast or unicast. IEEE Trans. Veh. Technol. **67**(5), 4401–4414 (2017)
62. Kyryk, M., Pleskanka, N., and M. Pleskanka. Content delivery network usage monitoring. In: 2017 14th International Conference the Experience of Designing and Application of CAD Systems in Microelectronics (CADSM), pp. 306–308. IEEE (2017)
63. Harahap, E., Fajar, Y., Ahmadi, D., Kudus, A., Ceha, R.: Modeling of request routing management on router for content delivery network. Int. J. Sci. Technol. Res. **9**, 308–314 (2020)
64. Cheng, Y.: Routing-independent Anycast for IPv6 content delivery networks. PhD diss., Case Western Reserve University (2019).
65. Huang, L., Guan, Y., Zhang, X., Guo, Z.: Consistent dynamic server assignment in content delivery network. CoRR (2019)
66. Shuai, Q., Wang, K., Miao, F., Jin, L.: A cost-based distributed algorithm for load balancing in content delivery network. In: 2017 9th International Conference on Intelligent Human-Machine Systems and Cybernetics (IHMSC), vol. 1, pp. 11–15. IEEE (2017)
67. Hu, H., Wen, Y., Chua, T., Huang, J., Zhu, W., Li, X.: Joint content replication and request routing for social video distribution over cloud CDN: a community clustering method. IEEE Trans. Circuits Syst. Video Technol. **26**(7), 1320–1333 (2016). https://doi.org/10.1109/TCSVT.2015.2455712
68. Fan, Q., Yin, H., Jiao, L., Lyu, Y., Huang, H., Zhang, X.: Towards optimal request mapping and response routing for content delivery networks. IEEE Trans. Serv. Comput. **14**(2), 606–613 (2021). https://doi.org/10.1109/TSC.2018.2796567

69. Chu, W., Dehghan, M., Lui, J.C.S., Towsley, D., Zhang, Z.-L.: Joint cache resource allocation and request routing for in-network caching services. Comput. Netw. **131**, 1–14 (2018)
70. Kalghoum, A., Saidane, L.A.: Fcr-ns: a novel caching and forwarding strategy for named data networking based on software-defined networking. Cluster Comput. **22**(3), 981–994 (2019)
71. Goudarzi, P., Mousavinejad, M.: A Lyapunov optimization approach to energy-efficient content delivery networks. SIMULATION **96**(3), 269–280 (2020)
72. Poirion, P.-L., Leguay, J., Ruosi, L.: Scalable request routing for VR-ready CDNs. In: 2018 21st Conference on Innovation in Clouds, Internet and Networks and Workshops (ICIN), pp. 1–6. IEEE (2018)
73. Omotunde, A.A., Okolie, S.O., Adekunle, Y.A., Izang, A.A., Ebiesuwa, O.O.: Request-routing for content delivery networks (CDN). Int. J. Adv. Res. Comput. Sci. **6**(6), 10–16 (2015)
74. Bilen, T., Kurnaz, D.S., Sevim, S., Canberk, B.: Parametric-decomposition based request routing in content delivery networks. In: Chowdhury, K.R., Di Felice, M., Matta, I., Sheng, Bo. (eds.) WWIC 2018. LNCS, vol. 10866, pp. 323–335. Springer, Cham (2018). https://doi.org/10.1007/978-3-030-02931-9_26
75. Papagianni, C., Leivadeas, A., Papavassiliou, S.: A cloud-oriented content delivery network paradigm: modeling and assessment. IEEE Trans. Dependable Secure Comput. **10**(5), 287–300 (2013)
76. Masa, M., Parravicini, E.: Impact of request routing algorithms on the delivery performance of content delivery networks. In: Conference Proceedings of the 2003 IEEE International Performance, Computing, and Communications Conference, 2003., pp. 5–12. IEEE (2003)
77. Presti, F.L., Petrioli, C., and Vicari, C.: Dynamic replica placement in content delivery networks. In: 13th IEEE International Symposium on Modeling, Analysis, and Simulation of Computer and Telecommunication Systems, pp. 351–360. IEEE (2005)
78. Mangili, M., Martignon, F., Capone, A.: Performance analysis of content-centric and content-delivery networks with evolving object popularity. Comput. Netw. **94**, 80–98 (2016)
79. Tang, S., Alnoman, A., Anpalagan, A., Woungang, I.: A user-centric cooperative edge caching scheme for minimizing delay in 5G content delivery networks. Trans. Emerg. Telecommun. Technol. **29**(8), e3461 (2018)
80. Ge, C., Sun, Z., Wang, N., Ke, X., Jinsong, W.: Energy management in cross-domain content delivery networks: a theoretical perspective. IEEE Trans. Netw. Serv. Manage. **11**(3), 264–277 (2014)
81. Pierre, G., Van Steen, M.: Globule: a collaborative content delivery network. IEEE Commun. Mag. **44**(8), 127–133 (2006)
82. coralcdn. https://www.coralcdn.org/overview/. Accessed 20 July 2021
83. COMODIN. https://cwatch.comodo.com/what-is-cdn.php. Accessed 20 July 2021
84. CoDeen. https://codeen.cs.princeton.edu/. Accessed 20 July 2021
85. Huang, Z., W. Wu, K. Nahrstedt, R. Rivas, and A. Arefin. SyncCast: synchronized dissemination in multi-site interactive 3D tele-immersion. In: Proceedings of the Second Annual ACM Conference on Multimedia Systems, pp. 69–80 (2011)
86. Pathan, A.-M.K., Buyya, R.: A taxonomy and survey of content delivery networks. Grid Computing and Distributed Systems Laboratory, University of Melbourne, Technical Report **4**, 70 (2007)
87. Mirror-Image. https://www.resizepixel.com/mirror-image/. Accessed 20 July 2021
88. Limelight. https://www.limelight.com/. Accessed 20 July 2021
89. Globix. http://www.globix.com/. Accessed 20 July 2021
90. Edgestreamlp. https://www.edgestreamlp.com/. Accessed 20 July 2021
91. Grégoire, J.-C., Hamel, A.M.: On scheduling live media streaming in the cloud—a study. In: Proceeding of IEEE International Symposium on a World of Wireless, Mobile and Multimedia Networks 2014, pp. 1–6. IEEE (2014)

92. Akamai. https://www.akamai.com/. Accessed 20 July 2021
93. Accellion. https://www.accellion.com/. Accessed 20 July 202
94. Tomarchio, O., Calcaterra, D., Di Modica, G.: Cloud resource orchestration in the multi-cloud landscape: a systematic review of existing frameworks. SpringerOpen-J. Cloud Comput. **9**, 1–24 (2020)
95. https://www.ibm.com/downloads/cas/DQXWN428
96. https://cloud.google.com/cdn
97. https://aws.amazon.com/cloudfront/
98. https://www.dailyhostnews.com/cdn-comparison-aws-google-cloud-azure-ibm-alibaba-cloud

Effectiveness of Patinformatics Research for Building IP Management Strategy in a Pharmaceutical Enterprise

Tatyana Maximova(✉), Andrei S. Nikolaev, Valentina V. Ivashchenko, Anastasia V. Sennikova, and Aleksandra A. Cherkashina

ITMO University, 11 Tchaikovsky Str., St. Petersburg, Russia
{nikand,a.cherk}@itmo.ru, vaflia@list.ru

Abstract. This article presents a patent information support approach focused on pharmaceutical R&D projects and the implementation of a full or partial development cycle in pharmaceutical companies. The paper presents a definition of patent information support, a description of the process and instruments for the realization of the approach in the pharmaceutical enterprise. Patent informatics is introduced as one of the key tools for the realization of the approach and shows the highest performance in the fulfillment of the implementation phases of the method. A key aspect in the implementation of patent information support is the provision of data for decision makers at all levels of management and the formation of an interoperable thread of pharmaceutical production and patent analytics to improve the efficiency of the process.

Keywords: Intellectual property management · Patinformatics · Intellectual property analysis

1 Introduction

The configuration of global markets is rapidly changing. Knowledge-based companies are taking center stage in establishing the global economy. The increased role of information as a driver of economic development is leading to changes in the organization of the innovation process, both within all companies and in the knowledge-based ones. Information and knowledge are the most important resources, and no company can do anything without them. Knowledge is now being concentrated in areas where there is a high demand for it. It is especially important to acquire the required knowledge in the lifecycle (LC) of innovative projects.

Based on the LC, today's authors highlight a series of approaches to the role and position of patent research. The requirements for strategic research have also changed. Today's companies want to be aware of what will exist, how it will exist, when it will exist and what its role will be, given the variability of different factors. Obviously, this expectation requires a comprehensive approach. Modern marketing research not only focuses on the product-consumer relationship, but also on the producer-provider relationship, considering the technical level. Both of their own and that of the industry.

R. Silhavy (Ed.): CSOC 2022, LNNS 501, pp. 244–261, 2022.
https://doi.org/10.1007/978-3-031-09070-7_21

In today's fast-changing worldview, we need to know the answer to the question of what the product of today should be and what it should be tomorrow. We can therefore highlight the main components that need to be considered in today's research:

- The state of the art (specify all existing solutions, including positive and negative aspects);
- Prospective level of technology, including existing innovations, which have not yet been implemented, as well as those in various stages of preparation
- Degree of readiness of society for innovation;
- Degree of innovation readiness of the market and competitors in particular, and our own readiness (business processes);
- The extent of expectations for innovation in different regions, both in terms of price and utility.

This means that today's market research has many potential beneficiaries, at various levels of the company's management. So, the number of people involved in the preparation of this level of research is also growing. The complexity of such studies is largely due to the increase in the amount of information that needs to be processed to produce a document. These sources can include both open sources and those derived from technological intelligence, which risk violating the law. Also, there is a low level of reliability and trustworthiness of the information obtained in this way.

All this leads to a search for a source of comprehensive, reliable, and verified information, about major technology market players and the nature of their project interests.

Leading experts refer to patent documents as such a source. Patent landscapes have been proven to be informative to date. Most authors (Ernst, Zhang) agree that patent analytics is the basis for building an intellectual property (IP) management strategy and a research and development (R&D) plan for enterprises. They also point to the role of such data in competitive intelligence and identifying and systematizing research objectives in relation to technology trends [4].

The authors of this paper operate on the assumption that this is a strategically attractive way of reasoning, however, it is not necessarily a potential economic one. This paper therefore hypothesizes that the initial results of patent analytics should form the basis of market research and that the output becomes a verified and more convincing basis for building intellectual property management strategies. This in turn increases the capacity of marketing departments of companies, forming a safe legislative field for marketing and promotion maneuvers.

2 Discussion

Modern pharmaceutical medical preparations (MPs) development projects increasingly require patent information support (PIS) to better meet their basic needs.

The concept of PIS is important for identifying the needs of knowledge-based projects. In this context, it is necessary to label, firstly, the scope of the process, secondly, the objectives and procedural methodology of its existence, as well as the evaluation of the effectiveness of the process.

The creation of innovative products in knowledge-intensive industries involves continuous knowledge engineering, which is built on a continuum of information processes: receiving, processing, creating, and disseminating.

PIS is a type of analytic-synthetic processing of patent data. Patent information is a general body of information which has developed its relevance through time [9].

E.V. Koroleva writes about patent and information support. In her articles, Elena reveals this concept through the network system of innovation process support centers and, correspondingly, as part of the innovation infrastructure [3].

She covers the concept of PIS support providing access of economic entities to information resources to address issues of protection, disposition, and use of rights to the results of intellectual activity (RIA). The author combines PIS of innovation activity with the need to develop a methodology of network development of patent information space. Such a combination gives a conceptual basis for the functioning and development of the patent information support system. This definition is much more expansive than the one presented in this paper. E.V. Koroleva's definition derives from the meaning of state support factors for innovation activities of the country as a whole, rather than to the activities of corporations.

This definition is not entirely appropriate in the concept of this paper, since it is the specifics of industry corporations, rather than the public sector in its totality of economic entities, that are being considered. Narrowing the concept of PIS to classify it by technology sectors and presenting it with specific types of patent analytics seems appropriate in relation to considering the effectiveness of marketing research based on patent information to form an IP management strategy for a pharmaceutical enterprise.

For the purposes of this article, we define PIS as a mechanism of action, a proprietary methodology for the internal interaction of functional departments, and the types of research received as part of the methodological approach.

Let us formulate the concept of PIS in a sector-specific way. PIS involves a process of communication between the patent analyst and the creator, whereby, through consultation or the provision of analytic-synthetic processing of patent information, the creator receives data for decision-making in his research area or a qualitative analysis of possible solutions in relation to the overall IP management strategy of the enterprise.

PIS products are created for the benefit of the end-user. The end user is expected to be able to use the information contained in such products in end-user's professional work, and thereby improve end-user's own efficiency and make decisions to optimize processes.

Risk management also plays an important role in all phases of the LC. The application of patent risk management tools will maximize value for the end-user. For knowledge-intensive production, risk assessments include a range of measures, including the hiring of patent professionals.

Chinese practitioners have written about the impact of patent research on market research. In their work, they identify three areas of impact of patent research results:

1. the commercial sector;
2. the technology sector;
3. the legal sector.

The analysis of the results of patent research, based on the authors' conclusions, has the greatest impact on the legal and technological sectors. For example, L. Zhang points out this circumstance in terms of decision-making and building a financing strategy for a project, especially if the project is aimed at raising external investments [12].

We also believe it is important to note that, according to the World Intellectual Property Organization (WIPO), effective exploitation of patent information can reduce research time by 60%, save research costs by 40% and 90–95% of R&D advances can be obtained from patent literature.

3 Processing

3.1 Integration of IPS Instruments in the Process of Shaping and Evolving R&D Programs of Pharmaceutical Companies

In order to create an innovative product, namely to create quality MPs, constant knowledge engineering is required at all stages of its LC. As already mentioned it is acquisition, processing, creation and distribution. Communication with all production departments is necessary at this point. When it comes to obtaining data, it is important to analyze patent documents and reach out to the secondary stream of documents. When it comes to information, without patent research reports, decisions made by senior management at this stage would not be justified. If major decisions are taken in a hurry that will seriously affect the future life of the drug, not only will the benefits be lost, but also other negative consequences, such as lengthy litigation, will result. Decisions made can be questioned or supported by a decision evaluation, which identifies the strengths and weaknesses of the strategic initiatives.

Communication between the departments accompanying the development of the MPs in all phases of the LC is extremely important. And it is worth paying attention to the fact that none of the stages should be overlooked. In the data phase, data verification is carried out by the functional specialists, in the information phase by the legal and marketing departments, and in the knowledge phase, the presentation to top management and decision makers (DM) is resisted (Fig. 1).

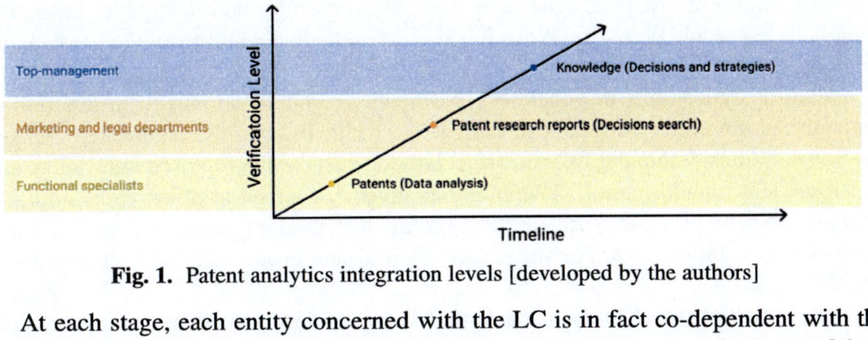

Fig. 1. Patent analytics integration levels [developed by the authors]

At each stage, each entity concerned with the LC is in fact co-dependent with the others without realizing this. Each actor here can and should request refinement of data, in-depth sampling, broadening the range of data for their own needs and processes from other actors whose specific activities make it possible to request the same.

Creating the above-mentioned interaction and establishing its frequency leads to continuous work of all departments and increases the speed of decision-making. Information environment in the company is formed in such work. A comfortable information environment, characterized by interactivity, flexibility, dynamism, selectivity, ability to interface with the existing body of knowledge and language of communication, is an indispensable link in the development of innovative products.

PIS of functional specialists in this context is presented as the process of bringing to the most readable formats of information presentation and verified data for the process of new knowledge generation within the project by specific specialists of different profiles. Information support is categorically part of information services.

If patent data changes constantly, the solutions that can be proposed will change. A patent research report provides information on those changes. Under similar circumstances, it is possible to classify PIS as monitoring patent data to generate ready-made solutions that are integrated into the policies, business processes and management practices of the organization. Thus, the importance of PIS in all phases of the LC development of MPs is highlighted as illustrated in Fig. 2.

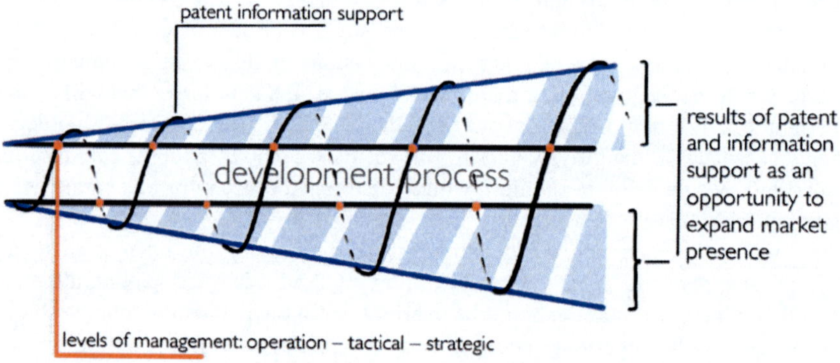

Fig. 2. The effect of PIS at all stages of the medicinal product LC [Developed by the authors]

PIS is closely linked not merely with the concept of information support, but also with that of patent analytics. The difference between information support as regular process in-house and patent analytics lies in the results that result from the use of these methods.

Patent analytics aims to generate new knowledge based on patent information in general, and therefore will result in knowledge itself. PIS, on the other hand, aims at shaping solutions within the LC of a project on the basis of the acquired knowledge and its subsequent transformation. The solutions given are aimed at effective planning of internal company processes, especially regarding the development of new products and the successful, and more importantly, their rightful and economically acceptable entry into the market.

In the availability of PIS, LC is constructed by considering the implementation of the necessary evidence-based decisions in the development phases. The fact that qualitative decisions are available as a result of PIS indicates the completeness of the research and the high degree of validation of the decisions taken, in the strategic concepts and business plans, but also in the operational activities and tactical decisions taken by specialists at different levels in the various departments.

Conceptually, this approach allows any employee involved in development, to formulate a request to the patent office of corporation to obtain additional details for making decisions as part of the workflow.

The iterative character of market research and patent research are interdependent, so the two cycles within development eventually come to converge.

The first step in the convergence is an analysis of the patent and non-patent situation in the markets. The results of the patent and non-patent situation balance the gap between the data on the number of market participants with the results of an analysis of the breadth of exclusive rights in RIAs and the IP objects of such participants. This approach is integrated into today's current strategy of augmented analytics, which allows double the value of investments in analytics, beyond the already typical funding in machine learning and big data machine processing.

Marketing follows the long-term economic growth trends of the industry. Patent prosecution will take into account long-term trends in strategies to protect subject matter analogues and contentious jurisprudence. This will enable the marketing department to immediately build a protection strategy for promotion and the patent department to build strong protection for all subject matter in the necessary markets.

The full convergence of marketing and patent research within the LC of product development is shown in Table 1.

Table 1. Relation of the stages of the organization's departments involved in the LC MPs process

Market research stages	PIS stages	LC stages of MPs/decision making stages
Identification of potential markets for sales and commercialisation of the technology Assess the impact of the external marketing environment on the commercialisation of the technology Evaluate the influence of the policy and legislative framework of the attractive markets Analysis of the commercial potential of the technology	Rapid patent landscapes for the targets involved in the selection, drawing up a map of the subject area, identifying the main market players Identification of prior art, analysis of prior art patents Verification of licenses for commercial and non-commercial cell/gene/protein banks, etc.	Research: idea generation and preparation for development: – Basic research (– Target identification and validation; – Translation of hits to leaders; – Lead optimization) – Chemical synthesis / biosynthesis (– Medicinal chemistry – selective synthesis) – In vitro research (– Target enzyme kinetics; – Receptor binding; – ARMV screening; – Toxicological screening) – Bioanalysis (– Development of bioanalytical techniques (non-GLP)) – In vivo studies (– Efficacy Assessment; – Preliminary bioavailability)

(continued)

Table 1. (*continued*)

Market research stages	PIS stages	LC stages of MPs/decision making stages
Market searches	Monitoring of competitors' patenting activities, their patenting strategies, and the development of an industry-specific patent landscape for the project Fulfilment of requests from functional specialists for narrow searches Analysis of in-house development for infringement of rights of third party rights holders	Pre-clinical development: – Production and quality (Synthesis of the active substance; – Characterization of the active ingredient; – preclinical development; – Specification analysis; – Analytical support) – Toxicology and Pharmacological Safety (– Animal safety assessment to initiate clinical development) – Bioanalysis (– Validation of bioanalytical techniques for toxicokinetic assessment) – Pharmacokinetics (Toxicokinetics and additional ARMV research)
Selection of markets Formulation of business plan, sales strategy, sales plan Branding Creation of brand Creation of promotion strategy	Formalization of service IP, monitoring of competitors' activities Forming the basis for a patent application Making own development patenting strategy Additional analysis of patent and non-patent situation in target markets	Clinical development – Production and quality (– Production of LMRs in compliance with GMP; – Preparation of production for registration) – Preclinical trials (– Safety assessment for further clinical development and registration) – Clinical trials (– Study planning and protocol development; – Study design and implementation; – Laboratory support; – Medical writing) – Bioanalysis (– Validated techniques for the analysis of human biological matrices)
	Patent application, registration of a trademark	Pre-registration clinical development Phase 1, 2, 3
	Obtaining a trademark Obtaining a patent license	Registration
Setting up and running advertisements	Transition of applications to the international phase	Post-registration period

Cooperation of the marketing department's work is most relevant for organizations which set up separate departments to commercialize the company's IP products, or where a specialized unit is set up within the marketing department. In this case, the creation of such complex studies or the conduct of work on the cooperation of their results will be the responsibility of a specialized department.

3.2 Concept

Among the existing approaches to patent analytics, the product approach is important for the purposes of this paper. The product approach describes the integration of patent research results into automated management systems and implies the generation of new service solutions in this area [5].

This approach can be done through the internal decision support system of (usually) heads of departments. In organizations where the process of analysis and decision making based on verified information is embedded in the corporate culture, it will be much easier to create such a product. In organizations where the functionality of patent departments is limited to meeting the point requirements of top management, it is important firstly to implement in stages and secondly to link with the types of work that are present in each project to move from the general format of PIS development to its particular applications adapted to each specific industry.

For companies in many countries where generic production dominates the market, the most relevant activity will be to monitor the patent failures of originator pharmaceutical firms. To correctly interpret and predict patent failures and build your own production on them, you need to be aware of the nuances. Patent informatics can play a key role in this respect.

4 The Practice of Applying Patent Analytics Tools to the Design of a Pharmaceutical Company's Development Strategy

4.1 Prerequisites for the Action Research

Let us form an analysis of patent data for decision-making and subsequent evaluation in terms of strategic planning, using the example of the pharmaceutical sector for generic companies in the Russian Federation.

Current research has characterized the pharmaceutical market through[1]:

- increased spending on R&D with a concomitant decline in efficiency;
- development of new technologies;
- integration of the industry with IT players for more effective development of new products;
- tightening state regulation of the circulation of MPs;
- growing the use of generics and biologics;

[1] Pharmaceuticals Market Research Reports & Industry Analysis // URL: https://www.marketresearch.com/Life-Sciences-c1594/Pharmaceuticals-c89/.

- lower pharmaceutical prices due to increased competition in many segments of the pharmaceutical market.

With the expiration of a patent license there is a company patent collapse, which is the loosening of the patent portfolio of the rightsholder's company. A patent collapse is an abandonment of a significant development. Each such collapse can be predicted during the basic patent search and examination of counterparties' patent portfolios.

For the pharmaceutical market, each such collapse is particularly relevant. A collapse can trigger the development of the generic market for certain types of medicines. For companies that are not developing original generic MPs, it is crucial to track this cycle. Tracking the cycle will enable start work on in-house R&D 3–4 years before a patent collapse and prepare the generic MP for release.

Patent monitoring here will predict the collapse of a MP of interest to a company within a few years with great probability.

In a highly competitive generic environment, the development of an IP object that affects the reduction of product costs and the addition of its own patent portfolio is of relevance. The creation of the most efficient ways to manufacture or use forms that increase the rate of delivery of the active substance into the body and modify any other pharmacokinetic and pharmacodynamic properties entail additional certain types of clinical trials.

The circumstances are instructive. They reflect the fact that the work of a company's patent department is not limited to basic patent searches to identify patent failures for its own generic MP. There needs to be constant monitoring of market players, generic companies. And strategic planning to promote a company's own generic MP to external markets requires not only market research, but also patent research reviews of existing patented solutions in each individual country, the market to which the company plans to enter based on the results of the market research and the patent system of each individual country.

Each of these and many other tasks will form the basis of a PIS plan for the departments within such a pharmaceutical company. This approach refers to the Quality-by-Design (QbD) concept whereby generic companies set up systematic quality management systems (QMS). Because of the systems that have been mentioned, the product is not copied in its entirety, but with adjustments for related innovations, which makes it possible to claim patent protection on its own. In this way, the PIS can significantly improve the efficiency of the quality management department, provided there is one in the company.

Establishing a system to integrate patent analytics into all levels of decision-making and into all phases of the LC development of the generic MP will enable the company to successfully build up its own R&D department in the future as well.

F. K. Suri и A. Banerji in their paper [10] have noted that the size of the generic market is built on such patent collapses. For example, for 2020, the monetary value of the patent collapse of originator drugs was $80 million USD, which would transfer to the generics market. In terms of the structure of the global pharmaceutical market by 2020, generics accounted for 35% along with 52% of the original MPs. According to Frost & Sullivan estimates, biologics with total sales of $70 billion will lose patent protection in the next 4 years. According to Evaluate Pharma in 2016–2022 the patent protection may be lost for drugs whose annual total sales are close to 250 billion $ [11].

On average, it takes 3–4 years to develop a generic MP and obtain approval, but for some drugs the entry threshold for equivalent development is much higher due to complex manufacturing technology and safety issues. For example, the patent for the original MP of fondaparinux [8] (Arixtra) came out of protection in 2002, and the first generic was approved only in 2011, although the application was filed back in 2009. So far, this part of the generic market is not highly competitive although a bioequivalence study is sufficient to release a generic MP and, in terms of the trials being conducted for release approval, this is the shortest route to market entry for an MP. Meanwhile, MPs with a high entry threshold may serve as a first step towards the development of originator MPs in generic companies.

These patents are most commonly found in affiliated generic companies of originator rights holders. They may be able to transfer the technology and bring the generic MP to market in a shorter period of time. Therefore, it is important to monitor M&A deals in order to anticipate such transfers. Technological patent intelligence on a particular MP plays a significant role here. By providing patent data, market research can also assess the risk of investing in such a project in terms of likely competition, as well as the return on investment in terms of estimated market share.

The work of full-service pharmaceutical companies is characterized by the same instruments of patent analytics used with a different focus. For example, patent intelligence would also include additional searches in the non-patent literature, looking at bioprospecting experiments as one of the relevant areas of the search for chemical entities.

4.2 Approbation of Patent Analytics Instruments

Let us work with the 2010–2021 International Patent Classification [2] index A61P:

A – Satisfaction of life human needs.

A61 – Medicine and veterinary medicine, hygiene.

A61P – Specific therapeutic activity of chemical compounds or MPs.

This subclass covers the therapeutic activity of chemical compounds or MPs already classified as such in subclasses A61K or C12N or in classes C01, C07 or C08.

In this subclass, the term "pharmaceutical products (FPP)" includes chemical compounds or compositions with therapeutic activity. Therapeutic activities in this case are classified in all relevant sites.

We consider it necessary to draw attention to cases where the subject matter of the invention relates only to the specific therapeutic activity of chemical compounds or MPs, and the chemical structure, compound, mixture, or composition of that subject matter is known. In such cases, the classification is made under both of subclasses A61K and A61P as information on the invention. In addition, if the chemical structure, compound, mixture or composition, or any single ingredient of a mixture or composition is considered to be of value for searching information, it may also be classified as additional information.

The classification symbols of the subclass are not indicated firstly when patent documents are assigned.

The primary sampling frame for the analysis formed by Lexis Nexis Patent Sight database. The analysis of the sample starts with the geography of the subject area (Fig. 3).

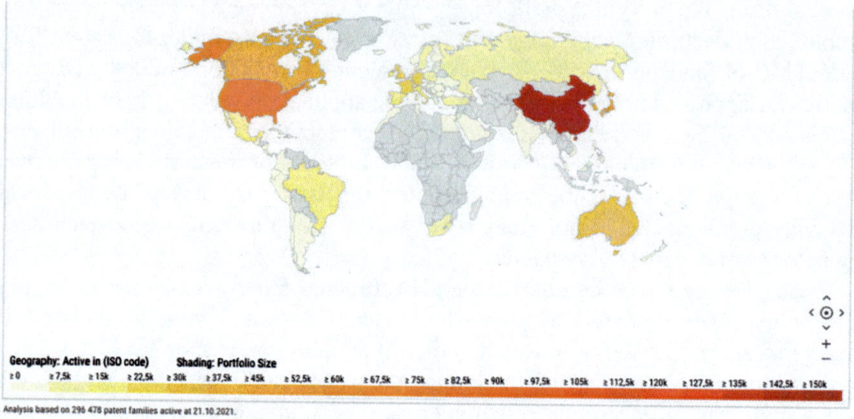

Fig. 3. Geography of patent activity by A61K index of IPC[2]

The geographical distribution of the sample is as follows: China has the largest number of patents filed in the last 10 years in the field of MPs. The US and Japan come second, followed by South Korea. Canada, Australia, and several European countries come next. Australia and Brazil are a large part of the sample, with higher numbers than even some European countries.

Most patent searches have considered China separately because of the factor of national originality as a criterion for patentability, which allows the copying of world achievements for local use.

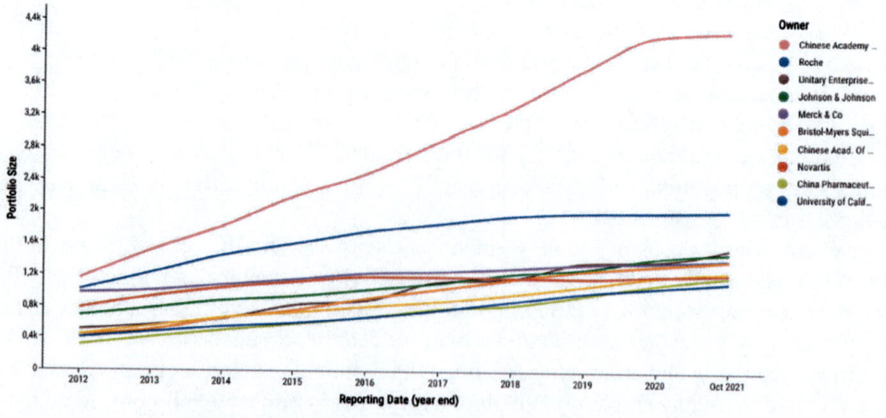

Fig. 4. A61P. Top companies by size of patent portfolio[3].

[2] Formed by using Lexis Nexis tools.

[3] Formed by using Lexis Nexis tools.

It is important to note that it is the Chinese Academy of Science that leads in terms of patent portfolio size with the index selected, indicating that IP work is encouraged in the academic environment. This is confirmed by the seventh-ranked Chinese Academy of Medical Science and the ninth-ranked Chinese University of Pharmacy. The size of the patent portfolio is only directly related to the patenting activity of the organization, but does not reflect the successful commercialization of the subject matter or its sustainability over time.

In this sample, China is highlighted because of the developments of the Chinese Academy of Science, which has the largest patent portfolio in the sample – 4,250 patents (Fig. 4). Most of the patents issued by the Chinese Academy of Science relate to pharmaceuticals through technological areas such as peptides, synthetic proteins, gene therapy, cancer treatment and compositions that regulate apoptosis.1/8 of the patents relate to the use of bacteria, antigens and antiviral agents, 1/8 to drug composition, antibiotic substances and biodegradable polymers (Fig. 5).

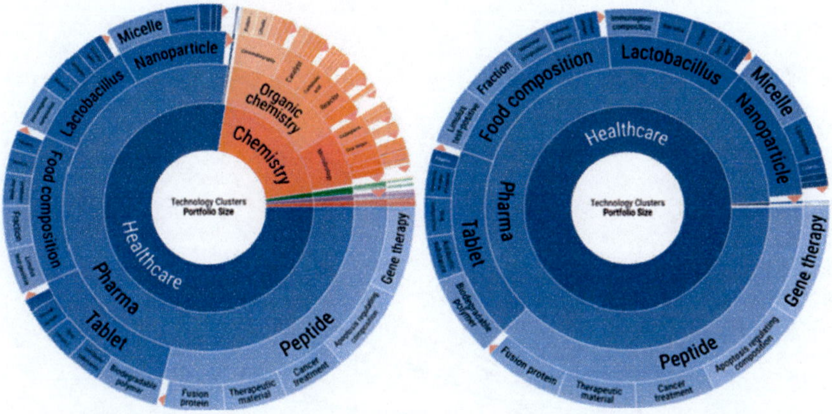

Fig. 5. Chinese Academy of Science subject area labelling: general and by healthcare cluster (A61P and A61K).[4]

The subject matter of the patent portfolio of the Chinese Academy of Science is narrowed down to patent families with classification class A61P (Therapeutic Activity of Chemical Compounds or Drugs). The most common subclasses in the patent portfolio are: A61P35 (Antineoplastic Agents), A61P25 (Drugs for the treatment of the nervous system) and A61P29 (Analgesics with non-central action, antipyretic or anti-inflammatory agents such as antirheumatic agents; non-steroidal anti-inflammatory drugs (NSAIDs)). Only subclass A61K31 (MPs containing organic active ingredients) occurs more frequently. Only 22 patent families (in green) refer to Nutrition, 12 patent families (in purple) to agrochemicals and 9 patent families (in red) to analyzing material.

If we classify the patent portfolio of the academy by Technology Relevance, the patent families with the highest numbers would be C01B31 – Carbon Compounds, C01B3 – Hydrogen, Gas Mixtures Containing Hydrogen, Hydrogen Separation from

[4] Formed by using Lexis Nexis tools.

Mixtures Containing Hydrogen, Hydrogen Purification and H01G9 – Electrolytic Capacitors, Rectifiers, Detectors, Switching Devices, Light Sensing or Heat Sensing Devices. The above indicates the weakness of the application in the field of pharmaceuticals and therapeutically active chemicals or compounds, despite their number.

The distribution in terms of Market Coverage shows that the patent families of class C08B – polysaccharides and their derivatives – have a wider geographical coverage of rights than patent families in other indices. These are followed by patent families relating to subclasses A61M31, devices for the introduction or retention of various nutrients, such as medicines, into body cavities, and A61M39, tubes, connectors or couplings for tubes, valves, access devices specifically designed for medical purposes.

To summarize these grounds, we believe that the low geographical coverage of patent rights for pharmaceutical ingredients is related to the lack of international novelty of such patents or that there is no provision for the production of exports for such products.

Let us present the sample companies with the strongest patent portfolios (Fig. 6).

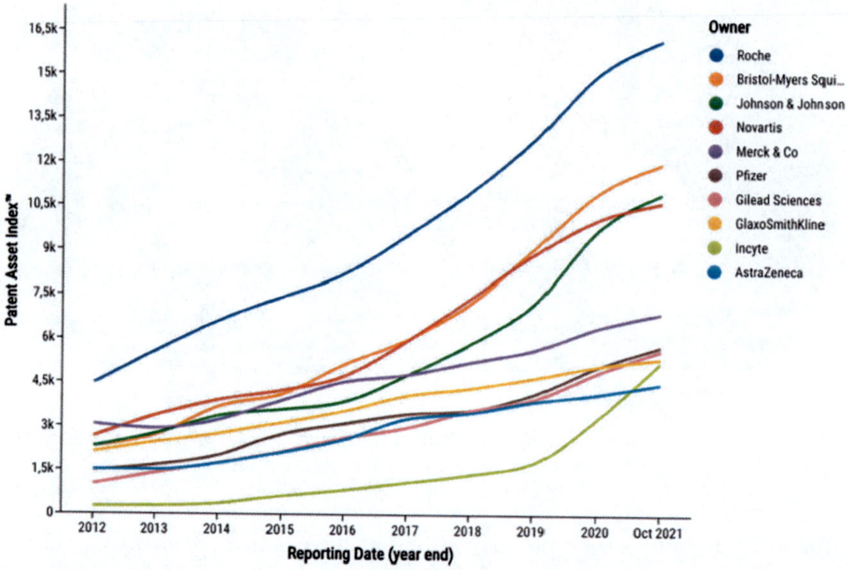

Fig. 6. Ranking of pharmaceutical companies by strength of patent portfolios[5]

The strongest patent portfolio is owned by a Swedish pharmaceutical company called Roche. We believe that this position is due to the age of the company. Roche was founded in 1896. The American pharmaceutical company Bristol-Myers Squibb ranks second and is almost the same age as Roche. Bristol-Myers Squibb was founded in 1887, and for the past 10 years has been on the same level as another much younger Swiss company, Novartis (1996). Only in the last three years has the company emerged as a leader, with the development of the rapid SARS-CoV-2 coronavirus test.

[5] Formed by using Lexis Nexis tools.

Most of the ranking positions are still occupied by US companies, which may indicate their interest in professional IP management and the use of patent informatics and patent analytics tools: Merck & Co (1891), Pfizer (1849), Gilead Sciences (1987), Incyte Corporation (1991 / 2002), Johnson & Johnson (1886). In addition to America and Switzerland, the British pharmaceutical companies GlaxoSmithKline (2000) and AstraZeneca (1999) are also in the top-10.

Several companies such as Incyte and Johnson & Johnson have shown strong growth just from 2019.

For the strongest patent portfolios, let's present the subject area markup (see Fig. 7).

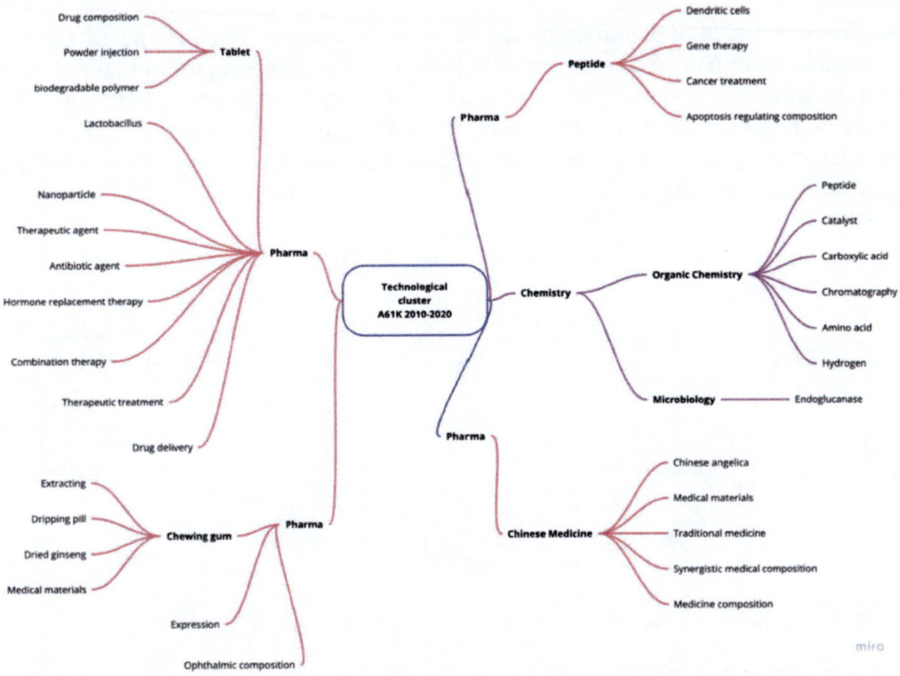

Fig. 7. Technological structure of the strongest patent portfolios [Developed by the authors]

The strongest patents relate to peptides, which are coded under the codes for pharmaceutical engineering and organic chemistry (C07K 1/13, C07K 5/12, C07K 19/00, A61K 38/12, C07K 7/04, C07K 7/64, C07K 14/785, C07K 1/00, C07K 1/107 and C07K 7/14). Because peptides were discovered at the beginning of the 20th century and scientists quickly found a way to synthesize them in the laboratory, much of today's science is based on this technology. The indices given contain ways of producing peptides in different ways, as well as different types of peptides with a classification according to the length of the amino acid sequence.

Therapeutic agents are coded as: A61K 47/55, A61K 47/64, C21D 1/60.Since therapeutic agents can be of biological origin or synthetic, having pharmacological or therapeutic activity they are also strongly related to peptides, proteins and polyamino acids.

Some indices describe MPs containing antigens, antibodies. In different types of agents, MPs are found in the group A61P, indices such as: A61P 11/14, A61P 21/06, A61P 39/04, A61P 27/06.This branch of development is the most attractive and allows the creation of unique MPs for existing therapeutic purposes.

Most of the patents classified through A61K have an additional index A61P. However, 78% of all the patents in the sample relate to antitumour MPs. 42% of MPs relate to anti-infective agents – antibiotics, antiseptics chemotherapeutic agents. 40% MPs for immunological or allergic diseases. 38% MPs for diseases of the nervous system. However, they have weak technological relevance and average market coverage. Patents for MPs with unique therapeutic active chemical activity are leading in this respect (13 positions).

The distribution of pharmaceutical companies on the average patent power relative to the number of patents in the portfolio shows that Roche is the largest in the market with the strongest patent portfolio only because of the total strength of each individual portfolio. However, Incyte has the strongest patent families, i.e. the average strength of each patent is significantly higher than the average patent strength of any other company's portfolio in the ranking (see Fig. 8).

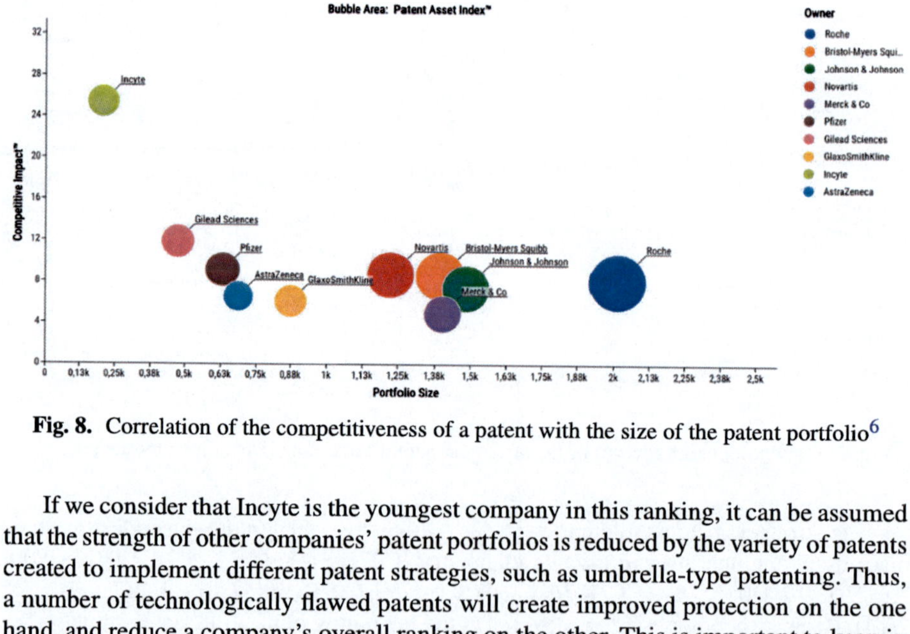

Fig. 8. Correlation of the competitiveness of a patent with the size of the patent portfolio[6]

If we consider that Incyte is the youngest company in this ranking, it can be assumed that the strength of other companies' patent portfolios is reduced by the variety of patents created to implement different patent strategies, such as umbrella-type patenting. Thus, a number of technologically flawed patents will create improved protection on the one hand, and reduce a company's overall ranking on the other. This is important to keep in mind when assessing the patent portfolios of competitor companies, including in order to monitor when a particular technology does fall out of protection. Also, Incyte's young age, combined with this high average patent strength, may point to a careful selection of subjects for patenting, a strong IP protection strategy, and a strong workforce within the organization.

[6] Formed by using Lexis Nexis tools.

Let's take a closer look at Incyte's patent portfolio (in terms of class A61P) to analyze how the high average patent strength will affect the exit from protection of their actual technologies. The company's main developments located in the technological sectors: substd, catalyst, optionally substituted, therapeutic treatment, carboxylic acid, therapeutic agent, peptide, composition, chromatography.

Incyte Corporation is an American biopharmaceutical company specializing in the development of therapeutic MPs and further commercialization of patented versions. The company's developments are mainly focused on the treatment of cancer and autoimmune diseases. To date, the company has 7 FDA-approved MPs and 23 MPs in clinical trials. The portfolio includes 20 molecular targets. The company's historical portfolio relates to genes and proteins. The non-exclusive license of this portfolio is owned by leading pharmaceutical players such as Phizer, Johnson&Johnson, Abbott Laboratories, Hoechst AG, Novo Nordisk Als and Pharmacia&Upjohn Inc.

Most of the patents were applied for internally or by employees. A few strong patents, apparently raising the overall average strength of patents in the portfolio, were bought from Fresh Express Incorporate.

Almost all patent families have a degree of geographical coverage of legal protection ranging from 2.5 to 3.5. More than half of the patents (180) are granted in the US, 82 in Taiwan, 80 in Canada, 77 in Japan, 74 in Australia, 69 in China and Korea, 67 in Argentina, 55 in the Philippines, 52 in Mexico and 49 in Peru. European countries have, on average, fewer than 50 patents registered (see Fig. 9).

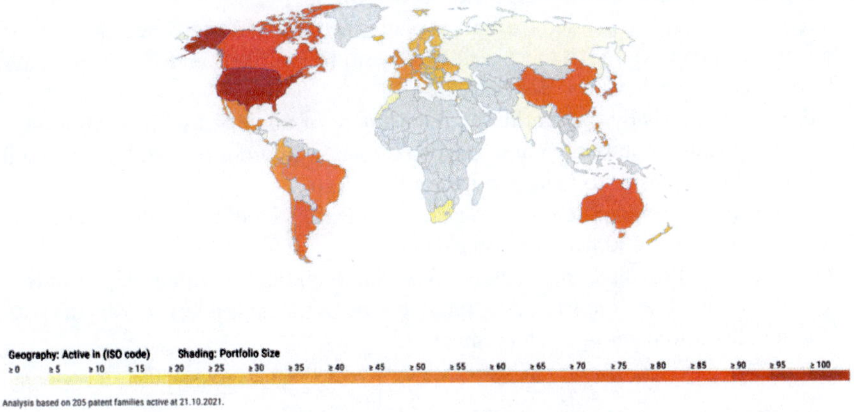

Fig. 9. Incyte's patent portfolio on the map.

For a company with such a short history in the market, this scope is quite wide, one thing being the structure of the corporation's subsidiaries. The company has no representative office in Russia and it is clear that Russia is not a target market for such a company, based on sectors.

Incyte's patent portfolio consists of various proteins, protein detection methods, DNA, protein homologues, antigens, antibodies and polynucleotides. To properly analyze the patent cliffs of such multi-faceted portfolios it is necessary to develop a tailor-made study design taking into account the specificities of the patent portfolio. In this case, it will consist mainly of inventions, with the company's MPs being mostly "targeted", which means targeting specific therapeutic target molecules (most often used in the treatment of cancer, malignant cells). This means that one MP can target specific markets rather than being created for the whole world at once. This can be seen in the significant number of MPs that have been introduced into the South American market.

Incyte's patent portfolio study design structure:

1. portfolio allocation by priority date, structuring of patent families according to their expected date of expiry of legal protection;
2. an analysis of the inactive part of the portfolio, to identify the average life span of a patent. The hypothesis for Incyte's portfolio would be that the majority of patents, 95% will maintain their status throughout the 20 years, due to the fact that they are prone to additional commercialization in the form of licensing. If a company does not have its own resources to carry out independent research in the clinical and post-registration phase, it can sell such an invention to other companies. Incyte has produced 4 partner MPs with other pharmaceutical companies;
3. using the resulting average patent lifespan, the maximum term of protection and prospective term of protection are plotted. The time gap between these dates is compared to predict the development of own MPs.
4. patent family map is retained – a market plan is formed to start marketing the product before the patent expires (if the 2-year PCT period has already ended and applications in those countries have not been published).
5. additional analysis by the marketing department of countries with legal protection to identify future competitiveness in such a market in order to avoid the possibility of technology transfer to local production;
6. an analysis of geographical coverage concludes by identifying the timeline of a competitor's filing and patenting strategy;
7. the list of inventors that are included in a patent application often helps to identify subsidiaries in other countries. And the structure of patent holders reveals the pattern of asset allocation among such companies.

5 Conclusion

Despite the indisputable role of knowledge in science-based companies, PIS is seen as an integral part of research in a broad sense. The broader view of PIS is not always justified by the specifics of the industry.

PIS (in narrow sense) involves a process of communication between the patent analyst and the creator, whereby, through consultation or the provision of analytic-synthetic processing of patent information, the creator receives data for decision-making in his research area or a qualitative analysis of possible solutions in relation to the overall IP management strategy of the enterprise.

Communication between the departments accompanying the development of the MPs in all phases of the LC is extremely important. And it is worth paying attention to the fact that none of the stages should be overlooked.

The low geographical coverage of patent rights in MPs is due to the lack of international novelty or to the fact that there is no provision for such products to be manufactured for export.

Patent informatics is one of the key instruments for the implementation of the approach and shows the highest performance in the implementation steps of the approach.

Acknowledgments. The research described in this paper is partially supported by the ITMO University Research program (№619403).

References

1. Damodharan, S.: Product development strategy definition using qualitative patent analysis. IEEE Technology & Engineering Management Conference (TEMSCON), IEEE (2020). www.ieeexplore.ieee.org/stamp/stamp.jsp?tp=&arnumber=9140130. Accessed 17 December 2021
2. International Patent Classification (IPC): World Intellectual Property Organization, Geneva, p. 7 (2020)
3. Koroleva, E.V.: Innovative process and its information support. Sci. J. NRU ITMO: Eco. Environ. Manage. 2, 24–34 (2015)
4. Kortov, S.V., Shulgin, D.B., Tolmachev, D.E., Egarmina, A.D.: Analysis of technological trends based on the construction of patent landscapes. Reg. Econ. 13(3), 935–947 (2017)
5. Nikolaev, A.S.: Key approaches for determining the role of patent analytics in innovation management. Innovations **12**(254), 119–123 (2019)
6. Fabry, B., Ernst, H., Langholz, J., Koster, M.: Patent portfolio analysis as a useful tool for identifying R&D and business opportunities – an empirical application in the nutrition and health industry. WPI **28**(3), 215–225 (2006)
7. Pharmaceuticals Market Research Reports & Industry Analysis: https://www.marketresearch.com/Life-Sciences-c1594/Pharmaceuticals-c89/
8. PubChem [Internet]:Bethesda (MD): National Library of Medicine (US), National Center for Biotechnology Information. PubChem Compound Summary for CID 5282448, Fondaparinux [cited 2021 Dec. 14] (2004). https://pubchem.ncbi.nlm.nih.gov/compound/Fondaparinux
9. Sick, N., Merigó, J.M. Krätzig, O., List, J.: Forty years of World Patent Information: a bibliometric overview. World Patent Info. 64 (2021). www.sciencedirect.com/science/article/pii/S0172219020301022. Accessed 16 December 2021
10. Suri, F.K., Banerji, A.: Super generics – First step of Indian Pharmaceutical Industry in the innovative space in US market. J. Health Manage. **18**(1), 161–171 (2016)
11. World Preview 2016: Outlook to 2022/EvaluatePharma, 9th edn., September 2016. https://info.evaluategroup.com/WP2016EPV.html
12. Zhang, L., Guo, Y., Sun, G.: How patent signals affect venture capital: the evidence of biopharmaceutical start-ups in China. Technol. Forecast. Soc. Change **145**, 93–104 (2019)
13. Yang, X., Yu, X.: Identifying patent risks in technological competition: a patent analysis of Artificial Intelligence industry. In: 2019 8th International Conference on Industrial Technology and Management (ICITM), pp. 333–337 (2019)

Correction of the Direction of the Pursuer's Movement with the Method of Parallel Approach to the Target in Space

Dubanov Alexander(✉)

Buryat State University, Ulan-Ude, Russian Federation
alandubanov@mail.ru

Abstract. In this article, a kinematic model of iterative calculation of the pursuer's trajectory is proposed for consideration. Based on the materials of the article, a mathematical model of constructing the pursuer's trajectory approaching the target in space is compiled. The speeds of the pursuer and the target at the beginning of the iterative process are directed arbitrarily. The point of the presented method is that the pursuer smoothly changes the movement direction with an acceptable angular velocity until it is aligned with the direction to a point belonging to the Apollonian circle, as in the parallel approach method. The permissible angular velocity is determined by the condition that the curvature radius of the pursuer's trajectory cannot be less than a certain value at a given movement velocity. The developed quasi-discrete model for constructing the pursuer's trajectory may be in demand by software developers for unmanned barrage devices with autonomous control.

Keywords: Trajectory · Angular velocity · Curvature restriction · Pursuer · Target · Interception · Apollonian circle

1 Introduction

Earlier, in the works of R. Isaacs [1], L. A. Petrosyan [4], L. S. Pontryagin [2], the method of parallel convergence was considered. The method of parallel approach on the plane is based on the fact that the directions of the pursuer P and the target T at a given time t intersect at point K (Fig. 1).

The point K belongs to the Apollonius circle. The Apollonius circle is the geometric location of points K, for which the ratio of distances to points in the plane P and T is unchanged $|KP|/|KT| = const$. In the works of the founders of game theory, it is shown that the strategy of parallel convergence on the plane is optimal for the pursuer that is, providing the shortest time to intercept the target.

R. Silhavy (Ed.): CSOC 2022, LNNS 501, pp. 262–269, 2022.
https://doi.org/10.1007/978-3-031-09070-7_22

In the case of parallel convergence on the plane, the ratio takes place to be constant:

$$\frac{|KP|}{|KT|} = \frac{\left|\vec{V}_T\right|}{\left|\vec{V}_T\right|} = k,$$

where V_P is the velocity vector of the pursuer P, and $\vec{V}_T$ is the velocity vector of the target T. The radius of the Apollonian circle is (Fig. 1):

$$R = \frac{k}{|k^2 - 1|} \cdot |T - P|.$$

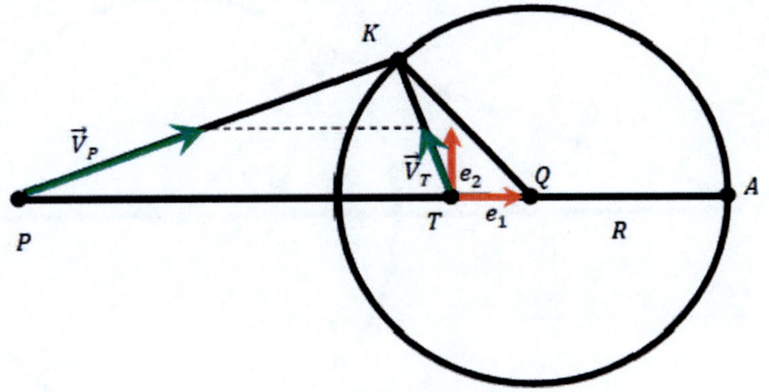

Fig. 1. Apollonian circle

The center of the Apollonian circle is calculated as follows (Fig. 1):

$$Q = P + \frac{k^2}{k^2 - 1} \cdot (T - K).$$

Since the directions of the pursuer's and the target's velocities, $\vec{V}_P$ and $\vec{V}_T$, respectively, are dependent on each other, or rather, they should be directed to the point K on the Apollonian circle at a given moment in time. By setting the value of the vector $\vec{V}_T$, we determine the point K and the value of the vector $\vec{V}_P$.

The point K (Fig. 1), depending on the vector V_T is calculated as follows, it is the solution of the system of Eq. (1):

$$\begin{cases} (K - Q)^2 = R^2 \\ K = T + \vec{V}_T \cdot \tau \end{cases}. \tag{1}$$

The system of Eq. (1) is transformed to a quadratic equation, with respect to the parameter τ:

$$\left(\vec{V}_T \cdot \vec{V}_T\right) \cdot \tau^2 + 2 \cdot (T - Q) \cdot \vec{V}_T \cdot \tau + (T - Q)^2 - R^2 = 0.$$

The resulting value t is substituted into the second equation of the system (1) and we get the coordinates of the point K on the Apollonius circle depending on the vector $\vec{V}_T$ of the target velocity T. The point K determines the direction of the vector $\vec{V}_P$ of the pursuer's speed P.

2 Problem Statement

The purpose of this article is to develop a quasi-discrete model for constructing the pursuer's trajectory. The pursuer and the target are moving in space. The speed vector of the pursuer is directed arbitrarily (Fig. 2).

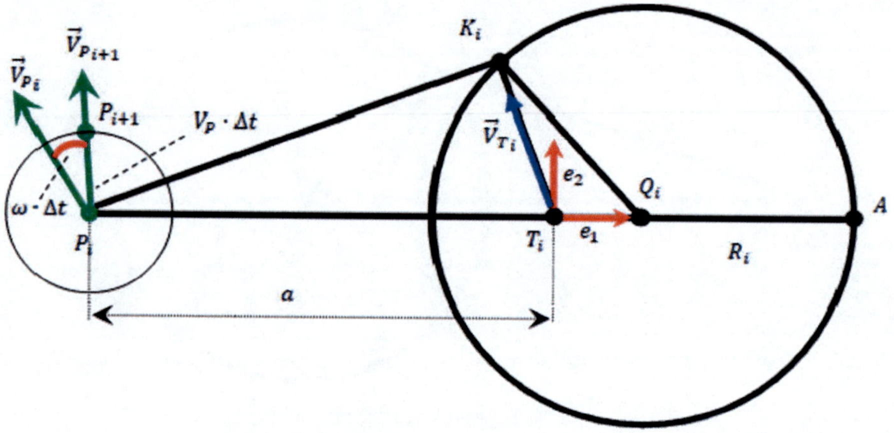

Fig. 2. Direction correction in parallel approach

At certain discrete points in time, the pursuer will make a turn at an acceptable angle and take a step at an acceptable distance. The plane is defined by the vector $(\overrightarrow{PK})$, the velocity vector $\vec{V_P}$ and the position of the pursuer P.

For certainty, since the test program is written in the MathCAD computer mathematics system, in the geometric model of this article it is assumed that the goal T moves rectilinearly and uniformly.

3 Theory

3.1 Step and Turn Formation

Let at some point t_i pursuer is in point of space P_i, while having a speed vector $\vec{V}_{P_i}$, and the target is at the point T_i at a speed $\vec{V}_{T_i}$ (Fig. 2). The pursuer can't change direction instantly, so in the model it is considered that the trajectory of the chaser restrictions on the curvature. The radius of curvature at the points of the trajectory cannot be less than a certain value R_{min}.

Accordingly, the instantaneous rotational speed cannot exceed the value of:

$$\omega = \frac{V_P}{R_{min}}.$$

V_P is the velocity module of the uniform movement of the pursuer.

During the time interval Δt, into which the quasi-discrete iterative pursuit process is evenly divided, the pursuer can make a turn by an angle $\alpha = \omega \cdot \Delta t$.

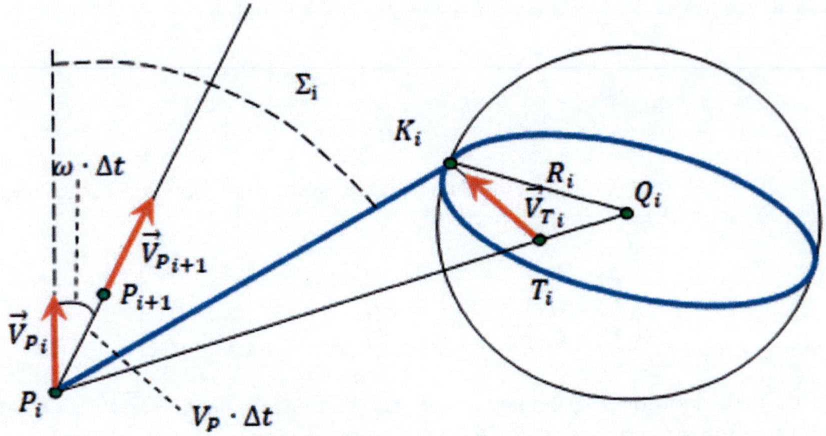

Fig. 3. Iterative process in space

The Apollonius circle in space (Fig. 3) at a certain moment of time t_i lies in the plane formed by the point P_i, the line of sight (P_iT_i) and the velocity vector $\vec{V}_{T_i}$.. To find the point K_i, it is necessary to solve the system of Eq. (2):

$$\begin{cases} (K_i - Q_i)^2 = R_i^2 \\ K_i = T_i + \vec{V}_{Ti} \cdot \tau \end{cases}. \quad (2)$$

The first equation of system (2), in contrast to system (1), shows that this is not a circle, but a sphere of radius R_i centered at point Q_i, and the problem is considered in space. The radius of the sphere containing the Apollonius circle: $R_i = \frac{k}{|k^2-1|}|T_i - P_i|$, where k is the ratio of the velocity modules of the pursuer and the target at time t_i. The center of the sphere is calculated by the formula:

$$Q_i = P_i + \frac{k^2}{k^2 - 1} \cdot (T_i - K_i).$$

The pursuer during a period of time Δt can make a turn at an angle ω × Δt. The rotation will be performed in the plane Σ_i. The plane Σ_i is formed by the line of sight (P_iK_i), the vector $\vec{V_i}$ of the pursuer's movement direction and the reference point P_i.

3.2 Building a Local Basis

In order to make a step and turn in the plane Σ_i, it is advisable to go to the local basis (E_1, E_2, E_3), where the basis vectors are calculated as follows (Fig. 4):

$$E_1 = \frac{K_i - P_i}{|K_i - P_i|}, E_3 = \frac{(K_i - P_i) \times \vec{V}_{P_i}}{\left|(K_i - P_i) \times \vec{V}_{P_i}\right|}, E_2 = \frac{E_3 \times E_1}{|E_3 \times E_1|}.$$

Figure 4 also shows the world coordinate system (H_1, H_2, H_3) centered at point O. The basic vectors of the world coordinate system look like this:

$$H_1 = \begin{bmatrix} 1 \\ 0 \\ 0 \end{bmatrix}, H_2 = \begin{bmatrix} 0 \\ 1 \\ 0 \end{bmatrix}, H_3 = \begin{bmatrix} 0 \\ 0 \\ 1 \end{bmatrix}.$$

The basis vectors of the world coordinate system in the coordinate system (E_1, E_2, E_3) look like this (h_1, h_2, h_3):

$$h_1 = \begin{bmatrix} E_1 \cdot H_1 \\ E_2 \cdot H_1 \\ E_3 \cdot H_1 \end{bmatrix}, h_2 = \begin{bmatrix} E_1 \cdot H_2 \\ E_2 \cdot H_2 \\ E_3 \cdot H_2 \end{bmatrix}, h_3 = \begin{bmatrix} E_1 \cdot H_3 \\ E_2 \cdot H_3 \\ E_3 \cdot H_3 \end{bmatrix}.$$

If the world coordinate system pursuer P_i takes a step in time Δt : $\overrightarrow{(\Delta P)} = P_i + \vec{V}_{P_i} \cdot \Delta t$ in the coordinate system (E_1, E_2, E_3) with center in the point P_i step will correspond to the point:

$$P^* = \begin{bmatrix} \left(\overrightarrow{\Delta P} - P_i\right) \cdot E_1 \\ \left(\overrightarrow{\Delta P} - P_i\right) \cdot E_2 \\ \left(\overrightarrow{\Delta P} - P_i\right) \cdot E_3 \end{bmatrix}.$$

The angle $\propto$ between the direction vector $\vec{V}_{T_i}$of the pursuer and the line of sight (P_iK_i) is equal to:

$$\propto = arccos\left(\frac{\vec{V}_{P_i} \cdot E_1}{\left|\vec{V}_{P_i}\right| \cdot |E_1|}\right).$$

The permissible angle of rotation $\propto_0$ is$\propto_0 = \omega \cdot \Delta t$. The position of the pursuer P^{**} after the step and rotation in the coordinate system (E_1, E_2, E_3) will be calculated as follows:

$$P^{**} = \begin{bmatrix} |P^*| \cdot cos(\propto_n) \\ |P^*| \cdot sin(\propto_n) \\ 0 \end{bmatrix}.$$

where $\propto_n = \propto - \propto_0$

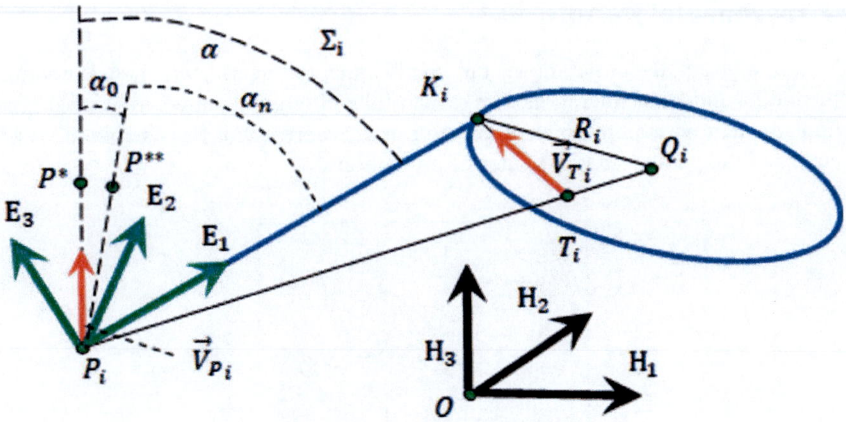

Fig. 4. Construction of a local basis

The transition of point P^{**} from the local coordinate system (E_1, E_2, E_3) centered at point P_i to the world coordinate system centered at point O (Fig. 4) is carried out as follows:

$$P_{i+1} = \begin{bmatrix} P^{**} \cdot h_1 \\ P^{**} \cdot h_2 \\ P^{**} \cdot h_3 \end{bmatrix} + P_i.$$

Accordingly, the new speed vector of the pursuer will be as follows:

$$\vec{V}_{i+1} = V_P \cdot \frac{P_{i+1} - P_i}{|P_{i+1} - P_i|}.$$

where V_P is the velocity modulus of the pursuer's uniform movement.

3.3 Condition for Switching to the Line of Sight

Based on the materials of the article, a test program was written in the MathCAD computer mathematics system, in which the target moves in space rectilinearly and uniformly. At the beginning of each cycle of this program, the angle α is calculated (Fig. 4). The angle ∝ is compared with the value of the angle of possible rotation ∝ _0 for a period of time Δt.

If the angle is $\propto > \propto_0$, then the algorithm calculates the step and rotation as indicated in formulas (3) and (4).

If the angle $\propto \leq \propto_0$, then the pursuer 's next step P_{i+1} belongs to the line of sight $(P_i K_i)$, i.e. $\propto \_n = 0$ (Fig. 4). Accordingly, the velocity in the following time $\vec{V}_{Pi+1}$ will be collinear with the line of sight $(P_i K_i)$.

4 Experimental Results

In the test program, the target moves uniformly and rectilinearly, which does not matter for the model under consideration. The result of the program is shown in Fig. 5. Figure 5 is supplemented with a link to an animated image where it will be possible to view the trajectory of the pursuer in two planes of projections.

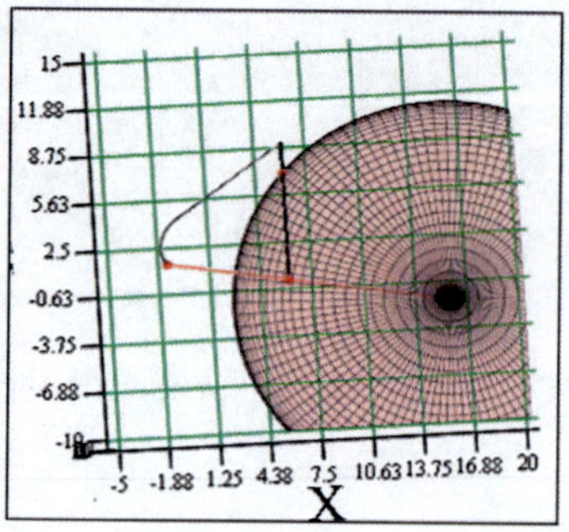

Fig. 5. Simulation of the pursuer's trajectory

5 Discussion of the Results

In the test program, the pursuit process takes place on a piece of space with dimensions [−10;20] × [−10;2] × [−10;20]. The pursuer's uniform movement velocity modulus is 16 m/s, the target speed is 12 m/s. The pursuer's direction vector at the moment of the beginning of the pursuit process is equal to:

$$\begin{bmatrix} -0.579 \\ 0.579 \\ 0.573 \end{bmatrix}.$$

The target moves uniformly and rectilinearly along the vector:

$$\begin{bmatrix} 0 \\ 1 \\ 0 \end{bmatrix}.$$

The radius of curvature of the pursuer's trajectory should not be less than two meters.

Figure 5 shows the trajectory of the pursuer's movement from the moment of the beginning of the pursuit to the interception of the target after a certain period of time, the trajectory of the pursuer moves to a straight section. While moving in a straight line, the speed of the pursuer is directed at a point on the Apollonian circle, which is located in the plane formed by the vectors of the pursuer's movements, the target and the location of the pursuer.

Until the moment of transition to a rectilinear section, the trajectory of the pursuer is a smooth curve that satisfies the curvature constraints.

6 Summary and Conclusion

The method for correcting the direction of the pursuer movement was developed for the case when the initial speeds of the pursuer and the target are arbitrarily directed. In this method, there is a gradual change in the direction of the pursuer until it is directed to a point on the circumference of Apollonius. The circle of Apollonius is given by the target's velocity vector, the pursuer's and target's coordinates, and the pursuer's velocity modulus.

This method can be useful for developers of unmanned aerial vehicles with autonomous control.

References

1. Isaacs, R.: Differential Games. Mir, Moscow (1967)
2. Pontryagin, L.S.: Linear Differential Game of Evasion. Tr. MIAN USSR., T. 112. pp. 30–63 (1971)
3. Krasovsky, N.N., Subbotin, A.I.: Positional Differential Games. M, Nauka (1974)
4. Petrosyan, L.A.: Differential Games Of Pursuit. Izd. LGU, 222 p (1977)
5. Vagin, D.A., Petrov, N.N.: The task of pursuing coordinated fugitives. In: Proceedings of the Russian Academy of Sciences. Theory and control systems, No. 5, pp. 75–79 (2001)
6. Bannikov, A.S.: Some non-stationary problems of group pursuit. Proc. Inst. Math. Inf. UdGU 1(41), 3–46 (2013)
7. Bannikov, A.S.: Non-stationary problem of group pursuit. In: Proceedings Lobachevsky Mathematical Center, vol. 34, pp. 26–28. Kazan Mathematical Publishing House Society, Kazan (2006)
8. Izmest'ev, I.V., Ukhobotov, V.I.: Acoust. The task of pursuing low-maneuverable objects with terminal set in the form of a ring. In: Proceedings of the international conference "Geometric Methods in Control Theory and Mathematical Physics: Differential equations, integrability, qualitative theory", Ryazan, September 15–18, 2016, Itogi science and technology. Subject. obz., 148, VINITI RAN, Moscow, pp. 25–31 (2018)
9. Video, the results of the simulation of the problem of pursuit in space. https://youtu.be/8nDUSi3ENB4

Methodological Problems Related to Big Data Processing and Analysis

Kamelia Shoilekova(✉)

Angel Kanchev University of Ruse, 8 Studentska Street, 7000 Ruse, Bulgaria
kshoylekova@uni-ruse.bg

Abstract. Big data, structured or unstructured, exceed the processing complexity of conventional databases. Information flows, coming from many separate and independent sources, pile up in complicated and evolving relations into continually growing big data sets. The availability of huge data volumes that are daily delivered to organizations and companies becomes a prerequisite for searching adequate processing methods allowing timely elimination and analysis of issues provoked by the big data sets. The paper analyzes the main problems connected with big data operations and proposes options for coping with them.

Keywords: Big Data · Data silos · Finding the signal in the noise · Inaccurate data · Software platforms

Big data is rapidly entering our everyday life. Companies seek and find ways to get the benefit from it. Proper processing and analysis of big data help decision-making and it leads to company optimization [7, 8], redirection of traffic to avoid data congestion, etc. Therefore, big data is in favor of business, medicine and other sectors [10]. Over 50% of companies declare that big data leads to an increase of their income because it provides a profound analysis of the information related to client's behavior. For a big part of companies, the possibility of getting such knowledge means a 360-degree view to their customers. Based on the above stated, it follows that the biggest issue for most organizations is to understand how they can get value from all the data processed and analyzed.

1 Big Data Processing Operation

Big data processing operations pass via the following stages:

1. Generation – incidentally or purposefully receiving data from a source:
   The generation stage is analogous to the observation stage meaning that the errors are in a sense analogous to the ones omitted during the process of collecting initial data, i.e. it may be mistaken or incomplete.
2. Extraction/Transformation/Loading (ETL) – combining all the data in a homogenous computing system in three steps:

R. Silhavy (Ed.): CSOC 2022, LNNS 501, pp. 270–275, 2022.
https://doi.org/10.1007/978-3-031-09070-7_23

a. Extraction – data is collected from relevant sources, "fragmented", validated and stored;
b. Transformation – data is translated, encoded, re-encoded, aggregated/disaggregated and/or edited;
c. Loading – data is integrated and stored.

3. Analytics – data is turned into information through:

   a. Filtration (sample release)/reduction – unwanted characteristics and content are cut out, some characteristics are combined to obtain new ones, data elements are removed or added for easier analysis and processing on the next steps;
   b. Computations/analytics/visualization – data is analyzed and/or presented for interpretation and information extraction.

Two basic methods are mainly applied for big data analysis:

- Descriptive analytics – aims to make statistics for the current and historical data and provides information for what has happened and why it has happened. It includes methods like statistical modeling, report generation for the occurring trends, visualization, association and correlation analysis;
- Predictive analytics – it focuses on familiar information (teaching data) that includes the input properties of data (attributes) and the values of responses (target models) to build a predictive model (solution) for making forecasts of the indiscernible data (test data). It uses methods based on vector machines, linear regression/classification, non-linear regression (generalized linear model), solution tree, Bayes theorem, neural networks, etc. [3].

On this stage, there is a risk of noise accumulation, false correlation and incidental endogeneity, which may be complicated by sampling and non-sampling errors. Concerning sampling errors, the data can be filtered, included in a sample, or reduced for obtaining representative and easily controllable data. These processes may need additional transformations demanding considerable efforts and time for preparing and combing data from different sources.

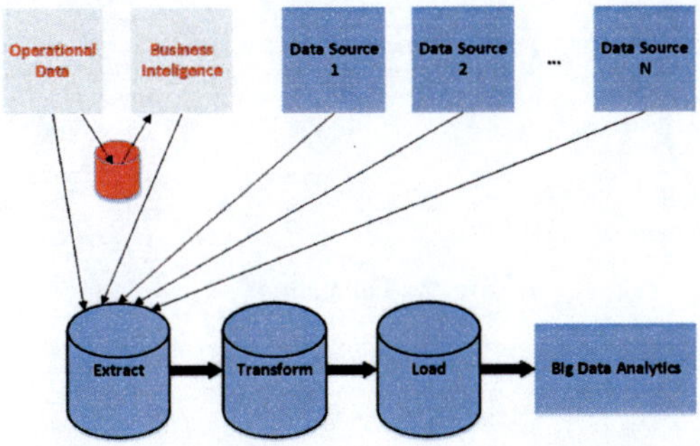

Fig. 1. Map of Big Data processing operation

Figure 1 represents Big Data processing operation and data flow movement as per the foregoing stages.

Big Data processing operation passes through several phases, which have a number of problems behind them. Problems elimination leads to more precise data and better outcomes.

2 Big Data Basic Problems

Only 27% of all Big Data project managers, who were involved in the CapGenimi surveying report, announced that they had successful initiatives [1]. This circumstance shows the existing gap between big data theoretical knowledge and the actual application of theory into practice [9]. The top five key problems related to big data can be defined as follows:

1. "Finding the signal in the noise"
   It is difficult to extract information from a huge data volume. Frequently, after data analysis, the conclusion admits the lack of insights as the investigated variable was incorrect. In result, a new analysis has to be conducted. The process goes on until a certain pattern in the analyzed data is actually detected and that is defined as "finding the signal in the noise".
2. Data silos
   Data silos are the weak point of big data processing [2]. They represent a collection of data that is not accessible at all levels of the organization hierarchy. They become a repository of all the data coming from multiple sources in various formats. Therefore, no insights can be gained from that data because it is plainly not integrated. The best method for coping with data silos is data integration. In other words, data silos create expensive and time-consuming problems for businesses but they are relatively easy to be coped with the help of data integration (Fig. 2).

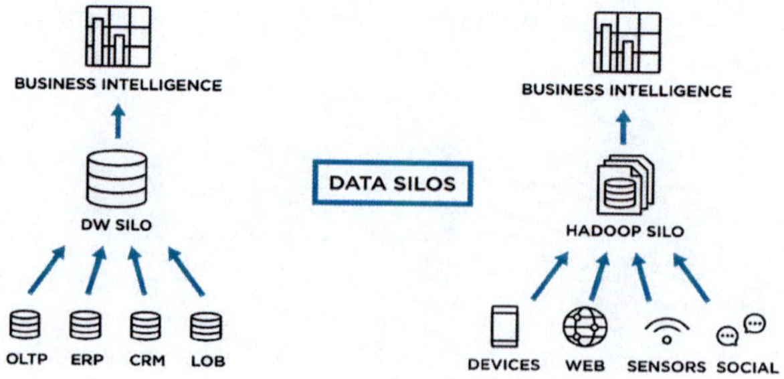

Fig. 2. Data silos [2]

3. Inaccurate data
   Besides being ineffective, the data silos are the reason for having a big data amount that proves inaccurate at a moment. The best way to combat with incorrect data is to eliminate silos by data integration.

   Figure 3 shows the stages of data assessment process. The fulfillment of these stages is of great importance for achieving data accuracy and reliability.

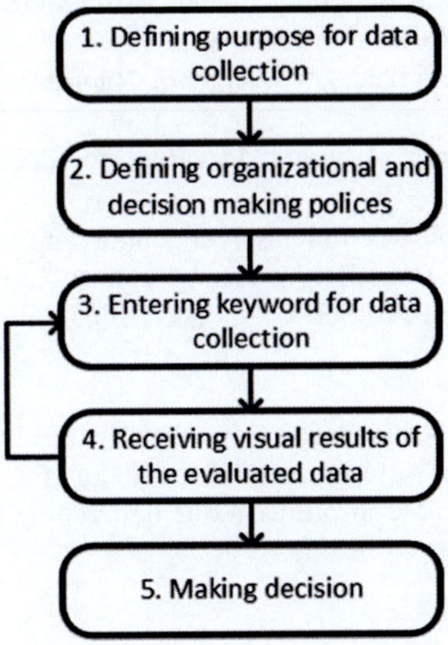

Fig. 3. Assessment process of data reliability

4. Technologies develop too fast
   Technologies are developing too fast over the recent years. Doubtless, they have an impact on the big data issues, as well. The trends in technologies development regarding big data processing are - to achieve the ability of a quick data analysis aiming to find and adequately use the target information as fast as possible. Hence, big data processing technologies aim to achieve the ability to catch things "in the act" so as to favorably effect the outcome, i.e. the ability to make quick decisions and act promptly in response to the insights gained in the process of big data analytics is of significant advantage.
5. Lack of skilled workers
   Many companies have troubles in finding skilled data analyzing specialists. The point is that employees, who are suitable for working with big data, must not only understand the data from a scientific perspective, but also they must be well acquainted with the business and its clients, placing the attention on how they can make use of the data findings [11]. Big data operations and problem solving require a team of experts including employees with expertise in the specific subject domain, researchers (methodologist), IT specialists, and system administrators.

3 Big Data Software Platforms and Processing Tools

Big Data problem solutions need adequate software platforms and tools for data processing and analysis. The software market of big data processing offers hundreds of tools and complex platforms.

The most efficient software tools for Big Data processing can be listed according to their destination and functions as it follows [6]:

- Data storage and control – Haddop, Cloudera, MongoDB, Talend;
- Data cleansing – OpenRefine, DataCleaner;
- Knowledge retrieval from data sets – Xplenty, RapidMiner, Orange, Weka, KNIME [4];
- Data analytics – Xplenty, Adverity, Dataddo, Apache Hadoop, CDH (Cloudera Distribution for Hadoop) [4];
- Data visualization – Tableau Public, Solver, Datawrapper [5];
- Data integration – Hevo Data, IRI Voracity, Xplenty, ZigiWave [4];
- Data languages – Python, JavaScript, Scala, R, SQL, Julia;
- Data gathering - Import.io.

The choice of a correct big data analysis platform normally depends on several important factors like data volume; platform velocity for real-time data processing; development model applicable in offline mode demanding a longer time, or in online mode displaying almost instant results.

4 Conclusion

Big Data analysis is based on complex technologies applied for processing huge data volumes, arranged in multiple sources of unstructured and structured data, some delivered as data streams in real-time mode. The successful analysis of Big Data provides the executives with new, recently unavailable information from external and internal sources that improves and speeds up the process of decision-making. The access to such kind of knowledge is determined by the preliminary elimination of the main key problems related to big data processing.

References

1. https://reports.capgemini.com/
2. https://www.tibco.com/reference-center/what-is-a-data-silo
3. Chaovalitwongse, W., Huang, S.: Computational Optimization and Statistical Methods for Big Data Analytics: Applications in Neuroimaging. http://faculty.washington.edu/artchao/INFORMS_Tutorials_2015-Web.pdf
4. www.softwaretestinghelp.com
5. https://bigdata-madesimple.com/top-30-big-data-tools-data-analysis/
6. All the best big data tools and how to use them. https://www.import.io/post/all-the-best-big-data-tools-and-how-to-use-them/

7. Krastev, G., Voinohovska, V.: Smart mobile application for public transport schedules – data organization and program realization. In: 2020 International Congress on human-computer interaction, optimization and robotic applications (HORA), pp. 1–4 (2020). https://doi.org/10.1109/HORA49412.2020.9152884
8. Krastev, G., Voinohovska, V.: Smart mobile application for public transport schedules - logical model. TEM J., ISSN 2217-8309 (2020) https://doi.org/10.18421/TEM92-15
9. Atanasova, G., Hristova, P.: The significance of software psychology knowledge in computer science related courses. In: INTED2020 Proceedings, pp. 3225–3229 (2020)
10. Petrov, V., Rusev, R. Rusev, I.: Anaesthesiology in the big data era - are we ready for that? Anaesthesiol. Int. Care, ISBN: 1310-4284 (2021)
11. Rusev, I., Rusev, R., Ivanova, B.: Training of medical staff for using data – usage of the full capability of smart devices. In: EDULEARN21 Proceedings, pp. 8804–8810 (2021)

Selection of Terms and Process Optimization of Loading and Unloading System of Viscous Oil Products

A. N. Qurbanov, Almaz M. Mehdiyeva(✉), and İ. Z. Sardarova

Azerbaijan State Oil and İndustry University, Baku, Azerbaijan
almaz.mehdiyeva@asoia.edu.az

Abstract. To perform technological operations for filling and draining petroleum products at oil tank farms, special structures are built and various technologies are applied depending on the viscosity of petroleum products. For highly viscous petroleum products for the purpose of erosion of solid residues, an upper erosion system with the use of special devices is recommended. The drainage system provides the possibility for erosion of solid residual fuel oil at the same time as the bottom circulation heating to use the system of high erosion. To optimize the drain operations, the automation system is applied. The applied technology provides an opportunity in winter to reduce the drain time of high-viscosity products by about half. The main advantage of the device of the upper erosion system is the provision of the planned volume of transshipment of highly viscous petroleum products.

Keywords: Fuel oil · Discharge operations · High-viscosity petroleum products · Solidification · Viscosity

1 Introduction

Oil products intake and release technology depends on the type of transport means for oil products delivery and shipment, weather conditions, the intensity of handling operations and the physical and chemical properties of petroleum products.

Petroleum products are transported by pipeline, railway, road, sea and river transport in accordance with the rules valid at each mode of transportation, approved in accordance with the established procedure.

Petroleum products should be accepted and released only through special handling facilities:

- in railway tanks-on special load racks, through separate filling pipes or discharge plants;
- to sea and river vessels-through berthing facilities or by a non-mooring method;
- in car tanks - at filling stations, car racks, through separate filling pipes;
- in barrels, cans and other containers-through filling facilities;
- through off takes of main oil product pipelines.

R. Silhavy (Ed.): CSOC 2022, LNNS 501, pp. 276–287, 2022.
https://doi.org/10.1007/978-3-031-09070-7_24

The temperature of the filling oil should not exceed 30 °C, the temperature of high-viscosity heated petroleum products during loading is determined in accordance with the shipping rules and safety requirements.

High viscosity heated petroleum products should be filled into railway tanks and tankers at temperatures specified in shipping rules, regulatory documents for petroleum products, taking into account the labor protection rules.

2 Materials and Methods

To facilitate the process of loading and unloading of petroleum products transported by rail, a railway loading/unloading rack is used [1, 2]. In order to carry out technological operations for loading and unloading tank wagons at oil depots, special facilities are built and various technologies are applied, depending on the viscosity of oil and petroleum products and the minree of their fire hazard and explosion hazard [3–6]. With regard to transshipment of petroleum products, including fuel oil, the operation is rather complex and responsible, requiring a reasonable approach to hydraulic and thermal calculations ensuring fuel preparation, and the availability of modern, certified equipment for draining tanks that can ensure the safety of the technological operations and the ability to bring the rheological properties of structured liquids to the required values [7].

Viscosity, density, pressure and temperature are the most important technical characteristics of viscous petroleum products, which determine both the methods and duration of dumping operations, the conditions of transportation and pumping, and hydraulic drags during transportation. Changes in these characteristics are described by the equations obtained in theoretical experimental studies [8].

Due to the fact that during draining, loading and transport operations, heavy fuel oil is repeatedly exposed to heat-transfer, this can cause a change in their viscosity and density. The dependence of the viscosity of fuel oil on temperature is approximately described by Walter's equation

$$lglg\left(\nu \cdot 10^{-6} \div 0.8\right) = A - B \cdot lgT \tag{1}$$

where: ν - magnitude coefficient of kinematic viscosity, m^2/c; A & B - empirical coefficient, which defines heavy fuel oil type; T - absolute temperature, ^{o}K.

Viscosity of heavy fuel oil changes under pressure. The more the pressure increases the more the viscosity goes up, aligned with the studies by P.V. Bridgman [8]. The viscosity increases ten and a hundred times under super-high pressure which is seemingly related to the deformation of deformation molecules, i.e., to the decrease in their volume. In temperature 175 °C and pressure variation between 0,1–14,72 MPa range, viscosity dependence on the pressure might be recognized linear:

$$\mu_p = \mu_0(I + a_t \cdot p) \tag{2}$$

where: μ_p - viscosity under pressure p, Pa·s; μ_0 - viscosity under atmospheric pressure Pa·s; a_t - invariable for current temperature (when t = 175 °C $a_t = 2, 65 \cdot 10^{-2}$ m/MN); p - manometric pressure, MPa.

There normally exists unique dependence between density and viscosity – once the density grows, the viscosity increases too. Relative density under thermal forces is calculated by the following formula:

$$\rho^{t} = \rho^{t=20} + \gamma(20 - t) \quad (3)$$

where: ρ_t - relative density of heavy fuel oil in the reservoir, kg/m^3; γ - average correction for temperature of density, 1/°C; t - temperature in the reservoir, °C.

For the properties of fuel oil in these conditions, in addition to the Newtonian viscosity, it is necessary to take into account their rheological properties. At temperatures close to the setting temperature, fuel oil is a structured system. Since during the filling-up of loading and handling operations it is necessary to deal with insufficiently heated fuel oil, when additional highly-viscous petroleum products are being drained, the oil bases are given additional time to warm up these products.

To determine the thermal conductivity of fuel oil, the calculated equation proposed by DF Smith [8]:

$$\lambda = 43 \cdot 10^{-9} c_p \rho^{4/3} M^{-1/3} \quad (4)$$

where: c_p - heating capacity J/kg·min, ρ - density, kg/m^3; M - molecular weight.

At a temperature of T = 30 °C, the thermal conductivity of fuel oil can be determined most accurately by the empirical formula proposed by V.B. Zenkiewicz [8]:

$$\lambda = \left[\frac{c_p - 1884}{20000} + \frac{\rho/M}{21,5} + \frac{\nu^{1/9}}{4,3} - 0,0188\right] \quad (5)$$

The formula allows to determine the thermal conductivity of petroleum products with a viscosity at 30 °C from 1 • 10^{-6} to 7 • 10^{-4} m^2/s with an inaccuracy of not more than 5%. Viscous and solidified petroleum products, when drained from railway tanks, are heated to reduce the viscosity by sharp (open) or closed steam with "deaf" mobile or stationary coils or electric heaters. The change in the temperature of fuel oil Δt during the time Δτ is determined from the equation of IG. Shukhov, obtained by separating variables and integrating the heat balance equation [10]:

$$\Delta t = t_k - t_0 = (t_H - t_0) \cdot e^{-m\Delta\tau} \quad (6)$$

where; $m = \frac{K \cdot F}{B_{jc}^{c}}$ - rate of temperature change, 1/c; B_{ij} - heavy fuel oil in the tank, kg; c - heavy fuel oil mean specific heat fuel oil, J/kg·min; K - heat transmission average ratio from heavy fuel oil to ambient medium, Vt/m^2·min; F - inner surface of the tank, m^2; t_H, t_K - accordingly, initial and final temperatures of heavy fuel oil, °C; t_0- surrounding temperature, °C.

Heat-transfer coefficient is defined by the following formula [10]:

$$K = \frac{1}{\frac{1}{\alpha_1} + \sum_{i=1}^{g} \frac{\delta_i}{\lambda_i} + \frac{1}{\alpha_2 + \alpha_B}} \quad (7)$$

α_1 - average coefficient of heat transfer from cooling heavy fuel oil to the tank wall, Vt/m^2·min;

δ_i - respectively, wall thickness of the tank, tank insulation and solidified heavy fuel oil layer, m;
λ_i - relevant coefficient of thermal conductivity (walls, insulation and heavy fuel oil, Vt/h·min;
α_2 - average coefficient of heat emission from the tank into atmosphere, Vt/m^2·min;
α_3 - radiant heat-transfer coefficient, Vt/m^2·min.

Value α_1 is defined by criteria equation:

$$Nu_1 = C(Gr \cdot Pr)_m^n \tag{8}$$

where;

$Nu_1 = \frac{\sigma_1 D}{\lambda}$ - heat transfer criterion of Nusselt (Nusselt number);
$Gr = \frac{g \cdot \beta \cdot D^3 \cdot \Delta t}{\nu^2}$ - Grashof natural convection criteria (Grashof number);
$P_Y = \frac{\nu}{\alpha}$ - criterion physical constants of Prandtl (Prandtl number);
D - tank diameter, m;
β - coefficient of cubic expansion of heavy fuel oil, 1/min;
$\Delta t = t_1 - t_2$ - mean temperature difference, °C;
t_1 - applied temperature, °C;
t_2- average temperature of heavy fuel oil, °C;
ν - kinematic viscosity coefficient, m^2/s;
$\alpha = \frac{\lambda}{\rho \cdot c}$ - thermal diffusion coefficient, m^2/s; g - gravitation acceleration, m/s^2.

Highly viscous and easily congealing oil products should be heated up to the temperature which provide for its kinematic viscosity not more than 600 • 10^{-6} m^2/s taking into account its physical and chemical properties and storage term of oil products [1, 2].

3 Result and Discussion

Heating temperature of viscous oil products (fuel oil) should not exceed 363K. When oil products in the tanks are heated by the heating units on the railway overpass, a water vapor collector must be provided with a tap to each tank and with a mandatory installed stop valve. Currently circulation heating is used to perform discharge operations of high-viscosity petroleum products. The essence of the circulation heating is that cold viscous oil product, being in the rail tank, is heated and washed away by a hot jet of the same oil product, preheated in any heating unit.

Two types of circulation heating have been examined:

1. with closed circulation;
2. with open circulation.

When heated with closed circulation the heated oil product which is pumped through a monitor installed in a railway tank turns into a cold product, and then a mixture of cold and hot oil is pumped from the tank to the heat exchanger and back to the tank.

Thus, closed oil circulation of takes place. After a while, part of the heated oil product is taken to the receiving tank. In case of open circulation part of the oil product, being washed, immediately merges through a tray or trough into tanks with dead point.

Compared to closed circulation, this method has the type of minus that the tanks may overflow during overheating and draining, however plus of this method of heating is that the oil product is not completely watered [3, 4]. The amount of heat necessary for heating petroleum products in tanks is determined by [5]:

$$Q = q_1 + q_2 + q_3 \tag{10}$$

where

$q_1 = M \bullet c_p \bullet (t_k - t_н)$ - the heat required to heat the entire mass of oil M from the initial t to the final temperature t_k; C_p – specific isobaric heat capacity, J/(kg•K); $q_2 = M_n \bullet x$ - the heat required to melt the frozen mineral oil (paraffin) in an amount of M_n. The heat required to melt the frozen mineral oil (paraffin) in an amount of Mn;
x – latent heat of melting of paraffin;
$q_3 = K \bullet F \bullet \tau \bullet (t_{cp} - t_0)$ - thermal losses to the environment.
K – coefficient of heat transfer from oil to the environment, W/(m^2•K);
F – heating surface, m^2;
τ – heat up time;
t_{cp} – average temperature of the oil product during the warm-up period, °C;
t_0 – ambient temperature, °C.

The heating time of fuel oil in the circulating installation τ is determined from the heat balance equation, by separating the variables and integrating:

$$\tau = \frac{B_{ij}c}{bc + k_{ij}F_{ij}} ln \frac{b \cdot c \cdot \Delta t + k_{ij} \cdot F_{ij}(t_K - t_Q)}{b \cdot c \cdot \Delta t + k_{ij} \cdot F_{ij}(t_K - t_Q)} \tag{11}$$

where: t, t_n, t_0 - respectively, the temperature of fuel oil in the tank, at the outlet of the heater and the surrounding medium, °C;

$$\Delta t = t_n - t$$

The formula (11) allows to determine the necessary heat capacity Q_n of the heater:

$$Q_n = b \cdot c \cdot \Delta t = K \cdot F \cdot \frac{(t_n - t_0)e^{\beta} - (t_K - t_0)}{e^{\beta} - 1} \tag{12}$$

$$\beta = \frac{K \cdot F \cdot \tau}{B_4 c}$$

Mass specific heat for petroleum products ranges from 1600 ÷ 2500 J/(kg•K). When calculated an average value of 2100 J/(kg•K) is used.

Table 1 shows the melting point and the required heat value for melting 1 kg of paraffin.

The coefficient of heat transfer K is calculated depending on the type of container, wall thickness, insulation according to the formula (7). With approximate calculations, the coefficient K is roughly assumed to be 5–7 W/(m^2•K).

Table 1. Melting heat of paraffin.

Density At 70 °C, kg/m^3	Melting temperature, K	Melting heat, KJ/kg
773	325	163
774	333	172
775	338	184

When calculating the heat loss through the tank wall within one hour, the heat transfer coefficient $K_ч$ will have a unit value of kJ/(m^2•h•°C). The approximate value $K_ч = 12 \div 25$. For railway tank-cars, the ambient temperature, °C is assumed equal to the open-air temperature. The time for heating and draining oil, for example, fuel oil depends on the ambient temperature. In warm weather it respectively equals 3 and 4 h and 6 and 10 h when it is cold (from September 15 to April 15).

The total time for heating and draining a rail tanker filled with oil depends on its kinematic viscosity, pour point, and is given in Table 2.

The heat expended directly on the heating is partially lost. Heat losses during heating can be calculated by formula (13) [6]

$$t_{cp.} = 0,5 \cdot (t_H + t_к) \tag{13}$$

when $\frac{t_k - t_o}{t_H - t_o} \leq 2$ where: $t_н$ and $t_к$ – initial and final temperature of the oil product, °C; t_O – ambient temperature, °C.

Table 2. Total time for heating and draining of viscous and solidified petroleum products

Oil product group	Kinematic viscosity at 323K, mm^2/s	Pout point, K	Warm up and drain time, hours
1	35–116	255–275	5
2	120–189	276–290	7
3	198–307	291–301	9
4	Above 307	Above 301	10

When heated with hot steam, the amount of hourly steam consumption

$$q = \frac{Q}{i_n - i_k} \text{ kg/h}, \tag{14}$$

i_n – thermal content of steam; i_k – thermal content of condensate.

When heated by closed stationary or portable coils, the heat transfer area is usually known, then the heating time is determined by the following formula:

$$\tau = \frac{Q}{KF\left(t_n - t_{cp.}\right)} \text{ hour.} \tag{15}$$

The diameter of the supply pipeline can be determined from formula (16):

$$f = \frac{d}{3600\omega\gamma} \text{ m}^2 \tag{16}$$

here, diameter

$$d = \sqrt{\frac{4f}{\pi}} \tag{17}$$

f - steam line section area, м^2; ω - pair velocity (accepted as 25–35 m/s); γ - relative weight of steam – kg/m^3.

For the discharge of high-viscosity petroleum products from railway tanks in winter, the method of circulation heating with the application of horizontal heat exchangers is mainly used. This current method significantly accelerates the drainage of fuel oil and other solidifying and high-viscosity products.

However, in the cold season, this method for draining of highly viscous petroleum products with a pour point up to 315 K is sometimes ineffective.

In this regard, an upper washout system with the use of special devices is recommended for high-viscosity petroleum products to wash out solid residues. These devices for discharging fuel oil and other heavy petroleum products from railway tanks include all the necessary components for discharging oil products during the cold season. For their connection, water vapor is required to operate the applied system, as well as to ensure the outlet of steam condensate. The drainage system provides the possibility to use an upper washout system with the application of appropriate devices for erosion of solid residues of fuel oil at the same time as the bottom circulation heating.

When using this method, the solid residue is washed out through the lower discharge device by jets of fuel oil heated in the heat exchanger, which is supplied under pressure to the tank through the washing device. In this case, the jets are directed obliquely upward, changing their height relative to the vertical axis of the tank during the washing and draining of the product. The principle of operation is based on combining the operations for heating and draining viscous oil products into one technological process using the heat energy recovery appearance. This device ensures the unchanged quality of drained petroleum products, as the heating and drainage takes place without watering the oil product, because the product does not have direct contact with water vapor.

This system of draining and circulating heating of high-viscosity fuel oil reduces their viscosity and ensures an increase in efficiency of the drainage system.

When using this method, the angle of inclination of jets of heated fuel oil can be changed only very slightly, because due to the design features of the tanks, the jets can leak out through the upper inspection hatch.

If the length of the transitional section c is zero, then for a jet symmetric about its axis, the distance h_0 from the nozzle cut to the pole of the jet, the length of the initial section s_0 and the lateral angle of the expansion w are determined from the formulas [10]:

$$\frac{h_0}{r_0} = \frac{0.29}{a}$$
$$\frac{s_0}{r_0} = \frac{0.29}{a} \tag{18}$$
$$tg(w) = 3.4a$$

here r_0 - nozzle radius, м; a - Coefficient, the value of which depends on the minree of unevenness of the velocity field in the outlet section of the nozzle.

When the Reynolds number is varied in the range Re = 2 • 10^4–4 • 10^4, the coefficient a does not depend on the Reynolds number.

For a perfectly uniform velocity field a = 0.006, for a fully established turbulent motion a = 0.076. When average value of a = 0.07, in this case:

$$h_0/r_0 = 4, 1,$$
$$s_0/\Gamma_0 = 10,$$
$$W = 0, 24\,rad$$

Experiment shows that fast fields are similar in all section of main jet zone. If total amount motion of second jet mass is constant in all sections, then law of variation of dimensionless axial velocity u0 has the appearance [9]:

$$\frac{w_m}{u_0} = \frac{13.7}{\frac{\varepsilon}{r_0} + 4.1} \tag{19}$$

where s - distance from section of main jet zone to nozzle section, *м;* u_m - speed on axis in this section, m/sec. Fluid consumption q in any section of main jet zone

$$q = 0.15 \cdot q \cdot \left(\frac{s}{r_0} + 4.1\right) \tag{20}$$

where q_0 - consumption in nozzle exit section, m^3/s.

By constancy of enthalpy law, excess temperature falling ΔT_m along axis of main jet zone is determined following [9]:

$$\frac{\Delta T_m}{\Delta T_0} = \frac{10}{\frac{s}{r_0} + 4.1} \tag{21}$$

where:

$$\Delta T_m = T_m - 7$$
$$\Delta T_0 = T_0 - t_0$$

T - average temperature of jet, $°K$; T_0 - initial temperature of jet, $°K$; t_o-environmental temperature, $°K$.

Kinetic jet energy directed obliquely upward is mostly consumed on vortex flows of liquid fuel oil, but washing of fixed residual is performed a long time due to gradual heating of its external layers by fuel oil jets through the hydraulic monitor of the lower washing device. The bottom unloading device has such advantages as long term operation reaching to ten or more years, exceptional usability and reliability.

Thus, efficiency upgrading of fixed residual unloading, as well as rundown time of high-viscosity oil-product (fuel oil) are provided. Given technology considers the possibility to reduce duration of high-viscosity oil-product discharge approximately two times in winter time. Application of given technology on railroad overpass is practical and economical, thus upper washing device only operates directly during the discharge operation execution. Operating costs are not required during down time, because product and water vapor supply are not performed. Presence of modern, high-technology equipment is a guarantee that the discharge and loading operation of oil products require minimum of time and labor, which, ultimately, will allow significantly improve the economy and safety of the whole cycle of operations. The main advantage of the upper washing system devices is the provision of the planned volume of high-viscosity petroleum products transshipment. In order to increase the efficiency of solidifying and highly viscous petroleum products transshipment, the method of circulation heating with an upper erosion system was applied at the oil terminal. The discharge of high-viscosity fuel oil was envisaged on the existing two-way railway overpass.

At the terminal for heating and discharge of highly viscous petroleum products from railway tanks, a method of circulating heating through heat exchangers was used.

However, in the winter, this method did not ensure the discharge of fuel oil with an increased pour point and, therefore, a reconstruction of the existing high-viscosity petroleum product discharge system was envisaged. Reconstruction of the drain system provides the possibility for erosion of solid residual fuel oil, simultaneously with the bottom circulation heating, to use the system of high erosion using the device УНЖ 6-100С-01. UNJ 6-100S-01 is pressure swinging balanced pipeline which is fixed to the support with tightening pin. The cover of the washing device closes the upper hatch of the tank in order to reduce the emission of fumes when fuel oil heating.

When this method is used, the solid residue washout is performed simultaneously through the lower discharge device USN-175g with the jet of fuel oil heated in the heat exchanger, which is supplied under pressure to the tank and the fuel oil heated in the same heat exchanger through the upper washout device to the surface of the solid residue. In such a way, efficiency of the solid residue unloading of high-viscosity fuel oil was ensured. Reconstruction actions were conducted on railroad overpass on the basis of development technology and the duration of discharge operations in winter time decreased about twofold after reconstruction of discharge system for viscous oil products. Circulation heating of fuel oil with an upper erosion system is shown in the schematic diagram. After discharging two-thirds of the fuel oil volume from the tank through the lower discharge device, it is possible to estimate the presence and location of solid residues of various shapes through the upper inspection neck of the tank. After this, without stopping the unloading process, an upper sighting hole is installed in the upper erosion device, which fuel oil is supplied from the tank through the heat exchanger and the upper pressure valve when the valve is opened, the flow in the upper part can be adjusted. Depending on this, flow control can optimize the circulation circuit in the tank.

With the use of automated process control systems, the process of discharge operations of high-viscosity petroleum products is improved. In laboratory conditions, analyzes of fuel oil samples obtained from Baku, Central Asian and West Siberian oils were carried out. The results of laboratory studies are shown in Tables 3, 4, 5 (Fig. 1).

Table 3. Characteristics of fuel oil produced from the oils of Baku deposits

Yield of oil, %	ρ_4^{20}	Coefficient of kinematic viscosity, cCт						Temperature, K		Coke – capacity, %
		323K	333K	343K	353K	363K	373K	Setting up	Flares	
Fuel oil >353-51	0,9100	62,55	51,53	42,01	35,0	32,0	18,5	306	503	3,56

Table 4. Characteristics of fuel oil produced from the oils of Central Asian deposits

Yield of oil, %	ρ_4^{20}	Coefficient of kinematic viscosity, cCт						Temperature, K		Coke capacity, %
		323K	333K	343K	353K	363K	373K	Setting up	Flares	
Fuel oil 45–50,2	0,9130	60,40	49,50	42,60	36,02	30,35	21,0	295	507	6,5

Table 5. Characteristics of fuel oil, derived from West Siberian oil

Yield of oil, %	ρ_4^{20}	Coefficient of kinematic viscosity, cCт						Temperature, K		Coke capacity, %
		323K	333K	343K	353K	363K	373K	Setting up	Flares	
Fuel oil 39	0,9520	151,50	125,08	103,36	88,24	77,32	37,24	299	479	4,9

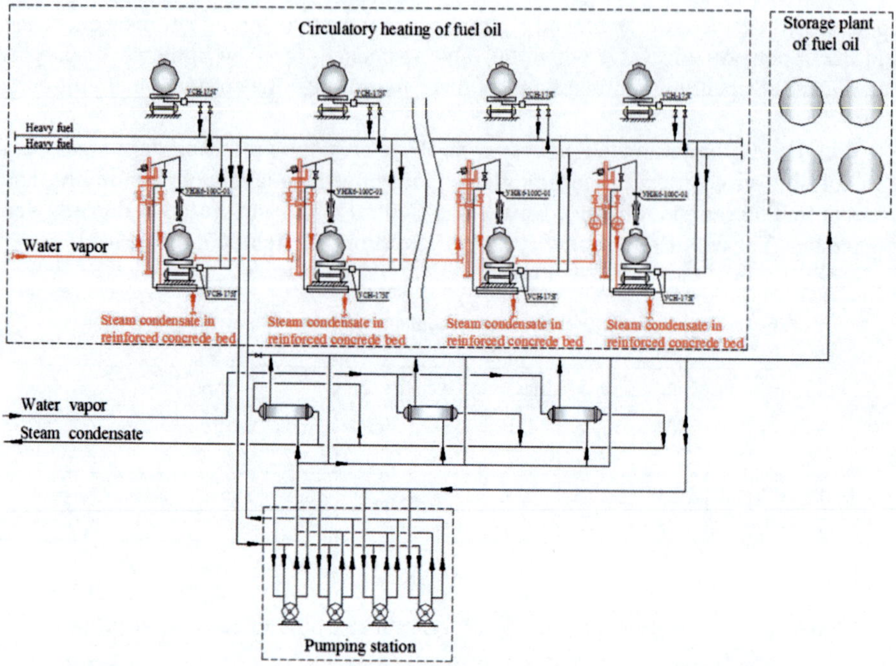

Fig. 1. Description of the scheme for heating high viscosity oil products

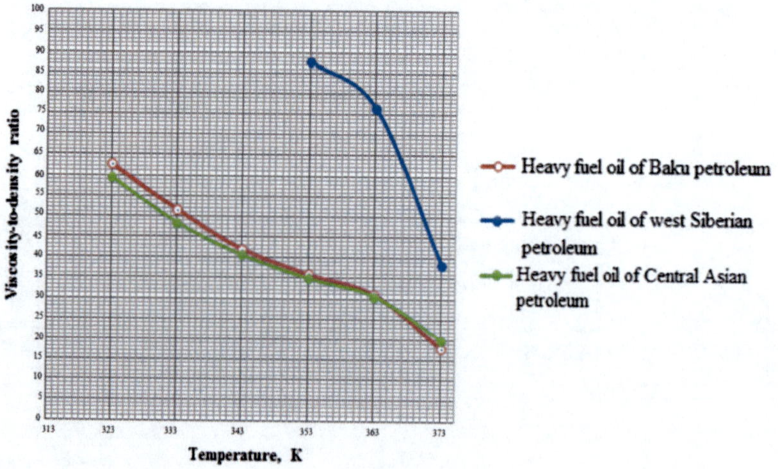

Fig. 2. Schedule of dependence of the fuel oil viscosity on temperature

Figure 2 shows the viscosity-temperature dependence for fuel oil, constructed by laboratory studies.

Viscosity is technical specification for dark oil products, including fuel oil. It determines the methods and duration of discharge-loading operations, transportation and pumping conditions, hydraulic resistance during transport of fuel through pipelines, and the efficiency of the process furnaces activity.

4 Conclusions

Heating is necessary to reduce viscosity for carrying out discharge-loading operations of viscous and highly viscous petroleum products. Depending on the nature of the raw materials and taking into account the physic-chemical properties of petroleum products, heating of highly viscous and easily solidifying oil products should be carried out to a temperature that prevents its solidification and ensures reliable performance of the discharge system. Both the pour point and the viscosity characterize the conditions for discharge and pumping of fuel. It depends on two main factors: the quality of the processed oil and the method of obtaining fuel. Depending on the physic-chemical properties of viscous oil products, viscous-temperature curves show that the heating temperature can range from 323K to 363K. The heating temperature depends on the viscosity and pour point of the product. How the solidification temperature is lower, the heating temperature of the oil is lower, i.e., below 363K, the total time of heating and drainage of oil products using modern high-tech equipment is correspondingly reduced. As a result, energy costs for heating fuel oil are significantly reduced.

References

1. Shalay, V.V., Makushev, Y.P.: Design and operation of petroleum depots and gas stations, 127 p. State Educational Establishment of Higher Professional Education, Omsk State Technical University (2010)
2. Technological design standards of oil product suppliers (petroleum depots). VNTP 5-95. Ministry of Energy of Russian Federation, Volgograd, 124 p. (1995)
3. Technological design standards of facilities for gathering, transport and treatment of oil, gas and water of oil fields (VNTP 3-85). Minnefteprom (1985)
4. Operation and maintenance of petroleum terminals, 168 p. Nedra (1986). Engineering instruction petroleum storage depot
5. Standard engineering solutions for design. Oil pumping stations with a tank farm in the system of trunk pipelines of OJSC "AK TRANSNEFT". Graphic materials, Moscow (2007). Book 1, 2
6. Zemenkov, Y.D.: Storage of Oil and Petroleum Products: Teaching Materials, 2nd edn., 536 p. Vector Book, Tyumen (2003). Revised and enlarged edition
7. Shishkin G.V.: Handbook on the Design of Tank Farms, 216 p. Nedra, Leningrad Dep., (1978)
8. Geller, Z.I.: Fuel Oil as Fuel, p. 498. Nedra, Moscow (1905)
9. See ref. in the book "Fuel oil as fuel"
10. Shageev, M.F.: Modeling and research of the circulation heating of fuel oil in separate 4-tank schemes of fuel oil facilities of TPS, Kazan (2002)

Addressing Covid-19 in Higher Education Institutions with a Blockchain-Based Mobile Payment System

Chimuka Moonde(✉) and Jackson Phiri

Computer Science Department, University of Zambia, P.O. Box 32379, Lusaka, Zambia
2018248898@student.unza.zm

Abstract. Looking at the higher learning institutions in developing countries, there is no question that the current methods for paying student fees are inefficient, inconvenient and wasteful of time. In addition, the rise in the number of students studying in higher learning institutions has led to long frustrating queues and overcrowding in most financial institutions during payment of student fees. With the advent of Covid-19, banknotes and contact-based payment methods should be avoided, according to the World Health Organization. Health experts have confirmed that the SARS-CoV2 virus could survive on surfaces such as banknotes for two to four days. This paper sought to design and implement a secure blockchain-based payment system for higher learning institutions in developing countries. The study adopted elliptic curve public-key encryption and signature scheme. Object-oriented software development methodology was used to implement the proposed payment system. The proposed system consists of an e-wallet mobile application, RESTful API and blockchain as the core component of the API. Students can use the proposed solution to address the difficulties mentioned when paying fees at a higher learning institution. The wallet can also be used to pay for goods and services offered by the institution as well as other merchants on the premise.

Keywords: Universities · Blockchain · Cryptography · Cryptocurrency

1 Introduction

Payment systems have become a vital constituent of the economic life of contemporary societies. The digital financial services ecosystem maps the overall Digital Financial Services (FDS) ecosystem by identifying all key stakeholders and examining the critical elements required to improve the ecosystem, promote and allow policies for financial inclusion [1]. In addition, the Digital Financial Services framework defines participants within the ecosystem and their functions. The players are users (consumers, companies, government agencies and non-profit groups), who need digital and interoperable financial products and services; the suppliers (banks, other licensed financial institutions and non-banks), who supply digital products and services; the financial, technological and other infrastructure that makes them possible; the governmental policies, laws and regulations

R. Silhavy (Ed.): CSOC 2022, LNNS 501, pp. 288–308, 2022.
https://doi.org/10.1007/978-3-031-09070-7_25

which enable them to be delivered in an accessible, affordable and safe manner [2]. The current coronavirus outbreak has highlighted the need of digital financial services. Advances in electronic wallets, financial technology services and online banking can all benefit consumers [3].

The expression Digital Financial Services refers to the use of an electronic device or mobile phone application to access financial services. Digital Financial Services include automated financial services among which are storing funds, conducting and receiving payments [4]. Digital Financial Services have become a viable way for the unbanked to access formal financial services. Where the service is available, users can securely receive funds, pay bills, make bank transactions, transfer funds as well as purchase goods and services. Most compelling evidence indicates that increasing access to formal financial services does not only reduce financial exclusion, but it has also become an important development goal for stimulating economic growth, increasing welfare and reducing poverty [4]. As such, the recent growth of mobile money has allowed millions of people who were financially excluded from the formal financial system to carry out financial transactions relatively cheaply, securely and reliably [5].

Money transfer and payment applications focused on mobile phones have been accepted as the means for bringing banking facilities to users beyond the traditional finance system. Mobile banking systems provide various advantages for the community, including improved profitability and capital transfers, helping to control cash balance and improving volatile income protection [6].

A sector that can highly benefit from the services provided by Digital Financial Services is the education sector. Most school systems in Africa rely significantly on cash transactions. This reliance on cash for financial transactions contributes to inefficiencies, cash leakage and carries security risks. Digital Financial Services will be able to provide unique opportunities in institutions of learning such as primary and secondary schools as well as universities. These opportunities include replacing cash, thus bringing efficiency to institutions of learning by improving management, increasing financial support from donor and non-governmental organizations, reducing financial costs, as well as laying a concrete way to developing more multifaceted systems that otherwise would not have been feasible in a cash environment [7].

Looking at Africa's higher learning institutions, there is no doubt that the existing student fees payment mechanisms are ineffective, inconvenient and time-wasting. Inefficient collection processes are especially troublesome for students in remote and hard-to-reach locations, and affect the productivity and efficiency of the workforce in accounting departments and administration. Moreover, the increase in the number of students studying in higher education institutions has led to long stressful queues and overcrowding in most financial institutions during payment of student fees.

Consumers should avoid banknotes and contact-based payments, according to the World Health Organization (WHO), and instead utilize digital payment solutions. The recommendations were based on information from health specialists who confirmed that the SARS-CoV-2 virus could survive for two to four days on surfaces such as cash and banknotes [3].

2 Literature Review

Relevant systems and technologies were analyzed to learn how common challenges and pitfalls were handled and to obtain a deeper understanding of the approaches currently in place. Best practices in the existing systems were learnt and shortcomings noted to seek and resolve them in the proposed system. The following are related existing systems and applications that were reviewed.

2.1 E-Wallet

M-Pesa is an innovative payment service for the unbanked. "Pesa" is the Swahili word for cash and the "M" is for mobile. Customers turn cash into e-money at Safaricom dealers and then follow simple instructions on their mobile phones to make payments through their M-PESA accounts; the system provides money transfers as banks do. The account is very secure, PIN-protected and supported with a 24/7 service provided by Safaricom and Vodafone Group. To implement it, Vodafone had to marry the incredibly divergent cultures of global telecommunications companies, banks and microfinance institutions–and cope with their massive and often contradictory regulatory requirements [8]. Using data preloaded on the SIM card, M-PESA utilizes a SMS-based interface to transmit money virtually to other mobile phones. To load money into one's virtual account, a customer visits one of Safaricom agents and exchanges cash for e-money, which is deposited into their account. Customers can transfer money to anyone who owns a mobile phone [9].

Paytm (earlier written as PayTM) began its journey as a recharge and bill payment platform which made its way into the e-commerce market. The Paytm wallet facilities ease of payment on almost all possible needs of daily life ranging from daily or monthly essentials, utilities (electricity bill, metro card, gas and water), entertainment, travel, mobile recharge, electronic gadgets, appliances, fashion, and many more [10]. A user can pay through scanning QR codes or entering a merchant's registered number. Users have the facility to add credit or debit cards, especially bank accounts to transfer money to Paytm wallets and transfer money from Paytm wallets to bank accounts [11].

PayPal is one of the most accepted payment services in the world. PayPal allows any business or individual to send and receive payments online, in-store or on mobile securely, conveniently and cost-effectively. PayPal deducts the funds from whatever bank account or credit card the consumer has linked [12]. PayPal uses public key cryptography to encrypt communication between the consumer and merchant. Public key cryptography is a popular technique used to encrypt data which are transmitted over the Internet from one location to another and to ensure that the sender's identity is guaranteed. It operates by using public keys and private keys, which are bits of data mathematically connected to each other by means of an algorithm [13].

Google Wallet lets a consumer save all their loyalty cards in once place. A consumer can also make payments online and make contactless payments with their mobile phone. Google Wallet offers tap-to-pay for near field communication (NFC) enabled devices. If a consumer needs to send money to someone, a consumer can do so without any transaction charges if they're using Google Wallet as well. Google Wallet allows users to store their debit cards, credit cards, gift cards and loyalty cards on a smartphone, transforming it into a virtual wallet [14]. It consists of an Android app with a user interface. The user interface is used to protect the wallet with a PIN code, to manage the payment, gift, and reward cards, to select the currently active card, to find specific offers, and to view the transaction history. The secure element of the wallet is used to store sensitive information of the payment, gift and reward cards and to interact with existing POS reader infrastructures [15].

Apple Pay is a mobile payment system that lets iPhone (iPhone 6 and higher versions) users pay for goods and services using Touch ID. As compared to entering or confirming payment card information (credit or debit card) every time they make a purchase, users can authorize payment for items securely by touching the home button of the iPhone. A payment token stores all the information needed to process the payment all the way from authorization to settlement. During an Apple Pay transaction, payment card information never leaves the user's iPhone; this information is stored securely in the device [16]. To set up Apple Pay, users simply add a debit or credit card to the wallet app. The card information can be imported from iTunes, entered manually, or added by taking a picture of a debit or credit card using an iPhone. To tap-and-pay in stores, users need to hold their iPhone close to an NFC reader. Users then place a finger on Touch ID (fingerprint scanner) to authenticate themselves and complete the payment process. The wallet app immediately notifies users about confirmed transactions [17].

Samsung Pay is a method to make purchases over the latest line of Samsung smartphones devices. Leveraging a new proprietary technology called magnetic secure transmission (MST) and NFC, Samsung Pay makes mobile payment more accessible to both merchants and customers. Samsung implemented a sophisticated alphanumeric algorithm called tokenization. In addition, Samsung partnered with payment card providers such as Visa and MasterCard, and embraced the VTS framework (Visa Token Service) to push its ambitious project. When a user adds a card to their Samsung Pay, the system generates a new "virtual random" CC (a new card number with some parameters) leveraging the framework which assigns a token to each card. That token is saved in a token vault relating the original primary account number (PAN) information. Therefore, in each transaction instead of using the original CC's data, the mobile device sends a tokenized number [18]. Samsung Pay works with traditional point-of-sale (POS) devices that allow the use of magnetic card swipes. The key technology that gives Samsung Pay such a superior position is called MST. Under this technology, devices using the Samsung Pay app can generate a magnetic signal that contains the same payment information as that generated by swiping a magnetic card in the POS card reader. Thus, the POS device can recognize the Samsung Pay signal when the mobile device is sufficiently close, even though nothing is swiped on the card reader. Unlike Apple Pay, Samsung Pay supports MST alongside NFC, which means it works with any payment terminal that accepts contactless payments or the more traditional method of swiping a card through the reader [19].

2.2 Blockchain Technologies

The concept of blockchain emerged initially from two people: Stuart Haber and W. Scott Stornetta. They did not necessarily coin or introduce the term "blockchain," because it came much later after Satoshi Nakamoto's original paper. Nevertheless, Haber and

Stornetta [20] introduced the concepts of what is now called the blockchain and actually all or most of the features and ideas behind the blockchain are present in the document.

A blockchain can be defined as an immutable transaction recording ledger, held within a distributed network of mutually untrusting peers. A copy of the ledger is maintained by any peer. A consensus protocol is used by the peers to verify transactions, group them into blocks, and construct a hash chain over the blocks [21]. A blockchain can be considered as a data structure, that is, a blockchain contains a connected list of blocks, each containing a collection of transactions. Furthermore, a blockchain can be considered as a network, that is, a blockchain is a combination of peer-to-peer networks [22]. Blockchain from a software engineering perspective enables the development of a new distributed and decentralized system software architecture, where a sensitive transaction or confidential transaction agreements may be made with untrusted individuals across the chain [23]. In addition, blockchain is a distributed system that operates without the use of a central authority or third-party intermediaries [24].

Public and private blockchains are the two primary types of blockchains. In a public or permission-less blockchain, without a particular identity, any device can participate. Usually, public blockchains provide a native cryptocurrency, use consensus based on Proof-of-Work (PoW) and financial rewards. Permissioned blockchains, on the other hand, run a blockchain among a collection of known, identified devices. A permissioned blockchain provides a way for a group of organizations that have a common purpose but do not fully trust each other to secure interact, such as businesses that share money, products, or information [21].

Blockchain as an emerging technology became popular in 2008. It was first used as a peer-to-peer ledger for registering the transactions of Bitcoin cryptocurrency [25]. The aim was to eliminate any third-party intermediary and allow users to make their transactions directly. Thus, blockchain was designed as a decentralized network of peer nodes. Each node in the network: (1) Holds a replica of the transaction's ledger; (2) writes an entry to its own ledger when it receives consensus from the other nodes in the network; (3) broadcasts any transaction made by its user to the other nodes in the network; (4) checks, on a regular basis, that the ledger it holds is identical to the ones across the network [26]. As Bitcoin continues to grow in popularity, researchers and practitioners realize the enormous potential of its underlying technology [27]. The unique capabilities of blockchain, which include immutability, transparency and trustworthiness, were found to be useful not only in cryptocurrencies, but also in many other fields. Therefore, an increasing number of blockchain-based applications have been developed in various fields [28]. Furthermore, the various features of blockchain technology, such as decentralization, immutability, robustness, privacy, consensus and cryptographic algorithms have the ability to alleviate reliance on a single centralized authority which is more susceptible to inaccuracy, insecurity, lack of complete trust and single point of failure [29]. This would lead to a reduction in the reliance on traditional central ledger managed by a trusted entity for holding and transferring funds [21].

According to Gatteschi et al. [30], the development of blockchain-based applications could be divided into three main stages: Blockchain 1.0, 2.0, and 3.0. Blockchain 1.0 was used for cryptocurrencies and its focus was to facilitate simple cash transactions. Subsequently, Blockchain 2.0 was introduced for properties and smart contracts. These smart contracts impose specific conditions and criteria to be met before registering them in the blockchain. Registration takes place without the intervention of a third party. In Blockchain 3.0, many applications were developed in various sectors, such as government, education, health and science. The application of blockchain in education is still in its early stages. Only a small number of educational institutions have started to utilize blockchain technology.

Since Diffie and Hellman introduced public-key cryptography in 1976, there has been awareness of the potential for the use of the discrete logarithm problem in public-key cryptosystems. Although the discrete logarithm problem as first used by Diffie and Hellman was specifically defined as the problem of finding logarithms in relation to a generator in the multiplicative group of the integer modulo a prime, this concept can be generalized to arbitrary groups and to elliptic curve groups. The resulting public-key systems provide relatively small block size, high speed and high security. Koblitz et al. [31] conducted a survey to investigate the development of elliptic curve cryptosystems from Koblitz and Miller's invention in 1985 to present-day implementations.

Singh and Singh [32] implemented text encryption using elliptic curve cryptography. They introduced a new technique, in which the authors omitted the classic technique of mapping characters to affine points in the elliptic curve. The respective plain text ASCII values are paired up. The paired values serve as input for the cryptography of the elliptic curve. This new technique eliminated the expensive mapping process and the need to share the standard lookup table between the sender and the recipient. The algorithm was built in such a way that any kind of script with specified ASCII values can be encrypted or decrypted.

In order to realize safe and fair payment of outsourcing services in general without depending on any third party, trusted or not, Zhang et al. [33] introduced a solution called BCPay. BCPay is a blockchain-based fair payment system for outsourcing services in cloud computing. Security review showed that BCPay achieved soundness and immunity to attacks of eavesdropping and malleability. Performance review showed that BCPay was very effective in terms of the number of transactions and the cost of computation.

Lusetti, Salsi, and Dallatana [34] proposed a blockchain-based solution for the custody of digital files in forensic medicine. The authors suggested addressing the problem through a hybrid operating model using a consensus framework to maintain a transparent access history and prevent unauthorized users from changing the access history. The digital evidence was encrypted and stored electronically, while the file properties were stored on a private blockchain implementation of the Hyperledger Fabric. The nodes to the blockchain would allow access to the data by means of a dynamic consensus process and any operations (uploads, images, or deletions) were registered on the blockchain, continuously and indefinitely. All information was accepted and exchanged amongst the blockchain nodes to prevent single points of failure and safe access to digital proof was achieved by combining cryptography and the blockchain consensus mechanism.

3 Methodology

In this study, the authors adopted an object-oriented software development methodology. It is a design strategy where system designers think in terms of "things" or real-life objects instead of operations or functions. The object-oriented software development methodology ensured that the system being developed was refined and transformed through phases of analysis, design, coding and testing [35]. The object-oriented software development life cycle is an iterative process that has five key phases. The following subsections outline some of the key phases the authors used in this study.

3.1 Requirements Specification

The stakeholder's expectations of the proposed system were reviewed and recorded in great detail. This was an iterative process which involved focus group meetings between the stakeholders and authors. In this study, main stakeholders included accountants and students from higher institutions of learning. The authors held discussions with the stakeholders to gather the desired system features and gain insight of the stakeholder's day-to-day processes [36].

The following list highlights the functional requirements for the e-wallet mobile application:

1. A user should be able to create a mobile wallet account using his/her student number, mobile number and PIN.
2. A user should be able to login his/her mobile wallet using his/her mobile number and PIN.
3. A user should be able to reset his/her forgotten PIN.
4. A user should be able to view his/her current mobile wallet balance and private and public keys.
5. A user should be able to load funds into his/her mobile wallet from either his/her bank account or mobile money account.
6. A user should be able to input necessary information in order to perform a transaction.
7. A user should be able to view payment information before committing a transaction.
8. A user should be able to receive feedback that relates to the transaction.
9. A user should be able to pay for various student fees and other merchant fees.
10. A user should be able to send funds from his/her mobile wallet to another registered mobile wallet.
11. A user should be able to request transfer of funds from another registered mobile e-wallet to his/her mobile e-wallet.
12. A user should be able to view his/her transaction history.

3.2 Design Specification

The design specification provided a description of the design that would help to realize the specified or identified requirements of the proposed system. It was a brief description of how the system would meet the user's expectations. It was the bridge between requirements and the implementation [36]. This study made use of Unified Modeling Language (UML) to visually represent the architecture, design, and implementation of the proposed system. The following highlights some of the UML diagrams used in this study.

Use Case Diagram. This study adopted use case diagrams as a precursor to use case specifications that captured the overall functionality of the proposed software system at very high-level. Use case diagrams often serve as a summary of all use cases in a software system. Figure 1 depicts the use case diagram used in this study.

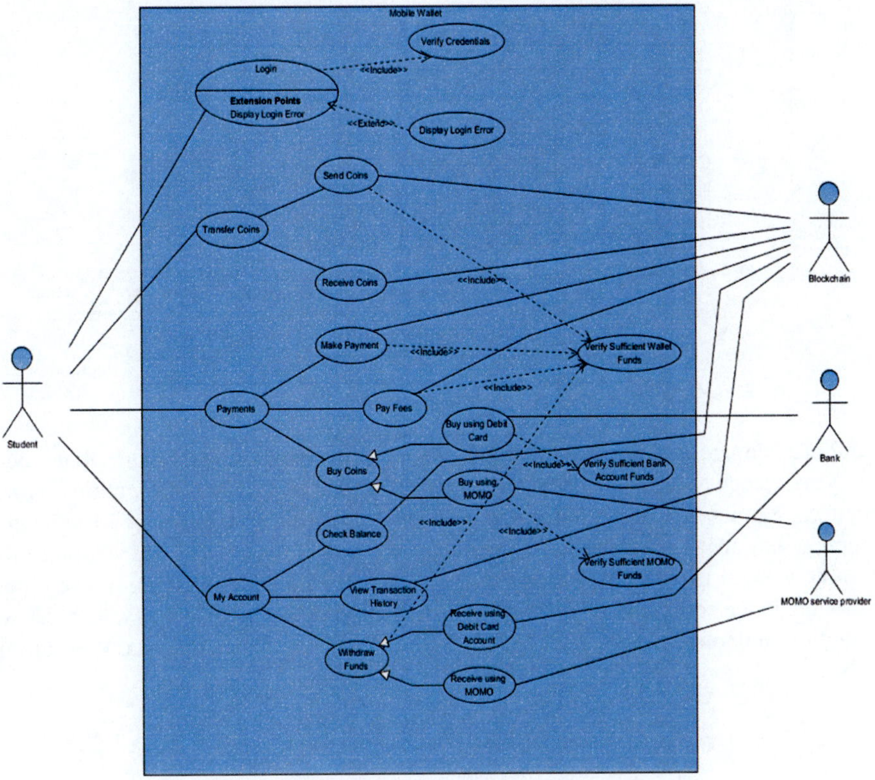

Fig. 1. Use case diagram

Class Diagram. In the Unified Modeling Language, a class diagram is a form of static structural diagram that depicts a system's structure. Class diagrams model the type of objects in the system, their attributes and operations along with the relationships among the objects. They are the most widely-used structural models, showing a static view of the system. Figure 2 shows a class diagram used in this study.

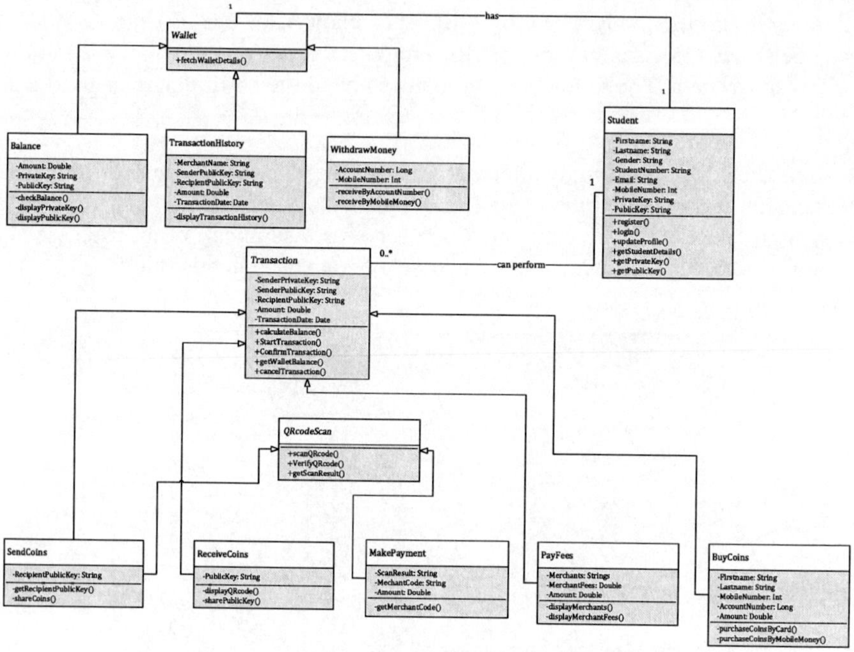

Fig. 2. Class diagram

Sequence Diagram. Sequence diagrams are one of the interaction diagrams that model the dynamic behavior of a system, focusing on interaction among the entities within a system. Sequence diagrams are not only one of the most widely-used interaction diagrams, but also one of the most popular UML diagrams as well. They capture the dynamic view of the system in terms of the sequence of interaction among its entities. They capture the communication among classes, through this, help identify the behavior needed to implement in the code. Figure 3 illustrates a sequence diagram used in this study.

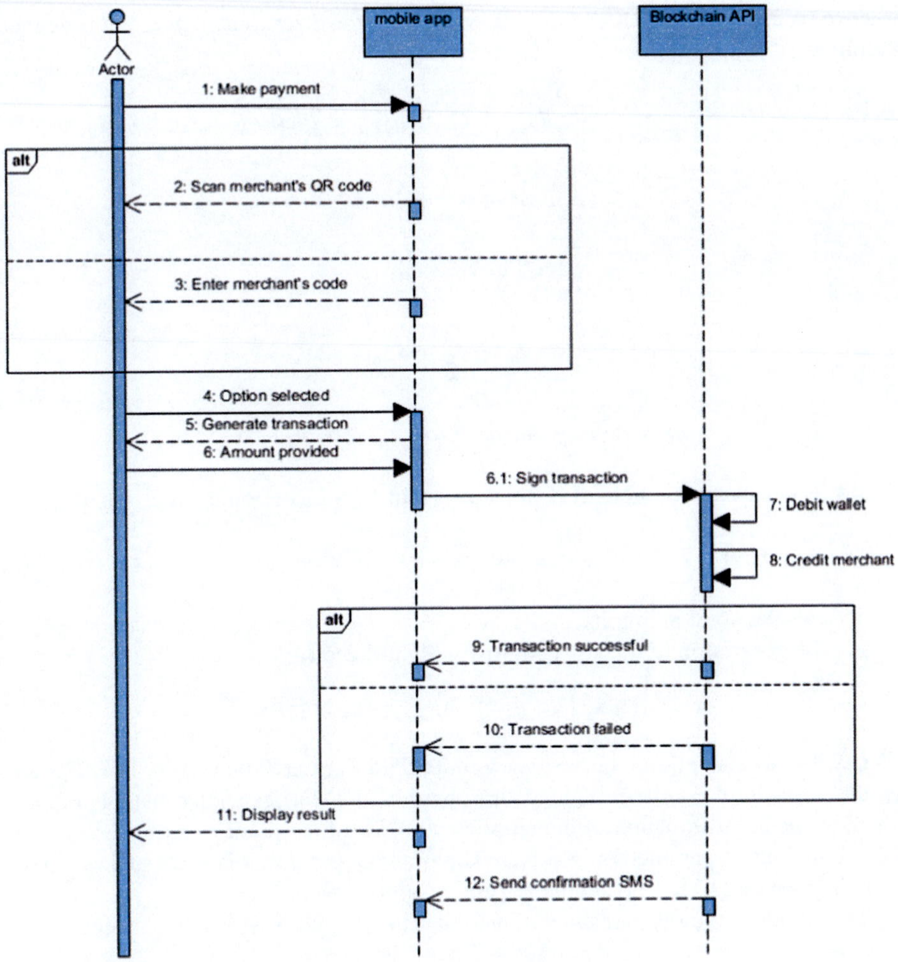

Fig. 3. Sequence diagram

3.3 Implementation

Fundamental in the implementation of the proposed blockchain-based payment system was the use of elliptic curve public-key encryption and signature scheme. Cryptography is the transformation of plain message into a different form that secure and immune from intruders. Elliptic curve cryptography is an independently developed public key cryptography by Victor Miller and Neal Koblitz [32]. This study adopted Eqs. (1), (2), (3) and (4) as outlined by Hankerson et al. [37]. An elliptic curve E is defined by an equation over a field K:

$$E: y^2 + a_1xy + a_3y = x^3 + a_2x^2 + a_4x + a_6 \tag{1}$$

where a1, a2, a3, a4, a6 ∈ K and Δ ≠ 0, where Δ is the discriminant of E and is defined as follows:

$$\Delta = -d_2^2 d_8 - 8d_4^3 - 27d_6^2 + 9d_2d_4d_6$$

$$d_2 = a_1^2 + 4a_2$$

$$d_4 = 2a_4 + a_1a_3 \tag{2}$$

$$d_6 = a_3^2 + 4a_6$$

$$d_8 = a_1^2 a_6 + 4a_2a_6 - a_1a_3a_4 + a_2a_3^2 - a_4^2$$

If L is any extension field of K, then the set of L-rational points on E is as follows:

$$E(L) = \left\{(x, y) \in LxL : y^2 + a_1xy + a_3y - x^3 - a_2x^2 - a_4x - a_6 = 0\right\} \cup \{\infty\} \tag{3}$$

where ∞ is the point at infinity.

For the generation of the key pair, this study utilized (4):

$$\langle P\rangle = \{\infty, P, 2P, 3P, \ldots, (n-1)P\} \tag{4}$$

Let E define an elliptic curve over a finite field F_p. Let P be a point in $E(\mathrm{F_p})$ and assume P has prime order n. The cyclic subgroup of $E(\mathrm{F_p})$ generated by P is Eq. (4), where prime p, the equation of the elliptic curve E and the point P and its order n are the public domain parameters. A private key is an integer d that is generated randomly from the interval $[1, n-1]$ and the respective public key is $Q = dP$.

The elliptic curve key pair generation used in the proposed system can be represented as follows:

INPUT: Elliptic curve domain parameters (p, E, P, n).

OUTPUT: Public key Q and private key d.

1. Select $d \in_R [1, n-1]$.
2. Compute $Q = dP$.
3. Return (Q, d).

For the encryption scheme, in this study the authors adopted the following algorithm:

INPUT: Plain transactional data denoted as m, public key Q and elliptic curve domain parameters (p, E, P, n).

OUTPUT: Ciphertext (C_1, C_2).

1. Represent the plain transactional data m as a point M in $E(\mathrm{F_p})$.
2. Select $k \in_R [1, n-1]$.
3. Compute $C_1 = kP$.

4. Compute $C_2 = M + kQ$.
5. Return (C_1, C_2).
6. Transmit C_1 and C_2.

Plain transactional data denoted m is first represented as point M, then encrypted by adding it to kQ where k is a randomly selected integer, and Q is the public key of the intended recipient of the plain transactional data. The sender transmits the points $C_1 = kP$ and $C_2 = M + kQ$ to the recipient.

For the decryption scheme, in this study the authors adopted the following algorithm:

INPUT: Ciphertext (C_1, C_2), private key d, and elliptic curve domain parameters (p, E, P, n).

OUT: Plain transactional data denoted as m.

1. Compute $dC_1 = d(kP) = k(dP) = kQ$.
2. Compute $M = C_2 - dC_1$.
3. Extract m from M.
4. Return (m).

The recipient uses his/her private key d to compute $dC_1 = d(kP) = k(dP) = kQ$ and thereafter recovers $M = C_2 - dC_1$.

The authors implemented a prototype of the proposed blockchain-based mobile payment system. The implementation stage of software development is the process of converting design specification to an executable software system. The authors implemented the blockchain mobile payment system using the following programming languages and tools: Flutter version 2.0, JavaScript Object Notation (JSON), Node.js version 12.18.1, Express.js version 4.17.1, Visual Studio Code version 1.48.2, Postman version 7.31.1 and GitHub.

Blockchain API. The authors implemented a RESTful API (application programming interface that conforms to the constraints of REST architectural style) using Express.js, which is a web framework for Node.js. The researchers used JavaScript, particularly Express.js to implement the API, while Visual Studio Code was the development tool the authors used to build the API. The API used JSON as the data-interchange format. The authors used Postman to test the API. Furthermore, Postman was used to view the various responses from the endpoints of the API. The complete API was hosted on Ubuntu 18.04 LTS powered virtual server provisioned from Amazon Web Services. At the core of the API, the authors implemented inherent characteristics of blockchain using JavaScript, particularly using Node.js and Express.js. Visual Studio Code was the development tool the authors used to implement the blockchain.

Mobile Application. The authors implemented an Android-based and iOS-based e-wallet mobile application using Flutter which is a tool that enabled the authors to build a native cross-platform (Android and iOS) application with one programming language and codebase. The user interface (UI), business and application logic were implemented using dart as the programming language. Overall, the mobile application was developed using Visual Studio Code as the development tool. The mobile wallet application was installed on an Android device running Android Marshmallow for testing purposes.

Code Version. The researchers used GitHub as a code version and backup tool while developing the proposed system.

4 Results

4.1 Blockchain API

The RESTful API was the communication and data sharing mechanism between the numerous nodes and e-wallet mobile application. Authentication between the RESTful API and e-wallet mobile application was achieved using JSON Web Tokens. The token was generated and signed by the API when a user of the e-wallet mobile application signed up. The JSON Web Token was used by the mobile application as an authentication token when fetching and sending data to the endpoints of the API.

The API leveraged the inherent characteristics of blockchain as the core component. Blockchain was the underlying technology and data structure of the API. As an immutable ledger, the blockchain served the purpose of storing transactional data. Shows the JSON formatted output obtained from consuming one the API's endpoint using Postman.


```
[
    {
        "timestamp": 1606461234754,
        "lastHash": "0000000000000000000000",
        "hash": "00b9924970ed999a43683c51cbda978f6a26c9e82c1e30c5d382c610c1bc5d71",
        "nonce": 0,
        "difficulty": 3,
        "data": []
    },
    {
        "timestamp": 1606461354922,
        "lastHash": "00b9924970ed999a43683c51cbda978f6a26c9e82c1e30c5d382c610c1bc5d71",
        "hash": "00a143df520611bbb56946d7e3978f8d65d1f249d68fe87e475c69643bb8c074",
        "nonce": 350,
        "difficulty": 2,
        "data": [
            {
                "id": "576f6f40-3080-11eb-8314-57fc37925ba4",
                "input": {
                    "timestamp": 1606461334582,
                    "sender_public_key": "0450ac88b4b69709c412ddc9ac7d4b05957697f114ca64c5d3d6f828a3b383c7054b066e9ab20afcdf01060ee887aa13a9e496c865a1e97d9e7743d0bc59268b5d",
                    "signature": "82ac15cdef3cb5c00d7c4a4c9f3ccacddc1880c2d52ae913c2d90adbf56625e0",
                    "amount_available": "7,000"
                },
                "outputs": [
                    {
                        "recipient_public_key": "043f7eab213d4999bd3b7ce2b61a62745262295472 8c48c84ee829c2db1fc07135a8b32be5d41bcde58c11e1222bd15fbee383ea970251ab4d96204256b4c4e6f3",
                        "fees": [
                            {
                                "fee_name": "Tuition fee",
                                "amount_paid": "5,907"
                            }
                        ]
                    },
                    {
                        "recipient_public_key": "0450ac88b4b69709c412ddc9ac7d4b05957697f114ca64c5d3d6f828a3b383c7054b066e9ab20afcdf01060ee887aa13a9e496c865a1e97d9e7743d0bc59268b5d",
                        "amount": "1,093"
                    }
                ]
            }
        ]
    }
]
```


Fig. 4. RESTful API output

A blockchain contains a connected list of blocks, each containing a collection of transactions and is considered one of the data structures [22]. The first JSON object in the JSON array shows the first block called the genesis block. This was the block that was generated after initializing the blockchain. The genesis block had meta-data such as the nonce which increased for every hash calculation, timestamp in seconds which referred to the time the block was created and the difficulty which was the existing hashing target. In addition, the genesis block did not have a previous or last hash. Conventionally, the last hash for the genesis block was represented with all zeros. Nevertheless, the genesis block had its own harsh. The second JSON object in the JSON array shows the second block that proceeded the genesis block. Like the genesis, the second block contained its own meta-data such as nonce, timestamp and difficulty. The last hash of the second block was identical to the hash for the genesis block. Therefore, the genesis block and second block were cryptographically linked with each other through the hashes. Furthermore, the second block had its own hash which was generated based on the block's meta-data and data to store. As shown in Fig. 4, transactions consisted of an input (where funds came from) and output (where funds were going) objects. Information about the sender was generated in the input of a transaction. Input information included a timestamp, sender's public key, the sender's balance as well as the digital signature of the sender which was generated using the sender's private key. Digital signatures, similar to unique handwritten signatures, but with significantly intrinsic security were used to overcome the problem of impersonation and tampering. The output object included recipient public key and a JSON array of fees being paid. The output object also included a transaction that specified the public key of the sender as the recipient key. This transaction is called Unspent Transaction Output (UTXO). UTXO refers to the amount of digital currency left after executing a transaction.

4.2 Mobile Application

Figure 5 shows the test results of the implemented e-wallet mobile app application where a student could make payments and view his or her transactions. The figure shows the sender's public key which was the student's public key and the recipient's public key which is the institution's public key. In addition, the figure shows the merchant's name, the coins spent and the date time the transaction occurred. Like a physical wallet, the e-wallet mobile application helped a student track how much digital currency he/she had. Furthermore, the e-wallet mobile application stored the student's private and public keys. The authentication token and keys were persisted in a local document-based NoSQL database called Sembast. Sembast supports encryption, hence the keys were stored in an encrypted form using the Salsa20 algorithm. The authentication token, private key, and public key were all encrypted so that if the mobile application was reverse engineered, the authentication token, private key, and public key could not be recovered and used in malicious actions or attacks against the proposed system.

The private key enabled the student to generate a unique digital signature to sign very exchange of digital currency between the student and the institution. The institution used the student's public key to verify the student's signature. A wallet's public key allowed other nodes or individuals to verify the digital signature generated by the private key of the wallet in the blockchain network. Using the student's public key, the institution decrypted the data presented by the signature. If the decrypted data did not match the data suggested, then the signature was deemed invalid. Likewise, the signature was valid if the decrypted data matched the proposed data. It was required that the signature undergoes verification to confirm whether the underlying transactional was sent by the allegedly student to prevent system fraud. The public key also served as the public address of the wallet. The public key was used as a wallet address and was what other nodes or fellow students would use to send digital currency to the wallet. The public key is a 2D point coordinate on an elliptic curve.

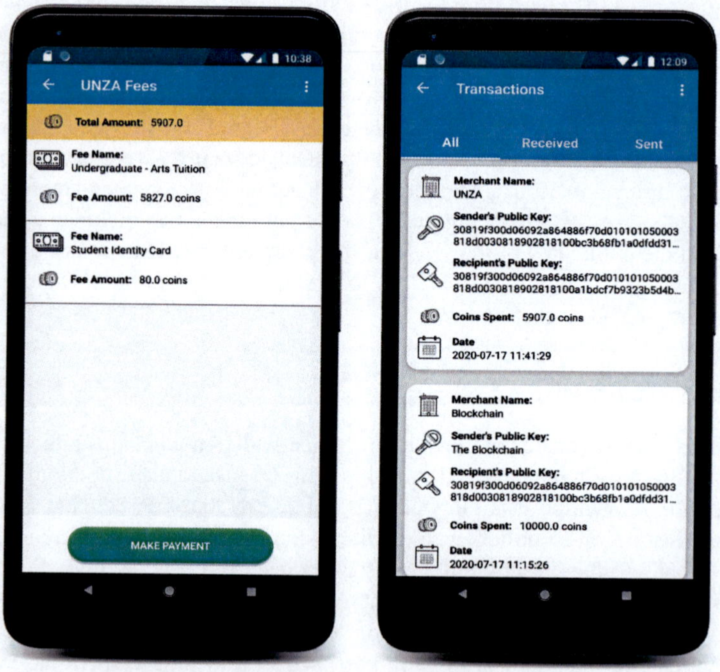

Fig. 5. Implemented e-wallet mobile application

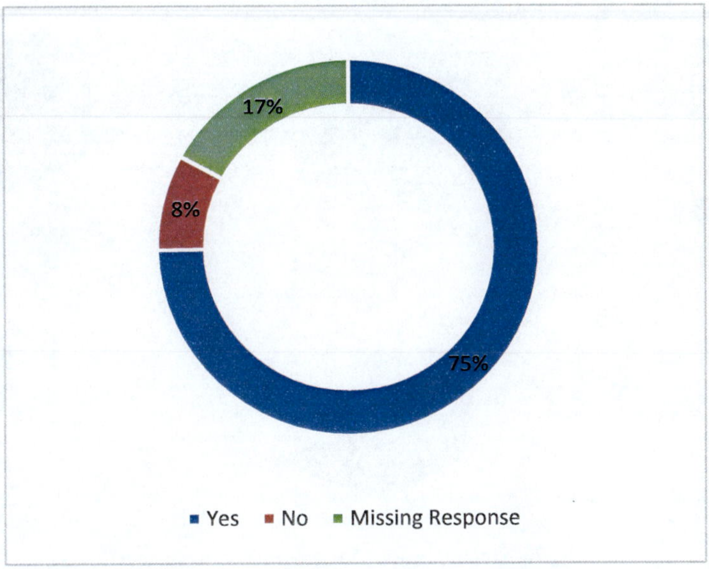

Fig. 6. Blockchain payment system as an avenue for paying tuition and other fees in Higher learning institutions

Figure 6 illustrates whether students believe the proposed Blockchain payment system should be used to pay for higher education tuition and other fees. The majority of respondents (n = 136) agreed that mobile payment systems should be used in higher education institutions when paying tuition and other fees, while 8% (n = 15) objected and the remaining 17% (n = 31) did not respond to the question.

5 Discussion

Figure 7 depicts the student and institution communication via an unsecured channel. In this study, the communication channel was the Internet. In an event that communication occurred in the presence of an attacker whose aim was to defeat any security services provided to the student and institution, the attacker could attempt to read the traffic from the student to the institution, therefore learning the student's personal identifiable information as well as bank card or account information. In addition, the attacker could attempt to impersonate either the student or institution.

Careful analysis of the aforementioned scenarios shows the following underlying factors that required to be addressed in the implementation of the proposed system:

1. Confidentiality: keeping data secret from everyone, but those authorized to view it. Transactional data sent by the student using the e-wallet mobile application to the intuition should not be readable by the attacker.
2. Data Integrity: ensuring that the transactional data sent through the unsecured channel is not altered by the attacker. The institution should be able to detect when the

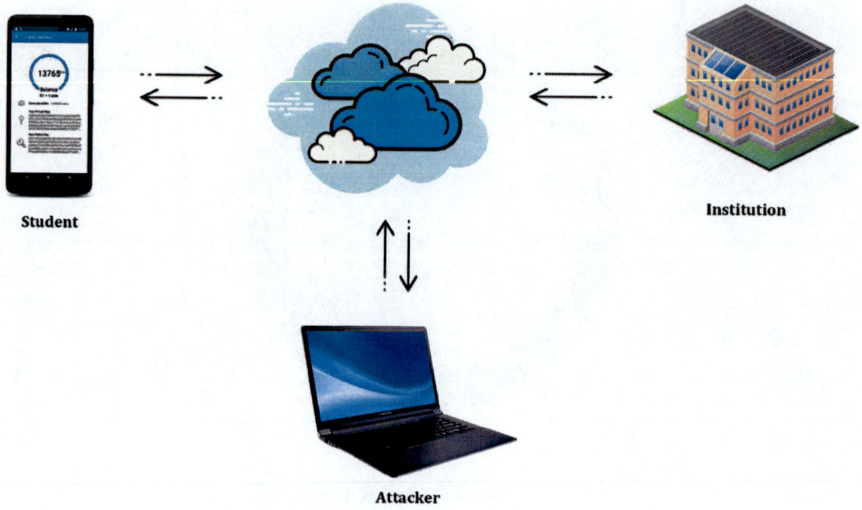

Fig. 7. Communication via an unsecured channel

transactional data sent by the student using the e-wallet mobile application have been modified by the attacker.

3. Data Origin Authentication: The institution should be able to check and verify that the transactional data supposedly sent by the student using the e-wallet mobile application did actually originate from the student.
4. Entity Authentication: The institution should be convinced of the identity of the student.
5. Non-Repudiation: Preventing the student from denying transactions previously made. When the institution supposedly receives transactional data from the student, it should not only be convinced that the transaction came from the student, but it should be able to convince a neutral third party that the transaction came from the student. Therefore, the student should not be able to deny having made the transaction.

The authors addressed the highlighted fundamental security factors by adopting elliptic curve public-key encryption and signature in the implementation of the proposed system. With the help of the blockchain, both the student and institution generated a pair of keys (e, d) consisting of a public key e and associated private key d. Each entity kept its private key a secret.

When a student wishes to send transactional data to the institution, the e-wallet mobile application obtains a copy of the institution's public key $e_{institution}$. The e-wallet mobile application uses the encryption function ENCRYPT of elliptic curve public-key encryption scheme to compute the ciphertext c = ENCRYPT $e_{institution}$ (transactional data). The e-wallet mobile application then sends c to the institution. The institution uses the decryption function DECRYPT and its private key $d_{institution}$ to recover the plaintext format of the transactional data, which are represented by m = DECRYPT $d_{institution}$ (c), where c is the ciphertext that was received. An attacker with access to the copy of the institution's public key cannot use the public key to decrypt the ciphertext sent from the student to the institution. The key pair has a property that is not computationally feasible to determine the private key only from the information of the public key.

A student using the e-wallet mobile application uses the signature generation algorithm SIGN of the digital signature scheme of the elliptic curve cryptography and their private key $d_{student}$ to compute the signature of the transactional data s = SIGN $d_{student}$ (transactional data). The e-wallet mobile application sends the signed transactional data in ciphertext format to the institution. The institution obtains a copy of the student's public key $e_{student}$ and uses a signature verification algorithm to confirm that *s* was indeed generated from m and $d_{student}$. Since $d_{student}$ is only known by the student, the institution is assured that the transactional data did indeed originate from a particular student. More importantly, since verification requires only the non-secret quantities m and $e_{student}$, the signature *s* for the transactional data can also be verified by a third party who could settle disputes if the student denies having signed the transactional data.

This study adopted blockchain technology because blockchain creatively blends cryptography with a decentralized database in a novel way, allowing a blockchain to be decentralized, immutable and transparent all at once [38]. Additionally, this study adopted blockchain technology because blockchain is a distributed and decentralized ledger that is tamper-proof and preserves the growing array of data such as transactions. There is no centralized approach and no master computer. The ledger comprises a chain of blocks linked to each other by cryptography [23]. Furthermore, the distributed and decentralized ledger is publicly shared across all the nodes of its network. This study adopted Proof-of-Work as the consensus protocol. By implementing the proof-of-work consensus protocol, each node of the network checks the validation of the block. The validated block is appended to the chain and the modified ledger is replicated through the permitted nodes of the chain. The consensus protocol replaces the requirement trusted third party or the central authority [29].

The blockchain consists of multiple blocks of data chained together like the links of a physical chain [39]. Once a block is added to the chain of blocks that precedes it, it becomes irreversible. Every block is given a unique value that looks like a random string of characters. The random string of characters is called the hash since it is generated from a hash function that creates a unique output for every unique input that it receives [25]. The unique hash is generated based on the metadata about the block and the unique data that are in the block. There is also one more very crucial data point used to generate the hash of a new block, namely the unique hash of the previous block. The unique hash of the previous block is also used as part of the input that the new block uses to generate its hash. This means that the previous block is an integral part of creating the hash value of the new block.

6 Conclusion

The purpose of this study was to design and implement a blockchain-based mobile payment system for higher learning institutions in developing countries. The initiative to implement a blockchain-based mobile payment system was prompted by the World

Health Organization's suggestion that consumers avoid banknotes and contact-based payments, as well as the realization of how the payment of student fees is undeniably ineffective, inconvenient and time-consuming. In addition, the increase in the number of students studying in higher education institutions has often resulted in long stressful queues and severe overcrowding when paying student fees in most financial institutions. Students can use the proposed system to pay their student fees. Finally, the proposed system can be used by students to pay for the goods and services offered by the higher learning institution and other vendors on the premises of the institution.

7 Limitation and Future Works

This is study did not look at design and development of an Unstructured Supplementary Service Data (USSD) application that would allow students to interact with the blockchain-based mobile payment via feature phones. Furthermore, the authors did not look into the design and development of API endpoints that would achieve seamless interoperability with commercial banks and other financial services such as mobile money and telegraphic money transfers. Therefore, the authors suggest that other scholars consider undertaking a similar study, but consider development of a USSD application and provisioning of desired features for clearing and settlement of account to account (A2A) interoperability transactions.

References

1. Benson, C.C., et al.: Financial inclusion. In: ITU-T Focus Group Digital Financial Services (2017)
2. Benson, C.C., et al.: The digital financial services ecosystem. In: ITU-T Focus Group on Digital Financial Services (2016)
3. Daragmeh, A., Sági, J., Zéman, Z.: Continuous intention to use E-wallet in the context of the COVID-19 pandemic: integrating the health belief model (HBM) and technology continuous theory (TCT). J. Open Innov. **7**, 2 (2021)
4. Martin, R., Mauree, V.: Commonly identified consumer protection themes for digital financial services. In: International Telecommunication Union (ITU) (2019)
5. Nandhi, M.A.: Effects of mobile banking on the savings practices of low income users. In: Institute for Money, Technology and Financial Inclusion (2012)
6. Deloitte, F.: Value of Connectivity: Economic and Social Benefits of Expanding Internet Access. UK Economic Consulting, Deloitte (2014)
7. Braniff, L.: Schools-africa-arent-taking-advantage-mobile-money-why. (2017). cgap.org/blog/schools-africa-arent-taking-advantage-mobile-money-why. Accessed 23 Aug 2020
8. Hughes, N., Lonie, S.: M-PESA: mobile money for the "unbanked" turning cellphones into 24-hour tellers in Kenya. Innov. Technol. Govern. Globaliz. **2**(1–2), 63–81 (2007)
9. Hinz, M.: M-PESA: The Best of Both Worlds, Banco Bilbao Vizcaya Argentaria (BBVA) (2014)
10. Singh, K., Mandal, S., Mahanti, A.: Opportunity analysis: the story of PayTm during demonetization. In: Sixth National Conference for Case Studies (COGNOSCO) (2017)

11. Joshi, T., Gupta, S.S., Rangaswamy, N.: Digital wallets 'turning a corner' for financial inclusion: a study of everyday PayTM practices in India. In: Nielsen, P., Kimaro, H.C. (eds.) ICT4D 2019. IAICT, vol. 552, pp. 280–293. Springer, Cham (2019). https://doi.org/10.1007/978-3-030-19115-3_23
12. Pandy, S.M., Crowe, M.: Choosing a Mobile Wallet: The Consumer Perspective. Federal Reserve Bank of Boston (2017)
13. Williams, D.: Encrypted Website Payments. In: Pro PayPal E-Commerce, pp. 55–57 (2007)
14. Ghag, O., Hegde, S.: A comprehensive study of google wallet as an NFC application. Int. J. Comput. Appl. **58**(16), 37–42 (2012)
15. Roland, M., Langer, J., Scharinger, J.: Applying relay attacks to google wallet. In: International Workshop on Near Field Communication (NFC), pp. 1–6 (2013)
16. Bruce, E.: Apple Pay Essentials. Packt Publishing Ltd. (2016)
17. Huh, J.H., Verma, S., Rayala, S.S.V., Bobba, R.B., Beznosov, K., Kim, H.: I don't use apple pay because it's less secure: perception of security and usability in mobile tap-and-pay. In: NDSS Symposium 2017 (2017)
18. Choi, D., Lee, Y.: Eavesdropping one-time tokens over magnetic secure transmission in Samsung pay. In: 10th USENIX Workshop on Offensive Technologies (WOOT 16). USENIX Association (2016)
19. Choi, D., Lee, Y.: Eavesdropping of magnetic secure transmission signals and its security implications for a mobile payment protocol. IEEE **6**, 42687–42701 (2018)
20. Haber, S., Stornetta, W.S.: How to time-stamp a digital document. J. Cryptol. **3**(2), 99–111 (1991)
21. Mvula, F., Tembo, S., Phiri, J.: A blockchain based mobile money interoperability scheme. Int. J. Adv. Comput. Sci. Appl. **11**, 1 (2020)
22. Hussein, D.M.E.-D.M., Taha, M.H.N., Khalifa, N.E.M.: A blockchain technology evolution between business process management (BPM) and Internet-of-Things (IoT). Int. J. Adv. Comput. Sci. Appl. **9**, 8 (2018)
23. Muhamad, W.N.W., et al.: Evaluation of blockchain-based data sharing acceptance among intelligence community. Int. J. Adv. Comput. Sci. Appl. **11**, 12 (2020)
24. Labrador, M., Hou, W.: Security mechanism for vehicle identification and transaction authentication in the internet of vehicle (IoV) scenario: a blockchain based model. J. Comput. Sci. **15**(2), 249–257 (2019)
25. Nakamoto, S.: Bitcoin: a peer-to-peer electronic cash system. Decentral. Bus. Rev. (2008)
26. Grech, A., Camilleri, A.F.: Blockchain in education. Publications Office of the European Union (2017)
27. Collins, R.: Blockchain-a-new-architecture-for-digital-content (2016). http://www.econtentmag.com/Articles/Editorial/Commentary/Blockchain-A-New-Architecture-for-Digital-Content-114161.htm. Accessed 1 Aug 2020
28. Memon, M., Hussain, S.S., Bajwa, U.A., Ikhlas, A.: Blockchain beyond bitcoin: blockchain technology challenges and real-world applications. In: International Conference on Computing, Electronics & Communications Engineering (iCCECE), pp. 29–34 (2018)
29. Yaqoob, S., et al.: Use of blockchain in healthcare: a systematic literature review. Int. J. Adv. Comput. Sci. Appl. **10**, 5 (2019)
30. Gatteschi, V., Lamberti, F., Demartini, C., Pranteda, C., Santamaría, V.: Blockchain and smart contracts for insurance: is the technology mature enough? Future Internet **10**, 20 (2018)
31. Koblitz, N., Menezes, A., Vanstone, S.: The state of elliptic curve cryptography. Des. Codes Crypt. **19**(2), 173–193 (2000)
32. Singh, L.D., Singh, K.M.: Implementation of text encryption using elliptic curve cryptography. Procedia Comput. Sci. **54**, 73–82 (2015)
33. Zhang, Y., Deng, R.H., Liu, X., Zheng, D.: Blockchain based efficient and robust fair payment for outsourcing services in cloud computing. Inf. Sci. **462**, 262–277 (2018)

34. Lusetti, M., Salsi, L., Dallatana, A.: A blockchain based solution for the custody of digital files in forensic medicine. Foren. Sci. Int. Digit. Investig. **35**, 301017 (2020)
35. Hevner, A.R.: Object-oriented system development methods. Adv. Comput. **35**, 135–198 (1992)
36. Maciaszek, L.A.: Requirements analysis and system design. Addison-Wesley (2007)
37. Hankerson, D., Menezes, A., Vanstone, S.: Guide to Elliptic Curve Cryptography. Springer, New York (2004). https://doi.org/10.1007/b97644
38. Werapun, W., Arpornthip, T., Sangiamkul, E., Wetprasit, R., Karode, T.: A blockchain-based renewable energy investment management platform: decentralized sustainable development (DeSDev). J. Comput. Sci. **16**(11), 1657–1668 (2020)
39. Atlam, H.F., Wills, G.B.: Technical aspects of blockchain and IoT. Adv. Comput. **115**, 1–39 (2019)

On the Design of Base Tables in the SQL Databases of Some Existing Software

Erki Eessaar(✉)

Department of Software Science, Tallinn University of Technology, Tallinn, Estonia
erki.eessaar@taltech.ee

Abstract. We investigated the quality of database conceptual schemas of PostgreSQL databases of four existing programs, three of which were open source. We automated the process with a set of system catalog-based queries that we have also published (https://github.com/erki77/database-design-queries). We presented 31 SQL database design problems about the data types, field sizes, and default values of base tables as well as sequence generators and the structure of base tables. All these problems appeared at least once in at least one of the databases. Most of the problems (55%; 17) were not mentioned in the reviewed literature. We explained different signs that these problems exist in a database. We explained how to detect the occurrences of the problems by using queries. System catalog-based queries can be efficiently used to continuously check the database design and thus reduce the delay in finding the problems and fixing these.

Keywords: SQL · Database design · Conceptual schema · Base table · PostgreSQL · Design smell · Automated quality assessment

1 Introduction

On the one hand, finding a good structure for a database, being consistent in the database design decisions as well as choosing the most suitable data types, field sizes, and default values increases the cost of system development. On the other hand, it improves usability of the database as well as maintainability and extensibility of the database and the software that uses the database. Moreover, it makes it easier for the human observers and various programs (including the database management system (DBMS) itself) to understand the meaning of data and apply this knowledge in their work processes.

Technical debt makes it gradually more difficult to understand, maintain, and extend software systems. In data-intensive software systems the debt can emerge from the software that uses data, data storage systems, and data [1]. Shortcuts and resulting increased debt in one part of a system (for instance, database structure) may lead to long-term problems in other parts of the system [1]. For example, insufficient constraint checking in databases decreases data quality and complicates application development.

If the DBMS implements a data model that foresees definition of the database schema at the database level (schema-on-write), then the system catalog part of the database contains information (metadata) about the structure of the database and about the behavior

R. Silhavy (Ed.): CSOC 2022, LNNS 501, pp. 309–324, 2022.
https://doi.org/10.1007/978-3-031-09070-7_26

that is implemented in the database. In our previous works, we introduced a catalog of system catalog-based database queries for finding the occurrences of database design problems in PostgreSQL databases [2] and applied some of the queries to evaluate the state of implementation of declarative integrity constraints (primary key, unique, not null, foreign key, and check) in the databases of four existing programs [3].

The set of potential SQL database design problems is large and the problems can appear in the schemas of any database level that is defined by the ANSI/SPARC layered database architecture [4]. In [3], we considered only a small portion of potential SQL database design problems that can appear in the database conceptual schema (also known as logical schema). The schema describes the data structures (base tables in case of SQL) of the entire database for all the potential users of the database. This paper continues the work. Firstly, we did not consider problems with the type and maximum field size (field size in short) of base table columns in [3]. Setting these is an act of specifying a constraint (an attribute constraint) [4]. Specifying constraints to the data that is permitted in base tables is only a part of designing the tables. Thus, we wanted to look other design issues in the database conceptual schema like the structuring of tables and the usage of default values. Due to the limited size of the paper, we leave out problems with external schemas, internal schema, and naming of database objects. We analyze the same databases and the same versions as in case of [3]. We describe database design problems and how to detect the occurrences of these by using system catalog-based queries. Many of the problems have different signs. Therefore, there are different approaches how to detect the problem occurrences and we explain these as well.

We have created queries for detecting many database design problems [2]. Many of the queries were inspired by the observation of the respective problems in the works of students, i.e., mostly novice developers. The literature that is named in Sect. 2 covers only a small subset of our compiled set of problems [2]. With the research in [3] and with this work, we want to find out as to what extent the problems from [2] exist in real, long time developed databases. Problems with integrity constraints do exist in such databases according to [3]. If the databases do have the problems, then we want to publish these in order to extend the body of knowledge about SQL database design problems and to stress the need to continuously monitor and refactor database schemas.

The rest of the paper is organized as follows. In Sect. 2, we give an overview of some related work. In Sect. 3, we explain the research method. Section 4 presents the analysis results and its subsections describe different classes of problems. In Sect. 5, we discuss the results. Finally, we conclude and point to the future work.

2 Related Works

Karwin's seminal book about SQL database design antipatterns [5] lists many design problems of base tables that are mentioned in other papers about the topic and will be also covered by this paper. There are works that report the analysis results of existing SQL databases [6–11]. Some of these are already quite old [6, 7] but still they mention design problems like similar fields do not have the same data type and length [6], "fractured objects" [7], and "multi-class tables" [7], which are also present in some of our analyzed databases. Some of the papers in this field report the results of analyzing SQL databases of open-source systems [8–11]. On the other hand, some of the studies of existing databases are fully or partially based on commercial or in-house developed systems [6, 7, 9, 10]. In the present study three databases are from open-source systems.

The number of covered design problems are usually quite small and some of the considered problems are database design antipatterns from [5] and some are about database design aspects (like naming, database external schemas, and integrity constraints) that are not covered by this paper. For instance, Al-Barak and Bahsoon [8] name 12 design problems (4 from [5]), Delplanque et al. [9] name 11 (2 from [5]), and Sharma et al. [10] name 13 (10 from [5]). Karwin [5] names 8 logical database design antipatterns and 4 physical design antipatterns. Some of these are covered in this research as well (see Sects. 4.1 and 4.4). Larger sets of problems are covered by Blaha and Premerlani [7], Dintyala et al. [11] (26 design problems), and Factor [12] (150 design problems). Balogh et al. [13] report a catalog of 53 rules, but they mention only some of these. Some of the analyses of existing databases are based on a large number of databases – 35 databases in [6], 357 industrial and 2568 open-source projects in [10], and 1406 open-source repositories that contain SQL statements in [11]. In this paper, we use a small set of databases but consider a large number of design problems.

3 The Method

We investigated occurrences of problems related with the base tables (tables in short), i.e., concentrated to the conceptual schemas of the databases. We did not investigate the programs that use the database. We did not use any documentation about the databases. To automate the work, we used our catalog of system catalog-based SQL SELECT statements (https://github.com/erki77/database-design-queries). The goal of the queries is to help identifying the possible occurrences of database design problems in PostgreSQL databases [2]. The problems could be design smells or outright mistakes.

The investigation involved queries from the categories about classifier tables, database design antipatterns, data types, default values, field sizes, sequence generators, implementations of state machines, and general structure of base tables. At the time of conducting the research the categories contained in total 213 queries, including 151 problem detection queries, which results directly point to the possible problem occurrences. We collected the queries that returned results that are presented in this paper to the collection "Find problems about base tables". It consists of 83 queries, 59 of which are problem detection queries. It means that we checked the databases against a larger set of problems and presented only these that appeared at least once.

We analyzed the same databases and versions as in [3]. We used PostgreSQL (13) DBMS. All the investigated programs and their databases have a long development history. FusionForge [14] is an open source development management and team collaboration software. The first commit in GitHub happened in August 13, 2001 and the analyzed version was committed in May 12, 2019. LedgerSMB [15] is an open source enterprise resource planning software. The first commit in GitHub happened in September 1, 2006 and the analyzed version was committed in November 24, 2020. OTRS Community Edition is an open source ticketing software, which can be used to track and manage issues that need resolving [16]. The first commit in GitHub happened in

December 1, 2001 and the analyzed version was committed in October 24, 2019. Stansoft is a Linux financial accounting software [17]. We analyzed the demo database of the version 7.20.

The method of the work was action design research [18]. We executed the queries based on the databases and continuously improved the queries, if needed. After each improvement, we executed the queries again against the databases. We found new ways of detecting possibly missing constraints, which complement the paper [3].

The work results would be useful for improving the existing databases. However, the main goal of the paper is to get a picture of the prevalence of various design problems in the real-world SQL databases. The main contribution of the work is systematic cataloging of SQL database design problems that are present in the existing databases. The number of covered problems is much larger than in the comparable studies.

4 Analysis Results

Table 1 gives a general overview about the base tables of the investigated databases.

Only LedgerSMB used table inheritance, which is an object-relational feature in PostgreSQL. It had 13% (21) tables that were defined by using inheritance. It had 2.5% (4) of tables (supertables) that were used as the basis for defining other tables by using inheritance. The subtables did not have themselves any subtables, i.e., the length of the longest inheritance path was 1. None of the databases used generated stored columns, which is a recent feature of PostgreSQL (from version 12) and had tables with zero columns, which could be used for storing truth values. Thus, we see a conservative approach of generally not using the most recent or more exotic features of the DBMS.

In the analysis results, we firstly show the percentage of elements that potentially have a problem and, in the brackets, we show the number of elements. The order of listed problems does not indicate the severity of problems. It depends on the context and should be assessed by the developers of the system in question.

Table 1. The numeric overview of the investigated base tables.

Software measure	FusionForge	LedgerSMB	OTRS	Stansoft
Base tables (tables)	206	162	116	174
Columns of the tables	1097	978	962	1931
Median number of columns	4	4	8	7
Maximum number of columns	39	32	32	142
Tables with one column	1	1	1	3
Different column data types	10	15	7	7
Different maximum character lengths of textual columns	22	10	23	36

4.1 Data Types of Columns

None of the analyzed databases used domains and only one (FusionForge) used a user-defined type (an enumeration type), which was used in case of one column. Despite the quite rich type system of PostgreSQL, the number of different column data types was rather small, especially in case of two databases. Only three types (Integer, Smallint, and Timestamp without time zone) were used in all four databases. The types Bytea, Interval, Name, Timestamp with time zone, and a user-defined type were only used in one of the databases. Surprisingly, Varchar type was not used in all the databases as a column type. The reason is that all the textual columns had type Character in Stansoft, meaning that all the registered textual values would be padded with spaces to the specified maximum length.

Selection of the column data type determines the storage size of data on disk and may influence performance. However, as importantly, the data types give information about the meaning of data and prevent recording some certainly incorrect propositions. The following is not a design problem in its own right but is a sign that *perhaps* the selection of types was suboptimal. In FusionForge 99.5% (802) and in LedgerSMB 96.4% (407) of all the columns for recording integer values had type Integer. However, in Stansoft the proportion of integer columns that had type Integer was 67.8% (367) and in OTRS 61.2% (268). Perhaps, in case of FusionForge and LedgerSMB, there has been less consideration about the potential set of integer values that could be registered in columns and because of that types Smallint and Bigint have been less used.

DT1: Using a floating-point type. Karwin [5] warns against possible rounding errors if columns with a float type are used to store fractional numbers. Whether it happens depends on the internal representation of the values, i.e., it is DBMS-specific. The PostgreSQL data types Real and Double precision are inexact, variable-precision numeric types that implement the IEEE Standard 754 for Binary Floating-Point Arithmetic. For the exact storage and calculations PostgreSQL suggests the usage of type Numeric. In FusionForge 1.8% (20) and in Stansoft 0.1% (1) of columns had a floating-point type.

DT2: Potentially incorrect data type or misleading column name. The queries guess the nature of expected data based on the column names. It can produce erroneous results due to the imprecision and inconsistency in human language. However, the names are a source of information for the human database users and should be precise as well.

At first, we searched columns that based on the column name could contain temporal data but the name is not consistent with the column type. Firstly, we searched the cases where the name of a column refers to date (for instance, *created_date*) but the type of the column is not Date. In FusionForge 5.4% (59), in OTRS 0.5% (5), in Stansoft 0.2% (3), and in LedgerSMB 0.2% (2) of columns were such. In FusionForge all the columns had type Integer. In LedgerSMB all the columns had type Timestamp without time zone. In OTRS types Timestamp without time zone, Integer, and Varchar were used. In Stansoft types Character and Timestamp without time zone were used.

Secondly, we searched the cases where the name of a column refers to time but the type of the column is not a time or timestamp type. In OTRS 1.4% (13), in Stansoft 0.5% (9), in FusionForge 0.5% (5), and in LedgerSMB 0.2% (2) of columns were such. All the columns had type Integer in OTRS and FusionForge. In Stansoft all the columns had type Character. In LedgerSMB types Date and Smallint were used. This search is prone to false positives because the columns could contain data about the length of time intervals. However, in this case an integer type is suitable, not Date or Character.

We searched columns that based on the column name probably do not contain textual values with a fixed length but still have type Character. For instance, codes or password hashes might have a fixed length but not names or descriptions. In Stansoft such columns made up at least 22.6% (437) of all the columns. For instance, in Stansoft 0.6% (11) of columns had a name that referred to the possibility that the column is meant for recording phone numbers but the type of the column was Character. However, the phone numbers have a variable length and the better type would be Varchar.

A convention is to use the phrase "id" in the names of surrogate key columns. We searched the columns that name has prefix or suffix "id" (for instance, *message_id*) or the name is "id" but the column does not have an integer type. In OTRS 3.4% (33), in Stansoft 0.7% (14), and in LedgerSMB 0.3% (3) of columns were such. The columns should have an integer type or the names of the columns should be changed.

We searched columns that based on the column name (it has prefix "is" or "has" – for instance, *is_deleted*) probably contain representations of truth values but the type of the column is not Boolean. In FusionForge 2.2% (24) and in OTRS 1.6% (15) of columns were such. In FusionForge all the columns had type Integer and in OTRS Smallint (better than Integer in terms of storage size). None of the columns had a check to permit only representations of truth values (0 and 1). The databases did not have any check constraints. Although Boolean would be the best suited type, a strategy for finding missing check constraints (in addition to the strategies in [3]) is to search columns without type Boolean that apparently (based on the column name) are meant for recording truth values but do not have a check constraint to restrict possible values in the column.

PostgreSQL supports Xml data type. Based on the column name, type, and field size 0.5% (5) of the columns in OTRS were meant for recording XML documents. However, columns with a textual type were used for the purpose instead of type Xml (for instance, column xml_content_raw with type Text). In this case the data recorder has to ensure the well-formedness of recorded documents and the DBMS does not check this.

In Stansoft 0.2% (3) of columns had a name that referred to the possibility that the column was meant for recording prices or costs (for instance, *an_print_price*) but the type of the column was Character, not Numeric or an integer type.

In LedgerSMB was a textual column with a default value that represented a number. Perhaps the columns should have type Numeric or an integer type.

In LedgerSMB was a column with type Text that had the default value expression that invoked a sequence generator. The column type should be an integer type.

DT3: Data type selection could be more expressive. In FusionForge 0.3% (3) of columns had type Varchar and field size 1. A symbol can be registered in this column and more expressive type would be Char. Factor [12] names it as a design problem on the grounds of data storage. However, this would depend on the used DBMS.

DT4: Inconsistencies in using data types. In Stansoft 92.9% (52) of columns that based on the name were probably meant for recording data about costs or prices had type Numeric. At the same time 5.4% (3) of such columns had type Character and 1.8% (1) had type Integer.

In LedgerSMB 95% (38) of columns that based on the name were probably meant for recording dates had type Date and 5% (2) had type Timestamp without time zone. In OTRS 40% (2) of such columns had type Timestamp without time zone, 40% (2) type Integer, and 20% (1) type Varchar. In Stansoft 96.8% (90) of such columns had type Date, 2.2% (2) type Char, and 1.1% (1) type Timestamp without time zone.

In LedgerSMB 91.4% (32) of columns with a timestamp type had type Timestamp without time zone and 8.6% (3) had type Timestamp with time zone. A question of system analysis is whether data about time zones is needed.

In FusionForge 92.3% (24) of columns for recording truth values had type Integer and 7.7% (2) of columns had type Boolean (see also DT2).

In case of the following inconsistencies, there is no difference in physical storage but such differences may cause confusion in understanding the meaning of data.

In FusionForge 99.5% (188) of the columns for registering long textual values had type Text and 0.5% (1) had type Varchar without specifying the maximum character length. In LedgerSMB the proportions of such columns were 96.3% (233) and 3.7% (9), respectively. In OTRS the proportions of such columns were 70.4% (38) and 29.6% (16), respectively.

In FusionForge 62.5% (5) of columns for recording one character had type Char and 37.5% (3) of the columns had type Varchar (see also DT3).

DT5: Overloaded column names in terms of data types. It was common that in different tables the columns with the same name had different types. This is also an inconsistency (see DT4). However, we listed it as a separate problem because some authors [6, 10, 12] name it as a database design problem. In LedgerSMB, we found 20, in FusionForge 13, in OTRS 8, and in Stansoft 6 such column names. For instance, in FusionForge column name *type* had 4 different data types.

4.2 Field Sizes

FS1: Too small field size of textual values. It is suspicious if a column for holding descriptions has less than twice as large field size than a column for holding names in the same table. In OTRS 16.4% (19), in Stansoft 0.6% (1), and in FusionForge 0.5% (1) of tables were such. We used column names for determining the intent of the columns.

In OTRS 2.6% (25), in Stansoft 1% (19), and in FusionForge 0.2% (2) of columns were textual columns that based on the name were meant for recording comments or descriptions. However, the field size was too small – in FusionForge 255, in OTRS 250, and in Stansoft between 12 and 50 characters (end points included).

Based on column names, in OTRS 0.4% (4), in Stansoft 0.2% (4), and in FusionForge 0.2% (2) of columns were meant for keeping components of personal names but had a too small field size for extreme name lengths. In OTRS the columns had the field size 100, in FusionForge 60, and in Stansoft 25 or 16, depending on the component.

Based on column names, in Stansoft 1.9% (37), in OTRS 0.2% (2), and in FusionFore 0.2% (2) of columns were meant for keeping various addresses but had a too small field size. For instance, in Stansoft the largest field size in case of e-mail addresses was 80. It was 40 in case of file paths. In case of some phone numbers it was 14, in case of postal addresses it was 90, in case of URLs it was 40, and in case zip codes it was 9.

FS2: Too large field size of textual values. Based on column names, in Stansoft 0.1% (2) and in OTRS 0.1% (1) of columns were meant for registering various addresses but had a too large field size. In all the cases the columns were meant for registering phone numbers and permitted up to 150 symbols in OTRS or 20 symbols in Stansoft.

FS3: Missing field size specification. In LedgerSMB 8.3% (81) of columns had type Numeric without the specification of precision and scale.

In OTRS 1.7% (16), in LedgerSMB 0.9% (9), and in FusionForge 0.1% (1) of columns had type Varchar without specifying the maximum character length.

FS4: Inconsistent field sizes. In Stansoft, we found (based on the column names and types) 5 different types of addresses (*address, mail, phone, url,* and *zip*) in case of which different columns that were meant for registering the addresses had different field sizes (including missing, i.e., unlimited size). For instance, 16 columns for registering postal addresses and 6 columns for registering e-mail addresses had both 3 different field sizes. In FusionForge, we found one such inconsistency – in case of registering URLs.

In Stansoft, we found 7 different numeric precision and scale combinations in case of columns that based on the name and data type were meant for registering data about sums of money. The most popular type declaration was Numeric (8,2) (27 columns).

FS5: Overloaded column names in terms of field sizes. In different tables the columns with the same name and type had different field sizes. In OTRS, we found 16, in Stansoft 7, in FusionForge 5, and in LedgerSMB 1 such name and type combination. For instance, in OTRS column *name* (type Varchar) had 5 different field sizes.

4.3 Default Values

Table 2 gives an overview of the usage of default values. In the brackets is the percentage of tables or columns that have the values.

Table 2. The numeric overview of the usage of default values.

Software measure	FusionForge	LedgerSMB	OTRS	Stansoft
Tables with defaults	146 (70.9%)	108 (66.7%)	80 (69%)	36 (20.7%)
Tables with a column where the default implements a surrogate key	97 (47.1%)	92 (56.8%)	80 (69%)	35 (20.1%)
Tables with a default on a non-surrogate key column	119 (57.8%)	65 (40.1%)	4 (3.4%)	1 (0.6%)
Different default values (excluding the use of sequence generators)	27	26	3	1
Columns with a default	584 (53.2%)	245 (25.1%)	84 (8.7%)	36 (1.9%)
Columns where the default is used to implement a surrogate key	97 (8.8%)	93 (9.5%)	80 (8.3%)	35 (1.8%)
Columns with a static default	487 (44.4%)	94 (9.6%)	4 (0.4%)	1 (0.1%)
Columns with a dynamic (returned by a function) default (excluding the use of sequence generators)	0 (0%)	58 (5.9%)	0 (0%)	0 (0%)

Not defining default values to columns is not a design mistake. These values may make database end users careless when they register new propositions. Programs may rely on the existence of the values that complicates their later removal. However, just like in case of any design aspect, one should try to be consistent. Moreover, the default values provide additional information about the expected values in columns and are thus helpful for getting a better understanding of the meaning of data. In case of two programs (OTRS and Stansoft), the default value mechanism was mainly used to implement surrogate keys where it was used for asking new values from a sequence generator. FusionForge and LedgerSMB used the default values more.

DF1: Perhaps an unnecessary default value. If all the table columns have a default value, then it is a sign of potentially excessive use of default values, which requires review. In FusionForge 24.3% (50) and in LedgerSMB 0.6% (1) of tables were such. In FusionForge 10.8% (63) and in LedgerSMB 0.4% (1) of columns with a default value had the empty string as the default value. In FusionForge 4 of the columns were optional (permitted NULLs) – all the others were mandatory (i.e., prohibited NULLs).

A more consistent approach, which also avoids encoding missing information in different places differently, is to make the columns optional, prohibit empty strings in these columns by using a check constraint, and remove the default value. In FusionForge 254 columns had a textual type. Thus, a question is why some of these had to have such default value, i.e., there was also an inconsistency.

DF2: An incorrect default value. The default value should reflect the type and field size of the column. In FusionForge one column with the Character type had a default value (the empty string) that was shorter than the character maximum length. However, values of type character are padded with spaces to the specified maximum length.

DF3: An imprecise default value. In LedgerSMB one of the columns had a default value expression that instead of invoking a function that returns precisely that is needed (*current_date*) invoked a function and then casted its result (*now()::date*).

In LedgerSMB 12.2% (30) of columns with a default value had the value expression that invoked a function that returns fractional seconds as a part of the value. A question of system analysis is whether data about fractional seconds is needed.

In LedgerSMB 11.8% (29) of columns with a default value had a value expression that returns a value that does not belong to the column type, forcing the system to conduct implicit type casting on the fly. The invoked functions returned current timestamp *with* time zone but the type of the column was Timestamp *without* time zone or even Date. In this case the problem may be an imprecise default value. However, it could also be that the default value reflects real requirements. In this case, one would have to change the column type to fix the problem. This is also a question of system analysis.

DF4: Perhaps a default value is missing. In LedgerSMB 20.3% (15) of columns with type Boolean did not have a default value. In addition, in OTRS 100% (15) and in FusionForge 4.2% (1) of columns that did not have type Boolean but apparently (based on the column name) were meant for recording truth values did not have a default value.

In OTRS, we found that 0.3% (3) of columns were foreign key columns that based on the column name apparently referenced to a state classifier table but the columns did not have a default value. The default value in this case should correspond to the state that the entities get when they firstly get the attention of the organization.

If the system has to register the creation time or the last modification time of rows, then the corresponding columns should have a default value that expression invokes a function that returns current date or timestamp. Based on column names and types, we found that in OTRS 16.4% (158) and in Stansoft 0.1% (2) of columns were meant for such data but none of these had a default value.

DF5: Inconsistencies in using default values. Firstly, we searched the names of non-key columns that have the same type in different tables but in some table have a default value and in some not. In FusionForge, we found 4, in LedgerSMB 2, and in Stansoft 1 such name and type combination. Secondly, we searched the names of non-key columns that have the same type but a different default value in different tables. In FusionForge, we found 13, in LedgerSMB 5, and in Stansoft 1 such name and type combination. In FusionForge the largest number of different default values (3) was in case of pair ("id", Integer) that in different tables had invocation of different sequence generators or "0" as the default value. Interestingly, such columns were not part of any key that is a sign of a design mistake in its own right (see [3]).

4.4 Structure of Base Tables

We start with the database design antipatterns in the Karwin's [5] book.

ST1: A textual column contains lists of values. In [5] it is in the set of logical database design antipatterns (Format Comma-Separated Lists). An indicator of the problem occurrence is a textual column that name refers to lists. Another way of finding the problem occurrences is to search textual columns that default value is a list of values where the values are separated with a separator like comma. In FusionForge 0.1% (1) of columns were such. Both search strategies found the same column.

ST2: A self-referencing table. In [5] it is in the set of logical database design antipatterns (Always Depend on One's Parent). We searched tables where a foreign key constraint references the same table. In LedgerSMB 3.1% (5), in OTRS 1.7% (2), and in FusionForge 0.5% (1) of tables had the design. A strategy for finding missing foreign key constraints (in addition to the strategies in [3]) as well as additional occurrences of the ST2 problem is to search non-key and non-foreign key columns that are not associated with a sequence generator and that name refers to the possibility that the column contains parent identifiers (contains the word "parent"), i.e., the table probably implements the adjacency list design. In FusionForge 0.5% (5) of columns were such.

We listed ST2 because it is mentioned as a design problem in the literature. However, in our opinion it is not a design problem any more in case of PostgreSQL because nowadays the DBMS supports recursive queries.

ST3: Multiple columns for the same attribute. In [5] it is in the set of logical database design antipatterns (Create Multiple Columns). We searched the tables with multiple columns with the same type and field size where the column names differ only by the digits in the name. We excluded the columns that probably end with a month or year number (two or four digits, respectively) because this is a sign of cloned columns. In Stansoft 7.5% (13), in OTRS 1.7% (2), and in FusionForge 1% (2) of tables had multiple columns for the same attribute. In Stansoft 4.6% (8) of tables had more than one such group. Stansoft had the problem more than the others. For instance, in one of its tables there were seven groups with 66 columns in total.

ST4: Cloned tables. In [5] it is in the set of logical database design antipatterns (Clone Tables or Columns). FusionForge had two tables with the identical set of columns that had the same name, type, field size, and ordinal position. The only difference of the table names was a digit in the end of one name.

ST5: Keeping referenced files outside the database. In [5] it is in the set of physical database design antipatterns (Assume You Must Use Files). We searched columns with a textual type that name refers to the possibility that the column is meant for recording file names or file paths. We found that in OTRS 6.9% (8), in LedgerSMB 5.6% (9), in FusionForge 5.3% (11), and in Stansoft 0.6% (1) of tables had such columns. On the other hand, based on column names and types (Text, Oid, or Byte) in OTRS 7.8% (9), in LedgerSMB 4.9% (8), and in FusionForge 1.9% (4) of tables contained file contents. In other words, there was a design inconsistency as well.

ST6: God tables. Large classes (also known as god classes) are considered design smells in object-oriented development. Base tables with a large of number of columns could be a side effect of using cloned columns or multiple columns for the same attribute (see ST3). A base table with a low normalization level, which is meant to hold data that corresponds to multiple entity types and relationship types, has typically also relatively large number of columns compared with other base tables. Thus, the normalization level of tables should be checked as well. Using SELECT * in queries, which is a query antipattern [5], could return a lot of unnecessary data in case of these tables.

Blaha and Premerlani [7] and Factor [12] name this database design problem but do not say the number of columns that would be the threshold for such tables. Sharma et al. [10] and Dintyala et al. [11] use the threshold 10 and Delplanque et al. [9] use the threshold 25. According to Table 1 there are some very large tables in all of the databases. The proportion of tables with more than 25 columns was 9.2% (16) in Stansoft, 1.9% (3) in LedgerSMB, 1.5% (3) in FusionForge, and 0.9% (1) in OTRS. The proportion of tables with 11 to 25 columns (end points included) was 22.4% (26) in OTRS, 21.8% (38) in Stansoft, 8.6% (14) in LedgerSMB, and 6.8% (14) in FusionForge.

ST7: Recording durations. In FusionForge a column had a name that referred to duration (time interval). The same table had columns for recording start and end date. Thus, there seems to be data redundancy and the duration column is not needed.

ST8: A redundant column. In LedgerSMB a table had two columns with an associated sequence generator. It is a sign that one of these columns may be redundant.

ST9: A missing classifier (reference data) table. We looked for potentially missing classifier tables by searching columns that instead of permitting free-form text (codes) could have a foreign constraint that refers to a table that currently is not in the database. More precisely, we searched columns that are not covered by any primary key, unique, and foreign key constraint but have a textual type with the field size 3 or less and where the column name does not refer to names and comments. Additional conditions were that there should not be any tables with a similar name to the column name and that there should be at least two tables that reference the potentially missing table. In Stansoft, we found 14, in FusionForge 2, and in LedgerSMB 1 such table. Examples of missing tables were *paytype* (Stansoft), *category* (LedgerSMB), and *owner_type* (FusionForge). If such classifier table does exist, then the search strategy would help us to find potentially missing foreign key constraints.

ST10: Inconsistency in recording the current state. Information systems have to know the current state of entities to correctly react to events. In case of "truth values columns" design, one or more columns for holding truth values are used, each of which corresponds to a certain state. To detect the occurrences of this design, we searched base tables that have at least one column that based on the name or type is meant for registering truth values and the name of the column possibly refers to a state. Another approach (we call it "timestamp columns" approach) is to use one or more columns for holding timestamp values. To detect this, we searched base tables that have at least two timestamp columns and the names of the columns possibly refer to a state. Yet another approach is to use state classifier tables with foreign keys that reference these. We used table names to search such tables. Finally, a possible approach is to use a column with an enumeration type to implement a state machine. We searched the occurrences of this approach by using type name and column name.

In FusionForge 5.8% (12) of tables had "truth values columns" design and 0.5% (1) of tables used a column with an enumeration type to implement a state machine. In LedgerSMB 12.3% (20) of tables had "truth values columns" design and 0.6% (1) of tables had "timestamp columns" design. In OTRS 62.1% (72) of tables had "timestamp columns" design. Most of the columns were for registering row creation and change times. In OTRS 4.3% (5) of tables had "truth values columns" design. In addition, the database also had 2 state classifier tables. We did not detect any variants in Stansoft.

The use of designs that require a separate column for each and every state complicates some queries and requires changing database structure if the set of states will change. The use of state classifier tables would be better in this regard and for recording the history of events (state changes) a separate table would be needed. We found only one example of using this design – table *ticket_history* in OTRS.

ST11: Short cycles between tables. In FusionForge were 3 and in OTRS 1 pair of tables where the tables referenced each other with a foreign key constraint. In both cases there was one pair where the foreign key columns were mandatory, meaning that for the registration of an A instance there has to be a B instance, and vice versa.

ST12: A table implements many-to-many relationship type without additional attributes. This is one of the information smells of domain modeling (many to many (missing information)), which points to a need to ask more questions about the model and its domain [19]. From such domain model construct, one would get a table that has two or more foreign keys and the table has only foreign key columns and possibly also a surrogate key column. It *could be* that behind such relationship type there is a separate entity type with yet to be identified attributes or relationship types. For instance, there may be a need to record the time when a relationship started and when it ended. In LedgerSMB 11.1% (18), in FusionForge 5.8% (12), and in OTRS 2.6% (3) of tables were such. An example that the same table can have multiple problems is that in OTRS 66.7% (2) and in FusionForge 8.3% (1) of such tables where without a key constraint.

Declaring foreign key constraints is a prerequisite for finding the occurrences of ST12. A strategy of finding missing foreign key constraints (in addition to the strategies in [3]) is to search columns that are covered by a composite key but not with a foreign key, that are not associated with a sequence generator, and the name of the column refers to the possibility that the column contains identifiers ("id" or "code" in the name). It would help us to find missing constraints in tables that implement many-to-many relationship types but which are not complete "islands" in terms of missing foreign key constraints. In FusionForge 1.4% (15), in LedgerSMB 1% (10), and in OTRS 0.2% (2) of columns were such, i.e., without a foreign key constraint.

ST13: The primary key columns are not the first in a table. Columns are ordered in SQL tables. For the better understandability of the table structure a convention is to have the primary key columns first in tables. In LedgerSMB 31.5% (51) and in FusionForge 2.4% (5) of tables did not follow it. In LedgerSMB 7 of these tables inherited from another table. In general, the parent table could be an abstract blueprint that is used to define common columns to different tables and subtables add additional columns, including the primary key columns. In this case the primary key columns are acceptably not the first in the subtables. However, in LedgerSMB both the supertable and its subtables had the same primary key columns and the position of these columns.

4.5 Sequence Generators

In PostgreSQL external sequence generators can be associated with a column by using default values. By analyzing default values, we discovered the following problems.

SG1: A sequence generator is not used. In FusionForge 2 and in LedgerSMB 1 sequence generator was not referenced from the default value expression of any table. Unused database elements are like dead code that should be removed for the clarity.

SG2: Potentially missing invocation of a sequence generator. We searched potential but not fully implemented surrogate key columns. These have an integer type and "id" in the column name, are covered with a simple primary key or unique constraint but not with a foreign key constraint, are not identity columns, and do not have the default value to invoke a sequence generator. In FusionForge 0.1% (1) of columns were such.

SG3: Potentially unnecessary use of a sequence generator. We assumed that codes should be registered by a human user, not generated by the system like "id" values. We searched columns with "code" in the column name that were covered with a simple primary key or unique constraint but not with a foreign key constraint and had an associated sequence generator. In FusionForge 0.1% (1) of columns were such.

5 Discussion

We listed 31 different SQL database design problems in the paper. We presented only the problems that appeared at least once in at least one of the analyzed databases. Many of the problems are generalizations of different more specific problems, i.e., there are different signs or smells of the same general problem. The collection "Find problems about base tables" contains queries that returned results that were presented in the analysis. The collection contains 59 problem detection queries. Each such query corresponds to a more specific problem. From the resources [5–13], we did not find any reference to 55% (17) of the problems listed in this paper. There were no references to the problems with default values and sequence generators. We also did not find references to the inconsistencies in using data types and field sizes, too small or large field sizes, and certain issues with table structure (see ST9–ST13). In the resources [5–13] the most mentioned problems that are also presented in this paper are incorrect data type (DT2) (9 references) and self-referencing table (ST2), cloned tables (ST4), and god tables (ST6) (all 5 references). We detected DT2 and ST6 in all the investigated databases. ST2 and ST4 were detected in three and one database, respectively.

Only a small subset of the identified problems can be characterized as mistakes that would cause errors in database applications. If the registration of a row in a table requires the existence of a row in another table, and vice versa (ST11), then the first rows to the tables can be added within a transaction, if the foreign key constraints in questions are deferrable. In two cases it was not this way and it is a definite mistake.

In addition, depending on the circumstances, one could see the following either as a mistake or as a technical debt, which could complicate maintenance and evolution in the future. DT1 can cause incorrect calculations, DT2 and FS1 may prevent recording permissible values. If a database user expects a default value (DF4) or the generation of surrogate key values (SG2) but the mechanism is not there, then the insertion of rows would fail. In case of most of the problems, their occurrences point to the need to re-check the database design and perhaps even revisit analysis to clarify requirements. Fixing a problem occurrence may fix other problem occurrences or provide information to the DBMS that makes it possible to discover new problem occurrences with a query. Thus, the use of the queries must be iterative. Any database management software that allows us to connect to a PostgreSQL database can be used to execute the queries.

Many of the problems had different signs. At the same time the problems can be classified under even higher-level generic problems. Six problems (DT4, DT5, FS4, FS5, DF5, ST10) are about *inconsistencies*. Developers have to check as to whether these are deliberate to satisfy user requirements or accidental. The fact that other problems appeared in only some tables is also a kind of inconsistency. Seven problems (FS3, DF4, ST1, ST5, ST9, ST12, SG2) are about something (field size specification, a default value, a column, etc.) that is *missing* from the database. Three problems (ST7, ST8, SG1) are about *unneeded* elements. Five problems (DT3, FS2, DF1, DF3, SG3) are about *imprecision*. The rest are about some kind of *incorrectness* in the design. Six of these (ST2–ST4, ST6, ST11, ST13) are about the *structure of base tables* and four (DT1, DT2, FS1, DF2) are about *column properties*. A problem with imprecisions as well as missing and incorrect elements is that database users may get imprecise or incorrect understanding of the meaning of columns as the result.

We detected the largest number of different problems in FusionForge (28), followed by LedgerSMB (18), OTRS (15), and Stansoft (14). However, we cannot compare the quality of the design of databases based on the numbers because these do not reflect the number of problem occurrences and the number of different subproblems in databases. Secondly, different problems have different importance that depends on the context where the database is used. Moreover, if a database does not extensively use a mechanism (like OTRS and Stansoft use relatively less default values), then the chances to misuse the mechanism are also lower.

There were seven problems (DT2, DT4, DT5, FS5, DF4, ST5, ST6) that occurrences were present in all the databases. Problems with data types were the most widespread. 60% (3) of the identified five datatype problems were present in all the databases. At the same time there were nine problems (DT3, DF2, DF3, ST1, ST4, ST7, ST8, SG2,

SG3) that occurrences were present in only one database. 66.7% (2) from the three identified sequence generator problems, 40% (2) from the five default value problems, and 30.7% (4) from the 13 table structure problems were present in only one database. Eight problems were present in three and seven problems in two investigated databases.

5.1 Weaknesses of the Research

A weakness is that 40% (34) of the queries in the collection of queries that gave the input to the paper have been marked as queries that rely on expressive table or column names. Good names are important. However, the analysis results that only rely on names may be misleading if the quality of names is not good and due to the imprecision of human language expressions. Constraint definitions are another important source of information about the meaning of data. Many of the used queries depend on the existence of declarative constraints. Unfortunately, many constraints were missing from the databases [3]. For instance, Stansoft does not have any declarative primary key, unique, foreign key, and check constraints [3].

Another weakness is that we considered only databases in one DBMS. On the other hand, it is a popular and technically capable system with a long development history. It follows quite well the SQL standard. Most of the queries that we used for the analysis are not solely based on the standardized information schema. On the other hand, the queries check design problems that in most part can occur in any SQL database. Most of the analyzed databases were a part of open source software. It could be that development processes of proprietary software will lead to different outcomes in terms of design quality. This will require further investigation.

6 Conclusion

We analyzed the SQL databases of four programs, three of which were open source. All the databases were implemented by using PostgreSQL DBMS. We used a set of system catalog-based queries for detecting occurrences of database design problems. We have worked out the queries and we continuously improved these as a part of this research. The collection of the used queries "Find problems about base tables" is available here https://github.com/erki77/database-design-queries. We found that all the databases had a lot of different design problems regarding base tables. In the paper, we listed only problems that had at least one occurrence in at least one of the databases. We covered much larger set of problems than the existing works in the field. Thus, the problems that we first observed in the works of students and that gave us the reason to produce the queries, are not limited with the work of students.

Future work should include investigating external schemas, internal schema, and naming of database objects in the databases of existing programs.

References

1. Foidl, H., Felderer, M., Biffl, S.: Technical debt in data-intensive software systems. In: 45th Euromicro Conference on Software Engineering and Advanced Applications, pp. 338–341. IEEE (2019). https://doi.org/10.1109/SEAA.2019.00058

2. Eessaar, E.: Automating detection of occurrences of PostgreSQL database design problems. In: Robal, T., Haav, H.-M., Penjam, J., Matulevičius, R. (eds.) DB&IS 2020. CCIS, vol. 1243, pp. 176–189. Springer, Cham (2020). https://doi.org/10.1007/978-3-030-57672-1_14
3. Eessaar, E.: The usage of declarative integrity constraints in the SQL databases of some existing software. In: Silhavy, R. (ed.) CSOC 2021. LNNS, vol. 230, pp. 375–390. Springer, Cham (2021). https://doi.org/10.1007/978-3-030-77442-4_33
4. Date, C.J.: An Introduction to Database Systems, 8th edn. Pearson, Addison Wesley, Boston (2003)
5. Karwin, B.: SQL Antipatterns. Avoiding the Pitfalls of Database Programming. The Pragmatic Bookshelf (2010)
6. Blaha, M.: A retrospective on industrial database reverse engineering projects - part 2. In: Eighth Working Conference on Reverse Engineering, pp. 147–153. IEEE (2001). https://doi.org/10.1109/WCRE.2001.957818
7. Blaha, M.R., Premerlani, W.J.: Observed idiosyncracies of relational database designs. In: 2nd Working Conference on Reverse Engineering, pp. 116–125. IEEE (1995). https://doi.org/10.1109/WCRE.1995.514700
8. Al-Barak, M., Bahsoon, R.: Database design debts through examining schema evolution. In: 8th International Workshop on Managing Technical Debt, pp. 17–23. IEEE (2016). https://doi.org/10.1109/MTD.2016.9
9. Delplanque, J., Etien, A., Auverlot, O., Mens, T., Anquetil, N., Ducasse, S.: CodeCritics applied to database schema: Challenges and first results. In: 24th International Conference on Software Analysis, Evolution and Reengineering, pp. 432–436. IEEE (2017). https://doi.org/10.1109/SANER.2017.7884648
10. Sharma, T., Fragkoulis, M., Rizou, S., Bruntink, M. and Spinellis, D.: Smelly relations: measuring and understanding database schema quality. In: 40th International Conference on Software Engineering: Software Engineering in Practice, pp. 55–64. ACM (2018). https://doi.org/10.1145/3183519.3183529
11. Dintyala, P., Narechania, A., Arulraj, J.: SQLCheck: automated detection and diagnosis of SQL anti-patterns. In: 2020 ACM SIGMOD International Conference on Management of Data, pp. 2331–2345 (2020). https://doi.org/10.1145/3318464.3389754
12. Factor, P.: SQL Code Smells. Redgate. http://assets.red-gate.com/community/books/sql-code-smells.pdf. Accessed 29 Dec 2019
13. Balogh, G., Gergely, T., Beszédes, Á., Szarka, A., Fábián, Z.: Capturing expert knowledge to guide data flow and structure analysis of large corporate databases. Acta Polytech. Hungar. **16**(4), 7–26 (2019)
14. FusionForge. https://github.com/fusionforge/fusionforge. Accessed 26 Oct 2020
15. LedgerSMB. https://github.com/ledgersmb/LedgerSMB. Accessed 26 Nov 2020
16. OTRS/otrs. https://github.com/OTRS/otrs/tree/rel-6_0. Accessed 04 Nov 2020
17. Stansoft. http://www.stansoft.org/download.html. Accessed 21 Nov 2020
18. Sein, M.K., Henfridsson, O., Purao, S., Rossi, M., Lindgren, R.: Action design research. MIS Quart. **35**, 37–56 (2011). https://doi.org/10.2307/23043488
19. McDonald, K., Matts, C.: The Seven Information Smells of Domain Modelling, InfoQ (2013). https://www.infoq.com/articles/seven-modelling-smells/. Accessed 17 Mar 2022

Safe Smart City

Michaela Kollárová(✉)

Fakulta bezpečnostného inžinierstva, Žilinská univerzita v Žiline, Katedra krízového manažmentu, 1. mája 32, 010 26 Žilina, Slovakia
misatko48@gmail.com

Abstract. The reason for implementing smart technology in cities is to solve various problems, such as transport, waste management and environmental protection, security and crime prevention. The concept of smart cities controls almost every aspect of how citizens' daily lives work. The smart city is closely linked to the Internet of Things (ICT). Within the smart city, ICTs are combined with traditional infrastructure, coordinated and integrated using new digital technologies. Ensuring physical security can be considered a key area in building smart cities. Supporting the construction of smart cities makes sense. Internet of Things (IoT) technology is helping to make it more common in countries around the world. These sensors of the Internet of Things interconnect and form an extensive network capable of collecting data that can then be analyzed to gain valuable insights on how to make modern cities more habitable. The scientific article will identify the security challenges facing the development of smart cities and then list their possible solutions.

Keywords: Safe city · Smart city · Internet of Things

1 Smart City

The concept of smart cities has increasingly become part of our daily lives in the world. A person who lives in an intelligent city becomes incredibly dependent on it, resp. already fully dependent on him. The concept of smart cities controls almost every aspect of the functioning of citizens' daily lives, such as the management of the transmission of engineering networks (e.g., water, electricity) (Dlodlo 2016). About 80% of Europeans live in urban areas, in several countries this proportion is 90% or more (Nikolov et al. n.d.).

R. Silhavy (Ed.): CSOC 2022, LNNS 501, pp. 325–330, 2022.
https://doi.org/10.1007/978-3-031-09070-7_27

1.1 Definition Smart City

Technological advances in hardware and software have recently taken place, resulting in a huge increase in information and communication technologies. Due to the use of information and communication technologies in the morning of cities through various activities, there was a relatively significant increase in efficiency within the operation of cities. This type of city in which advanced technologies are used can be called smart city (Mohanty 2016).

Despite the fact that the importance of a smart city is becoming more and more important, there is currently no generally clear and consistent definition of this term between the academic community and practitioners. The concept of smart city can be characterized as a place where traditional networks and services are more flexible, efficient and sustainable using digital information and telecommunications technologies, in order to improve its operations for the benefit of its citizens. One of the formal definitions of a smart city is the following: It is a city "bringing together physical infrastructure, information technology, infrastructure, social infrastructure and business infrastructure to use the city's collective intelligence". Another formal and comprehensive definition can be the following: "Smart sustainable the city is an innovative city that uses information and communication technologies (ICT) and other means to improve the quality of life, efficiency of urban operations and services and competitiveness in ensuring that they meet the needs of current and future generations with regard to economic, social and environmental aspects."

A smart city is defined as a city in which ICTs are combined with traditional infrastructure, coordinated and integrated using new digital technologies. A smart city is also a city that, through investment in human and social capital and traditional transport and modern ICT communication infrastructure, promotes sustainable economic growth and a high quality of life with prudent management of natural resources through participatory management. the center of the future, which is safe, secure, environmentally green and efficient, because all buildings, whether energy, water, transport, etc., are designed, constructed and maintained using advanced integrated materials, sensors. electronics and networks that are connected to computer systems composed of databases and monitoring and decryption algorithms.

From the application of complex information systems to the integration of urban infrastructure and service operations such as buildings, transport, electricity and water supply and public safety, the definition of a smart city has evolved to mean almost any form of technology-based innovation in urban planning, development and operation. The concept of a smart city has many forms, including smart cities, virtual cities, digital cities and information places (Fig. 1).

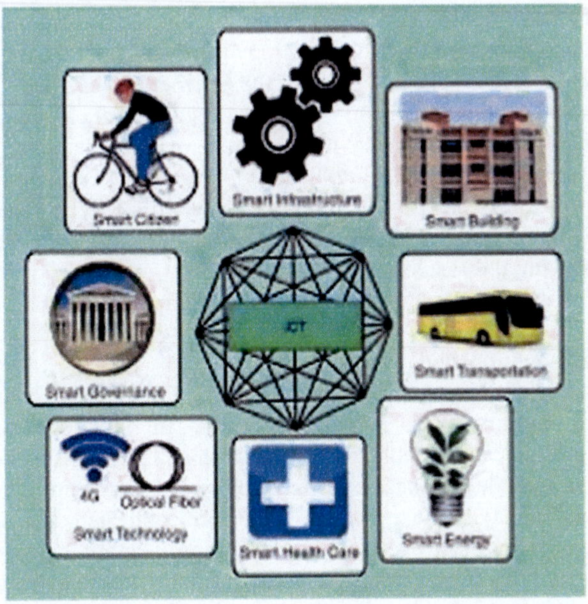

Fig. 1. A comprehensive overview of the various components required in a smart city (Mohanty 2016)

Any combination of these smart components can help make cities smart. In order for a city to be considered intelligent, it is not necessary for it to contain all the components of a smart city. The number of intelligent components itself depends on the financial costs and the available technology. More and more cities are becoming intelligent. Advanced intelligent technologies are increasingly becoming part of the everyday life of urban dwellers. Due to the growing dependence on intelligent technologies, more attention needs to be paid to increasing security measures. This is due to the impending cyber attack, which could paralyze the very operation of the city - the threat of failure of the entire network, loss of data, damage and property and the potential risk to life due to the disaster are staggering.

2 Safe City

Smart city security has become one of the issues in cyber security. Almost every developed country now has access to some type of facility that can access the Internet. This has not always been the case in earlier times. The security of smart cities has therefore become an important requirement.

2.1 Security Issues Within Smart City

Ensuring physical security can be seen as a key area in building smart cities. Although urban dwellers support their process of transformation into smart cities, it is essential to adequately explain the importance of security. Supporting the construction of smart cities makes sense. Internet of Things (IoT) technology is helping to make it more common in countries around the world. These sensors of the Internet of Things are interconnecting and form an extensive network capable of collecting data that can then be analyzed to gain valuable insights on how to make modern cities more habitable.

IoT and 5G technology, artificial intelligence (AI) and cloud computing technologies can be key elements of a smart city. The smart city network will be extremely sensitive, fast and reliable. Planners and developers need to consider how to protect citizens during emergencies and natural disasters.

Among the security challenges facing the development of smart cities:

1. Gradual phase-out of obsolete technologies
   Based on the existing technologies in modern cities to which we can assign: surveillance cameras and fiber-optic networks, it can be said with 100% certainty that the technology used to ensure public safety needs to be updated. Developers will need to ensure that a viable foundation for technology infrastructure is built. It is a real cloud application that is able to communicate with connected devices. Another type of technology that has been used for decades is closed circuit television (CCTV). By applying some of the new technologies used in smart cities, this network of cameras can interact at a deeper level and cover the entire city, which has not yet happened in major cities around the world. Planners and developers should have been able to anticipate the use of new networks and high-quality video capture systems to help stop the crime. The new CCTV network can be monitored from a single location and even connected to emergency services - for example, police stations.
2. Digital transformation of public security agencies
   Some of the vital services used to keep the city safe are fire and police stations, hospitals and government organizations. In order to work in a smart city, they need to democratize their data and improve their internal communication capabilities. These organizations must constantly communicate in real time, whether it's fighting a fire or caring for someone who has experienced a medical incident. Force work leads to inefficient communication, but new smart city technologies such as the Internet of Things and AI can help agencies work together in tandem.

   In addition to democratizing their data, agencies can work together to create effective emergency plans that will help keep the city safe. These are needed for everything from hurricanes and floods to pandemics and active shooter actions.
3. Change of police recruitment
   Police officers have to rely on vast amounts of data from a variety of sources as they work. Instead of gathering information during patrol routes, future police officers will patrol smart cities from a single station. Thanks to new technologies, they will be able to respond to crimes even faster than traditional methods. Thanks to remote monitoring, police officers in smart cities will have to be skilled at watching live video and refraining what they see. They currently rely on personal patrols. During the transition period until the introduction of smart city technologies, problems may arise with the recruitment of new police officers. Most current police officers lack the formal training needed to analyze data at a sophisticated level.
4. Creating a private broadband network

The smart city that has met this challenge is Nairobi, Kenya. The network relies on Huawei's own eLTE wireless network. The network connects command stations throughout the city with more than 1,500 high-definition cameras. There are more than 200 cameras and a number of wireless devices at the city's checkpoints.

5. Providing safe and habitable smart places
While these are just some of the physical security challenges facing the development of smart cities, they can be overcome with the right technology. People working on planning and implementing the Internet of Things, AI and 5G, should be aware of these challenges and plan accordingly. Residents of smart cities want to take advantage of all the benefits that an urban area has to offer without compromising their privacy. Smart cities will become the standard, so it is important to ensure that they live safely.

Another problem that is raised is the budget of the smart cities project. Smart cities are not paid for by one person or a group of people. The money they pay for these intelligent cities comes from the city itself. This is money from taxes that cities collect. The budget for the city is directly related to the smart city project, as it is decisive how much money the city will have to spend on its smart city security activities. The budget can prevent the implementation of a smart urban security initiative. Without the right budget, this can lead to situations such as a lack of trained and certified security professionals, as well as a lack of adequate resources for adequate smart city protection.

2.2 Security Solutions Within Smart City

One of the best things about solving smart city security issues is constant penetration testing. Permanent penetration testing is essential to keep every smart city operational. Another issue to consider is port security. Some cities offer free Wi-Fi, with the result that a lot of traffic passes through and to the ports on the network every day.

Hardware and software firewalls are one of the most important steps in protecting a smart city by increasing security. A firewall is a key factor to any network, but from a smart city and security perspective, a firewall is vital to everyday functions. Firewall and its proper configuration is one of the best ways to improve security around SCCI. However, even if we had all the hardware and software in the world for smart city cyber infrastructure (SCCI) protection, if there were no well-trained and certified individuals to secure SCCI, they would be useless. We believe that one of the most important ways to secure SCCI is to have trained and certified individuals who ensure the development, construction and maintenance of SCCI. SCCI is a very complex entity and having a team or individual trained to secure them is key to protecting them from security attacks on a daily basis. If the SCCI is down for only one day, events would lead to a catastrophic scenario.

Another solution is to vary the rules for using public Wi-Fi. Currently, some cities are free to offer Wi-Fi to the general public. Routers are located at street corners and Wi-Fi is used by the public. The introduction of a small paid Wi-Fi service could significantly reduce overall network traffic. Another significant problem with smart cities is the amount of traffic that the network can have at any given time. The introduction of a paid service would reduce the amount of traffic that passes through the network each day, with the result that security monitoring can be performed more closely with available resources. By introducing a paid service, the smart city administration can also solve the problem of reducing the budget by returning this money to the smart city security fund. The city already pays for the costs of installing routers and networks, and by charging a small monthly fee, the city should be able to recover money for this investment, as well as turn this revenue into new security, as well as reduce overall network traffic, which greatly facilitates the work of packet trackers, firewalls and other security measures. The introduction of paid Wi-Fi is a solution that has the potential to solve many of the problems that are usually associated with smart cities, and is therefore a solution worth thinking about smart city administrators and policy makers.

The last option we would like to explore is to evaluate existing security products on the market that could be adapted to SCCI's security needs. Preferably, the product should be pre-designed with smart city security requirements in mind. The system to do this is to combine security and safety solutions that help protect against crime, terrorism and civil unrest.

3 Conclusion

The importance of a smart city is becoming more and more important. The concept of smart cities is radically changing the world. Ensuring security can be considered a key area in building smart cities. City dwellers must be aware of the benefits of a smart city. Within the document, several security threats associated with the construction of intelligent cities were identified and subsequently security solutions within Smart City were identified.

References

Dlodlo, N.: The Internet of Things for the Safety and Security of Smart Cities (2016). https://www.researchgate.net/publication/303813523_THE_INTERNET_OF_THINGS_FOR_THE_SAFETY_AND_SECURITY_OF_SMART_CITIES

Lauf, J., Borion, H., Bradford, B.: Security and the Smart City: A Systematic Review (2020). https://www.researchgate.net/publication/338715824_Security_and_the_Smart_City_A_Systematic_Review

Mohanty, P.S.: Everything You wanted to Know about Smart Cities (2016). https://www.researchgate.net/publication306046857_Everything_You_Wanted_to_Know_About_Smart_Cities

Nikolov, R., Novakova, G., Shoikova, E.: Model for a Smart and Safe City (n.d.). https://hightechsocety.eu/sbornik/2016/1/27.MODEL%20FOR%20A%20SMART%20AND%20SAFE%20CITY.pdf

Partida, D.: The Physical Security Challenges of Smart Cities (2021). https://www.planetizen.com/blogs/114762-physical-security-challenges-smart-cities

Ralko, S., Kumar, S.: Smart City Security (2016). https://www.researchgate.net/publication/311574162_Smart_City_Security

Investigation of the Influence of the Particle Movement Velocity and the Light Flux Reflection Effect on the Conditions of Heat and Mass Transfer Processes in a Direct Absorption Solar Collector

Sergey Smagin and Ilya Manzhula(✉)

Computing Center FEB RAS, Kim Yu Chena Street, 65, 680000 Khabarovsk, Russia
manzhula1994@gmail.com

Abstract. The problem of developing alternative energy sources with a high efficiency was raised in the middle of the 20th century. One of the possible ways to solve it is the development of systems based on direct absorption solar collectors based on nanofluids. Systems of this kind are the most environmentally friendly and modern, the disadvantages of which include the complexity of maintenance and climatic restrictions of widespread use. The purpose of this paper is to determine ways to improve the efficiency of heat transfer in solar energy collection and storage systems by means of mathematical modeling of heat and mass transfer processes in direct absorption solar collectors based on nanofluid. When constructing a mathematical model, the contributions of the diffusion and electrostrictive components were taken into account. A mathematical model of heat and mass transfer in a liquid-phase medium with nanoparticles under the influence of a light field has been studied. It takes into account one-dimensional concentration convection in the form of a boundary value problem for a system of second-order partial differential equations. An algorithm for the numerical simulation of such processes has been developed and implemented in software, the main element of which is a conservative finite-difference scheme of the second order of approximation. The effect of particle motion velocity and the effect of reflection of the light flux on the conditions for the processes of heat and mass transfer in a direct absorption solar collector has been numerically studied. The reported study was funded by RFBR, project number 19-31-90070. To carry out the calculations, the computing resources of the Center for Collective Use "Data Center of the Far Eastern Branch of the Russian Academy of Sciences" were used.

Keywords: Direct absorption solar collector · Mathematical modeling · Heat and mass transfer process · Finite difference method · Approximation · Computational experiment

R. Silhavy (Ed.): CSOC 2022, LNNS 501, pp. 331–343, 2022.
https://doi.org/10.1007/978-3-031-09070-7_28

1 Introduction

The rapid development of technology in modern society imposes an increased requirement for the efficiency and environmental safety of energy supply systems. The main sources of energy supply at the moment are thermal, hydro, nuclear, and wind power plants. The problem of developing alternative energy sources with a high efficiency has been posed and has been actively studied since the middle of the 20th century. One of the possible ways to solve it can be the development of systems based on direct absorption solar collectors based on nanofluids [1–9].

Solar collector is a special heat exchanger that converts the energy of solar radiation either into thermal energy of the working fluid (in solar thermal applications) or into electrical energy (in photovoltaic applications). When using solar energy, solar radiation is absorbed by the solar collector in the form of heat, which is then transferred to its working fluid (air, water or oil). The heat carried by the working fluid can be used either to provide hot water/heating or to charge a thermal energy storage tank from which heat can be extracted for use later (at night or on cloudy days).

In photovoltaic applications, a photovoltaic module not only converts solar radiation directly into electrical energy (usually at a fairly low efficiency), but also produces a large amount of waste heat. It can be used by combining a photovoltaic panel with recuperative tubes filled with a carrier liquid.

The purpose of this paper is to develop a mathematical model that describes the processes of heat and mass transfer, as well as energy transfer in direct absorption solar collectors based on nanofluid. The developed mathematical model is planned to be further applied to study and improve the efficiency of heat transfer in solar energy collection and storage systems.

The paper is organized as follows. The Statement of the Problem section presents the statement of the boundary value problem for the nonlinear diffusion equation. The section "Finite difference algorithm for numerical simulation" describes the construction of a conservative finite-difference scheme of the second order of accuracy in the spatial coordinate. In the "Selection of initial data" section, the qualitative characteristics of the base liquid and nanoparticles used for the numerical experiment are given. The sections "Influence of fluid velocity on the process of heat and mass transfer" and "Accounting for the effect of light flux reflection on the process of heat and mass transfer" are devoted to numerical modeling of the evolution of the temperature field in a liquid with nanoparticles exposed to a light field, taking into account the thermal diffusion term, visualization of the results of the computational experiments. In conclusion, the results are discussed and conclusions are formulated.

2 Statement of the Problem

The paper considers a liquid-phase medium with nanoparticles irradiated with a light flux. We presume that the wavelength of light is much larger than the characteristic dimensions of nanoparticles. Thus, we do not consider the processes of diffraction and light scattering, we also exclude the processes of sedimentation of particles. For a more visual representation of how the studied physical processes proceed, Fig. 1 schematically shows a two-dimensional interpretation of the problem being solved.

From the point of view of this study, it is required to consider heat and mass transfer processes in which the characteristics do not change with time, i.e. with steady flow.

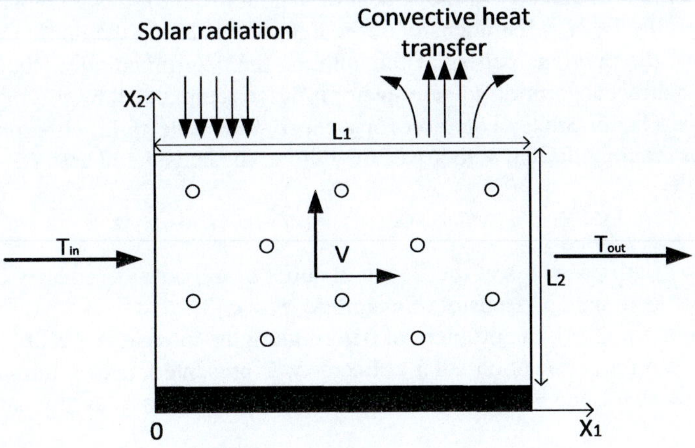

Fig. 1. Geometry of the problem.

As is known, as a result of the action of a light field, temperature gradients arise in the medium, which affect the processes of heat and mass transfer. These phenomena are described by a system of balance equations for the temperature of the medium T and the velocity $\mathbf{V}$ of fluid flow in the direct absorption solar collector, which in general look as follows [10–12]:

$$\nabla(\lambda\nabla T) - K\mathbf{V}\nabla T + q = 0, \quad x \in D, \tag{1}$$

$$\nu\Delta\mathbf{V} - (\mathbf{V}\nabla)\mathbf{V} - \rho^{-1}\nabla p = 0, \tag{2}$$

$$\nabla\mathbf{V} = 0. \tag{3}$$

Further on in this paper, the following designations are accepted: rectangular area $D = \{0 < x_1 < L_1,\ 0 < x_2 < L_2\}$, $x = (x_1, x_2)$ are spatial coordinates (m), $L = (L_1, L_2)$, L_1 is solar collector length (m), L_2 is solar collector height (m); $K = C_{nf}\rho_{nf}$, $C_{nf} = \left((1 - C)\rho_l C_l + C\rho_p C_p\right)/\rho_{nf}$ is the heat capacity of the nanofluid $\left(\mathrm{J}/(\mathrm{kg}\cdot\mathrm{K})\right)\mathrm{K}$, $\rho_{nf} = (1 - C)\rho_l + C\rho_p$ is the density of the nanofluid $\left(\mathrm{kg}/\mathrm{m}^3\right)$, C is the concentration of nanoparticles, C_l, ρ_l, C_p, ρ_p are heat capacity $\left(\mathrm{J}/(\mathrm{kg}\cdot\mathrm{K})\right)$ and density $(\mathrm{kg}/\mathrm{m}^3)$ of the enclosing liquid and nanoparticle material; λ is the thermal conductivity coefficient $\left(\mathrm{W}/(\mathrm{m}\cdot\mathrm{K})\right)$,

$$\lambda = \left(\frac{\lambda_p + 2\lambda_l + 2C\left(\lambda_p - \lambda_l\right)}{\lambda_p + 2\lambda_l - C\left(\lambda_p - \lambda_l\right)}\right)\lambda_l,$$

where λ_p is the thermal conductivity coefficient of nanoparticle material $(W/(m \cdot K))$, λ_l is the thermal conductivity coefficient of enclosing liquid $(W/(m \cdot K))$; T_{in} is the inlet temperature of the nanofluid in the solar collector (K), T_e is the temperature of ambient air (K); $\mathbf{V} = (V_1, V_2)$ is velocity of movement of nanofluid (m/s); ν is kinematic viscosity $(Pa \cdot c)$; $q = \partial I/\partial x_2$, I is the intensity of the light flux entering the liquid (W/m^2).

In one of the previous papers of the authors, the determination of the velocity of particles in a direct absorption solar collector is described in detail by solving a problem based on the Navier-Stokes equation for an incompressible fluid, an expression was obtained for the longitudinal velocity of the fluid in the considered area D

$$V_1(x_2) = (p_{in} - p_{out})x_2(L_2 - x_2)/2\mu L_1, \quad 0 \le x_2 \le L_2. \tag{4}$$

where p_{in}, p_{out} are pressures at the inlet and outlet of the considered area D, Pa; $\mu = (1 + 2.5C)\mu_l$ is dynamic viscosity of nanofluid, $Pa \cdot c$ [3].

In addition, in [1–3], the problem of determining the intensity of visible solar radiation inside a direct absorption solar collector was presented, taking into account the material, size, shape and volume fraction of nanoparticles, as well as the material of the base liquid.

$$I = \int_{\lambda_{\min}}^{\lambda_{\max}} \tau I_{b,\lambda} e^{(x_2 - L_2)K_{e\lambda,nf}} d\lambda, \quad q = \frac{\partial I}{\partial x_2} = \int_{\lambda_{\min}}^{\lambda_{\max}} \tau I_{b,\lambda} K_{e\lambda,nf} e^{(x_2 - L_2)K_{e\lambda,nf}} d\lambda, \tag{5}$$

where τ is the transmittance, $I_{b,\lambda}$ is the spectral intensity of the incident solar radiation on the upper part of the collector, (W/m^2), $K_{e\lambda,nf}$ is the spectral extinction coefficient of a nanofluid, $\lambda_{\min}$ and λ_{max} are the boundaries of the visible radiation spectrum and are equal to 380 and 780 nm, respectively. All relevant optical properties of bulk materials used to create nanofluids can be found in the literature on optical constants, for example [13].

To close the system (1)–(3), we formulate the initial and boundary conditions

$$T|_{x_1=0} = T_{in}, \quad \left.\frac{\partial T}{\partial x_2}\right|_{x_2=0} = 0, \quad \left.\left(\lambda \frac{\partial T}{\partial x_2} + \sigma(T - T_e)\right)\right|_{x_2=L_2} = 0, \tag{6}$$

$$p|_{x_1=0} = p_{in}, \quad p|_{x_1=L_2} = p_{out}, \quad V|_{x_2=0,L_2} = 0. \tag{7}$$

This formulation of the problem takes into account the reflection of the light flux passing through the colloidal nanosuspension from the lower wall of the direct absorption solar collector, as well as heat exchange with the external environment according to the Newton-Richman law.

Note that a similar problem of heat and mass transfer has already been solved in the papers of other authors [1–3]. However, the authors of the above papers considered the velocity of the nanofluid movement to be constant over the entire cross section of the direct absorption solar collector.

3 Finite Difference Algorithm for Numerical Simulation

For the purpose of a more detailed study of the processes of heat and mass transfer under study, let us consider the case when in the system of Eqs. (1)–(3) $L_2 >> L_1$ and λ is a constant value, i.e. unknown unknown function T depends only on the variable x_2, then the boundary value problem (1), (6) will look like

$$\lambda \frac{\partial^2 T}{\partial x_2^2} + q = 0, \tag{8}$$

$$\left.\frac{\partial T}{\partial x_2}\right|_{x_2=0} = 0, \quad \left.\left(\lambda \frac{\partial T}{\partial x_2} + \sigma(T - T_e)\right)\right|_{x_2=L_2} = 0. \tag{9}$$

In this formulation of the problem, $q = \alpha I$, α is the absorption coefficient of a light wave (m^{-1}), depending on the chemical nature of the medium and on the wavelength of the transmitted light, $I = I_0 \exp(\alpha(x_2 - L_2))$, I_0 is the initial intensity of the light flux which are entering the liquid.

We will solve this problem analytically. Due to the fact that the right side of the differential Eq. (8) depends only on the spatial variable x_2, then this differential equation is solved by successive integration of the right side, i.e.

$$T = -\frac{1}{\alpha\lambda} I_0 \exp(\alpha(x_2 - L_2)) + C_1 x_2 + C_2. \tag{10}$$

Further, using the boundary conditions (9) to find C_1 and C_2 we obtain

$$\begin{aligned} T = & -(\alpha\lambda)^{-1} I_0 \exp(\alpha(x_2 - L_2)) \\ & + \lambda^{-1} I_0 \exp(-\alpha L_2)\left(x_2 - \sigma^{-1}\lambda - L_2\right) \\ & + \left(\sigma^{-1} + (\alpha\lambda)^{-1}\right) I_0 + T_e. \end{aligned} \tag{11}$$

Analytical solution (11) will be used below to check the correctness of the numerical simulation results.

In order to build a more accurate mathematical model, we consider the boundary value problem (1), (6), when $L_1 >> L_2$ and λ is constant value, and there is no movement of particles in the vertical direction, i.e. $V_2 = 0$.

In this case, problem (1), (9) will take the following form

$$K V_1 \frac{\partial T}{\partial x_1} = \frac{\partial}{\partial x_2}\left(\lambda \frac{\partial T}{\partial x_2}\right) + q, \tag{12}$$

$$T|_{x_1=0} = T_{in}, \quad \left.\frac{\partial T}{\partial x_2}\right|_{x_2=0} = 0, \quad \left.\left(\lambda \frac{\partial T}{\partial x_2} + \sigma(T - T_e)\right)\right|_{x_2=L_2} = 0. \tag{13}$$

This boundary value problem does not have an analytical solution, therefore, to study it, it is necessary to use numerical methods, in particular, the finite difference method.

The method of finite differences physically means the transition from a continuous medium to some of its discrete model. The main requirement of such a transition is the preservation of the properties of the physical process. In this case, the difference schemes that express the conservation laws on the grid are called conservative [14, 15].

To obtain conservative difference schemes, it is natural to proceed from the balance equations written for the elementary volumes of the grid domain. The integrals and derivatives included in these equations should be replaced by approximate difference expressions.

We will assuming that $\omega_h = \{(x_{1,i}, x_{2,j}) : i = \overline{0, N_1}, j = \overline{0, N_2}\}$ is the set of nodal points of the grid area (grid), where $x_{1,i} = ih_1$, $x_{2,j} = jh_2$, $h = \{h_1, h_2\}$, $h_1 = L_1/N_1$, $h_2 = L_2/N_2$ - grid steps ω_h along x_1 and x_2 respectively, N_1, N_2 are natural numbers.

To simplify further writing, we will use the following notation for grid functions and expressions containing finite-difference derivatives

$$F_{i,\underline{x}} = (F_i - F_{i-1})/h,\ F_{i,\overline{x}} = (F_{i+1} - F_i)/h,\ F_{i,x} = (F_{i+1} - F_{i-1})/(2h),$$

$$\left(a_i F_{i,\underline{x}}\right)_{\overline{x}} = \left(a_{i+1} F_{i,\overline{x}} - a_i F_{i,\underline{x}}\right)/h,\ F_i = F(x_i).$$

Let us denote and assume that all the coefficients in Eq. (12) are equal to 0 outside the segment [0, L]:

$$v_{1,j} = 2KV_{1,j},\ \vartheta_j = 2q_j,$$

$$v_{1,\ 1/4} = h_2 KV_{1,\ 1/4},\ \psi_{1/4} = h_2 q_{1/4},$$

$$v_{1,N_2-1/4} = h_2 KV_{1,N_2-1/4},\ \psi_{N_2-1/4} = h_2 q_{N_2-1/4}.$$

For the numerical solution of the boundary value problem (12), (6), we will use the following conservative difference scheme

$$v_{1,j} T_{\overline{x}_1,ij} = \lambda\left(T_{\underline{x}_2\overline{x}_2,\,ij} + T_{\underline{x}_2\overline{x}_2,\,i+1j}\right) + \vartheta_j,\ i = 0, ..., N_1 - 1,\ j = 1, ..., N_2 - 1$$

$$T_{0,\,j} = T_{in},\ j = 0, ..., N_2,$$

$$0.25 v_{1,\ 1/4}\left(3T_{\overline{x}_1,i\,0} + T_{\overline{x}_1,i\,1}\right) = \lambda\left(T_{\overline{x}_2,\,i,\,0} + T_{\overline{x}_2,\,i+1,\,0}\right) + \psi_{1/4},\ i = 0, ..., N_1 - 1 \tag{14}$$

$$\begin{aligned}&0.25 v_{1,N_2-1/4}\left(T_{\overline{x}_1,i\,N_2-1} + 3T_{\overline{x}_1,i\,N_2}\right) = \\ &-\left(\sigma\left(T_{i,\,N_2} + T_{i+1,\,N_2} - 2T_e\right) + \lambda\left(T_{\underline{x}_2,i,\,N_2} + T_{\underline{x}_2,i+1,\,N_2}\right)\right) + \psi_{N_2-1/4}, \\ &i = 0, ..., N_1 - 1.\end{aligned}$$

The finite-difference scheme under consideration has an accuracy of the order $O\left(h_1^2 + h_2^2\right)$.

4 Selection of Initial Data

The choice of starting materials was due to a number of factors: practical interest; the presence of experimental data obtained in the course of a full-scale experiment, acting as test data to verify the correctness of the constructed mathematical model.

The thermophysical properties of the base fluid and aluminum particles are presented in Table 1. The particle size is 5 nm.

Table 1. The thermophysical properties of base fluid (water) and different types of nanoparticles at T = 298.15 K [16].

	Water	Al
Specific heat, $J/(kg \cdot K)$	4182	900
Density, kg/m^3	998.2	2700
Thermal conductivity, $W/(m \cdot K)$	0.607	247

Below are the values of other qualitative characteristics used in the framework of the numerical experiment: $T_{in} = 308.15$ K - the inlet temperature of the nanofluid in the solar collector; $T_e = 298.15$ K - the temperature of ambient air; $\alpha = 0.021\ m^{-1}$ - the light wave absorption coefficient; $I_0 = 635\ W/m^2$ - the initial intensity of the light flux entering the liquid; $f_v = 0.008\%$ - the particle volume fraction; $\tau = 0.9$- the transmittance; $\mu = 0.0009 \text{Pa} \cdot \text{c}$ - the dynamic viscosity of nanofluid; $n_{bf} = 1.33$ - the refractive index of base liquid; $\sigma = 6.43\ W/(m^2 \cdot K)$ - the heat transfer coefficient [1]. The values of the optical quantities: n is reflectance index of the nanoparticle material and κ is absorption index of the nanoparticle material were borrowed from [13].

In order to compare the obtained results with the authors of [1], we take the thickness of the solar collector L_2 and width L_1 equal to 0.0012 m and 1 m, respectively.

To carry out a numerical experiment, a numerical simulation algorithm was developed and implemented in software, and the results obtained using the supercomputer of the Computing Center of the Far Eastern Branch of the Russian Academy of Sciences [17] are presented below.

An argument in favor of the plausibility of the obtained numerical results is that at a speed equal to 0 m/s, the results of the numerical simulation algorithm completely converge to the results of the analytical solution (11).

5 Influence of Fluid Velocity on the Process of Heat and Mass Transfer

Let us determine the influence of the fluid velocity profile on the physical processes in a direct absorption solar collector. To do this, we construct the dependences of temperature on spatial coordinates at a constant speed equal to 1 m/s and a speed specified according to formula (4). The main condition for comparison will be that the amount of flow passing through the cross section of the direct absorption solar collector per unit time will be the same, Fig. 2.

For ease of interpretation of the results, the temperature is displayed in degrees Celsius. When converting from Kelvin to Celsius, a difference of 273.15 was taken into account.

Note that in the framework of this numerical experiment, in order to compare the results obtained with the authors of [1], the effect of particle reflection from the lower wall of the direct absorption solar collector was not taken into account. The results obtained at a speed of 1 m/s completely coincided with the results presented in [1].

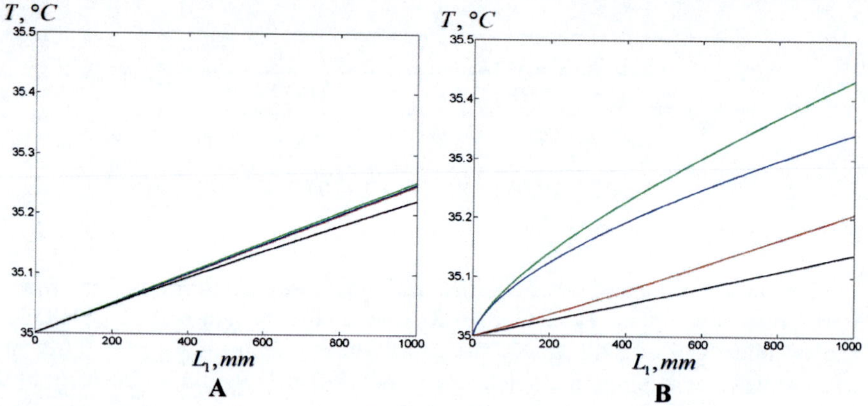

Fig. 2. Graph of temperature distribution along the length of the direct absorption solar collector ($x_2 = 0.25L_2$– black line, $x_2 = 0.5L_2$ - red line, $x_2 = 0.75L_2$– blue line, $x_2 = L_2$- green line). A is a constant flow rate equal to 1 m/s, B is a flow rate set according to formula (4).

From the graphs shown in Fig. 2, the following conclusions can be drawn: the velocity distribution profile has a significant impact on the conditions for the occurrence of physical processes. For a mathematical description of the fluid flow in the channel, in view of its geometry and features, it is better to use not a constant speed, but a speed given according to formula (4). The low increase in temperature values relative to the initial one is due to the small thickness of the collector, and also to the fact that, in order to reveal the influence of the velocity profile, the effect of reflection of the light flux passing through the colloidal nanosuspension from the bottom of the cuvette was not taken into account.

At the next stage, another numerical experiment was carried out, in which the thickness of the simulated direct absorption solar collector was increased by 10 times and amounted to 12 mm. For an adequate comparison of the obtained results with the results of the previous computational experiment, the constant flow velocity was reduced by a factor of 10 and amounted to 0.1 m/s. In this case, the speed given according to formula (4) was chosen in such a way that the value of the flow passing through the section of the direct absorption solar collector per unit time was the same.

The results of this numerical experiment are shown in Fig. 3.

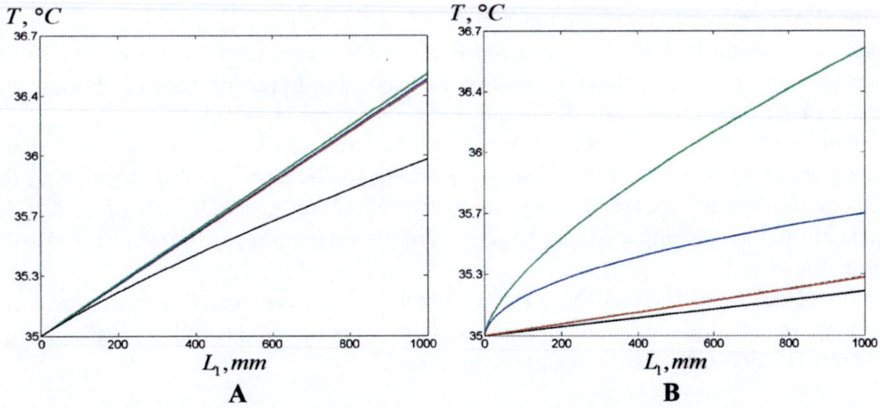

Fig. 3. Graph of temperature distribution along the length of the direct absorption solar collector ($x_2 = 0.25L_2$– black line, $x_2 = 0.5L_2$ - red line, $x_2 = 0.75L_2$– blue line, $x_2 = L_2$- green line). A is a constant flow rate equal to 1 m/s, B is a flow rate set according to formula (4).

Comparing the graphs shown in Figs. 3 with the graphs in Figs. 2, we can draw the following conclusions: there is a similar trend in the flow of heat and mass transfer processes inside the direct absorption solar collector, both at a constant flow rate and at a flow rate specified according to the formula (4).

6 Accounting for the Effect of Light Flux Reflection on the Process of Heat and Mass Transfer

When carrying out the numerical experiments described in the previous paragraph, it was found that in the case of a direct absorption collector thickness of 1.2 mm, the value of the light flux absorbed by the coolant liquid is about 0.3% of the initial one. In the case of a solar collector thickness of 12 mm - about 3%. In this regard, additional studies were carried out, the results of which are presented in this paragraph.

In order to reveal the influence of the effect of reflection of the light flux on the process of heat and mass transfer, it is necessary to change the expression responsible for determining the intensity of visible solar radiation, according to formulas (15).

$$I = \int_{\lambda_{\min}}^{\lambda_{\max}} \tau I_{b,\lambda}\left(e^{(x_2-L_2)K_{e\lambda,nf}} - e^{-(x_2+L_2)K_{e\lambda,nf}}\right)d\lambda, \tag{15}$$

$$q = \frac{\partial I}{\partial x_2} = \int_{\lambda_{\min}}^{\lambda_{\max}} \tau I_{b,\lambda} K_{e\lambda,nf}\left(e^{(x_2-L_2)K_{e\lambda,nf}} + e^{-(x_2+L_2)K_{e\lambda,nf}}\right)d\lambda.$$

From the point of view of physical theory, the differences in the case of taking into account and in the case of not taking into account the effect of reflection of the light flux are as follows: the first means that the lower wall of the direct absorption solar collector is made of a reflective material capable of reflecting the light flux behind limits of the solar collector through the heat transfer fluid. In the second case, the lower wall of the direct absorption solar collector is transparent and the light flux passing through the heat-transfer fluid also passes through it outside the solar collector.

For a qualitative comparison of the obtained results with the results presented in the previous paragraph, we set the value of the constant velocity equal to 1 m/s, and the value of the velocity specified according to formula (4) is chosen in such a way that the value of the flow passing through the section of the solar collector per unit time was the same, Fig. 4.

From the graphs shown in Fig. 4, the following conclusions can be drawn: due to the reflection effect, the temperature of the coolant in the direct absorption solar collector is significantly higher than in the previous case.

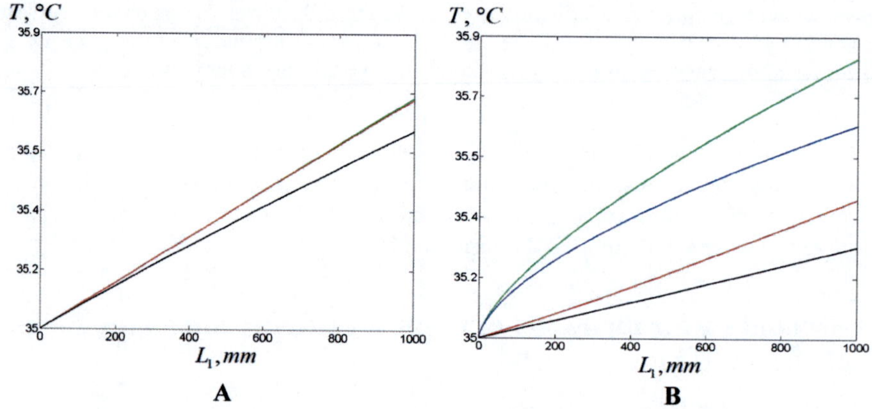

Fig. 4. Graph of temperature distribution along the length of the direct absorption solar collector ($x_2 = 0.25L_2$ - black line, $x_2 = 0.5L_2$ - red line, $x_2 = 0.75L_2$ - blue line, $x_2 = L_2$ - green line). A is a constant flow rate equal to 1 m/s, B is a flow rate set according to formula (4).

To identify a similar trend and formulate conclusions, another numerical experiment was carried out, taking into account the reflection effect, within which the thickness of the simulated direct absorption solar collector was increased by 10 times and amounted to 12 mm.

As in the previous case, the constant flow velocity was reduced by a factor of 10 and amounted to 0.1 m/s. The speed set according to formula (4) was chosen so that the value of the flow passing through the solar collector cross section per unit time was the same, Fig. 5.

Comparing the graphs shown in Fig. 5 with the graphs in Fig. 4, we can draw the following conclusions: there is a similar trend in the occurrence of heat and mass transfer processes inside the direct absorption solar collector.

Based on the results of the numerical experiments, some assumptions can be made regarding the efficiency of direct absorption solar collectors: a direct absorption solar collector, in which the lower wall is made of a reflective material, has a higher efficiency in relation to a direct absorption solar collector with a transparent bottom wall. This

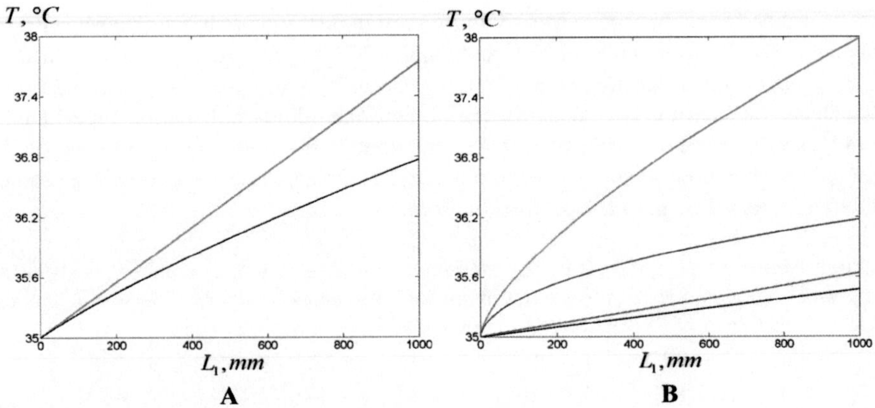

Fig. 5. Graph of temperature distribution along the length of the direct absorption solar collector ($x_2 = 0.25L_2$ - black line, $x_2 = 0.5L_2$ - red line, $x_2 = 0.75L_2$ - blue line, $x_2 = L_2$ - green line). A is a constant flow rate equal to 1 m/s, B is a flow rate set according to formula (4).

assumption is based on the fact that with the same thermophysical parameters of the heat transfer fluid, the geometry of the solar collector, as well as environmental conditions, the temperature of the heat transfer fluid is significantly higher. However, in order to form final conclusions regarding a more efficient type of direct absorption solar collector, it is necessary to conduct a number of numerical experiments, in which the absorbed amount of heat will be calculated and compared (in both types of solar collectors) for comparison.

7 Conclusion

A mathematical model has been constructed that describes the processes of heat and mass transfer, as well as energy transfer in direct solar collectors based on nanofluid. A numerical simulation algorithm has been developed and implemented in software, the main element of which is a conservative finite-difference scheme of the second order of approximation.

A mathematical model of heat and mass transfer in a liquid-phase medium with nanoparticles under the influence of a light field is studied numerically, taking into account one-dimensional concentration convection in the form of a boundary value problem for a system of second-order partial differential equations. The features of the temperature field are studied at a steady flow of a nano-fluid-coolant in a solar collector of direct absorption.

As a result of the numerical experiments, it was revealed that the velocity distribution profile has a significant effect on the conditions of the physical processes under consideration, a similar trend persists with varying direct absorption solar collector thickness; when the reflection effect is taken into account, the processes of heat and mass transfer proceed more uniformly, the temperature and heating of the coolant liquid in the central part of the collector are significantly higher.

A further vector of research is to study the efficiency of a direct absorption solar collector depending on the light-absorbing and reflective properties of the bottom wall. In addition, of particular interest is the influence of the geometry of the solar collector, as well as the thermophysical parameters of the nanofluid-coolant (the type of base liquid, as well as the material, size, shape, and volume fraction of nanoparticles) on the efficiency of the direct absorption solar collector. These and other aspects will be studied in detail in the following papers of the authors.

Acknowledgments. The reported study was funded by RFBR, project number 19-31-90070. This research was supported in through computational resources provided by the Shared Services Center «Data Center of FEB RAS» (Khabarovsk) [17].

References

1. Tyagi, H., Phelan, P., Prasher, R.: Predicted efficiency of a low-temperature nanofluid-based direct absorption solar collector. J. Sol. Energy Eng. **131** (2009)
2. Gorji, T.B., Ranjbar, A.: A numerical and experimental investigation on the performance of a low-flux direct absorption solar collector (dasc) using graphite, magnetite and silver nanofluids. Sol. Energy **135**, 493–505 (2016)
3. Vital, C., Farooq, S., de Araujo, R., Rativa, D., Malagón, G., Arturo, L.: Numerical assessment of transition metal nitrides nanofluids for improved performance of direct absorption solar collectors. Appl. Therm. Eng. **190**, 116799 (2021). https://doi.org/10.1016/j.applthermaleng.2021.116799
4. Kalogirou, S.A.: Solar thermal collectors and applications. Prog. Energy Combust. Sci. **30**, 231–295 (2004)
5. Raj, P., Subudhi, S.: A review of studies using nanofluids in flat-plate and direct absorption solar collectors. Renew. Sustain. Energy Rev. **84**, 54–74 (2018)
6. Chen, M., He, Y., Zhu, J., Kim, D.R.: Enhancement of photo-thermal conversion using gold nanofluids with different particle sizes. Energy Convers. Manage. **112**, 21–30 (2016)
7. Otanicar, T.P., Phelan, P.E., Prasher, R.S., Rosengarten, G., Taylor, R.A.: Nanofluidbased direct absorption solar collector. J. Renew. Sustain. Energy **2**, 033102 (2010)
8. Goel, N., Taylor, R.A., Otanicar, T.: A review of nanofluid-based direct absorption solar collectors: Design considerations and experiments with hybrid pv/thermal and direct steam generation collectors. Renew. Energy **145**, 903–913 (2020)
9. Farhana, K., et al.: Improvement in the performance of solar collectors with nanofluids–a state-of-the-art review. Nano-Struct. Nano-Objects **18**, 100276 (2019)
10. de Groot, S.R., Mazur, P.: Non-equilibrium thermo-dynamics. Courier Corporation, 510p. (1984)
11. Minakov, A.V., Rudyak, V.Ya., Guzey, D.V., Lobasov A.S.: Meas-urement of the heat transfer coefficient of a nanofluid based on water and particles of copper oxide. TVT **53**(2), 256–263 (2015). https://doi.org/10.7868/S0040364415020167
12. Incropera, F., Bergman, T.L., DeWitt, D., Lavine, A.S.: Fundamentals of Heat and Mass Transfer, 6th edn, pp. 260–261. Wiley (2007)
13. Hunter, W.R.: Optical constants of metals in the extreme ultraviolet. II. Optical constants of aluminum, magnesium, and indium at wavelengths shorter than their critical wavelengths*. J. Opt. Soc. Am. **54**, 208–212 (1964)

14. Matus, P.P., Hieu, L.M., Vulkov, L.G.: Analysis of second order dierence schemes on non-uniform grids for quasilinear parabolic equations. J. Comput. Appl. Math. **310**, 186–199 (2017)
15. Shi, D.: Numerical Methods in Problems of Heat Exchange, Moscow, 544p. (1988)
16. Cengel, Y.A.: Heat Transfer: A Practical Approach, 2nd edn. McGraw Hill, New York (2003)
17. Sorokin, A.A., Makogonov, S.V., Korolev, S.P.: The information infrastructure for collective scientific work in the Far East of Russia. Sci. Tech. Inf. Process. **44**(4), 302–304 (2017)

The Survey of Object-Oriented Software Programming Language from a Heterogeneous Cluster Programming Viewpoint

Tomas Brandejsky(✉) and Vaclav Hrbek

Faculty of Electrical Engineering and Informatics, University of Pardubice, Studentská 95, 53210 Pardubice, Czech Republic
tomas.brandejsky@upce.cz
http://www.upce.cz

Abstract. In this paper, the problem of programming language selection is presented from the position of large cluster with heterogeneous accelerators programming in the situations when it is need to apply object-oriented programming like in the case of heterogeneous multi-agent simulations or large data modelling using memetic algorithms. This work was inspired by experience with the Chapel language obtained during complicated conversion of hybrid evolutionary algorithm GPAes from a single node OpenMP C++ implementation onto HPC cluster with nodes equipped by both CPU and GPGPU.

The paper consists of discussion of many approaches to parallel programming including not only traditional ways such as OpenMP, MPI and Cuda and their combinations, but also modern extensions of C/C++ as OpenACC, Silk and CYCL. Emerging languages as Chapel and Julia are discussed too. The work concludes with an evaluation of the real state of parallel object-oriented programming on heterogeneous node HPC clusters.

Keywords: Object-Oriented Programming · C++ · OpenMP · MPI · SICL · Chapel · Julia · Parallel Programming · HPC cluster · Heterogeneous system · Memetic algorithm

1 Introduction

The total processor computing power continually increases, while the increase in single-thread performance of the processor is slowing down. For many years, their development detract from power improvement by parallelism increasing, because increasing of processor frequency is coupled with the quadratic increase of power consumption. In previous decades, many kinds of parallelism were established, especially

R. Silhavy (Ed.): CSOC 2022, LNNS 501, pp. 344–352, 2022.
https://doi.org/10.1007/978-3-031-09070-7_29

- pipelining
- superscalar (running more independent instruction in the same cycle, together with out-of-order execution)
- simd (e.g. vector instructions)
- etc.

With the incoming new hybrid architectures combining multi-thread many-core CPUs with accelerators, it is more and more significant to parallelize existing code. While in some cases, this way is relatively easy (e.g. some kinds of numerical calculations), in others it can be difficult and hard to automate. Problems frequently occur on clusters, heterogeneous architectures, non-uniform memory architecture systems, etc.

1.1 Background of the Publication

Within the frame of memetic algorithm research, the aim to implement GPAes [1] algorithm on heterogeneous computing cluster was formulated. On the one side, new computing accelerators like General-Purpose Graphic Processing Units (GPGPUs) from Intel, Nvidia and AMD), Field-Programmable Gate Arrays (FPGAs) and also vector processors as NEC Aurora are available now. On another side, memetic algorithms need a lot of computing power. E.g., following Flinn's taxonomy of processors [2], CPUs are now of Multiple Instruction Multiple Data (MIMD) architecture, but GPGPUs has Single Instruction Multiple Data (SIMD) one. Thus, a suitable programming language supporting many kinds of HW and allowing high programming productivity was searched.

Cluster or even HPC implementations of hybrid and memetic algorithms are essential for their computing complexity in the case of non-trivial training data size. The memetic algorithm GPAes consists of two parts - Genetic Programming Algorithm [3] and optimizer which can be represented e.g. by evolutionary strategy [4]. It is easier to process GPA on the CPU, while GPGPUs are better for optimization tasks. We can remember the pioneering works of Longdon. He used GPGPUs [5] for the fitness evaluation of structures produced by GPA and implemented two different approaches. The first was based on source code generating for all evaluated individuals, its compilation and running on the GPGPU; the second was based on the interpret implementation [6]. The work [7] distinguishes 3 different approaches to implementation of GPA on GPGPUs from the viewpoint of computing kernel management. These discussions are oriented to standard GPA implementations on GPGPUs, not to memetic algorithms adding optimization of parameters (constants) of each individual in each population by multiple instances of a nested optimizer. The implementation of many instances of nested optimizer increases the need for use of object-oriented programming (or difficult code and coding without it).

The complex structure of memetic algorithms and concluding computing complexity are the main reason for the search of the way how to implement them on heterogeneous clusters containing not only CPUs, but also computing accelerators like GPGPUs.

The memetic GPAes algorithm implementation uses the following data structures common for both GPA and ES parts. In the case of heterogeneous implementations, they must be present on both sides and synchronized. These structures are object arrays (representing array of optimizers or array of head nodes of trees representing individuals); tree structures of objects representing individuals and standard arrays containing fitness values and magnitudes of individual parameters (constants). If the interpret is used, there is also a need to store programs representing individuals as their alternative representations are processed by the interpret.

2 Hybrid Cluster Programming State of the Art

Together with the increase of accessible computing power, the improvement of many computationally extremely expensive parts of science, and applications are evolving nowadays, especially multi-physics simulations, to which implementations are now available tools suitable. Models described by extremely large sets of Ordinary Differential Equations (ODE) find application in military, nuclear science and also in industry. But, it is the question, how many new computing paradigms are prepared for the implementation of models based on models other than on solving of large sets of differential equations. For example, these alternative models are represented by Artificial Intelligence (AI) models which are not based on Artificial Neural Networks (ANNs), agent-oriented simulation models, etc., as well as the area of data analytic (if it is not based on the application of ANNs or statistic models).

Nowadays, we can observe intensive development of many different programming approaches and tools to simplify and thus speed up parallel program development for newly incoming heterogeneous computing systems, for increasing of programmers work efficiency in their programming. As heterogeneous computing systems are mainly labelled the computing clusters where standard compute nodes with CPU and local memory (thus Non-Uniform Memory Architecture systems) are extended by other computing accelerators, most frequently GPGPUs with separate local memory or FPGAs.

Data analytic and AI commonly may benefit from evolutionary algorithm application and especially from the memetic algorithm capability to provide fine symbolic regression. Also here, the problem is their computing expensiveness (the computing complexity of these algorithms) asking the need for implementation on large computing systems.

In the past, two basic architectures of parallel program design have expanded. The Message Passing Interface (MPI) is especially suitable for clusters composed of nodes interconnected by a data network; but it is capable to manage parallel programs on multi-core and multi-processor nodes too. The other one, OpenMP (Open Multi-Processing) is suitable for symmetric multiprocessor systems (and multi- or many-core processors) working with common shared memory. It allows one of the easiest programmings but it is constrained to a single node. In the past few years, clusters composed of hybrid nodes have begun to occur. These nodes do not contain only CPUs, but also computing accelerators specialized to some class of computing (typically vector ones), which contain their own separated memory like GPGPUs. These accelerators are highly efficient in some classes of tasks, like numerical solving systems of equations, but their programming brings different concepts.

2.1 MPI

MPI is now in version 4.0. Originally it was based on (NUMA) distributed memory model; since version 3 it introduces another approach to hybrid programming that uses the new MPI Shared Memory (SHM) model, see [12,13] introducing new basic principles.

For limited support of accelerator programming in the previous versions of OpenMP and lack of this support in MPI together with MPI programming difficulty and efficiency problems on multi-core processors, programmers still use complicated combinations MPI + CUDA, MPI + OpenMP, or even MPI + OpenMP + CUDA, MPI + OpenACC. The languages/techniques CUDA and OpenACC will be discussed below.

2.2 OpenMP

OpenMP is based on for-join model and parallelization of cycles, but the OpenMP standard 4.0 and later versions, have introduced pragmas for C, C++, and FORTRAN programming languages to offload work on general-purpose GPUs. It is possible to find information about the usage of OpenMP and GPU programming in the OpenMP specifications [8]. The following papers [9–11] explain the usage of GPU offloading pragmas. The problem of this application is in it's strong similarity to CUDA and OpenAcc because accelerator programming in OpenMP is also based on defining of compute kernels, parallelization of cycles. This style is similar to the frequently used combination OpenMP+CUDA. Worse is the need to solve the consistency of data in CPU and GPGPU memories manually.

2.3 OpenACC

OpenACC is an alternative to CUDA. OpenACC was defined by a broader consortium than CUDA, but its influence is significant and also OpenACC is based on C/C++ language and computer kernels, but it is a little higher level and hides some implementation details. It is available on both NVidia and AMD GPGPUs. The application limits are similar to CUDA.

2.4 CUDA

CUDA (Compute Unified Device Architecture) means both HW architecture and programming style. It was developed by NVidia company. The original support for Fortran and C was extended to C++ and OpenACC. Now, many programming languages and specialized tools support the programming of CUDA devices. The main problem of CUDA programming is given by its origin in graphic devices. GPGPUs are strongly parallel; their HW architecture is complicated. They contain many different memories (local, global, texture ones etc.). And CUDA does not hide this HW structure to support highly efficient programming. CUDA also allows a description of host code (code running on a host system, e.g. PC), but does not go far and does not allow to describe parallel computing on cluster level. It is possible to describe programs running on more graphic devices within the single node, but the limitation to single node remains.

Simply talking, while a modern CPU is a multi-core/multi-thread processor, GPGPU is vector one - there are many threads, but they process the same instructions (in CUDA within a so-called block). The basic programming concept in CUDA is the kernel function. These functions are executed in parallel as a block of threads. On CPU there is equivalent on some processors in the form of vector instructions. The use of vector (also called multimedia) instructions sometimes brings hardly predictable compute time on standard CPUs, as it will be discussed later.

CUDA as well as OpenMP, OpenACC, etc. offers concept of compute kernels to simplify pluralization of cycles. This fact limits its application in areas where pure multi-threading is required. Because GPGPU has its own memory, its use brings the need to solve data transfers between CPU's and GPGPU's RAM. Even if the newest versions of CUDA and NVidia's GPGPU allow direct access to host computer memory, this access is significantly slower than working with the device one. CUDA also allows to process complex structures, but it does not solve their synchronization between host and device memories implicitly. They must be organized by a programmer. This fact increases code complexity (and thus the probability of error occurrence). OOP on CUDA-like HW is possible since 2020, while on the host it was allowed since CUDA origin, see [14,15].

The remaining problem of OOP code on both host and CUDA accelerator is the need of two instances of object - the first on the host and the second on the device not only for the existence of two different memory spaces (this problem can be solved by unified memory), but also for different addresses of virtual methods for host and device. Their dual existence is caused by different machine codes for CPU and GPGPU. See also [16]. In the case of memetic algorithm implementation it is needed to distribute not only data with static structure but also trees (graphs) of objects - instances of classes, which may on the different subsystems (e.g. CPUs versus GPGPUs) different code, thus different pointers from virtual method tables, etc.

As it is mentioned above, standard Intel and AMD processors also offer vector instructions. It is possible to compile code using them, e.g. the option -fopenmp-simd of the GNU C++ compiler. When this possibility was tested on GPAes code, the resulting code was significantly slower, but such a result was expectable. Compiler options influence the whole code, but GPAes consist of two parts. While for es optimizer the vector instructions can bring advance, for operations with a tree-like representation of genes in GPA part, the vector processing brings the decrease of execution speed. The use of GNU C++ with option -fopenmp-simd brings also compatibility problems that not all functionalities of OpenMP library are in that moment available. For example, there is not available omp_get_wtime function to measure run time.

It is possible to conclude that "classic" methods of heterogeneous SW development represented by combinations of MPI, OpenMP, CUDA, and eventually OpenACC discussed above, are constrained by the need for a non-homogeneous programming model (e.g., the use of combinations MPI+CUDA, OpenMP+MPI+Cuda etc.). This style of programming is difficult for a combination of diverse programming paradigms, but it is frequently used on large (and heterogeneous) clusters now.

3 Modern Languages and Derivatives of C/C++

In the last years, these problems have caused formulations of new concepts of parallel software development occur and they stand at the origin of new extensions of standard languages (C++, Java) or even new languages (Chapel, Julia).

Cilk, Cilk ++ and Cilk Plus were developed as general-purpose languages for concurrent/multi-threaded parallel programming on MIT Laboratory For Computer Science. They are based on standard structured (procedural) programming and the basic parallel concept is similar to OpenMP fork—join. They extend C/C ++ by three constructs cilk_spawn, cilk_sync and cilk_for (definition of parallel thread, result synchronization and parallel loop). But these extensions to C/C++ does not offer direct support of accelerators.

3.1 Sycl

C/C++ language-based extension (or rather abstraction layer) Sycl, whose standard is developed by Kronos group (now in the version Sycl2020) exists in many implementations including Intel oneAPI DPC++. It also uses compute kernel model (like CUDA), but it solves memory sharing using accessors objects. This model allows to write versatile code capable to run on the wide spectrum of HW, but also this code is not pure original C/C++. Sycl is not a part of GNU compiler collection, but on the pages [17] it is possible to find the free accessible implementation of oneAPI DPC++ by Intel including Sycl. This implementation is based on LLVM free compiler architecture. Official support of NVidia and AMD GPGPUs is not, but it is possible using hipSYCL [18].

The Intel oneAPI DPC++ also contains OpenMP and especially MPI. This fact only concludes that neither Sycl is capable to describe parallelism on the level of heterogeneous clusters; it is limited to single heterogeneous node programming.

Also, Sycl will require a similar effort like previously described combinations based on CUDA or OpenACC, because Sycl also requires manual solving of complex structure consistence.

After this review of Cycl there is remaining an interesting group of new languages consisting of Chapel, X10, and Julia.

3.2 Chapel and X10

Because in this year the information about development of NVidia and AMD GPGPUs support in programming language Chapel [19] were publicized, this language become the main candidate for hybrid cluster implementation of GPAes code. At this moment, this support is still under construction, but it's hopeful for the future.

This language was formed within the above-cited project supported by Darpa. In this project also the second language origins. X10 was proposed by IBM. While Chapel was really a new language, X10 is Java-based. X10 is a programming language being developed by IBM at the Thomas J. Watson Research Center as part of above mentioned the Productive, Easy-to-use, Reliable Computing System (PERCS) project funded by DARPA's High Productivity Computing Systems (HPCS) program. X10 project at this moment seems to be stopped; the last upgrade was in 2019-01-07 [20]. Both languages (Chapel and X10) are from the so-called PGAS languages family. PGAS is an acronym for Partitioned Global Array Space [21]. Now there can be observed effort to include into this model also heterogeneous systems - clusters with nodes consisting of combination of CPUs and accelerators, as on the newest (and the most powerful) supercomputers.

3.3 Chapel - Practical Experience

With the work on the transformation of GPAes project into the Chapel language, strong problems were found. These problems are not in the support of cluster computing but in the actual state of OOP implementation incompatible with C++ one. Even if the language supports OOP, there is no possibility to use generic objects (polymorphism), especially arrays of generic objects and structures with more complex relations between them; it means structures referencing generic class objects. This unexpected step was done for inconsistency in the automatic garbage collector work. Without this ability accessible in other OOP languages, it is not possible at this moment to implement effective work with an array of genes in the GPA algorithm and thus no memetic algorithm with an nested optimizer, as GPAes. This problem in the Chapel language at this moment also affects such data structure implementations, as the implementation of stack, or join list.

This current state of Chapel implementation, non-looking to promising implementation of GPGPU support (based on compute kernels as in CUDA, e.g.) does not allow to use Chapel for reasoned purpose.

3.4 Julia

The Julia language was reasoned for its ability to describe parallel programs using OOP and functional programming. A practical problem is given by the fact, that Julia uses Java Virtual Machine (JVM) as e.g. X10 and thus at this moment cannot work on GPGPUs, because the instruction set of GPGPUs does not allow its work.

4 Conclusion

Presented paper points to the fact that if the solved problem tends to implementation of graph (or tree) of objects, forest of objects, array of graphs, etc., till now, there is no applicable progress in programming languages suitable to hybrid cluster programming. The situation still is not improving the standard approach based on combinations MPI+CUDA or MPI+OpenMP+CUDA used for many years with the known difficulties.

There are other approaches to solving the problem of hybrid single node programming as OpenACC or Sycl, but their use on clusters requires simultaneous application of MPI. On the opposite, there are languages optimized for problems where data are too big to fit a single node, to be processed on multiple nodes as Chapel or Julia, but there are other problems of accelerator programming (on heterogeneous nodes), and in the case of Chapel there are also still unsolved problems with OOP implementation.

Acknowledgments. The work was supported from ERDF/ESF "Cooperation in Applied Research between the University of Pardubice and companies, in the Field of Positioning, Detection and Simulation Technology for Transport Systems (PosiTrans)" (No. CZ.02.1.01/0.0/0.0/17_049/0008394).

References

1. Brandejsky, T., Zelinka, I.: Specific behaviour of GPA-ES evolutionary system observed in deterministic chaos regression. In: Zelinka, I., et al., (eds.) Nostradamus: Modern Methods of Prediction, Modeling and Analysis of Nonlinear Systems. Advances in Intelligent Systems and Computing, pp. 73–81. Springer, Heidelberg (2013). https://doi.org/10.1007/978-3-642-33227-2_10
2. Flynn, M.J.: Some computer organizations and their effectiveness. IEEE Trans. Comput. **C-21**(9), 948–960 (1972). https://doi.org/10.1109/TC.1972.5009071
3. Koza, J.R.: Genetic Programming: On the Programming of Computers by Means of Natural Selection. MIT Press, Cambridge (1992)
4. Beyer, H.-G., Schwefel, H.-P.: Evolution strategies - a comprehensive introduction. Nat. Comput. (2002)
5. Langdon, W.B.: A Many Threaded CUDA Interpreter for Genetic Programming (EuroGP) (2010)
6. Langdon, W.B.: Graphics processing units and genetic programming: an overview. Soft Comput. **15**, 1657–1669 (2011)
7. Kim, Jinhan, Kim, Junhwi, Yoo, Shin: GPGPGPU: Evaluation of Parallelisation of Genetic Programming Using GPGPU. In: Menzies, Tim, Petke, Justyna (eds.) SSBSE 2017. LNCS, vol. 10452, pp. 137–142. Springer, Cham (2017). https://doi.org/10.1007/978-3-319-66299-2_11
8. OpenMP Compilers and Tools. https://www.openmp.org/resources/openmp-compilers-tools/
9. Hayashi, A., Shirako, J., Tiotto, E., Ho, R., Sarkar, V.: Performance evaluation of OpenMP's target construct on GPUs - exploring compiler optimisations. Int. J. High Perform. Comput. Netw. **13**, 54–69 (2019)

10. CUDA C Best Practices Guide. https://docs.nvidia.com/cuda/cuda-c-best-practices-guide/index.html
11. OpenMP on GPUs First experiences and best practices. https://docs.nvidia.com/cuda/cuda-c-best-practices-guide/index.html
12. Brinskiy, M., Lubin, M.: An-introduction-to-MPI. https://www.intel.com/content/dam/develop/external/us/en/documents/an-introduction-to-mpi-3-597891.pdf
13. Intro to Parallel Programming. https://princetonuniversity.github.io/PUbootcamp/sessions/parallel-programming/Intro_PP_bootcamp_2018.pdf
14. CUDA C Programming Guide. https://docs.nvidia.com/cuda/cuda-c-programming-guide/index.html#c-cplusplus-language-support
15. Park, T.-J.: CUDA-based object oriented programming techniques for efficient parallel visualization of 3D content. J. Digit. Cont. Soc. **13**(2) (2012). https://doi.org/10.9728/dcs.2012.13.2.169
16. How to use class in CUDA C. https://forums.developer.nvidia.com/t/how-to-use-class-in-cuda-c/61761/2
17. Intel Compilers. https://github.com/intel/llvm
18. HipSYCL. https://github.com/illuhad/hipSYCL
19. Chapel Language. https://chapel-lang.org/
20. X10 Language. http://x10-lang.org
21. Almasi, G.: PGAS (Partitioned Global Address Space) Languages. In: Padua, D. (ed.) Encyclopedia of Parallel Computing, pp. 1539–1545, Springer, Boston (2011). https://doi.org/10.1007/978-0-387-09766-4_210

Toward Applying Agglomerative Hierarchical Clustering in Improving the Software Development Effort Estimation

Vo Van Hai(✉), Ho Le Le Le Nhung, and Roman Jasek

Department of Computer and Communication Systems, Tomas Bata University in Zlin, Nam. TGM 5555, 76001 Zlin, Czech Republic
{vo_van,lho,jasek}@utb.cz

Abstract. *Background*: There are many studies on the effect of data clustering on the effort estimation process. Most of them are on partitioning and density-based clustering, and some use hierarchical clustering but fewer details on the linkage methods. *Aim*: we concentrate on the aspect of the agglomerative hierarchical clustering algorithm's effectiveness on the accuracy of the effort estimation. *Method*: We used the agglomerative hierarchical clustering algorithm to group the data into clusters then performed the IFPUG FPA method for effort estimation. The ISBSG dataset was used in this study. The number of clusters is determined using the dendrogram's cut points. Different cut points and linkage methods were employed to cluster the dataset for the comparison. The estimated results of these clusters were compared with the result from the whole dataset without clustering. *Result*: with the selected number of clusters, results are consistently better than without clustering with all selected evaluation criteria. *Conclusion*: the accuracy of the effort estimation can be significantly improved when using agglomerative hierarchical clustering.

Keywords: Software effort estimation · Function Point Analysis · Hierarchical clustering · Machine learning

1 Introduction

Software effort estimation is one of the most challenging problems for project managers. This dramatically affects the success or failure of software development. We can view estimation techniques divided into three categories: algorithm-based, non-algorithm, and machine learning (ML) [1, 2]. Among the proposed solutions, applying machine learning techniques is a solution that has emerged recently and is seen as an alternative to the other two solutions [3]. However, the significant problem with ML is the data used to train the models. This dataset must be accurate and reflect the current software industry. Fortunately, the International Software Benchmarking Standards Group (ISBSG) [4] dataset is one of the most recognized globally, contributed by many companies around the globe. It partly solves the problem of data. However, with a large dataset like ISBSG, when we use the whole dataset to train for a model, the accuracy is not too feasible when

R. Silhavy (Ed.): CSOC 2022, LNNS 501, pp. 353–371, 2022.
https://doi.org/10.1007/978-3-031-09070-7_30

we use that model to estimate the project to be developed. There are many reasons for this inaccuracy; one important reason may be that the model's training data spread over a huge range, while the data to be estimated belongs to only a small segment. Therefore, the need to cluster the data before training the ML model plays an essential role in the model's accuracy.

Many studies have pointed out this problem [5, 7, 8], and many solutions have been proposed for clustering. Partitioning, density-based, and hierarchical clustering are commonly used solutions for data clustering. In this study, we use hierarchical clustering for our experimental implementation. The main objective of this study was to answer the following two research questions:

RQ1: How does the data clustering using hierarchical clustering affect FPA method estimation accuracy?

RQ2: Are there significant improvements when using hierarchical clustering compared to without clustering?

We did an experiment to solve these research questions. In addition, we performed a statistical pairwise t-test comparison to validate the accuracy of the proposed method. The hypotheses of this test are:

- H0: There is no significant difference in estimation capability between the proposed method and the FPA method. This statement means that the estimation accuracy of the proposed method is not significantly different from that of the FPA method.
- H1: There is a significant difference in estimation capability between the proposed method and the FPA method. This statement means that the estimation accuracy of the proposed method is significantly different from that of the FPA method.

The remainder of this work is divided into the following sections: Sect. 2 represents the related work. The background includes an overview of the FPA, the effort computation, and hierarchical clustering is present in Sect. 3. Section 4 describes the research methodology, which contains data preprocessing, experimental setup, and evaluation criteria. The result and discussion of this study are presented in Sect. 5. Section 6 demonstrates the conclusion and future work.

2 Related Work

Many solutions and models have been proposed before to improve the accuracy of software effort estimation. These studies have made significant contributions. However, there is no such thing as an ideal solution, as we all know. Therefore, research is still ongoing on the problem of improving estimation accuracy.

Prokopová et al. [5] performed a comparison among k-mean, density-based, and hierarchical clustering methods with three different distance metrics to find the most appropriate clustering technique. The results of their experiment showed that the k-mean is achieved the best result. The other conclusion from this study confirms the significant role of adequately selecting the clustering type and distance metric. The authors did not mention the linkage method used in their study, so they found that this method's result is not better than the other two methods.

Sehra et al. [6] used the k-means clustering algorithm and Ordinary Least Squared (OLS) Regression technique in data preprocessing, then estimated software effort using the COCOMO method. The experiment results have achieved shown that this combination was better than the method without using clustering in data preprocessing.

Wu and Keung [7] proposed a technique based on hierarchical clustering to construct a range of the values of k through various cluster quality criteria. The technique used in their study is the heuristic optimization technique, and the dataset is the PROMISE repository. The result shows that their proposed method can be used to determine an optimized set of k values for analogy-based prediction better than the expert's fixed k value. The authors also found that complete-linkage clustering is more fitting for large datasets, while single-linkage clustering is suitable for small datasets.

Silhavy et al. [8] proposed a novel model based on the dataset segmentation, called newCVS, in which Relative size was used as the parameter of segmentation. The authors also used the stepwise regression method to create the unique estimation model of each segment. The proposed model results were compared with the IFPUG FPA and clustering-based models (regression and spectral clustering). The experiment result showed that the newCVS model is outperformed the clustering-based models.

A challenging problem for managers in software estimation is determining the boundaries of the estimation. In the effort to solve this problem, Sarac and Duru [9] proposed the method that combined the COCOMO model with ANN and k-means algorithms to estimate effort and possible boundaries. The experiment results show that their model is acceptable over ANN and COCOMO.

In [10], the authors investigated to confirm the critical role of clustering in data mining. The authors also explored hierarchical clustering to improve the clusters' quality. Three parameters were implemented in the experiment are cohesion measurement, silhouette index, and elapsed time to find the relation between them. They found that a cluster with high cohesion, zero and positive values of silhouette index, and low elapsed time outcomes in a good quality cluster.

In the recent study [11], we proposed a new method to improve the FPA method. This study proposed a new complexity weight and combined it with the ensemble method to optimize the estimation result.

3 Background

3.1 The FPA Overview

Function Point Analysis (FPA) was proposed by Albrecht in 1979 [12]. FPA was presented as a metric to measure the functionality of the project. The International Function Point User Group (IFPUG) [13] has been the governing body for FPA since 1986, responsible for improving and developing counting rules and other related matters.

In FPA, two significant quantities should be computed before calculating the Adjusted Function Points (AFP), Unadjusted Function Points (UFP), and Value Adjustment Factor (VAF). The obtaining of UFP is based on the counting of five base functional components External Input (EI), External Output (EO), or External Inquiry (EQ), Internal Logic Files (ILF), and External Interface Files (EIF) of the developing system. Each of them is evaluated as simple, average, or complex and assigned a weight accordingly in Table 1. The formula to compute the UFP can be offered as Eq. (1).

$$UFP = \sum EI \times weight + \sum EO \times weight + \sum EQ \times weight + \sum ILF \times weight + \sum EIF \times weight \quad (1)$$

Table 1. Function complexity

Component	Functional complexity		
	Low	Average	High
EI	3	4	6
EO	4	5	7
EQ	3	4	6
EIF	5	7	10
ILF	7	10	15

The VAF value depends on the influence of the developing system on 14 factors (called general system characteristics – GSCs). These influences are judged from 0 to 5 depending on the level: no influence, incidental, moderate, average, significant, and strong. Equation (2) is the formula for computing the VAF.

$$VAF = 0.65 + 0.01 \times \sum_{i=1}^{14} (F_i \times rating) \quad (2)$$

The AFP can be computed using Eq. (3)

$$AFP = UFP \times VAF \quad (3)$$

3.2 From AFP to Effort

After counting AFP, we need an essential parameter to estimate the software effort: the productivity factor (PF). The PF was defined as the relationship between one function point and the number of hours needed for its development by one person. This factor was investigated in many studies, for example, [14–16].

Effort can be computed by using Eq. (4).

$$Effort = AFP \times PF \quad (4)$$

In this study, the ISBSG dataset was used, in which the Productivity Delivery Rate (PDR) is considered the metric for efficiency. This value uses the person-hours per FP as the measuring unit. PF and PDR differ, but we can derive the other when knowing the one [17].

3.3 Hierarchical Clustering

Hierarchical clustering is a famous clustering method that can be considered a set of flat clustering methods organized in a tree structure. These methods assemble the clusters by recursively partitioning the data in a top-down or bottom-up manner. Among hierarchical algorithms, bottom-up approaches tend to be more accurate but have higher computational costs than top-down approaches [18]. In this way, hierarchical clustering can be categorized into two types: Agglomerative hierarchical clustering (bottom-up approach) and Divisive hierarchical clustering (top-down approach).

Agglomerative hierarchical clustering (also known as Agglomerative Nesting - AGNES) begins by treating each observation as a distinct cluster. Then it repeats the following two steps: (1) identify the two clusters which are the closest together, and (2) combine the two clusters that are the most similar. This iterative procedure is iterated until all of the clusters have been merged.

Divisive hierarchical clustering supposes that all objects initially belong to a single root cluster and iteratively partitions existing clusters into sub-clusters [19].

There are two main issues in hierarchical clustering that we should solve: measuring the distance between two clusters (similarity) and selecting the linkage criteria (from where distance was computed). The typical solution for the first problem is using Euclidean distance. With the second problem, there are several linkage methods for cluster observation in hierarchical clusterings, such as Average, Complete, Single, Ward [20].

The distance between two clusters in the Average linkage method was computed as the distance between the two clusters' average values (the mean vectors or centroids).

$$d(X, Y) = \text{mean}(d(x_i \in X, y_j \in Y) \tag{5}$$

The Single linkage: distance between clusters is the distance between the closest points.

$$d(X, Y) = \min(d(x_i \in X, y_j \in Y) \tag{6}$$

The Complete linkage: The maximum distance between an observation in one cluster and an observation in the other cluster is the distance between two clusters.

$$d(X, Y) = \max(d(x_i \in X, y_j \in Y) \tag{7}$$

The Ward linkage: The total of squared deviations from points to centroids is the distance between two clusters:

$$d(X, Y) = \|x_i \in X, y_j \in Y\|^2 \tag{8}$$

where:

- $d(x, y)$ is the distance between clusters x and y ($x \in Xandy \in Y$).
- X, Y are two sets of objects (clusters).

This study uses the Agglomerative algorithm with three linkage methods: Ward, Complete, and Average. We choose different cut points (number of clusters – the k value), then perform the clustering based on this k value and evaluate the effort estimation accuracy.

4 Research Methodology

4.1 Experiment Setup

In this section, we explain the experiment process, which is graphically depicted in Fig. 1.

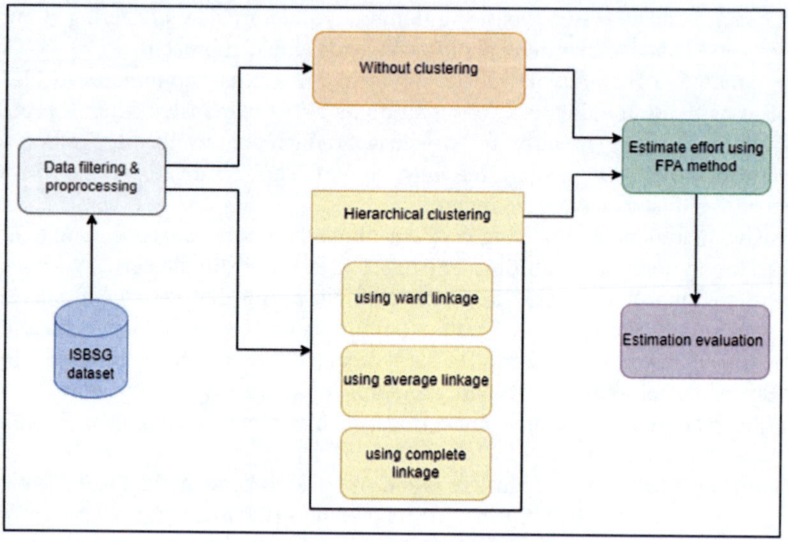

Fig. 1. Experiment process

We conduct two groups of experiments as described below:

1) Group 1: one experiment on the whole dataset

- Remove outliers for the whole dataset.
- Compute the effort using the FPA method

2) Group 2: three experiments correspondent with three linkage methods

- Remove outliers for the whole dataset.
- Draw dendrogram for dataset and determine the cut points (k values).
- For each k, clustering dataset into k clusters using the AGNES algorithm.
- Compute the effort of clusters using the FPA method.

Finally, we compare these clusters' effort estimation results versus the results from the dataset without clustering.

Data Filtering

The dataset we used in this experiment is the ISBSG repository August 2020 R1 [21]. In our study, the criteria for data filtering will be processed as follows:

1. Records with the counting approach of IFPUG (including IFPUG Old and IFPUG 4+) are selected.
2. Data quality of records A and B are chosen.
3. The development type should be the new development.
4. The rows with an empty value of Base Functional Components should be removed.
5. The rows with empty values in Normalized Productivity Delivery Rate and Summary Work Effort (SWE) were also eliminated.
6. Fill the blank cell of VAF by the value got from formula (3).

We observe that several AFP and SWE values are too far from the mean group, which suggests that the data could be noisy. To detect and eliminate outliers, we used the interquartile range (IQR) approach [22, 23] on these features, with the lower bound being 0.15 and the upper bound is 0.85. Figure 2 depicts the dataset's description before and after outlier reduction.

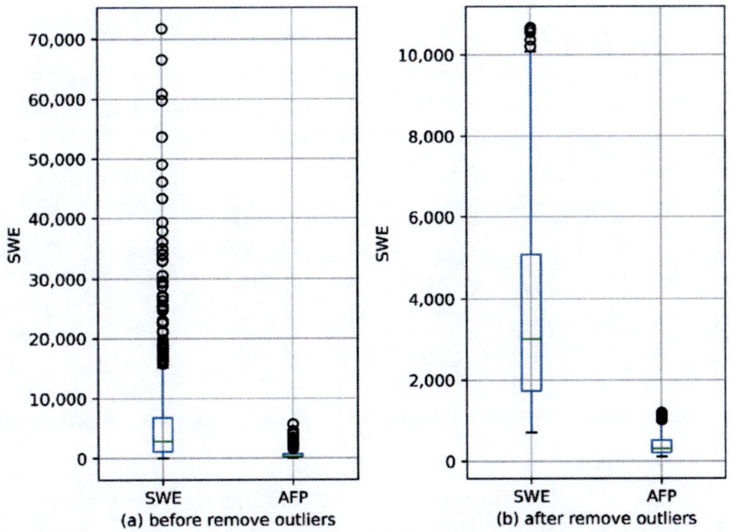

Fig. 2. Boxplot of the dataset

Data Clustering

In this study, we use the hierarchical clustering technique in clustering the dataset into clusters then apply the FPA method to estimate effort. Before proceeding with clustering, we use a dendrogram to determine the number of clusters. We use three linkage methods (ward, complete, and average) for clustering in this study. The following steps are the procedure we use to plot the dendrogram:

1. Data set normalization
2. Use the PCA algorithm to reduce the dimension of the dataset to 2.

3. Draw the dendrogram, determine cut points.
4. Clustering the dataset.

We performed a process using the linkage object and the Agglomerative Clustering algorithm in clustering to determine the cut points (k values) and clusters. These objects and their parameters are described in Table 2.

Table 2. Implementation and parameters of clustering algorithms

Implementation	Parameters
scipy.cluster.hierarchy.linkage	method = {ward, average, complete}
sklearn.cluster.AgglomerativeClustering	linkage = method, affinity = 'euclidean', method = {ward, average, complete}

The Ward Linkage Method

With the Ward linkage method, based on the observation of the dendrogram, we chose the number of clusters as 2, 3, 4, and 6. The dendrogram of these cut points is shown in Fig. 3. The illustration of these clusters can be found in Fig. 4.

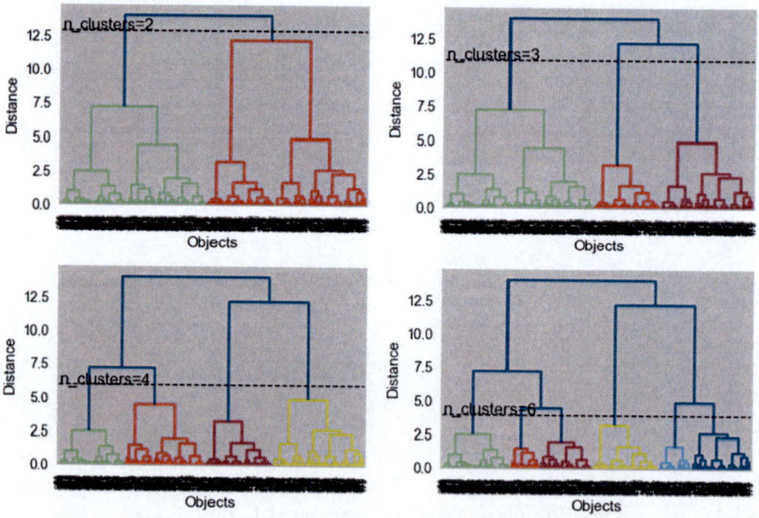

Fig. 3. Dendrogram using Ward linkage method

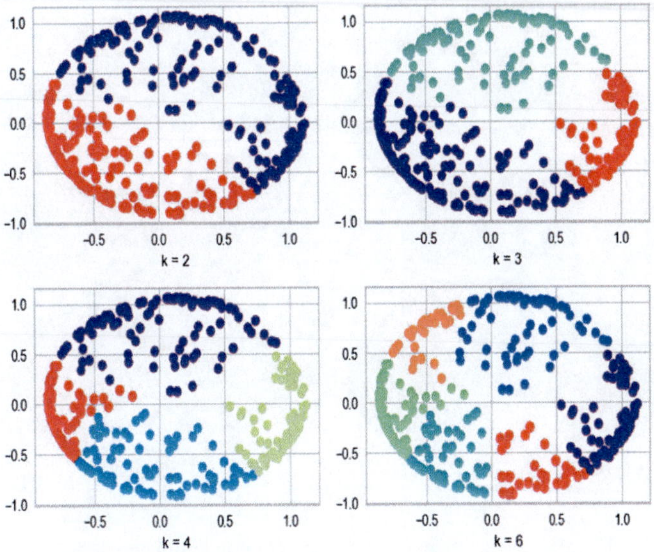

Fig. 4. Clusters distribution using Ward linkage method

The Complete Linkage Method

In this linkage method, the cut points we choose through observation are 2, 4, 6, and 9. The dendrogram and the distribution of the clusters can be seen in Fig. 5 and Fig. 6.

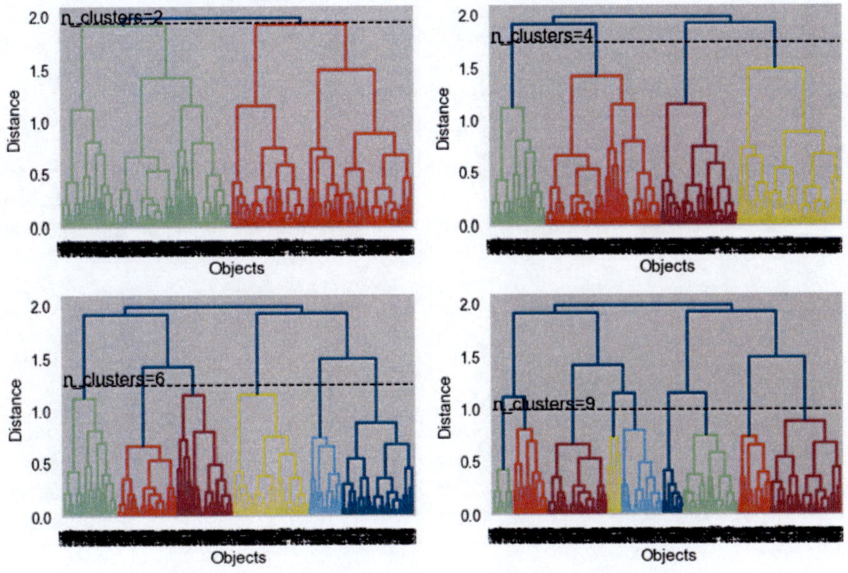

Fig. 5. Dendrogram using Complete linkage method

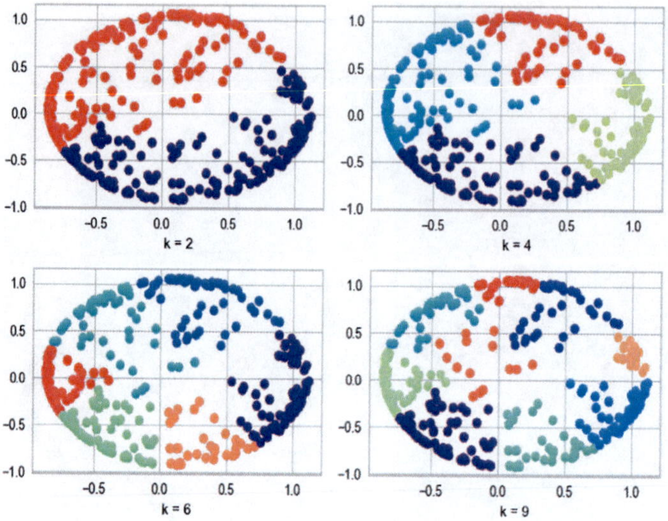

Fig. 6. Clusters distribution using Complete linkage method

The Average Linkage Method

The similarity of others, in this linkage method, 2, 3, 4, and 6 are the cut points that we choose. Figure 7 shows the dendrogram, and Fig. 8 presents the illustration of clustering.

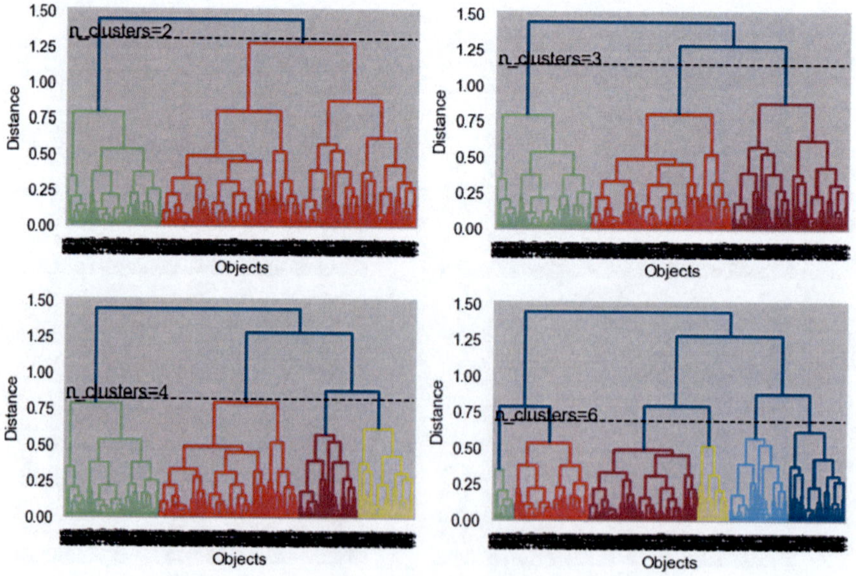

Fig. 7. Dendrogram using Average linkage method

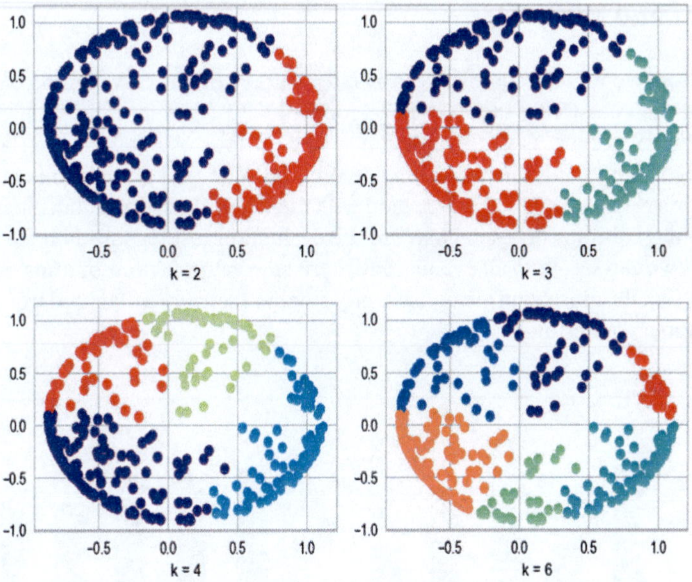

Fig. 8. Clusters distribution using Average linkage method

4.2 Evaluation Criteria

This study uses the evaluation criteria Mean Absolute Error (MAE), Root Mean Square Error (RMSE), Mean Balance Relative Error (MBRE), and Mean Absolute Percentage Error (MAPE). These evaluation criteria produce an unbiased and symmetric distribution [24–26].

$$MAE = \frac{1}{n}\sum_{i=1}^{n}|y_i - \hat{y}_i| \tag{9}$$

$$MAPE = \frac{1}{N}\sum_{i=0}^{n}\left|\frac{y_i - \hat{y}_i}{y_i}\right| * 100 \tag{10}$$

$$RMSE = \sqrt{\frac{\sum_{i=1}^{n}(y_i - \hat{y}_i)^2}{n}} \tag{11}$$

$$MBRE = \frac{1}{n}\sum_{i=1}^{n}\frac{|(y_i - \hat{y}_i)|}{min(y_i - \hat{y}_i)} \tag{12}$$

where y_i it the actual value, $\hat{y}_i$ is the estimated value, and n is the number of projects.

5 Result and Discussion

In this section, we will compare the estimated accuracy of the FPA method on the whole dataset and each cluster by the AGNES algorithm with different linkage methods.

First, Fig. 9 shows us the evaluation result on four evaluation criteria using the AGNES algorithm using the Ward linkage method. The first column on each graph is the evaluation result using the FPA method with the whole dataset without clustering. The four following columns on each chart are the evaluation results based on the number of clusters. As we can see, with all evaluation criteria, the values of the columns representing the results from the clustering are consistently smaller than those obtained from the whole dataset without clustering.

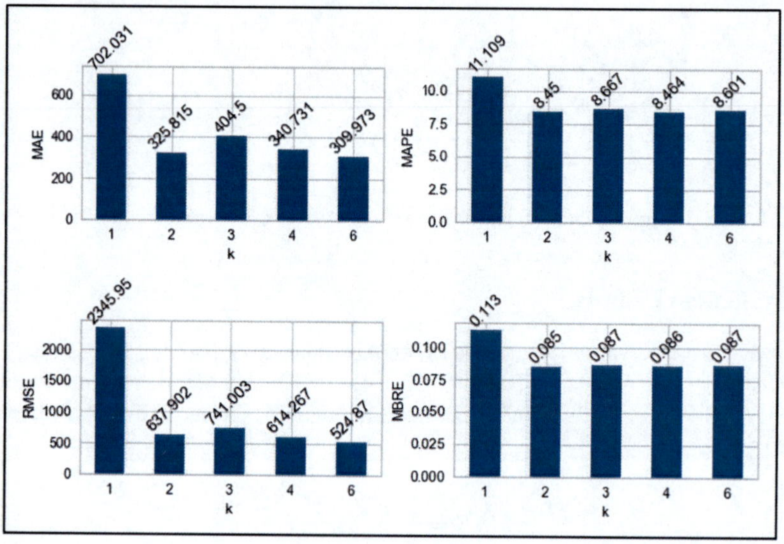

Fig. 9. The valuation results of hierarchical clustering using the Ward linkage method

The error value obtained from clustering is much smaller than that of no clustering with the MAE evaluation criteria. With the number of clusters being 2, the obtained value is 325,815 compared to 702,023 of the whole datasets; the improvement percentage reaches 53.59%. Same for the comparison with k = 3, the ratio is 42.83%, with k = 4, this rate is 51.46%. Especially for k = 6, this rate is 55.85%. The best number of clusters in the Ward linkage with the MAE evaluation criterion is 6. We can observe the percentage of improvement in Table 3 with other criteria.

Table 3. The improvement percentage of clusters in compared with without clustering in AGNES algorithm with the Ward linkage

k	MAE (%)	MAPE (%)	RMSE (%)	MBRE (%)
1	–	–	–	–
2	53.59	23.94	72.81	24.78
3	42.38	21.98	68.41	23.01
4	51.46	23.81	73.82	23.89
6	55.85	22.58	77.63	23.01

For the AGNES algorithm with the Complete linkage, Fig. 10 shows the evaluation results of the experiments. Same with the Ward linkage method result, all clusters have errors more minor than the whole dataset without clustering. The improvement percentage of clusters compared to without clusters on all evaluation criteria was shown in Table 4.

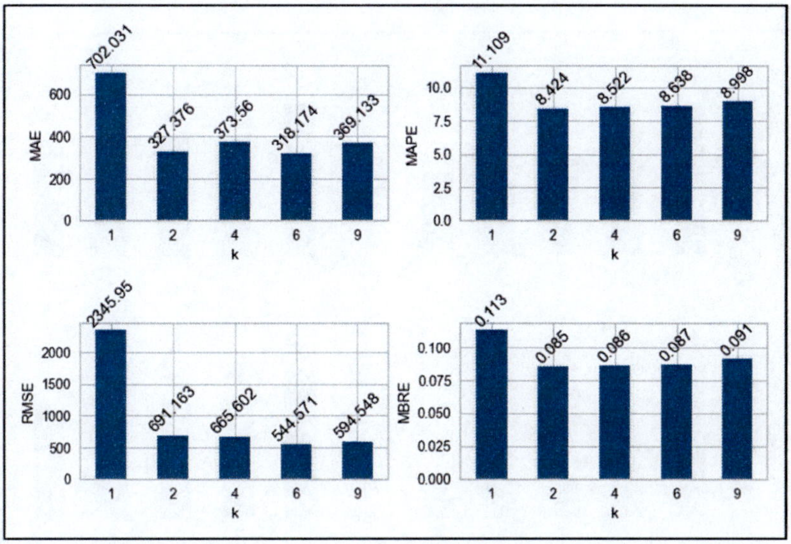

Fig. 10. The evaluation results of hierarchical clustering using the Complete linkage method

Table 4. The improvement percentage of clusters in compared with without clustering in AGNES algorithm with the Complete linkage

k	MAE (%)	MAPE (%)	RMSE (%)	MBRE (%)
1	–	–	–	–
2	53.37	24.17	70.54	24.78
4	46.79	23.29	71.63	23.89
6	54.68	22.24	76.79	23.01
9	47.42	19.00	74.66	19.47

With the Average linkage method in the AGNES algorithm, the evaluation results may not be satisfactory as the other two methods. But overall, the errors of clusters are smaller than without clustering on all evaluation criteria. Figure 11 shows the evaluation results of all evaluation criteria graphically. Table 5 shows the improvement percentage of clusters compared with without clustering.

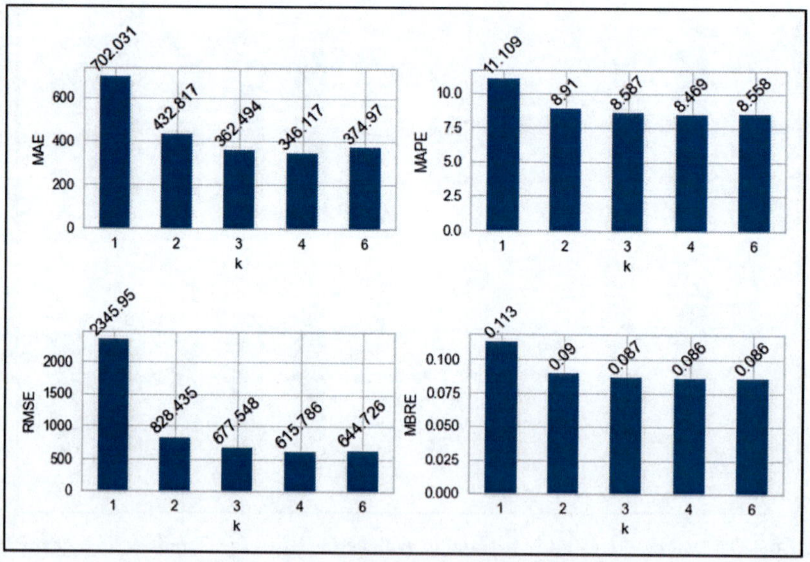

Fig. 11. The evaluation results of hierarchical clustering using the Average linkage method

Table 5. The improvement percentage of clusters in compared with without clustering in AGNES algorithm with the Average linkage

k	MAE (%)	MAPE (%)	RMSE (%)	MBRE (%)
1	–	–	–	–
2	38.35	19.79	64.69	20.35
4	48.36	22.70	71.12	23.01
6	50.70	23.76	73.75	23.89
9	46.59	22.96	72.52	23.89

We have an exciting statement to close this section: No matter what linkage methods are used in the group of methods studied; no matter how much k is used, the effort estimate using hierarchical clustering will consistently yield minor error values when compared to the results obtained from the dataset without clustering on any of the evaluation criteria.

Based on this statement and the assessment of the experimental results, we can move to answer the research questions.

RQ1: How does the data clustering using hierarchical clustering affect FPA method estimation accuracy?

Through experimentation in this study, we can easily see that clustering dramatically affects the accuracy of the estimation results. Figure 9, Fig. 10, Fig. 11, we can see that the estimation error of clusters on any of the evaluation criteria is smaller than on the whole dataset. It makes sense that clustering yields more accurate results.

RQ2: Are there significant improvements when using hierarchical clustering compared to without clustering?

To answer this research question, we used paired samples t-test to see if the proposed method is significantly different from the FPA method to confirm the evaluation conclusions. The notations $\gg$, $\ll$, and $\approx$ reflect the proposed approach's statistical superiority, inferiority, and similarity compared to the FPA method. The result of this t-test can be found in Table 6, Table 7, and Table 8. We can conclude that the difference in estimating accuracy between the proposed and the FPA approach is significant when the p-value is less than 0.05.

The t-test results on the above tables show the average p-value results and the MAE, MAPE, RMSE, and MBRE over different runs on linkage methods. This result confirms that the proposed method is statistically significant at the 95% confidence level compared to the FPA method applied on the whole dataset. Therefore, we are prone to accept the alternative hypothesis (H1), consistent with the results presented above.

Table 6. The statistical t-test based on evaluation results for the ward linkage method

Pairs of methods		Proposed method vs. FPA
MAE results	Avg. MAE	345.255 vs. 702.031
	Avg. p-value	0.00021
	Statistical conclusion	>>
MAPE results	Avg. MAPE	8.546 vs. 11.109
	Avg. p-value	0.00001
	Statistical conclusion	>>
RMSE results	Avg. RMSE	629.511 vs. 2345.948
	Avg. p-value	0.00002
	Statistical conclusion	>>
MBRE results	Avg. MBRE	0.086 vs. 0.113
	Avg. p-value	0.00001
	Statistical conclusion	>>

Table 7. The statistical t-test based on evaluation results for the complete linkage method

Pairs of methods		Proposed method vs. FPA
MAE results	Avg. MAE	347.061 vs. 702.031
	Avg. p-value	0.00007
	Statistical conclusion	>>
MAPE results	Avg. MAPE	8.646 vs. 11.109
	Avg. p-value	0.00014
	Statistical conclusion	>>
RMSE results	Avg. RMSE	623.971 vs. 2345.948
	Avg. p-value	0.00001
	Statistical conclusion	>>
MBRE results	Avg. MBRE	0.087 vs. 0.113
	Avg. p-value	0.00015
	Statistical conclusion	>>

Table 8. The statistical t-test based on evaluation results for the average linkage method

Pairs of methods		Proposed method vs. FPA
MAE results	Avg. MAE	379.1 vs. 702.031
	Avg. p-value	0.00022
	Statistical conclusion	>>
MAPE results	Avg. MAPE	8.631 vs. 11.109
	Avg. p-value	0.00006
	Statistical conclusion	>>
RMSE results	Avg. RMSE	691.624 vs. 2345.948
	Avg. p-value	0.00003
	Statistical conclusion	>>
MBRE results	Avg. MBRE	0.087 vs. 0.113
	Avg. p-value	0.00005
	Statistical conclusion	>>

6 Conclusion and Future Work

The standard IFPUG FPA method was used on the data clustering by the agglomerative hierarchical clustering algorithm was presented in this study. The paper desires to answer two research questions. In answer to RQ1, the agglomerative hierarchical clustering was used to verify that the effort estimation accuracy can be improved with data clustering.

In answer to RQ2, we performed a pairwise t-test and found that our proposed method is statistically significant at the 95% confidence level compared to the FPA method applied on the whole dataset. It means that our proposed method outperforms the FPA method applied on the entire dataset without clustering. Moreover, the improvement percentage varies from the evaluation criteria, linkage methods, and the number of clusters. In the Ward linkage method, with MAE and RMSE evaluation criteria, the optimal k is 6 with an improvement rate of 55.85% and 77.63%. With MAPE and MBRE, optimal k is 2 with an improvement rate of 23.94% and 24.78%, respectively. The same for the Complete linkage method, the number of clusters being 6 is the best for MAE and RMSE evaluation criteria while 2 is the k-optimal for two other evaluation criteria. The improvement percentage of this linkage method in their k-optimal with MAE = 54.68%, RMSE = 76.79%, MAPE = 24.17, and MBRE = 24.78. The best number of clusters is 6 for all evaluation criteria with the Average linkage method. The improvement percentage in this case is MAE = 50.70%, RMSE = 73.75%, MAPE = 23.76%, and MBRE = 23.89%.

Future work will investigate other clustering techniques combined with machine learning algorithms to improve the software effort estimation accuracy.

Acknowledgment. This work was supported by the Faculty of Applied Informatics, Tomas Bata University in Zlin, under project IGA/CebiaTech/2022/001.

References

1. Vera, T., Ochoa, S.F., Perovich, D.: Survey of software development effort estimation taxonomies. Technical Report, Computer Science Department, University of Chile, Chile (2017)
2. Khan, B., Khan, W., Arshad, M., Jan, N.: Software cost estimation: algorithmic and non-algorithmic approaches. Int. J. Data Sci. Adv. Anal. **2**, 1–5 (2020)
3. Sharma, P., Singh, J.: Systematic literature review on software effort estimation using machine learning approaches. In: 2017 International Conference on Next Generation Computing and Information Systems (ICNGCIS), pp. 43–47 (2017). https://doi.org/10.1109/ICNGCIS.2017.33
4. International Software Benchmarking Standards Group, ISBSG Repository August 2020 R1. https://www.isbsg.org/
5. Prokopova, Z., Silhavy, R., Silhavy, P.: The effects of clustering to software size estimation for the use case points methods. In: Silhavy, R., Silhavy, P., Prokopova, Z., Senkerik, R., Kominkova Oplatkova, Z. (eds.) CSOC 2017. AISC, vol. 575, pp. 479–490. Springer, Cham (2017). https://doi.org/10.1007/978-3-319-57141-6_51
6. Sehra, S.K., Kaur, J., Brar, Y.S., Kaur, N.: Analysis of data mining techniques for software effort estimation. In: 2014 11th International Conference on Information Technology: New Generations, pp. 633–638 (2014). https://doi.org/10.1109/ITNG.2014.116
7. Wu, J.H.C., Keung, J.W.: Utilizing cluster quality in hierarchical clustering for analogy-based software effort estimation. In: 2017 8th IEEE International Conference on Software Engineering and Service Science (ICSESS), pp. 1–4 (2017). https://doi.org/10.1109/ICSESS.2017.8342851
8. Silhavy, P., Silhavy, R., Prokopová, Z.: Categorical variable segmentation model for software development effort estimation. IEEE Access **7**, 9618–9626 (2019). https://doi.org/10.1109/ACCESS.2019.2891878
9. Saraç, Ö.F., Duru, N.: A novel method for software effort estimation: estimating with boundaries. In: 2013 IEEE INISTA, pp. 1–5 (2013). https://doi.org/10.1109/INISTA.2013.6577643
10. Nisha, Kaur, P.J.: Cluster quality-based performance evaluation of hierarchical clustering method. In: 2015 1st International Conference on Next Generation Computing Technologies (NGCT), pp. 649–653 (2015). https://doi.org/10.1109/NGCT.2015.7375201
11. Hai, V.V., Nhung, H.L.T.K., Prokopova, Z., Silhavy, R., Silhavy, P.: A new approach to calibrating functional complexity weight in software development effort estimation. Computers **11**, 15 (2022). https://doi.org/10.3390/computers11020015
12. Albrecht, A.J.: Measuring application development productivity. In: Proceedings of IBM Applications Development Symposium, p. 83 (1979)
13. International Function Point User Group (IFPUG). https://www.ifpug.org/. Accessed Sept 2021
14. Azzeh, M., Nassif, A.B.: Analyzing the relationship between project productivity and environment factors in the use case points method. J. Softw. Evol. Process **29**(9), e1882 (2017)
15. Prokopova, Z., Silhavy, P., Silhavy, R.: Influence analysis of selected factors in the function point work effort estimation. In: Silhavy, R., Silhavy, P., Prokopova, Z. (eds.) CoMeSySo 2018. AISC, vol. 860, pp. 112–124. Springer, Cham (2019). https://doi.org/10.1007/978-3-030-00184-1_11
16. Hai, V.V., Nhung, H.L.T.K., Hoc, H.T.: A productivity optimising model for improving software effort estimation. In: Silhavy, R., Silhavy, P., Prokopova, Z. (eds.) CoMeSySo 2020. AISC, vol. 1294, pp. 735–746. Springer, Cham (2020). https://doi.org/10.1007/978-3-030-63322-6_62

17. IFPUG, Function Point Counting Practices Manual, Release 4.3.1, International Function Point Users Group, Westerville, Ohio, USA (2010)
18. Masciari, E., Mazzeo, G.M., Zaniolo, C.: A new, fast and accurate algorithm for hierarchical clustering on euclidean distances. In: Pei, J., Tseng, V.S., Cao, L., Motoda, H., Xu, G. (eds.) PAKDD 2013. LNCS (LNAI), vol. 7819, pp. 111–122. Springer, Heidelberg (2013). https://doi.org/10.1007/978-3-642-37456-2_10
19. Rousseeuw, P.J., Kaufman, L.: Finding Groups in Data: An Introduction to Cluster Analysis. Wiley, Hoboken (2009)
20. Alpydin, E.: Introduction to Machine Learning, pp. 143–158. The MIT Press (2010)
21. International Software Benchmarking Standards Group, ISBSG Repository August 2020 R1. https://www.isbsg.org. Accessed 20 Sept2021)
22. Upton, G., Cook, I.: Understanding Statistics, p. 55. Oxford University Press (1996). ISBN 0-19-914391-9
23. Zwillinger, D., Kokoska, S.: CRC Standard Probability and Statistics Tables and Formulae, p. 18. CRC Press (2000). ISBN 1-58488-059-7
24. de Myttenaere, A., Golden, B., Le Grand, B., Rossi, F.: Mean absolute percentage error for regression models. Neurocomputing **192**, 38–48 (2016)
25. Nhung, H.L.T.K., Van Hai, V., Silhavy, R., Prokopova, Z., Silhavy, P.: Parametric software effort estimation based on optimizing correction factors and multiple linear regression. IEEE Access **10**, 2963–2986 (2022). https://doi.org/10.1109/ACCESS.2021.3139183
26. Boehm, B.: Software Engineering Economics. Prentice-Hall (1981). ISBN 0-13-822122-7

Analysis of Mimetic Muscle Rehabilitation: Data Fusion and Database Development

Karel Štícha[1(✉)], Jan Mareš[1], Jan Kohout[1], and Sanjeev Kumar[2]

[1] Department of Computing and Control Engineering, University of Chemistry and Technology in Prague, Technická 1905/5, 166 28 Praha 6, Czech Republic
{karel.sticha,jan.mares,jan.kohout}@vscht.cz

[2] Department of Mathematics, Indian Institute of Technology Roorkee, Roorkee 247667, India
sanjeev.kumar@ma.iitr.ac.in
http://optimcyb.vscht.cz , http://ma.iitr.ac.in/

Abstract. The paper deals with the software solution for facial data acquisition, which consists of a desktop and mobile application, relational database, and their mutual communication. Facial data in the form of facemarks, depth or RGB video are obtained during standardized mimetic muscles exercising of the patient with facial paralysis. Thanks to the collaboration with the *University Hospital Královské Vinohrady*, 228 measurement sessions of 118 patients were conducted so far.

Keywords: Facial paralysis · C# · .NET · Kinect v2 · Xamarin · Relational database · Entity framework · gRPC

1 Introduction

The objectification of the determination of mimetic muscle dysfunction represents a long-term goal in facial paralysis treatment. Until now, a large amount of subjective classification methods was invented, such as the *Botman-Jongkees* method, the *House-Brackmann* method and its modifications like the *Peitersen* grading system, the *Sunnybrook* or in Japan popular *Yanagihara's* grading system [6]. These methods and systems are usually based on evaluating the patient's face static symmetry, the effort to do some mimetic muscle movement or its amplitude and some of them also evaluating the side effects of improper muscle innervation.

Objective grading systems rest on statistical analysis which require many testing samples in the form of video, depth or facemarks records [1,3]. To acquire these samples, we developed two software solutions – desktop application for *Kinect v2* depth camera and the mobile application for patients' home rehabilitation, both using a central relational database.

R. Silhavy (Ed.): CSOC 2022, LNNS 501, pp. 372–381, 2022.
https://doi.org/10.1007/978-3-031-09070-7_31

1.1 State of the Art

One of the very hot topics in research in biomedical data analysis is focused on objectification and increasing accuracy and reliability. The key idea is a quick and accurate diagnosis of the patient. Software engineering and image processing methods are very helpful in this research and different software tools have been used for this purpose for ages.

Special case is the issue of data fusion. Mostly, papers in this field are focused on image fusion via a combining technique that merges useful information from two or more modalities of data captured through high definition scanning devices [5,7]. Some papers deal with image fusion from different scanners e.g. combination of data from CT (*Computed Tomography*), PET/CT (*Positron Emission Tomography*), SPECT/CT (*Single-Photon Emission Computed Tomography*), or MR (*Magnetic Resonance*) [2,4].

Very special case deals with data fusion, where the data are acquired in the hospital environment and meanwhile by the patient himself in the home conditions. There are almost no references dealing with this topic, although it is a key source of important information about the patient. This topic of data fusion is crucial from both the research and the clinical point of view because it can help with early diagnosis or rehabilitation.

Therefore, the aim of this paper is to introduce a complex system for data acquisition and fusion in the case of rehabilitation of patients after head surgery.

2 Materials and Methods

In the following text, the individual approaches to data measurement are described in more detail. Firstly, the acquisition of data in the clinical environment is described (see Fig. 1a), then the application for measurements in the home environment is introduced (see Fig. 1b), followed by a chapter on the relational database, in which this data are stored and merged.

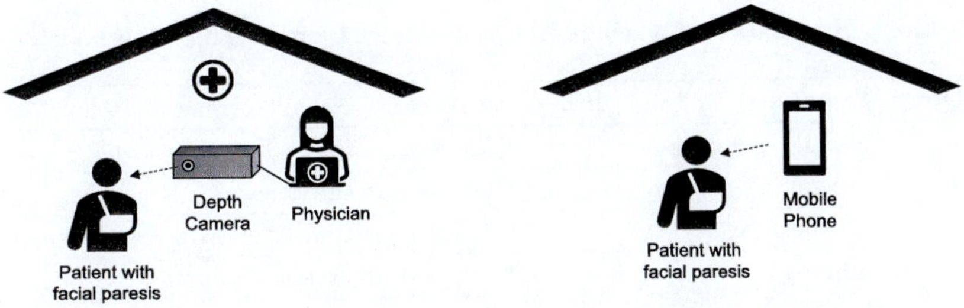

(a) Control session in the hospital. (b) Home rehabilitation.

Fig. 1. Individual approaches to data measurement.

Our interest is in the dynamic analysis of selected facemarks, which can be obtained from several sources – hardware with integrated online facemarks detection like *Kinect v2*, depth video (3D data) or RGB video (2D data) where additional facemarks detection is applied. The basic schema of data acquisition and storage is presented in the Fig. 2.

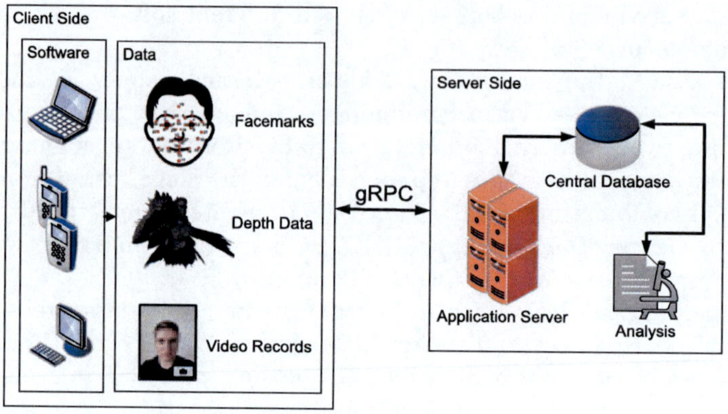

Fig. 2. Schema of the solution for data collection.

2.1 Desktop Application in Hospital

For the purpose to objectivise the physician evaluation of the patient after mimetic muscle rehabilitation, the desktop application for *Kinect v2* depth camera was developed. The application collects three-dimensional facemarks through *Kinect v2* camera during patient's standardised mimetic exercising in the hospital. The list of nine mimetic muscle exercises by prof. Vladimír Janda et al. [9] is showed in the Table 1. Collected facemarks are then processed by advanced statistical methods presented in [3].

Table 1. Hospital control rehabilitation exercises by prof. Janda methodology.

Order	Exercise	Order	Exercise
1	Raising eyebrows	6	Pursing the lips
2	Frowning	7	Blowing out the cheeks
3	Closing eyes tightly	8	Closing eyes tightly and baring the teeth
4	Smiling	9	Raising eyebrows and pursing the lips
5	Baring the teeth		

Technically, the application is based on *.NET Framework* and *Kinect for Windows SDK 2.0* and was created with WPF (*Windows Presentation Foundation*) graphical user interface (*GUI*).

Kinect v2 uses holistic facial deformation model from AAM (*Active Appearance Model*) family and detects 1,347 3D facemarks via *Kinect.FaceHD.x64* library. For analytic purposes, only 35 facemarks are used and are shown in the Fig. 3a.

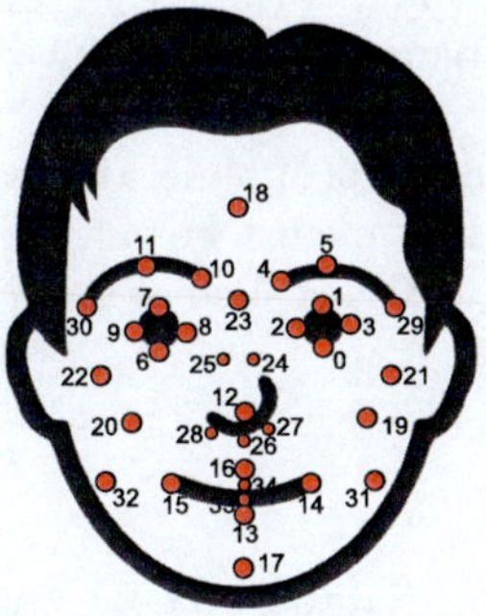

(a) Selected facemarks – Kinect v2.

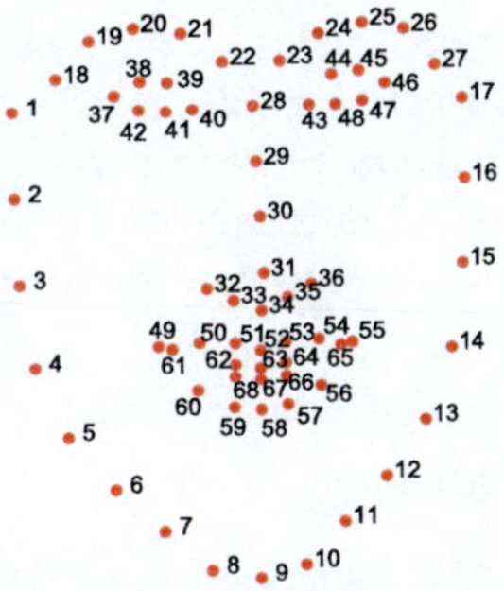

(b) Facemarks – OpenCV LBF algorithm.

Fig. 3. Detected facemarks used for further statistical analysis.

2.2 Mobile Application

As there is a lack of capability for observation of patients in the periods between hospital visits, a mobile application was created to fill this gap. The mobile application carries the patient through the home rehabilitation unit which consists of the mimetic exercises included in the Table 2.

Table 2. Home rehabilitation exercises.

Raising eyebrows and forehead crouching	Wrinkling of eyebrows
Nasal nostrils exercising	Winking
Eyes closing and opening	Eyes squinting
Lips pursing	Pressing teeth together
Lips pull out	Pulling lip over another lip
Baring the teeth	Smiling
Tongue exercising	Inflating/blowing out the cheeks
Pronunciation of vowels a, e, i, o, u	Pronunciation of labials b, p, m, f

The continuous observation of the patient through the whole rehabilitation process is crucial as nerves can innervate improper mimetic muscles and cause permanent consequences such as synkinesis, which, in the case of facial paralysis, is an involuntary muscle movement accompanying the voluntary one (for example eye closure accompanying smile).

The developed *Xamarin* mobile application not only motivates patients to exercise regularly but also supports facemarks detection from taken RGB video record. These two-dimensional facemarks data can be used for further statistical analysis. OpenCV LBF (*Linear Binary Features*) algorithm was chosen for facemarks detection. It recognises 68 facemarks as it is shown in the Fig. 3b.

According to [8], two-dimensional data underestimates the measured amplitudes up to 43%. That is the reason our next work focuses on new *LiDAR* sensors utilization in this mobile application.

The *Xamarin* platform was chosen for development of mobile application as it supports C# language and it is multi-platform, that means we can write code once and run our application on multiple operating systems such as *Android* and *iOS*.

2.3 Database

To collect the measured data, the central relational database (DB) was developed on the *Azure* platform. As a whole development of the solution stands on the programming language C# and .NET, also database was created and is maintained in C# language through *Entity Framework Core* (EF Core) and its code-first approach. EF is an *object-relational mapping* (ORM) tool, that means it enables to work with the relational model as if they were native objects in object-oriented language.

Two approaches of ORM are supported by EF – database-first: when DB already exists, classes are generated from a relational model, and code-first: classes are converted to the relational model. The access to the stored data in DB is provided by LINQ (*Language-Integrated Query*) to Entities and especially when it is used with method chaining and lambdas (anonymous functions), the written code is cleaner and better to understand than raw SQL (*Structured Query Language*).

Entity-relationship (ER) model of the created database consists of four modules: *Data, User, Measurement* and *Account.*

- **Data** module (see Fig. 4) stores facemarks and their underlying video or depth files with metadata about them and a source (camera).
- **User** module (see Fig. 5) saves the patient's or physician's personal data and important information about interventions, which caused facial paralysis.
- **Measurement** module (see Fig. 6) provides information about patient's measurement session in a hospital or via a mobile phone. Module also include classifications of sessions (where standard *House-Brackmann* classification is used) and exercises (for which advanced statistical analysis has been developed [1,3]).
- **Account** module (see Fig. 7) stores access information on personal accounts.

Client applications do not access DB directly but via the application server.

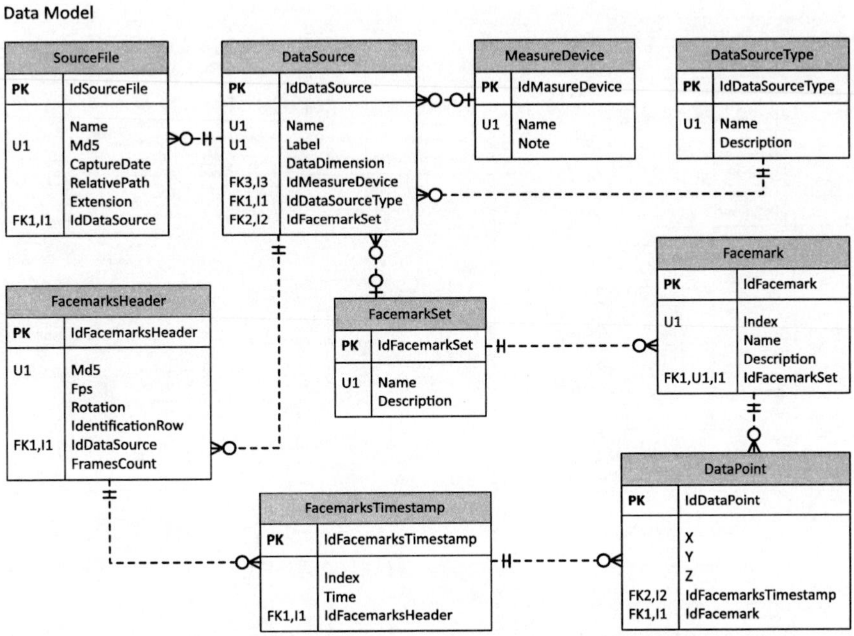

Fig. 4. ER diagram of data module of proposed relational database.

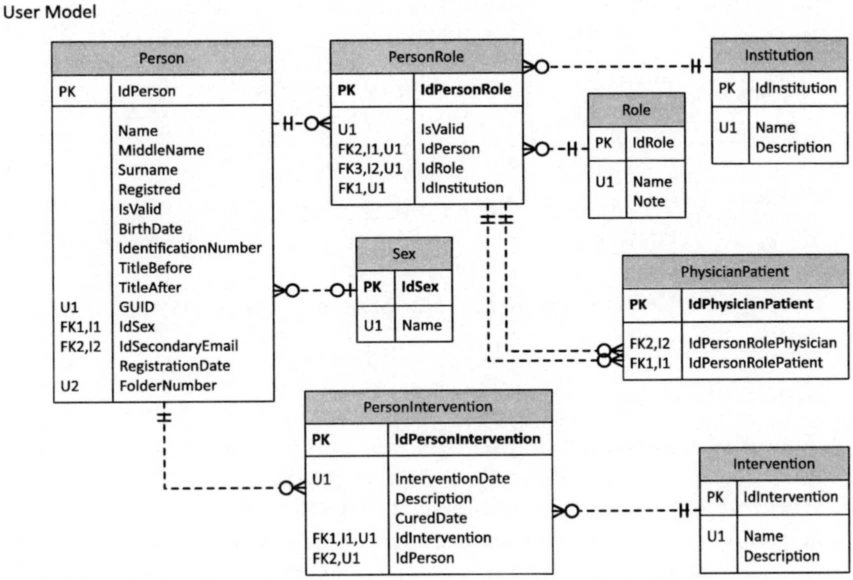

Fig. 5. ER diagram of user module of proposed relational database.

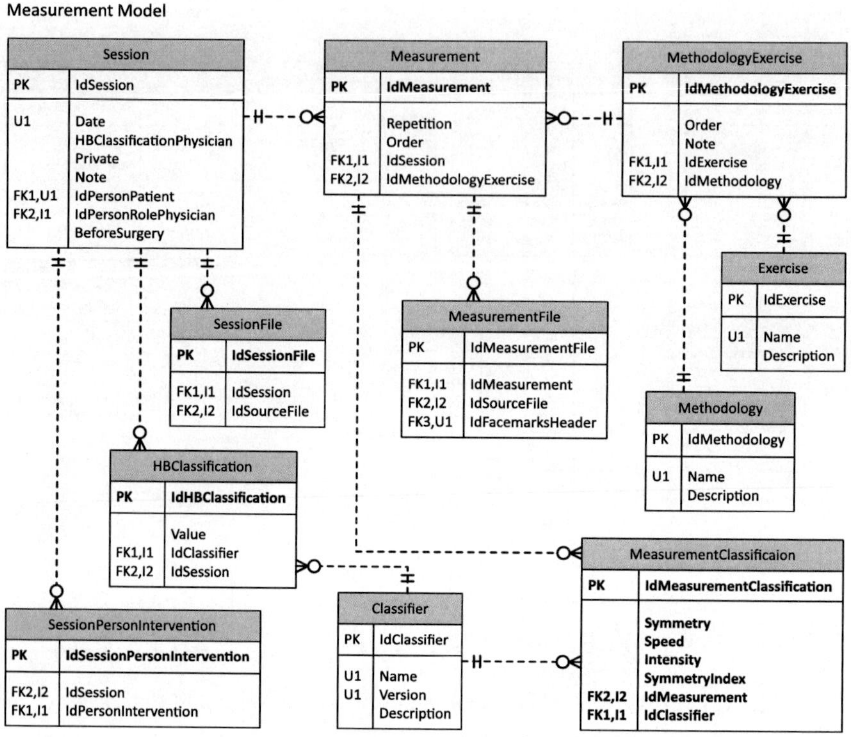

Fig. 6. ER diagram of measurement module of proposed relational database.

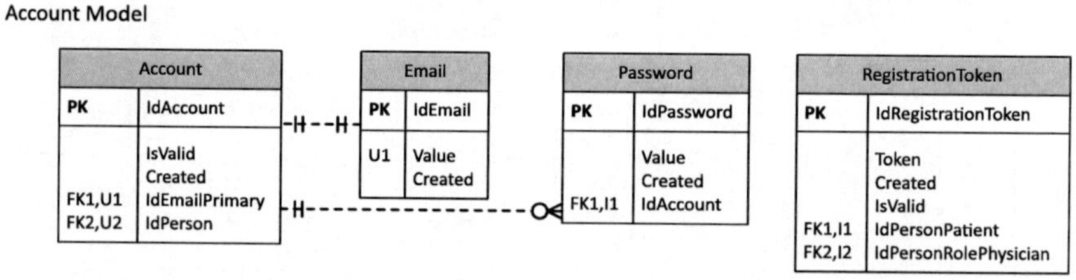

Fig. 7. ER diagram of account module of proposed relational database.

2.4 Communication

Communication between client applications (desktop or mobile application) and the application server is mediated by gRPC (*Remote Procedure Call*) API (*Application Programming Language*) protocol. Unlike SOAP (*Simple Object Access Protocol*) or REST (*Representational State Transfer*), gRPC supports *Protocol Buffers* for serializing structured data into binary format, which makes network communication more effective. Furthermore, gRPC offers *contract-first* API development (in contrast with SOAP and *code-first* API development) and uses

HTTP/2 (*Hypertext Transfer Protocol*) for transport which enables bidirectional streaming. gRPC currently supports (January 2022) eleven programming languages, including C#, in which it is possible to generate proper server or client classes from *.proto* contract files (see Fig. 8).

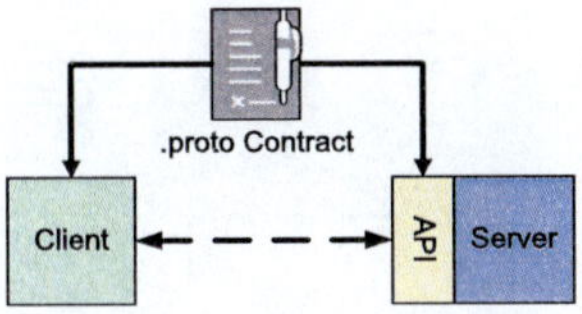

Fig. 8. Contract-first API development.

3 Results

The proposed software solution, represented by a triplet of desktop and mobile applications and the centralized relational database, provides a reliable source of data for future statistical analysis.

The desktop application is deployed in *University Hospital Královské Vinohrady* and more than 228 sessions of 118 patients were measured during three years. The screenshot of the user interface is shown in the Fig. 9. The left part of the window enables to control a session and its measurements. During each measurement, the patient do one of the mimetic muscle exercises. As a feedback for a physician, there are graphs in the middle part of the window that

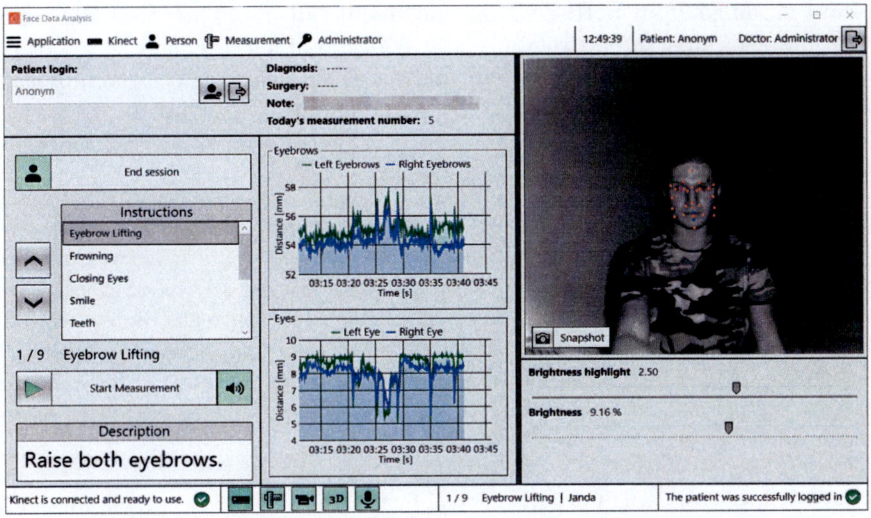

Fig. 9. Desktop application for kinect v2 used in hospital control sessions.

show the chosen distances of facemarks in time. The camera feedback is shown in the right part of the screenshot.

The screenshot of the developed mobile application is showed in the Fig. 10. As the desktop application, the mobile application provides camera feedback and description of rehabilitation exercises. The resulting video record can be sent by the user to our server for further analysis.

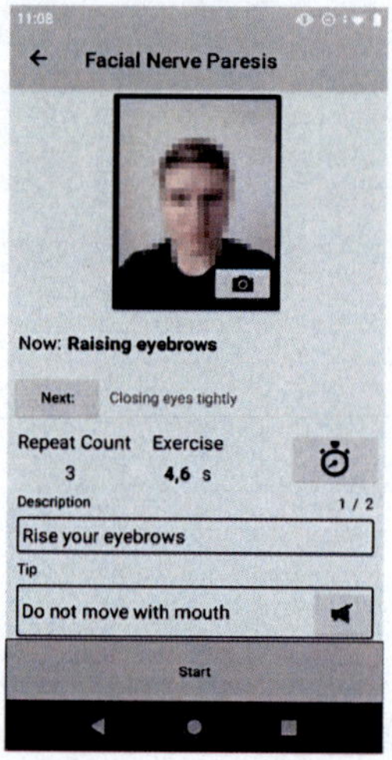

Fig. 10. Mobile application for home rehabilitation.

The relational database was developed for persistent and centralized storage of data obtained from both desktop and mobile clients. The relational representation of facemarks and related data like type of surgery, age or sex of the patient is advantageous for statistical analysis as data can be accessed uniformly by any qualified analyst or classifier via well-standardized SQL.

4 Conclusion

Objectification of the determination of mimetic muscle dysfunction represents a long-term goal in facial paralysis treatment. We introduced a two-software-solution. Firstly, we developed the desktop application for *Kinect v2* depth camera. Secondly, we developed the mobile application for patients' home rehabilitation. Owing to the fact mentioned above, it was necessary to create a central relational database for permanent facemarks and other patient-related personal information storage as qualified analysts or classifiers can consistently access needed data for their analyses.

Acknowledgement. The work was funded by the Ministry of Education, Youth and Sports by the grant 'Development of Advanced Computational Algorithms for evaluating post-surgery rehabilitation' number LTAIN19007.

References

1. Červená, L., et al.: Advanced statistical analysis of 3D Kinect data: a comparison of the classification methods. Appl. Sci. (Switzerland) **11**(10), 4572 (2021)
2. Juergens, K.U., et al.: Multi-detector row CT of left ventricular function with dedicated analysis software versus MR imaging: initial experience. Radiology **230**(2), 403–410 (2004). www.scopus.com
3. Kohout, J., et al.: Advanced statistical analysis of 3D Kinect data: mimetic muscle rehabilitation following head and neck surgeries causing facial paresis. Sensors (Switzerland) **21**(1), 1–21 (2021)
4. Pfluger, T., La Fougère, C., Stauss, J., Santos, R., Vollmar, C., Hahn, K.: Combined scanners (PET/CT, SPECT/CT) versus multimodality imaging with separated systems. Radiologe **44**(11), 1105–1112 (2004). www.scopus.com
5. Rani, V.A., Lalithakumari, S.: Efficient medical image fusion using 2-dimensional double density wavelet transform to improve quality metrics. IEEE Instrum. Meas. Mag. **24**(4), 35–41 (2021). www.scopus.com
6. Samsudin, W.S.W., Sundaraj, K.: Evaluation and grading systems of facial paralysis for facial rehabilitation. J. Phys. Ther. Sci. **25**(4), 515–519 (2013)
7. Tao, L., Qian, Z..: An improved medical image fusion algorithm based on wavelet transform. In: Proceedings - 2011 7th International Conference on Natural Computation, ICNC 2011, vol. 1, pp. 76–78 (2011). www.scopus.com
8. Tzou, C.H.J., et al.: Evolution of the 3-dimensional video system for facial motion analysis: ten years' experiences and recent developments. Ann. Plast. Surg. **69**(2), 173–185 (2012)
9. Zemanová, M., Janda, V., Ondráčková, Z.: Rehabilitace po obrně lícního nervu. Technical report, Státní zdravotní ústav, Praha (1998)

Solving Some Problems of Predictive Analytics for Time Series Data

Igor Botygin(✉), Anna Sherstneva, and Vladislav Sherstnev

Tomsk Polytechnic University, Tomsk, Russia
{bia,sherstneva,vss}@tpu.ru

Abstract. Approaches for constructing an additive regression model and a seasonally integrated autoregressive-moving average model were considered. Experiments were aimed at investigating and developing an algorithm for forecasting meteorological fields. In particular, a time series forecast of monthly precipitation totals for the next two years was built. In the software experiments for building forecast models, arrays containing data from eight-year observations of atmospheric precipitation were used. That is, sums of precipitation for the period between standard synoptic dates with an interval of three hours were used. The year 2016 was selected as the forecast year. The original data series was processed - emissions that were not attributed to seasonal variations were removed. Additionally, the allowable maximum amount of precipitation between dates for an observation station with synoptic index 29430 recommended by the All-Russian Research Institute for Hydrometeorological Information - World Data Centre (VNIIGMI-WDC) was controlled. The Python programming language was used as a tool to generate forecasts for the time series data. An additive regression model was created using the Prophet library from Facebook. A seasonally integrated autoregressive-moving average model was created using the StatsModels library. A DF-test (Dickey-Fuller basic test), which checks for the presence of a single "unit root", was used to test for stationarity. The testing and writing of the program code took place in the interactive Jupyter Notebook environment. The environment is a graphical web shell for Python and extends the idea of the console approach to interactive computing. Model forecast accuracy was assessed by calculating absolute and mean absolute prediction errors.

Keywords: Regression analysis · Additive regression model · Autoregressive integrated moving average model

1 Summary of Forecasting Models

Time series forecasting is a popular analytical task of data processing and analysis that helps organizations in planning, target setting and anomaly detection. However, there are serious problems associated with generating reliable and high quality forecasts, especially when there are multiple time series. The accuracy of time series forecasting is a value that depends on the qualitative characteristics of the time series, the selected forecasting models, the skill of the expert, the availability of software and hardware

R. Silhavy (Ed.): CSOC 2022, LNNS 501, pp. 382–391, 2022.
https://doi.org/10.1007/978-3-031-09070-7_32

tools to implement the model algorithms, and other factors. There are a large number of different models for time series forecasting: autoregressive, regression, exponential smoothing, Markov, neural network, etc. [1–10]. According to estimates of foreign and domestic researchers, there are already more than a hundred methods of forecasting.

Therefore, experts are faced with the task of selecting methods that would give adequate forecasts for the processes or systems under study.

Any time series can be viewed as a sampling realization from an infinite set of values that are generated by observations of selected processes. Therefore, of the formalized time series research methods, statistical prediction models are the most interesting. In statistical models, all functional relationships (between current and predicted values, external factors) are specified analytically.

One widely used approach in statistical modelling is regression analysis. There are linear and non-linear regression models of varying degrees of complexity, multiple regression models, additive regression models, etc. Most researchers note the transparency and flexibility of such models, the ease of their software implementation and, therefore, the speed at which the results are obtained. At the same time, autoregressive models and methods - AR(p) – are currently very popular and are widely used for solving time series forecasting problems in various subject areas [11].

An autoregressive model with conditional heteroscedasticity (GARCH) can be built on the residuals of AR(p) models with two parameters: p characterizes the order of autoregressive squares of the residuals; q is the number of prior estimates of the residuals. The most frequent application of this model is in the financial sector, where it is used to model volatility. Nowadays, there are a number of modifications of the model called NGARCH, IGARCH, EGARCH, GARCH-M and others [12, 13].

A moving average model of order q, often referred to in the literature as MA(q), is often used in conjunction with autoregression. The moving average model is a low-pass filter. There are simple, weighted, cumulative, exponential moving average models. Used together, there is more flexibility in model fitting. A common model is denoted ARMA(p, q), which combines a filter in the form of a moving average of order q and an autoregressive filtered process value of order p [14, 15].

If not the values of the time series themselves are used as input data, but their difference of order d, then the model is called autoregressive integrated moving average [16]. In the literature, this model is called ARIMA(p, d, q) (autoregression integrated moving average). If additional regressors of external factors are introduced into the autoregression model in order to predict the process, which is the result of the MA(q) model, then the ARIMAX model is used. Rather popular is the seasonally integrated autoregressive model - moving average SARIMA [17]. The input parameters of the SARIMA model are p, d, q, Ps, Ds, Qs. Basic parameters: p is the order of autoregression, d is the order of difference, q is the order of moving average. Seasonal parameters: Ps - order of seasonal autoregression, Ds - order of seasonal difference, Qs - seasonal moving average parameter.

Since the experiments use series of multi-year observations of atmospheric precipitation, the most significant component for model building that affects the accuracy of the forecast is the seasonal component. In this case, studied series have 12-link (annual) seasonality.

It is the construction and investigation of predictive models based on an additive regression model and a seasonally integrated autoregressive-sliding average model that this paper focuses on.

2 Research and Software Experiments on Building Predictive Models

The experiments were aimed at investigating and developing an algorithm for forecasting meteorological fields. In particular, a time series forecast of monthly precipitation totals for the next 2 years was made. The data files were taken from the FGBU VNIIGMI-MDC website [18]. The program experiments were carried out for data received from Tomsk station (station synoptic index - 29430). The observation periods were from 1966 to 2016. The programming language Python [19] and library statsmodels [20] were used. The testing and writing of the programming code took place in the interactive environment Jupyter Notebook [21]. The environment is a graphical web shell for Python and extends the idea of the console approach to interactive computing.

2.1 The Stationarity Test

When working with time series models, it is important to know if the series in question are stationary. This question is important because stationary and non-stationary series have different statistical characteristics and therefore must be evaluated in different ways.

The DF-test (Dickey-Fuller basic test), which checks for the presence of a single "unit root", was used to check for stationarity. The software code used:

```
sm.tsa.stattools.adfuller(data['num_data'])[1]
sm.tsa.seasonal_decompose(data['num_data']).plot()
print("Критерий Дики-Фуллера: p=%f" % sm.tsa.stattools.ad-
fuller(data['num_data'])[1])
```

The tested series is stationary, as the Dickey-Fuller test rejects the non-stationarity hypothesis. Figure 1 shows the graphs of the Dickey-Fuller test - they all indicate stationarity. In particular, there is no trend in the time series (trend), the values on the graphs fluctuate around zero, Dickey-Fuller criterion: p = 0.000000.

2.2 Additive Regression Model

It is a generalized linear model with a certain set of additional factors to calculate seasonality and to account for short-term important changes (anomalies):

$$y(t) = g(t) + s(t) + h(t) + \varepsilon t,$$

where: y(t) is the predicted value of the time series; g(t) is a piecewise linear or logistic function (trend); s(t) is a component reflecting modelling of periodic changes (seasonal component); h(t) is a component reflecting modelling of abnormal days and other irregular events; εt is a component not considered by the model (error).

Below (see Fig. 2) the result of the model on the time series of atmospheric precipitation for 2013–2016 is presented. In the experiments for construction of forecast models, the arrays containing data of eight-term observations of atmospheric precipitation were used. That is, the sums of precipitation for the period between standard synoptic dates with an interval of 3 h were used. The year 2016 was selected as the forecast year. The original data series was processed - emissions that are not related to seasonal fluctuations were removed. Additionally, the permissible maximum amount of precipitation between dates was controlled for the observation station with synoptic index 29430, recommended by the All-Russian Research Institute of Hydrometeorological Information - World Data Centre (VNIIGMI-WDC) [18].

The graph shows the initial data, predicted values and forecast boundaries (upper and lower).

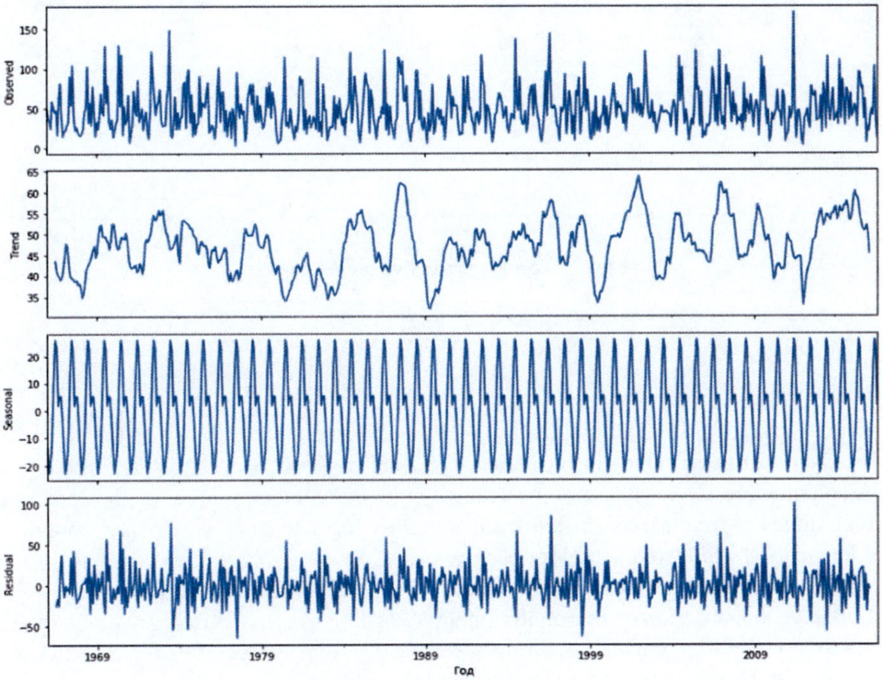

Fig. 1. Dickey-Fuller test result.

2.3 Seasonal Integrated Model

An autoregressive moving average (ARIMA) model of order p and moving average order q has the following specification:

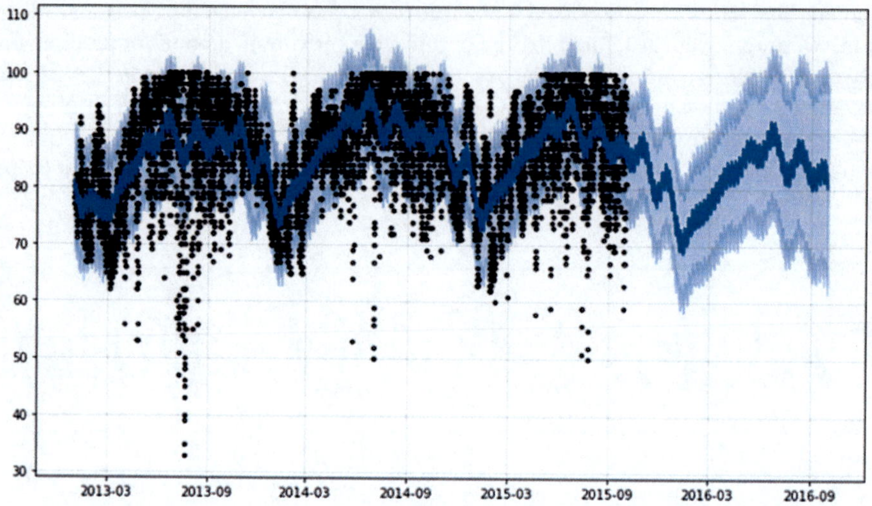

Fig. 2. Prophet predictive model chart.

$$y(t) = c + \alpha_1 y_{t-1} + \ldots + \alpha_p y_{t-p} + \beta_1 \varepsilon_{t-1} + \ldots + \beta_q \varepsilon_{t-q} + \varepsilon_t,$$

where: y(t) - forecast value of time series; c - model constant; $\alpha_1, \ldots, \alpha_p$ – autoregressive order parameters; $\beta_1, \ldots, \beta_q$ – moving average order parameters; $y_{t-1}, \ldots, y_{t-p}$ – previous actual values of time series shifted from actual by lag 1 to p; $\varepsilon_{t-1}, \ldots, \varepsilon_{t-q}$ – previous model errors shifted from actual by lag 1 to q.

If ARIMA modelling consists in correct fitting of ARIMA(p, d, q) model specification, then construction of seasonally integrated autoregressive moving average model (SARIMA modelling) consists in fitting of seasonally adjusted specification SARIMA(p, d, q) (Ps, Ds, Qs), where s is seasonality dimension.

The type of SARIMA model is determined by a correlogram. A correlogram is a graphical representation of the autocorrelation function. The functions plot_acf() and plot_pacf() of the StatsModels package were used to construct the corresponding correlograms.

```
ax=plt.subplot(211)
sm.graphics.tsa.plot_acf(data['num_data'][13:].values.squeeze(), lags=58, ax=ax)
ax=plt.subplot(212)
sm.graphics.tsa.plot_pacf(data['num_data'][13:].values.squeeze(), lags=58, ax=ax);
```

The blue colour on the autocorrelation plots (see Fig. 3) indicates the critical interval within which the ACF and PACF values are considered not to differ from zero. Outliers in the first few lags of the PACF partial autocorrelation function indicate that this is an autoregressive model. The outlier at the twelfth lag in the partial autocorrelation function indicates the presence of 12-lag seasonality.

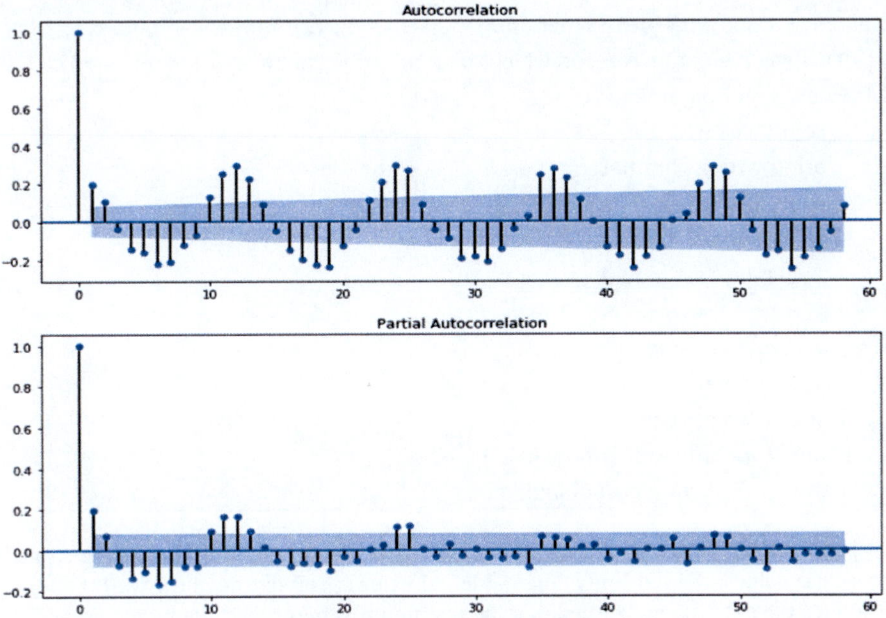

Fig. 3. ACF and PACF charts.

Thus, it is necessary to construct a seasonally integrated autoregressive-moving average model (SARIMA model) with the following input parameters. p – order of autoregression, d – order of differences of original series (order of integration), q – order of moving average (order of model), Ps – order of seasonal component of autoregression, Ds – order of seasonal difference (order of integration of seasonal component), Qs – seasonal parameter of moving average. We determine the initial approximations (according to the correlogram) for the construction of the model: Qs = 5, q = 12, Ps = 5, p = 2. The order of integration d and the order of integration of the seasonal component Ds are assumed to be 1.

```
%%time
results = []
best_aic = float("inf")
for param in parameters_list:
  try:
    model=sm.tsa.statespace.SARIMAX(data['num_data'], order=(param[0], d,
param[1]), seasonal_order=(param[2], D, param[3], 12)).fit(disp=-1)
  # display the parameters on which the model does not learn, and move on to
the next set of parameters
  except ValueError:
    print('wrong parameters:', param)
    continue
  aic = model.aic
  # saving the best model
  if aic < best_aic:
    best_model = model
    best_aic = aic
    best_param = param
  results.append([param, model.aic])
  warnings.filterwarnings('default')
```

Figure 4 shows the best predictive model over the whole period. The blue line is the actual values, the red line is the modelled values. You can see how well the model describes the data. It can be seen that the maximum values could not be predicted.

In order to plot the forecast for the next two years, for which no actual data is available, the following program code must be run.

```
from dateutil.relativedelta import relativedelta
data2 = data[['num_data']]
date_list = [pd.datetime.strptime("2016-01-01", "%Y-%m-%d") + relativedel-
ta(months=x) for x in range(0,24)]
future = pd.DataFrame(index=date_list, columns=data2.columns)
data2 = pd.concat([data2, future])
data2['forecast'] = best_model.predict(start=600, end=700)
data2['num_data'].plot(color='b')
data2['forecast'].plot(color='r')
plt.ylabel('осадки');
```

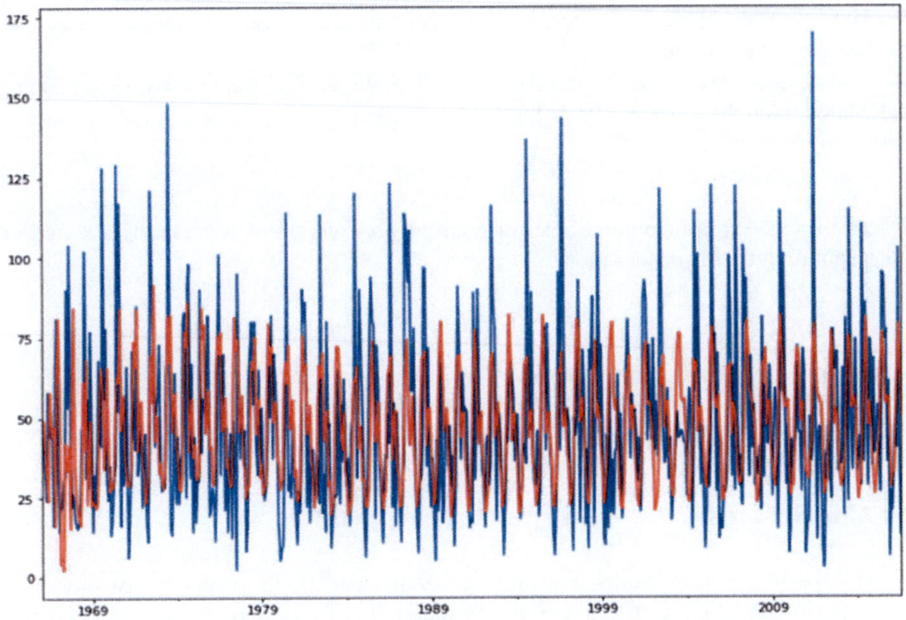

Fig. 4. Model SARIMA for the whole period.

The result of running the code is shown in Fig. 5.

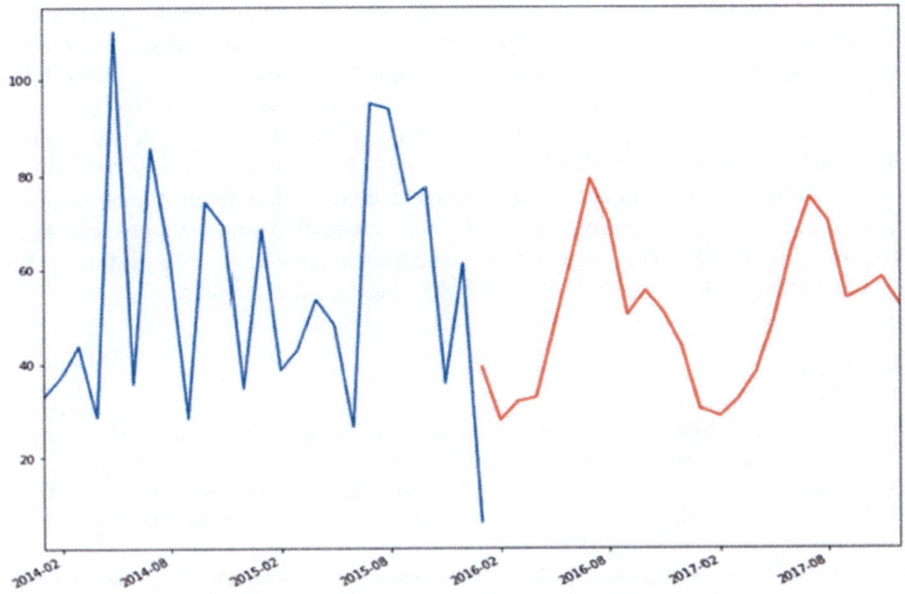

Fig. 5. Forecast for the next two years.

Here, as in the previous figures, it can be seen that the simulated values are in a close range of expected values.

The absolute error (MAE) and the mean absolute error of the forecast (MAPE) are calculated to analyze the results of the forecast calculation.

$$\mathrm{MAE} = \mathrm{mean}(|\mathrm{e_i}|),$$

where $\mathrm{e_i} = \mathrm{y_i} - \hat{\mathrm{y}}_\mathrm{i}$ is the prediction error; $\mathrm{y_i}$ is the observation; $\hat{\mathrm{y}}_\mathrm{i}$ is the model prediction corresponding to this observation.

$$\mathrm{MAPE} = \mathrm{mean}(|\mathrm{p_i}|),$$

where $p_i = \frac{\mathrm{e_i}}{y_i}$ – relative forecast error.

Final forecast errors of the SARIMA model time series: MAPE = 26.79%, MAE = 8.49. Prophet: MAPE = 37.35%, MAE = 10.62.

3 Conclusions

The SARIMA forecast model was more accurate than the Prophet forecast model in the experiments conducted on stationary time series. In view of the calculated error rates, the forecast models under study can only be used to make long-run forecasts, as at present their plausibility is only about 70%. When autoregressive models and their generalizations are used, additional difficulties arise in analyzing correlograms. In particular, it is necessary to distinguish between periodicity effects associated with the daily, weekly or other cycle, and intrinsic effects due to dependencies between row members. Increasing the averaging interval in this case does not lead to success, because the dependence observed in one sample may disappear in another of the same size, but referring to a different point in time. Averaging the correlogram over some time interval and switching to the average correlogram for the observation period increases the prediction inaccuracy even for a short time interval, and for a large interval it is not advisable to set such a task at all. The resulting predictions can be used as a new feature for other processing algorithms. The considered methods of time series forecasting may be used as an addition to existing methods (synoptic method, hydrodynamic method) of meteorological forecasting. This can help identify anomalies in the forecast models by comparing them with the SARIMA and Prophet mathematical models.

References

1. Khromov, S., Petrosyants, M.: Meteorology and Climatology: a Textbook for Universities. Moscow State University Press, Moscow (2001)
2. Draper, N., Smith, G.: Applied Regression Analysis. Finance and Statistics, Moscow (2016)
3. Shashkov, V.: Applied Regression Analysis. Multifactorial Regression. Orenburg State University Press, Orenburg (2003)
4. Sadovnikova, N., Shmoylova, R.: Time Series Analysis and Forecasting. Publishing house of the University "Synergy", Moscow (2016)

5. Terekhin, D., Botygin, I., Sherstneva, A., Sherstnev, V.: Research of GIS-services applicability for solution of spatial analysis tasks. In: IOP Conference Series: Journal of Physics: Conference Series, vol. 803, no. 012163, pp. 1–6 (2017)
6. Afanasyev, V., Yuzbashev, M.: Time Series Analysis and Forecasting. Finance and Statistics. Infra-M, Moscow (2010)
7. Lukashin, Y.: Adaptive methods of short-term forecasting of time series. Finance and Statistics, Moscow (2003)
8. Tartakovsky, V., Botygin, I., Sherstneva, A.: Synchronous analysis of statistical characteristics of natural climatic processes. In: Proceedings of 23rd International Symposium Atmospheric and Ocean Optics: Atmospheric Physics: Proceedings of SPIE (10466), pp. 104665J-1–104665J-5 (2017)
9. Sergienko, A.: Digital Signal Processing. Peter, Saint Petersburg (2006)
10. Kizbikenov, K.: Forecasting and Time Series: A Textbook. Altai State Pedagogical University Press, Barnaul (2017)
11. Time series models: AR (P), MA (Q), ARIMA (P, D, Q). http://economy.bsu.by/wp-content/uploads/2016/10/naft_ar_ma.pdf. Accessed 21 Nov 2018
12. Garcıa, F., Guijarro, F., Moya, I., Oliver, J.: Estimating returns and condicional volatility: a comparison between the ARMA-GARCH-M models and the backpropagation neural network. Int. J. Complex Syst. Sci. **2**(1), 21–26 (2012)
13. Model GARCH. https://basegroup.ru/community/glossary/garch-mod. Accessed 21 Nov 2021
14. ARMA Models. http://www.econ.ohio-state.edu/dejong/note2.pdf. Accessed 21 Nov 2021
15. Introduction to ARMA Models. http://www-stat.wharton.upenn.edu/~stine/stat910/lectures/08_intro_arma.pdf. Accessed 21 Nov 2021
16. Applied problems of data analysis. https://www.coursera.org/learn/data-analysis-applications/lecture/8yR4G/arima. Accessed 21 Nov 2021
17. Seasonal ARIMA with Python. http://www.seanabu.com/2016/03/22/time-series-seasonal-ARIMA-model-in-python/. Accessed 21 Nov 2021
18. All-Russian Research Institute of Hydrometeorological Information - World Data Center. http://meteo.ru/data. Accessed 21 Nov 2021
19. Python. https://www.python.org/. Accessed 21 Nov 2021
20. StatsModels. http://www.statsmodels.org/stable/. Accessed 21 Nov 2021
21. Project Jupyter. http://jupyter.org/. Accessed 21 Nov 2021

Novel and Simplified Predictive Computational Model for Clinical Diagnosis on Complex Medical Information

A. S. Chandru[1](✉) and K. Seetharam[2]

[1] Department of Computer Science and Engineering, Visvesvaraya Technological University (VTURRC), Belagavi, Karnataka, India
chandru.as.research@gmail.com

[2] Department of Computer Science and Engineering, Jnanavikas Institute of Technology, Bangalore, India

Abstract. The significance of adopting analytical operation towards extracting potential medical information for assisting towards an effective diagnosis is quite high. Existing review of literature suggest an enriched archives of Natural Language Processing (NLP) technique as well as machine learning techniques for this purpose; however, they are also characterized with essential drawbacks in various regards. Therefore, the proposed study presents a novel simplified computational model which is capable of performing a prediction of diagnosis on complex medical dataset by hybridized form of using Deep Neural Network, Long Short Term Memory, and Convolution Neural Network distinctly. The outcome of the proposed study shows that it offers better clinical and computational performs in comparison to existing analytical schemes of prediction and hence offer efficient predictive approach.

Keywords: Natural Language Processing · Machine learning · Long Short Term Memory · Convolution Neural Network

1 Introduction

The contribution of Natural Language Processing (NLP) as well as machine learning has been significantly high in the area of health section [1, 2]. Majority of the medical information obtained from Electronic Health Records and Electronic Medical Record system generate three types of data image-based (radiology), numerical based (pathology), and text-based [3]. Out of all these data, text-data is quite abundant in volume and generate all forms of challenges in the process of analysis [4]. At present, there are various text-mining algorithm [5–8], however, they are not applicable in case of complex form of medical data. Big data-based approach is just in beginning phase and there are various research contribution in this regards [9–11]; however, it still has a long way to go. The implication of NLP techniques on various text-related objects are witnessed to show promising result owing to its inclusion of various taxonomies of concept over text [12, 13]; however, when it comes towards medical dataset, identifying the proper

R. Silhavy (Ed.): CSOC 2022, LNNS 501, pp. 392–401, 2022.
https://doi.org/10.1007/978-3-031-09070-7_33

terms, extracting its features, as well as performing its inference is one of the most challenging task till data. There is also no denying the fact about the contribution of machine learning approach especially about deep learning techniques which has been contribution towards analyzing medical data [14, 15]. However, very less preference is offered towards preprocessing operation. This is because of adoption of normal medical dataset which is error free. However, such models encounters challenges towards accuracy which is achieved at the cost of computational efficiency and contextual outliers. Therefore, there is a need to carry out a study which can address this problem. This manuscript presents a novel and simplified computational modelling which is capable of performing predictive analysis of medical dataset using NLP and machine learning approach. Section 2 discusses about the existing research work followed by problem identification in Sect. 3. Section 4 discusses about proposed methodology followed by elaborated discussion of algorithm implementation in Sect. 5. Comparative analysis of accomplished result is discussed under Sect. 6 followed by conclusion in Sect. 7.

2 Related Work

This section discusses about various techniques being carried out towards analyzing medical data. Adoption of Convolution Neural network is reported in work of Ye et al. [16] is carried out to predict the mortality rate. Study towards preprocessing over sophisticated medical data is carried out by Ferrao et al. [17] which mainly emphasize on redundant and missing data for enhancing feature extraction. Further, work carried out by Kaur et al. [18] has presented a processing technique to rectify the errors present in clinical text data, where further machine learning (Support Vector Machine) has been used in order to classify the data. Significance of machine learning techniques was presented in discussion carried out by Mustafa and Azghadi [19] where various challenges are highlighted. Analysis of medical data is also subjected to variety of Natural Language Processing along with machine learning as seen in work of Spasic and Nenadic [20]. Adoption of K-means algorithm along with random forest for perform feature selection is seen in work of Mishra and Yadav [21] emphasizing on preprocessing operation of medical data. Significance of predictive modelling is reported in work of Hassler et al. [22] where semantic and statistical analysis is carried out toward extraction of significant features. The author has also used logistic regression in order to perform classification. The model implemented by Teo et al. [23] is reported to perform prediction of readmission of patient using various machine learning approach. Further comprehensive discussion about preprocessing and its demands in classification process using series of processing is highlighted in work of Kashima et al. [24]. Apart from this, there are various further models e.g. hybrid learning model (Prabhakar and Won [25], Moqurrab et al. [26]), Linguistic-driven framework using genetic programming (Poul et al. [27]), prediction of morbidity (Kumar et al. [28]), forecasting readmission using bidirectional encoder (Huang et al. [29]), unified supervised medical system (Weng et al. [30]), comprehensive investigation of rule-based NLP approaches for fall detection (Topaz et al. [31]). Therefore, there are different levels of contribution being carried out towards investigation and analysis of medical dataset using NLP and different variants of machine learning approaches. The next section outlines the significant research problems extracted from existing review work.

3 Problem Description

The research problems that are identified from prior section are as follows: i) existing machine learning approaches are highly focused on accuracy while doesn't much emphasize towards computational efficiency, ii) majority of existing approaches are reported to adopt normal mechanism of data preprocessing on overall dataset without identifying the significant and non-significant fields of corpus, iii) there are no exploratory analysis being carried out in existing medical dataset which fails to ensure better data preprocessing operation, iv) almost majority of studies are found to subject the dataset with normal preprocessing ignoring its implication towards the predicted outcome, and v) there is a need to hybridize machine learning and NLP operation in cost effective manner over highly standard as well as complex dataset.

4 Proposed Methodology

The prime purpose of the proposed study is to design a simplified computational framework that is capable of performing prediction of testimony of patient as well as diagnosis. An analytical research methodology is applied for this purpose.

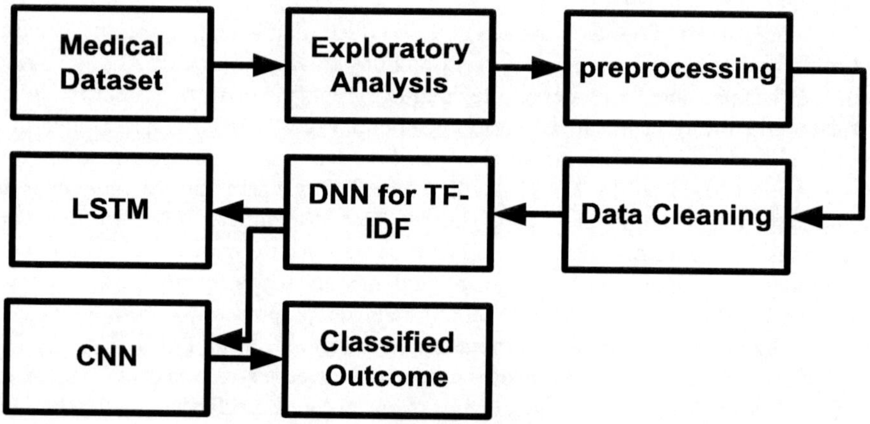

Fig. 1. Proposed architecture

According to proposed architecture exhibited in Fig. 1, the medical dataset is taken as an input of corpus which is followed by exploratory analysis on specific attributes of chosen fields in the dataset. Further preprocessing is applied where standards of ICD9 codes has been used for better clinical inference of the selected dataset. The next part of implementation concerns about performing data cleaning over the discharge summary of a patient followed by applying Deep Neural Network (DNN) for Term Frequency (TF) and Inverse Document Frequency (IDF), which yields cluster ID. Finally the system applies Long Short Term Memory (LSTM) and Convolution Neural Network (CNN), which yields the predicted classified outcome. The next section discusses about system implementation.

5 System Implementation

This section discusses about the flow of the system implementation which is carried out in order to perform the classification of the medical dataset. The novelty of proposed classification technique is that it doesn't subject the medical data directly to the machine learning approach directly as it is which is most the cases of implementation in existing approach. Inspite, various series of operation is carried out to fine tune the medical data prior to subjecting it to machine learning approach. The algorithmic steps of proposed study is as follows:

Algorithm for Classification
Input: d (dataset)
Output: q_2 (classified outcome)
Start
1. **For** i=1 : d
2. Extract $\eta_f \leftarrow d, \eta_f \subseteq d \,\&\, \eta_f < d$ fields
3. $\Omega = \sum_{n=1}^{max} (\eta_f)_n$
4. $q_1 = f(\Omega, d)$
5. $q_2 = f_1(q_1, d)$
6. **End**
End

The discussion of the steps of above algorithm is as follows: The above algorithm considers the medical dataset d (Line-1). It should be noted that the medical dataset d consists of mainly summary as well admission related information. The next part of the implementation consist of extracting specific number of fields η_f from the medical dataset d (Line-2). The extracted field η_f considered for the study are *csv* files associated with a_1 = event notes, a_2 = diagnosis information, a_3 = diagnosis with ICD codes, and a_4 = admission information i.e. $\eta_f = \{a_1, a_2, a_3, a_4\}$. Further, the sub-field a1 consists of more categories as exhibited in Fig. 2.

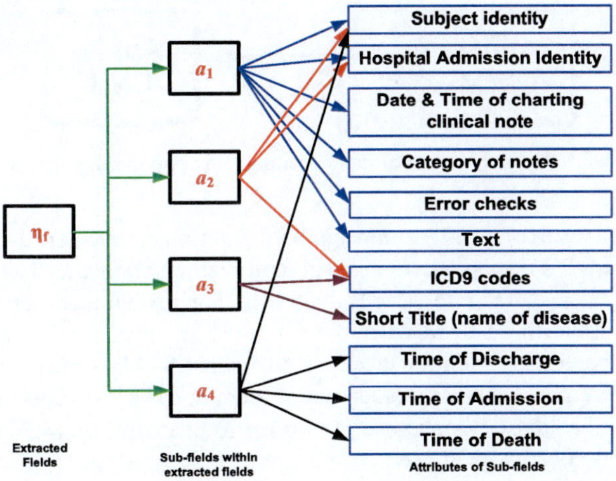

Fig. 2. System design of proposed exploratory analysis in medical dataset

It should be noted that proposed system considers maximum value max = 4 for number of the extracted fields n = 1, 2, 3, 4 (Line-3) and this leads to the generation of a master table Ω (Line-3) which is formed adopting inner join function on the 4 different sub-fields of η_f in order to carry out preprocessing. In this preprocessing step, the attributes for *subject identity* and *hospital admission identity* is eliminated as they mainly acts as primary key for joining the table in Ω. This step also consist of truncating in order to retain the first three character of ICD9 codes that represents the class of disease. This step basically contributes towards the identification of disease rather complete condition of it.

The next part of the algorithm implementation is about cleaning the column of text using *word2vec* algorithm using function $f(x)$ that takes the input argument of master matrix Ω and dataset *d* (Line-4). For this purpose, the complete text present in the corpus is converted to lower case in order to avoid any form of contradiction. Further all multiple whitespaces are replaced with a single whitespace. Further, the punctuation marks are eliminated with retaining the stop words. Further, the function f(x) performs TF-IDF vectorization followed by training using Deep Neural Network to the data cluster in order to recognize the text (Fig. 3).

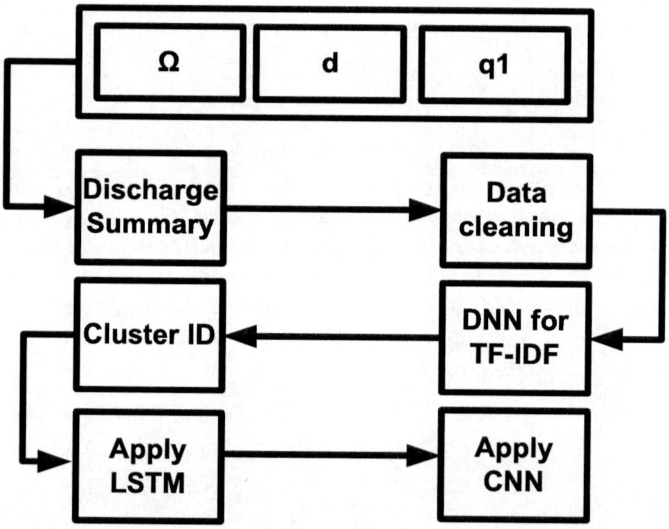

Fig. 3. System design of data cleaning and applying dual LSTM

Finally, the last part of the algorithm applies a function $f_1(x)$ where a dual Long Short Term Memory (LSTM) is applied. In this part of implementation, the discharge summary is used where it is classified into various diagnosis. The patient testimony is given more importance towards using LSTM when the patient is in out-patient department. On the basis of the cost center, the attention layer present in the LSTM permits either the primary set of biases or weights or the secondary set of them in order to process the data (Line-5). Therefore, the study will make use of two forms of neural network i.e. LSTM and Convolution Neural Network (CNN).

6 Results Discussion

The implementation of the proposed study is carried out in Python environment considering MIMIC III dataset [32]. The prime reason for selection of this dataset is because it has standard information about patient admitted in intensive care units and it offers all form of clinical information that is required in proposed study. The outcomes obtained are as follows.

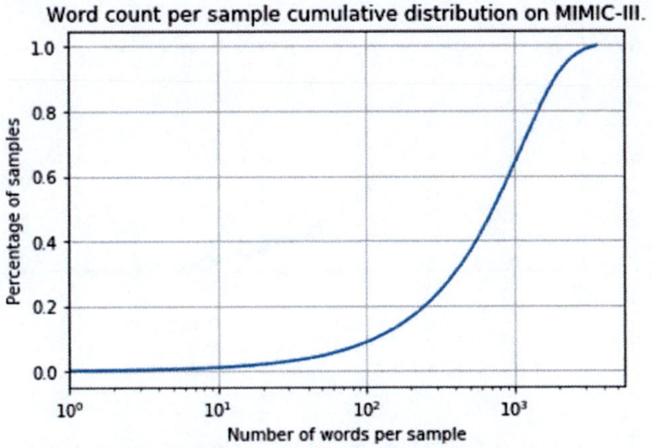

Fig. 4. Word count per sample

Figure 4 highlights the analysis being carried out towards number of words present in each sample with respect to percentage of sample. The trend of the graph shows that majority of the clinical information consists of 1000 or more words, which is inevitable owing to higher volume of documents in the form of discharge summary and all information of discharge to admission.

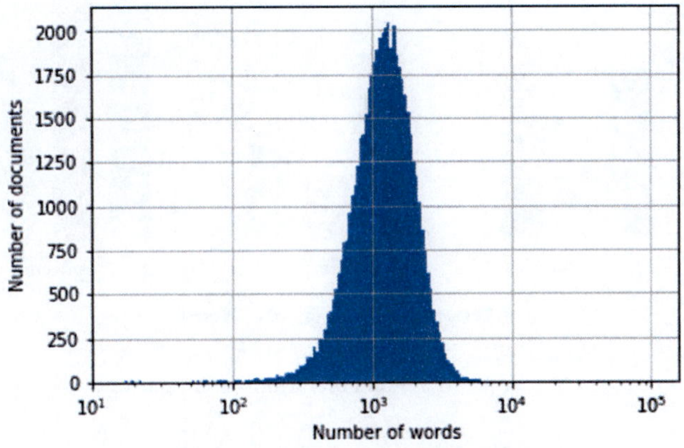

Fig. 5. Number of words histogram.

Figure 5 highlights the histogram for number of words with respect to number of documents. The trend of the graph shows the presence of document with maximum words; however 1000 word count is the average score observed.

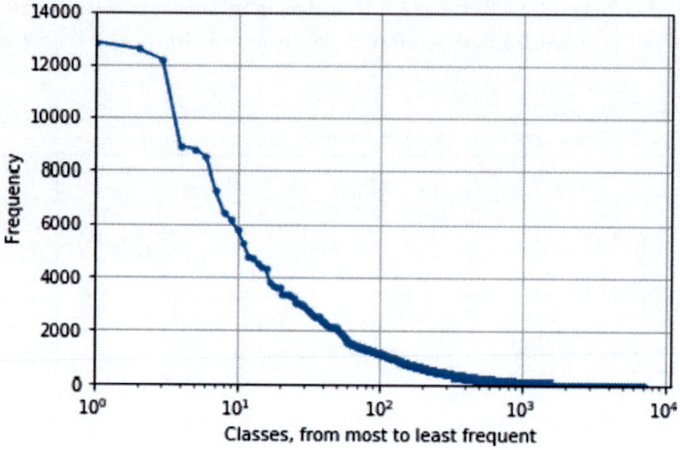

Fig. 6. Classes from most to least frequent

The above Fig. 6 shows the analysis regarding the number of classes from most to least occurrences versus frequency. The graph trend exhibits that the less than 100 classes have the highest frequency of occurrences. At the same time, most of the classes have fewer and more frequent occurrences.

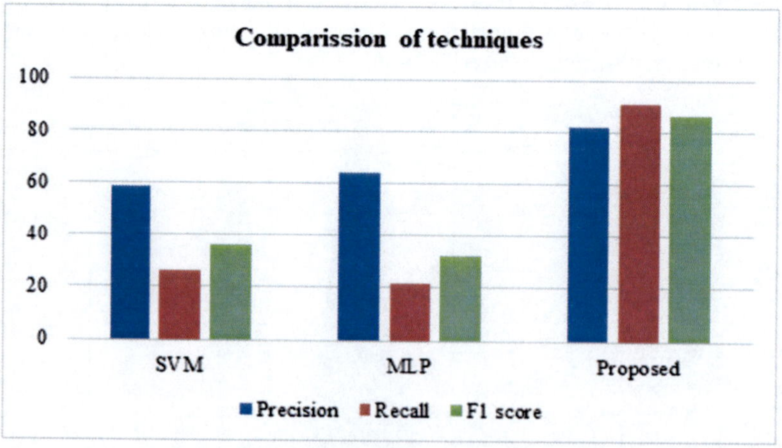

Fig. 7. Compression of various techniques

The analysis from Fig. 7 shows that the proposed method shows a better result than all other existing methods. This is due to the fact that we are using LSTM, and text data is a series data. LSTM works best for the series data. The proposed algorithm shows a better result than the other shallow learning and deep learning methods with improvements. As it can be observed, the recall rate has improved greatly. This is because in this study, the more precise and technical clinical notes are given preference; hence, the number of false negatives reduces greatly.

7 Conclusion

The prime contribution of the proposed paper is to present a novel and simplified learning approach which can be used for making prediction of diagnosis required for clinical decision making. The analytical modelling adopted in proposed system performs optimization in order to process the clinical notes that are written in rich text language. The novelty of the proposed study are as follows: i) the paper presents an appropriate and suitable data treatment and cleaning operation that is applied to clinical notes for the processing of NLP and learning mechanisms, unlike any existing system; ii) the proposed model also implements word2vec modeling in presence of autoencoder that is applied to perform clustering of the two distinct clinical classes for the predictive modeling and iii) the proposed model also constructs a dual LSTM on the basis of the joint operation of LSTM and CNN deep learning approach. The study outcome exhibited higher performance achieved by the proposed system compared to the shallow machine learning approaches.

References

1. Vajjala, S., Majumder, B., Gupta, A., Surana, H.: Practical Natural Language Processing-A Comprehensive Guide to Building Real-World NLP Systems. O'Reilly Media (2020). ISBN: 9781492054023, 149205402X
2. Saxena, A., Priya, C., Tanwar, P.: Deep Natural Language Processing and AI Applications for Industry 5.0. IGI Global (2021). ISBN: 9781799877288, 1799877280
3. Dekker, A., Dumontier, M., Kubben, P.: Fundamentals of Clinical Data Science. Springer, Cham (2018). https://doi.org/10.1007/978-3-319-99713-1. ISBN: 9783319997131, 3319997130
4. Chang, A.C.: Intelligence-Based Medicine- Artificial Intelligence and Human Cognition in Clinical Medicine and Healthcare. Elsevier Science (2020). ISBN: 9780128233382, 0128233389
5. Liu, S., et al.: Bridging text visualization and mining: a task-driven survey. IEEE Trans. Vis. Comput. Graph. **25**(7), 2482–2504 (2019). https://doi.org/10.1109/TVCG.2018.2834341
6. Buenaño-Fernandez, D., González, M., Gil, D., Luján-Mora, S.: Text mining of open-ended questions in self-assessment of university teachers: an LDA topic modeling approach. IEEE Access **8**, 35318–35330 (2020). https://doi.org/10.1109/ACCESS.2020.2974983
7. Karami, A., Lundy, M., Webb, F., Dwivedi, Y.K.: Twitter and research: a systematic literature review through text mining. IEEE Access **8**, 67698–67717 (2020). https://doi.org/10.1109/ACCESS.2020.2983656
8. Grönberg, N., Knutas, A., Hynninen, T., Hujala, M.: Palaute: an online text mining tool for analyzing written student course feedback. IEEE Access **9**, 134518–134529 (2021). https://doi.org/10.1109/ACCESS.2021.3116425

9. Dhayne, H., Haque, R., Kilany, R., Taher, Y.: In search of big medical data integration solutions - a comprehensive survey. IEEE Access **7**, 91265–91290 (2019). https://doi.org/10.1109/ACCESS.2019.2927491
10. Nazir, S., et al.: A comprehensive analysis of healthcare big data management, analytics and scientific programming. IEEE Access **8**, 95714–95733 (2020). https://doi.org/10.1109/ACCESS.2020.2995572
11. Bahri, S., Zoghlami, N., Abed, M., Tavares, J.M.R.S.: BIG DATA for healthcare: a survey. IEEE Access **7**, 7397–7408 (2019). https://doi.org/10.1109/ACCESS.2018.2889180
12. Otter, D.W., Medina, J.R., Kalita, J.K.: A survey of the usages of deep learning for natural language processing. IEEE Trans. Neural Netw. Learn. Syst. **32**(2), 604–624 (2021). https://doi.org/10.1109/TNNLS.2020.2979670
13. Zeng, Z., Deng, Y., Li, X., Naumann, T., Luo, Y.: Natural language processing for EHR-based computational phenotyping. IEEE/ACM Trans. Comput. Biol. Bioinform. **16**(1), 139–153 (2019). https://doi.org/10.1109/TCBB.2018.2849968
14. Rajabi Shishvan, O., Zois, D., Soyata, T.: Machine intelligence in healthcare and medical cyber physical systems: a survey. IEEE Access **6**, 46419–46494 (2018). https://doi.org/10.1109/ACCESS.2018.2866049
15. Shickel, B., Tighe, P.J., Bihorac, A., Rashidi, P.: Deep EHR: a survey of recent advances in deep learning techniques for Electronic Health Record (EHR) analysis. IEEE J. Biomed. Health Inform. **22**(5), 1589–1604 (2018). https://doi.org/10.1109/JBHI.2017.2767063
16. Ye, J., Yao, L., Shen, J., et al.: Predicting mortality in critically ill patients with diabetes using machine learning and clinical notes. BMC Med. Inform. Decis. Mak. **20**, 295 (2020). https://doi.org/10.1186/s12911-020-01318-4
17. Ferrão, J., Oliveira, M., Janela, F., Martins, H.: Preprocessing structured clinical data for predictive modeling and decision support: a roadmap to tackle the challenges. Appl. Clin. Inform. **7**, 1135–1153 (2016). https://doi.org/10.4338/ACI-2016-03-SOA-0035
18. Kaur, R.: A comparative analysis of selected set of natural language processing (NLP) and machine learning (ML) algorithms for clinical coding using clinical classification standards. Stud. Health Technol. Inform. **252**, 73–79 (2018)
19. Mustafa, A., Rahimi Azghadi, M.: Automated machine learning for healthcare and clinical notes analysis. Computers **10**(2), 24 (2021)
20. Spasic, I., Nenadic, G.: Clinical text data in machine learning: systematic review. JMIR Med. Inform. **8**(3), e17984 (2020)
21. Misra, P., Yadav, A.: Impact of preprocessing methods on healthcare predictions. SSRN Electron. J. (2019). https://doi.org/10.2139/ssrn.3349586
22. Hassler, A., Menasalvas, E., García-García, F., et al.: Importance of medical data preprocessing in predictive modeling and risk factor discovery for the frailty syndrome. BMC Med. Inform. Decis. Mak. **19**, 33 (2019). https://doi.org/10.1186/s12911-019-0747-6
23. Teo, K., Yong, C., Chuah, J.H., Murphy, B., Lai, K.W.: Discovering the predictive value of clinical notes: machine learning analysis with text representation. J. Med. Imaging Health Inform. **10**, 2869–2875 (2020). https://doi.org/10.1166/jmihi.2020.3291
24. Kashina, M., Lenivtceva, I.D., Kopanitsa, G.D.: Preprocessing of unstructured medical data: the impact of each preprocessing stage on classification. Procedia Comput. Sci. **178**, 284–290 (2020)
25. Prabhakar, S.K., Won, D.O.: Medical text classification using hybrid deep learning models with multihead attention. Comput. Intell. Neurosci. (2021)
26. Moqurrab, S.A., Ayub, U., Anjum, A., Asghar, S., Srivastava, G.: An accurate deep learning model for clinical entity recognition from clinical notes. IEEE J. Biomed. Health Inform. **25**(10), 3804–3811 (2021). https://doi.org/10.1109/JBHI.2021.3099755
27. Poulin, C., et al.: Predicting the risk of suicide by analyzing the text of clinical notes. PLoS ONE **9**(1), e85733 (2014)

28. Kumar, V., Recupero, D.R., Riboni, D., Helaoui, R.: Ensembling classical machine learning and deep learning approaches for morbidity identification from clinical notes. IEEE Access **9**, 7107–7126 (2021). https://doi.org/10.1109/ACCESS.2020.3043221
29. Huang, K., Altosaar, J., Ranganath, R.: ClinicalBERT: modeling clinical notes and predicting hospital readmission. arXiv preprint arXiv:1904.05342 (2019)
30. Weng, W.H., Wagholikar, K.B., McCray, A.T., et al.: Medical subdomain classification of clinical notes using a machine learning-based natural language processing approach. BMC Med. Inform. Decis. Mak. **17**, 155 (2017). https://doi.org/10.1186/s12911-017-0556-8
31. Topaz, M., et al.: Mining fall-related information in clinical notes: comparison of rule-based and novel word embedding-based machine learning approaches. J. Biomed. Inform. **90**, 103103 (2019)
32. https://physionet.org/content/mimiciii-demo/1.4/

Comparative Analysis of Mikado Method Approach to Refactor Legacy Code to Common Approach in Software Quality Measuring

Hansel Susanto(✉), Kelvin Alexander, Louis Vincent Sanjaya, and Alexander Agung Santoso Gunawan

Computer Science Department School of Computer Science, Bina Nusantara University, Jakarta 11480, Indonesia
{hansel.susanto001,kelvin.alexander001, louis.sanjaya}@binus.ac.id, aagung@binus.edu

Abstract. As a software engineer, facing legacy code can be challenging. Most of the time, software engineers hate legacy code. But at some point, to maintain our services, we are forced to refactor our legacy code because of many factors such as obsolete dependencies, version conflict, lousy performance, etc. Mikado Method is a way to simplify the basic needs of a software engineer to revamp complex code through a structured approach. Mikado Method has four basics: Set a Goal, Experiments, Visualizations, and Undo. These frameworks exist in Mikado Method to help software engineers to reconstruct a legacy code free of hassle. With the help of the Mikado Method, software engineers will be able to deal with legacy codes without fearing code smell. Additionally, by refactoring the Legacy Code, Software Engineers will achieve better-quality software than the old one. Thus, this paper concludes that Mikado Method will ensure engineers with a more consistent and planned refactoring of the codes compared with other common approaches with the same quality of codes or better.

Keywords: Legacy code · Code refactoring · Mikado Method

1 Introduction

In the process of dealing with a code that isn't well made, such as not using the SOLID Principles [1], not using Clean Code or even Clean Architecture, Software Engineers tend to find several problems when it comes to the time to refactor it, such as Obsolete Dependencies, Version Conflicts, and the worst of all is your code doesn't really do anything after you try to refactor it. The code has already become a Big Ball of Mud [2]. When it comes, you know that refactoring is one of the most annoying things to do as a software engineer. Most of the time software engineers would try their best to avoid refactoring legacy code for some reason. But in some cases, especially if we think that our code has a great value and is too cumbersome to recode from the ground, refactoring is inevitable. So, it is also important for a software engineer to understand and try to do some refactor if necessary.

R. Silhavy (Ed.): CSOC 2022, LNNS 501, pp. 402–412, 2022.
https://doi.org/10.1007/978-3-031-09070-7_34

From the explanations, we can see that legacy code is something that we will face someday when we want to maintain and improve them. The only way to deal with them is by doing refactoring, which is implementing greater techniques, greater architecture, and greater language skills to solve the problem that legacy code solves equally or even better. That's why we need a refactoring method to a legacy code that will ease our job as software engineers.

Currently, we already know several refactoring methods such as Red-Green Refactoring. But in this paper, we will discuss Mikado Method and how it will differentiate from the others. This method is a way to simplify the basic needs of a software engineer to revamp complex code such as legacy code through a structured approach. There are four basics in Mikado Method which are Set a Goal, Experiments, Visualizations, and Undo [3]. We will discuss these four basics step later in this paper.

These frameworks exist in Mikado Method to help Software Engineers to reconstruct a legacy code free of hassle. We will discuss these frameworks later. With the help of the Mikado Method, Software Engineers will be able to deal with legacy code without the fear of facing code smell. Additionally, by refactoring the legacy code, Software Engineers will achieve better-quality software compared to the old one. Thus, this paper will compare the Software Quality from Mikado Method Approach to the Common Approach.

2 Literature Review

2.1 Refactoring

Refactoring is the practice of altering a software system in such a manner that it improves its internal structure while leaving the code's exterior behaviors unchanged [4]. Refactoring is a technique used by programmers/developers to improve the functional aspects of the software in order to increase the code's readability and modifiability [5]. With refactoring, improving code quality is achieved by directly measuring software metrics of spoken software and indirectly the quality attributes of the spoken software [6].

Many refactoring methods have been proposed by researchers, including the extract method, inline method, move method, Mikado method, and so on. Each method has its own set of advantages and disadvantages. Software quality will be utilized as a criterion in comparing procedures in order to determine one is more effective than the other.

Software quality is defined as a set of attributes that must be present in software in order to increase code quality. It is characterized as a set of software characteristics that are used to express and assess quality. It is classified into two categories: internal software quality attributes and external software quality attributes [6].

Internal software quality attributes are concepts of code quality evaluation based on software architecture and clean code principles including cohesion, coupling, complexity, size, and inheritance. Whereas, External software quality attributes, such as refactoring, maintainability, readability, performance, security, dependability, and testability, are used to assess software systems [6]. With these attributes, We can determine whether the refactoring approach is effective in increasing the quality of the software code that is presently being worked on.

2.2 Legacy Code

There are several definitions of legacy code is. As mentioned by Michael Feathers, is simply a code without tests [7]. He said that usually without doing some tests on the code, it is quite hard to determine what a code can or cannot do. To understand what a code is fully capable of, we need to do some tests with the code, only then we will know what it can do, what it cannot do, and what it cannot do yet. Without doing some tests, we need to understand the full capability of the code by reading it carefully, line by line, then envision it in our head of all the possible scenarios that might happen within the code, which usually takes a lot of time and effort.

He also mentioned that several types of legacy code give the same nightmare to software engineers. Legacy code that has been tested but it is poorly written tests, which makes identifying what the code is supposed to be doing much harder than usual. It is more difficult to change the code when we do not know what the result of the test is supposed to mean. Unexperienced software engineers might find meddling with legacy code is the same as playing with an unfinished product that doesn't work as it is supposed to.

But not all legacy code is friendly to programmers and not all legacy code has an easy solution to refactoring it [8]. In our real-life case situation, codes that have tests in them also can become legacy code. It is because with the rapid growth of our codebase and with a software engineer that comes and goes very fast (changing), we could even call our well-tested last year's code to become a legacy code. So that's why some even suggest calling legacy code further than just a code without a test, they suggest a legacy code is a code that can be rewritten using better programming techniques, architecture, or languages.

2.3 Big Ball of Mud

Big Ball of Mud [1] is a term created by Brian Foote and Joseph Yoder, which implies an architecture that is dictated more by expediency than design. They stated it as a haphazardly structured, sprawling, sloppy, duct-tape and baling wire, spaghetti code jungle. The criteria of this architecture are the systems have significant growth that is very unregulated, which in our case is untested, and recurrent and expedient repair has always been done. Next, information passing is freely exchanged throughout system components, typically to the point where all of the critical data becomes global or replicated. It's possible that the system's general structure was never clearly defined. If it was, it might have deteriorated to the point of being unrecognizable.

Brian and Joseph stated that Big Ball of Mud emerges from Throwaway Codes [9], which is a code that is only intended for one use only and then discarded. But sometimes, this dirty code still exists in our repository, and by the time, it causes a problem because of a lack of documentation and understanding of the code.

The following Fig. 1 is an example of how Big Ball of Mud Architecture works. It has a lot of dependencies to other use cases that made it very complicated and very hard to trace each function will depend on what other functions.

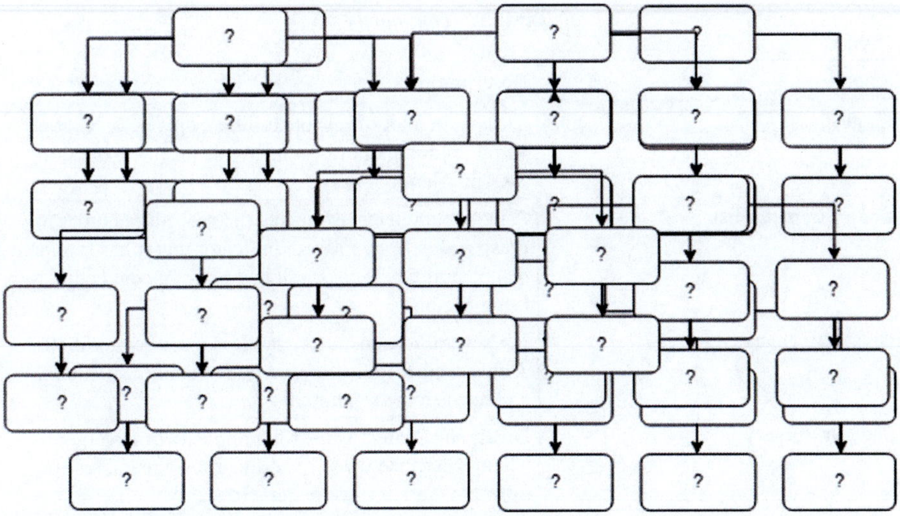

Fig. 1. The illustration of Big Ball of Mud Architecture [10]

2.4 Code Smell

Code Smell is a term that refers to an internal problem in the software in the lower level, which is the code level, and in the high level, which is the software architecture, languages, stacks, etc. It is very different from a code bug, a smell doesn't directly create an error in our system, but it may lead us to another problem that has a great relationship with how old our code is to be tested against the time. The problem also comes when a language doesn't implement a type-safe feature, which not every language does [11].

The idea of smells was initially used by the agile software development community to highlight a flaw or an area for improvement. As of now, the coding industry has also adopted this concept to represent anomalies in software.

Fowler et al. [4] identified 22 smells as well as refactoring sequences that might be used to lessen each one. Recently, Fowler introduced 6 more smells to the original list, making it 28 kinds of smells. Lacerda et al. created a table to arrange and provide a list of smells and their descriptions, which is shown in Table 1 [12].

Table 1. List of code smells according to Fowler

Smell	Description
Duplicated code	Consists of paragraphs in distinct segments of the same code base that are identical or nearly identical
Long method/long function	Large method/function, making it difficult to comprehend, expand, and alter. This method most certainly has too many responsibilities and violates one of the principles of excellent OO design which is SRP (Single Responsibility Principle)

(continued)

Table 1. (*continued*)

Smell	Description
Large class	Class with a lot of responsibilities, thus it has a lot of variables and methods in it. In this instance, the same SRP problem applies
Long parameter list	A large parameter list makes the procedure harder to grasp and is frequently an indicator that it has too many responsibilities. This smell has a strong connection to Long Method
Divergent change	For several reasons, a single class must be modified. This is a clear indicator that it is not coherent enough and needs to be separated
Shotgun surgery	Divergent Change is the total opposite of Divergent Change because when a modification occurs, several different classes must be altered
Feature envy	When a method is more concerned with members of other classes than with its own, it is evident that it is in the incorrect class
Data clumps	When one of the pieces in data structure is missing, the entire set loses its meaning
Primitive obsession	It shows a scenario in which primitive types are utilized instead of light classes
Switch statements/repeated switches	It is not necessary smells, but they are frequently a symptom of difficulties when they are extensively utilized, especially when used to identify the behavior of an object based on its type
Parallel inheritance hierarchies	Presence of two fully linked hierarchies of classes, such that adding a subclass in one hierarchy requires the creation of a corresponding subclass in the other
Lazy class	Classes that lack sufficient responsibilities and, as a result, should be eliminated
Speculative generality	Code snippets are intended to facilitate non-essential behavior in the future
Temporary field	Member-only is a term that is only used in specified contexts and has no significance outside of it
Message chains	One item interacts with another, which in turn interacts with yet another object belonging to the second, resulting in a high level of coupling between classes
Middle man	Identified how a class had essentially no logic since it delegated almost all its responsibilities to another class
Inappropriate intimacy	A situation in which two classes are also known, indicating a high level of connection, which create a tight coupling class

2.5 Red-Green Refactoring

Red-Green Refactor is an engineering pattern that uses Test-Driven Development [13]. It uses very short development cycles to minimize leap-of-faith assumptions. This method works as follows before any implementation work is done, there are several conditions to be met. Since no work has yet been done to satisfy the test, it may reasonably be expected that the test would fail were it to be run. However, it must not be assumed that such a test will in fact fail.

The missing feature should be implemented so that the test will now pass (green). Just enough code should be written to satisfy the test and no more. It's also important to execute all other tests to make sure that the implementation has not caused existing tests to fail [14]. The code should be revised until all tests execute successfully. This is the "Green" step in Red-Green-Refactor.

Now that the functionality is in place, it can be optimized for improved readability and maintainability [15]. Any expedient design decisions should be revised so that technical debt is not incurred. Better design patterns may be leveraged, and the quality of the code improved as per the team's coding standards. After the code has been refactored in this way, the tests should be re-run. They should still come up green, thereby indicating that refactoring has not compromised the functional correctness of the implementation [16].

There is a possibility that we will write so many tests during a test-first development initiative. So it's important to make sure the test keeps applying the clean architecture pattern, such as keeping tests small and focused and executing them regularly. The following Fig. 2 is showing the three main processes of Red-Green Refactoring which is a Red process, when we write a test that fails, a Green process, when we make the test pass, and the final process which is doing the refactor.

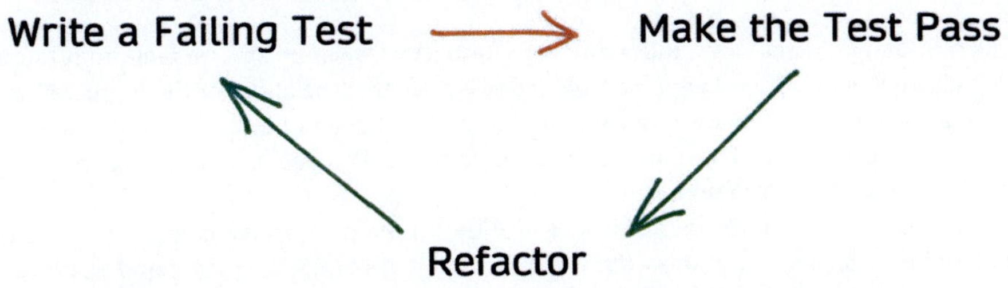

Fig. 2. The three main processes of the Red-Green refactoring method

2.6 Mikado Method

The Mikado Method is a structured methodological way to make changes to a complex code [3]. The framework that the Method provides can help individuals and whole teams to morph a system into a new desired shape. The Method itself is straightforward and simple and can be used by anyone at any time.

There are four basic and well-known concepts that summarize the "process" of the Mikado Method: Set a goal, Experiments, Visualization, and Undo [3]. To set a goal, think about what you want for the future. After a goal has been established, write it. The goal serves two purposes, which is first, represents a starting point for the change, and second, determines if the method has achieved success or not. An experiment is a procedure that makes a discovery or establishes the validity of a hypothesis. In the context of the Mikado Method, you use experiments to change the code so that you can see what parts of the system break. Visualization is when we write down the goal and the prerequisites to that goal. When an experiment for implementing a goal or a prerequisite has broken your system, and you have visualized what you need to change in the system to avoid it breaking, you want to restore a previously undone working state. The undo process is the most important step in Mikado Method because it gives us knowledge of what is possible and not possible to refactor.

3 Methodology

This research paper will try to show the difference between several methods of refactoring legacy codes, which are the Mikado Method and Red-Green Refactoring, or some know it as a TDD approach refactoring. We analyzed several legacy codes which were obtained from several sources, such as the internet, old projects, etc. From there we create a scenario that is commonly happening in daily life as a software engineer. We refactor each code with all methods to see which one is the most efficient.

We discovered via our research that using the Mikado Method to do code refactoring on old codes is far more efficient than using the previously mentioned approaches. The findings conclude that the Mikado Method, as one of the refactoring methods available, is a viable option for code refactoring.

4 Result and Discussion

This section will discuss the following case, as software engineers, we want to refactor a legacy code in our mobile application code. We see that the code for the login authentication system in our code still uses a singleton global class object to store the token in our mobile application. The code has been there for several years, and it works properly and doesn't have any problem so far.

The current system for login works as follows if the user opens the login page, then the mobile application will show the login form that asks for a username and password. After the user fills the form, then it will validate those inputs, and then send them to the server. There, the server will validate the user input and compare it with the data in the database. If the input is okay then, the server will generate token access and will send it to the mobile application. After the mobile application receives the token access, it will try to save it into the singleton global class object, so every time the token is needed it can be accessed anywhere. After that, the login is successful, and the mobile application will show a success popup.

Besides that, regarding the token storage system, the current system also does the checking for the token as follows, if the user tries to access the sign-in only page, then the mobile application will try to access the token that exists in the singleton global class object. If the token exists, then, it will try to validate those to the server, then it will return the validation result. If the token is validated then it will show the required sign-in page, otherwise it will show the login page again. The following Fig. 3 is the swim lane diagram for this login scenario, and the Fig. 4 is the swim lane diagram for the check token scenario.

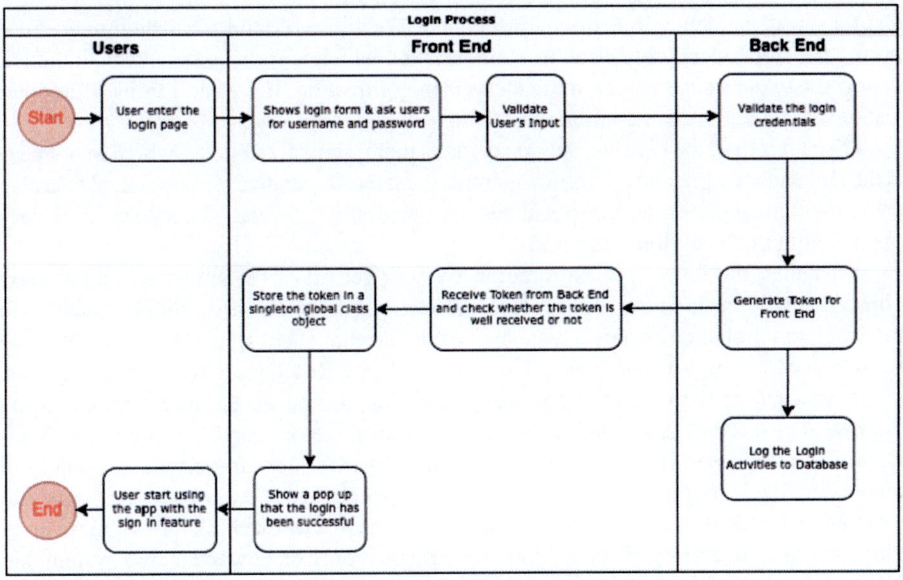

Fig. 3. The current system for login process

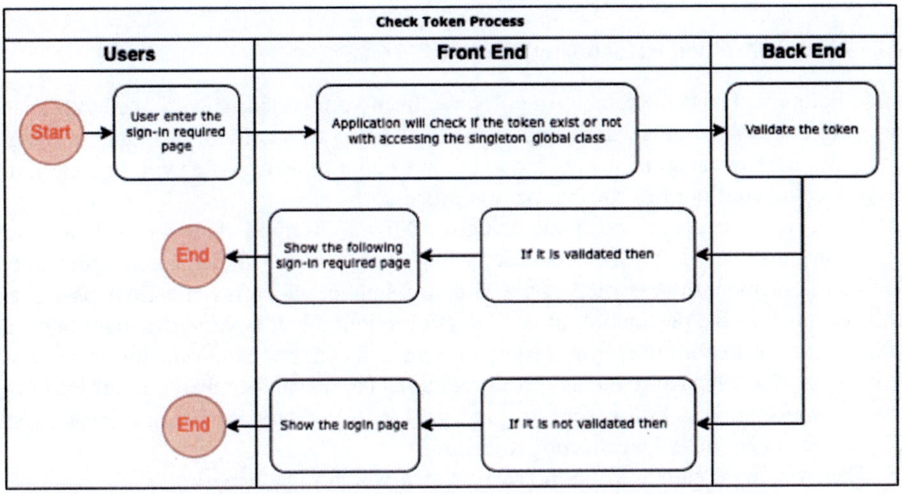

Fig. 4. The current system for the check token process

This code is not a good one, because it smells. Our singleton object is a global one, so every part of the code can easily access the code, which is vulnerable if there's some unauthorized access from some codes. So, we would like to refactor that part to use a Dependency Injection to access the token. We will try to do the refactor in Red-Green Refactor and Mikado Method Refactor.

4.1 Red-Green Refactoring Process

If we want to refactor our code using Red-Green Refactoring, first, we will create a test for our dependency injection pattern class. We will write the test which may cause during the Dependency Injection implementation, which is if the token exists in the DI, then if the token doesn't exist in the Dependency Injection. If we can't think of another test to write, then we can continue to the next step.

Then after that, the next step suggests for us to implement our code, which in our case is the Dependency Injection. While you implement the Dependency Injection, please also try to fix our test code, and try to make it run successfully (green). This iteration is very important in the Red-Green Method.

After that, the last step is to refactor, we will get rid of the singleton global class object that still exists in our global class. We also get rid of all depending on features to the singleton global class object, especially the part that checks the token whenever we want to access a sign-in-only page. This concludes the Red-Green Refactoring process.

If we analyze the refactoring process that is suggested by the Red-Green method then we can see that this method is only trying to refactor a code by doing the Test-Driven Development. It lacks a refactoring plan, which is very important, especially in refactoring the legacy code because sometimes, we can mess up our code. It is essential to do an experiment of our refactoring code and to have a rollback plan if we mess up in the process of refactoring. This method's sole focus is only on how to create a system that has a test on it, but because the lack of the other point makes there is another refactoring method that can do better than this one.

4.2 Mikado Method Refactoring Process

First, before you start the refactoring process, from your codebase, you need to branch out your code. We didn't want to risk your code being overwritten, so branching your code into a new branch is always a good idea to do a refactor. After you branch your code, we will start our refactoring method process.

Next, we will refactor our code using the Mikado Method. This method has four main steps that we will do which are Set a Goal, Experiment, Visualization, and Undo. We will implement the first step of the Mikado Method, which is Set a Goal. The goal will be drawn as the destination of our refactoring journey. It is very important to set a goal before we do our refactoring kick-off because, it can scope our refactoring process to only do what we need to do, any irrelevant things to do can be taken aside if we have created a visible goal. In this scenario, our goal is to Refactor Singleton Global Class Object for Token into a Dependency Injection.

Then we can continue to do the second step which is an experiment. Because we have already created a new branch of our code that isn't affecting our main branch of the code, then we can safely experiment. This step is important because not every refactor always comes smoothly. There are maybe some obstacles in our way to refactor. The obstacle may come from different sources such as the new library that we are going to use in the refactoring process, the old library that still has a strong coupling with our existing codes, the broken unit test, etc. For example, if we have a new library that we want to use, then this experiment step is the best way to try understanding the library because we can freely try to implement it into our system without having any issues so far. In our scenario, we will try to implement several libraries to implement dependency injection and try to find a library that fits our use cases.

Then the next step is to do a visualization. In this step, it is essential to create a Mikado Graph. This graph will map us the steps that we already explored in the experiment part and will be our guide in doing this refactor. The Mikado Graph also shows us which prerequisites and single dependency that we need to do first before we go to a bigger, and finally finish the goal. In this scenario, we will create a Mikado Graph that looks like the following Fig. 5

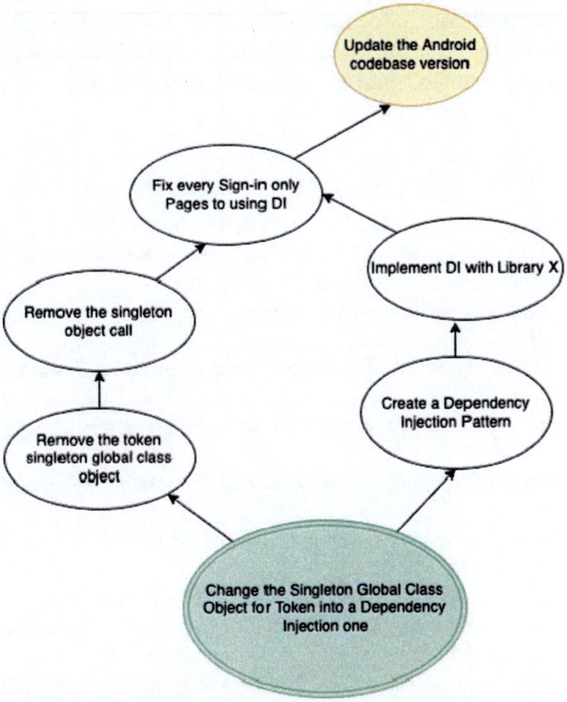

Fig. 5. Mikado Graph for the refactor

After we have already done all the above steps, the final step is to undo. This step is a step to make sure we undo all your irrelevant code, especially the code that can create the smell and the code that part of your "experiment code" which we have also known as the throwaway code. This step has also encouraged us if we think that some parts maybe can be refactored better, then undo our progress and try to do it from scratch again with better refactoring.

One thing that we can look at from Mikado Method is this method has great planning. It has a great rollback plan. So, Mikado Method is a better way to do a refactor. But keep in mind this process has a bigger iteration, so it takes much time compared to the other refactoring method. Then this method works great for a large codebase, especially the one that has a lot of tight dependencies on each other.

5 Conclusion

So, from the result, we can see that using the Mikado Method approach and compared to the common way, which is the Red-Green Refactoring, to refactor a code, doesn't change the quality of the code. But the process of how we do the Mikado method forces us to refactor the code with safety and with great planning. Using the Mikado method, we can make ourselves more consistent and planned in refactoring the code. So, in the end, if our code is a small codebase and still not sure how to refactor it anyway, then the Mikado method is not for you. But if our code is big and has a lot of dependencies, and it is prone to breaking changes, then the Mikado Method might come in handy for us.

References

1. Martin, C.R.: Design Principles and Design Patterns (2000)
2. Foote, B., Yoder, J.: Big ball of mud. Pattern Lang. Prog. Des. **4**, 654–692 (1997)
3. Ellnestam, O., Brolund, D.: The Mikado Method (2014)
4. Fowler, M., Beck, K., Brant J., et al.: Refactoring: Improving the Design of Existing Code (2018)
5. Kaur, A., Kaur, M.: Analysis of code refactoring impact on software quality. In: MATEC Web of Conferences, vol. 57 (2016). https://doi.org/10.1051/MATECCONF/20165702012
6. Draz, A.M., Farhan, M.S., Eldefrawi M.M.: A Survey of Refactoring Impact on Code Quality (2021)
7. Feathers, M.: Working Effectively with Legacy Code, 1st edn. Pearson Education, London (2004)
8. Madni, A., Madni, C., Lucero, S.: Leveraging digital twin technology in model-based systems engineering. Systems **7**(1), 7 (2019)
9. Foote, B., Yoder, J.: Big Ball of Mud (1999). http://www.laputan.org/mud/mud.html
10. The Domain-Driven Design: Big Ball of Mud (2019). https://thedomaindrivendesign.io/big-ball-of-mud/. Accessed 14 Mar 2022
11. Samadhiya, D., Wang, S.-H., Chen, D.: Quality models: role and value in software engineering. In: Proceedings of the 2nd International Conference on Software Technology and Engineering (ICSTE), pp. 320–324 (2010)
12. Lacerda, G., Petrillo, F., Pimenta, M.: Code smells and refactoring: a tertiary systematic review of challenges and observations. J. Syst. Softw. **167**, 110610 (2020). https://doi.org/10.1016/j.jss.2020.110610
13. Martin, R.: Red green refactor. In: Pattern of the Month (2017)
14. Johnson, B.: Refactoring code. In: Essential Visual Studio 2019. Apress, Berkeley (2020). https://doi.org/10.1007/978-1-4842-5719-7_4
15. Beyls, K., D'Hollander, E.H.: Refactoring for data locality. Computer **42**(2), 62–71 (2009). https://doi.org/10.1109/MC.2009.57
16. Embury, S.M., Borizanov, M., Jay, C.: Red-green-go! A self-organising game for teaching test-driven development. In: Parsons, D., MacCallum, K. (eds.) Agile and Lean Concepts for Teaching and Learning, pp. 415–441. Springer, Singapore (2019). https://doi.org/10.1007/978-981-13-2751-3_19

Towards a Correction Factors-Based Software Productivity Using Ensemble Approach for Early Software Development Effort Estimation

Ho Le Thi Kim Nhung(✉), Vo Van Hai, and Roman Jašek

Faculty of Applied Informatics, Tomas Bata University in Zlin, Nad Stranemi 4511, 76001 Zlin, Czech Republic

{lho,vo_van,jasek}@utb.cz

Abstract. Accuracy of effort estimation is one of the necessary conditions for efficiently managing software development projects. Since the information available in the early stages of software development is insufficient, software sizing metrics are considered critical factors for effort estimation. However, there is no consistent method for converting software sizing into the corresponding effort. Previous estimation methods have not considered software productivity a critical factor in estimating effort based on software sizing. This paper proposes a software productivity model based on correction factors in the Optimizing Correction Factors method through an ensemble construction mechanism of three popular machine learning techniques. The results show that using the proposed software productivity minimizes the estimation error of the methods compared to using fixed productivity metrics.

Keywords: Software Development Effort Estimation · Optimizing Correction Factors · Software productivity

1 Introduction

Software Development Effort Estimation (SDEE) is one of the biggest challenges in the early stages of bidding on a software project and allocating human resources. It helps to minimize the project's risks or reduce the risk of surprises during the project to the lowest value [1]. Since the essential data is not known at the initial stage, most project managers focus on functional size metrics that can be used to calculate the effort of a software project [2]. The functional size metric - Use Case Points (UCP) [3] - can be helpful to measure the estimated effort. More than 50% of software projects have used use cases in the requirements elicitation, description, and modeling phase [4]. Recently, most researchers have evaluated the UCP method as a possible method for such a problem [5, 6] by multiplying the calculated UCP by a given productivity ratio to estimate the effort.

However, the original UCP method has been criticized for its low precision. First, there are biases and uncertainties in evaluating the components of UCP. In particular,

R. Silhavy (Ed.): CSOC 2022, LNNS 501, pp. 413–425, 2022.
https://doi.org/10.1007/978-3-031-09070-7_35

it is difficult to assign appropriate value to the correction factors - thirteen technical complexity factors (TCF) and eight environmental complexity factors (ECF) - due to a lack of relevant information. Therefore, the correction factors need to be reevaluated [7, 8] to simplify the original principles of UCP. We solved this problem by developing the Optimization Correction Factors (OCF) method [9]. The method has studied the Least Absolute Shrinkage and Selection Operator (LASSO) method [10, 11] to determine the best correction factors, which significantly affect the estimation accuracy of the UCP method. Second, it is risky if project managers do not have the experience to evaluate software productivity before estimating effort [12]. Many researchers consider that the software productivity factor, i.e., the amount of software produced relative to effort, is essential in effort estimation [13]. This term is also interpreted as the ratio of effort to size, known as the productivity factor (PF). Some researchers suggested using predetermined software productivity values to estimate effort based on UCP. For example, Karner's method [3] determines the corresponding effort by multiplying the calculated UCP by the PF, corresponding to 20 person-hours/UCP. Schneider and Winters (SW) recommended using three levels of software productivity (fair, low, and very low) based on the analysis of environmental complexity factors [14]. Concretely, 20 person-hours/UCP for the fair group, 28 person-hours/UCP for the low group, and 36 person-hours/UCP for the very low group. However, each software project has different characteristics and is carried out in different environments. Therefore, a fixed PF for all software projects is not a specific solution.

In this study, we aim to propose a more radical method to improve the estimation accuracy of the OCF method. Our idea is to suggest a new approach for computing the PF based on correction factors through an ensemble construction mechanism of machine learning models (PF_{CFE}) instead of using a fixed PF as in the OCF method. The research questions are answered as follows:

- RQ1: Does the OCF using the proposed software productivity approach outperform the OCF using a fixed PF, base method (UCP), and other tested methods?
- RQ2: Is the difference in the estimation accuracy using various methods statistically significant?

To answer the research questions, we designed experimental studies to investigate the estimation accuracy of the methods. Specifically, we made the following contributions:

- A project productivity model is constructed based on correction factors through an ensemble construction mechanism of three machine learning (ML) techniques, such as Support Vector Regression (SVR) [15], Multiple Linear Regression (MLR), and Decision Tree (DT). We select these techniques due to their fast-learning speed, good generalization ability, and more straightforward design [16, 17]. The voting ensemble is used as an ensemble ML model.
- With the dataset on different areas of government, healthcare, and business [18], we conducted experiments to evaluate the estimation accuracy of the methods in terms of various unbiased evaluation criteria (8–12) and the statistical pairwise t-test comparisons [19, 20].

The statistical hypothesis was tested as follows:

- H0: There is no significant difference in estimation capability between the OCF method using the proposed PF_{CFE} model and other methods.
- H1: There is a significant difference in estimation capability between the OCF method using the proposed PF_{CFE} model and other methods.

The rest of this paper is organized as follows: Sect. 2 presents the overview of the OCF method. Section 3 presented the research methodology, including the process of the experiments, the dataset description, the evaluation criteria/metrics, and the methodology used. Section 4 emphasizes the achieved results. Section 5 presents the conclusions and future works.

2 An Overview of Optimization Correction Factors

The Optimization Correction Factors (OCF) method [9] investigated the Least Absolute Shrinkage and Selection Operator method (LASSO) [10, 11] to determine the best technical and environmental complexity factors. Our method aimed to reduce the risks in evaluating the correction factors of the UCP method and obtain accurate results of the estimates. The phases for the OCF method take place as follows:

Phase 1: Determination of correction factors in regression analysis by the LASSO regression model. The LASSO estimate denoted $\hat{\beta}(\lambda)$ is presented as follows:

$$\hat{\beta}(\lambda) = argmin_{\beta}\left(\frac{\|Y - X\beta\|_2^2}{n} + \lambda\|\beta\|_1\right)$$
$$subject\ to\ \sum_{j=1}^{k}|\beta_j| < t \tag{1}$$

where

$$\|Y - X\beta\|_2^2 = \sum_{i=0}^{n}(Y_i - (X\beta)_i)^2 \tag{2}$$

$$\|\beta\|_1 = \sum_{j=1}^{k}|\beta_j| \tag{3}$$

where $\lambda \geq 0$ is the lasso parameter that controls the strength of the penalty determined by the Leave One Out Cross-Validation (LOOCV) method [21, 22]. The choice of the lasso parameter is adjusted based on the lowest possible estimation error and a lack of bias against the correction factors of the observations in the training set. The lasso parameter relates directly to the number of correction factors selected via non-zero β.

Phase 2: The Least Squares Regression (LSR) is used to obtain the regression coefficients for the selected technical and environmental variable sets. The n-selected technical factors (LaTF) and m-environmental factors (LaEF) based on LASSO are respectively calculated using Eqs. (4) and (5).

$$LaTF = \alpha_0 + \sum_{i=1}^{n} \alpha_i \times LaT_i \times WLt_i \tag{4}$$

$$LaEF = \beta_0 + \sum\nolimits_{i=1}^{m} \beta_i \times LaE_i \times WLe_i \tag{5}$$

where LaT_i, LaE_j are the technical and environmental factors that take values from the interval [0, 5]. WLt_i, WLe_j are the weight of factor i, j, respectively. $\alpha_0, \alpha_i, \beta_0, \beta_1$ are regression coefficient parameters from the LSR model.

Phase 3: The OCF is computed using Eq. (6) as follows:

$$OCF = (UAW + UUCW) \times LaTF \times LaEF \tag{6}$$

Phase 4: The obtained OCF is multiplied by the PF to give the final effort.

$$Effort = Size \times PF \tag{7}$$

where Effort is measured in person-hours and Size is measured in OCF. A standard of 20 person-hours to develop each OCF is used [3].

3 Research Methodology

3.1 Experimental Setup

This section presents the experimental process to evaluate the accuracy of effort estimation methods. Our proposed method and related methods are summarized in Table 1.

Table 1. Methods implemented for experiments

Effort estimation method	Summary	Notation
Use Case Points (UCP)	– Size is measured by UCP variables (UAW, UUCW, TCF, and ECF) – 20 person-hours to develop each UCP (PF = 20) – The effort is computed by multiplying Size by the PF	UCP [3]
Schneider and Winters (SW)	– Size is measured by UCP variables (UAW, UUCW, TCF, and ECF) – PF is computed from the UCP environmental complexity factors – The effort is computed by multiplying Size by the PF	SW [14]
Algorithmic Optimization Method (AOM)	– Size is based on linear regression on UCP variables (UAW, UUCW, TCF, and ECF) – 20 person-hours to develop each UCP (PF = 20) – The effort is computed by multiplying Size by the PF	AOM [28]

(continued)

Table 1. *(continued)*

Effort estimation method	Summary	Notation
Optimization Correction Factors (OCF)	– Correction factors are determined in regression analysis by the LASSO regression model – Size is measured in UCP size variables (UAW, UUCW) and correction factors (LaTF, LaEF) – 20 person-hours to develop each UCP (PF = 20) – The effort is computed by multiplying Size by the PF	OCF [9] (see Sect. 2)
The OCF method using the proposed PF_{CFE} model	– Size is measured in OCF size variables (UAW, UUCW) and correction factors (LaTF, LaEF) – A proposed PF_{CFE} model is constructed based on correction factors through an ensemble construction mechanism of three machine learning models (SVR, MLR, and DT) – The effort is computed by multiplying Size by the PF_{CFE}	OCF(PF_{CFE}) (proposed in Sect. 3.4)

The experimental methods are performed with five different runs (5-fold cross-validation). The comparisons of each method are based on the average results of these five runs, which are presented in Sect. 4. The results were evaluated using some evaluation criteria MAE, RMSE, MBRE, MIBRE, and PRED (0.25), as shown in Eqs. (8–12). A paired t-test (with a significance level of 5%) was also used to validate the accuracy of the methods.

3.2 Data Description

The experiments were conducted with 70 observations [18], a benchmark dataset for evaluating UCP-based effort estimation methods. The statistical properties of the dataset are described in Table 2, where Real_P20 is the actual effort in person-hours divided by PF. The dataset is based on three data donors with different sizes (measured in UCP). Figure 1 illustrates the boxplot of Real_P20 in the dataset.

Table 2. Statistical characteristics of the dataset

	Median man-hours	Median Real_P20	Standard deviation	Minimum Real_P20	Maximum Real_P20	n
Dataset	6406	320.3	33.21	288.75	398.5	70

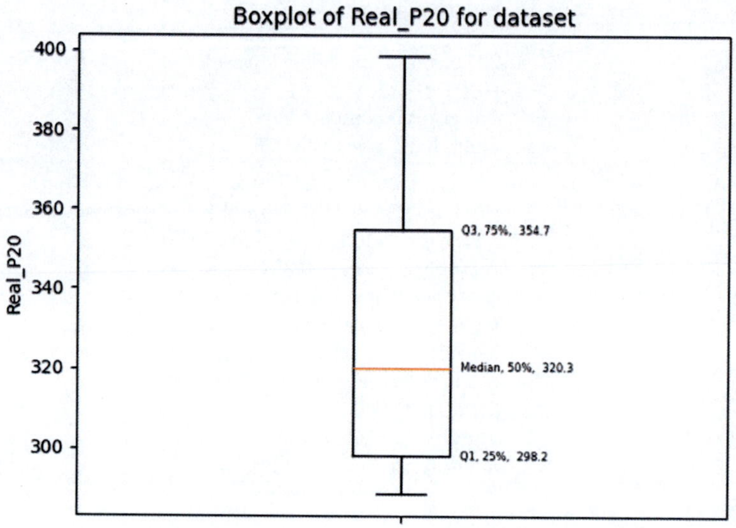

Fig. 1. Boxplot of Real_P20 in the dataset

3.3 Evaluation Criteria

In this study, we use five measurement criteria that yield an unbiased and symmetric distribution to evaluate the estimation methods [26, 27], such as (8) Mean Absolute Error (MAE), (9) Root Mean Square Error (RMSE), (10) Mean Balance Relative Error (MBRE), (11) Mean Inverted Balance Relative Error (MIBRE), and (12) Percentage of Prediction within x% (PRED(x)).

$$MAE = \frac{1}{n}\sum\nolimits_{i=1}^{n}|y_i - \hat{y}_i| \tag{8}$$

$$RMSE = \sqrt{\frac{\sum_{i=1}^{n}(y_i - \hat{y}_i)^2}{n}} \tag{9}$$

$$MBRE = \frac{1}{n}\sum\nolimits_{i=1}^{n}\frac{|(y_i - \hat{y}_{i)}|}{min(y_i - \hat{y}_i)} \tag{10}$$

$$MIBRE = \frac{1}{n}\sum\nolimits_{i=1}^{n}\frac{|(y_i - \hat{y}_{i)}|}{max(y_i - \hat{y}_i)} \tag{11}$$

$$PRED(x) = \frac{1}{n}\sum_{i=1}^{n}\begin{cases} 1\ if\ \frac{|y_i - \hat{y}_i|}{y_i} \leq x \\ 0\ otherwise \end{cases} \quad (12)$$

where n is the number of observations, y_i is the known real value, $\hat{y}_i$ is the estimated value.

3.4 Methodology Used

Figure 2 illustrates the proposed software productivity model. The following methodology was used:

1. Correction factors from OCF are used as input.
2. Built the voting regressor algorithm consisting of three base estimators. Details of these estimators are described in Table 3.
3. The project probability model through the voting ensemble model is called PF_{CFE}.

Table 3. Base estimators of the voting regressor algorithm

Base estimator	Implementation
Support Vector Regression	sklearn.svm.SVR
Multiple Linear Regression	sklearn.linear_model.LinearRegression
Decision Tree Regression	sklearn.tree.DecisionTreeRegressor

4. The convergence of the optimizations of the objective functions in the experimental methods is established based on the smallest Mean Absolute Error, or the maximum number of iterations reaches 10000 [23]. The parameters of the experimental models are tuned on from 20% of the training set. Based on grid search method [24, 25], we find the optimal configuration parameter sets for each experimental method on the dataset. Details of the optimal parameter sets after tuning are shown in Table 4.

5. Estimated effort is obtained by multiplying OCF by PF, as follows Eq. (13).

$$Effort_{OCF(PF_{CFE})} = OCF \times PF_{CFE} \quad (13)$$

Table 4. The optimal configuration parameters in the experiments

Machine learning technique	Optimal configuration parameters
Support Vector Regression	
• The regularization term (C) searched from { 0.1, 1, 10, 100, 1000} • The kernel coefficient (gamma) searched from { 1, 0.1, 0.01, 0.001, 0.0001} • Tolerance for stopping criterion (tol) searched from {0.1, 0.001, 0.001}	C = 1 gamma = 1 tol = 0.001
Decision Tree Regression	
• Maximum depth of the tree (max_depth) searched from {1, 3, 5, 7, 9, 11, 12} • Minimum number of samples (min_samples_leaf) searched from { 1, 2, 3, 4, 5, 6, 7, 8, 9, 10} • Minimum weighted fraction of the total of weights (min_weight_fraction_leaf) searched from { 0.1, 0.2, 0.3, 0.4, 0.5, 0.6, 0.7, 0.8, 0.9 }	max_depth = 7 min_samples_leaf = 2 min_weight_fraction_leaf = 0.2

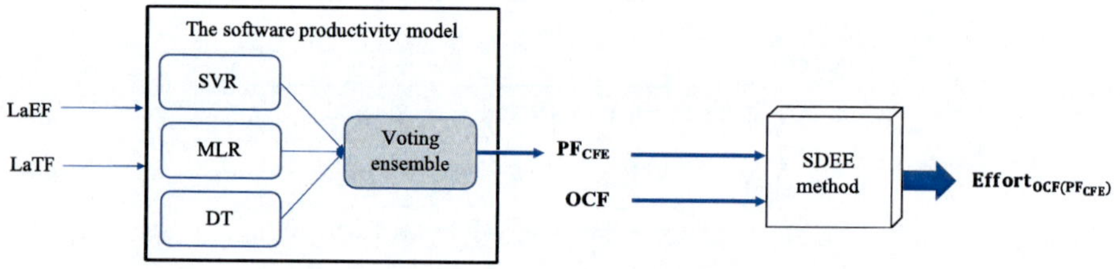

Fig. 2. The proposed software productivity model

4 Results

In this section, we evaluate the empirical results of the conducted experiment and answer our research questions.

4.1 RQ1

Does the estimation accuracy of OCF using the proposed software productivity approach outperform the estimation accuracy of OCF using a fixed PF, base method (UCP), and other tested methods?

To answer this research question, we constructed an ensemble model from three ML methods, SVR, MLR, and DT, to establish the relationship between project productivity and the independent variables LaTF and LaEF. Then, the predicted project productivity with OCF size is used to estimate the effort using Eq. (7). The experimental results of the constructed methods are shown in Fig. 3.

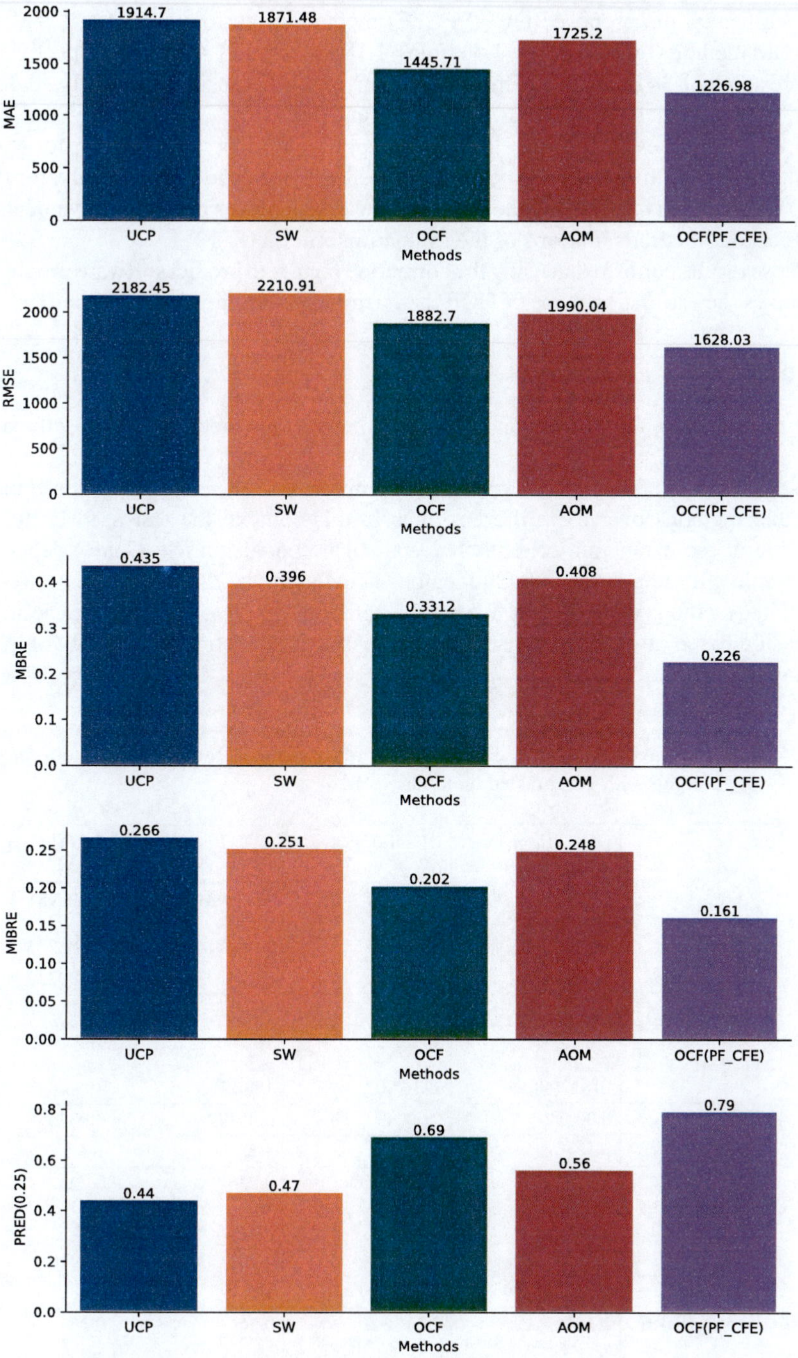

Fig. 3. The results of the proposed CFP method and other methods

Specifically, the proposed OCF(PFCFE) method outperforms the UCP, SW, OCF, and AOM methods by 1.56 times, 1.53 times, 1.18 times, and 1.41 times, respectively, for MAE result, by 1.34 times, 1.36 times, 1.16 times, and 1.22 times, respectively, for RMSE result, by 1.92 times, 1.75 times, 1.46 times, and 1.80 times, respectively, for MBRE result, by 1.66 times, 1.56 times, 1.25 times, and 1.54 times, respectively, for MIBRE result, and by 1.77 times, 1.67 times, 1.15 times, and 1.41 times, respectively, for PRED (0.25) result. It can be seen that the proposed method achieves better improvements than the other four methods in terms of the evaluation criteria (8–12).

These results confirm that using the correction factors to predict software productivity minimizes the estimation error of OCF than using the fixed productivity metrics.

4.2 RQ2

Is the difference in the estimation accuracy using various methods statistically significant?

We conducted statistical pairwise t-test comparisons between our proposed method and other methods over five different runs. In this context, the test used in the t-test consisted of the evaluation criteria. The statistical conclusion (St. Conc.) depends on their p-value. If the p-value ≤ 0.05, it can be concluded that the difference between the two related methods is significant. We use the notation $A \gg B$ to explain the experimental conclusion based on their p-value, which expresses the statistical superiority of method A compared with method B.

Table 5. The t-test results of five different runs for statistical comparison of our proposed OCF(PF_{CFE}) method with other tested methods

Pairs of methods		OCF(PF_{CFE}) vs. UCP	OCF(PF_{CFE}) vs. SW	OCF(PF_{CFE}) vs. OCF	OCF(PF_{CFE}) vs. AOM
MAE results	Avg. MAE	1226.981 vs. 1914.697	1226.981 vs. 1871.477	1226.981 vs. 1445.714	1226.981 vs. 1725.202
	Avg. p-value	0.00024	0.00114	0.02459	0.00023
	St. Conc.	>>	>>	>>	>>
RMSE results	Avg. RMSE	1628.033 vs. 2182.448	1628.033 vs. 2210.907	1628.033 vs. 1882.705	1628.033 vs. 1990.043
	Avg. p-value	0.00032	0.00049	0.00477	0.00267
	St. Conc.	>>	>>	>>	>>
MBRE results	Avg. MBRE	0.226 vs. 0.435	0.226 vs. 0.396	0.226 vs. 0.331	0.226 vs. 0.408
	Avg. p-value	0.00067	0.00162	0.0011	0.00105
	St. Conc.	>>	>>	>>	>>
MIBRE results	Avg. MIBRE	0.161 vs. 0.266	0.161 vs. 0.251	0.161 vs. 0.202	0.161 vs. 0.248
	Avg. p-value	0.00054	0.00189	0.01422	0.00054
	St. Conc.	>>	>>	>>	>>

Table 5 shows the average results of MAE, RMSE, MBRE, and MIBRE over five different runs and final statistical conclusions. The OCF using the proposed software productivity approach is statistically better than other methods, as the obtained p-values are all below 0.05. All these statistical conclusions strongly confirm the answer to our RQ2.

5 Conclusions

In this work, our goal is to improve the estimation accuracy of the OCF method. The challenge is to convert the OCF size into the corresponding effort to minimize estimation errors. We solved this problem by developing the software productivity model based on correction factors through the voting ensemble approach, consisting of three ML techniques.

As mentioned in Sect. 1, this paper analyzed two research questions related to the proposed approach. Regarding RQ1, the obtained results allow us to confidently conclude that the OCF using the proposed software productivity approach achieves better improvements than the previous related methods (UCP, SW, OCF, and AOM) using fixed productivity metrics, concerning all accuracy measures involving MAE, RMSE, MBRE, MIBRE, and PRED (0.25). This conclusion is confirmed by statistical t-test comparisons for each corresponding method. For RQ2, it can be concluded that the proposed method is statistically significantly superior to the other methods.

In the future, we continue to improve the estimation accuracy of the proposed method by applying clustering approaches. In addition, we will test more datasets in the future.

Acknowledgment. This work was supported by the Faculty of Applied Informatics, Tomas Bata University in Zlín, under Project IGA/CebiaTech/2022/001.

References

1. Boehm, B., Abts, C., Chulani, S.: Software development cost estimation approaches - a survey. Ann. Softw. Eng. **10**, 177–205 (2000). https://doi.org/10.1023/A:1018991717352
2. Trendowicz, A., Jeffery, R.: Software Project Effort Estimation. Foundations and Best Practice Guidelines of Success. Springer, Cham (2014). https://doi.org/10.1007/978-3-319-03629-8
3. Karner, G.: Resource Estimation for Objector Projects. Objective Systems SFAB (1993)
4. Jorgensen, M., Shepperd, M.: A systematic review of software development cost estimation studies. IEEE Trans. Software Eng. **33**(1), 33–53 (2007). https://doi.org/10.1109/TSE.2007.256943
5. Nhung, H.L.T.K., Hoc, H.T., Hai, V.V.: A review of Use Case-based development effort estimation methods in the system development context. In: Silhavy, R., Silhavy, P., Prokopova, Z. (eds.) CoMeSySo 2019 2019. AISC, vol. 1046, pp. 484–499. Springer, Cham (2019). https://doi.org/10.1007/978-3-030-30329-7_44
6. Mahmood, Y., Kama, N., Azmi, A.: A systematic review of studies on Use Case Points and experts-based estimation of software development effort. J. Softw. Evol. Process **32**(7), e2245 (2019)

7. Huanca, L.M., Oré, S.B.: Factors affecting the accuracy of use case points. In: Mejia, J., Muñoz, M., Rocha, Á., San Feliu, T., Peña, A. (eds.) Trends and Applications in Software Engineering, CIMPS 2016. Advances in Intelligent Systems and Computing, vol. 537, pp. 133–142. Springer, Cham (2016). https://doi.org/10.1007/978-3-319-48523-2_13
8. Azzeh, M.: Software cost estimation based on use case points for global software development. In: 5th International Conference on Computer Science and Information Technology (CSIT), Amman, Jordan, pp. 214–218 (2013)
9. Nhung, H.L.T.K., Hoc, H.T., Hai, V.V.: An evaluation of technical and environmental complexity factors for improving Use Case Points estimation. In: Silhavy, R., Silhavy, P., Prokopova, Z. (eds.) Software Engineering Perspectives in Intelligent Systems. CoMeSySo 2020. Advances in Intelligent Systems and Computing, vol. 1294. Springer, Cham (2020). https://doi.org/10.1007/978-3-030-63322-6_64
10. Fonti, V.: Feature selection using LASSO. Research paper in Business Analysis (2017)
11. Muthukrishnan, R., Rohini, R.: LASSO: a feature selection technique in predictive modeling for machine learning. In: IEEE International Conference on Advances in Computer Applications (ICACA), pp. 18–20 (2016). https://doi.org/10.1109/ICACA.2016.7887916
12. Heidrich, J., Oivo, M., Jedlitschka, A.: Software productivity and effort estimation. J. Softw. Evol. Process **27**(7), 465–466 (2015)
13. Rodriguez, D., Sicilia, M., Garcia, E., Harrison, R.: Empirical findings on team size and productivity in software development. J. Syst. Softw. **85**(3), 562–570 (2012)
14. Schneider, G., Winters, J.P.: Applied Use Cases, A Practical Guide, 2nd edn. Addison-Wesley, Boston (2001)
15. Oliveira, A.L.I.: Estimation of software project effort with support vector regression. Neurocomputing **69**(13–15), 1749–1753 (2006). https://doi.org/10.1016/j.neucom.2005.12.119
16. Elish, M.O.: Improved estimation of software project effort using multiple additive regression trees. Expert Syst. Appl. **36**(7), 10774–10778 (2009). https://doi.org/10.1016/j.eswa.2009.02.013
17. Hai, V.V., Nhung, H.L.T.K., Prokopova, Z., Silhavy, R., Silhavy, P.: A new approach to calibrating functional complexity weight in software development effort estimation. Computers **11**(2), 15 (2022). https://doi.org/10.3390/computers11020015
18. Silhavy, R., Silhavy, P., Prokopova, Z.: Analysis and selection of a regression model for the Use Case Points method using a stepwise approach. J. Syst. Softw. **125**, 1–14 (2017). https://doi.org/10.1016/j.jss.2016.11.029
19. Anderson, D.R., Sweeney, D.J., William, T.A.: Statistics for Business and Economics. Thomson South-Western, Ohio (2009)
20. Ross, A., Willson, V.L.: Paired sample t-test. In: Basic and Advanced Statistical Tests, pp. 17–19 (2017)
21. van der Vaart, A.W., Dudoit, S., van der Laan, M.J.: Oracle inequalities for multi-fold cross-validation. Stat. Decis. **24**(3), 351–371 (2006). https://doi.org/10.1524/stnd.2006.24.3.351
22. Van der Lann, M.J., Dudoit, S.: Unified cross-validation methodology for selection among estimators and a general cross-validated adaptive Epsilon-Net estimator: finite sample oracle inequalities and examples. U. C. Berkeley Division of Biostatistics Working Paper Series (2003)
23. Srinivasan, K., Fisher, D.: Machine learning approaches to estimating software development effort. IEEE Trans. Softw. Eng. **21**(2), 126–137 (1995)
24. Ma, X., Zhang, Y., Wang, Y.: Performance evaluation of kernel functions based on grid search for support vector regression. In: IEEE Conference on Robotics Automation and Mechatronics (2015)
25. Song, L., Minku, L.L., Yao, X.: The impact of parameter tuning on software effort estimation using learning machines. In: 9th International Conference on Predictive Models in Software Engineering (2013)

26. Silhavy, R., Silhavy, P., Prokopova, Z.: Using actors and use cases for software size estimation. Electronic **10**, 592 (2021)
27. Nhung, H.LT.K., Hai, V.V., Silhavy, R., Prokopova, Z., Silhavy, P.: Parametric software effort estimation based on optimizing correction factors and multiple linear regression. IEEE Access **10**, 2963–2986 (2022). https://doi.org/10.1109/ACCESS.2021.3139183
28. Silhavy, R., Silhavy, P., Prokopova, Z.: algorithmic optimisation method for improving use case points estimation. PLoS ONE **10**, e0141887 (2015)

Express Identification of the Frame-Based Signals Produced According to the Narrowband Digital Communication Standards

Leila Golpaiegany[1], Aleksandr Litvinov[2], Anton Tokarev[2,3], Alexandra Salnikova[4](✉), Dmitry Shepelev[4], and Alexey Shinyaev[2,3]

[1] University of Ayatollah Ozma Boroujerdi, 69199-69411 Boroujerd, Iran
[2] JSC "IRCOS", Rabochiy Avenue 101B, 394049 Voronezh, Russia
[3] Voronezh State Technical University, Moscow Avenue 14, 394026 Voronezh, Russia
[4] National Research University "MPEI", Krasnokazarmennaya str. 14, 111250 Moscow, Russia
al.vl.salnikova@mail.ru

Abstract. An important task included in the complex of radio monitoring measures is to search among the huge set of registered radio signals for such signals that violate the established procedure of radio frequency spectrum usage. A sign indicating that the observed radio signal is nonnormative can be the appearance in a frequency band of a signal using a communication standard uncharacteristic for this band. Detailed studies of the structure and service parameters of radio signals are, as a rule, laborious. Therefore, it is of considerable interest to develop simple, computationally efficient procedures that, without the need to analyze the "fine structure" of each signal, allow deciding on the type of the observed radio signals in terms of one or another communication standard. The paper considers the possibility of determining whether the particular narrowband signals belong to this or that communication standard employing frame-based signals. It is shown that the signal identification can be carried out based on a unified approach that does not require a priori information about the signals operating in the analyzed frequency range and without the need for precise tuning to the carrier oscillation frequencies. The use of sample sequences obtained at the output of the FM demodulator ensures reliable identification of signals of the dPMR, NXDN, TETRA standards, while direct processing of the samples of complex envelopes of oscillations does not guarantee reliable recognition of these signals, even with a noticeable increase in the signal-to-noise ratio. The transition from analyzing the sequences of samples of the complex envelopes of radio signals to the study of the samples generated at the output of the FM demodulator results in a decrease in the resources required for the implementation of the proposed processing algorithm by means of the FPGA. Thus, the recommendation for employing demodulated samples when implementing the express signal identification procedure is confirmed. #CSOC1120.

Keywords: Radio monitoring · Signal search · Express identification · Communication standard · Narrowband frame-based signal · Frequency modulation/demodulation · Correlation analysis · RTL-SDR device

R. Silhavy (Ed.): CSOC 2022, LNNS 501, pp. 426–442, 2022.
https://doi.org/10.1007/978-3-031-09070-7_36

1 Introduction

The observed rapidly growing amounts of transmitted information entail the active development of communication and data transmission systems based on multiple standards. As a result, there is an increasing need for radio environment analysis tools capable of recognizing and distinguishing radio signals observed on the air. The identification of radio signals that are not characteristic or permitted for a specific frequency and time point may indicate an unauthorized radio transmission, information leakage or some other events important in the context of the spectrum usage monitoring. Thus, a prompt decision on the correspondence of the observed radio signals to specific data transmission standards is a task in demand in practice.

It is well known [1] that in order to distinguish between the signals matching certain sets of "standards", a correlation processing is often effective (and therefore is actively used). With its help, a short auxiliary signal (for example, a preamble, a synchronization word or a pilot code) is searched for in a signal of long duration. However, given the abundance of potentially possible types and modulation parameters, it is unpromising to use correlators as the key tools for recognizing an extensive list of communication standards (since it has an extremely negative effect on hardware resources and a recognition time).

Commonly known [2, 3] and generally applied are the methods for recognizing the modulation formats, based on a comparison of the averaged spectra of the recorded radio signals with a set of reference radio signal spectral masks. However, when monitoring the radio environment within wide frequency ranges, these methods are effective only in the tasks of identification by the modulation formats of wideband signals. Among narrowband radio signals, there are many different oscillations of the very similar spectrum shapes, and that makes their recognition ineffective (at least under the conditions of an insufficient detailed spectral resolution). For example (Fig. 1), among the five common narrowband radio communication standards applicable in the frequency range from 136 to 860 MHz, APCO and DMR signals have the same spectrum width of 12.5 kHz, for TETRA it is 25 kHz, and for dPMR – 6.25 kHz, while NXDN can have a bandwidth of 6.25, and 12.5 kHz [4].

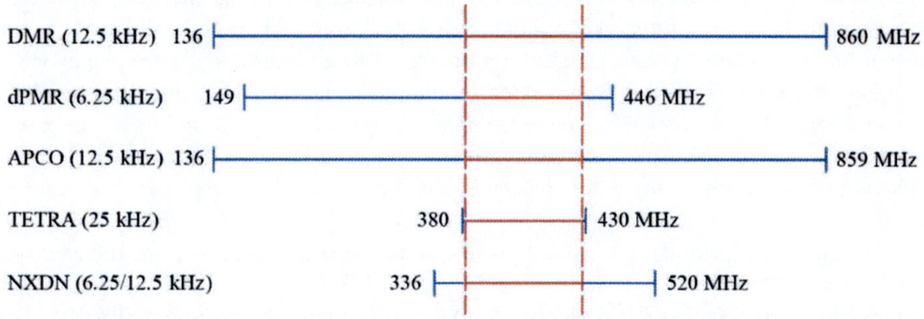

Fig. 1. The comparison of the wavebands of the operating frequencies and the spectral widths of the signals corresponding to the DMR, dPMR, APCO, TETRA и NXDN standards

Obviously, based on the spectral data only, it is extremely problematic to distinguish between the similar signals located nearby by frequency and having the same (and close in width) bandwidth and approximately the same type of power spectral density (PSD). Thus, it is of interest to find the ways to quickly decide on the correspondence of the recorded oscillations to the specific communication standards, and, first of all, the so called "universal" methods are needed, i.e. the ones for which, with the expansion of the number of tested hypotheses, only a slight increase in labor intensity is required.

It is useful to take into account the fact that among modern narrowband communication standards there is an extensive set of standards based on time-division multiple access (TDMA). Of course, to receive information and/or estimate the parameters of radio signals generated within time slots, it is necessary to take into account many technical details defined by a specific standard. However, the very fact of dividing the time axis into separate slots with time boundaries, albeit different in nature, but equally spaced from each other, generates a certain periodicity of radio signals, and the very existence of such a periodicity can serve as the basis for the development of signal identification algorithms. To test this assumption, one should make sure that the influence of the periodicity generated by the TDMA technology is confidently and clearly manifested even in the absence of a priori information about the features of the received radio signals as well as in the case of gross errors in a tuning on the observed radio signals.

2 Time Sampling Based Search for Latent Periodicities of the Recorded Radio Signals

Modern radio equipment and, in particular, radio monitoring systems are based, as a rule, on the software-defined radio (SDR) technology, allowing flexible adjustment of the algorithms and parameters of the reception of the oscillations through a firmware update. A typical stage in the reception and processing of signals in SDR systems is the transfer of the signal spectrum to the low-frequency region, which implies the sampling of the complex envelope of oscillations. For this, a downconversion is used, which is a quadrature transfer of the spectrum of a real signal to a zero frequency.

The principle of downconversion of the received signal in a digital receiver is simplified in Fig. 2 [5]. Here the notations are the following: BF is the band filter; IFF is the intermediate frequency filter; ADC is the analog-to-digital converter; DLPF is the low-pass filter. The received analog signal is transferred to an intermediate frequency (IF) using a heterodyning, then it is subjected to analog-to-digital conversion, and, finally, it is multiplied by the numerical sequence $\exp\left(-j2\pi f_{IF} n / f_S\right)$ resulting in the complex-valued samples being shifted to a zero frequency, generating a complex envelope of a radio signal in the frequency band determined by the intermediate frequency path of the radio receiver.

Algorithms for express identification of communication standards are focused on the search for "atypical" signals in wide frequency bands and, as a rule, involve their implementation by means of relatively wideband radio monitoring systems (RMS). The bandwidth of the IF paths of such systems often includes tens (and hundreds) of narrowband radio channels, and therefore the mandatory components of the processing procedure are energy search, translation to zero frequency and narrowband filtering of radio signals corresponding to individual radio channels. Studies have shown that for identification purposes, filters with bandwidths of 12.5 and/or 50 kHz, designed for processing complex samples, can be recommended.

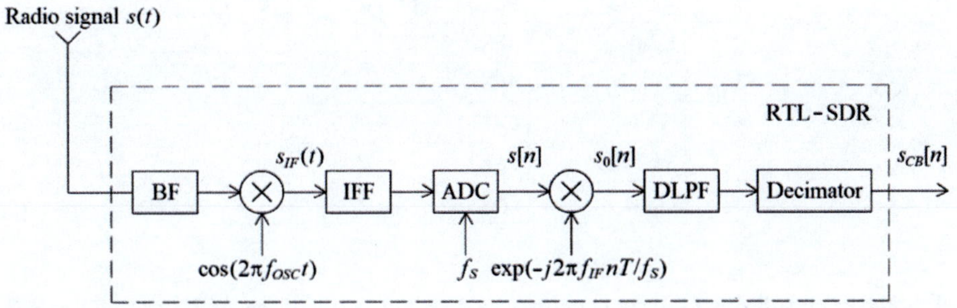

Fig. 2. The algorithm of generating a complex envelope of a radio signal in the frequency band determined by the IF path

It is noteworthy that the analysis of the most actively used modern narrowband communication standards presented in Table 1 shows that almost all of them use one or another version of frequency modulation for data transmission, and the periodicity corresponding to the complex envelopes of such oscillations

$$s_{CB}[n] = A\exp(j\theta_{FM}[n]),$$

also occur during the analysis of their demodulated samples [6].

$$s_d[n] = angle\left(s^*_{CB}s_{CB}[n-1]\right) = -(\theta_{FM}[n] - \theta_{FM}[n-1]) = \mathrm{d}\theta_{FM}[n]/\mathrm{d}t = K_{FM}\, s_i[n].$$

Here $angle(\cdot)$ is an operation of determining the phase of a complex signal argument; $s_i[n]$ is the sequence of the information samples that forms the basis for the modulated oscillation phase change law

$$\theta_{FM}[n] = K_{FM}\sum_{k=1}^{n} s_i[k].$$

The "tool" for identifying periodicities in the analyzed narrowband radio signals is their autocorrelation function (ACF)

$$\rho_k = \sum_{i=0}^{N} Y_i Y^*_{k-i}. \tag{1}$$

Here k is the index that determines the mutual bias of the sample Y_i and its complex conjugate copy, and * is the sign of the complex conjugation operation. Since to reveal hidden periodicities it is required to accumulate and process sampling made of tens of thousands of samples, to increase the computational efficiency of calculations, instead

Table 1. Characteristic features of the main types of the modern narrowband standards.

No.	Name of the standard	Modulation format	Use of TDMA
1	APCO	PSK, C4FM	✓
2	DMR	4FSK	✓
3	dPMR	4FSK	
4	NXDN	C4FM	✓
5	TETRA	PSK	✓
6	GSM	GMSK	✓

of directly determining (1), one usually uses the calculation of the ACF through the FFT in the form

$$\rho = IFFT\left(FFT(Y) \cdot FFT(Y)^{*}\right),$$

where FFT is Fast Fourier Transform and IFFT is Inverse Fast Fourier Transform. The samples Y_i can be both the samples of the complex envelopes of narrowband oscillations and the FM-demodulated samples of these signals.

In general, the stages of processing can be represented in the form of a block diagram shown in Fig. 3.

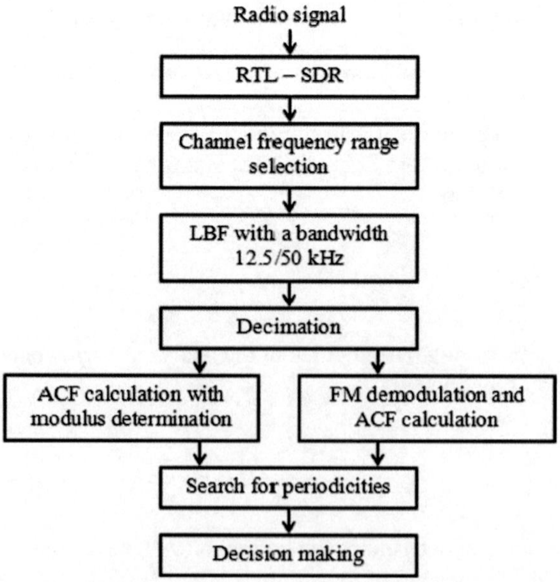

Fig. 3. The block diagram for the ACF based radio signal identification

Although the ultimate goal of the research is to develop a tool for automatic identification of radio signals observed on the air, the current phase of research is focused on identifying and analyzing the recommended decision determining statistics, and therefore an active personal participation of the operator-researcher in developing solutions is assumed. As a consequence, the last stage of the block diagram in Fig. 3 is implemented "manually" within the current research phase.

3 Hardware Research Apparatus

For the practical implementation of the above-described sampling mechanism, the RTL-SDR TERRATEC receiver is used, the appearance of which is shown in Fig. 4. TERRATEC receivers are based on the RTL2832 chipset – a microcircuit containing two 8-bit ADCs with a sampling rate of up to 3.2 MHz and using a USB interface for a communication with a computer. The range of permissible sampling rates that is between 0.25 and 3.2 MHz allows applying such a receiver for generating samples of the complex envelope of oscillations with a spectrum width from 250 kHz to approximately 2 MHz. In frequency bands with a width of 250 kHz, several narrowband radio signals can be simultaneously observed (acted upon), and the sampling rate corresponding to such a band is redundant to represent narrowband oscillations. Accordingly, the selection of the signals to be processed assumed the use of a digital low-pass filter with a subsequent decimation of the samples (see the starting stages of processing in Fig. 3).

The interaction with the RTL-SDR receiver and the preprocessing of the analyzed signals is carried out in the Simulink graphical programming environment, with the final data processing performed in the MATLAB application package. In addition to receiving and processing real signals "from the air" to speed up the operations and to expand them to the signals that are not registered at the point of the test bench, as part of the study, the complex samples from the dedicated library of radio signals have also been processed in MATLAB.

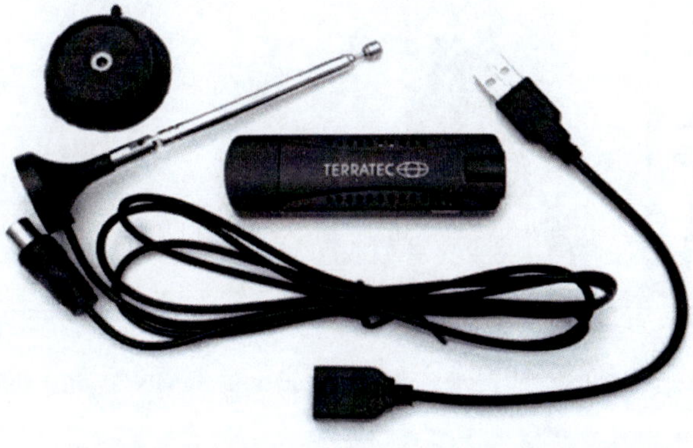

Fig. 4. The look of the set of receiving RTL-SDR device

4 Testing the Possibility of Identifying the Periodicity of the Structure of Radio Signals During "Blind" Tuning to the Frequency

4.1 Analysis of the GSM Signal Features

The analysis of radio signals with a frame structure one starts with considering the GSM standard [7], based on TDMA technology. According to this standard, the data stream is divided into frames with a duration of 4.615 ms, that, in their turn, are divided into 8 equal timeslots with a duration of one eighth of the frame duration, that is, 577 μs.

The first stage of the analysis of the correlation properties of this radiation is carried out based on the processing filtered samples of radio signals operating at a frequency of 950.8 MHz (the central frequency of one of the GSM base stations (BS)). The duration of the samples is 0.5 s. The objects of processing are the samples obtained from a standard spectrum with a bandwidth of 200 kHz due to low-frequency filtering in a band of 50 kHz. The such narrowband fragments of the signal spectrum are selected to substantiate the possibility of using a "blind" unified approach in the future, which does not require binding to a specific standard and is suitable for processing various narrowband signals of different structures.

The research is carried out in the facilities at the outskirts of the city. The receiving antenna is placed at the height of the 6th floor, in the absence of high buildings and structures nearby. According to SDRSharp software, the received signal level is from minus 60 to minus 20 dB full scale (dB FS). In this case, the difference between the level of the analyzed signal and the level of noises (interferences) appears to be approximately 25 dB FS. The PSD graph of the 950–952 MHz spectrum section, with the PSD of GSM BS signals obtained from SDRSharp is shown in Fig. 5.

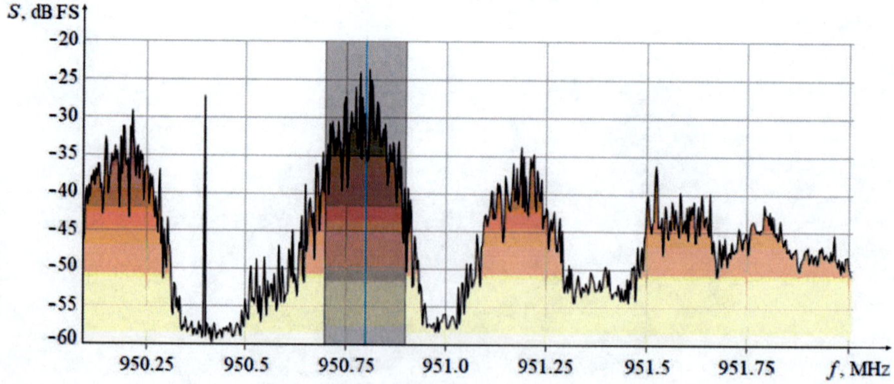

Fig. 5. The power spectral density of GSM base station signals against the background of SDRSharp noises

In Fig. 6 on the left, the ACF diagram of narrowband samples after the preliminary use of FM demodulation is shown, while to the right of it one can see the very same diagram but without FM demodulation involved. As it can be seen, both ACF diagrams clearly demonstrate the bursts at a frequency of 577 μs and less obvious bursts with the periods of 4.615 ms.

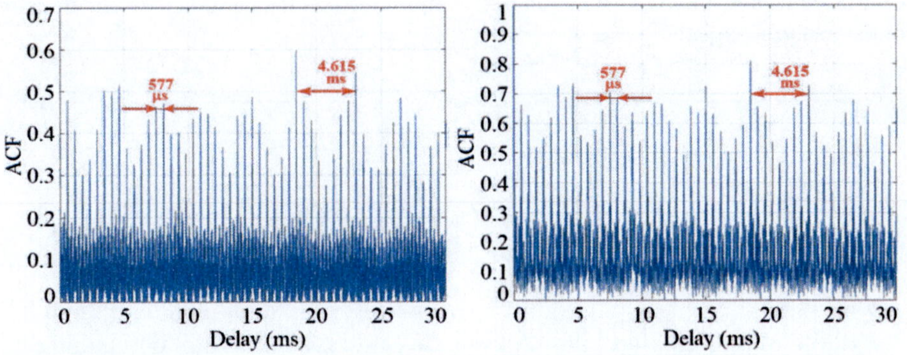

Fig. 6. GSM ACF at the frequency of 950.8 MHz with the FM demodulation involved (left) and not involved (right)

The second stage of a study of GSM signals is concerned with the possibility of fixing the periodicity of the structure of radio signal with noticeable errors in frequency tuning. At this stage, the receiver is tuned to the frequency below the center by 60 kHz (the rest of the recording parameters remain the same). A shift of 60 kHz is chosen as a fairly significant detuning, which still guarantees, however, a high enough signal-to-noise ratio (SNR) within the spectrum fragment that is allocated due to an additional narrowband filtering and that does not include the center frequency.

The ACF diagrams of the sampling obtained with and without FM demodulation are shown in Fig. 7.

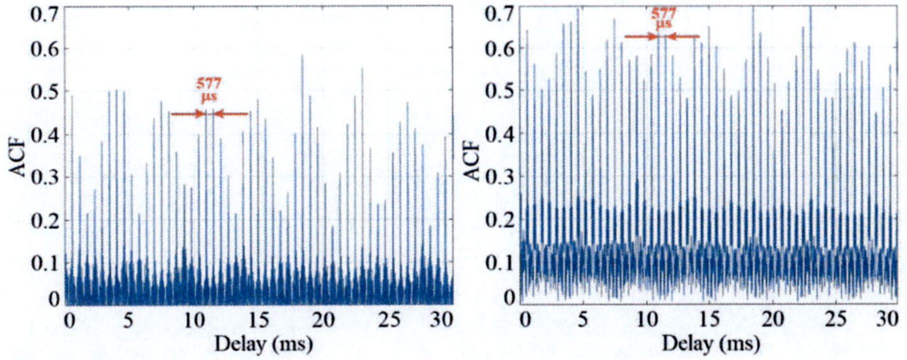

Fig. 7. GSM ACF at the frequency of 950.740 MHz with the FM demodulation involved (left) and not involved (right)

It can be seen from them that the results have slightly deteriorated, namely, the amplitudes of the peaks have slightly decreased, but the presence of a periodicity of 577 μs is still.

In Fig. 8, one can see the ACF diagrams with and without FM demodulation involved for the case of tuning the receiver to the frequency of 60 kHz above the central one. As well as the previously cited diagrams, they prove that even when processing a relatively insignificant fragment of the GPS radio signal spectrum, which does not include the central frequency, the ACF of the filtered samples reflects the periodicity characteristic of the analyzed signals.

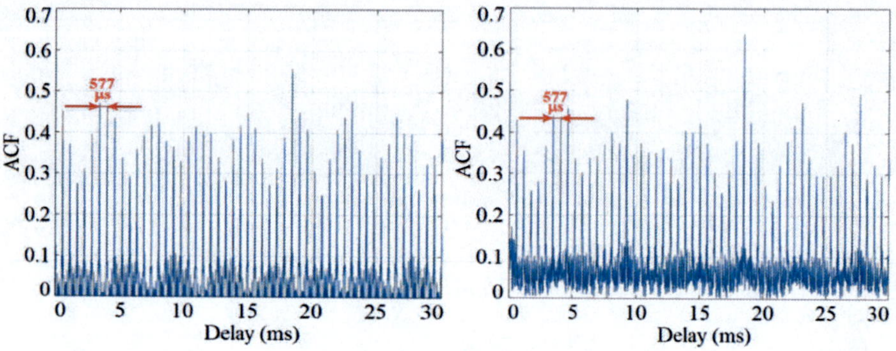

Fig. 8. GSM ACF at the frequency of 950.860 MHz with FM demodulation involved (left) and not involved (right)

4.2 Analysis of the APCO-25 Signal Features

The width of the APCO-25 signal spectrum is only 12.5 kHz [4], and when such narrow-band signals are detected on the frequency axis, the whole spectrum naturally becomes the object of analysis. The ACF sampling diagrams obtained by fine tuning the radio control equipment to the band center of the spectrum for the samples with a duration of 0.8 s are shown in Fig. 9.

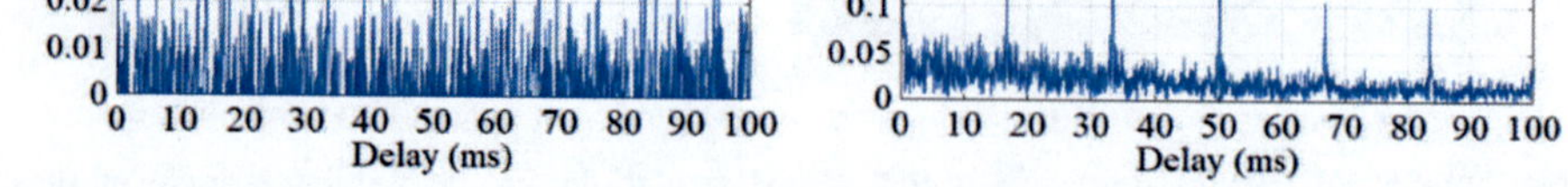

Fig. 9. APCO-25 ACF at the record duration of 0.8 s with FM demodulation involved (left) and not involved (right)

As it can be seen, the ACF bursts corresponding to the periodicity of 16.7 ms are traced for the samples both with the use of FM demodulation and without it.

The results of studying the ACF of the samples obtained with the spectrum shifted downward in frequency by 6.25 kHz, when the radio monitoring equipment captured only the upper half of the signal spectrum and the 6.25 kHz band containing noise and/or other signals, are shown in Fig. 10. It can be seen that the ACF for the samples without FM demodulation makes it possible to see the periods characteristic of APCO-25 even when the allocated frequency band and the band occupied by the signal spectrum do not coincide.

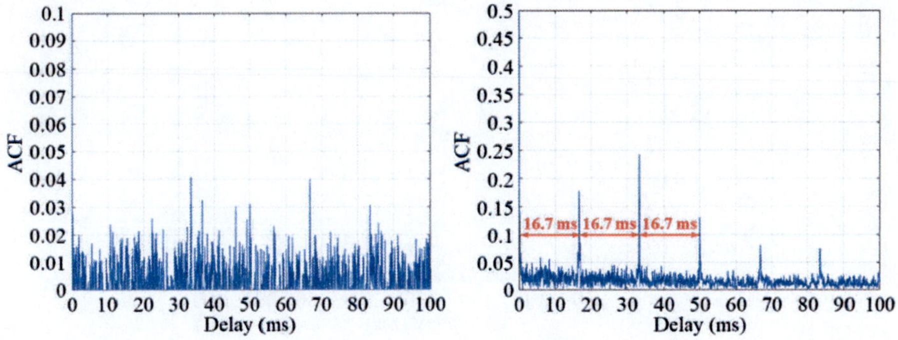

Fig. 10. APCO-25 ACF at the record taken not at the spectrum band center and having the duration of 0.8 s with FM demodulation involved (left) and not involved (right)

4.3 DMR Sampling Correlation Analysis Results

The DMR standard, like the APCO-25, uses signals with a spectrum width of 12.5 kHz [4], but it is characterized by a different frequency indicator. The ACF sampling diagram for this standard with fine tuning, analysis of the entire signal spectrum and 0.8 s record duration is shown in Fig. 11 on the left, and the results corresponding to a frequency shift up by 6.25 kHz are on the right.

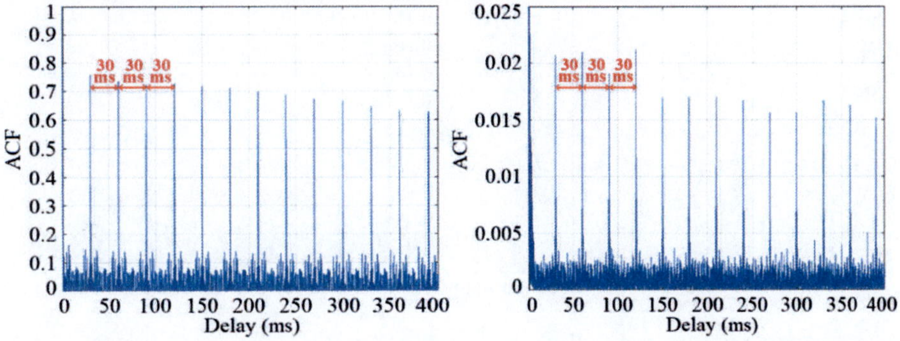

Fig. 11. DMR ACF with FM demodulation involved on band center (left) and at 6.25 kHz shift up, with the record duration of 0.8 s

In both cases, the diagrams clearly show significant amplitude bursts with a period of 30 ms. If the duration of the records is increased up to 20 s or more, then it will be possible to distinguish the maxima with a period of 60 ms on the ACF diagram (Fig. 12). Analyzing the DMR standard, one can find out that 30 ms is the duration of a single time slot, and 60 ms is the duration of a time frame [7].

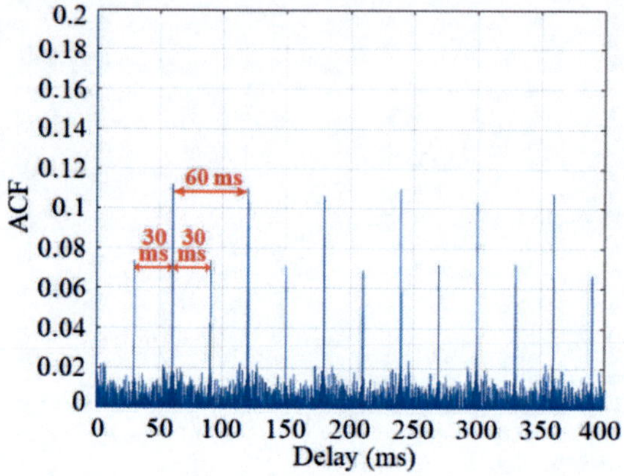

Fig. 12. DMR ACF with FM demodulation and fine tuning involved, the record duration is 20 s

4.4 dPMR Sampling Correlation Analysis Results

The narrowband dPMR standard has a spectrum width of 6.25 kHz [4]. The ACF diagram presenting the results of the 12.5 kHz band analysis (which is permissible in the absence of signals in the adjacent radio channel) with the record duration of 0.4 s is shown in Fig. 13 on the left. Similar results, although somewhat worse in quality, for the case of the frequency detuning at 3.125 kHz, i.e. by the half of the dPMR spectrum, are shown in Fig. 13 on the right. In both Figures, the frequency of the ACF maxima corresponding to the duration of the transmitted data frame of this standard is 80 ms.

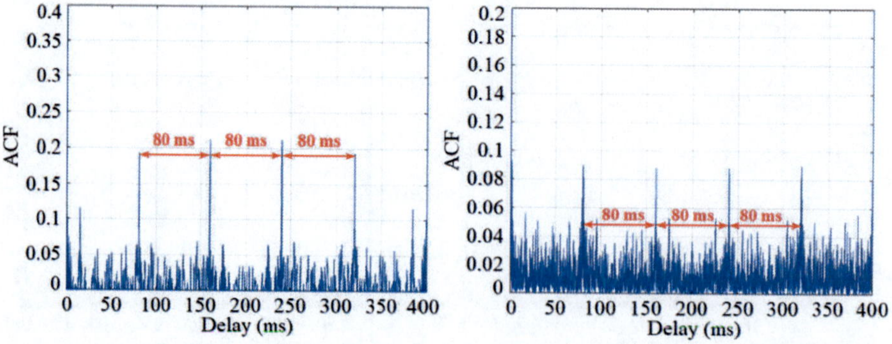

Fig. 13. dPMR ACF with FM demodulation involved on the band center (left) and at 3.125 kHz shift up, with the record duration of 0.4 s.

4.5 Analysis of the NXDN Signal Periodicity

The NXDN digital standard uses the signals with the bandwidth of 6.25 kHz like dPMR standard and the signals with the bandwidth of 12.5 kHz like APCO-25 and DMR standards [4]. The ACF diagrams obtained while analyzing the recordings of the signals with a bandwidth of 12.5 kHz for the case of fine tuning and FM demodulation involved are presented in Fig. 14 on the left. And on the right there, one can see the ACF diagram for the case of the 6.25 kHz frequency shift up and the modulo operation application after the ACF. It can be seen from the Figures that the ACF displays bursts with a period of 40 ms, one in two of them having an increased amplitude, i.e. one can say that on the ACF diagram there is a second burst periodicity present, and its value is 80 ms. The fact that this period coincides with the value characteristic of dPMR creates the problem of distinguishing between the NXDN and dPMR signals. To distinguish these standards, and to increase the system performance, it would be reasonable to start from the lower periodicities (40 ms), neglecting the higher ones, though they are characterized by the ACF peaks that are more pronounced by their amplitude.

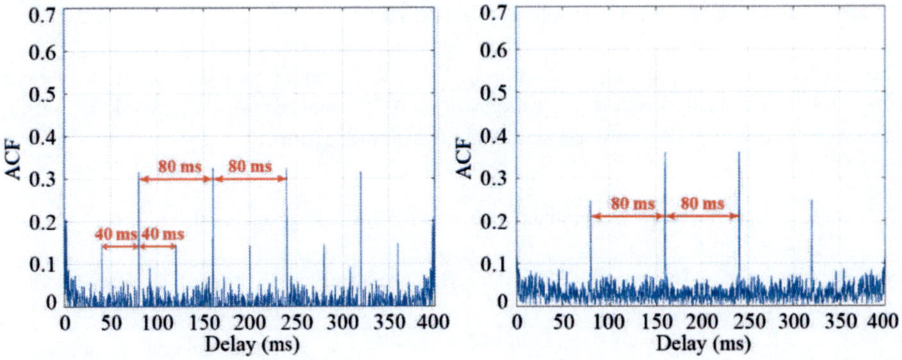

Fig. 14. NXDN AFC on the band center and with FM demodulation involved (left) and with the frequency shift up and without FM demodulation involved, in both cases record duration is 0.4 s

4.6 TETRA Signal Correlation Analysis Results

TETRA signals have a bandwidth of 25 kHz [4]. Applying, by analogy with the cases considered above, the filtering of the registered samples using a filter with a 12.5 kHz bandwidth, and further calculating the ACF, one can obtain the diagrams similar to that shown in Fig. 15 on the left. The Figure shows the periodicity of the ACF maxima of 14.167 and 56.67 ms. The first value of these values corresponds to the duration of a time slot, and the second to the duration of a TETRA frame [8]. The ACF diagram for the case of the spectrum frequency shift downward by 12.5 kHz is shown in Fig. 15 on the right. As it can be seen, in the analysis of a part of the spectrum, the periodicity of the ACF bursts is retained, although their amplitudes noticeably decrease.

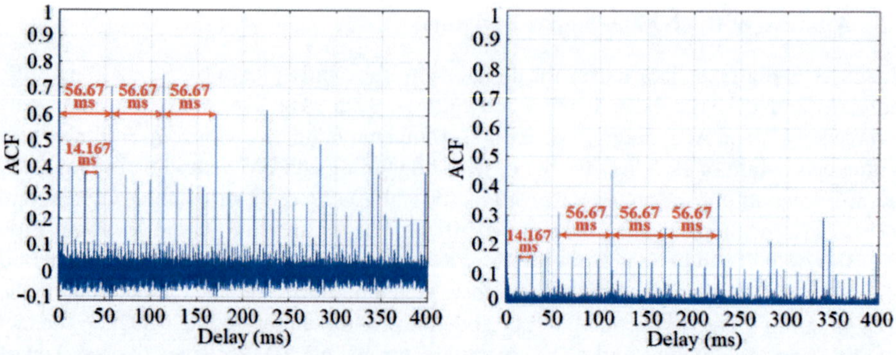

Fig. 15. TETRA ACF with FM demodulation at the band center (left) and in the case of the mismatch between the receiver and the spectrum band center, record duration being 0.8 s

5 Analysis of the Possibility of Identifying Signals by the ACF of Their Filtered Samples and Recommendations for Developing an Automatic Identification Procedure

The results of the research are presented in Table 2. It can be seen that practically each of the standards studied with the application of ACF demonstrates a periodicity with a unique period, which can be used as a basis for identification.

Table 2. Generalized results of the correlation analysis.

No.	Name of the standard	The periods of ACF bursts, ms	The duration of the sampling, s	Analysis bandwidth, kHz	Minimum allowable SNR during processing of	
					demodulated samples	complex envelopes
1	APCO	16.7	0.8	12.5	15[a]	7
2	DMR	30, 60	0.8		10	10
3	dPMR	80, 160	0.4		17	–
4	NXDN	40, 80	0.4		15	16–18
5	TETRA	14.67, 56.67	0.8		15	17–19
6	GSM	0.577, 4.615	0.5	50	5	1–3

[a] for the reliable recognition, data samples are required that exceed the sample sizes for the other standards by 4…8 times

The studies have shown that for a reliable manifestation of periodicity in the structure of the autocorrelation functions, the samples with a duration of 0.4 to 0.8 s should be used. An increase in the sampling duration has a positive effect on the results of the correlation analysis for all standards, however, an excessive increase in the duration is accompanied by a sharp increase in the complexity of calculations, therefore, as a relatively universal value, it can be recommended to use a sample that is about 0.8 s long.

The possibility of the correct recognition to what standard a particular oscillation belongs is preserved in the case of the narrowband signals even in the presence of a significant frequency detuning, i.e. when the samples correspond to only less than 50% of the spectrum of the processed oscillation. The studies have also shown that, as a rule, the capture and processing of about 40% of the spectrum bandwidth is sufficient to ensure the correct recognition of the signal belonging to its communication standard.

It is necessary to dwell separately on the estimation of signal intensity, at which the frequency of the signal structure is reliably recognized by the ACF. Since both the samples obtained from the RTL-SDR receiver and the samples from the dedicated library have been obtained at the different SNR values, an additional study of the effect of SNR on the reliability of recognition for each standard has been conducted. The essence of the study is the artificial addition of wide-band noise realizations of various intensities to complex signal samples. Examples of ACF for TETRA and NXDN standards obtained at a 15 dB SNR and shown in Figs. 16 and 17, respectively, demonstrate the differences in the quality of signal identification when the SNR is lowered.

The outcome of the study demonstrates that, at the significant SNR, both presented approaches lead to similar results, but the quality of identification ensured by different processing methods differs at low SNRs:

1) in the case of dPMR, NXDN, TETRA standards, when relying on demodulated signal samples, the ACF periodicity is reliably recognized, but when calculating the ACF by the sequences of complex envelope samples, the periodicity distinctness is noticeably reduced;
2) in the case of GSM and DMR standards, both processing methods ensure comparable identification quality;
3) in the case of the APCO standard, the use of demodulated samples results in a loss in the required SNR, but, at the same time, a considerable increase in the amounts of the samples to be processed is necessary to maintain the identification reliability.

The minimum SNR values, at which the signals of the analyzed standards are correctly identified by ACF peaks with a probability of at least 85%, are presented in the last two columns of Table 2.

It should also be taken into account that when identifying signals using FM demodulation, the ACF calculation applying the real samplings requires less resources when it is FPGA-based implemented than when a similar calculation based on complex-valued data is involved.

The established facts can be used as a starting point for the development of a procedure for automatic identification of the received signals by their belonging to communication standards. Taking into account that the communication standards analyzed in the paper are applied, as a rule, at frequencies not higher than 900 MHz, when performing radio monitoring that involves assessing the compliance of the detected narrowband radio signals with specific communication standards, the following algorithm can be recommended:

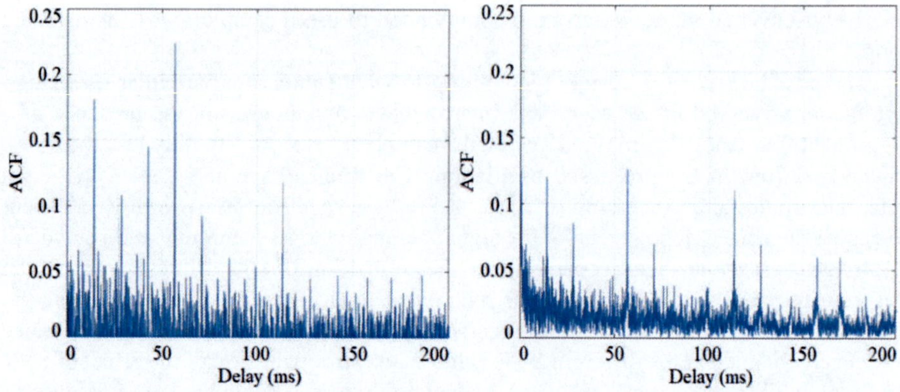

Fig. 16. TETRA signal ACF diagram while tuning to the band center, at 15 dB SNR, and with FM demodulation involved (left), and while complex envelope sampling processing (right)

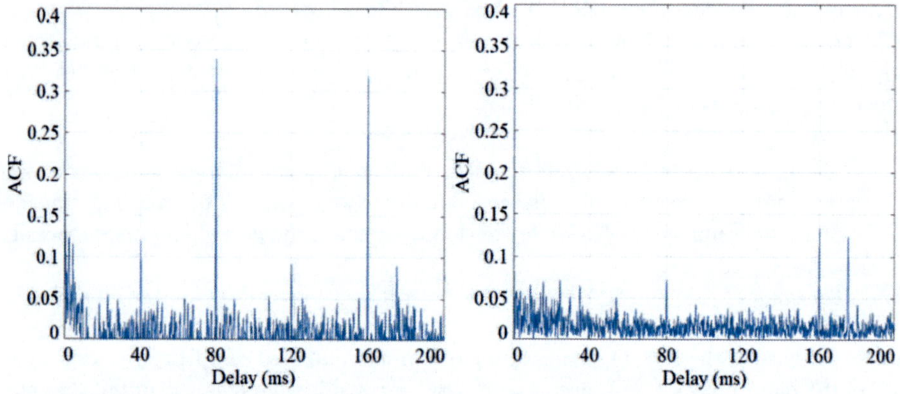

Fig. 17. NXDN signal ACF diagram while tuning to the band center, at 15 dB SNR, and with FM demodulation involved (left), and while complex envelope sampling processing (right)

a) the analyzed frequency range is divided into the bands with a width of 50 kHz and spectral bursts are searched for in each band using an energy detector. Bands in which no energy activity is detected are skipped, and steps "b"–"c" of the current algorithm are performed for the intervals occupied by radio signals;
b) for an "occupied" frequency interval, using a digital filter with a bandwidth of 50 kHz, a narrowband "single-channel" sampling is allocated, which is subjected to FM demodulation, and then the ACF is calculated for the demodulated sample sequence. If the calculated ACF displays a clear periodicity characteristic of any wideband communication standard, this fact is recorded, the analysis of the current frequency interval is completed and the transition to the analysis of the next "occupied" interval is carried out;
c) if it is not possible to determine the belonging of the signal to any standard in the 50 kHz wide band with the recorded spectral activity, then this band is additionally

divided into 4 sub-bands 12.5 kHz wide. For each of the sub-bands with an energy activity, the ACF of the corresponding demodulated samples is calculated and the search for the periodicities characteristic for narrowband communication standards is performed again. If the periodicities corresponding to the above considered standards are not yet found, then the signal non-conformity with the frame-based signal standards is recorded and a decision is made to address the "third-party" approaches to the signal identification that are not covered by the current study.

In conclusion, it is noteworthy that the use of multiple bands with a width of 12.5 or 50 kHz means that the multiple sampling frequencies can be employed, and thus the resources necessary for the recommended FPGA-based implementation procedure will be significantly reduced.

6 Conclusion

The research carried out has shown that a relatively universal algorithm can be built employing the correlation analysis of the samples of the observed radio signal complex envelopes and allowing detecting the periodicities uniquely identifying a specific standard among the set of modern narrowband communication standards. An important feature of the algorithm is its low sensitivity to receiver detuning relative to the spectrum center frequency, which makes it possible to process signals "blindly", i.e. in the absence of a priori information about their frequency "binding" and without the need for precise tuning to the frequency.

The results produced during processing real signals by the RTL-SDR receiver TERRATEC, as well as using the digital samples from the signal library of the research and production company RPC JSC IRCOS demonstrate high reliability of radio signal identification at SNRs in the signal band being no less than 5 dB in the case of GSM standard and no less than 10 dB in the case of DMR standard. For the recognition of the signals of other narrowband standards, SNR exceeding 15 dB is required, and for dPMR, NXDN, TETRA standards, at low SNR, it is advisable to use as source data the samples obtained as a result of FM demodulation of the filtered oscillations, instead of the very samples of the radio signal complex envelopes.

The version of the algorithm described in the paper does not yet provide automatic processing, but assumes decision-making with the participation of the operator. Thus, it is not allowed to recommend the given algorithm for the "as is" implementation. At the same time, the obtained results clearly indicate the possibility of obtaining a sufficiently universal algorithm for identifying radio signals with a frame structure, provided that the procedure for detecting ACF periodicities is effectively automated, i.e. at reasonable time and computational costs.

Acknowledgements. The work was financially supported by the Ministry of Education and Science of the Russian Federation (research project No. FSWF-2020-0022).

References

1. Recommendation ITU-R SM.1600-3: Technical identification of digital signals. https://www.itu.int/dms_pubrec/itu-r/rec/sm/R-REC-SM.1600-3-201709-I!!PDF-E.pdf. Accessed 15 Feb 2022
2. Spectral shape detector. https://scdn.rohde-schwarz.com/ur/pws/dl_downloads/dl_common_library/dl_news_from_rs/210/NEWS_210_english_Spectral_Shape_Detector.pdf. Accessed 15 Feb 2022
3. Studenikin, A.G., Kryzhko, I.B., Tokarev, A.B., Ashihmin, A.V., Fateev, A.A.: Algorithm for preliminary identification of radio signals with spectral masks. Syst. Control Commun. Secur. **4**, 10–39 (2021). (in Russian)
4. Signal identification guide. https://www.sigidwiki.com/wiki/Signal_Identification_Guide. Accessed 15 Feb 2022
5. Lyons, R.: Understanding Digital Signal Processing, 3rd edn. Pearson, London (2010)
6. Stewart, R.W., Barlee, K.W., Atkinson, D.S.W., Crockett, L.H.: Software-defined radio using MATLAB, Simulink, and the RTL-SDR. Strathclyde Academic Media, Glasgow (2015)
7. Rembovsky, A.M., Ashikhmin, A.V., Kozmin, V.A., Smolskiy, S.M.: Radio Monitoring: Problems, Methods and Equipment. Springer, Cham (2009)

Models and Algorithms for Analysis the Software Quality of the System of Automatic Segmentation and Pathology Analysis of the Lumbar Spine MRI Images

Alexander D. Selyutin[4](✉), V. A. Kushnikov[1,2], A. S. Bogomolov[1,2], A. F. Rezchikov[3], V. A. Ivashchenko[1], M. M. Kotyga[5], O. A. Toropova[4], S. V. Kumova[4], E. M. Kulakova[4], T. Y. Petrova[4], and M. A. Bolshelapov[4]

[1] Federal State Budgetary Institution Federal Research Centre «Saratov Scientific Centre of the Russian Academy of Sciences», 24 Rabochaya Street, Saratov 410028, Russia

[2] Saratov State University, 83 Astrakhanskaya Street, Saratov 410012, Russia

[3] V. A. Trapeznikov Institute of Control Sciences of RAS, 65 Profsoyuznaya Street, Moscow 117997, Russia

[4] Yuri Gagarin State Technical University of Saratov, 77 Politechnicheskaya Street, Saratov 410054, Russia

aseliutin@ya99.ru

[5] Russian State Social University, Wilhelm Peak Street, Moscow 129226, Russia

Abstract. This article describes a method to improve the work efficiency of the software complex for automatic segmentation and analysis of lumbar spine MRI images in conditions of dynamic changes. For the modules of the software complex, an event tree and a state graph developed by authors, which will allow taking into account the conditions leading to software complex failures. These conditions with a set of certain factors at different time intervals can break down all modules of the software complex. The developed models can reduce the program errors, debugging time and recovery of a non-functioning system. A system of Kolmogorov-Chapman differential equations based on the state graph. For the system of differential equations, a utility software developed via the Python programming language and SciPy library, which allows estimating the level of probability of failure of system modules over a certain time. The behavior of the software complex modelled under various initial conditions. As result of the solution of the differential equations system, the influence of the intensities of recovery and failure of elements on the operability of the system modules investigated. For the various initial states of the system, an analysis, based on which recommendation rules created for the management personnel of the software complex, which will avoid a critical combination of events leading to a system failure.

Keywords: Web services · Fault tree · Convolutional neural network · State graph · Kolmogorov-Chapman equation

R. Silhavy (Ed.): CSOC 2022, LNNS 501, pp. 443–453, 2022.
https://doi.org/10.1007/978-3-031-09070-7_37

1 Introduction

Intervertebral disc degeneration is an age-related disease associated with chronic back pain and functional disability. Pathology affects a significant part of the population and entails negative socio-economic consequences. The expenses of the Russian Ministry of Health for the treatment of degenerative diseases of the spine estimated at 2.6 billion rubles per year. In a study [2] devoted to the correlation between changes displayed on MRI in the sagittal plane and degeneration of intervertebral discs, it was revealed that the progression of the degeneration process includes both structural disorders and cell-mediated changes in the composition of intervertebral discs.

To build a model of the spine, for its further analysis for the presence of pathologies, various methods used, while magnetic resonance imaging (MRI) is the most common method for assessing degeneration of intervertebral discs.

Today, in most polyclinics, the work process of a radiologist analyzing MRI images not optimized. Radiologist make manually analyze many images for the presence of pathologies in patients. The main problem that arises during a manual analysis of MRI images is the human factor, which does not allow taking into account all the existing pathologies in patients, which presented in the article devoted to medical errors of radiologists based on statistical analysis. Manual analysis is time-consuming and interferes with the patient's prompt decision for surgery or the appointment of a course of treatment. In addition, the radiologist performs many other duties, including creating a report with a diagnosis of the condition of the patient's spine.

Another problem is the variability of images. MRI images taken from different devices do not have the same characteristics, which leads to a difference in the projections of the images, as described in [5] devoted to the analysis of degenerative intervertebral discs based on T2 weighted MRI images. These differences can lead to an incorrect diagnosis. For example, an MRI image using a single device can show the presence of deformation of the upper boundaries of the intervertebral discs. Meanwhile, there is no guarantee that this deviation will be present when scanned by another MRI machine.

Based on the above-described problems of the subject area, we automated the work processes of radiologists by developing the lumbar-spine information system, which performs the following functions:

- keeping records of patients and their medical history;
- automatic segmentation of T2 weighted MRI images of the lumbar spine;
- analysis of segmented anatomical bodies by determining spatial parameters and forming probabilistic characteristics of the presence of hernias for each intervertebral disc;
- generation of reports and statistics on patient examinations;
- editing of MRI images with interactive visualization tools;
- personal blogging by radiologists;
- training of novice radiologists.

It follows from the description of the system that for the full functioning of the software, it is necessary to ensure an appropriate level of continuous functioning of the software solution. Thus, the work aims to estimate the probabilities of failure of the project's web services by modelling a tree of failure events and implementing a probability estimate of a 5-element cross-section for a parallel recovery strategy on a state graph.

2 Subsystems of the Software Suite

To ensure independence from the equipment, as well as the possibility of functioning, including on insufficiently reliable equipment, the system under study designed and implemented using distributed technologies. The software solution is a set of interconnected web applications with a parent service for post-processing MRI images. To ensure the reliability of patient data storage and guarantee the efficiency of calculations, the software solution works in the infrastructure of the AWS public cloud provider.

The software suite includes the subsystems:

- patient data accounting module and automatic post-processing of MRI images;
- radiologist personal blogging subsystem;
- aspiring radiologists e-learning system;
- interactive MRI images editing system.

Figure 1 shows the diagram of the subsystems.

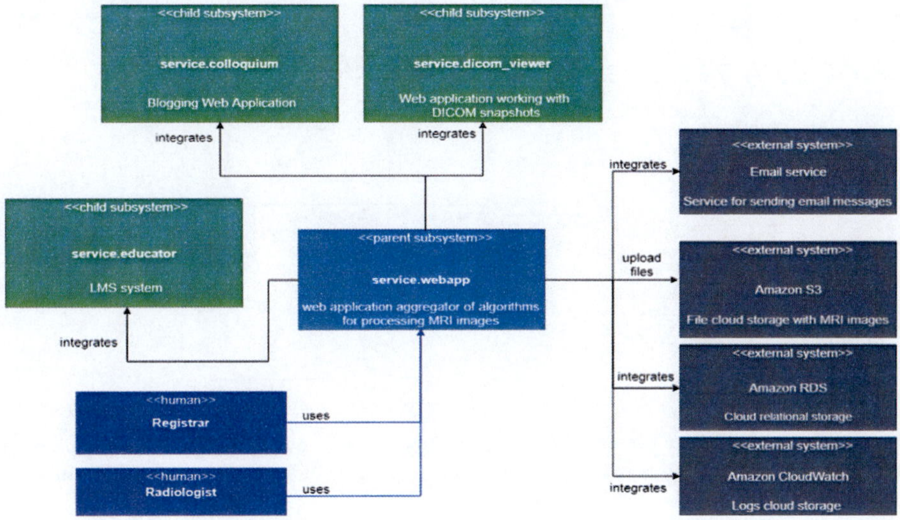

Fig. 1. Context diagram of the software solution

3 Probability Estimation of Failure

Failure trees use to analyze factors that may contribute to the onset of system failure [4]. The reasons that may affect the occurrence of an undesirable situation are determined deductively and presented graphically in the form of a tree diagram that depicts the relationship of the causes with the main event.

The failure event tree for each of the lumbar-spine subsystems described above shown in Fig. 2.

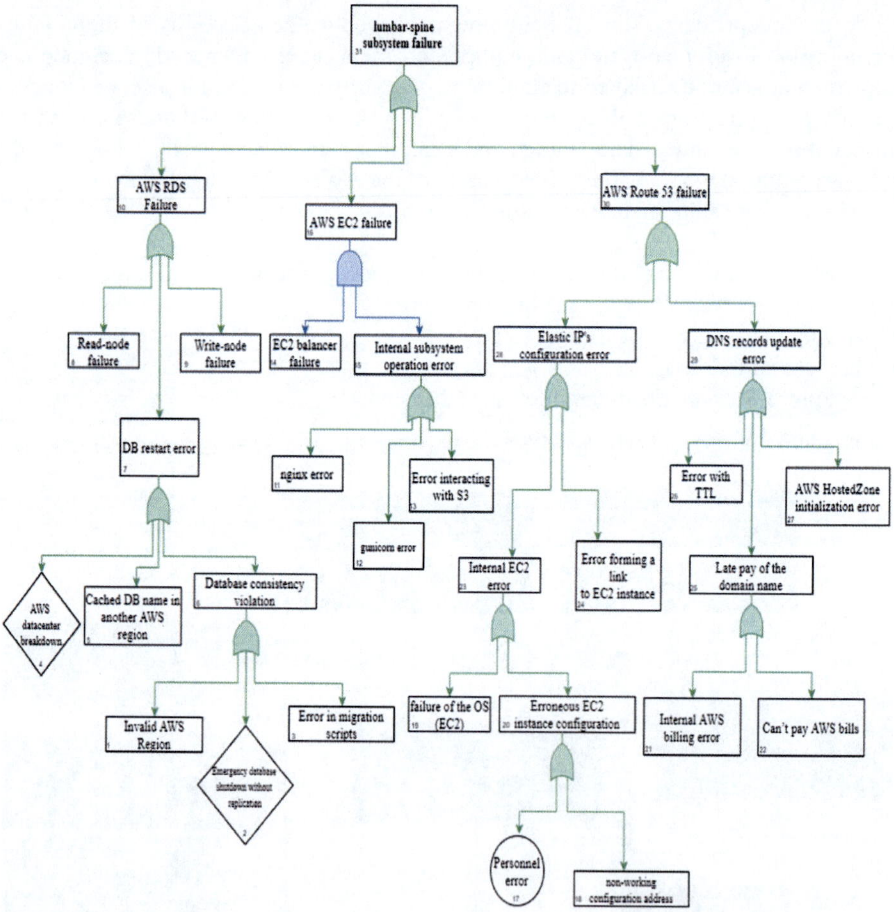

Fig. 2. The tree of failure events for the lumbar-spine system (logical disjunction is indicated in green; logical conjunction is indicated in blue; the rectangles indicate the nodes of the software package that can cause the system to fail)

When analyzing changes in the behavior of a system during its operation, it is also convenient to use a state graph. The state graph is a directed graph whose vertices represent various states of the system, and the edges reflect possible transitions of the system from one state to another with parameters of failure rates and recoveries [3].

From the analysis of the fault tree, it concluded that the state graph contain the following vertices:

- 0 - the subsystem is working correct;
- 1 - invalid AWS Region;
- 2 - emergency shutdown of the database without replication;
- 3 - error in migration scripts;
- 4 - AWS datacenter breakdown;
- 5 - cached DB name in another AWS region;
- 6 - database consistency violation;
- 7 - DB restart error;
- 8 - read-node failure;
- 9 - write-node failure;
- 10 - AWS RDS Failure;
- 11 - nginx error;
- 12 - gunicorn error;
- 13 - error interacting with S3;
- 14 - EC2 balancer failure;
- 15 - internal subsystem operation error;
- 16 - AWS EC2 failure;
- 17 - management personnel error;
- 18 - non-working configuration address;
- 19 - failure of the OS (EC2);
- 20 - erroneous EC2 instance configuration;
- 21 - internal AWS Billing error;
- 22 - can't pay AWS bills;
- 23 - internal EC2 error;
- 24 - error forming a link to EC2 instance;
- 25 - late payment of the domain name;
- 26 - error in TTL;
- 27 - AWS HostedZone initialization error;
- 28 - elastic IP's configuration error;
- 29 - DNS records update error;
- 30 - AWS Route 53 failure;
- 31 - lumbar-spine subsystem failure.

For the 5-element cross section, the state graph shown in Fig. 3. Therefore λ_i – a failure rate of i-th node, μ_i - recovery intensity of i-th node.

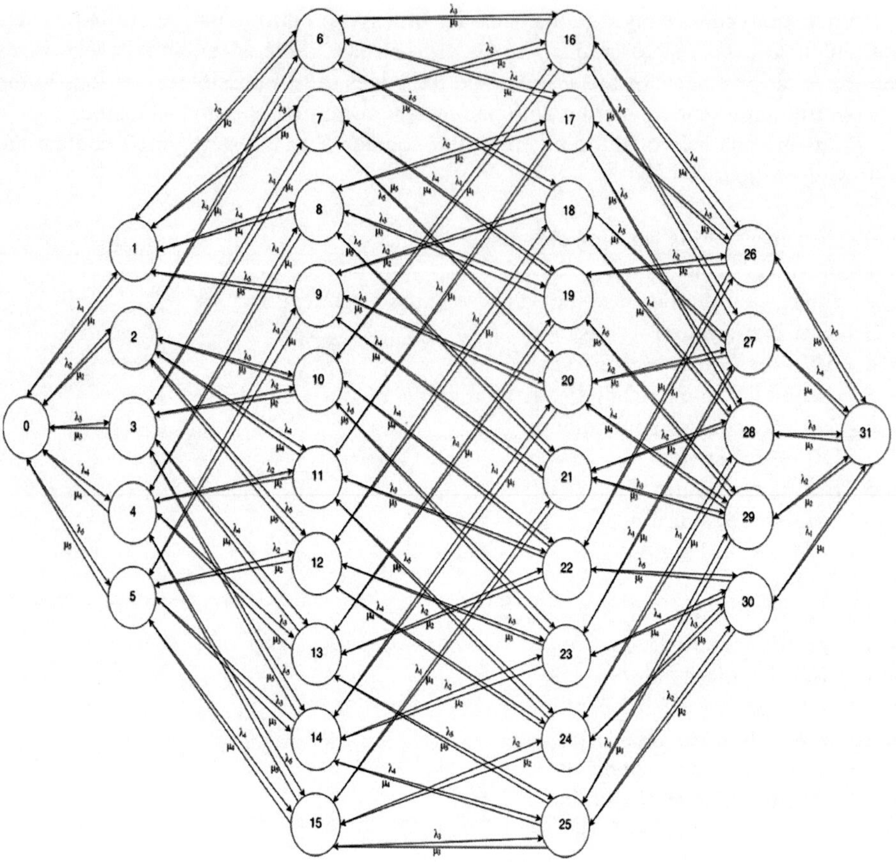

Fig. 3. State graph

Based on the graph under consideration, a system of Kolmogorov-Chapman differential equations compiled, having the form [1]:

$$\begin{cases} \dfrac{dP_0(t)}{d(t)} = -(\lambda_1 + \lambda_2 + \lambda_3 + \lambda_4 + \lambda_5) \cdot P_0(t) + \mu_1 P_1(t) + \mu_2 P_2(t) + \mu_3 P_3(t) \\ \qquad + \mu_4 P_4(t) + \mu_5 P_5(t); \\ \frac{dP_1(t)}{d(t)} = \lambda_1 P_0(t) - (\lambda_2 + \lambda_3 + \lambda_4 + \lambda_5 + \mu_1) \cdot P_1(t) + \mu_2 P_6(t) + \mu_3 P_7(t) \\ \qquad + \mu_4 P_8(t) + \mu_5 P_9(t); \\ \vdots \\ \frac{dP_{31}(t)}{d(t)} = \lambda_1 P_{30}(t) + \lambda_2 P_{29}(t) + \lambda_3 P_{28}(t) + \lambda_4 P_{27}(t) + \lambda_5 P_{26}(t) \\ \qquad - (\mu_1 + \mu_2 + \mu_3 + \mu_4 + \mu_5) \cdot P_{31}(t), \end{cases} \tag{1}$$

4 Software for Solving a System of Kolmogorov-Chapman Differential Equations

System of differential equations solve using the Python programming language using the SciPy library [10, 11]. For the developed script for solving a system of equations, a utility with a graphical interface using the PyQt5 library was developed [9]. Figure 4 shows a screenshot of the developed utility.

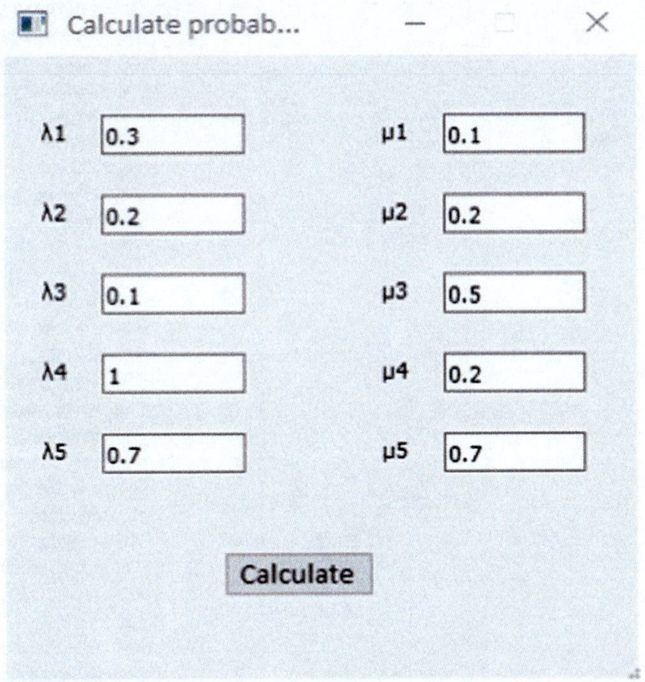

Fig. 4. Screenshot of the utility for estimating changes in the probabilities of implementing a five-element section for a parallel recovery strategy

5 System States Modelling

Solving the system under the conditions:

$$\text{nodes/day}\begin{cases}\lambda_1 = 0.3,\ \mu_1 = 0.2;\\ \lambda_2 = 1.1,\ \mu_2 = 1;\\ \lambda_3 = 0.4,\ \mu_3 = 0.2;\\ \lambda_4 = 2.34,\ \mu_4 = 2.65;\\ \lambda_5 = 2.7,\ \mu_5 = 2.86,\end{cases} \tag{2}$$

We obtain a graph of solutions to a system of differential equations (Fig. 5), where the probability of finding the system in a state of full operability is higher than the probability of any other state under mixed conditions of the ratio of recovery intensities to node failure rates.

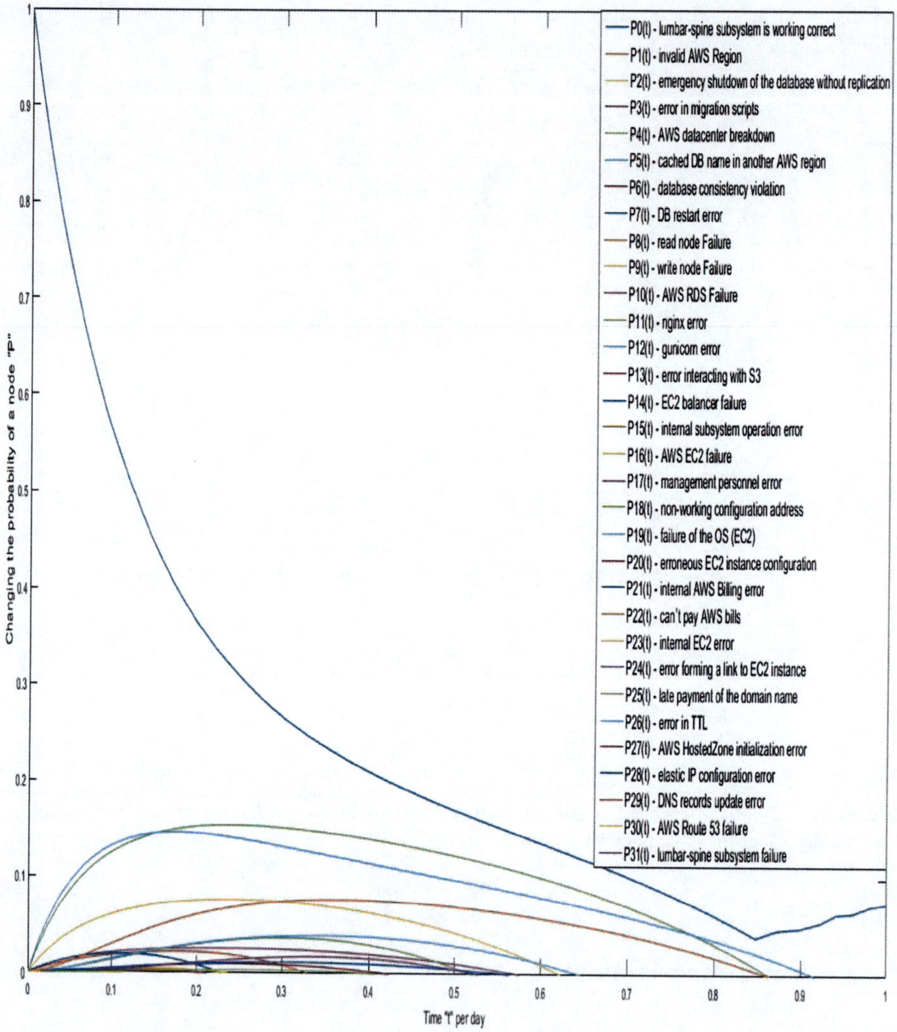

Fig. 5. A graph of changes in the probabilities of implementing a five-element cross-section for a parallel recovery strategy

Similarly, modelling the behavior of the system for different cases of lumbar-spine failure, Table 1 compiling, which presents the parameters of the probabilities of system nodes failure during t.

Table 1. Simulated values of failure/recovery of system nodes per day

	λ_1	μ_1	λ_2	μ_2	λ_3	μ_3	λ_4	μ_4	λ_5	μ_5
Case 1	0.20	9.00	0.10	9.00	0.30	10.00	0.30	7.00	0.30	8.00
Case 2	9.00	2.00	6.30	3.42	8.34	2.20	6.50	2.56	3.30	3.50
Case 3	2.20	1.70	2.10	1.70	2.00	1.80	2.00	1.90	2.00	1.90
Case 4	3.20	3.10	3.30	3.50	2.90	3.50	3.00	3.10	3.00	3.30
Case 5	1.00	0.90	1.50	1.10	0.90	1.00	2.80	2.50	0.50	0.70
Case 6	0.20	1.20	0.10	1.80	0.20	1.20	0.50	1.35	0.80	1.77
Case 7	2.20	4.50	2.30	4.50	2.40	4.50	2.50	4.50	2.40	4.50
Case 8	0.35	2.00	0.35	2.00	0.35	2.00	0.35	2.00	0.35	2.00
Case 9	5.70	3.10	5.10	3.60	5.50	3.30	5.35	3.70	5.00	3.58
Case 10	4.60	1.11	4.60	2.22	4.60	1.12	4.60	2.24	4.60	2.25

A petal diagram forming, shown in Fig. 6, which describes the probability of a system failure due to a critical combination of events, based on data from Table 1.

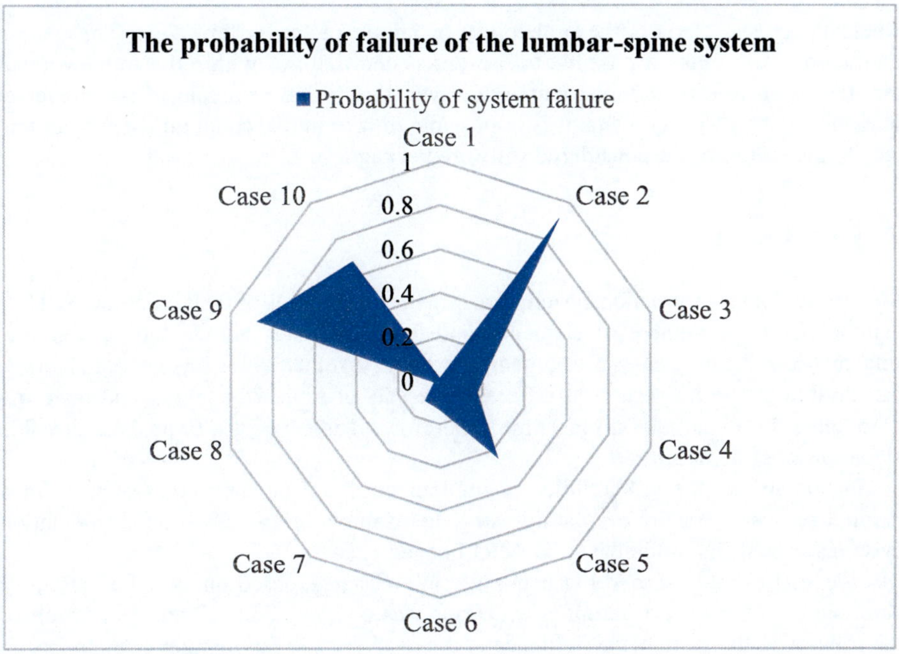

Fig. 6. Graph of the probability of system failure due to a critical combination of events

6 Analysis of System Failure Probabilities

Based on the above data on the probabilities of system failure due to a critical combination of events, an analysis and production rules were developed [6] for the management personnel of the lumbar-spine software complex. They will allow system administrators to respond promptly and promptly to changes in the state of the complex, thereby avoiding failures in applications. Production rules compiling for the 20 most frequently occurring states of the system (the state of the system means the ratio of the values of the failure/recovery of system nodes per day). The frequency of states was determined during the operation of the system.

Due to the flexibility of the AWS resource configuration (starting from the number of DB replicas [7, 8], ending with the choice of the processor architecture of the webserver), exceptional situations can be easily prevented, thanks to the friendly graphical web interface provided by the cloud provider.

For example, for the values of the output/recovery of system nodes, which present in (2), such production rules as:

$$\begin{cases} IF\ P_1 > 0.5\ AND\ t > 0.7, THEN\ \text{"reset S3 bucket cache";} \\ IF\ P_2 > 0.75, THEN\ \text{"reduce the GCD by 1 in EC2";} \\ \vdots \\ IF\ P_{30} > 0.43, THEN\ \text{"restart AWS RDS DB manually",} \end{cases} \tag{3}$$

where P_i $(0 < i < 31)$ is the probability of failure of the system node. Thus, these production rules make it possible to prevent system failures within the framework of the developed tree of failure events. However, due to the principle of the universal interconnection of phenomena, it is impossible to take into account all the factors that lead to the failure of the considered software package.

7 Conclusion

Modern software information systems are large-scale in terms of their complexity. They consist of a large number of elements, and these elements include subsystems, etc. Information systems necessarily contain errors that, together with a set of certain factors, can disable the entire system at different intervals of time. Debugging and restoring a broken system can take a long time. Therefore, it is necessary to analyze possible situations of system failure.

In this article, the probabilities of implementing a 5-element cross-section for a parallel recovery strategy evaluating using the example of an automatic segmentation system and analysis of lumbar spine MRI images.

The analysis was carried out using a graph of states, based on which a system of Kolmogorov-Chapman differential equations was obtained and solved. As a result of the solution of the system, the influence of the recovery and failure rates of elements on the operability of the system modules was investigated, and a utility was developed that allows analyzing changes in the probabilities of the system.

Production rules compiling for the management personnel of the software complex, which will allow responding promptly to changes in the state of the complex. Because avoiding failures in applications within the framework of the developed tree of events-failures, but due to the principle of the universal interconnection of phenomena, it would be impossible to take into account all the factors that lead the system to fail.

References

1. Ali, M., et al.: A data-driven knowledge acquisition system: an end-to-end knowledge engineering process for generating production rules. IEEE Access **6**, 15587–15607 (2018)
2. Berven, S., Wadhwa, R.: Sagittal alignment of the lumbar spine. Neurosurg. Clin. N. Am. **29**, 331–339 (2018)
3. Fominykh, D.S., Bogomolov, A.S., Ivashchenko, V.A., Kushnikov, V.A., Rezchikov, A.F., Filimonyuk, L.Y.: The problem of minimizing the probability of accidents at welding in robotized technological complexes. Vestnik Astrakhan State Tech. Univ. Ser. Manage. Comput. Sci. Inform. 21–30 (2017)
4. Helmer, G., Wong, J., Slagell, M., Honavar, V., Miller, L., Lutz, R.: A software fault tree approach to requirements analysis of an intrusion detection system. Requirements Eng. **7**, 207–220 (2002)
5. Kumar, M., Baklanov, A., Chopin, D.: Correlation between sagittal plane changes and adjacent segment degeneration following lumbar spine fusion. Eur. Spine J. **10**(4), 314–319 (2001). https://doi.org/10.1007/s005860000239
6. Nagamoto, Y., Okuda, S., Matsumoto, T., Sugiura, T., Takahashi, Y., Iwasaki, M.: Multiple-repeated adjacent segment disease after posterior lumbar interbody fusion. World Neurosurg. **121**, e808–e816 (2019)
7. Oliinyk, A., Skrupsky, S., Subbotin, S.: Experimental research and analysis of complexity of parallel method for production rules extraction. Autom. Control. Comput. Sci. **52**, 89–99 (2018)
8. Shcherbakov, M., Kushnikov, V.: Models and algorithms of the emergency management system in glass production. Saratov State Tech. Univ. Vestnik **1**(55), 196–200 (2011)
9. Veshneva, I., Chernyshova, G., Bolshakov, A.: Regional competitiveness research based on digital models using Kolmogorov-Chapman equations. In: Kravets, A.G., Bolshakov, A.A., Shcherbakov, M. (eds.) Society 5.0: Cyberspace for Advanced Human-Centered Society. SSDC, vol. 333, pp. 141–154. Springer, Cham (2021). https://doi.org/10.1007/978-3-030-63563-3_12
10. Yandybayeva, N., Kozhanova, E., Kushnirov, V.: Development of a software product for determining the effectiveness of higher education institutions. Saratov State Tech. Univ. Vestnik **1**(75), 214–219 (2014)
11. Yazdi, M., Nikfar, F., Nasrabadi, M.: Failure probability analysis by employing fuzzy fault tree analysis. Int. J. Syst. Assur. Eng. Manage. **8**, 1177–1193 (2017). https://doi.org/10.1007/s13198-017-0583-y

Coherent Demodulation of the Two-Level Amplitude Phase-Shift Keyed Signals: Noise Immunity Issue

Oleg Chernoyarov[1,2,3](✉), Alexey Glushkov[4], Vladimir Litvinenko[5], Maksim Maksimov[6], and Alexandra Salnikova[1]

[1] National Research University "MPEI", Krasnokazarmennaya Street 14, 111250 Moscow, Russia
chernoyarovov@mpei.ru
[2] National Research Tomsk State University, Lenin Avenue 36, 634050 Tomsk, Russia
[3] Maikop State Technological University, Pervomaiskaya Street 191, 385000 Maikop, Russia
[4] Zhukovsky-Gagarin Air Force Academy, Marshal Nedelin Street 133, 394052 Voronezh, Russia
[5] Voronezh State Technical University, Moscow Avenue 14, 394026 Voronezh, Russia
[6] Plekhanov Russian University of Economics, Stremyanny Lane, 117997 Moscow, Russia

Abstract. There are considered the fast algorithms for the coherent demodulation of the two-level amplitude-phase shift keyed signals (APSK signals) and for the same operation but accompanied by the estimation of the received symbol amplitude via its relative comparison with the preceding symbol amplitude (APSK-RA signals). A great advantage of these algorithms is that they require a minimum number of arithmetic operations per signal period. The phase demodulation is implemented based on the phase detector of multi-level phase keyed signals. The symbol amplitudes are determined by the quadrature channels responses. It is shown that the introduced demodulation algorithms can be implemented in both analog and digital versions. Then one more benefit from the proposed procedure is demonstrated: there is no need for the received signal normalization by its constellation during the APSK-RA signal demodulation implementation. The probabilities of the erroneous demodulation of the received symbol phase and amplitude are estimated. The comparative analysis of the noise immunity of the two-level APSK and APSK-RA signals for the two variants of signal constellations is performed. Statistical simulation of the proposed demodulation algorithms is carried out. #CSOC1120.

Keywords: Amplitude-differential phase shift keying · Signal demodulation · Coherent processing · Bit-error probability · Noise immunity · Statistical simulation

R. Silhavy (Ed.): CSOC 2022, LNNS 501, pp. 454–468, 2022.
https://doi.org/10.1007/978-3-031-09070-7_38

1 Introduction

The amplitude phase-shift keyed (APSK) signals [1] are widely used in mobile radio networks [2], in digital TV [3] and in optical cable communication lines [4, 5]. They are characterized by a high noise immunity and information delivery speed, but they also require a demodulator phase locking and a strict control of the received signal amplitude in order to be able to compare it with the preset threshold values. Signal positions are presented as the "constellation" in the form of the concentric circles where the dots are the images of the initial phases of the received symbols, while the circle radius corresponds to their amplitude. In the case of the two-level amplitude phase-shift keying, the constellation is composed of the two circles. On the first of them with the amplitude (radius) U_1, M_1 signal positions are situated, while on the second, with the greater amplitude - $U_2 > U_1$ - one finds M_2 signal positions, so that the general number of the signal positions is expressed as

$$M = M_1 + M_2, \quad (1)$$

To describe the signal, one uses the following notations: MAPSK (8APSK, 16APSK) or $M_1 + M_2$ APSK (that are 4+4APSK, 8+8APSK, 4+12APSK). The choice of the constellation is made in terms of noise immunity, power efficiency and signal modulation/demodulation comfort. The 4+12APSK signal is utilized in digital TV [3], while the 4+4APSK and 8+8APSK signals are very suitable for use in optic cable communication lines [4, 5]. The different corresponding constellation choices possible are discussed in [6–8]. The symbol amplitude ratio in the two-level APSK is $U_2/U_1 \approx 3$, and it depends on the internal coding speed [9].

APSK signal demodulation presupposes determining the normalized amplitude and phase of the received signal and their comparison with the coordinates of the decision making regions for the corresponding constellation, see, for example, [10, 11]. In the simpler case [6, 8], when one deals with the 8+8APSK signal, one should separately compare the two possible values of the symbol amplitude with the threshold value, and only then the signal phase is determined. For this purpose, the received signal level should be normalized using the automatic gain control of the receiver. If the noise is present, symbol amplitude determination errors may result in decreasing noise immunity.

In practice, when determining symbol amplitude, one should find out, if it belongs to the lower or the upper level of the APSK signal (the numerical values of these levels are then neglected). In this case, precise normalization of the received signal gain is not required. Below, the fast and technically simple algorithm for the APSK signal demodulation and the received symbol amplitude estimation via its relative comparison with the amplitude of the preceding symbol (APSK-RA demodulation) is introduced. It is shown that it provides high noise immunity and can be effectively implemented on modern element base.

2 Amplitude Phase-Shift Keyed Signal

The received element of two-level APSK signal can be presented as

$$s(t) = S_m \cos(2\pi f_0 t + \psi_{mn} + \psi_0). \tag{2}$$

In (2), the notations are the following: S_m is the amplitude with $m = 1$ or $m = 2$ being the numbers of the signal gains, it can take on the values S_1 or S_2 (for example, $S_2 = 3S_1$), f_0 is the carrying frequency, ψ_0 is the initial phase. And ψ_{mn} is the signal phase:

$$\psi_{mn} = a_{mn} 2\pi / M_m, \quad a_{mn} = 0,\ 1,\ \ldots,\ (M - 1),$$

where n is the number of the signal value of the phase for the m-th level, and the general number of signal positions is determined according to (1).

Under the coherent demodulation, the phase-locking device in an operation mode provides the condition $\psi_0 = 0$.

3 Demodulation Algorithms

For the demodulation of the APSK signal, it is necessary to determine the amplitude S_k and the initial phase ψ_k of the received k-th symbol and to compare them with the decision making regions of the preset constellation. In Fig. 1a, one can see the 4+12APSK signal constellation ($M_1 = 4$, $M_2 = 12$), while in Fig. 1b the 8+8APSK signal constellation is shown ($M_1 = M_2 = 8$). In both cases, U_1 and U_2 are the normalized symbol amplitudes and U_0 is the threshold level. Signal positions are presented as dots; the thickening lines stand for the decision making regions for the received positions that can be described within the quadrature coordinates I, Q or by polar coordinates U, ψ for which the amplitudes of the received symbols of the APSK signals should be normalized.

APSK-RA signal demodulator is designed based on the phase [12] and amplitude [13] detectors, and its block diagram is demonstrated in Fig. 2. The input signal $s(t)$ arrives to the quadrature processing unit (QPU) [1], by the signal synchronous with $s(t)$ that is $s_0(t) = S_0 \cos(2\pi f_0 t)$ coming from the reference generator (RG) and, with the help of 90° phase shifter (PS), the multipliers and the integrators, the responses $y_0(t)$ and $y_1(t)$ are calculated at the T symbol length interval as follows:

$$y_0(t) = \int_{t-T}^{t} S_m \cos(2\pi f_0 t' + \psi_{mn}) S_0 \cos(2\pi f_0 t')\, dt', \tag{3}$$

$$y_1(t) = \int_{t-T}^{t} S_m \cos(2\pi f_0 t' + \psi_{mn}) S_0 \sin(2\pi f_0 t')\, dt'. \tag{4}$$

phase shifter PS in the quadrature channel.

After the reception of the current symbol is completed, the responses (3) and (4) of the quadrature processing unit undergoing integration at the interval $T = NT_0$ (there $T_0 = 1/f_0$ is the carrier period) are equal to

$$y_0(kT) = y_0 = (S_m S_0/2) \cos \psi_{mn}, \quad y_1(kT) = y_1 = (S_m S_0/2) \sin \psi_{mn}. \tag{5}$$

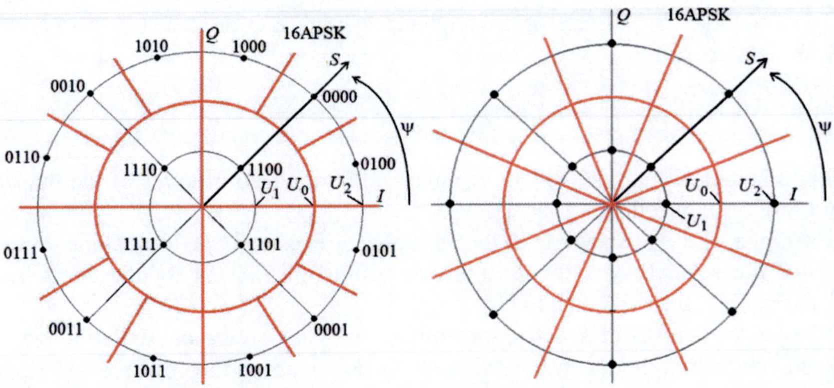

a) b)

Fig. 1. The signal constellations: a) 4+12APSK; b) 8+8APSK

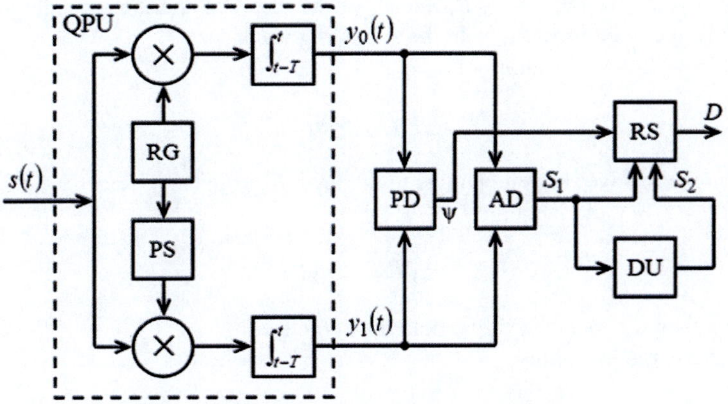

Fig. 2. The block diagram of the demodulator

Under the digital demodulation implementation, according to [12, 13], the corresponding responses of the quadrature channels are determined as

$$y_0(kT) = y_0 = 2NS_m \cos\psi_{mn}, \quad y_1(kT) = y_1 = 2NS_m \sin\psi_{mn}. \tag{6}$$

In (5), (6), k is the number of the received symbol.

By the values found in (5) or (6), at the output of the phase detector (PD), the estimate of the initial phase ψ of the input APSK signal symbol is generated according to rule:

$$\psi_i = \begin{cases} -\arctan(y_1/y_0), & \text{if } y_0 \geq 0, \\ \pi - \arctan(y_1/y_0), & \text{if } y_0 < 0. \end{cases} \tag{7}$$

At the same time, the estimate of the symbol amplitude S is produced in the amplitude detector (AD) as follows:

$$S = \sqrt{y_0^2 + y_1^2}. \tag{8}$$

It should be noted that the estimate (8) must be normalized in terms of the threshold level U_0 of the constellation (Fig. 1).

Examples of the realizations of the normalized responses (3), (4) and the time diagrams of the normalized symbol amplitude estimate S/S_1 (7) for the constellation depicted in Fig. 1b can be seen in [14].

The values (7), (8) of S and ψ determine the point on the constellation, and the decision concerning the received position is made regarding the constellation element closest to this point.

As such a kind of an algorithm of decision-making concerning the level of the received symbol that does not require this symbol to be normalized and compared with the threshold (i.e., the demodulator implementation is simplified considerably) but uses the ratio between the amplitudes S_1 and S_2 of the received and the preceding symbols, respectively, it can be formulated in the following way:

– if at $a_1 \approx 2$ the inequality

$$S_1/S_2 \geq a_1 \tag{9}$$

holds, then the received signal belongs to the level U_2 (Fig. 1);

– if at $a_2 \approx 1/2$ the inequality

$$S_1/S_2 \leq a_2, \tag{10}$$

is satisfied, then the received signal belongs to the level U_1;

– if there is the inequality

$$a_2 < S_1/S_2 < a_1, \tag{11}$$

then the level of the received signal coincides with the level of the preceding signal.

The values of the coefficients a_1 and a_2 can undergo optimization in the vicinity of the specified values according the recommendations specified in [9].

It is practical to rewrite the conditions (9)–(11) in the following way:

$$S_1 - a_1 S_2 \geq 0, \;\; S_1 - a_2 S_2 \leq 0, \;\; S_1 - a_1 S_2 < 0 \,\text{and}\, S_1 - a_2 S_2 > 0. \tag{12}$$

Thereat, it should be noted that in the digital devices, the multiplication by $a_1 = 2$ and $a_2 = 1/2$ can be carried out by the simple shift in word size of the binary codes to the left or right, respectively.

To implement the algorithm (12), the preceding value of the amplitude S_2 has to be saved into the delay unit (DU) (Fig. 2), while the values of S_1 and S_2 together with the value of ψ should be transferred to the resolver (RS), which applying (12) generates the code D for the received symbol.

4 Phase Demodulation Error Probability

If the signal (2) is distorted by the Gaussian white noise, then, for the received symbol with the minimal amplitude S_1 (that is U_1 in the constellations in Fig. 1), the ratio between the signal power and the noise power at the output of the demodulator presented in Fig. 2 (the output signal-to-noise ratio (SNR)) is the following:

$$h_0^2 = BS_1^2 \Big/ \sigma_n^2. \tag{13}$$

Here σ_n^2 is the dispersion (mean power) of the Gaussian noise within the signal bandwidth Ω_0 while $B = T\Omega_0$ is the signal base. The digital implementation of the demodulator according to [12, 13] presupposes the generation of the two samples in each quadrature channel during the signal period. Thus, in this case, the output SNR is determined as [12, 13]

$$h_0^2 = NS_1^2 \Big/ \sigma_n^2, \tag{14}$$

For the symbols with the higher amplitude (when $U_2 = 3U_1$ as in Fig. 1), the SNR is higher than 9.5 dB, and the phase demodulation error probability is substantially lower in this case. That is why, its approximate estimate may be carried out using the expression describing the multi-level PSK signal reception error probability, the number of phase values being M_1.

In [1], one can find the expression for the symbol error probability during the optimal coherent demodulation of the M-level PSK (MPSK) signals in the channel with Gaussian noise and at the SNR denoted as h:

$$P_{PSK}(h, M) = 2Q\Big(\sqrt{2}h \sin(\pi/M)\Big),$$

where $Q(x) = \int_x^{\infty} \exp(-t^2/2)\, dt \Big/ \sqrt{2\pi}$. Consequently, the approximate expression for the phase demodulation error when receiving the $M_1 + M_2$ APSK signal is

$$P_P(h) \approx M_1 P_{PSK}(h, M_1) \big/ (M_1 + M_2). \tag{15}$$

5 Symbol Amplitude Determination Errors

It is presupposed that the condition of high a posteriori accuracy [15] is satisfied, i.e. $h_0 >> 1$ (13), (14). Then, at the end of the symbol, one can hold that the received symbol amplitude S, while its real value is S_m, is the Gaussian random value with the probability densities

$$w_m(S) = \exp\Big[-(S - S_{mmean})^2 \Big/ 2\sigma_n^2\Big] \Big/ \sqrt{2\pi}\sigma_n, \quad m = 1, 2,$$

where $S_{1mean} = S_1$, $S_{2mean} = S_2$. Taking into account (14), one gets

$$w_m(S) = h_0 \exp\big[-h_0^2 (S - S_{mmean})^2 / 2N^2 S_{1mean}^2\big] \Big/ NS_{1mean}\sqrt{2\pi}.$$

When the amplitude of the received symbols is ideally normalized, then, for the constellations in Figs. 1a and 1b, it can be taken as $S_{1mean} = U_1$, $S_{2mean} = U_2$ with the threshold $U_0 = (U_1 + U_2)/2$. A a result, the symbol amplitude demodulation error probability in the case of Fig. 1a is

$$P_{A0} = \frac{1}{4}\int_{U_0}^{\infty} w_1(S)\,\mathrm{d}S + \frac{3}{4}\int_{-\infty}^{U_0} w_2(S)\,\mathrm{d}S, \tag{16}$$

while in the case of Fig. 1b one gets

$$P_{A0} = \frac{1}{2}\int_{U_0}^{\infty} w_1(S)\,\mathrm{d}S + \frac{1}{2}\int_{-\infty}^{U_0} w_2(S)\,\mathrm{d}S. \tag{17}$$

With a relative estimate of the symbol amplitude, according to (9)–(11), the decision is made based on the ratio

$$V = S_1/S_2, \tag{18}$$

between the estimates of the amplitudes of the received and the preceding symbols, that are S_1 and S_2, respectively. If they are independent, the probability density of the random variable V (18) can be defined by the expression from [16] as follows:

$$u(V) = \int_{-\infty}^{\infty} u_2(y)\, u_1(Vy)\, |y|\, \mathrm{d}y, \tag{19}$$

where $u_1(y)$ and $u_2(y)$ are the probability densities of the numerator and the denominator, respectively.

There are 4 possible variants of the amplitude values S_1 and S_2:

1) the amplitude value S_2 corresponds to the level U_1, and the amplitude value S_1 – to the level U_2 so that the probability density (19) is

$$u(V) = u_{12}(V) = \int_{-\infty}^{\infty} w_1(y)\, w_2(Vy)\, |y|\, \mathrm{d}y;$$

2) the amplitude value S_2 corresponds to the level U_2, and the amplitude value S_1 – to the level U_1 and therefore

$$u(V) = u_{21}(V) = \int_{-\infty}^{\infty} w_2(y)\, w_1(Vy)\, |y|\, \mathrm{d}y;$$

3) the amplitude values S_1 and S_2 corresponds to the level U_1 and so

$$u(V) = u_{11}(V) = \int_{-\infty}^{\infty} w_1(y)\, w_1(Vy)\, |y|\, \mathrm{d}y;$$

4) the amplitude values S_1 and S_2 corresponds to the level U_2, transforming from (19) into

$$u(V) = u_{22}(V) = \int_{-\infty}^{\infty} w_2(y)\, w_2(Vy)\, |y|\, dy.$$

Thus, for the symbol amplitude determination error probabilities, the following expressions are valid:

1) the preceding symbol corresponds to the level U_1, while the received – to the level U_2:

$$p_{12} = \int_{-\infty}^{a_1} u_{12}(V)\, dV;$$

2) the preceding symbol corresponds to the level U_2, while the received – to the level U_1:

$$p_{21} = \int_{a_2}^{\infty} u_{21}(V)\, dV;$$

3) both the preceding and the received symbols correspond to the level U_1:

$$p_{11} = 1 - \int_{a_2}^{a_1} u_{11}(V)\, dV;$$

4) both the preceding and the received symbols correspond to the level U_2:

$$p_{22} = 1 - \int_{a_2}^{a_1} u_{22}(V)\, dV,$$

where a_1 and a_2 are determined according to (9) and (10).

The received symbol amplitude determination error probability is

$$P_{AR}(h) = q_{12}p_{12} + q_{21}p_{21} + q_{11}p_{11} + q_{22}p_{22}, \tag{20}$$

where the probabilities q_{mn}, $m, n = 1, 2$ of the appearance of $U_m U_n$ pairs depend on the constellation according to the Table 1.

A general demodulation error probability can be estimated using the expression

$$P_{err}(h) = 1 - (1 - P_P(h))\,(1 - P_A(h)), \tag{21}$$

where $P_P(h)$ is determined according to (15) and $P_A(h)$ – according to (16), (17) or (20).

Table 1. The probabilities of joint appearance of possible pairs of amplitude values of the preceding and the received symbols.

Constellation	q_{12}	q_{21}	q_{11}	q_{22}
Figure 1a	3/16	3/16	1/16	9/16
Figure 1b	1/4	1/4	1/4	1/4

6 8+8APSK Signal Demodulation Errors

8+8APSK signal (Fig. 1b) undergoes 8-level phase shift keying (8PSK) and binary amplitude shift keying (2ASK). In the 2-level 8+8APSK signals, the symbols with the amplitudes S_1 and S_2 have the same phase change step $\pi/4$ ($M_1 = 8$ positions), so that the signal phase determination error probability P_P mainly depends on the symbols with the amplitude S_1. Thus, from (15), one gets:

$$P_P(h) \approx M_1 P_{PSK}(h, M_1)/(M_1 + M_2) = P_{PSK}(h, 8)/2, \tag{22}$$

In the case of the ideally normalized received symbol amplitude the probability P_{A0} of the amplitude demodulation error is estimated using (17).

In Fig. 3, by dashed and solid curves, there are presented the theoretical dependencies (17) and (21) upon the SNR defined as h_0 (14). One can also see the results of the statistical simulation there: the general demodulation error probability P_{err} is denoted

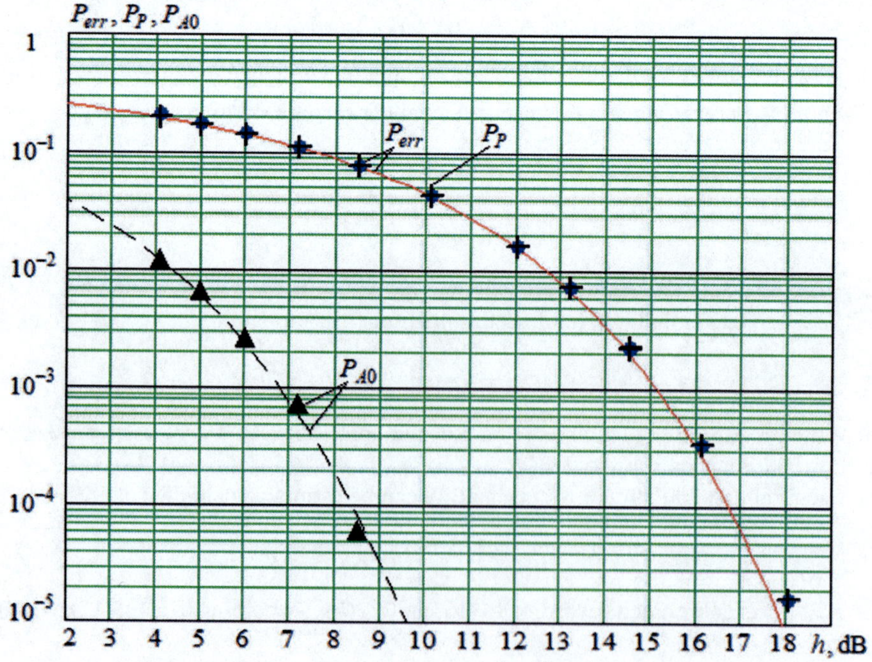

Fig. 3. The error probabilities when receiving a 8+8APSK signal

by blue dots; the phase estimate error probability P_P – by pluses; and the amplitude estimate error probability P_{A0} – by triangles. It follows from Fig. 3 that $P_{A0} << P_P$, and thus, in the case of the 8+8APSK signal, a noise immunity mainly depends on the phase demodulation errors: $P_{err}(h) \approx P_P(h)$. It means that one can neglect the probability P_{A0} in a relative estimation of the symbol amplitude.

7 4+12APSK Signal Demodulation Errors

In the 4+12APSK signal (Fig. 1a), the symbols with the amplitude S_1 are subjected to phase shift keying with the step $\pi/2$, while the symbols with the amplitude S_2 – with the shorter step of $\pi/6$. In this case, as it follows from (15)

$$P_P(h) \approx 4P_{PSK}(h, 4)/16 = P_{PSK}(h, 4)/4. \tag{23}$$

while the probability P_{A0} of the amplitude demodulation error is determined using (16).

In Fig. 4, by dashed and thin solid curves, there are shown the theoretical dependencies (16) and (23) upon the SNR h_0 (14). One can also see the results of the statistical simulation there: the general demodulation error probability is denoted by blue dots; the phase estimate error probability P_{P1} for the symbol with the amplitude S_1 – by squares; the phase estimate error probability P_{P2} for the symbol with the amplitude S_2 – by triangles. The symbol amplitude estimate error probability P_{A1} when the amplitude is lower

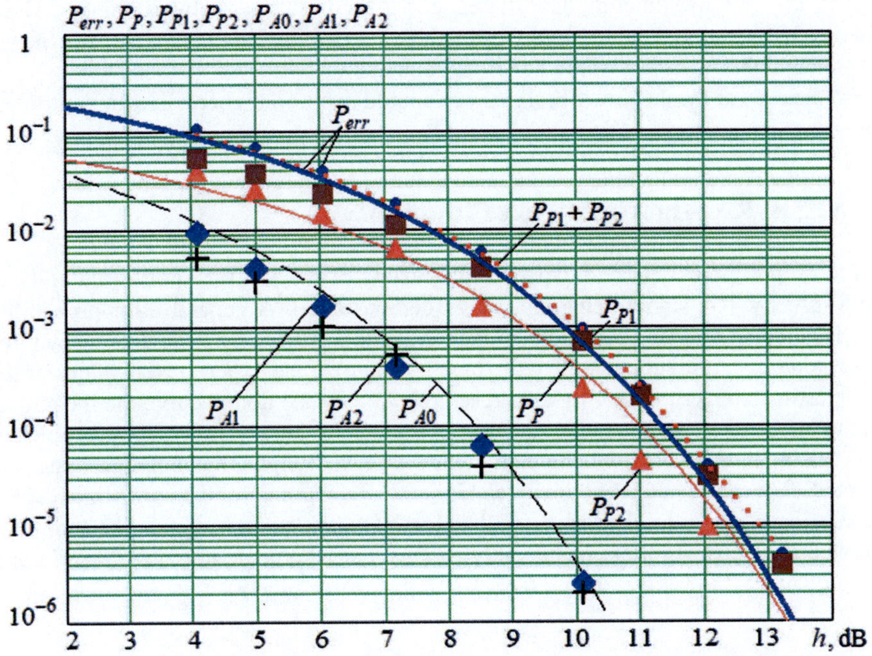

Fig. 4. The error probabilities when receiving a 4+12APSK signal

is presented as rhombuses, and when the amplitude is higher the similar probability P_{A2} is presented as pluses.

As it follows from Figs. 3 and 4, the noise immunity when using 4+12APSK signal is considerably higher compared to the case when 8+8APSK signal is used. The general error probability is underestimated compared to its real values. An analysis shows that it still practically depends upon the phase estimate errors only, so based on the simulation results it can be taken that

$$P_{err} = P_{P1} + P_{P2}. \tag{24}$$

For illustration purposes, the experimental values (24) are connected by dotted line in Fig. 4.

From here, after applying (15) the theoretical values of the general error probability P_{err} can be calculated using the formula

$$\begin{aligned} P_{err} &\approx P_P(h) = 4P_{PSK}(h, 4)/16 + 12P_{PSK}(3h, 12)/16 \\ &= P_{PSK}(h, 4)/4 + 3P_{PSK}(3h, 12)/4. \end{aligned} \tag{25}$$

The corresponding dependence is demonstrated by the heavy solid line in Fig. 4, and it fits in well with the simulation results.

It should also be noted that the symbol amplitude determination error probabilities when receiving the 4+12APSK signal are not too low compared with the phase demodulation error probabilities as is the case when receiving the 8+8APSK signal (see Fig. 3). But, when estimating the difference in noise immunity between the 8+8APSK and 4+12APSK signals, one should remember that the mean 4+12APSK signal power is by 3dB higher than that of the 8+8APSK signal.

In the case of the 4+12APSK signal one should expect a higher noise immunity loss when transferring to the APSK-RA.

8 APSK-RA Algorithm Noise Immunity

In Fig. 5, the theoretical and experimental results of the study of noise immunity of the 8+8APSK-RA signal demodulator are demonstrated. A general error probability P_{err} is calculated using (21) and denoted by the solid curve. The red dots display the corresponding simulation data. By the blue rhombuses, the experimental values of the phase estimate error probability are presented. And by the dotted curve the theoretical estimate (22) is shown. By squares и triangles the experimental values of the probabilities of the demodulation error when dealing with lower and higher symbol amplitudes are denoted. One can see that they are much higher than in the case of the normalized APSK signals (Fig. 3). It is also demonstrated that, for the 8+8APSK-RA signals, the amplitude estimate error probabilities considerably increase, but the noise immunity deteriorates insignificantly (the solid and dotted lines in Fig. 5, respectively).

In Fig. 6, the similar results for the 4+12APSK-RA signal are introduced (its constellation is in Fig. 1a). A general error probability is determined as in (21) and is defined by the solid line, while by red dots the corresponding experimental values obtained by means of statistical simulation are presented. The squares и rhombuses denote the experimental values of the phase estimate error probabilities for the symbols with the amplitudes S_1 and S_2, respectively. The dashed line stands for the theoretical estimate (23) for the phase demodulation error. The experimental dependences of the probabilities of the demodulation error when dealing with the higher and lower symbol amplitudes are defined as triangles and pluses, respectively. As in the previous case, these probabilities are much higher than the ones for the normalized 4+12APSK signals (Fig. 4). The dotted line describes the dependence $P_{AR}(h)$ (20).

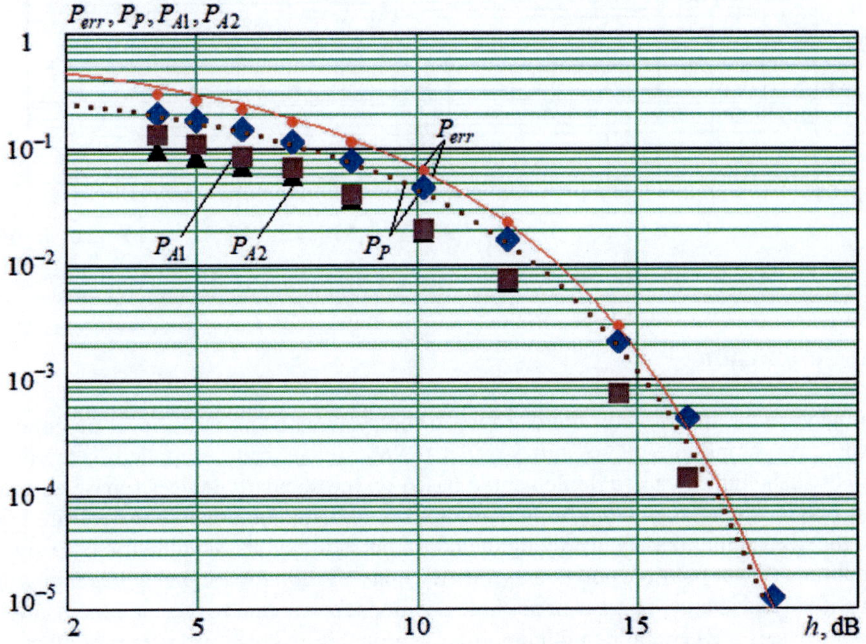

Fig. 5. The error probabilities when receiving a 8+8APSK-RA signal

The Figs. 4 and 6 prove that for the 4+12APSK-RA signal the transition to the relative estimate of the symbol amplitude results in the 3–4 dB decrease in the noise immunity.

The relative demodulation of the symbol amplitudes does not require a calibration of the decision-making threshold levels and a corresponding complex equipment, thus making it especially effective under the signal fading conditions. Still a decreased noise immunity remains a drawback of this approach.

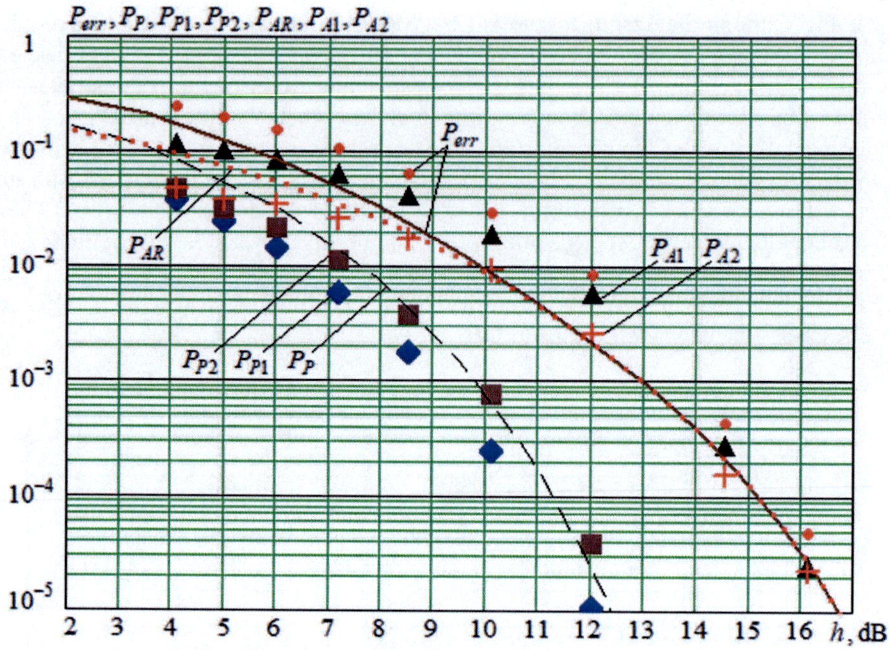

Fig. 6. The error probabilities when receiving a 4+12APSK-RA signal

9 Conclusion

To increase the information transfer rate in the systems using multi-level modulation techniques, amplitude-phase shift keyed (APSK) signals can be applied. Demodulators of such signals can be implemented based on the standard devices (phase detector and envelope detector) with a minimum number of arithmetic operations required per the received signal period. Applying the constellations that are asymmetric by a signal amplitude (fewer possible positions within the constellation for a signal with a minimum amplitude and a larger number of possible positions for a signal with a maximum amplitude) reduces the general demodulation error probability, but is more energy intensive (for example, the mean power of 4+12APSK signal is 3 dB higher than the mean power of 8+8APSK signal). With a relative estimate of the amplitude of the received symbol (based on a relative comparison of the amplitudes of the received and the previous symbols), i.e. when transferring from APSK signals to APSK-RA signals, the noise immunity of the communication system decreases, and when using asymmetrical constellations, the losses in a demodulation quality are greater compared to the case of the symmetrical constellations. In particular, when 4+12APSK signals and a relative symbol amplitude demodulation are applied, the SNR decrease at the output of the demodulator is 3–4 dB, while when applying 8+8APSK signals in a similar case SNR change is insignificant (less than minus 1 dB). However, the relative demodulation of the symbol amplitudes does not require the amplitude calibration of the decision-making threshold levels. That means dealing more effectively with the issues arising under the signal fading conditions, and the structure of the receiving device can also be simplified considerably. Both the operability and efficiency of the proposed demodulation algorithms are confirmed using statistical computer simulation of their application. It is established that the found analytical formulas for the probabilities of the phase demodulation error, amplitude demodulation error, and general demodulation error are in satisfactory agreement with the corresponding experimental data within a wide range of SNR values.

Acknowledgements. The work was financially supported by the Ministry of Education and Science of the Russian Federation (research project No. FSWF-2020-0022).

References

1. Sklar, B.: Digital Communications: Fundamentals and Applications. Prentice-Hall, Hoboken (2001)
2. Baldi, M., Chiaraluce, F., De Angelis, A., Marchesani, R., Schillaci, S.: A comparison between APSK and QAM in wireless tactical scenarios for land mobile systems. EURASIP J. Wirel. Commun. Netw. **2012**, 1–14 (2012)
3. ETSI EN 302 583 V1.2.1: Digital video broadcasting (DVB); Framing structure, channel coding and modulation for satellite services to handheld devices (SH) below 3 GHz. https://www.etsi.org/deliver/etsi_en/302500_302599/302583/01.02.01_40/en_302583v010201o.pdf. Accessed 21 Feb 2022
4. Häger, C., Graell I Amat, A., Alvarado, A., Agrell, E.: Design of APSK constellations for coherent optical channels with nonlinear phase noise. IEEE Trans. Commun **61**(8), 3362–3373 (2013)
5. Prat, J., Velásquez, J.C., Tabares, J.: Direct PSK-ASK modulation for coherent udWDM. In: 2019 21st International Conference on Transparent Optical Networks, ICTON, pp. 1–4. IEEE, Angers (2019)
6. Liu, Z., Xie, Q., Peng, K., Yang, Z.: APSK constellation with Gray mapping. IEEE Commun. Lett. **15**(15), 1271–1273 (2011)
7. Peng, S., Liu, A., Wang, K., Pan, X.: A modified design of APSK constellations for AWGN channel. In: 2016 IEEE/CIC International Conference on Communications in China, ICCC, pp. 1–6. IEEE, Chengdu (2016)
8. Ferrand, P., Maso, M., Bioglio, V.: High-rate regular APSK constellations. IEEE Trans. Commun. **67**(3), 2015–2023 (2019)
9. Recommendation ITU-R BO.2098-0: Transmission system for UHDTV satellite broadcasting. https://www.itu.int/dms_pubrec/itu-r/rec/bo/R-REC-BO.2098-0-201612-I!!PDF-E.pdf. Accessed 21 Feb 2022
10. Bao, J., Xu, D., Zhang, X.: A rapid de-mapping method for 64-APSK. In: 6th International Conference on Communications and Broadband Networking, ICCBN, pp. 27–31. Association for Computing Machinery, Singapore (2018)
11. Yao, E., Yang, S., Jiang, W.: A simplified soft decision demapping algorithm of 16-APSK signals in AWGN channels. In: 2nd International Conference on Networks Security, Wireless Communications and Trusted Computing, NSWCTC, vol. 1, pp. 103–106. IEEE, Wuhan (2010)
12. Chernoyarov, O.V., Glushkov, A.N., Golikov, A.A., Litvinenko, V.P., Mironov, V.A.: On digital phase detector. J. Phys: Conf. Ser. **2094**(2), 1–6 (2021)

13. Glushkov, A.N., Litvinenko, V.P., Bukharin, S.V., Rostrov, N.A.: Fast digital algorithms for detecting narrow-band signals. Telecommun. Radio Eng. **64**(3), 211–218 (2005)
14. Chernoyarov, O.V., Glushkov, A.N., Litvinenko, V.P., Mironov, V.A., Salnikova, A.V.: Coherent demodulation of the two-level APSK signals with the symbol amplitude estimation. J. Phys: Conf. Ser. **2094**(2), 1–6 (2021)
15. Chernoyarov, O.V., Zakharov, A.V., Trifonov, A.P., Salnikova, A.V., Makarov, A.A.: The common approach to calculating the characteristics of the signal parameters joint estimates under the violation of the decision statistics regularity conditions. Eng. Lett. **28**(2), 492–503 (2020)
16. Springer, M.D.: The Algebra of Random Variables. Wiley, New York (1979)

Water Availability in the Ilave Basin, Puno, Peru: SWAT Hydrological Model

Katerin Quispe Chahuara[1], Celenia Huanca Quispe[1], Hugo Fernández[1](✉), and Efrain Lujano Laura[2]

[1] Facultad de Ingeniería y Arquitectura, Escuela Profesional de Ingeniería Ambiental, Universidad Peruana Unión, Lima, Peru
{katerin.qc,celenia.hq,hugof}@upeu.edu.pe

[2] Facultad de Agrícola, Universidad Nacional del Altiplano Puno, Puno, Peru

Abstract. The drastic changes in the climate have been threatening the availability of water in quantity and quality at a global level. The objective of this study was to assess future water availability in the Ilave river basin by using the Soil and Water Assessment Tool (SWAT) model. The calibration and validation of the model was carried out over a period of 30 years (1986–2016) using climatic data and daily flows. For the future simulation (2025–2050), data from global climate models (GCM) MIROC6 and MPI-ESM1-2-LR of the Coupled Model Intercomparison Project Phase 6 (CMIP6) were adopted in the trajectory scenarios SSP245 and SSP585. The bias correction linear scaling method was used to correct the climatic data and then to simulate future streamflows in the basin were entered into SWAT. The results of the investigation project a decrease in precipitation (−38.08 mm to −0.01 mm) in the rainy season and an increase (0.62 mm to 4.30 mm) in the dry season. The maximum temperature increases by +0.66 °C, while the minimum temperature decreases by −0.29 °C. The research found that the future climate could decrease water availability by −41.2% to 40.0% in the rainy season, with a −33.8% to −0.1% decrease in the dry season. The findings of this study would be beneficial for water resource managers to develop adaptation strategies to climate change in the Ilave river basin.

Keywords: Ilave basin · Water availability · Global climate models · SWAT model · Climate change

1 Introduction

One of the main current problems facing humanity due to global climate changes are alterations in the hydrological system, variations in temperature and precipitation directly impact evapotranspiration, soil moisture, runoff, and the volume of the water. Surface water flow. The Ilave river basin has already been facing various extreme events such as droughts and floods that cause the loss of hectares of crops, destroyed homes, affected educational centers and a large part of the affected rural population, this due to the variations in the hydrological process that is a result climate change (CC) and negatively affect the population, economic growth, agriculture, biodiversity and water availability.

R. Silhavy (Ed.): CSOC 2022, LNNS 501, pp. 469–483, 2022.
https://doi.org/10.1007/978-3-031-09070-7_39

According to the fifth report of the Intergovernmental Group of Experts on Climate Change (IPCC), CC is one of the main challenges of the 21st century [1], it is the most relevant environmental problem worldwide due to the effects on the increase in global mean temperature, rainfall variations and sea level rise [2], the global mean temperature of the land surface observed in the years 2006–2015 was 0.87 °C and 0.99 °C [3], there is scientific research that proves that from 1880 to 2017 the global average temperature has risen by 1 °C and is increasing at a rate of 0.2 °C per decade [4], if greenhouse gas emissions are not reduced, the earth's surface temperature is likely to be above 1.5 °C in all climate scenarios by the end of the 21st century [5]. These changes are altering hydrological conditions, presenting various effects on global water resource systems [6]. Various scientific investigations affirm that in the future there will be a reduction in the availability of water resources due to the effects of CC, since in the long term there are forecasts of water scarcity [7], however, with warming 1.5 °C versus 2 °C could reduce the proportion of the population vulnerable to water scarcity by more than 50%, due to CC [8].

Peru is vulnerable to global climate changes denoted by variations in global average temperature, precipitation, humidity and runoff, this leads to the possibility of changes in the hydrological cycle, consequently there are significant effects on water such as floods and droughts. [9], this mainly affects the economic development of the country.

Hydrological models are an important tool to assess the impacts of CC on the water resources of basins under CC scenarios based on future greenhouse gas emissions [10]. To evaluate the hydrological behavior of a basin in the long term in relation to CC, models based on processes such as the Soil and Water Assessment Tool (SWAT) are used to represent the processes of generation of runoff, the influence of water use and feedback mechanisms between water availability and use [11]. The SWAT tool has established its applicability for hydrological impact studies, it has been widely used to simulate and evaluate the effects of changes in the hydrology of basins and land use [10].

The climate change scenarios represent in a simplified way the climate of the future and are constructed to be used in simulation investigations of the impacts of anthropogenic CC [12]. Global climate model simulations (GCM) archived by the Coupled Model Intercomparison Project (CMIP) are one of the most important sources for quantitative projection of climate for the 21st century [13], these GCM are used as the main tool to analyze extreme climate changes of the past and future [14].

The Ilave river basin is characterized by being an area of agricultural and livestock production and population growth, these are affected by climate change [15]. According to the inventory of water sources prepared by the National Water Authority (ANA) in the Ilave basin there are 1836 springs, 1300 streams, 67 rivers, 27 wetlands and 14 lagoons [15], these water sources face the threat of the loss of available water volume due to the increase in temperature and the decrease in rainfall [16].

In this context, it is essential to evaluate future water availability in order to propose action plans for stable and sustainable water management, for which there are tools such as the SWAT model that allow simulating and evaluating the variations of the water system in a basin and the CGM that allows us to investigate the causes and effects of climate change.

Therefore, the objective of this research work is to evaluate the availability of the water resource of the Ilave basin under climate change scenarios of CMIP6 of routes SSP 4.5 and SSP 8.5 of the global climate models MIROC6 and MPI-ESM1-2- LR using the SWAT model for sustainable planning and management.

2 Materials and Methods

2.1 Study Area

The Ilave River Basin (CRI) is located within the hydrographic region of Titicaca (Peru). The main river of the CRI has a length of 192.8 km from its source to its mouth and is one of the important tributaries of Lake Titicaca (Fig. 1). The CRI covers approximately 7,743 km^2, with an elevation range of 3,813–5,414 m.a.s.l. (average 4326.7; Fig. 2a). Most of the basin is characterized by flat topography, with slopes less than 10% (Fig. 2d).

According to the climatic classification of Peru, the CRI has a predominantly rainy climate type in summer, with dry autumn and winter [17], it has an average annual rainfall of 636 mm, most of it occurs in summer (December-January-February, 60%). The average annual temperature is 6.4 °C and the lowest temperatures occur in winter with values below 0 °C. The CRI has an annual average flow of 72.4 m^3/s.

The CRI consists of six types of land use (Fig. 2b). The main use is Andean grasslands (PAST, 73.0% of the basin area), followed by high Andean areas with little or no vegetation (SWRN, 14.7%), coastal and Andean agriculture (AGRL, 7.3%), wetlands (WETL, 4.4%), lakes, lagoons, lakes, rivers, glaciers (WATR, 0.5%) and urban area (URBN, 0.1%). In addition, the study area has seven soil classes (Fig. 2c)

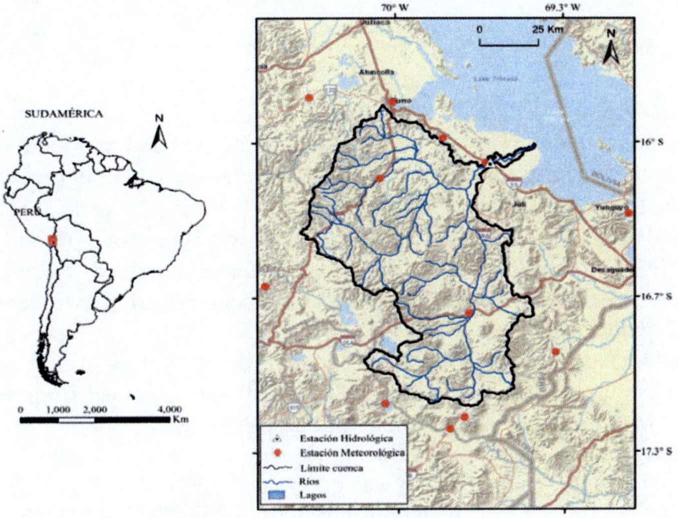

Fig. 1. Location of the study area.

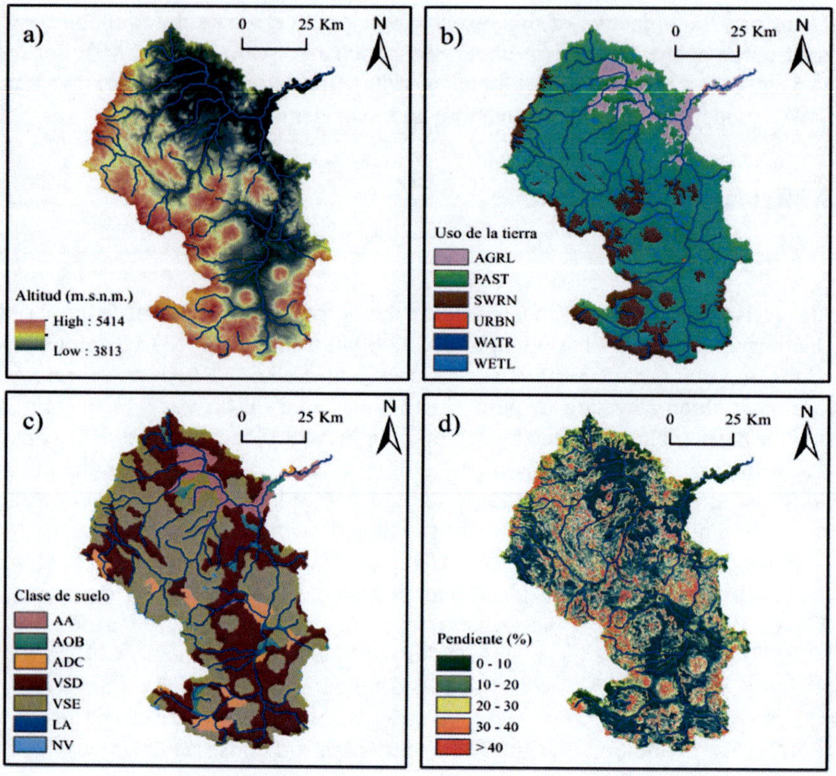

Fig. 2. Characteristics of a) DEM, b) land use, c) soil class, d) slope.

2.2 Data

- **Hydrometeorological data**
  The daily precipitation (Pr), maximum temperature (Tx) and minimum temperature (Tn) observation data for the period 1986–2016, were obtained from Peruvian Interpolated data of the SENAMHI's Climatological and hydrological (PISCO) [18]. PISCO is a gridded data set of Pr, Tx and Tn, with a spatial resolution of (~10 km) and a daily temporal resolution of the National Service of Meteorology and Hydrology of Peru (SENAMHI). Relative humidity, wind speed and solar radiation data were used from the SWAT global database.

  To calibrate and validate the SWAT model, the daily flow data from the Ilave bridge measurement station (period: 1986–2016) were obtained from the National Service of Meteorology and Hydrology of Peru (SENAMHI).
- **Topography, land use and soil class**
  The digital elevation model (DEM) was generated by NASA's Shuttle Radar Topography Mission (SRTM) at a spatial resolution of ~90 m. The DEM was provided by site

  The land use data in vector format were obtained from the map of major use capacity of the Ministry of the Environment MINAM, while the soil class was obtained from the physiographic map of Peru of the National Office for the Evaluation of Natural Resources ONERN (Table 1).

Table 1. Information supplied to the model.

Information	Source	Description	Format
DEM	SRTM/NASA	Digital elevation model (~90 m spatial resolution)	Raster
Vegetation cover and land use	MINAM	Greater land use capacity map	Shapefile
Soil Units (Edaphology)	ONERN	Physiographic map of Peru	Shapefile
Hydrography	ANA	Delimitation and hydrographic drainage network of the Ilave basin	Shapefile
Hydrography	ANA SENAMHI	Hydrometeorological stations	Shapefile
Basic cartography	IGN	Ilave basin national charts	Shapefile

2.3 SWAT Hydrological Model

The SWAT model is a continuous time distributed hydrological model that was developed to predict the impact of soil management practices on the generation of water from hydrographic basins, it requires specific information on meteorological data (precipitation, maximum and minimum temperature, relative humidity, solar radiation and wind direction) and space to simulate the hydrological cycle [19]. The hydrologic cycle simulated by SWAT is based on the following hydrologic equilibrium equation.

$$SW_t = SW_0 + \sum_{i=1}^{t} (R_{day} - Q_{surf} - E_a - W_{scep} - Q_{qw})i \tag{1}$$

where: SW_t is the final water content in the soil (mm); SW_0: is the initial water content on the day (mm); t: is the time (day); R_{day}: is the amount of precipitation in the day (mm); Q_{surf}: is the amount of surface runoff per day (mm); Ea: is the amount of evapotranspiration on day i (Mm H2O); W_{scep}: is the amount of water accumulated in the unsaturated zone of the soil profile in the day (mm); Q_{qw}: is the amount of return flow in the day (mm).

The SWAT model divides the hydrographic basin into sub-basins, to transform them into hydrological response units (HRUs), which consist of homogeneously combining the soil, vegetation cover and slope in the basin [19].

2.4 Future Projection of Climate Change

Two GCMs based on the Coupled Model Intercomparison Project Phase 6 ensemble (Table 2) were used in this study. Future climate scenarios were used for the time period 2025–2050. The MCGs namely MIROC6 and MPI-ESM1-2-LR in trajectory scenarios SSP245 and SSP585.

Table 2. Description of CMIP6 global climate models.

GCM	Origin	Institution	Resolution Long × Lat
MIROC6	Japan	Japan Agency for Marine and Earth Sciences JAMSTEC and the National Institute for Environmental Studies NIES	1.9° × 1.9°
MPI-ESM1-2-LR	Germany		1.9° × 1.9°

2.5 Methods

PISCO SENAMHI was used for the study. Consequently, the PISCO Pr, Tx and Tn data were extracted from the pixel grid from the coordinate points of 12 meteorological stations.

For the land use data, a classification was established between the vegetation cover map of the Ministry of the Environment (MINAM) and the SWAT land use codes. Likewise, from the ONERN Physiography map of Peru, the polygons of the soil units were extracted, assigning to each one the information of the soil properties of the Ilave basin. Taking advantage of the SOILGRIDS database (https://soilgrids.org/), the values of the soil characteristics were entered (depth, texture, bulk density, organic carbon content, clay, silt, sand, among others.), the available water capacity and soil hydraulic conductivity was calculated with the SPAW Soil Water Characteristics software, some missing values were completed with the harmonized soil database (HWSD) prepared by FAO and UNESCO (http://webarchive.iiasa.ac.at/Research/LUC/ExternalWorld-soildatabase/HTML/). It is necessary to indicate that the land use was used in one phase for the year 2014, remaining unchanged during the simulation process with SWAT.

For the hydrological modeling with SWAT, the QSWAT interface available for QGIS was used, introducing terrain data, hydrometric station point, land use, soil class and climatic data.

Prior to the sensitivity analysis, a comparison was made between the simulated and observed flows for the period 1986–2016. In order to reduce the uncertainty about the parameters that are used in the calibration, the sensitivity analysis was applied using generally recommended parameters such as are: CN2, GW_DELAY, GW_REVAP, REVAPMN, GWQMN, SURLAG, ESCO, EPCO, CANMX, CH_N2, RCHRG_DP, SOL_AWC). According to Arnold et al., 2012); The "t-Stat" that expresses the sensitivity of the parameter and "p-value" that represents the importance of the sensitivity must be evaluated and the parameters that have a "p-value" factor greater than 0.2 must be included. To carry out the calibration process, the SWAT - CUP (Calibration and Uncertainty Programs For SWAT) program and the SUFI - 2 (Sequential Uncertainty Fitting 2) algorithm were used.

The calibration and validation of SWAT was carried out with daily data for the period 1986–2016, considering 3 years as the warming period (1986–1988), 18 years for the calibration stage (1989–2007) and 9 years for the warming stage of validation (2008–2016). The calibration consisted of performing multiple simulations in order to find the optimal values for each parameter. Model calibration requires identifying parameters that are more sensitive to changes in their values, so that a minimal variation would generate significant changes in the final results [20]. Consequently, to evaluate the accuracy of the model with data outside the calibration stage, it was passed to the validation stage where the optimal values of the calibrated parameters were used.

To evaluate the performance of the model in the calibration and validation stage, 3 metrics of hydrological ability were used, namely, the coefficient of determination (R2), the Nash-Sutcliffe efficiency (NSE) and the percentage bias (PBIAS) [21] (Table 3).

Finally, for the future simulation (2025–2050), MCG data (MIROC6 and MPI-ESM1-2-LR) from CMIP6 in trajectory scenarios SSP245 and SSP585 were adopted. The bias correction linear scale method was used to correct the climatic data, for this process we used the quantile mapping library available for statistical r and the code adapted by Gudmundsson [22], the climatic base was considered for the period 1986–2016, while future climate 1986–2050. Accordingly, bias-corrected data were entered into SWAT to simulate future basin flows.

Table 3. Hydrologic ability metrics to assess SWAT model performance.

Metrics	Equation	Range	Optimal value
Nash-Sutcliffe Efficiency	$NSE = 1 - \left[\frac{\sum_{i=1}^{n}\left(Y_i^{obs} - Y_i^{sim}\right)^2}{\sum_{i=1}^{n}\left(Y_i^{obs} - \underline{Y}^{obs}\right)^2}\right]$	$-\infty$–1	1
Determination coefficient	$R^2 = \left[\frac{\sum_{i=1}^{n}\left(Y_i^{obs} - \underline{Y}^{obs}\right)\left(Y_i^{sim} - \underline{Y}^{sim}\right)}{\sqrt{\sum_{i=1}^{n}\left(Y_i^{obs} - \underline{Y}^{obs}\right)^2}\sqrt{\sum_{i=1}^{n}\left(Y_i^{sim} - \underline{Y}^{sim}\right)^2}}\right]^2$	0–1	1
Bias percentage	$PBIAS = \left[\frac{\sum_{i=1}^{n}\left(Y_i^{obs} - Y_i^{sim}\right)}{\sum_{i=1}^{n}\left(Y_i^{obs}\right)}\right] \times 100$	$-\infty$–∞	0

Note: Y_i^{obs} y Y_i^{sim} denote the observed and simulated flows at time i; $\underline{Y}^{obs}$ y $\underline{Y}^{sim}$ the average of the observed and simulated flows; n is the number of observations.

3 Results

3.1 Calibration and Validation

Twelve parameters were identified as the most sensitive (CN2, GW_DELAY, GW_REVAP, REVAPMN, GWQMN, SURLAG, ESCO, EPCO, CANMX, CH_N2, RCHRG_DP, SOL_AWC) (Table 4), the same ones that were used in the calibration and validation stage.

Table 4. SWAT model calibrated parameter values.

Parámeter	Definition of parameters	Rango	Valor ajustado
r_CN2.mgt	Initial runoff curve number	0–0.1	0.055
v_GW_DELAY.gw	Groundwater delay time (days)	30–100	61.50
v_GW_REVAP.gw	Groundwater coefficient	0.01–0.5	0.0835
v_REVAPMN.gw	Water depth threshold in shallow aquifer (mm)	30–100	33.50
v_GWQMN.gw	Shallow aquifer water depth threshold	4000–5000	4950.00
v_SURLAG.hru	Delay coefficient of surface rainwater	1–15	14.30
v_ESCO.hru	Soil evaporation compensation factor	0.01–1	0.4555
v_EPCO.hru	Plant absorption compensation factor	0.01–0.5	0.3285
v_CANMX.hru	Maximum storage by retention due to vegetation cover	30–100	40.50
v_CH_N2.rte	Manning's n value for the main channel	0.01–0.3	0.1985
v_RCHRG_DP.gw	Percolation coefficient to deep aquifer	0.01–0.5	0.2795
r_SOL_AWC.sol	Available water capacity (mm water/mm soil)	0.01–0.5	0.3775

Note: The qualifier (r_) indicates that the parameter is multiplied by 1 plus the value; (v_) refers to the substitution of the parameter by the value.

The calibration and validation stage of the SWAT model (1989–2007) (2008–2016) present satisfactory results for the Ilave basin, with an NSE = 0.81, R2 = 0.86 and PBIAS = −10.5 for calibration and an NSE = 0.79, R2 = 0.82 and PBIAS = −1.6 for validation (Fig. 3) (Table 5).

Table 5. SWAT model performance.

Stage	Daily		
	NSE	R^2	PBIAS
Calibration	0.81	0.86	−10.5
Validation	0.79	0.82	−1.6

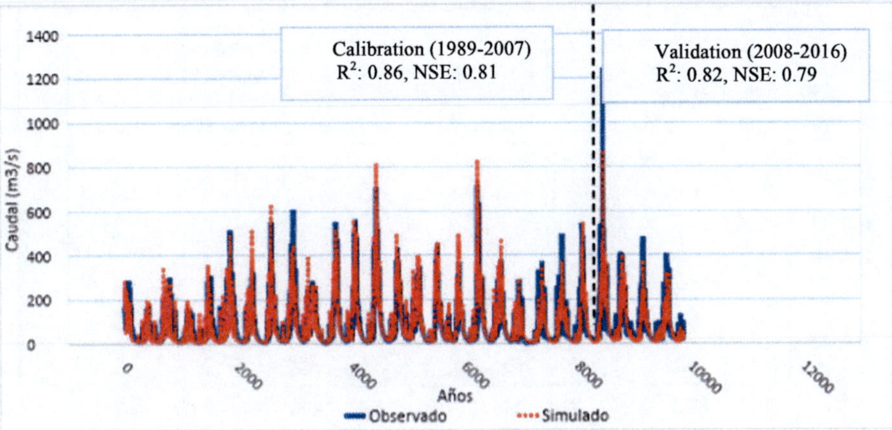

Fig. 3. Observed and simulated flows for the calibration and validation period.

3.2 Projected Climate Scenaries

The global climate models of good fit to represent the future climate (2025–2050) for the Ilave basin were MIROC6 and MPI-ESM1-2-LR and the climate scenarios correspond to the shared socioeconomic trajectories (SSP4.5 and SSP8.5). The selection of the models to be used was made by comparing the multi-monthly average total precipitation variable, where they show a "good" coefficient of determination (R2) (Table 6).

Table 6. Validation of global climate models (CMIP6)

MCG	R^2
MIROC6	0.61
MTI-ESM1-2-LR	0.62

Figure 4 shows the results of multimonthly average precipitation for the period (1986–2016) of the Ilave basin.

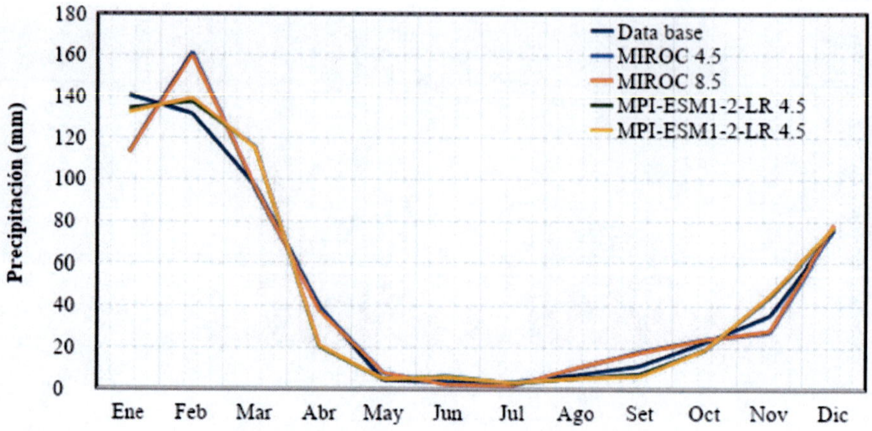

Fig. 4. Multi-monthly average total precipitation

3.3 Evaluation of Changes in Precipitation and Temperature

The percentage variation of monthly precipitation for the future (2025–2050) compared to the base period for the MIROC6 model for routes SSP 4.5 and SSP 8.5 increased by 0.07% and 0.02% during the dry season (May–September) and decreased by −0.16% and −0.07% during the rainy season (October–April), for the MPI-ESM1-2-LR model of routes SSP 4.5 and SSP 8.5 they decrease by −0.15% and −0.18% in the dry season and −0.18% during the rainy season.

The monthly average maximum temperature variation between the MIROC6 and MPI-ESM1-2-LR models for routes SSP 4.5 and SSP 8.5 project an increase of 0.8 °C during the dry season and 0.6 °C during the rainy season and the percentage variation of the minimum temperature decreases by −0.35% in the dry season and −0.23% during the rainy season.

3.4 Variation of Future Water Availability

Table 7 shows the monthly average flows of the Ilave basin of the base scenario (1986–2016) and the future scenarios (2025–2050), it can be seen that in the four future scenarios the projected maximum flows are presented in the month February and the minimum flow occurs in the month of September.

Table 8 indicates the variation of the water availability of the Ilave basin between the base scenario and the future scenarios, in the dry season variations can be identified from +221.0% to −33.8% and in the rainy season variations from +132.7% to −41.2%, the highest variations occur between the months of April to June.

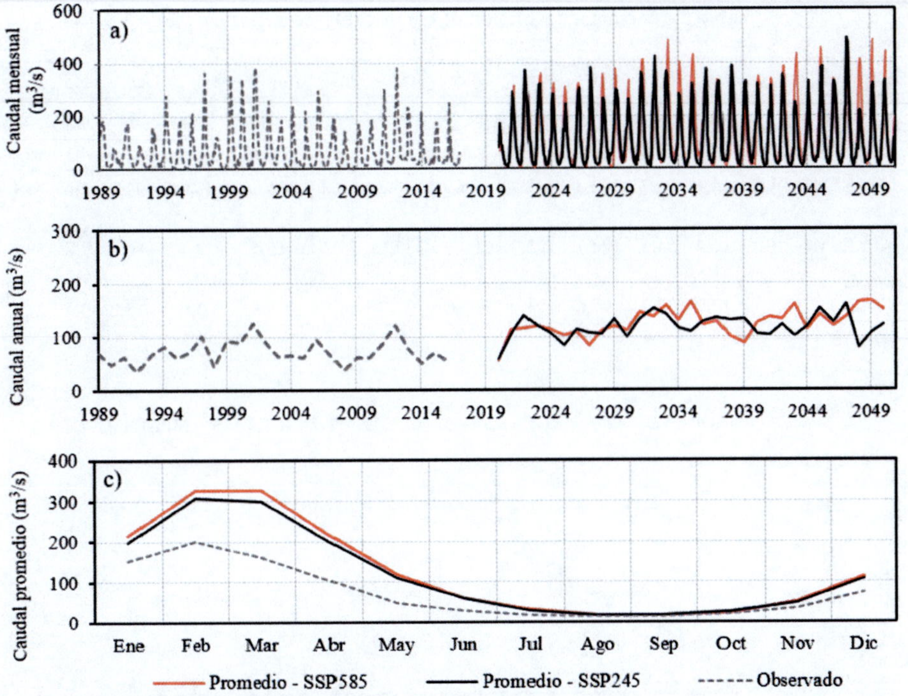

Fig. 5. Future hydrological projections a) monthly variations in flows, b) interannual variations and c) changes in the seasonal cycle

Table 7. Monthly average flow of the scenarios

Scenaries	Jan	Feb	Mar	Apr	May	Ju	Jul	Aug	Sep	Oct	Nov	Dec
Scenary Base	183.9	208.1	158.5	94.6	37.8	25.8	18.6	19.9	16.6	28.3	47.3	91.6
MIROC6 4.5	186.2	300.2	277.4	196.4	111.1	58.0	29.7	18.8	22.7	34.0	47.4	96.9
MIROC6 8.5	215.2	334.7	301.1	220.2	121.5	61.3	30.4	19.8	21.3	30.5	49.3	106.2
MPI-ESM1-2-LR 4.5	209.8	316.8	322.2	206.9	111.7	59.4	28.3	14.6	11.0	16.6	48.0	120.6
MPI-ESM1-2-LR 8.5	214.1	321.5	352.1	213.2	117.3	60.0	29.7	16.8	11.9	17.0	51.0	122.6

Figure 5 shows the percentage variation of the monthly average flow for the horizon (2025–2050) of the Ilave basin (Fig. 6).

Table 8. Percentage variation of the monthly average flow (%).

Scenarios	Jan	Feb	Mar	Apr	May	Ju	Jul	Aug	Sep	Oct	No	Dec
MIROC6 4.5	1.3	44.2	75.0	107.5	193.6	124.9	59.7	−5.3	36.5	20.3	0.1	5.7
MIROC6 8.5	17.0	60.8	90.0	132.7	221.0	137.7	63.1	−0.1	28.1	8.0	4.1	15.9
MPI-ESM1-2-LR 4.5	14.1	52.2	103.3	118.7	195.2	130.2	52.0	−26.3	−33.8	−41.2	1.4	31.6
MPI-ESM1-2-LR 8.5	16.4	54.5	122.1	125.3	209.9	132.6	59.3	−15.5	−28.4	−40.0	7.8	33.8
Mean	12.2	52.9	97.6	121.0	205.0	131.3	58.5	−11.8	0.6	−13.2	3.4	21.8

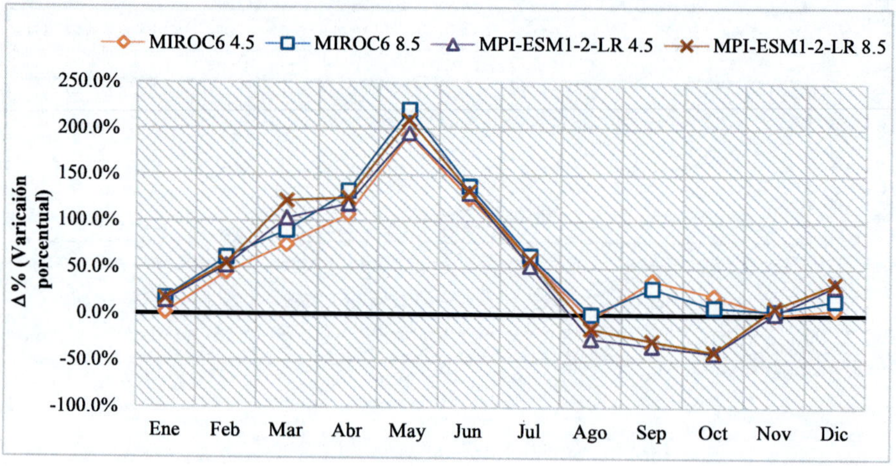

Fig. 6. Percentage change in the monthly average flow projected into the future (2025–2050).

4 Discussions

The SWAT model calibration results for the period (1989–2007) showed an NSE efficiency of 81% indicating that the model fit is "very good", on the other hand, the validation results for the period (2008–2016) showed a "good" fit with an NSE of 79%, results that are considered an adequate pattern to simulate the flow of rivers according to Moriasi [21]. The satisfactory results of calibration and validation for the Ilave basin indicate that they can be used as a management and planning tool in hydrographic basins [23].

The analysis of future precipitation variations for the Peruvian highlands (2025–2050) shows a decrease of up to −38.08 mm in an extreme scenario of SSP 8.5 and −0.29 mm in a scenario of SSP 4.5, these results coincide with Lujano [24] that affirms the decrease in precipitation by −9.7% for the horizon 2071–2100 in the Ilave basin, in the same way Moya [25] affirms the decrease in precipitation in the southern Peruvian highlands. It should be noted that the greatest decrease in precipitation occurs in the rainy season and increases in the dry season, likewise the maximum temperature increases by 0.66 °C and the minimum temperature decreases by −0.29%, analyzing these investigations the increase in the temperature and the decrease in precipitation in the rainy months could have a significant effect on the volume of the flows. According to Jodar [26], in the coming decades precipitation, surface runoff and aquifer recharge will decrease, while the temperature increases between 1.5 °C and 1.7 °C, which is evident that the management of water resources It will be a challenge for local authorities, likewise according to the sixth evaluation report on climate change 2021 [27], alterations in the hydrological cycle entail a greater intensity of rainfall, floods and droughts in various regions, with a greater increase precipitation in high latitudes and decrease in subtropical regions.

The results of the volume of the future flow for the year 2050 in the SSP 4.5 and SSP 8.5 scenarios of the MIROC6 and MPI-ESM1-2-LR models of the CMIP6 indicate that there will be a decrease in water resources more frequently in the rainy season. Varying between −41.2% to 40.0%, while for the dry season a decrease of −33.8% to −0.1% is projected. The greatest decrease in flow in the Ilave basin will occur in the months of August, September and October and will increase in the months of April, May and June. These results corroborate the investigations carried out by Lujano [24] and Lujano [28] where they mention a decrease in the annual water resource of −29.3% and −48% for the 2071–2100 horizon in the Ilave basin. Likewise Laura [29] indicates a decrease in water resources of up to −24% for the future 2020–2100 in an extreme scenario, it is true that the decrease in water availability in the dry season would increase the demand for water in use agricultural and population, which would put at risk the ecological flow and possibly the viability of future hydraulic infrastructure projects [30], according to Felipe [31] the impacts of climate change will affect the availability of water in the Titicaca slope with a decrease of −28% by the year 2050, likewise the reduction in the volume of surface water captured by the natural drainage network in the dry season implies great challenges to reduce the water gap and the socioeconomic development of the inhabitants [32], these results show us that in the near future a decrease in water resources awaits us with an impact of greater or lesser magnitude.

5 Conclusion

The SWAT semi-distributed mathematical model is an adequate tool to analyze periods of water excesses and deficits in hydrographic basins. The calibration and validation of the SWAT model was evaluated using 3 statistical indices: R2, NSE, PBIAS, where the results present an efficiency between "good" and "very good" with an NSE of 81% and 79% at the output, for the series (1986–2016).

The projected precipitations of the GCM indicate a decrease of −38.08 mm for an extreme scenario, the decrease in precipitation occurs in the rainy season and increases in the dry season. The results of the simulations of the MIROC6 and MPI-ESM1-2-LR models in the SSP 4.5 and SSP 8.5 scenarios for the horizon (2025–2050) show a percentage decrease in flow of −41.2% to 40% in the months of rain and a less frequent decrease from −33.8% to −0.1% in the dry season, these results predict the impact of climate change in the Ilave basin where flow decreases are predisposed in the near future, so it is deduces that it will be an area affected by water availability.

These results can be considered for the preparation of water resource management plans, implement early warning systems, identify areas prone to droughts and floods in the Ilave basin in relation to climate change in development programs, thus guaranteeing the management sustainability of the water resource.

References

1. Romero Cuéllar, J., Buitrago Vargas, A., Quintero Ruiz, T., Francés, F.: Simulación hidrológica de los impactos potenciales del cambio climático en la cuenca hidrográfica del río Aipe, en Huila, Colombia. Ribagua **5**(1), 63–78 (2018). https://doi.org/10.1080/23863781.2018.1454574
2. Bando Murrieta, U., Garrido, A., Altamirano del Carmen, M.Á., Martínez Fernández, J.: Proyecciones de cambio climático para las cuencas de méxico. Las cuencas hidrográficas México **2100**, 138–141 (2007)
3. Stocker, T.F., et al.: Contribucion del Grupo de trabajo I al Quinto Informe de Evaluacion del Grupo Intergubernamental de Expertos sobre el Cambio Climático. In: IPCC (2013)
4. Leung, J.Y.S., Russell, B.D., Connell, S.D.: Adaptive responses of marine gastropods to heatwaves. One Earth **1**(3), 374–381 (2019). https://doi.org/10.1016/j.oneear.2019.10.025
5. Jarraud, M., Steiner, A.: Climate Change 2014 Synthesis report, vol. 9781107025 (2012)
6. Yang, N., Men, B.H., Lin, C.K.: Impact analysis of climate change on water resources. Procedia Eng. **24**, 643–648 (2011). https://doi.org/10.1016/j.proeng.2011.11.2710
7. Chávez Jiménez, A., González Zeas, D.: El impacto de los caudales medioambientales en la satisfacción de la demanda de agua bajo escenarios de cambio climático. Ribagua **2**(1), 3–13 (2015). https://doi.org/10.1016/j.riba.2015.04.001
8. Masson Delmotte, V., Panmao, Z., Maycock, T., Moufouma Okia, W., Mattews, R.: Impactos del Calentamiento Global de 1,5 °C con respecto a los niveles preindustriales y las trayectorias correspondientes que deberian seguir las emisiones mundiales de gases de efecto invernadero, en el contexto del reforzamiento de la respuesta mundial (2019)
9. Alarcón Hincapié, J., Zafra Mejía, C., Echeverri Prieto, L.: Cambio climático y recursos hídricos en Colombia Climate change and water resources in Colombia. Rev. U.D.C.A. Actual. Divulg. Científica **2**, 1–10 (2019). https://doi.org/10.31910/rudca.v22.n2.2019.1368
10. Anand, J., Gosain, A.K., Khosa, R.: Prediction of land use changes based on Land Change Modeler and attribution of changes in the water balance of Ganga basin to land use change using the SWAT model. Sci. Total Environ. **644**, 503–519 (2018). https://doi.org/10.1016/j.scitotenv.2018.07.017.
11. Guug, S.S., Abdul Ganiyu, S., Kasei, R.A.: Application of SWAT hydrological model for assessing water availability at the Sherigu catchment of Ghana and Southern Burkina Faso. HydroResearch **3**, 124–133 (2020). https://doi.org/10.1016/j.hydres.2020.10.002
12. Rios Patrón, E.: Manejo De Cuencas Y Adaptación Al Cambio Climático. Secr. Medio Ambient. y Recur. Nat. (SEMARNAT), Red Mex. Cuencas Hidrográficas, August 2015 (2016)
13. Baker, N.C., Huang, H.P.: A comparative study of precipitation and evaporation between CMIP3 and CMIP5 climate model ensembles in semiarid regions. J. Clim. **27**(10), 3731–3749 (2014). https://doi.org/10.1175/JCLI-D-13-00398.1
14. Kim, Y.H., Min, S.K., Zhang, X., Sillmann, J., Sandstad, M.: Evaluation of the CMIP6 multi-model ensemble for climate extreme indices. Weather Clim. Extrem. **29**, 100269 (2020). https://doi.org/10.1016/j.wace.2020.100269
15. Ancco Carita, C., et al.: Evaluacion De Los Recursos Hidricos En La Cuenca Del Rio Ilave (2009)

16. Aguilar Lima, J.C.: Analisis de tendencias de temperatura y precipitacion pluvial en la cuenca del rio Ilave. Universidad Nacional del Altiplano, p. 87 (2017)
17. SENAMHI, "CLIMAS DEL PERÚ – Mapa de Clasificación Climática Nacional. Servicio Nacional de Meteorología e Hidrología del Perú (SENAMHI)," CLIMAS DEL PERÚ. Dirección de Meteorología y Evaluación Ambiental Atmosférica, Lima, p. 100 (2020)
18. Dataset: SENAMHI HSR PISCO. http://iridl.ldeo.columbia.edu/SOURCES/.SENAMHI/.HSR/.PISCO/index.html?Set-Language=es. Accessed 22 Feb 2022
19. Uribe, N.: Conceptos básicos y guía rápida para el usuario. Versión SWAT2005. SWAT Soil Water Assessment Tool, pp. 1–47 (2010)
20. Perez Valdivia, C., Cade Menun, B., Mc Martin, D.W.: Hydrological modeling of the pipestone creek watershed using the Soil Water Assessment Tool (SWAT): assessing impacts of wetland drainage on hydrology. J. Hydrol. Reg. Stud. **14**, 109–129 (2017). https://doi.org/10.1016/j.ejrh.2017.10.004
21. Moriasi, D.N., Gitau, M.W., Pai, N., Daggupati, P.: Hydrologic and water quality models: performance measures and evaluation criteria. Trans. ASABE **58**(6), 1763–1785 (2015). https://doi.org/10.13031/trans.58.10715
22. Gudmundsson, L.: Package 'qmap' statistical transformations for post-processing climate model output. Hydrol. Earth Syst. Sci. (2016). https://doi.org/10.5194/hess-16-3383-2012.bernexp
23. Guevara, O.C., Cazenave, G., Vazquez, A.G., Collazos, G., Vives, L.: Empleo del modelo hidrológico SWAT en regiones de llanuras. Aplicación en una cuenca rural, arroyo Santa Catalina, Provincia de Buenos Aires. II Congr. Int. Hidrol. Llanuras 10 (2014)
24. Lujano Laura, E., Diaz Aguilar, R.D., Lujano Laura, A., Quispe Aragon, J.P.: Evaluacion De La Disponibilidad Hídrica Actual Y Futura En El Contexto Del Cambio Climatico En Las Cuencas Ilave Y Ramis. Rev. Investig. Altoandinas - J. High Andean Investig. **16**(2) (2014). https://doi.org/10.18271/ria.2014.61
25. Moya Alvarez, A.S., Ortega León, J.M., Jurado Pacheco, X.: Evaluación del Modelo Climático Global MIROC5 y estimaciones de temperatura y precipitaciones para las zonas sur y norte del Perú. Apunt. Cienc. Soc. **5**(2), 1–8 (2015). https://doi.org/10.18259/acs.2015028
26. Jodar-Abellan, A., Ruiz, M., Melgarejo, J.: Evaluación del impacto del cambio climático sobre una cuenca hidrológica en régimen natural (SE, España) usando un modelo SWAT. Rev. Mex. Ciencias Geol. **35**(3), 240–253 (2018). https://doi.org/10.22201/cgeo.20072902e.2018.3.564
27. IPCC, Sexto Informe De Evaluación Sobre Cambio Climático 2021, Grup. Intergub. Expert. sobre el Cambio Climático (2021)
28. Lujano-Laura, E., Hidalgo-Sanchez, L.S., Tapia-Aguilar, B., Lujano-Laura, A.: Cambios proyectados de los recursos hídricos bajo escenarios de emisiones RCP4.5 y 8.5 de modelos climáticos globales del CMIP5 en el Altiplano Peruano. Rev. Investig. Altoandinas - J. High Andean Res. **18**(2), 195–204 (2016). https://doi.org/10.18271/ria.2016.200
29. Laura Santos, J.W.: Efecto del Cambio Climaático en la Oferta Hidrica Superficial de la Cuenca del Rio Ilave – Puno. Univ. altiplano Puno **051**, 363543 (2019)
30. Leon, R., Portuguez, D., Chávarri, E.: Modelación de la disponibilidad hídrica del rio Piura - Perú, considerando la incidencia del cambio climático. Rev. Investig. Altoandinas **21**(3), 182–193 (2019). https://doi.org/10.18271/ria.2019.476
31. Felipe Obando, O., Sánchez Bernardo, T., Astorayme Valenzuela, M.: Actualización de los escenarios de disponibilidad hídrica en el Perú en contexto de cambio climático. Serv. Nac. Meteorol. e Hidrol. del Perú (SENAMHI). Dir. Hidrol. 89 (2015)
32. Vázquez-Ochoa, L.A., Correa-Sandoval, A., Vargas-Castilleja, R.D.C., Vázquez-Sauceda, M.D.L.L., Rodríguez-Castro, J.H.: Modelo hidrológico, calidad del agua y cambio climático: soporte para la gestión hídrica de la cuenca del río Soto la Marina. CienciaUAT **16**(1), 20–41 (2021). https://doi.org/10.29059/cienciauat.v16i1.1498

Convergence of Metropolis-Hastings Algorithm with Optimal Mixing Parameter in Bipartite Regular Graphs

Martin Kenyeres[1](✉) and Jozef Kenyeres[2]

[1] Institute of Informatics, Slovak Academy of Sciences, Dubravska cesta 9, 845 07 Bratislava, Slovakia
martin.kenyeres@savba.sk

[2] Frequentis AG, Innovationsstraße 1, 1100 Vienna, Austria
jozef.kenyeres@frequentis.com
http://www.ui.sav.sk/w/en/dep/mcdp/

Abstract. The Metropolis-Hastings algorithm, a multi-functional iterative procedure applicable in various technical/non-technical industries, has been attracting great attention of the scientific community since its definition. In this paper, we address the generalized Metropolis-Hastings algorithm for distributed averaging in bipartite regular graphs, where this algorithm with the optimal mixing parameter diverges. We identify that the convergence of the algorithm in bipartite regular graphs is achieved (in spite of the initial divergence), provided that a distributed mechanism for ensuring the convergence of the average consensus algorithm with the Maximum-degree weights in this mentioned critical graph topology is applied. Moreover, we examine in the paper how to reconfigure the algorithm the most efficiently after detecting a bipartite regular topology in a distributed way. The experimental part is concerned with the convergence rate of bounded Metropolis-Hastings algorithm with the optimal mixing parameter in bipartite regular graphs of various degrees when the mentioned mechanism with different rounding accuracy is used.

Keywords: Bipartite regular graphs · Consensus · Data aggregation · Data fusion · Distributed algorithms · Information fusion · Metropolis-Hastings algorithm · Multi-sensor data fusion · Sensor fusion

1 Introduction

Multi-agent systems (MASs) are generally understood as a loosely interrelated computerized collection of autonomous intelligent components (referred to as agents of MAS) that regularly interact with their peers and adjacent environment [1–4]. MASs often consist of numerous agents, whose goal is to efficiently solve complicated issues (impractical, hard, or even impossible to solve by individuals or monolithic systems) [5,6]. The agents of MAS, flexible hardware/software/hybrid entities, are required to quickly react to environmental

R. Silhavy (Ed.): CSOC 2022, LNNS 501, pp. 484–498, 2022.
https://doi.org/10.1007/978-3-031-09070-7_40

impulses, regularly exchange information with their peers/people, reliably operate with no supervision, and act independently [7]. They cooperate or compete with each other in order to collectively solve their design objectives [8]. Among the requirements placed on topical MASs, we can include simple and time-undemanding implementation, high performance, financial affordability, high availability, efficiency, easy maintenance, the flexibility of architecture, etc. [9–12]. As a literature review shows, MASs have found application in various fields over the past decades (e.g., grid computing, computational biology, military, monitoring, etc.) [13–17]. The general architecture of an agent in MAS is depicted in Fig. 1.

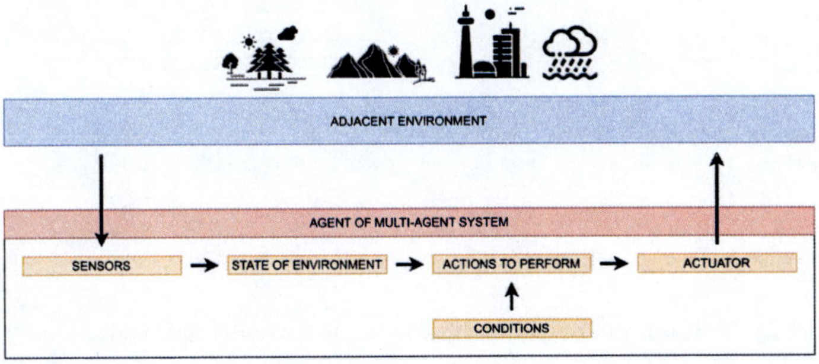

Fig. 1. General architecture of agent forming multi-agent systems

As the data volume is drastically growing nowadays, it is fundamental to find ways to effectively express a mass of data in a transparent compiled form for further uncomplicated analyzing/processing [18]. For this purpose, algorithms for data aggregation (sometimes referred to as data aggregation tools) are used as a complement of many modern MAS applications, as shown in [19–22]. Generally, the term "data aggregation" is used to refer to any integration process of extensive raw information amount into a comprehensive, high-level summary form representing overall statistics of the processed data set [18]. Simply said, data aggregation is applied to summarize a numeric/non-numeric data set from multiple, independent, and disparate sources of information [18]. The whole process of data aggregation can be generally divided into three phases, namely [23, 24]:

- ☐ *Collecting data* – extracting data from multiple information sources and storing it.
- ☐ *Processing data* – correcting damaged data and removing garbage information from the collected data.
- ☐ *Presenting data* – expressing the data in a readable and transparent form.

As shown in the literature, one of the most frequently applied classes of data aggregation tools is consensus-based algorithms for estimating/determining aggregate functions (e.g., the arithmetic mean, the sum, the extreme, the count, etc.) [25–28]. A consensus problem (sometimes referred to as an agreement problem) is a scientific discipline addressing how to enable multiple independent entities to agree on some value, finding various applications in modern MASs (e.g., leader election, data aggregation, prediction of flocking behavior, coordination of robotic/sensor systems, community management decisions, etc.) [29–31]. Typically, consensus-based algorithms are based on local interactions of adjacent entities (either selfish or cooperative) and regular exchanges of information among these entities. As stated in [29], a consensus problem can be understood as a process of convergence to a common value and is divided into three challenges: operating, negotiating, and reaching the agreement. See Fig. 2 for an example of a people group that is/is not in the consensus.

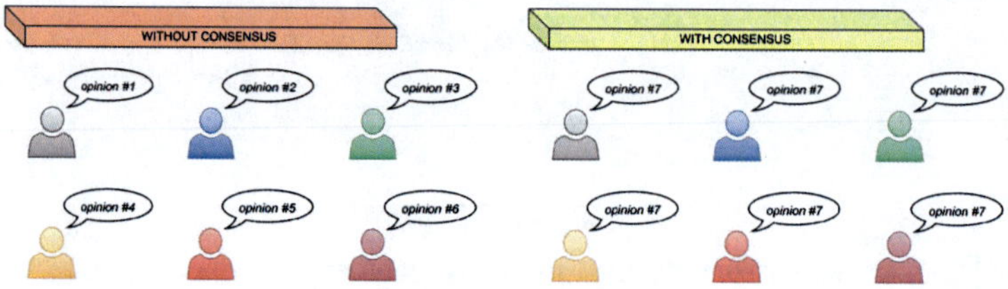

Fig. 2. Example of people group with and without consensus

In this paper, we address the Metropolis-Hastings algorithm (sometimes referred to as the Hastings-Metropolis algorithm), an iterative scheme defined by Nicolas C. Metropolis, Charles S. Hastings, and their scientific teams [32]. More specifically, we focus our attention on its generalized variant (GMH) defined by Valentin Schwarz et al. in [33]. The Metropolis-Hastings algorithms, one of the most common Markov chain Monte Carlo methods, was originally designed to generate samples from complicated probability distributions that are difficult to sample [32]. Since its definition, it has found the application in various fields such as statistics, computer science, financial sector, etc. [34–36]. As mentioned earlier in this paper, our goal in this paper is to address the consensus-based GMH with the optimal mixing parameter for distributed averaging over bipartite regular graphs. More specifically, we apply the distributed mechanism (defined in [37]) for detecting the average consensus algorithm with the Maximum-degree weights (AC MD) in these graphs in order to identify whether the convergence of GMH with the optimal mixing parameter in this critical topology can be ensured despite the initial divergence of the algorithm. Also, we analyze its convergence rate, how to optimally reconfigure the algorithm, and examine the impact of the graph degree on the algorithm performance. In our experiments, GMH has bounded execution and is examined under three different scenarios.

The content of the paper is organized as follows: Sect. 2 provides a theoretical insight into the topic. In Sect. 3, we introduce the applied research methodology and metric to evaluate the performance of the examined algorithm. Section 4 is focused on our numerical experiments executed in Matlab2018a and a discussion about observed phenomena. Section 5 briefly summarizes the main contribution presented in this paper.

2 Theoretical Background into Topic

This section, which is divided into two subsections, provides an insight into GMH, bipartite regular graphs, and the applied distributed mechanism for detecting whether a graph is biparite regular.

2.1 Generalized Metropolis-Hastings Algorithm in Bipartite Regular Graphs

As mentioned earlier in this paper, we focus on the Metropolis-Hastings algorithm (more specifically, its generalized variant) over bipartite regular graphs, where it diverges if the optimal mixing parameter is used. In this subsection, we define the graphs of our interest, GMH, and provide other relevant information. At first, let us turn our attention to the definition of bipartite regular graphs [37]:

Definition 1. *An undirected simple finite graph G*[1] *is considered to be bipartite regular, provided that*

- ☐ *its vertex set* **V** *can be split into two disjoint subsets (i.e., two arbitrary vertices are adjacent unless they are in the same disjoint subset).*[2]
- ☐ *the degrees of all the vertices from* **V** *are equal to each other.*[3] *The label "d–regular" indicates that the degree of each vertex is equal to d.*

In Fig. 3, we show four graph topologies with the graph order $n = 4$ in order to demonstrate the difference between four graph types.

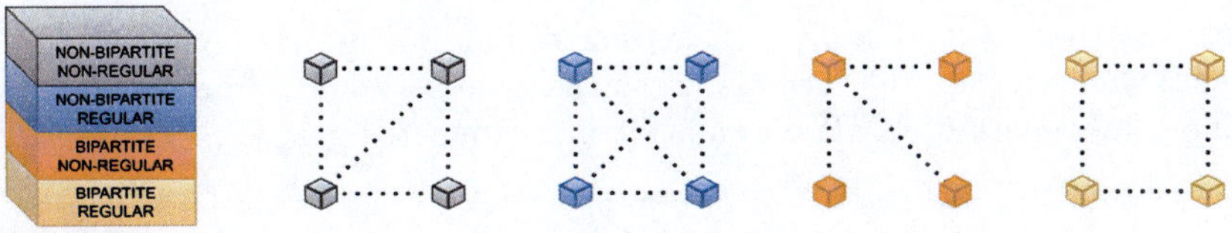

Fig. 3. Example of four graph types with graph order $n = 4$

Now, we draw the readers' attention to GMH for distributed averaging. As mentioned earlier, it is a distributed iterative consensus-based algorithm whose update rule can be expressed as the following difference Eq. [38]:

[1] determined by the vertex set **V** (v_i is the label of a vertex) and the edge set **E** (e_{ij} is the label of an edge linking v_i and v_j).

[2] Definition of bipartite graphs.

[3] Definition of regular graphs.

$$\mathbf{x}(k+1) = \mathbf{W}\mathbf{x}(k) \tag{1}$$

Here, $\mathbf{x}(k)$ is a variant column vector whose entries are equal to the inner states of all the agents at a kth iteration. Note that $\mathbf{x}(0)$ represents the initial inner states. The symbol $\mathbf{W}$ represents the weight matrix of the algorithm, which is defined for GMH in bipartite regular graphs as follows [33]:

$$[W]_{ij} = \begin{cases} \frac{1}{d+\epsilon}, & \text{if } e_{ij} \in \mathbf{E} \\ 1 - \sum_{i \neq l} [W]_{il}, & \text{if } i = j \\ 0, & \text{otherwise} \end{cases} \tag{2}$$

Note that the generalized variant differs from its original definition in adding the mixing parameter ϵ to the formula whereby the entries of the weight matrix $\mathbf{W}$ can be easily reconfigured (in its original definition, the weight matrix is unconfigurable, and the mixing parameter ϵ is substituted by one). In the algorithm, the inner state of each agent asymptotically converges to the value of the arithmetic mean, which can be expressed as follows [37]:

$$\lim_{k\to\infty} \mathbf{x}(k) = \lim_{k\to\infty} \mathbf{W}^k \mathbf{x}(0) = \frac{1}{n}\mathbf{1}\mathbf{1}^{\mathrm{T}}\mathbf{x}(0) \tag{3}$$

Here, $\mathbf{1}$ is an all-ones column vector, and $\mathbf{1}^{\mathrm{T}}$ represents its transpose. As stated in [33], the limit defined in (3) exists if all these three convergence conditions depending on the entries of the weight matrix $\mathbf{W}$ are met:

$$\mathbf{1}^{\mathrm{T}}\mathbf{W} = \mathbf{1}^{\mathrm{T}} \tag{4}$$

$$\mathbf{W}\mathbf{1} = \mathbf{1} \tag{5}$$

$$\rho\left(\mathbf{W} - \frac{1}{n}\mathbf{1}\mathbf{1}^{\mathrm{T}}\right) < 1 \tag{6}$$

Here, ρ is the spectral radius of the corresponding matrix. As identified by Schwarz et al. in [33], the formula (7) is the necessary and sufficient conditions for the convergence of GMH on an arbitrary connected graph.

$$\epsilon > 0 \tag{7}$$

However, as further stated in [33], the convergence is also obtained for $\epsilon = 0$, provided that a graph is not bipartite regular (see Fig. 4 and Fig. 5 for a comparison of GMH with $\epsilon = 0$ and $\epsilon = 1$ in a bipartite regular graph). Besides, it is identified that the algorithm with $\epsilon = 0$ performs better than with greater values of ϵ; therefore, we refer to this initial configuration as the optimal mixing parameter [33]. As further identified in [33,37], the weight matrices $\mathbf{W}$ of GMH with $\epsilon = 0$ and AC MD are identical in bipartite regular graphs and determinable as:

$$\mathbf{W} = \frac{1}{d}\mathbf{A} = \frac{1}{d}\begin{bmatrix} 0 & \mathbf{A}_0 \\ \mathbf{A}_0^{\mathrm{T}} & 0 \end{bmatrix} \tag{8}$$

Here, $\mathbf{A}$ is the adjacency matrix of the analyzed graph. As stated in [37], the smallest eigenvalue of the weight matrix $\lambda_n(\mathbf{W})$ is always equal to minus one when the weight matrix $\mathbf{W}$ of AC MD is applied in bipartite regular graphs (let us recall that its weight matrix is identical to the weight matrix of GMH with $\epsilon = 0$ in bipartite regular graphs). Thus, the third convergence condition of GMH with $\epsilon = 0$ is broken, as identified in [37]:

$$\begin{aligned}&\lambda_n(\mathbf{W}) = -1 \wedge \lambda_n(\mathbf{W}) < \lambda_2(\mathbf{W}) < 1 \implies \\ &max\{\lambda_2(\mathbf{W}), -\lambda_n(\mathbf{W})\} = 1 \implies \rho\left(\mathbf{W} - \frac{1}{n}\mathbf{1}\mathbf{1}^{\mathrm{T}}\right) = 1\end{aligned} \tag{9}$$

Here, $\lambda_2(\mathbf{W})$ is the second-largest eigenvalue of the weight matrix $\mathbf{W}$. Thus, the formula (9) causes that GMH with $\epsilon = 0$ diverges in bipartite regular graphs [37]. See our previous paper [37] for further details about AC MD in bipartite regular graphs.

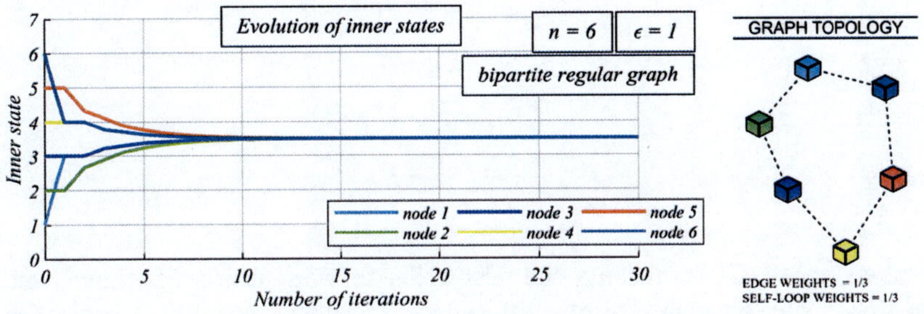

Fig. 4. Evolution of inner states in bipartite regular graph – algorithm converges as mixing parameter is equal to one

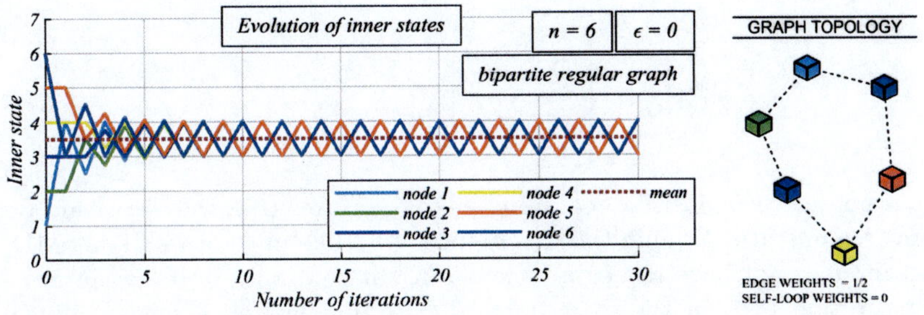

Fig. 5. Evolution of inner states in bipartite regular graph – algorithm diverges as mixing parameter is equal to zero

2.2 Distributed Mechanism for Detecting Metropolis-Hastings Algorithm with Optimal Mixing Parameter in Bipartite Regular Graphs

In this subsection, we turn our attention to the mechanism for detecting the algorithm in bipartite regular graphs defined in [37]. It is a very accurate distributed approach based on the first-order forward and backward finite-difference of the inner states. As discussed in [33,37], it is important to find out whether GMH with $\epsilon = 0$/AC MD is executed over bipartite regular graphs since these two algorithms diverge in this critical topology. Thus, the algorithms are unable to fulfill their design functionality, i.e., estimating the desired aggregate function [37]. Thus, we apply the mechanism determined by five conditions provided in (10)–(14) to overcome this issue. This mechanism classifies a graph as bipartite regular at an iteration if all these five conditions are met at this iteration. Note that each agent in MAS executes this classification independently from the other agents and only according to distributedly obtainable information.

$$\nabla_2^1 \, x_i(k+1) = 0 \tag{10}$$

$$\nabla_2^1 \, x_i(k) = 0 \tag{11}$$

$$\triangle^1 \, x_i(k) = \nabla^1 x_i(k-1) \tag{12}$$

$$\triangle^1 \, x_i(k) \neq 0 \tag{13}$$

$$\nabla^1 \, x_i(k-1) \neq 0 \tag{14}$$

Here, the symbol $\triangle^1$ represents the first-order forward finite-difference, and ∇^1 is the first-order backward finite-difference. The lower symbol 2 indicates the difference between inner states that are two iterations away from one another. When no lower symbol is provided refers to the difference between two subsequent inner states. Once all the agents detect that a graph is bipartite regular, the mixing parameter ϵ can be reconfigured to a value that ensures the convergence of the algorithm [37]. Thus, after detection is completed, the weight matrix $\mathbf{W}$ is required to be reconfigured as follows:

$$[W]_{ij} = \begin{cases} \frac{1}{d+0+c}, & \text{if } e_{ij} \in \mathbf{E} \\ 1 - \sum_{i \neq l} [W]_{il}, & \text{if } i = j \\ 0, & \text{otherwise} \end{cases} \tag{15}$$

The parameter c is a positive real number that ensures that the algorithm begins to converge in spite of the initial divergence. Just like in the case of AC MD, the algorithm does not have to begin once again but continues in its execution after reconfiguration [37]. In the experimental part, we deal with how to optimally change this parameter. See Fig. 6 for the evolution of the inner states after GMH is reconfigured.

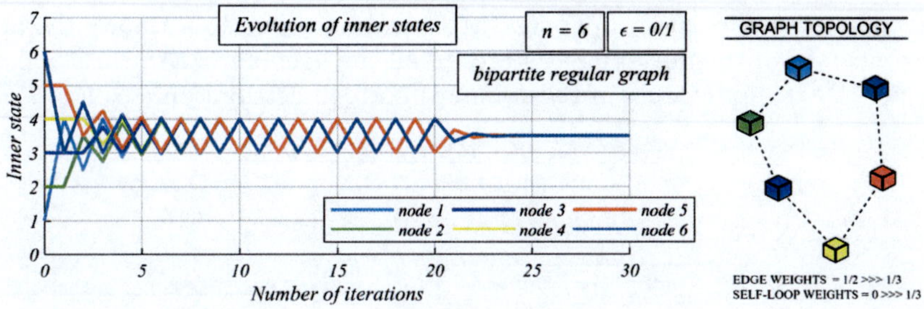

Fig. 6. Evolution of inner states in bipartite regular graph – algorithm is reconfigured

3 Applied Research Methodology

In this section, we introduce the applied research methodology and metric to evaluate the performance of the examined algorithm. As mentioned earlier, the convergence rate (expressed as the number of iteration required for the consensus achievement) of GMH with $\epsilon = 0$ reconfigured by c taking the values from the following interval (the step is equal to 0.1) is examined in this paper.

- $c \in [0.1, 6]$

As already mentioned above, the execution of the algorithm is bounded by a stopping criterion - in this paper, we apply the one determined by the parameter P (its value conditions the precision of the final estimates) and defined as:

$$\left| x_i(k) - \frac{\mathbf{1}^{\mathrm{T}}\mathbf{x}(0)}{n} \right| < P, \ for \ \forall v_i \in \mathbf{V} \tag{16}$$

In order to evaluate the performance of the algorithm in our experiments, we measure the number of iterations required for the consensus achievement, i.e., the iteration number when (16) is met for the first time.

All the experiments presented in this paper are carried out in randomly generated bipartite d-regular graphs, all with the graph order $n = 30$. For each d, we generate 100 unique graphs (in order to ensure high credibility of the presented research), and only the average (over 100 graphs) of the applied metric is shown in the figures provided in the experimental section. We assume an error-free execution in all the performed experiments, i.e., the execution of GMH is affected by no potential failures such as dead nodes, Gaussian noise, quantized noise, link failures, etc. The value of the vertex degree takes the following four values in our experiments:

- $d = \{2, 3, 4, 5\}$

As mentioned earlier in this paper, GMH can be applied to estimate the arithmetic mean from the initial inner states of all the agents in MAS. In our paper, they are IID random values of the standard normal distribution (see Fig. 7), i.e.,

$$x_i(0) \sim N(0,1),\ for\ \forall v_i \in \mathbf{V} \tag{17}$$

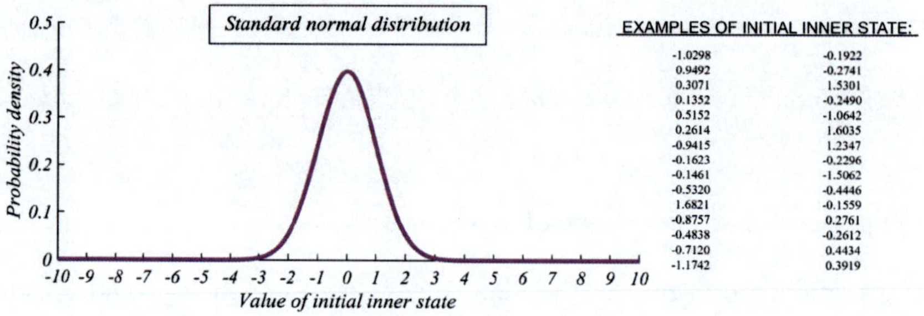

Fig. 7. Probability density of standard normal distribution

As stated in [37], it is necessary to round the inner states processed by the mechanism; therefore, we analyze three scenarios differing from each other in the precision of rounding. Namely, we analyze these scenarios in the paper:

- ☐ *Scenario 1* (S1): the inner states are rounded to decimals.
- ☐ *Scenario 2* (S2): the inner states are rounded to hundreds.
- ☐ *Scenario 3* (S3): the inner states are rounded to thousands.

Note that the inner states utilized by GMH are not rounded the ways provided above.

4 Experiments and Discussion About Observed Phenomena

This section consists of the results obtained in Matlab2018a, where our experiments are performed using either Matlab built-in functions or software designed by the authors of this paper. As mentioned above in this paper, we apply the distributed mechanism determined by (10)–(14) and analyze the convergence/the convergence rate of bounded GMH with the optimal mixing parameter ϵ (GMH is bounded according to the stopping criterion defined in (16) with $P = 0.0001$) over bipartite regular graphs with various d under three different scenarios. Moreover, we examine the impact of various values of c on the performance of the algorithm after its reconfiguration in order to identify the optimal way of reconfiguring GMH in bipartite regular graphs.

First of all, let us focus our attention on the results for $d = 2$, which are shown in Fig. 8. From the figure, we can see that the algorithm converges (proved by the fact that the number of iterations can be measured since the inner states converge to the arithmetic mean after reconfiguration), and the best performance is observed to be achieved in Scenario 3, while, the worst one in Scenario 1. Also, as seen from the figure, the difference between the analyzed scenarios grows as the value of c is increased. Besides, it is seen that an increase in the value of c results in an increase in the number of iterations for the consensus achievement (i.e., the convergence rate is decreased). Thus, the highest performance is achieved in all the three scenarios, provided that $c = 0.1$ (the minimal analyzed value of this parameter). As further seen from the figure, the highest convergence rate is equal to 367.7 it. in Scenario 1, 366.1 it. in Scenario 2, and 361.1 it. in Scenario 3.

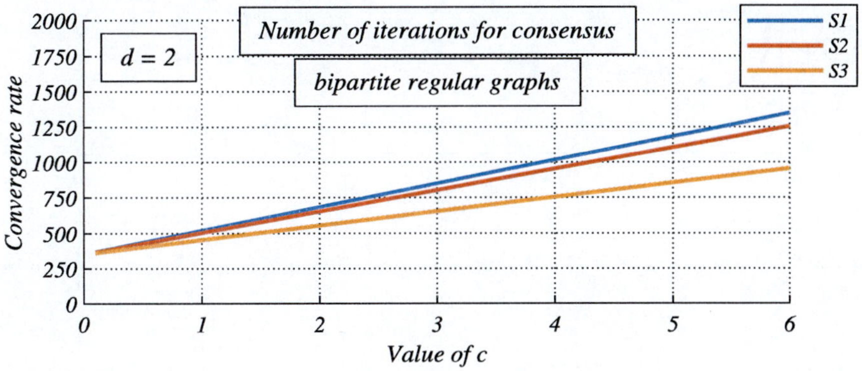

Fig. 8. Convergence rate of GMH (averaged over 100 unique graphs) in all three scenarios - $d = 2$

Now, let us turn our attention to Fig. 9, where the results for $d = 3$ are depicted. We can see from this figure that the algorithm converges again. However, the highest performance is not achieved for the lowest applied value of c (in contrast to the previous analysis) but for $c = 0.3$ in Scenario 1, 0.4 in Scenario 2, and 0.7 in Scenario 3. Thus, we can see that an increase in the rounding accuracy causes that this value of c is greater. Furthermore, it is seen that the convergence rate increases as c grows, reaches its maximum, and begins to decline with a further increase in c. This character of the functions is also presumed for very low values of c in the previous analysis, but we cannot experimentally verify this statement due to computational constraints of our simulation technique. For lower values of c, the highest performance is achieved in Scenario 1 and the worst one in Scenario 3 (unlike the previous analysis). In contrast, the maximum convergence rate is achieved in Scenario 3 and the worst one in Scenario 1 (like in the previous analysis), however for a different value of c, as mentioned earlier. The highest convergence rate is 73.77 it. in Scenario 1, 73.75 it. in Scenario 2, and 73.68 it. in Scenario 3 - so, we see that the difference in this aspect is just marginal among the examined scenarios. Then, as c is increased, the convergence rate decreases, and the difference among the examined scenarios increases like in previous analysis. For higher values of c, the best performance is reached in Scenario 3, meanwhile, the worst one is observed in Scenario 1 (like in the previous analysis). Also, it is identified that both the maximal and the average convergence rate are higher than in the previous analysis (i.e., $d = 3$ causes better performance of GMH than $d = 2$).

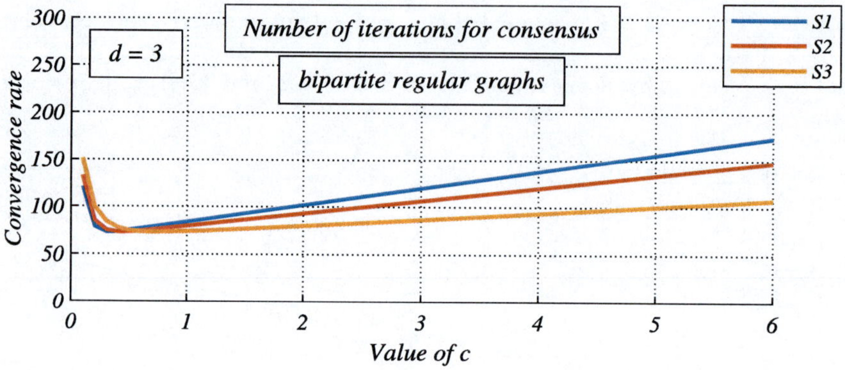

Fig. 9. Convergence rate of GMH (averaged over 100 unique graphs) in all three scenarios - $d = 3$

In the two last figures (Fig. 10 and Fig. 11), we depict the results for $d = 4$ and $d = 5$, respectively. From the results, one can see that the character of the functions is the same as in the previous analysis, and the algorithm converges in both cases again. One of the differences is that the value of c with which GMH achieves the maximal convergence rate varies. For $d = 4$, we can see that this value takes the following values: $c = 0.9$ in Scenario 1, 1.2 in Scenario 2, and 2.8 in Scenario 3. For $d = 5$, these values of c are greater compared to the previous case and equal to 1.7 in Scenario 1, 2.6 in Scenario 2, and 5 in Scenario 3. Also, the maximal convergence rate differs and takes the following values for $d = 4$ in the examined scenarios: 36.99 it. in Scenario 1, 36.78 it. in Scenario 2, and 36.12 it. in Scenario 3. For $d = 5$, its value equals 24.88 it. in Scenario 1, 24.49 it. in Scenario 2, and 22.46 it. in Scenario 3, as seen from the figure. Thus, a marginal difference among the scenarios is only observed in both cases just like in the previous analysis. As further observed from Fig. 9 – Fig. 11, the performance of the algorithm is higher for smaller d, provided that c takes lower values. On the other hand, GMH performs better for higher values of d if c is equaled to higher values. Also, the maximal convergence rate is higher when d is greater. Furthermore, we observe that the difference in c with which the maximal convergence rate is achieved is more significant among the analyzed scenarios with an increase in d.

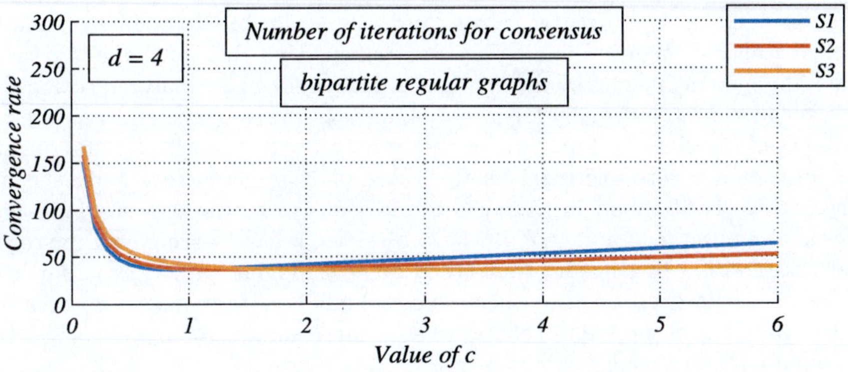

Fig. 10. Convergence rate of GMH (averaged over 100 unique graphs) in all three scenarios - $d = 4$

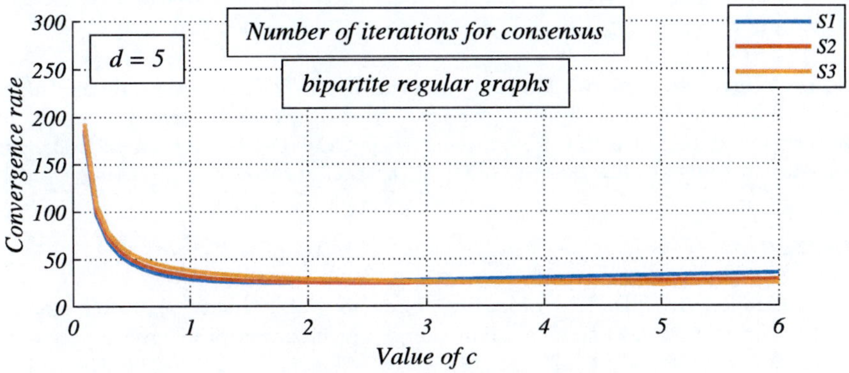

Fig. 11. Convergence rate of GMH (averaged over 100 unique graphs) in all three scenarios - $d = 5$

5 Conclusion

In this paper, we apply a distributed mechanism for detecting whether or not a graph is bipartite regular to bounded GMH with the optimal mixing parameter. We experimentally verify that GMH achieves the converge in bipartite regular graphs of various d, provided that the algorithm is reconfigured after successful detection. From the experimentally obtained results, it can be seen that an increase in the value of c results in a decrease in the number of iterations for the consensus achievement for lower values of this parameter (except for $d = 2$, where the convergence rate declines on the whole interval of the examined values of c). On this interval, the best performance is achieved in Scenario 1, while, the worst one is in Scenario 3. It is then observed that there is the value of c that ensures the maximal convergence rate (in the case of $d = 2$, this maximum is achieved for the lowest analyzed c in all the three examined scenarios) - its value is conditioned by d (its greater values causes this value of c to be higher) and the applied scenario (higher rounding accuracy causes that this value of c is greater). The difference in the maximal convergence rate among the analyzed scenarios is however marginal, especially for $d = 3$, 4, and 5. The maximal convergence rate is observed in Scenario 3 and worst one in Scenario 1 for each d. Subsequently, the convergence rate decreases as the value of c is increased - here, the best performance is observed in Scenario 3 and the worst one in Scenario 1 again (this is also observed for each c if $d = 2$). Besides, it can be seen from the results that lower values of c ensure that GMH achieves higher performance for lower values of d. However, on the other hand, a higher performance is observed for higher values of d, provided that c takes greater values. Also, we can see that the maximal convergence rate is increased as d grows.

The most important observable fact is that GMH with the optimal mixing parameter achieves the convergence in bipartite regular graphs (despite the initial divergence in this critical topology), provided that the mentioned mechanism is applied, and the algorithm is subsequently reconfigured the way it is identified in this paper.

Acknowledgment. This work was supported by the VEGA agency under the contract No. 2/0155/19 and by the project Cyber-Physical System for smart telemonitoring and tele-medicine for patients with COVID-19 (BAS-SAS-21-03). Since 2019, Martin Kenyeres has been a holder of the Stefan Schwarz Supporting Fund.

References

1. Lez-Briones, A.G., De La Prieta, F., Mohamad, M.S., Omatu, S., Corchado, J.M.: Multi-agent systems applications in energy optimization problems: a state-of-the-art review. Energies **11**, 1928 (2018). https://doi.org/10.3390/en11081928
2. Dominguez, R., Cannella, S.: Insights on multi-agent systems applications for supply chain management. Sustainability **12**, 1935 (2020). https://doi.org/10.3390/su12051935
3. Liang, C., et al.: Intrusion detection system for the internet of things based on blockchain and multi-agent systems. Electronics **9**, 1120 (2020). https://doi.org/10.3390/electronics9071120
4. Pokorny, J., Seda, P., Seda, M., Hosek, J.: Modeling optimal location distribution for deployment of flying base stations as on-demand connectivity enablers in real-world scenarios. Sensors **21**, 5580 (2021). https://doi.org/10.3390/s21165580
5. Dorri, A., Kanhere, S.S., Jurdak, R.: Multi-agent systems: a survey. IEEE Access **6**, 28573–28593 (2018). https://doi.org/10.1109/ACCESS.2018.2831228
6. Dong, Y., Lin, Z.: An event-triggered observer and its applications in cooperative control of multi-agent systems. IEEE Trans. Automat. Contr. (2021). https://doi.org/10.1109/TAC.2021.3105322
7. Kenyeres, M., Kenyeres, J.: Comparative study of distributed consensus gossip algorithms for network size estimation in multi-agent systems. Future Internet **13**, 134 (2021). https://doi.org/10.3390/fi13050134
8. Peng, Z., et al.: A novel optimal bipartite consensus control scheme for unknown multi-agent systems via model-free reinforcement learning. Appl. Math. Comput. **369**, 124821 (2020). https://doi.org/10.1016/j.amc.2019.124821

9. Fiore, D., Russo, G.: Resilient consensus for multi-agent systems subject to differential privacy requirements. Automatica (Oxf.) **106**, 18–26 (2019). https://doi.org/10.1016/j.automatica.2019.04.029
10. Guedes, G.T.A.: MASRML-a domain-specific modeling language for multi-agent systems requirements. Int. J. Softw. Eng. Appl. **11**, 25–45 (2020). https://doi.org/10.5121/ijsea.2020.11503
11. Törsleff, S., Hildebrandt, C., Fay, A.: Development of ontologies for reasoning and communication in multi-agent systems. In: 11th International Joint Conference on Knowledge Discovery, Knowledge Engineering and Knowledge Management (IC3K 2019), pp. 450–457. Springer Publishing, New York (2019). https://doi.org/10.5220/0008355804500457
12. Khan, M.W., Wang, J., Ma, M., Xiong, L., Li, P., Wu, F.: Optimal energy management and control aspects of distributed microgrid using multi-agent systems. Sustain. Cities Soc. **44**, 855–870 (2019). https://doi.org/10.1016/j.scs.2018.11.009
13. Nair, A.S., Hossen, T., Campion, M., Selvaraj, D.F., Goveas, N., Kaabouch, N., Ranganathan, P.: Multi-agent systems for resource allocation and scheduling in a smart grid. Technol. Econ.Smart Grids Sust. Energy **3**(1), 1–15 (2018). https://doi.org/10.1007/s40866-018-0052-y
14. Tianfield, H., Unland, R.: Towards self-organization in multi-agent systems and grid computing. Multiagent Grid Syst. **1**, 89–95 (2005). https://doi.org/10.3233/MGS-2005-1203
15. Seda, M., Seda, P.: Computational geometry data structures in logistics and navigation tasks. In: 12th International Congress on Ultra Modern Telecommunications and Control Systems and Workshops, (ICUMT 2020), pp. 179–184. IEEE Press, New York (2020). https://doi.org/10.1109/ICUMT51630.2020.9222453
16. Cil, I., Mala, M.: A multi-agent architecture for modelling and simulation of small military unit combat in asymmetric warfare. Expert Syst. Appl. **37**, 1331–1343 (2010). https://doi.org/10.1016/j.eswa.2009.06.024
17. Villarrubia, G., De Paz, J.F., De La Iglesia, D.H., Bajo, J.: Combining multi-agent systems and wireless sensor networks for monitoring crop irrigation. Sensors **17**, 1775 (2017). https://doi.org/10.3390/s17081775
18. Kenyeres, M., Kenyeres, J.: Average consensus over mobile wireless sensor networks: weight matrix guaranteeing convergence without reconfiguration of edge weights. Sensors **20**, 3677 (2020). https://doi.org/10.3390/s20133677
19. Zou, Y., Zuo, Z., Xia, K.: Sampled-data distributed protocol for coordinated aggregation of multi-agent systems subject to communication delays. Nonlinear Anal. Hybri. **43**, 101108 (2021). https://doi.org/10.1016/j.nahs.2021.101108
20. Sabo, R., Krammer, P., Mojzis, J., Kvassay, M.: Identification of spontaneous spoken texts in slovak. Jazykoved. čas. **70**, 481–490 (2019). https://doi.org/10.2478/jazcas-2019-0076
21. Sutagundar, A.V., Manvi, S.S.: Fish bone structure based data aggregation and routing in wireless sensor network: multi-agent based approach. Telecommun. Syst. **56**(4), 493–508 (2013). https://doi.org/10.1007/s11235-013-9769-z
22. Mojzis, J., Balogh, S.: Breaking microsoft azure information protection viewer using memory dump. Adv. Intell. Syst. Comput. **1294**, 913–920 (2020). https://doi.org/10.1007/978-3-030-63322-6_78
23. Ullah, A., Azeem, M., Ashraf, H., Alaboudi, A.A., Humayun, M., Jhanjhi, N.Z.: Secure healthcare data aggregation and transmission in IoT - a survey. IEEE Access **9**, 16849–16865 (2021). https://doi.org/10.1109/ACCESS.2021.3052850

24. Pourghebleh, B., Navimipour, N.J.: Data aggregation mechanisms in the Internet of things: a systematic review of the literature and recommendations for future research. J. Netw. Comput. Appl. **97**, 23–34 (2017). https://doi.org/10.1016/j.jnca.2017.08.006
25. He, J., Cai, L., Cheng, P., Pan, J., Shi, L.: Consensus-based data-privacy preserving data aggregation. IEEE Trans. Automat. Contr. **64**, 5222–5229 (2019). https://doi.org/10.1109/TAC.2019.2910171
26. Chang, J., Liu, F.: A byzantine sensing network based on majority-consensus data aggregation mechanism. Sensors **21**, 248 (2021). https://doi.org/10.3390/s21010248
27. He, J., Cai, L., Cheng, P., Pan, J., Shi, L.: Distributed privacy-preserving data aggregation against dishonest nodes in network systems. IEEE Internet Things J. **6**, 1462–1470 (2021). https://doi.org/10.1109/JIOT.2018.2834544
28. Wang, Y.C., Chen, T.C.T.: A partial-consensus posterior-aggregation FAHP method-supplier selection problem as an example. Mathematics **7**, 179 (2021). https://doi.org/10.3390/math7020179
29. Abdulghafor, R., Abdullah, S.S., Turaev, S., Othman, M.: An overview of the consensus problem in the control of multi-agent systems. Automatika **59**, 143–157 (2018). https://doi.org/10.1080/00051144.2018.1492688
30. Olfati-Saber, R., Fax, J.A., Murray, R.M.: Consensus and cooperation in networked multi-agent systems. Proc. IEEE **95**, 215–233 (2007). https://doi.org/10.1109/JPROC.2006.887293
31. Ni, W., Cheng, D.: Leader-following consensus of multi-agent systems under fixed and switching topologies. Syst. Control. Lett. **59**, 209–217 (2010). https://doi.org/10.1016/j.sysconle.2010.01.006
32. Hastings, W.K.: Monte Carlo sampling methods using Markov chains and their applications. Biometrika **57**, 97–109 (1970). https://doi.org/10.1093/biomet/57.1.97
33. Schwarz, V., Hannak, G., Matz, G.: On the convergence of average consensus with generalized metropolis-hasting weights. In: 2014 IEEE International Conference on Acoustics, Speech, and Signal Processing, (ICASSP 2014), pp. 5442–5446. IEEE Press, New York (2019). https://doi.org/10.1109/ICASSP.2014.6854643
34. Chib, S., Greenberg, E.: Understanding the metropolis-hastings algorithm. Am. Stat. **49**, 327–335 (1995). https://doi.org/10.1080/00031305.1995.10476177
35. El Chammie, M., Açıkmeşe, B.: Safe Metropolis-Hastings algorithm and its application to swarm control. Syst. Control. Lett. **111**, 40–48 (2018). https://doi.org/10.1016/j.sysconle.2017.10.006
36. Chen, P., Xu, R.X.: Metropolis-hastings adaptive algorithm and its application. Syst. Eng. **1**, 100–108 (2008)
37. Kenyeres, M., Kenyeres, J.: Distributed mechanism for detecting average consensus with maximum-degree weights in bipartite regular graphs. Mathematics **9**, 3020 (2021). https://doi.org/10.3390/math9233020
38. Xiao, L., Boyd, S.: Fast linear iterations for distributed averaging. Syst. Control. Lett. **53**, 65–78 (2004). https://doi.org/10.1016/j.sysconle.2004.02.022

A Concept of Nowcasting of Convective Precipitation Using an X-band Radar for the Territory of the Zlín Region (Czech Republic)

Irina Malkin Ondík[1], Lukáš Ivica[1], Peter Šišan[1], Ivan Martynovskyi[1], David Šaur[2](✉), and Ladislav Gaál[1]

[1] MicroStep-MIS, Čavojského 1, 841 04 Bratislava, Slovakia
[2] Faculty of Applied Informatics, Tomáš Bat'a University in Zlín, Nad Stráněmi 4511, 760 01 Zlín, Czech Republic
saur@utb.cz

Abstract. The paper presents the first results of precipitation nowcasting that is targeted to a smaller administrative unit in the Czech Republic (Zlín region), to aid better preparedness of the staff of crisis management of the region. The methodology is based on the scans of the X-band radar located at Holešov (240 m above the sea level). More specifically, the CAPPI 2000 data were used, which, in the first step, were converted into a 4-category system of warning classes (0: no event; 1/2/3: weak/moderate/strong events) for all the 13 smaller administrative districts of the region. Besides the radar echo extrapolation method TITAN that is operationally available in the region, a further extrapolation-based nowcasting method (COTREC) was selected and compared with a third approach (Persistence) that supposes no changes in the observed field of precipitation. The predictions of the CAPPI warning classes within the 13 districts with the lead times ranging from 5 to 30 min with 5 min intervals were compared with the observed CAPPI warning classes. Since the analysis was carried out in the framework of imbalanced ordinary regression, several issue-specific binary metrics had to be developed. Our first results indicate that the COTREC model outperformed the TITAN method, both in terms of the POD (probability of detection) and FAR (false alarm ratio) statistics, for all the lead times and the relevant binary metrics. Further we found that the reliability of the predictions sharply decreases with increasing lead times, and they are of low practical usability beyond the horizon of 30–40 min.

Keywords: Radar · CAPPI · Nowcasting · Class conversion · Imbalanced ordinal regression · TITAN · COTREC · Persistence · Performance metric

1 Introduction

Anthropogenic-driven climate change is expected to change (predominantly to increase) the frequency and the intensity of various types of extreme weather events, including the extreme rainfall, all around the world. Climate models project changes in the distribution

R. Silhavy (Ed.): CSOC 2022, LNNS 501, pp. 499–514, 2022.
https://doi.org/10.1007/978-3-031-09070-7_41

of the precipitation amounts: the frequency of occurrence extremes (both heavy precipitation and no precipitation, i.e., dry spells) is generally expected to increase, whereas the frequency of occurrence of moderate rainfall intensities is expected to decrease [1, 2]. Hosseinzadehtalaei et al. [3] concluded that the future intensity-duration-frequency relationships will be uplifted and steepened, i.e., one can expect 'an intensification of extreme precipitation with a larger intensification of rarer extreme events compared to less rare events'. From the practical point of view, it is likely that the extreme rainfall will be more extreme in the future, the torrential rainfall events may result in even more fatale disasters (flash floods, debris flow etc.). But how can people be better prepared for such unexpected and spatio-temporally localized weather phenomena? A helping hand should come from the specific field of weather forecast called nowcasting.

According to the definition of the World Meteorological Organization, the nowcasting is a detailed analysis and description of the current weather parameters and their forecasting for a period from 0 to 2 h ahead [4]. It is not easy to follow the development of nowcasting systems in the world and compare their pros and cons, because on the one hand there is a plenty of forecast models and on the other hand, these models are extremely complex, based on many-year efforts of scientific teams from large forecast centers worldwide [5, 6]. Some overviews, however have been published during the recent decades, especially in connection with Summer Olympic Games in Sydney and Beijing, and the associated specific research campaigns, from which the reader can get an overview on the performance of different nowcasting systems from the leading national meteorological services, their strengths and weaknesses and scientific challenges of nowcasting at the time of the campaigns [7, 8].

Instead, our focus was set on nowcasting approaches that can be implemented/developed within realistic temporal scales and using reasonable human resources, i.e., within a horizon of 1–2 scientific projects and with an inclusion of a smaller group of experts. The candidate methodologies include the so called extrapolation-based methods, and statistical-advective approaches, respectively. The first group of methods can be characterized by the fact that they purely rely on the radar data, and attempt to generate forecasts on the basis of tracking centroids or boundaries of storms cells (such as the TITAN method, [9]) or the nowcasts are derived by means of a cross-correlation analysis between consecutive radar fields (such as TREC or COTREC methods, [10, 11]). On the other hand, the second subset of the methods, the statistical-advective ones are primarily built on a multivariate statistical model (e.g., multilinear regression, neural networks, machine learning etc.) of input variables, which can be outputs of some forecast model, and also variables from radars, satellites and/or measurements from the network of ground meteorological stations, etc. [12–14].

The current study is an inherent part of the ongoing scientific project focusing on a development of more accurate methods of prediction of convective rainfall events in the Czech Republic. The project consists of two tasks, which differ in their methodological, spatial and temporal scopes; however, they are inter-connected within the framework of convective rainfall and efforts towards a better preparedness for their occurrence. The first subpart of the project focuses on the whole territory of the country (in a spatial granulation at the level of 206 administrative units called municipalities with extended competence), and attempts to forecast the occurrence of convective rainfall with lead times reaching up to 24 h, based on statistical methods of multivariate analysis. These approaches rely on synoptic analysis of past convective events and their potential triggering factors, and an analysis of a huge, related dataset of state and forecast variables from a number of regional weather forecast models. The second subpart of the pilot project is targeted at a smaller area (Zlín region, covering 13 municipalities with extended power) and at shorter time horizons in the future (tens of minutes, up to lead time of 1 h), and is aimed at providing nowcasts by making use the volume scans of the X-band radar located within the target region.

The current paper aims at presenting the basic concept and the first results of the attempts of the project staff to develop the nowcasting subpart of the scientific project. The paper is structured as follows. The Data and Methods section summarizes the most important information on the data and the scientific approaches adopted, including the geographical settings of the area, the radar data and their conversion into the so called warning classes, the selected nowcasting methods, and finally, the platform (metrics and statistics) for the inter-comparison of the performance of the forecast models. The third section summarizes the results and evaluates the behavior of the nowcasting models, whereas the last section discusses the results and further features of the analysis, and describes some potential pathways to a further development of the ongoing research.

2 Data and Methods

In the following subsections, we briefly characterize the target area, the radar data and the adopted approach to their conversion, the selected methodologies of rainfall nowcasting, and also the methods of the evaluation of the predictions vs. the observed measurements.

2.1 Target Area

The target area of the analysis is the Zlín region, one of the 14 major administrative regions of the Czech Republic. It is located in the eastern parts of the country, in the historical land called Moravia, near to the borders with Slovakia. The region itself is further divided into 13 smaller administrative units, officially termed as Administrative Districts of Municipalities with Extended Competence. Herein, for the sake of convenience, we will term them simply as administrative districts (abbr. AD). Table 1 presents a brief overview of the individual ADs. Figure 1 shows the boundaries between the ADs, with the location of the radar and some major Czech and Slovak cities nearby. The terrain of the target area is generally hilly, but as Fig. 1 indicates, the vertical division is more emphasized in the northeastern districts of the region.

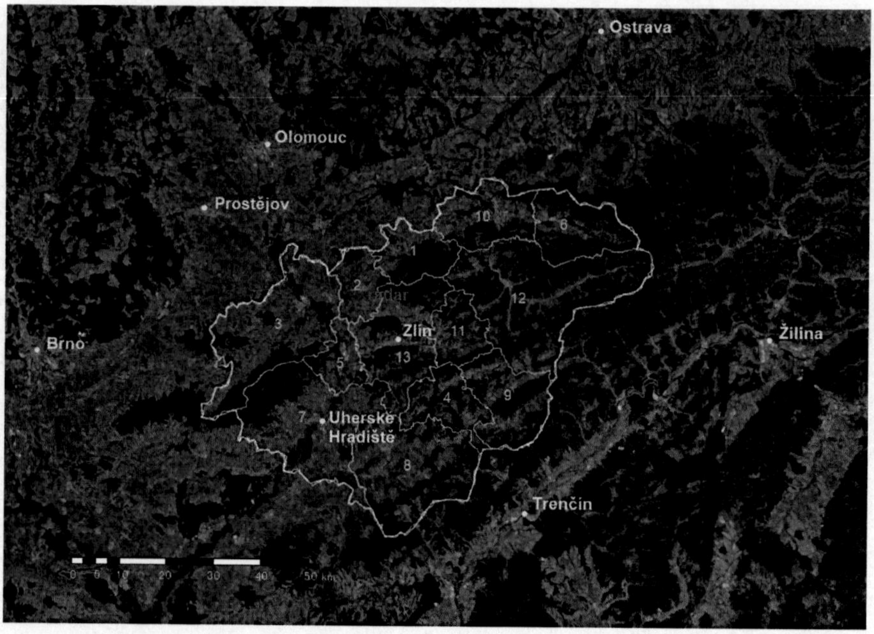

Fig. 1. Administrative division of the target area (Zlín region, Czech Republic) into 13 districts. The red dot indicates the location of the X-band radar at Holešov.

Table 1. Basic characteristics of the 13 administrative districts (AD) of the Zlín region.

AD #	Name	Area [km^2]
1	Bystřice pod Hostýnem	164
2	Holešov	133
3	Kroměříž	498
4	Luhačovice	178
5	Otrokovice	112
6	Rožnov pod Radhoštěm	239
7	Uherské Hradiště	518
8	Uherský Brod	473
9	Valašské Klobouky	258
10	Valašské Meziříčí	230
11	Vizovice	146
12	Vsetín	662
13	Zlín	350
Total	**Zlín region**	**3963**

2.2 Radar Data

The radar from the manufacturer MicroStep-MIS (type MMR-50) is installed near Holešov, at the elevation 240 m above the sea level. It is an X-band radar, working with a signal wavelength of 3.19 cm, and producing full volume scans with a frequency of about 5–5.5 min. The standard range of the radar is over 150 km.

The radar, operated by the Regional Municipality of the Zlín Region, has been working in a continuous regime since the year 2015. Due to the questionable reliability related to the magnetron of the radar during the first months of its operation, the whole set of measurements from the year 2015 was ignored in the current analysis, and only the data from the period from 2016 till 2020 were taken into consideration. Moreover, the analysis was restricted to the warm season, i.e., to the period from April to September, due to the low probability of occurrence of convective precipitation in the cold season (October–March). The entire dataset was further split into two parts: the training (2016–2018) and the validation datasets (2019–2020).

From the radar volume scans, the Pseudo CAPPI 2000 data were retrieved. CAPPI (Constant Altitude Plan Position Indicator) is a standard and widely used radar product, based on a number of PPI (Plan Position Indicator) scans. PPI is obtained from a complete 360-degree rotation of radar with a fixed elevation angle. In a PPI scan, data from a cone surface in 3D are projected to a surface of a circle in 2D. CAPPI is a slice through the data volume (i.e., several PPI scans at different elevation angles) at a user defined altitude or height above the Earth's surface. Pseudo CAPPI is then a further extension of the CAPPI data: in the regions where the horizontal slice does not cross any PPI, the reflectivity values from the lowest available PPI are taken. Usually, the elevation of 1500 m or 2000 m is used, mostly due to the fact that these are the elevations where the most significant cloud-forming processes take place. According to this, we adopted the hypothesis that the higher the CAPPI, the more intensive are the convective processes. We selected the elevation 2000 m, and in the remaining parts of the paper, we will refer to this radar product as CAPPI, instead of Pseudo CAPPI 2000.

Table 2. Conversion table of CAPPI values into warning classes, and their interpretation in terms of strength of convective processes.

CAPPI [dBZ]	Warning class	Interpretation
$\langle 0;44)$	0	No convective event
$\langle 44;52)$	0.5	Light intensity
$\langle 52;56)$	1	Moderate intensity
$\langle 56;60)$	2	Heavy intensity
$\langle 60;\infty)$	3	Extreme intensity

In the current work, however, we did not work with the CAPPI values directly. Instead, we converted them into a 4-class system of qualitative classification, termed as "warning classes" (Table 2) where '0' yields no convective event, and '3' indicates extremely intense convective events. Such a classification system is in line with the generally accepted approach of the units of crisis management as well as the Czech Hydrometeorological Institute where warnings on different meteorological, hydrological or other phenomena are expressed in terms of codes from 0 to 3, and are also visualized by green, yellow, orange and red colors. Note that the class of '0.5' was only introduced for practical reasons; in the end, these values are rounded up to 1, and treated as the class '1' [15].

The conversion of radar reflectivity values into the warning classes may be better understood using the Marshall-Palmer relation, which defines the relationship between the observed radar reflectivity factor Z [mm^6/m^3, or often expressed in logarithmic scale in dBZ] and the rainfall rate R [mm/h]

$$Z = aR^b \tag{1}$$

where a and b are empirical constants, and their common values are $a = 200$ and $b = 1.6$. Figure 2 shows this relationship, with the boundaries of the individual warning classes converted into rainfall intensity values.

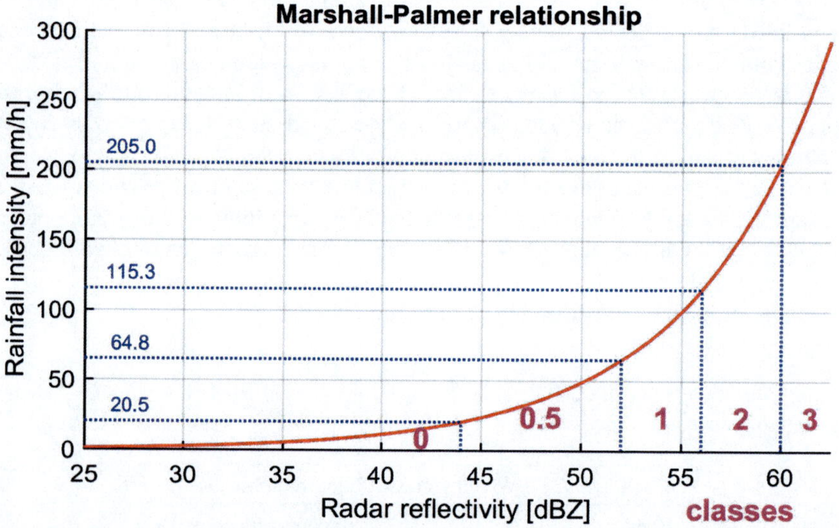

Fig. 2. Marshall-Palmer's relationship between radar reflectivity [dBZ] and rainfall intensity [mm/h], with the values indicating the boundaries between the individual warning classes, corresponding to the classification defined in Table 2.

The conversion of the CAPPI values into the warning classes was carried out for each radar pixel within each radar measurement. Consequently, the operation to derive a single value of warning class for the individual ADs followed. The set of warning class values that belong to the given AD was selected on the basis of pre-defined geographical masks of ADs, and then, the method of MaxPixels(N), with the parameter N = 2 was adopted [16], i.e., the final value of warning class within the given AD was estimated as the average of the two largest values of warning class observed for the radar pixels within the AD, which then was rounded up to the nearest integer 1, 2 or 3.

Table 3 presents the distribution of the entire set of available radar measurements (157 997 scans) for the whole training period (warm seasons of the years 2016–2018) within the 4 warning classes, stratified by the ADs. Note that if within the whole AD no CAPPI values (NaN) were assigned to the radar pixels (e.g., in the case of the sunny weather), the conversion for the particular AD was set to '0'. On the other hand, the per cent ratio of the measurements with non-zero CAPPI values is presented in the last column of Table 3. Table 3 indicates two important features of the conversion: (i) a low frequency of occurrence of non-zero warning classes, and (ii) extremely low numbers of heavy and extreme convective events. Such a biased distribution represented a further serious issue of the data processing procedure (see Sect. 2.4 for more detail). The odd behavior of the AD #6 (with very low number of non-zero warnings) can be explained by the fact that this AD is practically invisible for the radar at the elevation of 2000 m due to the blockage of the radar beams by the terrain.

Table 3. Number of cases with the given warning class assigned to each ~5-min time interval, and the ratio of the measured CAPPI values, for the entire training period of 2016–2018, stratified by administrative districts (AD).

AD #	0	1	2	3	Measured CAPPI [%]
1	157420	572	5	0	19
2	157379	616	2	0	21
3	156455	1502	39	1	20
4	157403	589	5	0	26
5	157537	458	2	0	10
6	157982	15	0	0	9
7	156685	1296	16	0	21
8	157126	857	14	0	22
9	157166	822	9	0	17
10	157566	421	10	0	15
11	157395	587	15	0	15
12	156667	1310	20	0	19
13	156977	986	34	0	27

2.3 Nowcasting Methods

In the current study, we considered three nowcasting methods:

- TITAN,

- COTREC and
- Persistence.

TITAN (Thunderstorm Identification, Tracking And Nowcasting; [9]) is a real-time algorithm for automatic identification, tracking and short-term storm prediction by extrapolation, based on meteorological radar volume data. For each discrete time step, TITAN identifies the storm as a continuous area that is characterized by a certain above-threshold value of radar reflectivity (e.g., 35 or 40 dBZ) and size (either storm area or volume). An optimization scheme is used to match storms identified at a given time with those at a subsequent time, taking into account the mergers or splits of storms. The short-term forecast of both storm position and size is based on a weighted linear fit of data on previous storm positions [17]. TITAN outputs are operationally available to the users of the products of the radar at Holešov.

TREC (Tracking of Radar Echoes by Correlation; [10]) is a nowcasting method that is based on a comparison of two consecutive images of radar reflectivity. For each block of radar pixels, the TREC method identifies a motion vector by maximizing the cross-correlation coefficient between the two consecutive reflectivity images. While TREC is successful in tracking the movements of individual radar echoes, in practice it usually captures the direction of individual rain cells instead of moving the entire meteorological system on a larger scale. Due to this drawback of the TREC method, various alternatives such as COTREC (Continuity Tracking Radar Echoes by Correlation [11, 18]) have been developed.

Persistence is a trivial methodology that assumes that the radar field does not change in the near future, i.e., the Persistence method extrapolates the observed state to the future without change. It is a commonly adopted baseline approach in any inter-comparison of different forecast models.

In line with the concept of conversion, the analysis of the current observation (say, at time T_0) and the forecasts (at time T_{0+X}, X = 5 min, 10 min, ...) produced by the selected nowcasting methods had to be also converted into the defined warning classes. Such a task was a bit complicated for the TITAN model since this method does not directly work with the CAPPI product. To accomplish this, we made use of a TITAN model output that we simply term as the 'mask of convective rainfall'. This mask, along with the mask of boundaries of ADs was applied to the picture of CAPPI at time T_0. For the nowcasts at time T_{0+X}, however, an additional step, namely a semi-Lagrangian backward scheme [19] had to be included to derive the picture of CAPPI at time T_{0+X}. As a result, we obtained the distribution of convective rainfall at each time (for the current observation at T_0 as well as for the nowcasts at T_{0+X}) in each AD, expressed in terms of CAPPI, and consequently, it was possible to apply the conversion into warning classes, described in Sect. 2.2.

Since the COTREC method provides complete CAPPI fields in its nowcasts, the conversion to warning classes can be applied directly to these fields and there is no need to deal with any mask of convective rainfall.

The nature of the examined convective phenomena as well as the experience of nowcasts specialists implies that the reliability of nowcast products sharply decrease beyond the lead times of 30–40 min [e.g., 14]. In line with these findings, we have chosen the 30 min interval as the maximum lead time. Additionally, since the radar volume scans are produced with approximately 5-min steps, we have defined this interval as the prediction step to facilitate the evaluation of the nowcast results.

2.4 Metrics and Performance Statistics

After completing the sequences described above, the radar observations, and TITAN/COTREC nowcasts were all converted into warning classes {0}... {3}. Due to this transformation, however, the statistical evaluation of the nowcasts was not straightforward. We thus had to deal with unexpected problems of the maximum likelihood methodology stemming from two major features of the analysis: (i) the probability of occurrence of non-zero warning classes is relatively low (Table 3), and (ii) classes {1},{2},{3} are ordered. These facts imply that we had to deal with the warning classes in the framework of the imbalanced ordinal regression. The strategy for tackling the problem includes operations such as resampling the training dataset (so that the rare {1},{2},{3} and the common {0} events are statistically equally distributed [20, 21]), selection of an adequate penalty function (to introduce a higher misclassification cost for wrongly classified rare events [22, 23]) or anomaly detection [24].

Nowcasts of warning classes may be evaluated from different aspects. We focused on three scientific questions:

1. 'How accurately can we predict convective events?' In terms of warning classes, the question is equivalent with the binary classification of the class {0} vs. {1,2,3}.
2. 'How accurately can we predict dangerous convective events?' This question corresponds with the binary classification of the classes {0,1} vs. {2,3}.
3. 'How accurately can we predict increases in warning classes?' We call this as the 'Change-up' metric, since it expresses the ratio of the correctly identified increases (tendencies to rise) to the total number of increases in warning classes.

The first two metrics are expressed in terms of the traditional statistics POD (Probability of Detection) and FAR (False Alarm Ratio) derived on the basis of a 2×2 contingency table where.

- TP (true positives, hits) – correctly identified convective events,
- FP (false positives, false alarms) – wrongly labeled non-convective events,
- TN (true negatives) – correctly identified non-convective events, and
- FN (false negatives, misses) – wrongly labeled convective events,

and consequently

$$FAR = \frac{FP}{FP + TP} \tag{2}$$

$$POD = \frac{TP}{TP + FN} \tag{3}$$

The third, 'Change Up' metric seems to be advantageous in operational planning of intervention of rescue units during a given convective rainfall event.

In accordance with the radar data availability every 5 min, we decided to evaluate the nowcasts in 5-min steps.

3 Results

The evaluation of the nowcasting methods is presented in Figs. 3 and 4. Figure 3 summarizes the results based on the training dataset from the period 2016–2018, whereas a similar panel is shown in Fig. 4 for the verification dataset from the period 2019–2020. Even though the methodology to derive the results for both datasets does not differ, we kept them separated with a vision of future potential benefits from this splitting, due to a planned extension of our analysis with machine learning algorithms.

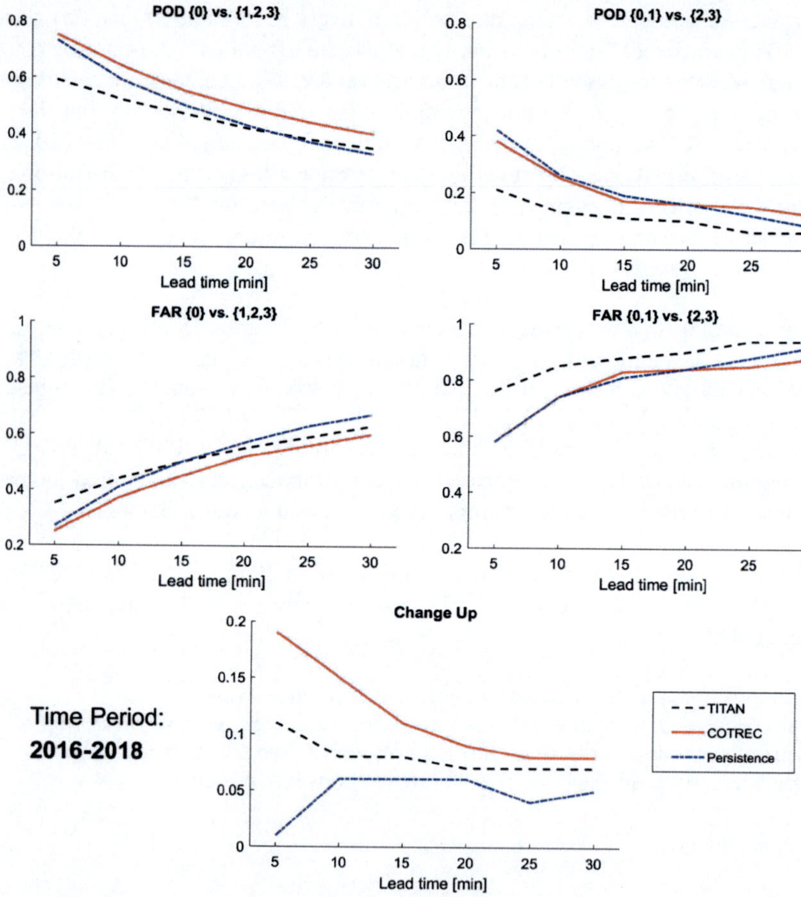

Fig. 3. Results of the comparison of the nowcasted warning classes with the observed truth, evaluated by different metrics and statistics, for lead times ranging from 5 to 30 min, and for the training dataset from the period 2016–2018. FAR – false alarm ratio, POD – probability of detection.

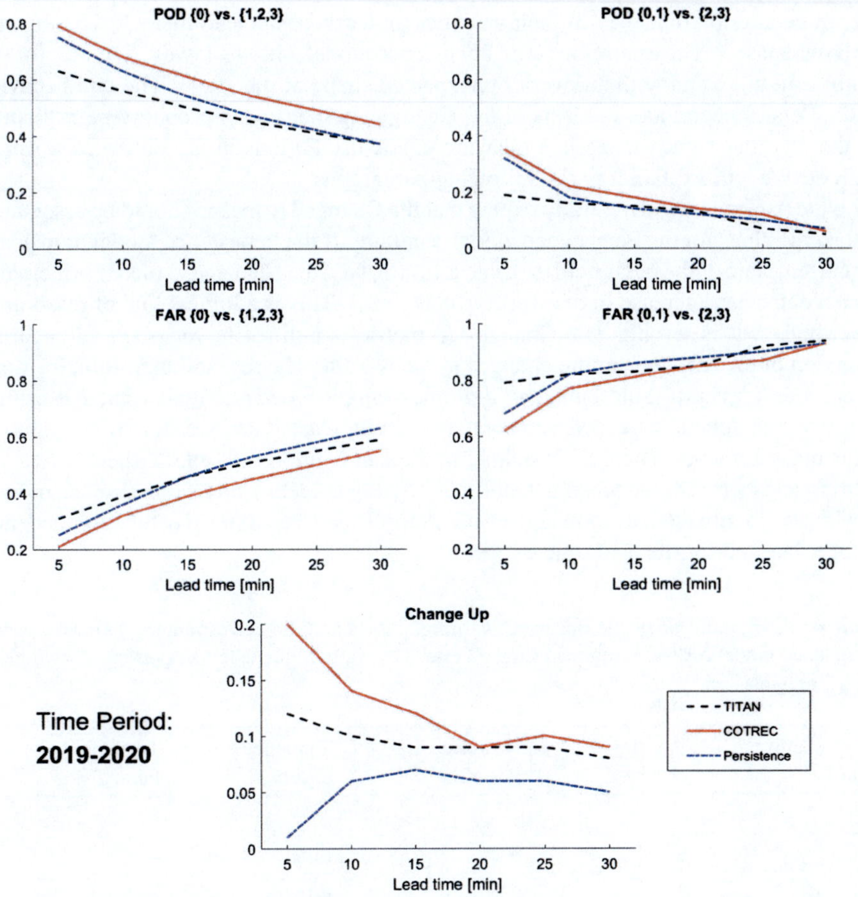

Fig. 4. The same as Fig. 3, but for the validation dataset from the period of 2019–2020.

The overall features of both of the panels in Figs. 3 and 4 are similar and are in line with the expectations:

- the statistics POD shows decreasing tendency with lead time; i.e., it is harder to forecast convective storms for a longer period of time;
- the statistics FAR shows increasing tendency with lead time; i.e., with the increase of the nowcast time horizon, the probability of erroneous nowcast also increases;
- it is obviously harder to distinguish between the strong vs. weak convective events (metric {0,1} vs. {2,3}) than to predict convective vs. non-convective events (metric {0} vs. {1,2,3}), as indicated by the difference between the left and the right sides of the panels in Figs. 3 and 4.

The behavior of the Change Up metric at the bottom of the panels in Figs. 3 and 4, however, needs a further explanation. Generally, all the nowcasting models show an overall decreasing tendency in their statistics, at least for the lead times ≥ 15 min, and for both datasets. The exception is the Persistence model with the lowest statistics for the 5 min lead time, and with an overall unexpected shape of the graph. The odd behavior of the Persistence model in terms of the Change Up metric can probably be explained by the fact that what we see is a measure of the oscillations in the observed warning degrees. We will explain it in the following paragraphs.

Note that one would naturally expect that the Change Up metric should be associated with zero values for the Persistence model, thinking 'If the Persistence model transposes the current state of the radar field to the near future, then the Change Up metric principally cannot catch any increase in the warning classes…'. This is a logical line of reasoning; nevertheless, it is wrong. The Change Up metric is defined to focus on an accurate detection of the *tendency* in the changes in the warning classes, and therefore, the non-zero figures are right. Table 4 presents a simple example based on a real event, to illustrate it. It is a rare special case when the actual warning class increases in two consecutive radar measurements. Then, as indicated in the last column of Table 4, the Change Up metric yields a true value when detecting increasing tendency by the Persistence model. Obviously, 15 min later, as soon as the time stamp $T_0 + 15$ becomes T_0, the 'up' tendency will be detected for the lead time of 5 min.

Table 4. A special case of the observed warning classes and the corresponding behavior of the Persistence model for the 5 min lead time ('Pers.' T + 5), to indicate tendencies in the change in the warning classes.

Time stamp [min]	Warning observed	Nowcast Pers. T + 5	Tendency observed	Tendency Pers. T + 5
T_0	0	–	–	–
$T_0 + 5$	0	0	Equal	–
$T_0 + 10$	0	0	Equal	Equal
$T_0 + 15$	1	0	Up	Equal
$T_0 + 20$	2	1	*Up*	*Up*

Table 5 presents a further example of the oscillations in the observed warning classes. The effect of these fluctuations typically between the warning classes {0} and {1} then manifests itself in the ability of the Change Up metric to identify increases in the Persistence model. The question for which lead time the Persistence model would identify the 'up' tendency depends on the distance between the sequences {0-1} in the observations. For instance, if one detects oscillations like {0-1-0-1}, the Persistence model will match the 'up' tendency for the lead time of 10 min; if the observations are like {0-1-0-0-1} or {0-1-1-0-1}, the lead time of 15 min will yield the 'up' tendency (exactly as in Table 5).

Table 5. A special case of the observed warning classes and the corresponding behavior of the Persistence model for the 15 min lead time ('Pers.' T + 15), to indicate tendencies in the change in the warning classes.

Time stamp [min]	Warning observed	Nowcast Pers. T + 15	Tendency observed	Tendency Pers. T + 15
T_0	0	–	–	–
$T_0 + 5$	1	–	Up	–
$T_0 + 10$	1	–	Equal	–
$T_0 + 15$	1	0	Equal	–
$T_0 + 20$	0	1	Down	Up
$T_0 + 25$	1	1	Up	Equal
$T_0 + 30$	1	1	Equal	Equal
$T_0 + 35$	0	0	Down	Down
$T_0 + 40$	1	1	*Up*	*Up*
$T_0 + 45$	0	1	Down	Equal
$T_0 + 50$	0	0	Equal	Down

From the inter-comparison of the individual nowcasting methods, we can derive the following conclusions:

- the COTREC model always outperforms TITAN, regardless the metrics, lead times and the datasets;
- the COTREC model also outperforms Persistence for the verification dataset; in the case of the training dataset, the comparison of these models is ambivalent;
- the COTREC (Persistence) model shows the best (poorest) performance in terms of the Change Up metric.

Based on the dominance of the COTREC model in its evaluation by the Change Up metric, we can conclude that the COTREC model is the nowcasting method with the best performance among the analyzed methodologies.

4 Discussion and Conclusions

In the current paper, we tried to summarize the first results and our experiences with implementing and developing different nowcasting methods based the principles of radar echo extrapolation. The target area was the Zlín region (a major administrative unit in the Czech Republic), the data (the CAPPI 2000 product) came from an X-band radar at Holešov, and the most specific feature of the analysis was the fact that the nowcasts were based on warning classes labelled from {0} to {3}, which were supposed to express the strength of the convective rainfall, and thus, they were to indicate the degree of preparedness of the staff of crisis management of the region.

Besides the TITAN nowcasting model that is currently being used in the operational practice in the Zlín region, we focused on nowcasts by the COTREC method, and we also compared them with the baseline Persistence model. The inter-comparison of these three models indicated the dominance of the COTREC model: it outperformed the TITAN method in all aspects, and, at the same time, it showed a clearly better performance than the Persistence model in terms of the Change Up metric, which evaluates the ability of the models to predict increasing trends in warning classes, i.e., deteriorating weather situations. We think that the unique behavior of the COTREC model can be attributed to the fact that COTREC is able to directly handle the observed CAPPI field and make the nowcasts also in terms of this radar product. These outputs can then easily be converted into warning classes. On the contrary, the TITAN model lacks the ability to directly utilize the fields of CAPPI. Therefore, we had to make use one of the TITAN outputs, the areas of convective rainfall, to get CAPPI data into the actual run of the model, and additionally, to implement a semi-Lagrangian backward scheme, to obtain the TITAN nowcasts in terms of the CAPPI product. These steps might have to lead to loss of some information, and thus, to poorer performance of the TITAN model in comparison with the COTREC.

The relatively good performance of the Persistence model in the case of the process of distinction between the strong and weak convection (metric {0,1} vs. {2,3}), and especially for the training dataset is a bit surprising. This feature might be explained by the very low number of cases with the warning classes {2} and {3}; however, further experiments are needed to verify this hypothesis.

The analysis further confirmed poorer and poorer performance and, thus, decreasing usability of the nowcast models with increasing lead times. We have also got test results of the nowcast models for lead times beyond 30 min (up to 60 min), but we decided not to include them into our summary since they do not bring added value. These findings are in line with the general features of the convective events, which can be characterized by large spatial and temporal variability, in magnitudes of tens of minutes and tens of kilometers.

It should be emphasized again that the current study is a publication of the very first results of our attempts to develop a useful nowcasting model for the target region. We intend to extend the scope of the current research in several ways, such as:

- To introduce a voting function that would enable to apply different functions (such as average or maximum) to the individual nowcasting models as well as different (pre-defined or optimized) weighting factors between them. For instance, selecting the maximum from the nowcasted warning classes by all the three (or more) models corresponds to a 'worst case scenario' for the crisis management.
- To incorporate other radar products than CAPPI in the current analysis. One of the potential candidates is VIL (vertically integrated liquid), which expresses the total amount of rain that would fall if all the liquid water in a column (over a radar pixel) within a rain cloud reached the surface of the Earth. This radar product might be useful in characterizing the flooding potential of the heavy (torrential) rainfall, obviously beyond the hydrological, hydrostatical and geomorphological settings of the target area.

- To extend the range of the input variables beyond those of the radar products. These may be a set of outputs of some regional weather forecast models. Nevertheless, one of the toughest issues of such a task is to match the spatial and temporal resolutions of the regional models that are generally coarser than those required by the radar echo based nowcast for the lead times up to 30–40 min with a step of 5 min. This plan naturally indicates a movement towards implementing statistical-advective approaches of nowcasting.
- To extend the spatial range of the analysis, i.e., to have a look at the neighboring administrative districts and take into consideration the information on the preceding position and the movements of convective systems to the target area 'from outside'.

Acknowledgments. This work was supported by the Ministry of Interior of the Czech Republic within the program 'Security Research Programme of the Czech Republic in 2015–2022' (project no. VI20192022134 'System of more accurate prediction of convective precipitation over the regional territorial unit').

References

1. Coumou, D., Rahmstorf, S.: A decade of weather extremes. Nat. Clim. Chang. **2**, 491–496 (2012). https://doi.org/10.1038/nclimate1452
2. Sharma, C., Ojha, C.S.P.: Changes of annual precipitation and probability distributions for different climate types of the world. Water **11**, 2092 (2019). https://doi.org/10.3390/w11102092
3. Hosseinzadehtalaei, P., Tabari, H., Willems, P.: Climate change impact on short-duration extreme precipitation and intensity–duration–frequency curves over Europe. J. Hydrol. **590**, 125249 (2020). https://doi.org/10.1016/j.jhydrol.2020.125249
4. World Meteorological Organization: Manual on the Global Data-processing and Forecasting System - Annex IV to the WMO Technical Regulations (WMO-No. 485), Geneva (2019)
5. Joe, P., et al.: Automated processing of Doppler radar data for severe weather warnings. In: Bech, J., Chau, J.L. (eds.) Doppler Radar Observations - Weather Radar, Wind Profiler, Ionospheric Radar, and Other Advanced Applications, IntechOpen (2012). https://doi.org/10.5772/39058
6. Wapler, K., de Coning, E., Buzzi, M.: Nowcasting. In: Reference Module in Earth Systems and Environmental Sciences (2019). https://doi.org/10.1016/b978-0-12-409548-9
7. Wilson, J.W., et al.: Sydney 2000 forecast demonstration project: convective storm nowcasting. Weather Forecast. **19**, 131–150 (2004). https://doi.org/10.1175/1520-0434(2004)019%3c0131:SFDPCS%3e2.0.CO;2
8. Wilson, J.W., Feng, Y., Chen, M., Roberts, R.D.: Nowcasting challenges during the Beijing Olympics: successes, failures, and implications for future nowcasting systems. Weather Forecast. **25**, 1691–1714 (2010). https://doi.org/10.1175/2010WAF2222417
9. Dixon, M., Wiener, G.: TITAN: thunderstorm identification, tracking, analysis, and nowcasting–a radar-based methodology. J. Atmos. Oceanic Tech. **10**, 785–797 (1993). https://doi.org/10.1175/1520-0426(1993)010%3c0785:TTITAA%3e2.0.CO;2
10. Rinehart, R.E., Garvey, E.T.: Three-dimensional storm motion detection by conventional weather radar. Nature **273**(5660), 287–289 (1978). https://doi.org/10.1038/273287a0

11. Li, L., Schmid, W., Joss, J.: Nowcasting of motion and growth of precipitation with radar over a complex orography. J. Appl. Meteorol. **34**, 1286–1300 (1995). https://doi.org/10.1175/1520-0450(1995)034%3c1286:NOMAGO%3e2.0.CO;2
12. Sokol, Z., Kitzmiller, D., Pesice, P., Mejsnar, J.: Comparison of precipitation nowcasting by extrapolation and statistical-advection methods. Atmos. Res. **123**, 17–30 (2013). https://doi.org/10.1016/j.atmosres.2012.10.013
13. Foresti, L., Sideris, I.O., Nerini, D., Beusch, L., Germann, U.: Using a 10-year radar archive for nowcasting precipitation growth and decay: a probabilistic machine learning approach. Weather Forecast. **34**, 1547–1569 (2019). https://doi.org/10.1175/WAF-D-18-0206.1
14. Hamann, U., et al.: Nowcasting of thunderstorm severity with machine learning in the Alpine region. In: 3rd European Nowcasting Conference, Madrid, 24–26 April 2019
15. Šaur, D., Kuliushina, A., Gaál, L.: Radar and station measurement thresholds for more accurate forecast of convective precipitation. In: Proceedings of 8th International Conference on Military Technologies, ICMT 2021, 8–11 June 2021, Brno, Czech Republic, p. 7 (2021). https://doi.org/10.1109/ICMT52455.2021.9502811
16. Šaur, D., Švejda, J.: Conversion methods of data mining analysis in algorithms of statistical and nowcasting forecast of convective precipitation. In: Silhavy, R. (ed.) CSOC 2021. LNNS, vol. 230, pp. 437–450. Springer, Cham (2021). https://doi.org/10.1007/978-3-030-77442-4_38
17. Thunderstorm Identification, Tracking, Analysis, and Nowcasting (TITAN). https://ral.ucar.edu/solutions/products/thunderstorm-identification-tracking-analysis-and-nowcasting-titan. Accessed 23 Jan 2022
18. Novák, P.: The Czech hydrometeorological institute's severe storm nowcasting system. Atmos. Res. **83**(2–4), 450–457 (2007). https://doi.org/10.1016/j.atmosres.2005.09.014
19. Germann, U., Zawadzki, I.: Scale-dependence of the predictability of precipitation from continental radar images. Part I Description Methodol. Monthly Weather Rev. **130**(12), 2859–2873 (2002). https://doi.org/10.1175/1520-0493(2002)130<2859:SDOTPO>2.0.CO;2
20. Chawla, N.V., Bowyer, K.W., Hall, L.O., Kegelmeyer, W.P.: SMOTE: synthetic minority over-sampling technique. J. Artif. Intell. Res. **16**, 321–357 (2002). https://doi.org/10.1613/jair.953
21. Barua, S., Islam, M.M., Yao, X., Murase, K.: MWMOTE – majority weighted minority oversampling technique for imbalanced dataset learning. IEEE Trans. Knowl. Data Eng. **26**(2), 405–425 (2014). https://doi.org/10.1109/TKDE.2012.232
22. Ting, K.M.: A comparative study of cost-sensitive boosting algorithms. In: Proceedings of 17th International Conference on Machine Learning, June 2000, pp. 983–990 (2000)
23. Jiang, L., Li, C., Wang, S.: Cost-sensitive Bayesian network classifiers. Pattern Recogn. Lett. **45**, 211–216 (2014). https://doi.org/10.1016/j.patrec.2014.04.017
24. Goldstein, M., Uchida, S.: A comparative evaluation of unsupervised anomaly detection algorithms for multivariate data. PLoS ONE **11**(4), e0152173 (2016). https://doi.org/10.1371/journal.pone.0152173

Framework of a Platform for Information and Analytical Support of Educational and Methodological Work

Vladimir A. Sudakov and Tatiana V. Sivakova(✉)

Keldysh Institute of Applied Mathematics, Miusskaya Square 4, 125047 Moscow, Russia
sivakova15@mail.ru

Abstract. The article proposes a toolkit for information and analytical support of educational and methodological work based on web technologies. The platform is built on the basis of the Ruby framework from Rails, on the MVC architectural pattern. This toolkit is implemented on the portal https://ws-dss.com [1]. The proposed technologies are in demand in organizing the learning process and make it possible to form the necessary training courses for students. A unique feature of the proposed solution is the ability to integrate models developed by scientists into the educational process. The possibility of launching and analyzing the simulation results is available to students in the browser and through the call to the Restful API.

Keywords: Platform · Algorithm · Information and analytical support · Educational and methodical work · Teaching

1 Introduction

The COVID-19 pandemic that has swept the world has affected more than just the health sector. Its consequences today are manifested in the political, economic, social spheres of society, including education [2, 3]. Recently, distance learning has been actively introduced, developed and used in education, implemented with the help of an information and educational system, including information, technical and methodological support. The backbone of the educational information system is the distance learning platform, which is an area for the delivery and management of training courses. It is built on tools that support online course creation, maintenance and delivery, student enrollment and management, administration of learning, and reporting of student progress. Distance learning platforms can be divided into learning management systems and curriculum management systems [4]. These systems are divided into commercial systems and open access systems.

The disadvantages of these platforms include the inability to provide the necessary control at the intermediate stages of the learning process, and the ability to make the necessary adjustments during the course and take into account the category of users.

Therefore, the authors have developed a platform for information and analytical support of educational and methodological work of specialists, implemented on the portal https://ws-dss.com [1].

R. Silhavy (Ed.): CSOC 2022, LNNS 501, pp. 515–523, 2022.
https://doi.org/10.1007/978-3-031-09070-7_42

2 Literature Review

Modern information technologies have made changes in all spheres of human activity, including education. Networked computer technologies and knowledge management systems can speed up and simplify access to the learning process.

Learning tools based on Internet platforms have been widespread for a long time. The first learning systems fall on the 50–60 s of the last century, professor B.F. Skinerr in 1954 proposed the term program learning [5]. The next stage was associated with the automation of program learning, the development of artificial intelligence and the development of training systems, in the 60–70 s [6].

Currently, a large number of information and analytical platforms for training are being developed.

Of the recent works, one can single out the work devoted to the learning algorithm in the field of building modeling, implemented at the University of Slovenia [7]. And also a number of other works [8, 9].

As a rule, the developed platforms are mainly intended for teaching schoolchildren and students, but there is a need to train specialists who have already received certain qualifications. In addition, there is a need to make changes and courses in the learning process. Therefore, the authors have developed a program designed to provide information support for forecasting and expertise of management decisions and support educational and methodological work in training specialists in the field of transport, but it can also be used in other areas. The program provides: analysis of the results of transport modeling; analysis of the adequacy of modeling; ranking of feasible solutions; selection of preferred forecasts and solutions; examination of the infrastructure of the transport system, including: topology of the road network, route network of public transport, traffic organization, schedules of passenger and freight traffic. The program implements the weighted sum method, selection of Pareto-optimal solutions and a hybrid method for constructing preference functions [10–13].

Let us consider in more detail the features of the implementation of the information model of educational activity, Fig. 1.

To apply the information model of the differentiated learning system via the Internet database server, it must perform the following functions: work with the SQL query language; support the storage of complex objects, not just the atomic values of entity attributes; ensure the reliability of storage and data integrity in accordance with the rules laid down in the database itself; provide access to information about the object for the entire period of its existence. In addition, an important quality for teaching technology is openness and accessibility for potential users of the system.

Let us dwell in more detail on the development of an algorithmic solution for information and analytical support of educational and methodological work.

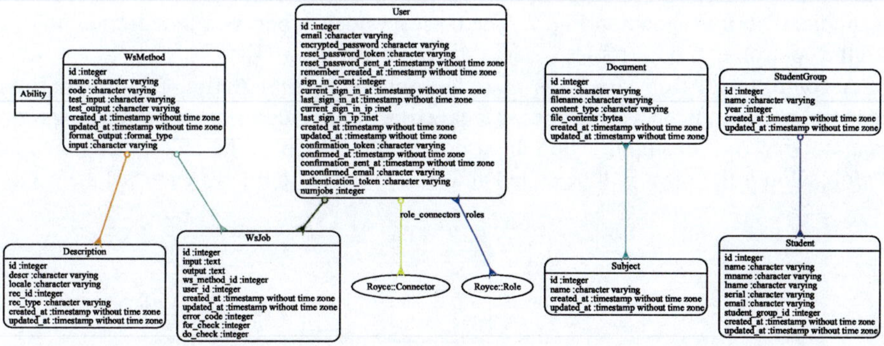

Fig. 1. Information model of educational and methodical work support

3 Formalization of the Method

For the effective development of an algorithmic solution for information and analytical support of educational and methodological work, a convenient platform is needed that provides tools for developing and testing the technology's performance. The platform was built on top of the Ruby on Rails framework [14]. Ruby on Rails is a framework written in the Ruby programming language. Ruby on Rails provides a Model-View-Controller architectural pattern for web applications, as well as integrates them with a web server and a database server. Ruby on Rails is open source software and is licensed under the MIT license. The framework is based on the MVC architectural pattern. Ruby on Rails defines the following principles for application development:

- provides mechanisms for reuse, allowing to minimize duplication of code in applications (the principle of Don't repeat yourself);
- by default, configuration conventions are used that are typical for most applications (Convention over configuration principle). An explicit configuration specification is required only in non-standard cases.

The main components of Ruby on Rails applications are Model, View, and Controller. Ruby on Rails uses the REST style of building web applications.

The model provides the rest of the application with an object-oriented display of data. Model objects can load and save data in a relational database, and also implement business logic. The ActiveRecord library is used to store model objects in a relational DMS. In the problem under consideration, the class diagram of the model corresponds to the information model shown in Fig. 1.

The view creates a user interface using the data it receives from the controller. The view also passes user requests for data manipulation to the controller (as a rule, the view does not directly modify the model). Fragments of views are shown in the figures in the previous section.

A controller in Rails is a set of logic that runs after the server receives an HTTP request. The controller is responsible for calling the methods of the model and starts the formation of the view. The class diagram of controllers of the portal of educational and methodical work is shown in Fig. 2. The correspondence between the Internet address and the controller action (route) is set in the config/routes.rb file.

A controller in Ruby on Rails is a class inherited from Action-Controller: Base. The public methods of a controller are called actions. Action often corresponds to a separate view. For example, when the user requests ws_jobs, the index method of the WsJob-sController class will be called and then the index.html.erb view will be used.

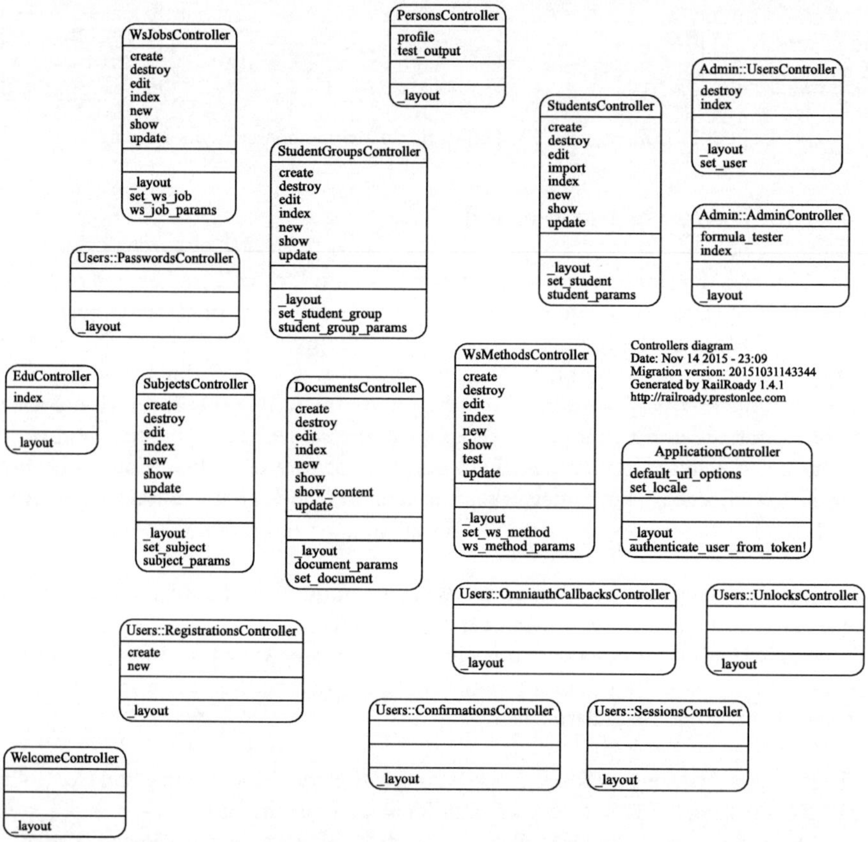

Fig. 2. Class diagram of controllers of the portal of educational and methodical work

There is a large ecosystem of plugins around Rails - pluggable "gems", some of which have been included in the base Rails distribution over time, such as Sass and CoffeeScript, while others, although not included in the base distribution, are the de facto standard for most developers. For example, the RSpec unit test tool. In this project, gems are used: Devise (for authentication), CanCatCan (for authorization), re-captcha (to protect against attacks when registering by robots), capistrano (for deployment on the server).

The user's work begins with registration in the system. Registered users are matched by e-mail with the students of the groups loaded into the system. The registration procedure requires a mandatory confirmation of the email address.

Teachers form educational and research modules in the system. Some of them are shown in Fig. 3.

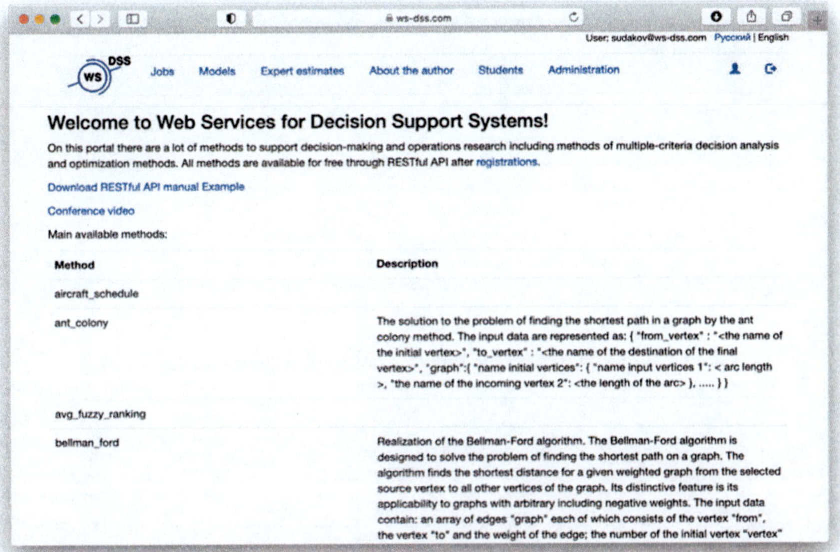

Fig. 3. Models and methods implemented on the ws-dss.com portal

Training modules usually contain a generator of a set of tasks. Tasks are generated using a pseudo-random number generator. The pseudo-random number generator is initialized with a unique student number, which includes the group number, year of study, and the sequence number in the journal.

Students run the problem generators twice: the first time to get their own unique version of the problem, and the second time to check the entered solution of these problems. The system reports in which points of the tasks errors were made, and after that you can try to find the error. In this mode, for example, tasks for generating functions of the algebra of logic and generators of random graphs work (see Figs. 4 and 5).

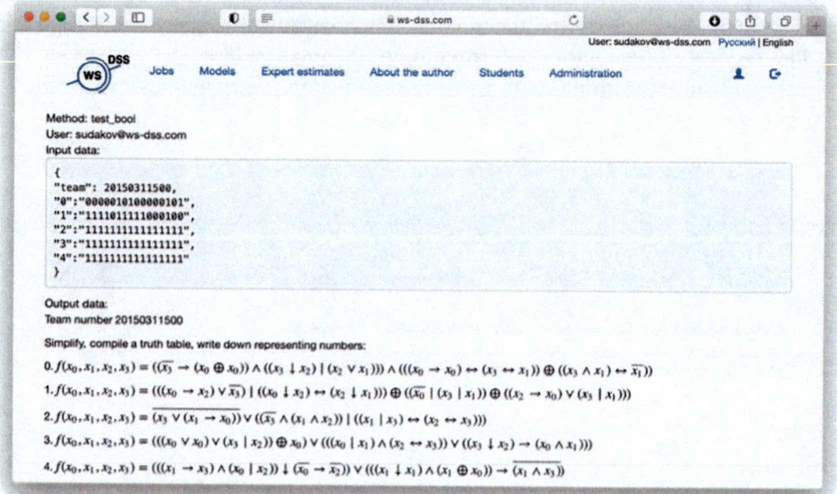

Fig. 4. Tutorial tasks on the functions of logic algebra on the portal

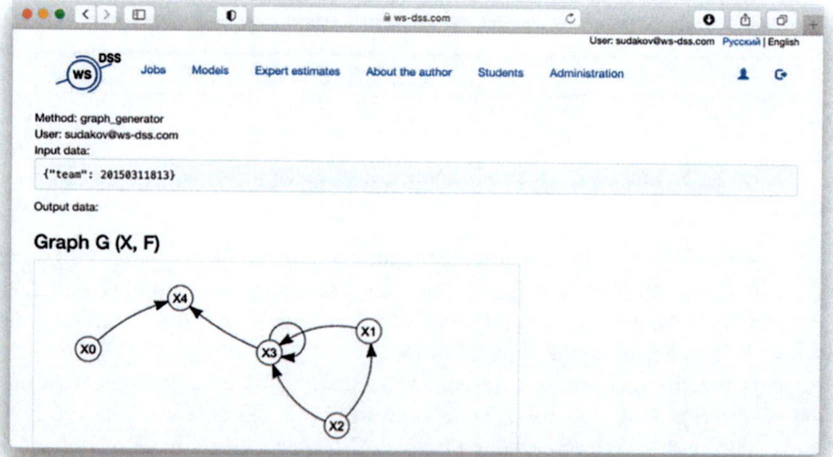

Fig. 5. Working with graphs on the portal

There is another type of models - these are modules "solvers", they allow you to load a formalized description of the problem and the system helps to find a solution. In this form, for example, the modules for searching for Hamiltonian and Euler paths work, students formalize the description of the graph, and then look for the corresponding route.

4 Results

A generalized scheme of the algorithm for information and analytical support of educational and methodological work is shown in Fig. 6.

Students print out the results of solving problems and give them to the teacher for checking. It generates a QR code in the printout of the systems, it allows the teacher to quickly find the result of the solution on the website. The teacher scans the QR and gets to the page in the solution of the problem. Thus, it is possible to avoid forgeries in reports and attempts to submit the same work by different students. If someone has already approached this work, then the teacher will register a rescanning.

At the output of the algorithm, a unified database of work results is formed. The teacher always sees through the site who did not their own versions of the assignments, who and when started the work, and who did not start at all.

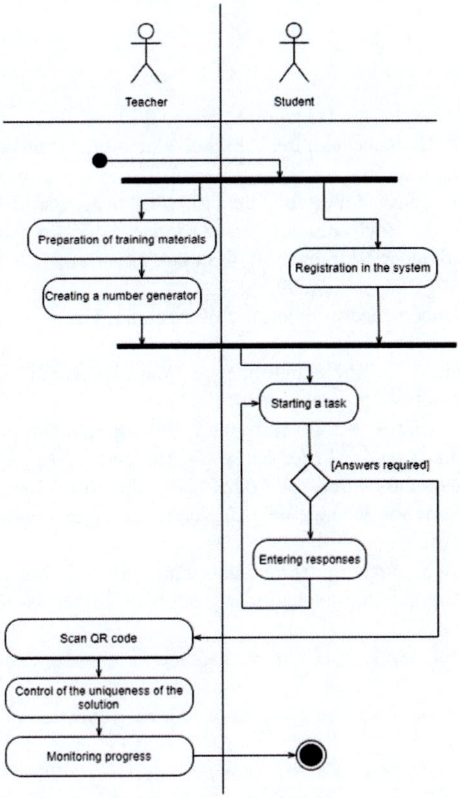

Fig. 6. Algorithm for information and analytical support of educational and methodological work

5 Conclusion

The developed algorithm for information and analytical support of the process of educational and methodological work in the training of specialists provides:

- formation of information resources in the form of a knowledge base in the subject area and a database of solutions to typical problems;
- remote network access of users to information resources of the educational module;
- formation of general and individual training courses in the field of transport modeling;
- remote control of the knowledge of learners.

The advantages of this algorithm include the ability to implement models prepared by scientists into the learning process. Students have the ability to run and analyze simulation results using the Restful API call.

In addition, the developed toolkit allows you to support the organization of the process of educational and methodological work and form the required number of courses. You can get acquainted with the implementation of this technology on the portal wss-dss.com [1].

References

1. WS-DSS Homepage. https://ws-dss.com. Accessed 20 Feb 2022
2. Dubey, P., Pandey, D.: Distance learning in higher education during pandemic: challenges and opportunities. Int. J. Indian Psychol. **8**(2), 43–46 (2020). https://doi.org/10.25215/0802.204
3. COL Homepage. http://oasis.col.org/handle/11599/3576. Accessed 1 Mar 2021
4. Ritke-Jones, W.: Virtual Environments for Corporate Education: Employee Learning and Solutions. Business Science Reference. IGI Global, Hershey, New York (2010). https://doi.org/10.4018/978-1-60960-195-9.ch609
5. Skinner, B.: The science of learning and art of teaching. Harward Educ. Rev. **24**(2), 86–97 (1954)
6. Poole, D., Mackworth, A.: Artificial Intelligence: Foundations of Computational Agents, 2nd edn. Cambridge University Press, New York (2017)
7. Pučko, Z., Šuman, N., Uroš, K.: Teaching and lifelong education on BIM at FCETEA in Maribor, Slovenia. In: Završki, I., Cerić, A., Vukomanović, M., Huemann, M., Ronggui, D. (eds.) 14th International conference OTMC 2019 and 7th IPMA, pp. 465–473. Croatian Association for Construction Management University of Zagreb, Faculty of Civil Engineering (2019)
8. Cyberlinka Homepage. https://cyberleninka.ru/article/n/information-and-methodological-support-of-the-continuity-of-the-educational-process-on-the-basis-of-electronic-support. Accessed 1 Mar 2021
9. Tait, A.: Education for development: from distance to open education. J. Learn. Dev. **5**(2), 101–115 (2018)
10. Sudakov, V.A., Sivakova, T.V.: The problem of rational allocation of resources for replacing aircraft. In: Silhavy, R., Silhavy, P., Prokopova, Z. (eds.) CoMeSySo 2020. AISC, vol. 1294, pp. 1042–1050. Springer, Cham (2020). https://doi.org/10.1007/978-3-030-63322-6_90
11. Sudakov, V., Sivakova, T.: Use of fuzzy areas of preference in tasks of multi-criteria evaluation of freight air transport. In: Bogdan, A. (eds.) Conference 2020, AMMAI, vol. 927, p. 012060. IOP Publishing (2020). https://doi.org/10.1088/1757-899X/927/1/012060

12. Sudakov, V., Sivakova, T.: Methodology for processing expert judgments based on fuzzy weighted summation. In: Bogdan, A. (eds.) Conference 2020, AMMAI, vol. 927, p. 012041. IOP Publishing (2020). https://doi.org/10.1088/1757-899X/927/1/012041
13. Sudakov, V., Sivakova, T.: Features of aggregation of fuzzy expert judgments. In: Bogdan, A. (eds.) Conference 2020, AMMAI, vol. 927, p. 012034. IOP Publishing (2020). https://doi.org/10.1088/1757-899X/927/1/012034
14. Ruby, S., Thomas, D., Hansson, D.: Agile Web Development with Rails, 6th edn. The Pragmatic Bookshelf, Raleigh North Carolina (2020)

Management of Home Electricity Costs with a Simple Computer Application

Ramadan Beqiri[1,2] and Valmir Hoti[3](✉)

[1] Faculty of Medicine, University "Fehmi Agani", Gjakova, Kosovo
ramadan.beqiri@uni-gjk.org
[2] Department of Management and Finance, Private College "Biznesi", Prishtina, Kosovo
ramadan.beqiri@kolegji.biznesi.com
[3] University "Hasan Prishtina", Prishtina, Kosovo
valmir.hoti5@student.uni-pr.edu

Abstract. Environmental pollution is significantly increasing global warming and damaging the ozone layer. We can reduce these damages in the external environment by using electricity rationally. Saving it has a good effect on households' incomes too. The purpose of this study is to present the effects of electricity use in the environment through a review of the literature, as well as presenting the use of a simple application, created in Microsoft Excel, to calculate the monthly electricity costs in a simple way in Republic of Kosovo. We have done this by seeking scientific research in the journals of Scopus and WOS, as well as on the websites of the Kosovo Energy Corporation (KESCO), and the Energy Regulatory Office (ERO). The use of this simple computer application, greatly simplifies the calculation of monthly electricity consumption and consumer expectations for the electricity bill.

Keywords: Environmental pollution · Electricity consumption management · Electric Bill calculator

1 Introduction

The nature that we have today is not like the one that was many centuries ago. It was much cleaner and much better. Many factors have contributed in its pollution, and still, are influencing it to be polluted even more. But what exactly is global warming, what are the factors that influence its growth and the consequences that affect the community?

Global warming is the extra heat inside the Earth's atmosphere which continues to cause global warming to rise. Global warming is causing and continuing to cause climate change [1]. Climate change can cause sea level rise, devastation of communities, as well as extreme weather conditions. There are many factors that affect global warming. The rapid development of industry has made human life easier and less costly, but its rapid development has greatly affected environmental pollution [2]. Industrialization is harmful in various ways. The waste that this industry produces ends up all on Earth or in our environment. The chemicals and materials used during industrialization are not only polluting the atmosphere but also the earth beneath it [3].

R. Silhavy (Ed.): CSOC 2022, LNNS 501, pp. 524–535, 2022.
https://doi.org/10.1007/978-3-031-09070-7_43

Another factor of global warming is the uncontrolled and the production of non-qualitative products which are not sustainable enough and the same during the dump create waste more than ever before. Such products need to be transported by vehicles which consume fuel and which release gases which pollute the air. Therefore, transport by various means through land, air and water transport, pollutes the air, water and land, creating the "greenhouse" effect [4]. Replacing the energy source for moving transport vehicles from fuel to electric ones, would reduce global warming. Oil and Gas are used almost all the time in many industries. They are also used in vehicles, buildings, production and for the production of electricity.

The production of electricity, through thermal power plants has made it about 46%, to affect global warming. When coal, oil, and other gases are burned, climate change greatly increases, endangering the lives of plants, animals, and humans themselves, leaving such areas uninhabited.

Kosovo, as well as the surrounding countries, produce electricity using coal combustion, which releases large amounts of carbon (C), and oxygen-related gases (O), such as CO_2 (carbon dioxide), CO (carbon monoxide), etc. Such gases affect the "greenhouse" effect, which also has the effect of creating "holes" in the ozone. Ozone is the Earth's gaseous envelope, which eliminates the carcinogenic radiation of sunlight that can bring serious consequences for life on Earth. Ozone (O3) [3] in the stratosphere absorbs ultraviolet radiation entering from the Sun (unlike other trace gases, it absorbs both infrared and ultraviolet radiation) [3].

A team of researchers from the British Antarctic Survey, working at Halley Bay, in Antarctica, in 1982, first noticed the hole in the sky. Regular observations have been made at the same site, throughout the months from October to March, every year since 1957, the International Geophysical Year [3].

The size of these holes has increased and they present a permanent threat to life on Earth. This growth trend must be stopped, in order for nature to use its power to repair the damage caused by the human factor. Such a problem has affected the need to reduce electricity production, which has increased the demand for its purchase. The increase in demand for electricity has increased the price of electricity, not only in Kosovo and the region, but also beyond, which in the market is calculated in megawatt hour (MWh) (Fig. 2).

Many details have been published about this, conferences and decisions have been taken to influence the reduction of various global changes [5–10].

To raise consumer awareness, they are notified through meetings, publications, TV appearances and other forms of the need to reduce global warming. For this, it is necessary to know that every household electrical appliance consumes electricity, which performs electrical work with its power, during the time it is used, which is expressed in kWh (kilowatt hour). Electricity consumption during 24 h is not the same. The demand for electricity consumption during the day is greater. In the evening hours, respectively at night, the demand for its expenditure drops significantly. Since Kosovo produces electricity day and night, in order to avoid accidents and even to make the most rational distribution of its consumption, two tariffs have been applied. Tariff A1 is known as the high tariff, which has the highest price compared to the low tariff, which has the lowest price and which is expressed by: A2. In order to reduce electricity consumption, it has now been decided in Kosovo to resume the application of two tariff blocks [11]. The first block with consumption per consumer up to 800 kWh, which will maintain the current price per kWh and the second tariff block with consumption over 800 kWh, which will have a significantly higher price. In order to manage better the electrical costs incurred by household electrical appliances, all are discussed in detail in the following parts. Also provided is a simple application (Table 4) which completely calculates the total household electricity costs by entering the costs recorded by the electricity meter as A1 (high tariff) and A2 (low tariff). All calculations are displayed immediately which are easy to read and clearly explained.

2 Literature Review

The need for electricity is undisputed everywhere. Local and international institutions are committed with great dedication to alleviate this deep crisis and these high prices. While the need to spend household electricity is ever increasing. Recently, high demands are on the use of electricity for district heating. This requires a large power supply because heating boilers have high electrical power which reaches up to 30 kWh for a family, and this can be higher or lower, it depends on the area of the settlement that will be heated by the district heating system [10].

To see how much concerned are consumers, civil society, scientists, researchers and all those who think that they can contribute in anyway in this case, it is enough to do just simple research in internet browser and you can find a lot of various portals, social networks, research platforms, that examine the problematics of the crisis and the necessity of rising electricity prices. Within the Scopus or Web of Science database, we found researches and authors from all over the world who have never resorted to scientific writing to find solutions to this reasonable request [12–14]. Electricity cost management is diverse. However, the global electricity crisis and the enormous prices that gripped Kosovo at the end of 2021 (Fig. 2), compared to a year ago passed 200% even 300%. Therefore, this would be one of the main reasons for quick access to research and find a solution on this phenomenon.

The proposal of a home energy management scheme to attenuate the energy cost was the objective of this scientific paper. In a study Aslam with colleagues [9] elaborate on the need to connect smart devices which will create the electric microjet in a house which would reduce the cost of paying for electricity. Such a house would be called as a "smart home". Based on the same paper the simulation results have proved that the proposed approach efficiently plans the intelligent devices, as well as the charging/unloading of EV according to the model of electricity generation in microgrids. Consequently, applying the scheme presented [10] in (Fig. 1. The developed energy-efficient model), would reduce the cost of electricity by integrating the household electrical grid.

The increase in demand for electricity has led many countries to try to reduce household electricity costs. One of the steps taken was the need to set up a blockchain system with at least two price levels for a certain limit of electricity consumption. In Kosovo, such a two-level system is restored as BT1 (block tariff 1) up to 800 kWh and BT2 (block tariff 2) over 800 kWh.

Due to environmental pollution, during the production of electricity in power plants, where the burning of coal greatly releases CO_2 and other harmful gases, there is a need to look for other alternatives. In the study of Çakar et al. [12] it was shown that renewal at nuclear power plants has a positive effect and reduces the release of harmful gases to the environment, in this case carbon. An advanced technology, not only increases energy efficiency but also reduces the level of air pollution. The need for electricity is growing day to day, but along with it, increases the risk for our safety and health. So now to reduce both of the above factors (the need for electricity and the risk of environmental pollution) much has been invested in finding other alternatives for generating electricity without polluting the environment. Many countries have developed to a considerable extent the production of electricity through wind and solar energy. We see such a good practice made by Germany, which to avoid the continuous rise of electricity prices, it has increased the electricity production through the solar and wind energy [10]. This would reduce the production of electricity in power plants as well as the gradual abolition of nuclear energy [15].

3 Material and Methods

This paper is a review type of the existing literature. The literature search was conducted in scientific journals, part of Scopus and Web of Science, on the websites of the Kosovo Energy Corporation (KESCO), and the Energy Regulatory Office (ERO) in order to collect as much data as possible about the use of electricity in Kosovo and in the world, and its impact on the environment. In this paper we also present an example of using a simple computer application, created in Microsoft Excel, that helps calculate monthly electricity costs. We elaborate it and we tell a simple calculation that can be made by it.

4 Results and Discussion

Works that in the past were done mechanically, today are replaced and still being replaced by electrical work. Almost in every corner of our house we have electrical appliances which make our life easy and comfortable. Everywhere, all around us, using the computer to write, analyze, comment and quote this work, we are doing it by utilizing electricity through its source, the battery, which is charged with an adapter, from the electricity grid, and a lot more electrical appliances which, more or less, consume electricity. All in one word are electricity consumers. Each of them has its own power and depending on the time of use of the electricity network, increases the cost of monthly expenses. Each of them does its own work. This work is known as electrical work and according to the laws of physics, it has different definitions, as it depends on many other physical quantities [16, 17]. Based on Joule's Law, the operation of electricity is proportional to the multiplication of electrical power and time.

$$E = P \times t \tag{1}$$

where: E – Electric energy and its international unit is Joule (J).

P - Power of electricity and its international unit is Watt (W).

t - Time and its international unit is second (s).

As seen from formula (1), electric energy (E) is a dependent magnitude on electrical power (P) and time (t). So electric energy (E) is a dependent variable, while power (P) and time (t), are independent variables. Electric energy depends on electric power and time. From formula (1), we read that electric energy is in direct proportion to power, because high electric power performs higher electrical work and vice versa: low electric power performs low electrical work.

Table 1. Relation of electric energy (E) to power (P) and time (t).

Situation	Constant	Variables	
I	t = const.	E ↑	P ↑
II	t = const.	E ↓	P ↓
III	P = const.	E ↑	t ↑
IV	P = const.	E ↓	t ↓

As seen in Table 1 in first situation, electric energy (E) is directly proportional to electrical power (P), when time is constant (const.). As the electric power of the electrical appliances connected to the home electrical network increases, so does the electric energy (E) and consequently the electricity costs increase. Electric energy also depends on time (t). The electricity bill at the end of the month will increase depending on the time of use of household electrical appliances, when the power of electrical equipment connected to the city grid remains unchanged (P = const.), as shown in third and fourth situation. Therefore, when the electrical equipment connected to the electricity grid stays longer, the use of electric energy is higher and along with it the cost of the electric bill increases.

We now validate formula (1), to obtain the international unit of electric energy.

$$E = P \times t(=)W \times s = \frac{J}{s} \times s = J \qquad (2)$$

From the formula above (2), we define joule as the equivalent of one watt of power radiated or dissipated for one second.

Since Watt and second are small units for measuring electricity consumption, we use: for electricity – “kilowatt (kW)” and for time – “hours (60 min)”. In this way we reach the expression of measuring electricity consumption, in kWh (kilowatt hour). This cost is then multiplied by the price of 1kWh and then we get the total amount of electricity use to which is added the VAT (valued added tax) and the fixed tariff of the electric meter.

Each electrical device has its own power. Thus, the cost of electricity when using the electrical appliance depends directly on the length of its connection to the electric grid. The power of electricity for electrical equipment is found on its ticket which is expressed in the international unit Watt (W). If the device has greater electrical power, it usually is expressed in kilowatt (kW).

Time management, and the use of electrical equipment according to their power, affects the reduction of electrical costs. According to data from KESCO (Kosovo Energy

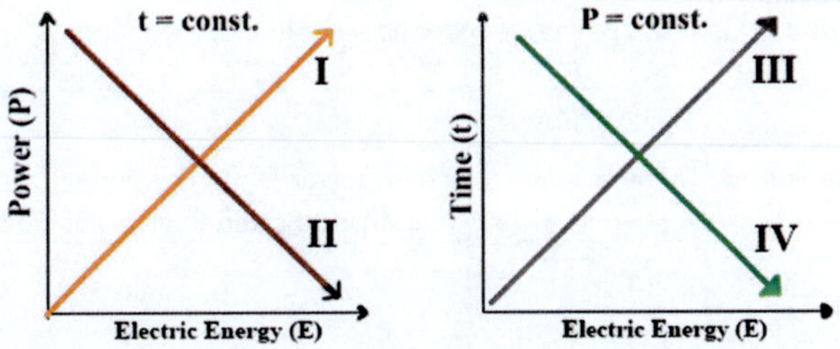

Fig. 1. Relation of electric energy (E) to power (P) and time (t) illustrated [Source: Author].

Corporation), the increase in electricity import has increased "inexplicably" as a result of changes in the price of rising global electricity [18].

Table 2. The most used multiple units of Watt (W) [16].

Multiple unit	Symbol	Value in Watt
Kilowatt	kW	1000 W
Megawatt	MW	1 000 000 W
Gigawatt	GW	1 000 000 000 W

Table 2 shows the three most commonly used multiple units of the power of electricity, Watt (W). Such numerical values can also be expressed in terms of power based on 10, such as: 1 kW = 1 000 W = 10^3 W, and so on.

To save electricity we must be careful in the time we leave them connected to the electric grid. Each device consumes electricity. Therefore, the management of electricity costs depends on our care and the ability to use them in the shortest possible time. Proper management enables the electricity bill to be acceptable and reasonable for the operation of the consumed electricity. Even if the cost of the bill at the end of the month is affordable for our budget, we are obligated to save it if possible. Saving electricity consumption is an obligation and a civilization. By saving electricity we reduce environmental pollution.

As is shown in Table 3, the boiler heater has a power of 2 kW and if it stays on for 5 h, within a day it will consume 10 kWh, and for a month (30 days) it will consume 300 kWh. If from the monthly household expenses we have a total of 1000 kWh, then almost 30% of the expenses were made only by the boiler. To reduce electricity costs, if we cannot save on time (hours during the month), then we can save by turning on (activating) the boiler during the night hours when the electricity tariff is with lower price.

Extra care should be taken in the use of the electric stove. Depending on how many heaters will be connected, it ranges from 1 kWh to 5 kWh. As it is known he has the electric oven with two heaters as well as the plates which also have their own heaters each. If an electric stove is used for 60 h/month (on average 2 h a day), it will consume a total of 60 kWh to 300 kWh. Consider also the power of electric boilers for central heating. Their power reaches up to 24 kW. Managing expenses requires extremely rigorous discipline.

Table 3. Electric power of some household electrical appliances.

Device	Electric Power	Note
Mobile phone adapter	30 W	Their power is not the same, it depends on the manufacturer and the type of electrical appliance
Laptop adapter	230 W	
Efficient electric bulbs	4–10 W	
Boiler heater	2 kW	
Washing machine heater	3 kW	
Electric stove heater	1–5 kW	
Electric heater	1–9 kW	

Reducing expenses would greatly help in the successful management of family incomes. Kosovo produces electricity but it is not enough for the consumption that has increased. Electricity import have increased rapidly. According to KESCO declarations, in 2020 1 MWh was purchased for € 60, while in 2021 the price for 1 MWh reached the amount of € 274 [18].

Fig. 2. Price per MWh in 2020 and 2021 [18].

There are 630,000 electric meters in Kosovo. If every household would manage electricity costs in the best possible way, the need to import electricity would be reduced and no block tariff II would be applied at all, which has significantly higher prices, as described in this research.

How the countries around Kosovo stand, such as: Albania, Macedonia, Montenegro and Serbia, is seen in Fig. 3.

So far, based on the statistics provided by KESCO, which are presented in Fig. 3, for 800 kWh spent consumers pay: in Kosovo - 47.12 €, Serbia - 60.31 €, Albania 76.63 €, Macedonia 83.10 € and Montenegro 89.49 €.

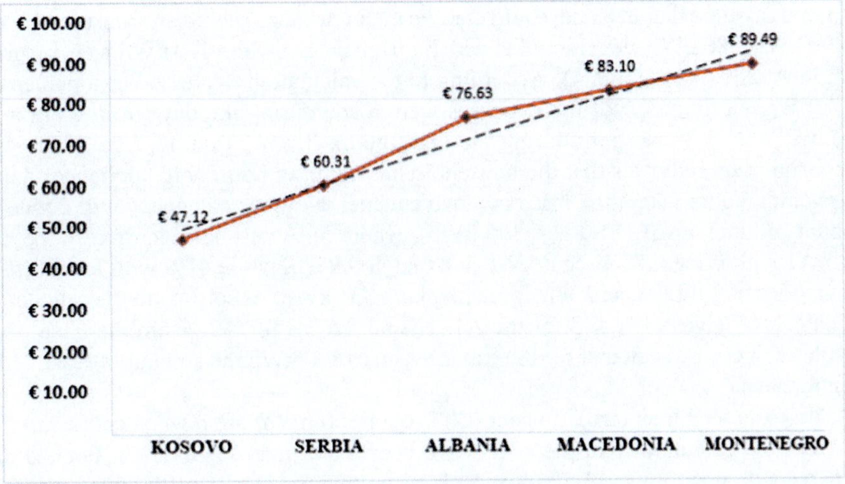

Fig. 3. Monthly electricity consumption of countries around Kosovo for 800 kWh [18].

Based on these data, Montenegro, compared to Kosovo, pays almost twice as much for electricity. Macedonia is the second, followed by Albania. Serbia also pays more for electricity than Kosovo.

An awareness campaign for its customers, KESCO on 2021/11/30, through personal pronouns connects the necessity for saving electricity, justifying why every kWh saved is valuable. So we can start saving a little, for example, I stop heating in the room where I do not stay; YOU stop a light; HE/SHE uses household appliances after 22:00; WE turn off office equipment after we finish working hours; YOU tell about the importance of saving; THEY believe in saving [18].

4.1 Calculation of Electricity Costs

Based on the decision of ERO [11] for the increase of electricity prices, the system with two-tariff block is applied. Also, electricity costs vary for the day and night. During the day applies: high tariff. This tariff during the winter season is from 7 a.m. to 10 p.m., and is known as tariff A1, while from 10 p.m. to 7 a.m., is applied the low tariff known as tariff A2. Tariffs A1 and A2 vary in price per kWh. Even block tariffs vary based on the price per kWh. According to ERO, the price for expenses up to 800 kWh is not increased, while above 800 kWh the higher prices are applied, and this counts for both tariffs A1 and A2. To manage easier, through a simple calculator we demonstrate the total amount of electricity that accumulates for a month for an electric meter.

The price per kWh is given in €-cent. Thus, for the high tariff A1 - the price is 6.75 €-cent/kWh, while for the low tariff A2 the price is 2.89 €-cent/kWh. This applies to block tariff I (consumption up to 800 kWh). For block tariff II (consumption over 800 kWh) the payment for expenses increases according to this description: high tariff A1 - the price is 12.52 €-cent/kWh, while for low tariff A2 the price is 5.90 €-cent/kWh.

Let us assume that in a household electric meter during a month are consumed a total of 990 kWh, 580 kWh were recorded with high tariff (A1) while 410 kWh were recorded with low tariff (A2) (Table 4). According to the calculation expressed as a percentage (%), it is seen that 59% of the expenses were made during the day (from 7 am to 10 pm), while 41% were spent during the night hours (from 10 pm to 7 am). This type of consumption indicates that the household has not used household appliances during the evening hours where the price is much cheaper. Now the calculator will divide the amount of block tariff I (BT1 ≤ 800 kWh), where 59% will be according to the high tariff A1 with price 6.75 €-cent/kWh and that 469 kWh, while 41% with low tariff A2 with a price of 2.89 €-cent/kWh, including here 331 kWh. According to the calculation, the calculator gives: 31.64 € for the A1 fee and 9.57 € for A2. From the table of the calculator, it can be seen that for the expenses up to 800 kWh the amount of € 41.21 has been reached.

The costs for block tariff II spent (BT2, over 800 kWh) are now calculated. As we see from the calculator that the costs have been exceeded by 190 kWh. For the high tariff A1 with price now (12.52 €-cent/kWh) are 59% or 111.31 kWh with the amount of € 13.94, and for the tariff A2 with price (5.90 €-cent/kWh) are 41% or 78.69 kWh with the amount of € 4.64. The amount per block tariff II (costs over 800 kWh) are:

Table 4. Household electricity consumption calculator

Consumption	High tariff (A1)	Low tariff (A2)	Total
Enter the data (kWh)	580	410	990
Allocation Coefficient	59%	41%	
High and Low tariff	A1	A2	Unit
Block-tariff 1 (BT1) ≤ 800 kWh	6.75	2.89	€-cent/kWh
Block-tariff 2 (BT2) > 800 kWh	12.52	5.90	€-cent/kWh
Fix tariff		1.74	€/month
kWh ≤ 800	469	331	800.00
Euro ≤ 800 kWh	31.64	9.57	41.21
kWh > 800	111.31	78.69	190.00
Euro > 800 kWh	13.94	4.64	18.58
Total kWh	**580**	**410**	**990**

Results:	Euro (€)
Fix tariff (€)	1.74
Total of energy (€)	59.79
Neto (€)	61.53
VAT (8%)	4.92
Bill amount for payment (€)	**66.45 €**

€ 18.58. In the following are collected the amounts of tariff blocks I and II, which is a total amount: € 59.79. A fixed fee of € 1.74 has been added to the bill, which is for each electricity meter regardless of costs. This fee applies even if there were no expenses at all during the measurement interval by the official person of the corporation. The bill is also charged with tax (VAT).

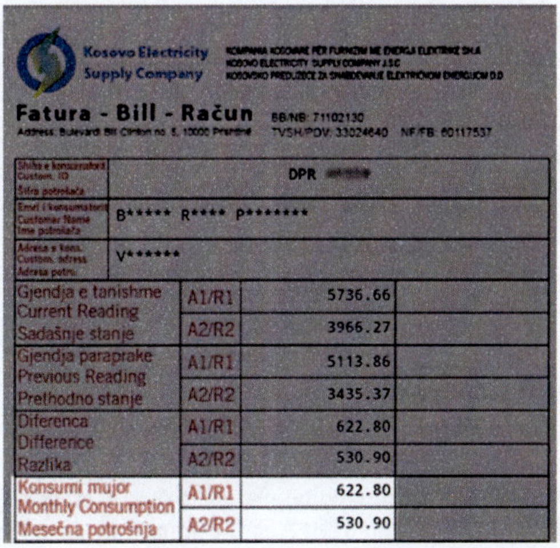

Kosovo Electricity Supply Company

KOSOVO ELECTRICITY SUPPLY COMPANY J.S.C

Fatura - Bill - Račun

BB/NB: 71102130

TVSH/PDV: 33024640 NF/FB: 60117537

Custom. ID	DPR		
Customer Name	B***** R**** P*******		
Adresa e kons. Custom. adress Adresa potr.	V******		
Gjendja e tanishme Current Reading Sadašnje stanje	A1/R1	5736.66	
	A2/R2	3966.27	
Gjendja paraprake Previous Reading Prethodno stanje	A1/R1	5113.86	
	A2/R2	3435.37	
Diferenca Difference Razlika	A1/R1	622.80	
	A2/R2	530.90	
Konsumi mujor Monthly Consumption Mesečna potrošnja	A1/R1	622.80	
	A2/R2	530.90	

Fig. 4. Electricity bill [18].

Figure 4, shows an example of a customer electricity bill issued by the Kosovo Electricity Supply Company (KESCO) [18]. Each customer has his unique ID number, where the first three digits identify the customer from which municipality of Kosovo he is, such as DPR (Pristina Division), followed by the unique number which is protected by law and does not appear in this picture. Other information on this bill is the name and surname of the customer and his address. The amount of kWh consumed is found as the difference between the current state (CS) and the previous state (PS), i.e.: D = CS − PS. Since there are two types of cost calculation, as high tariff (A1) and low tariff (A2), they are calculated separately. This bill is before the new rules came out, for applying the first and second block tariffs.

To calculate the electricity costs and to manage the electricity costs of household appliances, a simple calculator developed in MS Excel has been prepared, which is available for free and is very easy to use. All that is required from the consumer is to give the sum of kWh read in the electric meter as high tariff (A1) and low tariff (A2). The calculator (Table 4) then calculates the percentage of the amount as expense A1 and A2. If the total amount exceeds 800 kWh, during the two readings, it applies the two-level block tariff. As seen in Fig. 4, in the electric bill, the expenses are expressed as difference between current state and previous state: D = CS − PS. Numerical value means the unit kWh (kilowatt hour).

$$D = CS - PS \tag{3}$$

where: D – Difference of states; CS – Current state; PS – Previous state

5 Conclusions

Electricity and its use, as much as it is useful, it can also cause damage. We can reduce these damages in the external environment by using electricity rationally and taking care not to misuse it. Education and discipline in this aspect is important not only for Kosovo, but for the whole world. In addition to the environment, saving electricity has a positive effect on households and maintaining their income. The rise in electricity prices in Kosovo, and the imposition of block tariffs, have caused confusion among the general population. The use of this simple computer application, created through Microsoft Excel, greatly simplifies the calculation of monthly electricity consumption and consumer expectations for the electricity bill.

References

1. NASA's Scientific Visualization Studio. Graphic: Global Warming from 1880 to 2018. http://climate.nasa.gov/climate_resources/139/graphic-global-warming-from-1880-to-2018 Updated 12 February 2019. Accessed 15 Dec 2021
2. Bhattacharjee, P.K.: Environmental pollution control and prevention system. In: Indian Engineering Congress, 2008 at Warangal, A.P, India, pp. 148–150, December 2008
3. Gribbin, J.: The Hole in the Sky: Man's Threat to the Ozone Layer. Bantam Books, Great Britain (1988)
4. Thangavel, S., Reddy, K.K.S.: Ozone layer depletion and its effects: a review. Int. J. Environ. Sci. Dev. **2**, 30–37 (2011). https://doi.org/10.7763/IJESD.2011.V2.93
5. Seppänen, J., Ilin, A.: An LSTM model for power grid loss prediction. Electr. Power Syst. Res. **189**, 106823 (2020). https://doi.org/10.1016/j.epsr.2020.106823
6. Mott, S., Ihonen, J., Kiviluoma, J.: A new method for analysing financial damages caused by grid faults on individual customers. Electr. Power Syst. Res. **207**, 107839 (2022). https://doi.org/10.1016/j.epsr.2022.107839
7. Lorca, Á., Prina, J.: Power portfolio optimization considering locational electricity prices and risk management. Electr. Power Syst. Res. **109**, 80–89 (2014). https://doi.org/10.1016/j.epsr.2013.12.004
8. Sousa, T., Soares, T., Morais, H., Castro, R., Vale, Z.: Simulated annealing to handle energy and ancillary services joint management considering electric vehicles. Electr. Power Syst. Res. **136**, 383–397 (2016). https://doi.org/10.1016/j.epsr.2016.03.031
9. Aslama, S., Khalid, A., Javaid, N.: Towards efficient energy management in smart grids considering microgrids with day-ahead energy forecasting. Electr. Power Syst. Res. **182**, 106232 (2020). https://doi.org/10.1016/j.epsr.2020.106232
10. Merdanoğlu, H., Yakıcı, E., Doğan, O.T., Duran, S., Karatas, M.: Finding optimal schedules in a home energy management system. Electr. Power Syst. Res. **182**, 106229 (2020). https://doi.org/10.1016/j.epsr.2020.106229
11. ERO Homepage. https://www.ero-ks.org/zrre/sq/ballina. Accessed 21 Feb 2022
12. Çakar, N.D., Erdoğan, S., Gedikli, A., Akif Öncüa, M.: Nuclear energy consumption, nuclear fusion reactors and environmental quality: the case of G7 countries. Nucl. Eng. Technol. **54**, 1301–1311 (2021). https://doi.org/10.1016/j.net.2021.10.015

13. Oviedo-Gómez, A., Londoño-Hernández, S.M., Manotas-Duque, D.F.: Effects of the COVID-19 pandemic on the spot price of Colombian electricity. Energies **14**, 6989 (2021). https://doi.org/10.3390/en14216989
14. Kirli, D., Parzen, M., Kiprakis, A.: Impact of the COVID-19 lockdown on the electricity system of Great Britain: a study on energy demand, generation pricing grid stability. Energies **14**, 635 (2021). https://doi.org/10.3390/en14030635
15. Beucherie, P.: Selective absorbant surfaces for high temperature solar collectors. Electr. Power Syst. Res. **3**(1–2), 125–137 (1980). https://doi.org/10.1016/0378-7796(80)90028-0
16. Beqiri, R.: Detyra të zgjidhura. Gjeniu. Prishtinë, Kosovë (2002)
17. Beqiri, R.: Formulat e fizikës. Gjeniu. Prishtinë, Kosovë (1999)
18. KESCO Homepage. https://www.kesco-energy.com/?fbclid=IwAR2wdC2zPpwT1KrR44mpPy72fgWZMJLKD4B164voPbzisAoYFUtT5QXjLas. Accessed 26 Jan 2022

Gestural Interaction Design Model to Support Car Drivers Interacting with Smartphone

Jayasankari Ganasan[1(✉)], Ahmad Sobri Hashim[1], and Norshahila Ibrahim[2]

[1] Computer and Information Sciences Department, Universiti Teknologi PETRONAS (UTP), Seri Iskandar, Perak, Malaysia
{jayasankari_20000248,sobri.hashim}@utp.edu.my

[2] Faculty of Art, Computing and Creative Industries, Universiti Pendidikan Sultan Idris (UPSI), Tanjong Malim, Perak, Malaysia
shahila@fskik.upsi.edu.my

Abstract. The number of collisions on the road is alarming, and one of the major contributing factors is drivers' habit of using smartphones while driving. The issue is attributed to the driver's state when interacting with smartphone user interfaces, particularly when driving in a confined space with minimal distraction. The current method of interacting between the driver and smartphone user interfaces is not ergonomic or safe. The gestural interface is seen as a potential solution to this issue. However, it needs to be further investigated to support drivers' driving conditions. Therefore, this paper presents a gestural interaction design model which consists of four key components: user interface design, cognitive, ergonomic and safety, and distraction. These components are derived from a thorough literature review and theoretical background analysis. These four components are seen to be the key elements to be included in the development of the conceptual model to address the current challenges faced by drivers. The components are addressed based on three age group drivers: 18–35 years old, 36–50 years old and more than 50 years old. The outcome of the developed model could assist the drivers in engaging with the smartphone safely and contribute to the reduction in car crash cases in the long run. This will guide future smartphone developers to build a more user-centred smartphone user interface that will support drivers' interaction based on age category.

Keywords: Gestural interaction · Gestural interface · Car drivers · Smartphone user interface · Conceptual model · Car-crash cases · Human-computer interaction

1 Introduction

Smartphones have growingly become valuable devices to people for their portability and ease of access to information anytime anywhere. Several applications have been developed in different areas such as communication, education, finance, entertainment, and transportation [1] and [2]. Since these applications reside on smartphones, users constantly interact with the smartphones for their daily-routine activities, even while driving cars. This behaviour, unfortunately, leads to catastrophic incidents like car crashes.

R. Silhavy (Ed.): CSOC 2022, LNNS 501, pp. 536–545, 2022.
https://doi.org/10.1007/978-3-031-09070-7_44

Recent statistics have shown a high percentage of car-crash cases which are caused using smartphones while driving [3] and [4]. This report has contributed to the adaptation of a new lifestyle of avoiding smartphones across many users while driving as strongly recommended. However, some situations require drivers to use smartphones while driving such as in the case of an emergency which may require a quick response.

There are some technological solutions available to support the use of smartphones while driving [5]. However, the outcome of the implementation of these solutions did not translate to the required decrease in the number of car crash cases today. Studies have been conducted on gestural interactions with different groups of people for various reasons. The available solutions have certain drawbacks and drivers are prone to ergonomic and safety issues while interacting with smartphones.

Thus, concerning the current dilemma car drivers face, this paper presents a conceptual model consisting of four key components. These components are merely identified through an extensive literature review and theoretical background analysis which have been discussed in Sect. 2 and Sect. 3, respectively. The model is developed concerning three different age groups i.e., 18–35 years old, 36–50 years old and more than 50 years old. The components are further elaborated based on the age category which has been addressed in Sect. 4.

In the long run, this model is targeted to support the road safety industry by minimizing distractions and easing the use of smartphones while driving; thereby reducing car crash cases. The best option is always to avoid using smartphones while driving. However, some situations require them to use smartphones such as in the case of emergency or any important things that require a quick response. Hence, this paper aims to address the issues raised above by providing a solution and improving the user experience with smartphone interaction while driving.

This paper is divided into sections and subsections. Section 2 discusses the literature review, while Sect. 3 describes the theoretical background. Section 4 is about the development of the model, and Sect. 5 is on the conclusion and future work.

2 Literature Review

2.1 Statistics of Road Crash Cases

Based on the Ministry of Transport Malaysia statistics, 521,466 cases of road crashes were recorded in 2016 and 7,152 of them caused death. This number has increased 6.50% from the previous year for general road crash cases, and 6.65% for death cases. Every time a road crash happens, it results in traffic jams especially on highways, no matter the crash is minor or major. 28.48% of death cases occurred due to drivers' carelessness especially using smartphones while driving.

Research conducted by the Malaysia Institute of Road Safety Research (MIROS) to 300 random respondents in 2017, 43.4% of them used smartphones while driving and riding, while 61.7% of them accepted or made calls while driving [6]. Besides, it was also revealed that 53.6% of them used smartphones when they were trapped in a traffic jam and 53% of them texted while driving.

Similar findings were recorded in the studies conducted in the United States of America and Australia, respectively. A study conducted by America [7] involving 1211 US drivers revealed that 48% of them read the text and 33% of them write text while driving. Meanwhile, the study in Australia [8] reported that 45% of total respondents text while driving.

This finding is alarming, and it seems that drivers are unable to obey the traffic rule which is not to use mobile phones while driving. This becomes worse with the current trend and social lifestyle to heavily rely on a smartphone for most of the daily routines [9].

2.2 Related Work

Smartphone interface should be designed in such a way that it can minimize distraction and does not disturb drivers' driving performance [10]. A study was conducted by [11] to evaluate multi-touch gestures using kinetic scrolling in interacting with the smartphone user interface. An experiment was performed involving 29 participants who were instructed to perform certain tasks using smartphones while driving in a driving simulation environment. The analysis was done in terms of task completion, eye gaze and driving performance. It was found that kinetic scrolling affects sufficient visual distraction to the participants which disturbed their driving performance. This technique required better finger-pointing accuracy to improve the efficiency of the interface.

Another study conducted by [12], where the authors proposed a method for designing a user interface that will display sufficient and important information such as driver, vehicle and environment data to drivers. The aim is to enhance drivers' driving performance. Different from the study conducted by [11], the interface was designed by incorporating other communication components including vehicle-to-vehicle communication, sensory information and monitoring capability. The information received from the three communication input devices were displayed on the driver's smartphones. Based on the experiment conducted, it was found that information on the driver, vehicle and environment conditions are three sufficient and important information that needs to be displayed to the drivers to ensure safe driving.

Distraction during driving would be harmful to the drivers. As a result, gestural interaction is viewed as a viable solution to driver ergonomic and safety constraints. However, it has not been comprehensively explored and understood for interaction with smartphone user interfaces [13]. Thus, there is a need to critically examine and analyse existing gestural interactions to establish a more sophisticated gestural interaction design model. The aim is to make it easier for drivers to engage with smartphone user interfaces while driving, resulting in a safer driving environment.

3 Theoretical Background

Concerning the current challenges faced by drivers, there is a need to develop a user-centred gestural interaction design model. As the need of getting a solution increases day by day, the literature study and related work have been revised several times. But the question is, what are the components that need to include in the conceptual model?

Based on the analysis conducted, four components are highly recommended to be placed as part of the conceptual model which is closely supported by the literature review. Thus, the subsections below thoroughly describe those components and the importance of including those in the development of the conceptual model to support drivers in engaging with smartphones while driving.

3.1 User Interface Design (UID)

As stated by [14] user interface design is an important factor that is paid much attention to these days. This acts as a bridge between the user and smartphone for interaction. Thus, it is important to know what are the gestures that suit the user and how a more user-friendly interface should be developed.

Any bodily movement that can be recognised by a computer system and responded to without the need of a standard pointing device such as a mouse or pen is known as a "gesture" [15]. A wave, head nod, touch, toe tap, or brow lift can all be considered gestures depending on the device. Gestural interfaces are a part of a digital system interface that uses gestures to complete certain activities [16].

The ease of use in user experiences is always considered in the process of interface designation, especially for touchscreen [17]. The ability to manipulate digital things on the screen without using command-line commands is known as direct manipulation [18]. For example, moving an undesired file or application to the desktop bin without having to write *'del'* on the command line is now widely used in smartphones and other devices. Whereas indirect manipulation is the indirect control over their physical surrounding through an interactive gesture. An example that can present the use of indirect manipulation is "The Clapper" where it functions to turn on or off with the clap of the user.

In the world of gestures, the main difference that exists between traditional interfaces is gestural interfaces contains a wide scope of actions with which to manipulate a system. Moreover, gestural interfaces can have befitted on the whole body in the process of triggering system behaviours. Thus, it is dominant to design the user interface which could cater for all three age categories in the case of our research. This proves the importance of including user interface design in the development of a conceptual model to further assist the drivers in safe driving while engaging with a smartphone.

3.2 Cognitive

Impairments to driving arise from various interference and cognitive workload is one of them which has a large impact on the user experience on smartphone user interface [19]. Memorization is an important cognitive process in the brain since it underpins the majority of human intelligence capabilities. So, what is memorization or cognitive process is done with the development of the gestural interaction design model?

As the ageing process occurs, humans' motor skills decline [20]. As a result, when a senior driver interacts with a smartphone while driving, he/she is unable to recall information from the preceding screen. As this is the case, the process of settling things on smartphones takes a longer duration than it should be.

Keeping this in mind, it is important to design a user interface that concurrently works well with various age category drivers. It is required to study the mental workload of drivers as they are interacting with smartphones and accommodate their needs based on the age category.

However, memorizing, as a dynamic cognitive process that manipulates information across memories in the brain, has to be fully researched, especially for different age groups of drivers. Memorization, or the power of motor abilities, is a process of retaining and retrieving acquired knowledge and prior incidents, based on the physiological basis of memories.

Thus, there is a need to thoroughly study and analyze the mechanism of motor skills of drivers based on the age group. The cognitive component plays a vital role in assisting drivers while interacting with a smartphone user interface to avoid longer duration is taken to complete smartphone tasks.

3.3 Ergonomic and Safety

Ergonomics is the study of the physical features of interaction, such as how the controls are constructed, the physical space in which the interaction takes place, and the screen layout and physical attributes [21]. The interface is primarily concerned with user performance and how it promotes or detracts from it.

The user should be able to quickly access sets of controls and elements of the display by grouping them logically. When dealing with a single user, this may not seem relevant, but it becomes critical when dealing with safety-sensitive applications. The physical layout of things must be proper in this case. Returning to the smartphone user interface, improper control and display layout can result in inefficiency and annoyance. Thus, the user should be able to reach all necessary functions and view all displays without making superfluous gestures. Important displays should be placed at eye level. The lighting should be set up to prevent glare and reflections from distorting the image. Controls should be spaced such that the user has enough area to manoeuvre.

Knowing that our research presents the conceptual model based on the age categories, thus it is important to know who the users are. This leads to the question, "Which age category do users fall into?" By identifying the users, the smartphone developers can easily design the smartphone user interface by considering how ergonomics and safety can be assured for certain age categories.

Although we may not instantly consider smartphone use to be a risky activity, it is important to consider the potential repercussions of a smartphone's badly designed user interface. Leaving aside the obvious safety risks of the smartphone user interface will harm the users.

Ergonomics is closely related to human psychology since it is concerned with human perceptual limits. The use of colour in displays, for example, is an ergonomics issue. The visual system has some colour restrictions, such as the limited number of colours that can be distinguished and the low blue acuity. Colour vision insufficiency affects a sizable section of the population. The colours used in the smartphone user interface display should be as distinct as possible, with variations in contrast not affecting the differentiation. Colours should also adhere to established conventions and user expectations. These rules should not be violated unless there is a compelling reason to do so.

3.4 Distraction

According to the American Automobile Association Foundation for Traffic Safety (AAAFTS), distraction occurs when "a driver is delayed in the recognition of information needed to safely accomplish the driving task due to some event, activity, object, or person within or outside the vehicle compelled or tended to induce the driver's shifting attention away from the driving task" [22]. When a driver's gaze is diverted from the road and engages themselves in any secondary activities such as communicating with passengers or manually interacting with any devices their focus goes away. This situation is known as "structural interference" where the drivers are unable to focus at two disparate locations at the same time.

In this case, the secondary task that we are looking into is the interaction with smartphones while driving. For instance, answering a phone call while driving requires mental resources to interact with the next person. Performing a conversion task while driving proves that it diverts the attention of drivers from the primary task. This being a sample scenario, there are many ways and reasons for drivers to engage in a smartphone while driving. There are various types of distraction that co-relates with the use of smartphones when driving.

Distraction may be classified into several types, and the four main distraction that plays a huge role are visual distraction, cognitive distraction, manual distraction, and auditory distraction. The action that needs drivers to glance away from the road to visually acquire particular information is referred to as visual distraction, while any task that arises during driving that diverts the driver's mental focus is referred to as a cognitive distraction. Manual distraction on the other hand is when the driver is doing any physical tasks that are not related to driving such as engaging in the smartphone. Auditory distraction is where the driver is distracted via an auditory source that is not related to driving [23].

The study conducted by [24] proves that while engaging with a smartphone, drivers somehow got distracted in either way. Thus, the reason for involving the distraction component in the conceptual model is that, with any sort of secondary task, especially smartphone usage, drivers are somehow prone to a dangerous situation. Thus, by incorporating distraction elements in developing the model, it could minimize eyes off road duration, reduce the need for mental workload and could quickly settle the task perform using a smartphone. Thus, a well-developed and the designed smartphone user interface is essential to accommodate the user task while prioritizing drivers' safety.

Finally, the four key components discussed above summarize the conceptual model elements in brief. The importance and the need of implementing these components will rather provide a more advanced gestural interaction design model and support drivers' interaction with the smartphone. The complete developed model is further discussed in Sect. 4.

4 Development of Gestural Interaction Design Model

4.1 Implementation of 4 Key Components

The gestural interaction design model has been developed by incorporating all those four key components discussed in Sect. 3. The model is further extended with a summary by addressing each age category of drivers. The data is gathered from a literature review for each of the age category drivers. The developed model is depicted as shown in Fig. 1 below:

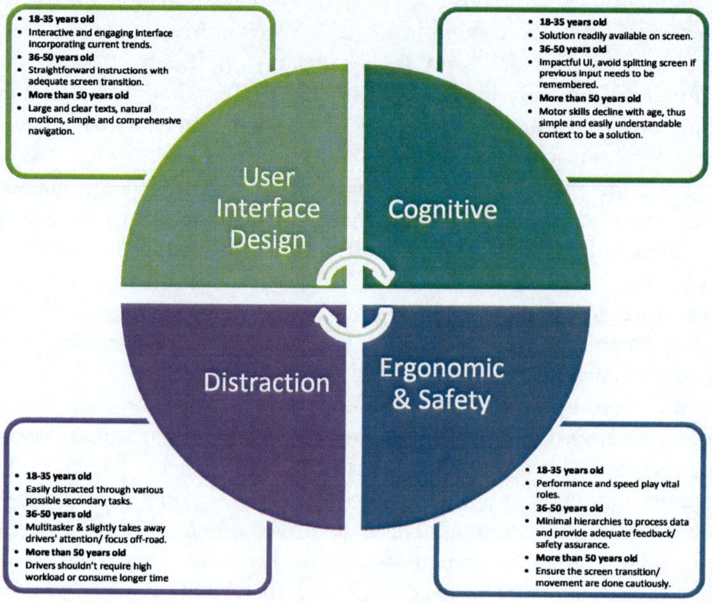

Fig. 1. Gestural interaction design model

User interface design is one of the most crucial components that need to be included in the model development. Various age group drivers would prefer the user interface to be seen differently. Drivers of age group 18–35 years old prefer to have a more interactive and engaging interface by incorporating current trends. Whereas drivers of age group 36–50 years old would like to have straightforward instruction with adequate screen transition. They do not prefer cluttered information dumped on a single screen. The drivers of the age group 50 years and above, would like to have large and clear texts, natural motions, and simple and comprehensive navigations across the screen. The colour contrast plays a huge role for this age group of drivers.

Furthermore, cognitive is seen to be a crucial element to be included in the development of the conceptual model. The drivers of the age group 18–35 years old would prefer to have the solution or guide readily available on the screen. These drivers lay lesser emphasis on searching and exploring for a solution. Drivers of the age group 36–50 years old would like to spot an impactful user interface with adequate information.

They do not prefer to have a split-screen if the previous input needs to be remembered to perform the following task. Drivers of age group 50 years and above on the other hand are conscious of their motor skill declination. As this is their major consent, they do prefer a simple language and easily understandable context. This will help with their motor control issues such as memory, concentration, and speed to process information.

Moreover, ergonomics and safety are considered important components in the conceptual model. As much as ergonomics and safety are paid attention to in Human Safety and Health (HSE) field, it is important too in addressing drivers' safety while interacting with a smartphone. User interface performance and speed play a vital role for the drivers of the age category 18–35 years old. Simplicity is seen to be the key aspect as they are frequent smartphone users while driving. Whereas drivers of age group 36–50 years old desire to have minimal hierarchies in data processing and constantly provide safety assurance. Drivers of age group 50 years and above are more cautious on their gestures or movements on screen. Thus, providing a safe and simple interface would ease the use of smartphones for drivers of this age group. They prefer to avoid the use of unambiguous language to avoid misinterpretation of information.

Concerning the use of a smartphone while driving brings distraction component in the picture. It is seen to be one of the most crucial elements that we need to investigate. Drivers of age group 18–35 years old get easily distracted through the various possible secondary task, in particular engaging with a smartphone. This takes away their focus off-road. Whereas drivers of age group 36–50 years old does multitasking most of the time. This simply means that they engage themselves in a smartphone while driving more often. Drivers of age group 50 years and above prefer not to focus on a secondary task for a longer duration. They are very much concerned on the road and tries to avoid any sort of distraction while driving. Thus, it is important to address the challenge faced by drivers in minimizing the duration of eyes off-road as much as possible.

In summary, these four key components that have been discussed above summarize our conceptual model. The explanation for each component based on the age group is supported by the literature review and theoretical analysis discussed in Sect. 2 and Sect. 3, respectively. Avoiding the use of smartphones while driving is always the best approach. However, in specific cases, such as an emergency or other critical matter that requires a quick response, drivers must use smartphones. Hence, the goal of the conceptual model is to assist the drivers in interacting with smartphone user interfaces in a more safe and secure environment.

5 Conclusion and Future Work

We proposed a gestural interaction design model that consists of four core components: user interface design, cognitive, ergonomic and safety, and distraction. This is further elaborated for each of the age categories: 18–35 years old, 36–50 years old and more than 50 years old. A thorough literature review has been conducted to gather all those data presented in Sect. 4 above. This is to allow the drivers to engage with smartphone user interfaces efficiently without commencing any danger on road.

For future work, we will be verifying the gestural interaction design model presented in Sect. 4 with the authorized bodies, Malaysian Institute of Road Safety Research (MIROS) experts and Human-Computer Interaction experts. We aim to build a gestural interaction design model which could support drivers' interaction with the smartphone user interface. In the long run, we ensure the developed model could reduce the car-crash cases.

Acknowledgement. The authors acknowledge the support of the Ministry of Higher Education (MOHE), Malaysia to provide financial assistance under Fundamental Research Grant Scheme (FRGS). Reference number: FRGS/1/2019/ICT04/UTP/02/1 and Universiti Teknologi PETRONAS to provide the Lab facilities to conduct this research work.

References

1. Ettema, D.: Apps, activities and travel: an conceptual exploration based on activity theory. Transportation **45**(2), 273–290 (2017). https://doi.org/10.1007/s11116-017-9844-5
2. Wang, D., Xiang, Z., Fesenmaier, D.R.: Smartphone use in everyday life and travel. J. Travel Res. **55**(1), 52–63 (2016). https://doi.org/10.1177/0047287514535847
3. Bergmark, R.W., Gliklich, E., Guo, R., Gliklich, R.E.: Texting while driving: the development and validation of the distracted driving survey and risk score among young adults. Inj. Epidemiol. **3**(1), 10 (2016). https://doi.org/10.1186/s40621-016-0073-8
4. Oviedo-trespalacios, O., King, M., Haque, M., Washington, S.: Risk factors of mobile phone use while driving in Queensland: prevalence, attitudes, crash risk perception, and task-management strategies. PLoS ONE **12**(9), e0183361 (2017)
5. Bennakhi, A., Safar, M.: Ambient technology in vehicles: the benefits and risks. Procedia Comput. Sci. **83**(FAMS), 1056–1063 (2016). https://doi.org/10.1016/j.procs.2016.04.223
6. Burhan, F.H., Abashah, A., Ibrahim, Z., Hashim, M.: Mobile phone distraction factors among Malaysian youth driver (2018). 335908193_Mobile_Phone_Distraction_Factors_Among_Malaysian_Youth_Driver
7. Gliklich, E., Guo, R., Bergmark, R.W.: Texting while driving: a study of 1211 U.S. adults with the distracted driving survey. Prev. Med. Rep. **4**, 486–489 (2016). https://doi.org/10.1016/j.pmedr.2016.09.003
8. King, M., Legge, M., Oviedo-Trespalacios, O., Regan, M., Rakotonirainy, A.: Scoping Study of Mobile Phone Use, p. 76 (2017)
9. Damyanov, I., Tsankov, N.: Mobile apps in daily learning activities. Int. J. Interact. Mob. Technol. **12**(6), 133–140 (2018). https://doi.org/10.3991/ijim.v12i6.9659
10. Guo, F., et al.: The effects of age on crash risk associated with driver distraction. Int. J. Epidemiol. **46**(1), 258–265 (2017). https://doi.org/10.1093/ije/dyw234
11. Louveton, N., McCall, R., Koenig, V., Avanesov, T., Engel, T.: Driving while using a smartphone-based mobility application: evaluating the impact of three multi-choice user interfaces on visual-manual distraction. Appl. Ergon. **54**, 196–204 (2016). https://doi.org/10.1016/j.apergo.2015.11.012
12. Sadigh, D., Driggs-Campbell, K., Bajcsy, R., Sastry, S.S., Seshia, S.A.: User interface design and verification for semi-autonomous driving. In: HiCoNS 2014 – Proceedings of 3rd International Conference High Confidence Networked Systems (Part CPS Week), pp. 63–64 (2014). https://doi.org/10.1145/2566468.2576851
13. Grossman, M.: Next-generation user interface technologies for mobile and consumer devices. Report, no. March, p. 71 (2014)

14. Hsiao, S.W., Lee, C.H., Yang, M.H., Chen, R.Q.: User interface based on natural interaction design for seniors. Comput. Human Behav. **75**, 147–159 (2017). https://doi.org/10.1016/j.chb.2017.05.011
15. Saffer, D.: Designing Gestural Interfaces. O'Reilly Media, Inc., Newton (2009)
16. Khan, M.I.U.: 毕业论文 Hand gesture detection & recognition system. Master's Thesis, Computer Engineering, Dalrana University, Sweden (2011)
17. Oulasvirta, A., Feit, A., Lähteenlahti, P., Karrenbauer, A.: Computational support for functionality selection in interaction design. ACM Trans. Comput. Interact. **24**(5), 1–30 (2017). https://doi.org/10.1145/3131608
18. Taralle, F., Paljic, A., Manitsaris, S., Grenier, J., Guettier, C.: Is symbolic gestural interaction better for the visual attention? Procedia Manuf. **3**(Ahfe), 1060–1065 (2015). https://doi.org/10.1016/j.promfg.2015.07.172
19. Strayer, D.L., Cooper, J.M., Turrill, J., Coleman, J.R., Hopman, R.J.: The smartphone and the driver's cognitive workload: a comparison of Apple, Google, and Microsoft's intelligent personal assistants. Can. J. Exp. Psychol. **71**(2), 93–110 (2017). https://doi.org/10.1037/cep0000104
20. Angelini, L., et al.: Comparing gesture, speech and touch interaction modalities for in-vehicle infotainment systems. In: IHM 2016 - Actes la 28ieme Conférence Francophone sur l'Interaction Homme-Machine, pp. 188–196 (2016). https://doi.org/10.1145/3004107.3004118
21. Rozanski, E.P., Haake, A.R.: Human–computer interaction (2017)
22. Burnett, G.E., Stevens, A.: Defining Driver Distraction, no. February (2018)
23. Spyropoulou, I., Linardou, M.: Modelling the effect of mobile phone use on driving behaviour considering different use modes. J. Adv. Transp. **2019** (2019). https://doi.org/10.1155/2019/2196431
24. Lyngsie, K., Pedersen, M.S., Stage, J., Vestergaard, K.F.: Don't text while driving: the effect of smartphone text messaging on road safety during simulated driving. In: Kotzé, P., Marsden, G., Lindgaard, G., Wesson, J., Winckler, M. (eds.) INTERACT 2013. LNCS, vol. 8119, pp. 546–563. Springer, Heidelberg (2013). https://doi.org/10.1007/978-3-642-40477-1_35

Framework of Critical Infrastructure Resilience in Tasks of Regional Development

D. N. Khaliullina[1(✉)], V. V. Bystrov[1], and S. N. Malygina[1,2]

[1] Institute for Informatics and Mathematical Modeling, Kola Science Centre of the Russian Academy of Sciences, Apatity, Murmansk Region, Russia
{khaliullina,bystrov,malygina}@iimm.ru

[2] Apatity Branch of Murmansk Arctic State University, Apatity, Murmansk Region, Russia

Abstract. The research is aimed to improve the methodology of regional development management by designing of problem-oriented decision support systems. The article deals with the problem of managing regional socio-economic systems from the perspective of critical infrastructures resilience. The authors propose a formalized conceptual description of the research task and the subject area being studied, as well as the developed means of information and analytical management support. The conceptual model makes it possible to ensure the interoperability of knowledge representation about the object of research by fixing the most significant components of the regional development system and the relations between them.

Keywords: Critical infrastructures resilience · Decision-making system · Socio-economic systems · Regional development

1 Introduction

Despite a large number of research and scientific and practical developments in the field of regional management, the problem of designing information and analytical tools to support decision-making does not cease to be relevant. This is due to the fact that regional development is associated with the functioning of complex dynamic socio-economic systems. This type of system has a weakly formalized structure, in which there are implicit feedbacks that complicate the behavior of the system.

Currently, there are various approaches to the study and management of regional socio-economic systems. In the last century, the most popular concept for the study of such systems was the theory of sustainable development [1]. But in the last decade, this theory has been replaced by a new concept of systems resilience [2]. In particular, the methodology of managing of critical infrastructure resilience in various areas of human activity (for example, works [3–6]) is actively developing.

The EU Council Directive defines the concept of critical infrastructure as "an asset, system or part thereof located in Member States which is essential for the maintenance of vital societal functions, health, safety, security, economic or social well-being of people,

R. Silhavy (Ed.): CSOC 2022, LNNS 501, pp. 546–556, 2022.
https://doi.org/10.1007/978-3-031-09070-7_45

and the disruption or destruction of which would have a significant impact in a Member State as a result of the failure to maintain those functions" [7].

Considering the above, the authors of the current study suggested it is possible to create means of information and analytical support for regional development management based on the concept of critical infrastructure resilience (CIR). The main idea of the study is to generalize and systematize existing methods in the field of CIR and selectively adapt them to the management of regional socio-economic systems.

2 Background

The key concept of this study is "resilience". The following definitions of this concept are recorded in the world scientific community:

"System resilience is the intrinsic ability of a system to adjust its functioning prior to, during, or following changes and disturbances, so that it can sustain required operations under conditions created by both expected and unexpected causes, such as human errors, system faults, adversarial attacks, and natural disasters" [8].

"Resilience is defined as the ability of a system, community or society exposed to hazards to resist, absorb, accommodate, adapt to, transform and recover from the effects of a hazard in a timely and efficient manner, including through the preservation and restoration of its essential basic structures and functions through risk management" [9].

To date, analytical papers have already been published on the comparative analysis of critical infrastructure resilience methodologies in various cross-sections and for different subject areas. For example, the paper [10] provides several definitions of resilience in different domains and focused on qualitative and quantitative assessment approaches and their subcategories. A critical review of the hazards/threats affecting the building of CIR is given in the work [11]. The research [12] is devoted to the consideration of the existing approaches to modeling and simulation of CIR. As a separate direction in the field of CIR, we can single out research focused on the formation of a system for measuring resilience. For example, the work [13] provides a modern overview of existing systems for assessing CIR. Within the framework of the works [14, 15], the issues of transition from the theory of management of complex systems to the concept of critical infrastructure resilience in the development of management systems are raised, in particular, conceptual descriptions of information management systems of regional security are proposed.

In the current study, the works related to the creation of software to support the management of critical infrastructure resilience are of the greatest interest. In particular, the paper [16] describes CRISADMIN project. The project examines the consequences of a critical event in the context of three critical infrastructure sectors: transport (private and public), energy (distribution and consumption of electricity) and telecommunications (mobile and fixed). The paper [17] describes a multi-agent model of public water supply from the perspective of critical infrastructure resilience. The authors of the article [18] have developed software for analyzing the critical infrastructure safety based on a risk matrix.

These publications show that the development of software for managing critical infrastructure resilience remains relevant.

The first step in software development is the formation of a conceptual understanding of the subject area, structure and functionality of the software solution.

3 Methods and Tools

Within the framework of the initial stage of the study, the following methods are used:

- Search and analysis of information from open scientific and practical publications and web resources to systematize knowledge about CIR.
- Conceptual modeling in terms of set-theoretic relations to represent the subject area of research in a formalized form.

In the future, it is planned to use the following set of approaches, methods and tools to develop a system to support the management of CIR of regional socio-economic systems:

- Process and project approaches;
- Principles of network-centric management;
- The concept of situational awareness;
- Elements of the theory of risk management and crisis management;
- Simulation modeling (agent-based models, system dynamics);
- Technologies to develop multi-agent software systems (including intelligent agents).

According to the authors, the integration of the above methodological and software tools will create an effective decision support system in the field of regional development.

4 Results

At the stage of initiating the design and implementation of a software system to support the management of CIR of regional socio-economic systems, it is necessary to present the concept of the software product being developed. To this end, it is rational to present the software product and the scope of its use in the form of a formalized conceptual description.

If a software product is considered as an information management system, it is logical to consider it from the perspective of the traditional theory of systems management. That is, to distinguish in it the object of control, the subject of control, control influences and disturbances.

In particular, critical infrastructures will act as an object of control, decision makers and a set of software and hardware tools – as a subject of control, developed measures and solutions to change the state of critical infrastructures – as control actions, external and internal threats and challenges affecting the resilience of regional socio-economic systems – as disturbances.

Thus, in terms of set-theoretic relations, it is possible to present a formalized conceptual model of a system to support management of CIR of regional socio-economic systems as following:

$$DSS_{RCI} = \langle O_{CI}, S_{RCI}, CA_{RCI}, D_{RCI} \rangle,$$

where

O_{CI} – object of control,
S_{RCI} – subject of control,
CA_{RCI} – control actions,
D_{RCI} – disturbances,
CI – critical infrastructure,
RCI – resilience of critical infrastructure.

It should be noted that the global goal of the software product being developed is to ensure the risk-resistant development of the socio-economic systems of the region. The stated goal will be achieved by the development of control actions. The control actions will be formed in an automated mode by adaptation of methods from the field of CIR.

Object of Control

Within the framework of the current study, the authors are interested in the issues of ensuring regional security of the socio-economic systems development.

In modern publications on the management of complex socio-economic systems, regional security (RS) is traditionally considered from two main positions:

1. Regional security is a state of protection of the macro system from internal and external threats or disturbances;
2. Regional security is a set of means (administrative, technical, informational, etc.) that are involved in the transition to and subsequent maintenance of the state of security of the macro system.

Within the framework of this study, the authors pay more attention to the second of these approaches, as they consider the problems of information and analytical support for the management of CIR of regional security.

Regional security is a complex system of interdependent subsystems, which can be considered as separate components of RS (Fig. 1). It is worth noting that from the entire set of RS subsystems, a subset of subsystems with signs of critical infrastructures (in their classical sense) can be distinguished.

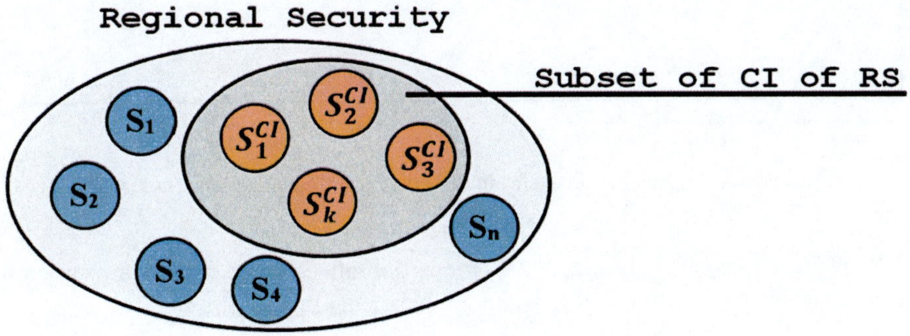

Fig. 1. The set of regional security subsystems

Thus, taking into account the above information, it is possible to formally describe the object of control of the decision support system being developed in the following form:

$$O_{CI} = <S^{CI}, R^{CI}>,$$

where

$S^{CI} = \{S_i^{CI}\}, i = \overline{1, N^{CI}}$ – set of critical infrastructures (S_i^{CI}) included in the regional socio-economic macro system.

R^{CI} – set of relationships between critical infrastructures.

Each critical infrastructure is described as follows:

$$S_i^{CI} = \langle Str_i^{CI}, St_i^{CI}, Prop_i^{RCI} \rangle,$$

where

$Str_i^{CI} = \langle E_i^{CI}, L_i^{CI} \rangle$ – CI structure, which is the set of critical infrastructure elements (E^{CI}) and links between them (L^{CI}).

$St_i^{CI} = \{st_k^{CI}\}, k = \overline{1, N^{St}}$ – set of states of critical infrastructure. The following critical infrastructure states (understand risk, anticipate/prepare, absorb/withstand, respond/recover, adapt/learn) are indicated in the article [19] (Fig. 2).

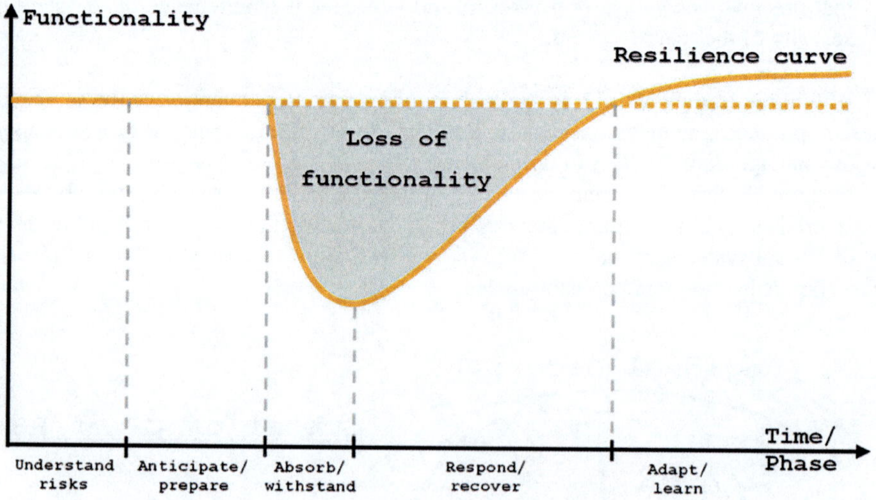

Fig. 2. Critical infrastructure states in accordance with the resilience cycle [19]

$Prop_i^{RCI} = \{prop_l^{RCI}\}, l = \overline{1, N^{Prop}}$ – properties reflecting the critical infrastructure resilience (robustness, redundancy, resourcefulness, and rapidity) [20].

Relationships Between Critical Infrastructures

Traditionally, there are four main classes of critical infrastructure interdependencies [18]:

- Physical Interdependency – two infrastructures are physically interdependent if the state of each depends upon the material output(s) of the other.
- Geographic Interdependency – infrastructures are geographically interdependent if a local environmental event can create state changes in all of them. This implies close spatial proximity of elements of different infrastructures, such as collocated elements of different infrastructures in a common right-of-way.
- Cyber Interdependency – an infrastructure has a cyber interdependency if its state depends on information transmitted through the information infrastructure. The computerization and automation of modern infrastructures and widespread use of supervisory control and data acquisition systems have led to pervasive cyber interdependencies.
- Logical Interdependency – two infrastructures are logically interdependent if the state of each depends upon the state of the other via some mechanism that is not a physical, cyber, or geographic connection.

Thus, the set of relationships between critical infrastructures can be formally described as:

$$R^{CI} = \left\{ r_{ph}^{CI}, r_{geo}^{CI}, r_{cyb}^{CI}, r_{log}^{CI} \right\},$$

where

r_{ph}^{CI} - subset of material nature relationships (for example, the supply chain of certain products or the movement of vehicles);

r_{geo}^{CI} - subset of geolocation relations (for example, the relation of geographical proximity of critical infrastructure objects);

r_{cyb}^{CI} – subset of cyber relationships (for example, information flows between critical infrastructures);

r_{log}^{CI} – subset of logical relationships (for example, changes in one CI may be caused by management decisions in another CI. In particular, summer vacationers may flock to the highways when gasoline prices are low, resulting in increased traffic congestion. In this case, the logical interdependency between the petroleum and transportation infrastructures is due to human decisions and actions and is not the result of a physical process [18]).

Example of Critical Regional Security Infrastructures

The following RS components can be considered as critical infrastructures in the regional security system, they are: the subsystem of economic security (CI_1); the subsystem of environmental security (CI_2); the subsystem of industrial security (CI_3); the subsystem of personnel security (CI_4) and other subsystems ($CI_i \ldots CI_f$). It is worth noting that the regional security system (RSS) can have a network, hierarchical or hybrid structure.

From the point of view of system analysis, in order to ensure the resilience of a set of regional security CIs, it is necessary to take into account the task that is being solved at the moment by the management support system.

It is worth noting that when developing information and analytical management support tools, each of the critical infrastructures can be considered at different levels: for example, at the informational, logical, physical and other levels of system description (Fig. 3) [21].

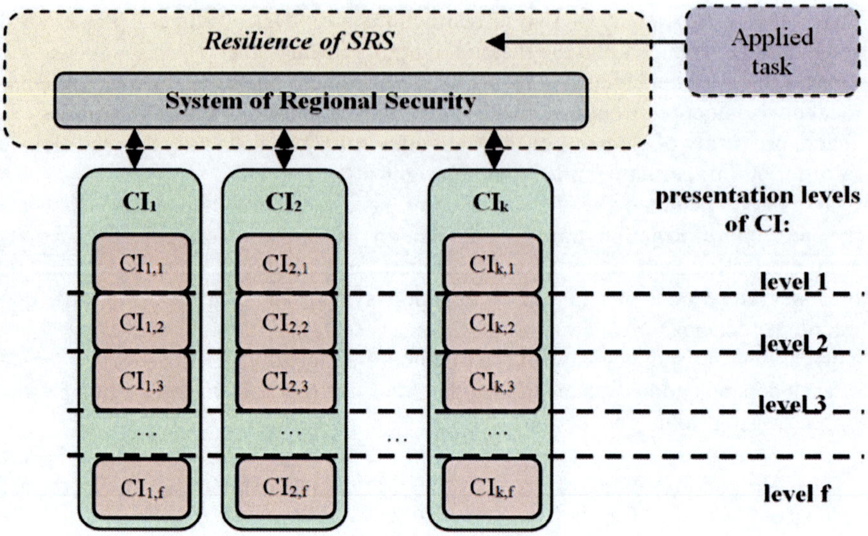

Fig. 3. Multilevel representation of the regional security system.

Subject of Control

Decision makers/organizational structure and the set of software and hardware can act as the subject of control.

$$S_{RCI} = \langle ORG^{S_{RCI}}, SH^{S_{RCI}}, R^{S_{RCI}} \rangle,$$

where

$ORG^{S_{RCI}} = \left\{ org_j^{S_{RCI}} \right\}, j = \overline{1, N^{S_{RCI}}}$ – set of organizational structures that manage critical infrastructures;

$SH^{S_{RCI}}$ – set of software and hardware tools by which information and analytical support of management is carried out.

$$SH^{S_{RCI}} = \left\{ prog^{SH}, tech^{SH} \right\},$$

where $prog^{SH}$ – set of software tools,

$tech^{SH}$ – set of technical means.

$R^{S_{RCI}}$ – set of relations between the elements of the subject of control:

$$R^{S_{RCI}} = \left\{ r^{ORG}, r^{SH}, r^{S_{RCI}} \right\},$$

where r^{ORG} – reflects the relationship between organizational structures:

$$r^{ORG} : ORG^{S_{RCI}} \rightarrow ORG^{S_{RCI}},$$

r^{SH} – relations between software and hardware:

$$r^{SH} : SH^{S_{RCI}} \rightarrow SH^{S_{RCI}},$$

$r^{S_{RCI}}$ – relationship between organizational structures and the set of software and hardware:

$$r^{S_{RCI}} : SH^{S_{RCI}} \rightarrow ORG^{S_{RCI}}.$$

Control Actions
The control actions include the measures and decisions developed to change the state of critical infrastructures. Formally, the control actions can be represented as the set,

$$CA_{RCI} = \left\{ ca_r^{RCI} \right\}, r = \overline{1, N^{CA}},$$

the elements of which represent the relationship between the subject and the control object:

$$ca_r^{RCI} : S_{RCI} \rightarrow O_{CI}.$$

As a result of the control action, a logical relationship between critical infrastructures can work.

Disturbances
Disturbances, which can be external and internal threats and challenges affecting the resilience of regional socio-economic systems, are formally described as the set of:

$$D_{RCI} = \left\{ D_p^{RCI} \right\}, p = \overline{1, N^{D_{RCI}}},$$

the elements of which can be represented as follows:

$$D_p^{RCI} = \langle Name, Cat, Loc, I_{hazard} \rangle,$$

where

Name – name of the disturbance;

Cat – category of disturbance, for example, natural disaster, emergency, crisis, etc.;

Loc – location of occurrence. This parameter indicates which critical infrastructure was affected, as well as whether it was caused by internal or external causes;

I_{hazard} – set of hazard indicators. Different hazard indicators can be included in the set. For example, they include: the degree of danger, the level of threat, the probability of occurrence, the expected damage, the duration of exposure, etc.

The analysis of the above formulations allows us to conclude that the critical infrastructure is a multicomponent distributed biogeotechnical system consisting of many interdependent subsystems (System of systems), a malfunction of at least one of which can lead to a significant deterioration in the life safety of the population included in the CI, and have a significant impact on the functioning of other subsystems and on the viability of the CI as a whole [21].

5 Conclusion

In modern conditions of digitalization of all spheres of human activity, the development of information and analytical support tools for managing complex management objects remains an important and urgent task. Such complex objects include regional socio-economic systems that are susceptible to various types of threats and dangers. In particular, the ever-changing modern world makes it more and more attentively and judiciously treat the decisions taken in the field of regional governance, including socio-economic ones. One of the effective ways to ensure the correctness of decision-making in conditions of high uncertainty and complexity of the object is the use of modern information technologies (including those from the field of artificial intelligence).

The main purpose of the current research is to develop methodological and software tools for information and analytical support of decision-making in the field of regional development. As a cross-cutting research methodology, it is proposed to use approaches and methods from the field of management of CIR adapted to the research/study of resilience to crisis events of regional socio-economic systems.

Within the framework of this article, the formalized conceptual model of a decision support system in the field of managing CIR of regional socio-economic systems is proposed. The conceptual model allows researchers to fix their points of view on the key components of the software being developed and the subject area being studied. The presentation of the conceptual model in the form of a mathematical description allows structuring information about the key elements of the object under study and establishing relationships between them in an explicit form. The proposed formalized conceptual model will be directly used to build an applied ontology of CIR of regional development. It is assumed that the constructed applied ontology will serve as the main means of presenting and storing data and knowledge in the decision support system being developed.

The article makes the first step of the research, such as formulation of the problem. The authors face a whole range of open questions that have yet to be answered as part of the implementation of the study. Such questions are:

- How to assess the state of CIR of regional socio-economic systems?
- What factors and their indicators should be taken into account when describing the CIR of regional development?
- What architecture of the software system to be developed should be chosen?
- How to use simulation to assess and predict CIR?

The list of open questions is not complete and will continue to expand as researchers immerse themselves in the subject area and the problems of research.

Acknowledgements. The research was supported by the Ministry of Science and Higher Education of the Russian Federation (project No. FMEZ-2022-0023 «Development of theoretical and organizational and technical foundations of information support for the management of the resilience of regional critical infrastructures of the Arctic zone of the Russian Federation»).

References

1. Meadows, D.H., Meadows, D.L., Randers, J., Behrens, W.W., III: The Limits to Growth, 205 p. Universe Books, New York (1972)
2. Holling, C.S.: Resilience and stability of ecological systems. Annu. Rev. Ecol. Syst. **4**, 1–23 (1973)
3. Bruneau, M.A., et al.: Framework to quantitatively assess and enhance the seismic resilience of communities. Earthq. Spectra **19**, 733–752 (2003)
4. Liu, H.J., et al.: Conceptual framework of life-cycle performance measurement: ensuring the resilience of transport infrastructure assets. Transp. Res. Part D Transp. Environ. **77**, 615–626 (2019)
5. Aldrich, D.P., Meyer, M.A.: Social capital and community resilience. Am. Behav. Sci. **59**, 254–269 (2014)
6. Adger, W.N.: Social and ecological resilience: are they related? Progress in Human Geography. Prog. Hum. Geogr. **24**, 347–364 (2000)
7. Council directive 2008/114/EC of 8 December 2008 on the identification and designation of European critical infrastructures and the assessment of the need to improve their protection. Off. J. Eur. Union (2008). https://eur-lex.europa.eu/legal-content/EN/TXT/PDF/?uri=CELEX:32008L0114&qid=1635332328283&from=EN. Accessed 2 Mar 2022
8. Zhu, Q., Wei, D., Ji, K.: Hierarchical Architectures of Resilient Control Systems: Concepts, Metrics, and Design Principles, 32 p. (2016). https://www.taylorfrancis.com/chapters/edit/10.1201/b19629-11/hierarchical-architectures-resilient-control-systems-concepts-metrics-design-principles-quanyan-zhu-dong-wei-kun-ji/. Accessed 2 Mar 2022
9. Pursiainen, C.: Critical infrastructure resilience: a Nordic model in the making? Int. J. Disaster Risk Reduct. **27**, 632–641 (2018)
10. Hosseini, S., Barker, K., Ramirez-Marquez, J.E.: A review of definitions and measures of system resilience. Reliab. Eng. Syst. Saf. **145**, 47–61 (2016)
11. Osei-Kyei, R., Tam, V., Ma, M., Mashiri, F.: Critical review of the threats affecting the building of critical infrastructure resilience. Int. J. Disaster Risk Reduct. **60**, 102316 (2021)
12. Ouyang, M.: Review on modeling and simulation of interdependent critical infrastructure systems. Reliab. Eng. Syst. Saf. **121**, 43–60 (2014)
13. Guo, D., Shan, M., Owusu, E.K.: Resilience assessment frameworks of critical infrastructures: state-of-the-art review. Buildings **11**(10), 464 (2021). https://doi.org/10.3390/BUILDINGS11100464
14. Masloboev, A.V.: Towards a theory of regional critical infrastructure security and resilience. Reliab. Qual. Complex Syst. **4**(32), 115–130 (2020). https://doi.org/10.21685/2307-4205-2020-4-13
15. Masloboev, A.V., Bystrov, V.V.: Conceptual model of critical infrastructures resilience in the context of modern theory of complex system security. Econ. Inf. Technol. **47**(3), 555–572 (2020). https://doi.org/10.18413/2687-0932-2020-47-3-555-572(InRussian)
16. Cavallini, S., et al.: A system dynamics framework for modeling critical infrastructure resilience. In: Butts, J., Shenoi, S. (eds.) ICCIP 2014. IAICT, vol. 441, pp. 141–154. Springer, Heidelberg (2014). https://doi.org/10.1007/978-3-662-45355-1_10
17. Rasoulkhani, K., Mostafavi, A.: Resilience as an emergent property of human infrastructure dynamics: a multi-agent simulation model for characterizing regime shifts and tipping point behaviors in infrastructure systems. PLoS ONE **13**(11), e0207674 (2018). https://doi.org/10.1371/JOURNAL.PONE.0207674
18. Brezhnev, E., Chernetskiy, B.: Risk matrix-based method for critical infrastructure safety assessment taking into account interdependencies. In: MATEC Web Conferences, vol. 76, p. 02050 (2016). https://doi.org/10.1051/matecconf/20167602050

19. Øien, K., Bodsberg, L., Jovanovic, A.: Resilience assessment of smart critical infrastructures based on indicators. In: Haugen, et al. (eds.) Safety and Reliability – Safe Societies in a Changing World. Taylor & Francis Group, London (2018)
20. Fang, Y.: Critical infrastructure protection by advanced modelling, simulation and optimization for cascading failure mitigation and resilience. Electric power. Ecole Centrale Paris, English (2015). fftel-01144504f
21. Khaliullina, D.N., Bystrov, V.V.: Resilience of critical infrastructure of regional security. Inf. Technol. **12**(5), 104–116 (2021). https://doi.org/10.37614/2307-5252.2021.5.12.009. Transactions of the Kola Science Centre. (in Russian)

Security Audit of a University Information System: A Functional Model

Andrei Kan and Ahmad Albatsha(✉)

HSE University, Moscow, Russian Federation
avkan@edu.hse.ru, aalbatsha@hse.ru

Abstract. In the context of transition to online education worldwide, it is of utmost importance to provide the protection of confidential data from information security threats. The present study intends to evaluate the level of security of the HSE University information system. The security audit of the system is carried out using the methodology of penetration testing which allows identifying vulnerabilities and assessing risks. The paper describes the stages of preliminary information gathering, preliminary vulnerability analysis. The exploration of methods and gathered information enabled developing a functional penetration testing model. Although some of the discussed (sub-)stages were not completed due to the legislative aspect, we were able to collect detailed information about the information system, determine providers and server locations, scan ports, determine file structures, describe the current state of the technologies used in the HSE University information system and compile a list of potential vulnerabilities. The proposed functional model and the completed phases will serve as a strong basis for conducting a full-fledged information security audit.

Keywords: Security audit · Penetration test · Information system

1 Introduction

The use of information technology in all spheres of life has caused numerous positive changes in society, but at the same time, it has given rise to numerous new threats to the security of confidential data. With the global transition to online education caused by COVID-19 [1], protecting the security of distance learning systems is of particular importance [2]. The information system of the National Research University Higher School of Economics handles different types of sensitive information including personal data of students and employees, curricula, course information, payment data, information for official use. This information can be exposed to threats during processing, storage and transmission, which will lead to a violation of information security: availability, integrity, confidentiality. In this regard, an urgent scientific task is to develop a functional model for auditing the information security of the University system.

The purpose of the present study is to assess the level of security of the University information system using penetration testing. To achieve this aim, it is necessary to gather information about the University system, analyze its security level based on

R. Silhavy (Ed.): CSOC 2022, LNNS 501, pp. 557–564, 2022.
https://doi.org/10.1007/978-3-031-09070-7_46

the collected information and develop a functional model for testing the system via penetration based on the work done. The proposed model for penetration testing of the University information system is aimed at identifying and exploiting the maximum number of vulnerabilities in order to compromise confidential data, disrupt systems, which will serve as the basis for the plan of the transition to the target state of university information systems.

2 Methodology

The security audit of the University information system is carried out using penetration testing. Penetration testing involves simulating the actions of a real attacker and comprises the following steps: gathering information, analysis of system vulnerabilities (scanning, building an attack vector), exploitation of vulnerabilities and reporting [3–5]. The result of penetration testing is a report containing a description of the attacks, identified vulnerabilities and recommendations for their elimination. The advantages of this method are repeatability and visual demonstration of vulnerabilities and consequences. Penetration testing enables detailing the vulnerabilities found, more accurately assessing risks and the real state of information security.

3 Results

3.1 Analysis of Open Sources and Preliminary Collection of Information

The official website of the National Research University Higher School of Economics contains a list of functioning systems: Single Personal Account (lk.hse.ru), Process Automation System (bpm.hse.ru), Electronic Document Management System based on 1C (confluence.hse.ru), Project Management System (pmo.hse.ru), Mail Host Provider MX Record (mx2.hse.ru), Online Support System for the Educational Process (lms.hse.ru), HSE Online Education Distance Learning System based on Moodle (online.hse.ru), Class Schedule Planning System (ruz.hse.ru), Cloud Storage Service (disk.hse.ru), Mail Server MX record (mx1.hse.ru), Mail Server MX record (mx.hse.ru). The systems having a web interface are under the domain name hse.ru that was registered to the National Research University Higher School of Economics in 1997. The IP address 186.2.163.228 is located in Belize and is owned by DDOS-GUARD (AS262254 DDOS-GUARD CORP., BZ). This IP address hosts at least 16 other domain names including pages of HSE employees, pages of outdated and current events, templates and one-pagers (econprize.ru, mclog.ru, energyleadership.ru, miem.edu.ru, uurok.ru, rosro.ru, opec.ru, etc.).

As a result of port scanning using nmap it was found that open ports of the IP address 183.2.163.228 are 80 (http) and 443 (https), and the rest of the ports are filtered. 179 (bgs) is an open port for the Single Personal Account system (45.89.226.3) and the Project Management System (45.89.226.154); 179 (bgs) and 53 (domain) are open for the Mail Server (45.89.226.137). Regarding the Online Education Distance Learning System (130.193.37.151), open ports are 22 (ssh), 80 (http), 443 (ssl/http), 9100 (see Fig. 1).

```
Nmap scan report for online.hse.ru (130.193.37.151)
Host is up (0.053s latency).
Not shown: 996 closed ports
PORT     STATE SERVICE    VERSION
22/tcp   open  ssh        OpenSSH 7.6p1 Ubuntu 4ubuntu0.3 (Ubuntu Linux; protocol 2.0)
80/tcp   open  http       nginx 1.14.0 (Ubuntu)
443/tcp  open  ssl/http   nginx 1.14.0 (Ubuntu)
9100/tcp open  jetdirect?
Service Info: OS: Linux; CPE: cpe:/o:linux:linux_kernel
```

Fig. 1. Scanning of 130.193.37.151.

3.2 Preliminary Analysis of Vulnerabilities

For the Single Personal Account System, Process Automation System, Online Support System for the Educational Process, Electronic Document Management System, Project Management System, Class Schedule Planning System, Cloud Storage Service, host site, ports 80 and 443 turned out to be open. These are standard ports for web applications.

Port 80 is a port for transferring document markup, also known as Hypertext Transfer Protocol (HTTP), which is normally used for web services. Attacks on this port can be divided into 2 types: (1) using a network connection. Cross-site scripting, phishing; (2) using application behavior, provided that the application is listening on port 80/443. Cross-site scripting, SQL injection. With regard to HSE University system providers, the use of the first type of attack does not make sense, since an appropriate level of protection is provided against network attacks, denial of service attacks. In the case of using the second type, the success of penetration depends on applications and vulnerabilities in them.

The Online Education Distance Learning System is based on Moodle and is deployed on the nginx web service version 1.14.0. As of 2021/05/04, 421 vulnerabilities of varying complexity are known in the Moodle system. Table 1 lists four examples of Moodle vulnerabilities related to both XSS vulnerabilities, SQL injection, and privilege escalation vulnerabilities [6]. For version 1.14.0 of the nginx web service, three vulnerabilities on the CVE-DETAILS resource are relevant: a vulnerability in the ngx_http_mp4_module module, which can lead to an infinite loop of the worker process, cause the worker process to fail, or expose the memory of the worker process; vulnerability in http/2 with ngx_http_v2_module present, could allow excessive CPU usage; the same ngx_v2_module can allow excessive memory consumption [7].

Port 22 is open on the server with the Online Education Distance Learning System. Port 22 (SSH) is a network protocol that allows the remote control of the operating system. Attack scenarios to gain unauthorized access might include (1) brute-force password attack (framework tools such as metasploit, hydra, medusa, ncrack and a password dictionary are needed); (2) exploitation of the RCE vulnerability CVE-2018-10933, if the libssh library version 6.0 and higher is available, the authentication vulnerability can be exploited in the server code. It was possible to get access to the file structure and source code of the web interface of the Online Education System.

Table 1. Vulnerabilities of Moodle.

Vulnerability	Description	Published on	Updated on	Complexity of use	Vulnerability type
CVE-2021-20283	The web service responsible for fetching other users' courses did not check that the requesting user had permission to view this information in each course in Moodle prior to 3.10.2, 3.9.5, 3.8.8, 3.5.17	2021/03/15	2021/03/23	Low	
CVE-2021-20282	In Moodle prior to version 3.10.1, some search inputs were found to be vulnerable to reflected XSS due to insufficient rendering of search queries	2021/01/28	2021/02/01	Medium	XSS
CVE-2020-25700	Some database module web services allowed students to add entries to groups they did not belong to. Versions affected: 3.9 to 3.9.2, 3.8 to 3.8.5, 3.7 to 3.7.8, 3.5 to 3.5.14 and earlier unsupported versions	2020/11/19	2020/12/03	Low	SQL
	Some users without permission could view the full names of other users using the online users block in Moodle prior to 3.10.2	2021/03/15	2021/03/23	Low	

Figure 2 illustrates the file structure of the HSE Online Education System. Based on the directory structure and some of the found files, it is possible to draw conclusions about the technologies used in the system. The root directory and such directories as "files", "auth", "blocks", "files", "backup" etc. contain.php files. When examining the source code, it is possible to determine the entry points of a web application, draw up an application map. It is important to note that all actions described in this section were carried out strictly within the law. None of the scanners or analyzers violated the integrity, performance and structure of the systems.

```
==> DIRECTORY: https://online.hse.ru/blocks/

==> DIRECTORY: https://online.hse.ru/blog/

==> DIRECTORY: https://online.hse.ru/cache/

==> DIRECTORY: https://online.hse.ru/calendar/

==> DIRECTORY: https://online.hse.ru/comment/

==> DIRECTORY: https://online.hse.ru/course/

==> DIRECTORY: https://online.hse.ru/error/
+ https://online.hse.ru/favicon.ico (CODE:200|SIZE:1150)

==> DIRECTORY: https://online.hse.ru/files/

==> DIRECTORY: https://online.hse.ru/filter/

==> DIRECTORY: https://online.hse.ru/group/
+ https://online.hse.ru/index.php (CODE:200|SIZE:51760)

==> DIRECTORY: https://online.hse.ru/install/

==> DIRECTORY: https://online.hse.ru/lang/

==> DIRECTORY: https://online.hse.ru/lib/

==> DIRECTORY: https://online.hse.ru/local/

==> DIRECTORY: https://online.hse.ru/login/
+ https://online.hse.ru/master (CODE:200|SIZE:48152)
+ https://online.hse.ru/master.passwd (CODE:200|SIZE:48152)
+ https://online.hse.ru/masterpages (CODE:200|SIZE:48152)
+ https://online.hse.ru/masters (CODE:200|SIZE:48152)

==> DIRECTORY: https://online.hse.ru/media/

==> DIRECTORY: https://online.hse.ru/message/

==> DIRECTORY: https://online.hse.ru/mod/

==> DIRECTORY: https://online.hse.ru/my/

==> DIRECTORY: https://online.hse.ru/notes/
+ https://online.hse.ru/online (CODE:200|SIZE:47523)
+ https://online.hse.ru/partners (CODE:200|SIZE:44151)
```

Fig. 2. File structure of the Online Education System (fragment).

3.3 Developing a Functional Penetration Testing Model Based on the Analysis of Methods and Collected Information

Based on the stages of preliminary collection and analysis of information, a high-level structure of university web applications can be distinguished. The HSE Online Education System infrastructure is a set of web applications that are located on servers from various providers. The interaction between them is based on data transmission via the Internet.

By functionality, the system can be divided into 3 modules: web interface, main functional mode, database. The web interface is an interface designed to gain access to the content, functionality of a web application. The main functional module is responsible for authentication and authorization management, the implementation of hierarchy and differentiation of access rights, system administration, monitoring and collection of analytics, learning materials management, interaction with other infrastructure components. The database allows storing training courses, assessment materials, student progress.

Physically, these modules can be located both on one server and distributed over several severs. From the point of view of information security, it is optimal to distribute such parts of systems, since having gained access to a system that includes all functional modules, an attacker gains access to all parts of the system.

As it was determined, information security breaches are implemented on vulnerabilities that are built on the basis of asset weaknesses. These assets should be taken into account when compiling the model and threats. The assets of the Online Education System are as follows:

- personal data of teachers, employees, students, applicants;
- user authentication and authorization data;
- teaching materials, manuals, tests, courses;
- score sheets;
- lists of courses, students, groups;
- student payment details;
- educational materials;
- curricula.

When compiling a list of potential threats, it is necessary to take into consideration the objects of protection, the logical and physical boundaries of the system. Therefore, the threat model includes four aspects of the Online Education System: web interface, main functional module, database and payment system. The list of potential threats for the web interface includes PHP injections; SQL injection in the authorization form; cross-site scripting, cross-site request forgery; attacks on the client side; RCE (remote code execution); DDOS attacks. The list of potential threats for the main functional module includes brute force attacks (password brute force, attack on the authentication and authorization system), privilege escalation, DDOS attacks, throwing exceptions. Among potential threats for the database are SQL injection, identity theft, deletion of data (accidental or intentional deletion of information from the database).

Regarding the payment system, it should be noted that the HSE Online Education Distance Learning System infrastructure does not use or does not find any payment system modules. Payment for training and courses is made according to the details in partner banks. Among potential threats are attacks on the client side and phishing.

Based on the evaluation of the HSE Online Education System, we suggest using a practical approach to conducting an information security audit, which consists of five stages: planning, collecting information, analyzing vulnerabilities, exploiting and reporting. Table 2 summarizes the stages and sub-stages of the security audit through penetration testing.

Table 2. Security audit model for the online educational system.

Stage	Description
1. Planning	1.1. Definition of aims and means 1.2. Coordination with responsible persons
2. Information collection/System scanning	2.1. Collecting data on the HSE Online Education System infrastructure 2.2. Formation of the topology of infrastructure systems 2.3. Scanning the ports of systems 2.4. Collecting data on the file structure of systems, searching for source code
3. Vulnerability analysis	3.1. Summarizing the results of the collection 3.2. Identification of potential threats 3.3. Construction of attack vectors
4. Exploitation	4.1. Implementation of attacks 4.2. Exploitation of vulnerabilities 4.3. Fixing the presence of vulnerabilities and exploiting them
5. Report	5.1. Summarizing the results of the operation phase 5.2. Forming a list of recommendations for improving the state of the information system

4 Discussion

In this paper, a study of the HSE information systems infrastructure was carried out. As part of the study, information was gathered about the servers and providers. All the servers were also scanned for open data transfer ports. A preliminary analysis of vulnerabilities was carried out, during which a list of potential vulnerabilities for the HSE Online Education System was compiled based on intelligence data: web service version, educational platform, file structure.

A functional model of penetration testing for HSE systems was developed based on the world practices of these methods, as well as collected and analyzed data on systems. The proposed model is based on the assessment of the University information system by conducting system penetration tests, which allows focusing on practical security, thereby reflecting the objective state of security.

The functional model comprises five stages. In this study, planning was not performed fully because the audit had not been coordinated with the University authorities. At the second stage, all tasks were completed: detailed information about HSE systems was collected, providers and server locations were determined, ports of all systems were scanned, and systems were scanned to determine file structures. At the vulnerability analysis stage, an analysis of the data obtained was carried out allowing for the analysis of the current state of the technologies used in the HSE University information system and the compilation of lists of potential vulnerabilities. The operational phase was not performed due to the legislative aspect. Attacks can be carried out only upon agreement with the administration and the provision of special conditions that do not violate the integrity and performance of the system. As a continuation of the present study, it is planned to obtain agreement on the conduct of active testing phases and, accordingly, complete the audit by compiling the final report.

The developed model can be applied as part of an external audit ("black box"), internal audit with limited access rights ("grey box") or internal audit with unlimited access rights ("white box"). The completed phases according to the functional model were carried out as part of an external audit; also, not all data was taken into account at the stage of vulnerability analysis and threat modeling. For example, files with the source code of the web interfaces of the systems were obtained, but the analysis of the source code can be included in the exploitation or attack phase. The developed functional model and the completed phases will serve as a basis for conducting a full-fledged information security audit.

References

1. Paudel, P.: Online education: benefits, challenges and strategies during and after COVID-19 in higher education. Int. J. Stud. Educ. **3**(2), 70–85 (2021)
2. Salvador, L.C.R., Llerena, C.L.A., Nguyen, H.P.D.: Digital education: security challenges and best practices. Secur. Sci. J. **2**(2), 65–76 (2021)
3. Santhi, V., Raja Kumar, K., Vinay Kumar, B.L.V.: Penetration testing using Linux tools: attacks and defense strategies. Int. J. Eng. Res. Technol. **5**(12), 153–158 (2016)
4. Denis, M., Zena, C., Hayajneh, T.: Penetration testing: concepts, attack methods, and defense strategies. In: 2016 IEEE Long Island Systems, Applications and Technology Conference (LISAT), pp. 1–6 (2016)
5. Khera, Y., Kumar, D., Sujay, Garg, N.: Analysis and impact of vulnerability assessment and penetration testing. In: 2019 International Conference on Machine Learning, Big Data, Cbud and Parallel Computing (Com-IT-Con), pp. 525–530 (2019)
6. Security Vulnerabilities. https://www.cvedetails.com/vulnerability-list/vendor_id-2105/product_id-3590/Moodle-Moodle.html. Accessed 17 Feb 2022
7. Security Vulnerabilities. Nginx. https://www.cvedetails.com/vulnerability-list/vendor_id-10048/Nginx.html. Accessed 17 Feb 2022

The Management of Examination Malpractice Using Blockchain Technology

Mweemba Sikuyuba(✉) and Jackson Phiri

Department of Computer Science, School of Natural Science, University of Zambia, Lusaka, Zambia
mweembasikuyuba2016@gmail.com

Abstract. The recent increase in post-examination malpractice at the grade twelve (12) level at marking centers has sparked widespread concern among education stakeholders. The management process at a marking center involves scoring, transfer of grades from hardcopy scripts onto the softcopy system provided by the Examination Council of Zambia and conducted by examiners from various subject panels. However, pockets of post-examination malpractice have been recorded, which include the change of grades initially assigned to a particular candidate's script at the data entry point. The research is based on finding measures to combat examination malpractice through the use of Blockchain Technology, to stop the post modification of entered results on the Electronic Results Management System of grade 12 results. A baseline study was conducted using purposive sampling of different respondents, i.e., Chief-Examiners, Team Leaders, and Markers, and a prototype using Blockhcain Technology was developed to address this need by the current ERMS. The technology would employ the SHA-256 hashing technique. The final hash document would be shared from a marking center to a central point at the examination council for the verification process before publication of the results. The proposed system showed improved service delivery at marking centers through making it impossible to experience grade modification.

Keywords: Blockchain technology · Marking center · SHA-256 · Examination malpractice · Hashing · Management

1 Introduction

The management of examination malpractice is every nation's wish as this vice impedes the development of the education sector through the production of personal that are not well vetted. Examination malpractice could possibly be managed well with blockchain technology. A technology uses a hash system to enable the secure transmission of data from one point to another. "Researchers and practitioners have started to recognize the value of Blockchain technology for addressing data sharing challenges" [1] and this value would help the education system in managing examination malpractice in schools in developing countries. The management process at a marking center involves scoring and transferring grades from hardcopy scripts onto the Examination Electronic Entry System

R. Silhavy (Ed.): CSOC 2022, LNNS 501, pp. 565–583, 2022.
https://doi.org/10.1007/978-3-031-09070-7_47

(EEES) provided by the Examination Council of Zambia and conducted by examiners from various subject panels. However, pockets of post-examination malpractice in some centers have been recorded, and that includes changes in grades initially assigned to a particular candidate's script by some data entry officers using the Examination Electronic Entry System (EEES). This study proposes the use of blockchain technology in trying to manage examinations and enhance the security of the examination process.

The main aim of the study was to assess the challenges that are posed by the current trends in the management of post examination malpractice at a marking center and develop a model for a prototype to manage examination malpractice using blockchain technology.

2 Literature Review

This literature review is based on various sources such as conference papers, journals, ministerial statements, etc. The following subtopics were considered: Post-examination malpractice in the education system, unified theory of acceptance and use of technology, types of blockchain technologies, EduCTX is a blockchain-based higher education credit platform, and EduRSS is a blockchain-based education record storage and sharing scheme.

2.1 Post Examination Malpractice

Post-examination malpractice is any form of malpractice that happens outside the examination room and after the exam. This may include entering the wrong results for a candidate. In certain cases, examiners have been able to aid candidates by putting scripts written outside the examination room into main envelopes before sealing them, as observed by [2], a supervisor left the envelope containing examination scripts open on a previous arrangement, so that scripts written outside the class could be included in the envelope before sealing and submitting it to the examination body.

Cases of post-examination malpractice involve the stealing of already written and submitted scripts that are due to be taken to centers for marking, "Eastern Province commissioner of police, Luckson Sakala, said the thieves drilled a hole from one of the classrooms into the deputy head teacher's office strong room where the papers were kept. Sakala said the incident occurred between November 23, 2018 at 17:00 h and November 25, 2018 at 07:30 h at Ncheka Primary School. He said the thieves went away with all the written grade nine examination papers, including maths and answer sheets, which were written on Friday, November 23, 2018"[1].

[1] https://zambiareports.com/2018/11/28/thieves-steal-written-grade-nine-exampapers-answer-sheets/.

Data entry officers are also sources of examination malpractice as they are able to change the grades of particular candidates. "Some data entry operators change the correct figures as presented by the team leaders. This occurs mostly after the counter checking sheet has already been produced and given to the team leader"[2]. This may not be with data entry officers, but could be with team leaders that are responsible for belt marking on their own belts. "Some team leaders do not only total up the marks after marking but also carry the marked papers to the data entry operators. Therefore, some of these people deliberately enter inflated marks for some papers, and data entry operators just enter the marks that they see on the mark sheet".[3]

Dr. Gbenga Adewale also noted that changing of results on the computer for the storage system is a form of malpractice that is also experienced for post examination malpractice as can be seen from the study by [3], another form of post examination malpractice has to do with officials in the computer department of an examination body changing a candidate's result on a computer storage facility.

Other forms Post-examination malpractice involves candidates finding a way of conniving with examiners at the marking center so that they can pay them some money in order for them to mark their scripts and award marks that are unnecessary, [3]. According to [3], explains that post examination malpractice could take the form of lobbying the examiners by begging and sending friends and senior colleagues to the examiner to be lenient while marking.

Certain scholars have conducted studies that focus on examiners' inherent malpractice habits, as evidenced by a study by [4].

2.2 Unified Theory of Acceptance and Use of Technology Model (UTAUT)

The Unified theory of acceptance and use of technology (UTAUT) in this context will focuses on the four main determinates: performance expectancy, effort expectancy, social influence, and facilitating conditions to show how they relate to the demographic data for any given research and have an influence on the outcome decision, [5] The model suggests that when users are presented with a new technology, a number of factors influence their decision.

Performance expectancy is the factor that looks at the perceived usefulness of the user in an organization [6]. It makes sense that the more a user uses a technology to improve their performance, the more their intent to use it increases. While effort expectancy focuses on the amount of effort required by the user to effect some significant change in behavior [7], effort expectancy is associated with the efficiency of technology.

The social influence factor is how individual users view the opinions of other players on a particular new product or technology, [8, 9]. Social influence is defined as the extent to which an individual perceives that important others believe the new system should be used. Facilitating conditions is the support that is made available to a user so that the outcome is useful [10], which refers to consumers' perceptions of the resources and support available to perform a behavior [10] (Fig. 1).

2 https://www.zambiawatchdog.com/how-examination-malpractices-are-carried-outin-zambia/.

3 https://www.zambiawatchdog.com/how-examination-malpractices-are-carried-outin-zambia/.

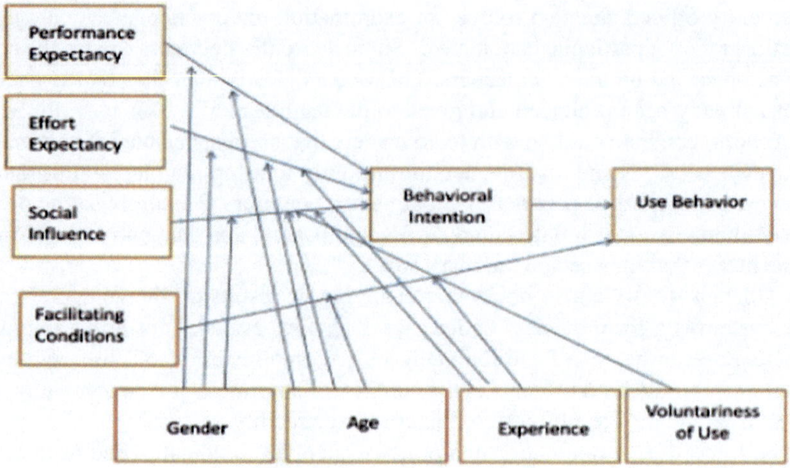

Fig. 1. Utaut model [10].

2.3 Blockchain Technology

The distributed database ensures that the ledger maintains transactions in hashed blocks, so as to make it difficult for someone to tamper with the transaction [11], A cryptographic hash function is a mathematical algorithm that takes an input and transforms it into an output. A cryptographic hash function is characterized by its extreme difficulty to revert, in other words, to recreate the input data from its hash value alone. The architecture of a block is made up of the blocks that are attached together, with each successive block having the hash tag for the preceding block in order to keep a record of the transactions.

Blockchain in the field of education is also used for keeping records in a more secure form. Records such as results for learners are kept for this purpose, to make sure it is difficult for people to manipulate records [12]. Blockchain technology allows the storage of a digital identity that can provide proof of education for learners (Fig. 2).

2.4 Smart Contracts

Any computerized transaction is known as a "smart contract." The transactions on a smart contract are triggered by way of a unique address that is assigned to it by blockchain technology and are securely validated before transmission is done [13]. These nodes are credited with the responsibility of first validating the transactions prior to broadcasting them further to other peers in the network (Fig. 3).

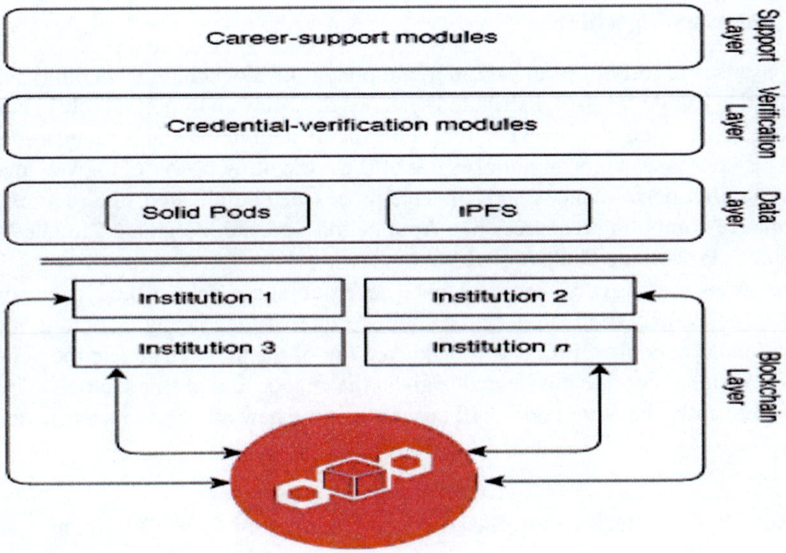

Fig. 2. The layered architecture of the Blockchain-based proposed decentralized educational landscape [12]

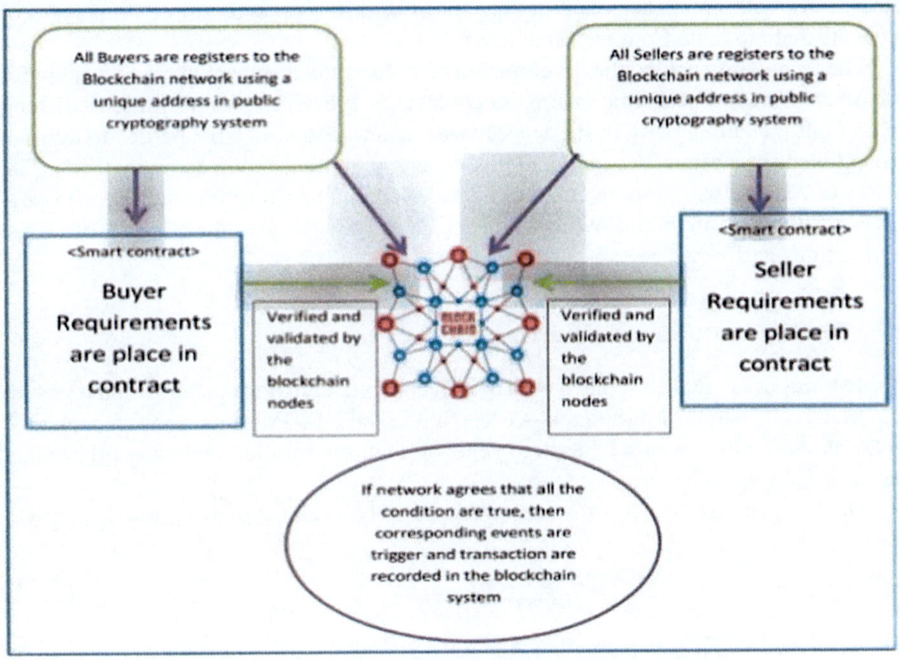

Fig. 3. A smart contract based supply chain system [14]

2.5 Consensus Algorithms

The consensus algorithm is an essential component of blockchain technology and can be applied to a wide range of distributed systems; they play an important role in ensuring that nodes connected to a blockchain have the same data items and format as shown in [15]. This consensus algorithm makes use of the computing power of a given machine, which is a high-performance one [16]. This process is complicated, and requires high-performance computing resources like Application Specific Integrated Circuits (ASIC) or Graphics Processing Units (GPU).

The proof-of-stake (PoS) requires that a node puts some stake in the blockchain; this provides for security of the nodes on a chain because if they is any malicious act done by a particular node involved, it will be kicked out of the blockchain and the node with the highest stake wins the puzzle and is given the chance to add a new block to its node and consequently the other nodes will take note of the new addition to a particular node [17]–[21].

2.6 EduCTX: A Block Chain Based Higher Education Credit Platform

This blockchain that is found in [22] was arrived at based on the concept of the European Credit Transfer and Accumulation System (ECTS), a credit platform that could be used for a global peer-to-peer network comprising of higher education institutions (HEI) plus other users such as students and organizations with the potential to be employers of students that graduate from the same institution.

The technology behind the development of a block chain education credit-system for the transmission of academic records of students' achievements to institutions wishing to find out the authenticity of the records was meant to enable large parties to transact using tokens in a secure manner, and this is largely appreciated in so many fields. The safety of records with a more enhanced way of conducting business on a peer-to-peer basis, as observed in the following articles [12, 22, 23], ensures the constant of all the transactions performed over the network.

2.7 The Key Features and Security EduCTX

The features of the EduCTX that would help in understanding the operation of the system are, firstly, that each institution that wishes to join the HEI-EduCTX network would be given an API, which would be used to generate their own wallet, including private and public keys.

The figure below displays the wallet that could be used for such a purpose (Fig. 4).

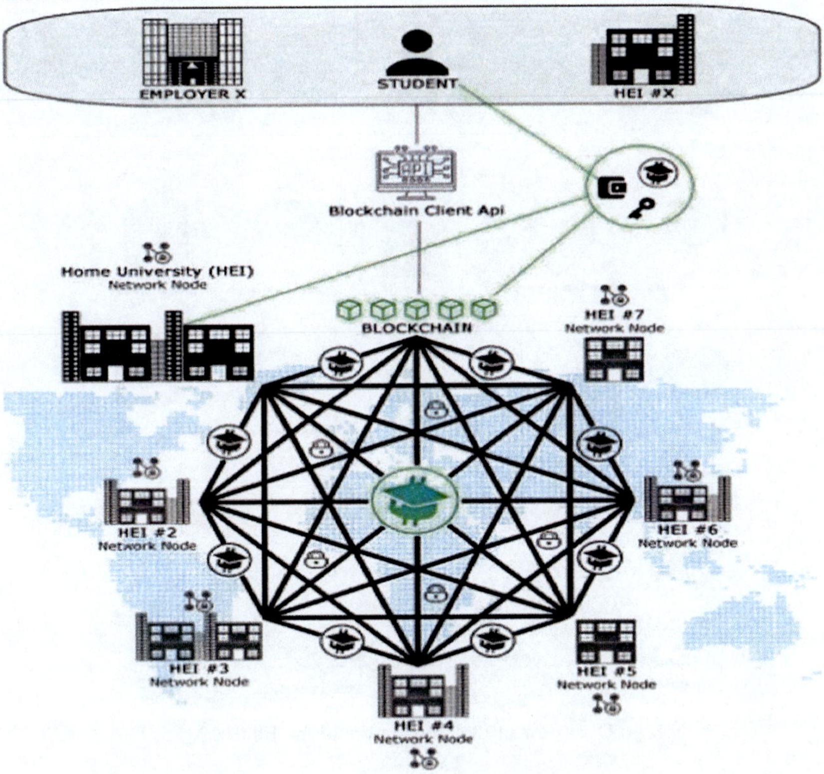

Fig. 4. A high-level depiction of the proposed EduCTX platform [22]

2.8 EduRSS: A Blockchain-Based Educational Record Secure Storage and Sharing Scheme

To meet some of the challenges that are encountered with dealing with a system that highly depends on physical handling of the record system, as provided in the article here [24], it is proposed to achieve an optimal solution through an efficient way of record keeping using blockchain that would include quick-access to the records as and when required, certification etc (Fig. 5).

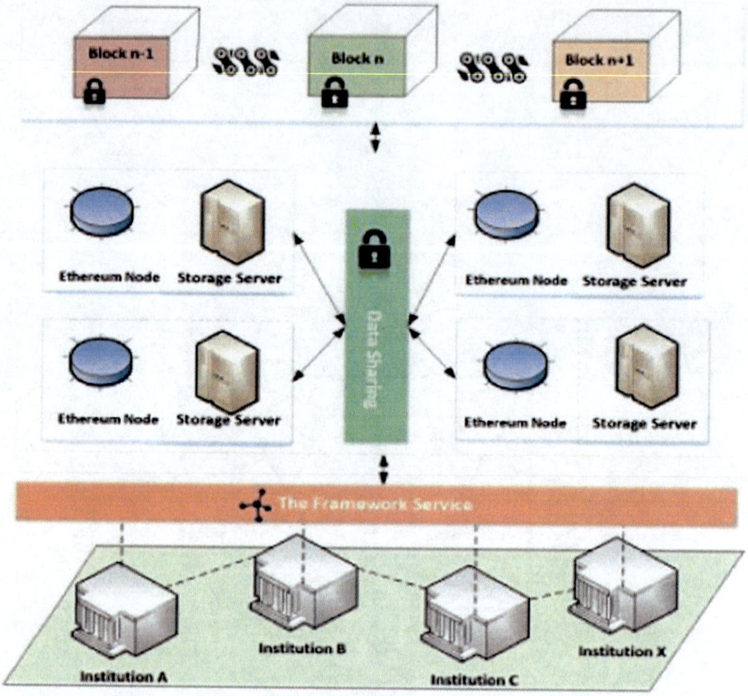

Fig. 5. Overview of the architecture of the EduRSS [25]

2.9 Key Features of EduRSS: A Blockchain-Based Educational Records Secure Storage and Sharing Scheme

The Educational Records Secure Storage (EduRSS) had features that were meant for the performance of the system; First s to take care of the individual enrolment of institutions as they join a HEI through the steps as proposed system is to have a three phase system to enable the establishment of a link i.e. the request phase, the vote to join phase and the set up member to join. This process involves the use of a certificate to enable the smooth running of the transactions and establishing what is called the (x) for each institution.

Its smart contracts had interfaces with IISC to enable the smooth transfer of information on a blockchain, as observed in the below with the two algorithms (Table 1).

Table 1. The description of the two functional operations to be done when establishing a link by a node [24].

Algorithm 1 initInstitutionMap (*InsitutionInfo*, *PublicKey*)

Input: *InsitutionInfo* is the object of institution information, *PublicKey* is the public key of an institution.
Output: boolean.
1: **if** *msg.sender* is not the *BlockAddress* of the institution **then**
2:     **return** false;
3: **end if**
4: *PubKeyInstitutionInfoMap* ← *InstitutionInfo*
5: **return** true;

Algorithm 2 isExistInstitution (*PublicKey*)

Input: *PublicKey*
Output: boolean.
1: **if** *PublicKey* is NULL **then**
2:     **return** false;
3: **end if**
4: **For**(skey **in** keys(*PubKeyInstitutionInfoMap*)) **do**
5:     **if** *PublicKey* == *skey* **then**
6:         **return** true;
7:     **end if**
8: **End For**
9: **return** false;

The system was developed under the premise of meeting secure data storage needs through block chain-enabled secure data storage that is based on blockchain techniques and storage servers that have a distributed system. The process of storing data is provided by the equation below [24]:

$$(X) = \{(X), (now), Sig\} \tag{1}$$

The last feature was an anti-tampering inspection based on blockchain that provides for educational records that are held by the institution and can only be tempered with during the process of proofreading of a blockchain [24]. The diagram below provides an example of a complete structural mechanism of the anti-checking mechanism for the anti-tempering mechanism of the EduRSS (Fig. 6).

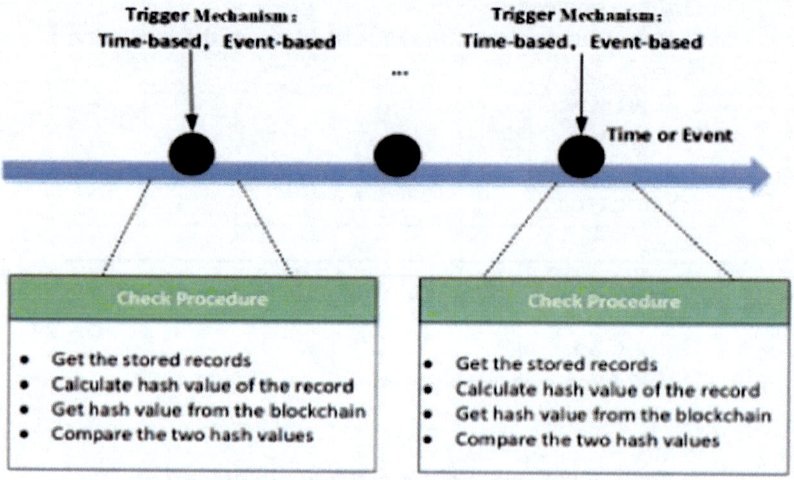

Fig. 6. An overview of the anti-tampering check mechanism [24]

3 Methodology

The proposed system based on blockchain is meant to change the current form of doing business by using the blockchain system to digitally sign on the document sheets for the raw results of candidates using some algorithms that would enhance the security of the entered results during the processing from the marking center. This same proposed system seeks to address many of the weaknesses that are presented by the current system.

3.1 Web Based System

This web application would be a great improvement in trying to do away with the manual system of confirming the results and transporting the same results to the Examination Council headquarters in Lusaka as hardcopies and storing them on the temporal local database (Personal Computer) with the view of transferring them onto the main server at the Examination Council when aggregating the results from various marking centers (Fig. 7).

Fig. 7. Web based blockchain system interface.[4]

3.2 Digital Signing Phase

The digital signing phase is a final confirmation by the examiners, who are required to provide a digital signature to the documents being transferred. The digital signature is meant to detect the unauthorized tempering or modification of the grades, as provided for by the federal information processing standards publication 186-4 (2013) in its definition of digital signature as follows "The result of a cryptographic transformation of data that, when properly implemented, provides a mechanism for verifying origin authentication, data integrity and signatory non-repudiation"[26].

3.3 Group Signature

The group signature was the one that was proposed for the development of the model, and it was suitable since the number of examiners involved per panel was more than one person and highly recommended because of its security features. [27] "A group signature scheme allows a group member to sign messages anonymously on behalf of the group." The group signature that was under consideration (Fig. 8).

When signing on the message, we use an algorithm where we let the generators of a key $g_1 and$ g_2 be members of a group of key generators G_q, this follows that the secret key P_i is $(x_{i1}, x_{i2}) \in Z_q^2$ for every $i = 1, 2$ and its public key is given by $h_i = g_1^{x_{i1}} g_2^{x_{i2}}$, this also assumes that no two persons from the same marking panel will have the same public key, therefore it shows that $h_1 \neq h_2$ (Fig. 9 and Fig. 10).

4 https://www.digisigner.com/free-electronic-signature/sign-document-online.

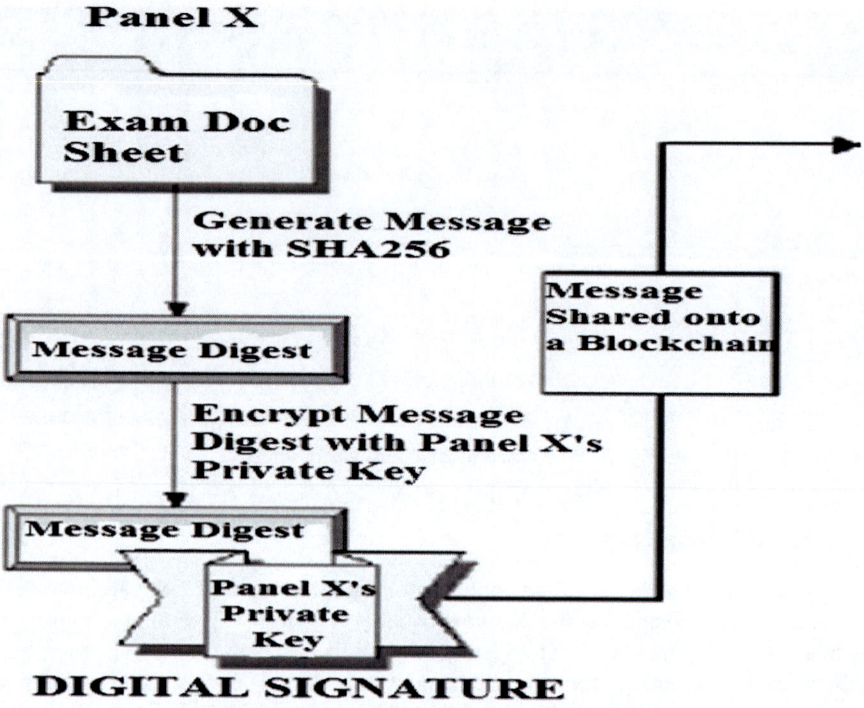

Fig. 8. Digital signing process [**Adopted Image**] [28]

P **V**

$s_1, s_2, t_1, t_2, d_2, \in_R, Z_q^*$

$a_1 \leftarrow g_1^{s_1} g_2^{s_2}$

$b_1 \leftarrow m_1^{s_1} m_2^{s_2}$

$a_2 \leftarrow g_1^{t_1} g_2^{t_2} h_2^{-d_2}$

$b_2 \leftarrow m_1^{t_1} m_2^{t_2} z^{-d_2}$

$\xrightarrow{(a_1, a_2, b_1, b_2)}$

$c \in_R Z_q^*$

Fig. 9. Proving that z is constructed by correctly with respect to h_1

3.4 Verification Phase

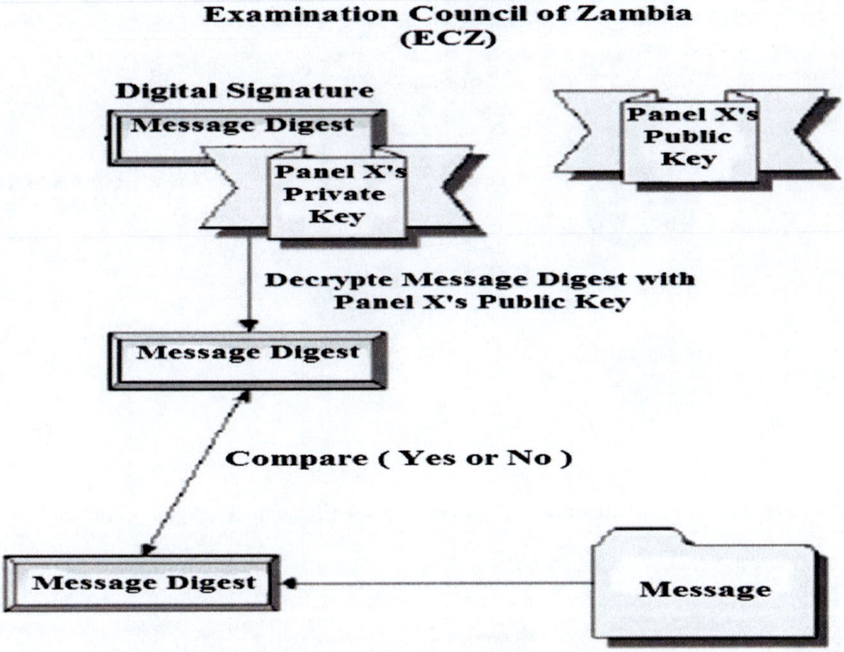

Fig. 10. Verification process of the document [**Adopted Image**] [28]

3.5 Use Case Diagram

The use case diagrams that are discussed in this section bring out the different interactions the examiners and other users of the system conduct (Fig. 11).

3.6 The Sequence Diagrams

The sequence diagram depicted below was used by Keiko Hashizume and Eduardo B. Fernandez in their article in titled digital Signature with Hashing and XML Signature patterns in [30]. The sequence diagram shows the process of signing the message as well as that of verifying the signature of the message in both Fig. 12 & Fig. 13 was used in the article to show the digital signature would be developed and used to enhance data security while is this article the same sequence is being suggested to be used to encrypt data being transmitted from a marking center to the ECZ servers (Fig. 12 and Fig. 13).

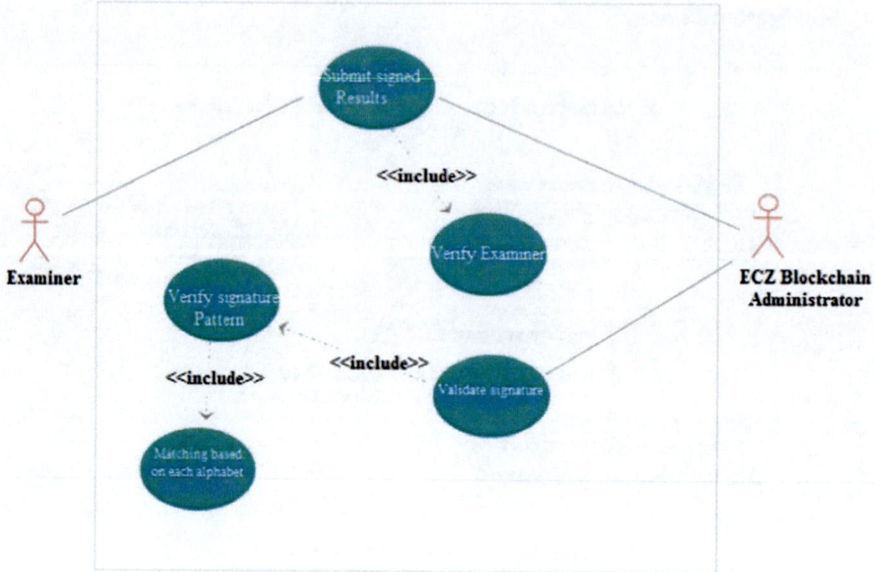

Fig. 11. Use case diagram on the signing and verification of a digital signature

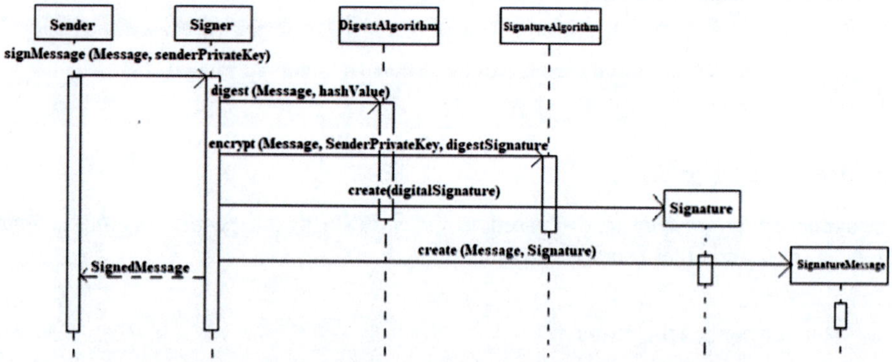

Fig. 12. The sequence diagram for signing a digital signature [30].

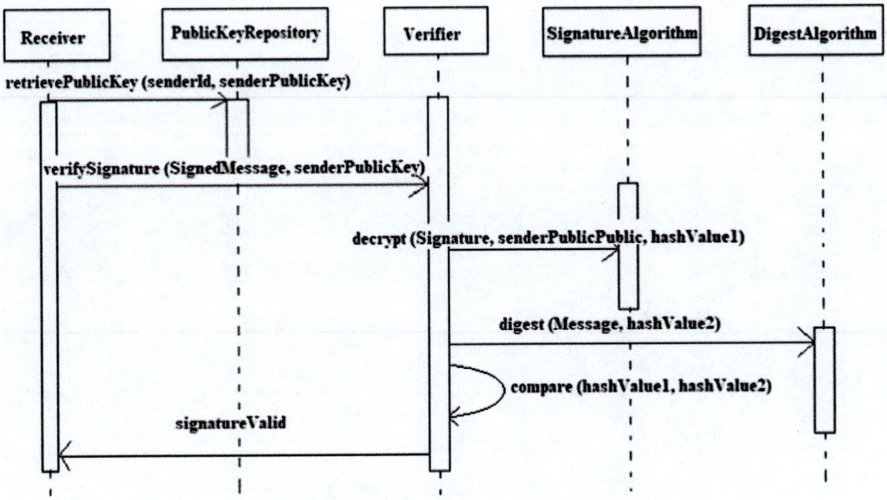

Fig. 13. The sequence diagram for verifying a digital signature [30]

4 Results and Discussion

The results for the code in the figure below was obtained in Java using NetBeans, a programming platform that was created by Oracle (Fig. 14 and Fig. 15).

```

    private void SignActionPerformed(java.awt.event.ActionEvent evt) {
        // TODO add your handling code here:
        File file =new File (Input.getText());


        try {

            //Use SHA-1 algorithm
            shaDigest = MessageDigest.getInstance("SHA-256");
        } catch (NoSuchAlgorithmException ex) {
            Logger.getLogger(Upload.class.getName()).log(Level.SEVERE, null, ex);
        }


        try {

            //SHA-1 checksum
            String shaChecksum = getFileChecksum(shaDigest, file);
            //see checksum

            Output.setText(shaChecksum);
        } catch (IOException ex) {
            Logger.getLogger(Upload.class.getName()).log(Level.SEVERE, null, ex);
        }


    }

    private void InputActionPerformed(java.awt.event.ActionEvent evt) {
```

Fig. 14. Shows the output obtained from the digital signature created using SHA256.

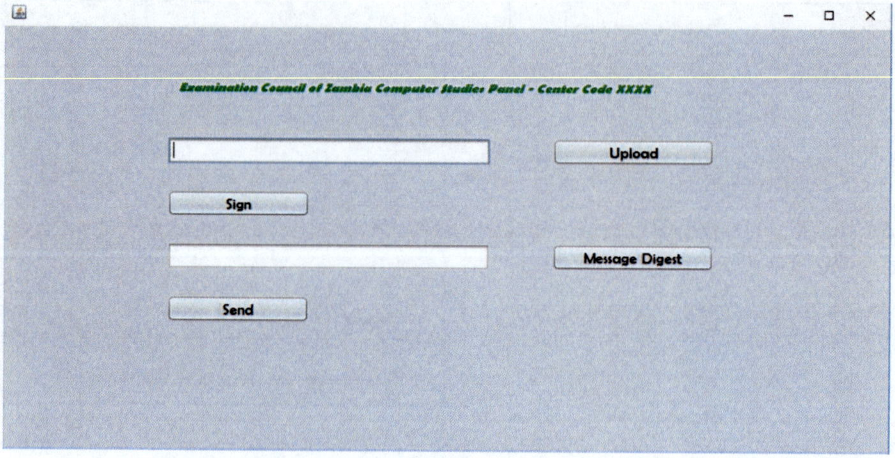

Fig. 15. The message upload/sign page

The above feature of the proposed system will provide users with the ability to upload their documents onto a webbased system in order to create a message digest that is shared on a blockchain in a secure way using the encryption algorithm of SHA 256 (Fig. 16).

Fig. 16. The document verification page for the blockchain system.

The above feature will provide a verification panel for the content to be verified in order to establish the authenticity of the content as well as the source of the block.

4.1 Encryption Mechanism

This feature of a blockchain helps in locking up each transaction with a code to secure the content of each transaction only to authorized users, as observed through [31], in which the article outlines the security benefits of having such a feature SHA256 as an encryption mechanism. SHA 256 provides for an almost impossible way of guessing the correct security combination of the encryption key through "In order to guess the correct secret key, A needs to guess the sequence of 256." For 256 bits, there are possible sequences, and among them, only one can be the Question Cloud (QC) secret key. The probability of guessing the secret key, which is 256-bit long, is $\frac{1}{2^{256}} = 2^{-256}$, which is practically not feasible." This feature will bring a novel to the management of post-examination malpractice that is attributed to the modification of already entered results in a web-based system in the transmission of results from a marking center to the examination council.

4.2 Time Stamping

Another feature of a blockchain that guarantees the transactions to be recorded with the time at which they were performed is the time stamp. This enables the hash that is generated by the algorithm to be easily shared onto the blockchain so that any tempering of a particular transaction is recorded and a new hash is created as observed through "A hash is the identity of a block that is generated from the aforementioned properties. If any value in the transaction changes, the value of the Merkle tree also changes. As the Merkle tree value changes, the hash of the block also changes, as the hash is totally dependent on the value of each property. In this way, if any illegal changes occur, they can be detected very easily in the blockchain," [31]. This feature will help the proposed blockchain on the management of post-examination malpractice keep a record of the transactions made on a shared network.

4.3 Immutability

This feature provides for data that has been shared not to be modified since a copy of the specific transaction has been shared on a network. "As for immutability, the Blockchain technology itself guarantees integrity of data. If a user wants to modify some piece of information, his or her acts will be registered in the ledger, being very easy to detect. This makes this technology trustworthy." [32]. This feature will allow for the safe sharing of results in order to avoid post-examination malpractice.

5 Conclusion

The study's novelty lies in the management of post examination malpractice using BlokChain technology, which provides for various features that can be used in the transmission of results from a marking center to a central point such as the examination council. The web-based transmission system makes use of blockchain features such as the SHA 256 algorithm, immutability of blockchain, timestamping, etc.

References

1. Rifi, N., Rachkidi, E., Agoulmine, N., Taher, N.C.: Towards using blockchain technology for eHealth data access management. In: 2017 Fourth International Conference on Advances in Biomedical Engineering (ICABME), pp. 1–4 (2017)
2. Karafiloski, E., Mishev, A.: blockchain solutions for big data challenges: a literature review. In: IEEE EUROCON 2017–17th International Conference on Smart Technologies, pp. 763–768 (2017)
3. Adewale, J.G.: Examination malpractice: a stigma on school effectiveness in Nigeria (2004)
4. Adeleke, J.O., Oluwatayo, G.K.: Examiners' malpractice dispositions: paradigm of assessment quality (2011)
5. Soneka, P.N.: A model for improving e-tax systems adoption in Zambia: a case study of solwezi town of Zambia. In: PhD Thesis, University of Zambia (2019)
6. Francisco, K., Swanson, D.: The supply chain has no clothes: technology adoption of blockchain for supply chain transparency. Logistics **2**(1), (2018)
7. Khazaei, H.: Integrating cognitive antecedents to UTAUT model to explain adoption of blockchain technology among Malaysian SMEs. JOIV Int. J. Inform. Vis **4**(2), 85–90 (2020)
8. Attuquayefio, S., Addo, H.: Using the UTAUT model to analyze students' ICT adoption. Int. J. Educ. Dev. Using ICT, **10**(3), (2014)
9. Escobar-Rodríguez, T., Carvajal-Trujillo, E., Monge-Lozano, P.: Factors that influence the perceived advantages and relevance of Facebook as a learning tool: an extension of the UTAUT. Australas. J. Educ. Technol, **30**(2). https://doi.org/10.14742/ajet.585 (May 2014)
10. Venkatesh, V., Thong, J.Y., Chan, F.K., Hu, P.J.-H., Brown, S.A.: Extending the two-stage information systems continuance model: Incorporating UTAUT predictors and the role of context. Inf. Syst. J **21**(6), 527–555 (2011)
11. Pilkington, M.: Blockchain technology: principles and applications. Edward Elgar Publishing, Research handbook on digital transformations (2016)
12. Lévy, W.S., Stumpf-Wollersheim, J., Welpe, I.M.: Disrupting education through blockchain-based education technology? (2018)
13. Allam, Z.: On smart contracts and organisational performance: a review of smart contracts through the blockchain technology. Rev. Econ. Bus. Stud **11**(2), 137–156 (2018)
14. Mohanta, B.K., Panda, S.S., Jena, D.: An overview of smart contract and use cases in blockchain technology In: 2018 9th International Conference on Computing, Communication and Networking Technologies (ICCCNT), pp. 1–4 (2018)
15. Chaudhry, N., Yousaf, M.M.: Consensus algorithms in blockchain: comparative analysis, challenges and opportunities. In: 2018 12th International Conference on Open Source Systems and Technologies (ICOSST), pp. 54–63 (2018)
16. Jalal, I., Shukur, Z., Bakar, K.A.A.: A study on public blockchain consensus algorithms: a systematic literature review (2020)
17. De Angelis, S.: Assessing security and performances of consensus algorithms for permissioned blockchains (2018)
18. Nguyen, G.T. Kim, K.: A survey about consensus algorithms used in blockchain. J. Inf. Process. Syst **14**(1) (2018)
19. Thin, W.Y.M.M., Dong, N., Bai, G., Dong, J.S.: Formal analysis of a proof-of-stake blockchain. In: 2018 23rd International Conference on Engineering of Complex Computer Systems (ICECCS), pp. 197–200, (2018)
20. Vashchuk, O., Shuwar, R.: Pros and cons of consensus algorithm proof of stake. Difference in the network safety in proof of work and proof of stake. Electron. Inf. Technol. **9**(9), pp. 106–112 (2018)

21. Yang, F., Zhou, W., Wu, Q., Long, R., Xiong, N.N., Zhou, M.: Delegated proof of stake with downgrade: a secure and efficient blockchain consensus algorithm with downgrade mechanism. IEEE Access **7**, 118541–118555 (2019)
22. Turkanović, M., Hölbl, M., Košič, K., Heričko, M., Kamišalić, A.: EduCTX: a blockchain-based higher education credit platform. IEEE Access **6**, 5112–5127 (2018)
23. Aste, T., Tasca, P., Di Matteo, T.: Blockchain technologies: the foreseeable impact on society and industry. Computer **50**(9), 18–28 (2017)
24. Gräther, W., Kolvenbach, S., Ruland, R., Schütte, J., Torres, C., Wendland F.: Blockchain for education: lifelong learning passport (2018)
25. Li, H., Han, D.: EduRSS: a blockchain-based educational records secure storage and sharing scheme. IEEE Access **7**, 179273–179289 (2019)
26. Gallagher, P.: Federal information processing standards publication digital signature standard (DSS). Fips Pub 186–3 (2009)
27. Camenisch, J., Michels M.: A group signature scheme with improved efficiency. In: International Conference on the Theory and Application of Cryptology and Information Security, pp. 160–174 (1998)
28. Sheldon, F.T., Jerath, K., Pilskalns, O.: Case study: B2B e-commerce system specification and implementation employing use-case diagrams, digital signatures and XML. In: Fourth International Symposium on Multimedia Software Engineering, 2002. Proceedings., pp. 106–113 (2002)
29. Chen, L., Pedersen, T.P.: New group signature schemes. In: Workshop on the Theory and Application of of Cryptographic Techniques, pp. 171–181 (1994)
30. Hashizume, K., Fernandez, E.B., Huang, S.: Digital signature with hashing and XML signature patterns (2009)
31. Islam, A., Kader, M., Shin, S.Y.: BSSSQS: A blockchain based smart and secured scheme for question sharing in the smart education system (2018)
32. Ramos-Sosa, M. del P., Cabrera, D., Moreno, B.: Blockchain and smart contracts for education (2020)

Verification of Convection Predictors for the Algorithm of Statistical Prediction of Convective Precipitation

David Šaur[1](✉) and Michal Žák[2]

[1] Faculty of Applied Informatics, Tomas Bata University in Zlin, 4511 Nad Stranemi, Czech Republic
saur@utb.cz

[2] Faculty of Mathematics and Physics, Charles University, Ke Karlovu 3, 121 16 Praha 2, Czech Republic
michal.zak@mff.cuni.cz

Abstract. This paper focuses on the description of the methodology of verification of convection predictors, which are implemented in the combined prediction in the Algorithm for statistical prediction of convective precipitation. The methodological part describes the algorithm of the statistical prediction of convective precipitation and the procedure for evaluating the verification of predictors. The resulting part contains the outputs from the verification of convection predictors for three main prediction categories - Convection triggering, Storm intensity and Convective cloud type. The outputs of this article will be intended for the design of a combined forecast, which is part of the Algorithm for the statistical prediction of convective precipitation. This algorithm can be used as one of the sources of more accurate predictions of convective storms, convective precipitation and dangerous phenomena as a support in the decision-making of the crisis management bodies of the region.

Keywords: Weather forecast · Severe storm · Convective precipitation · Algorithm · Forecast verification · Crisis management

1 Introduction

One of the effects of current climate change is extreme weather events. These phenomena occur unevenly and locally, and therefore it is very difficult to predict them accurately and well in advance. For this reason, they are one of the most significant natural threats, which has a number of negative impacts on our society [1].

In this article, we will deal with the prediction of extreme phenomena that occur mainly in the summer. These are strong convective storms, which are accompanied by heavy convective precipitation (torrential rain), hail, strong wind gusts and, in rare cases, tornadoes. A characteristic feature of the occurrence of these phenomena is their local occurrence in the order of several km^2, occasionally hundreds of km^2, and their relatively short duration. The typical impact is the occurrence of very intense precipitation, which

R. Silhavy (Ed.): CSOC 2022, LNNS 501, pp. 584–593, 2022.
https://doi.org/10.1007/978-3-031-09070-7_48

can raise the levels of smaller watercourses in a few tens of minutes and occasionally hours and cause local flooding in a very small area. And it is precisely these reasons in the form of the nature of these extreme events that have led to the design of a convective precipitation statistical prediction algorithm based on such data sources to reflect the physical causes of convective storms as much as possible in order to achieve a more successful convective storm prediction and their accompanying phenomena Prediction of convective storms and their accompanying dangerous phenomena is one of the most difficult disciplines in contemporary meteorology [2, 3].

The fundamental problems of prediction tools stem from the nature of these convective phenomena, which have already been mentioned above. At present, convective storm prediction is realized by three approaches: 1) Numerical models of weather forecasting, especially using a non-hydrostatic core designed for modeling convective processes in the atmosphere [4–6]; 2) Nowcasting systems using data from radar precipitation measurements to calculate the extrapolated echo for several tens of minutes [7–9]; 3) Expert meteorological systems that combine data from various forecast sources, e.g. numerical models, now-casting systems, conceptual models, local orography analysis, including statistical analysis of historical storm events [10–12]. And the currently proposed Algorithm for the statistical prediction of convective precipitation is closest to these expert systems aimed at predicting the conditions of convection. This algorithm combines data from numerical models, data from spatial analysis of GIS relief characteristics with data from radar and station measurements, which form a database of historical convective events for the years 2003 to 2021. In this article we will deal with input data from numerical models on the basis of which the so-called combined convective precipitation forecast is calculated. This combined prediction also includes an input source of GIS data, from which a spatial analysis of the relief characteristics is calculated, which characterizes the conditions of the initial initiation of convection at the local level [2].

The aim of this paper is to provide information on the evaluation of the success of convection predictors from numerical models that were used to calculate the combined prediction of convective precipitation. These convection predictors were selected as suitable input data sources that best describe the conditions for atmospheric convection. These are, for example, temperature, humidity and wind conditions, which affect the occurrence and occurrence of convective storms and at the same time dangerous accompanying phenomena. By calculating the results of the combined forecast and statistical analysis from the database of historical convective events, a more accurate forecast of convective precipitation is created, which can significantly help as decision support of flood authorities and crisis management authorities of the region and municipalities in warning of these dangerous phenomena.

2 Methods

This chapter focuses on the description of the methodology of verification of convective precipitation predictors, the results of which were used to design an algorithm for statistical prediction of convective precipitation.

The algorithm of statistical prediction of convective precipitation calculates the prediction of the probability of occurrence of convective precipitation and torrential rainfall for 206 municipalities with extended powers throughout the Czech Republic with the length of the forecast at 3 to 24 h [2].

This forecast is calculated on the basis of:

- Statistical analyzes of historical storm events.
- Combined convective precipitation forecasts.

Statistical analysis of historical storm events is calculated on the basis of five forecasting attributes such as wind direction forecast at 500 hPa, storm tracks, time of occurrence, type of convection and convective storm movement speed. The aim of this analysis is to select one or more historical storm events that best match the predicted situation.

The combined prediction of convective precipitation includes the prediction of temperature, humidity and wind atmospheric conditions for the occurrence and development of convective storms. This prediction also includes local surface conditions, especially orographic and other conditions that are very important for the initial initiation of convection.

The resulting predictions of convective and torrential rainfall are calculated by combining the outputs of the statistical analysis and the combined forecast of convective precipitation.

2.1 Input Parameters of Combined Forecast of Convective Precipitation

This paper focuses on the input parameters of combined forecasting, which are the outputs from numerical weather forecasting models. These parameters are referred to as convection predictors and are one of the most important input parameters affecting the accuracy of convective precipitation predictions.

Conversion predictor verification outputs are calculated based on the following main categories of combined convective precipitation prediction:

1. Convection triggers.
2. Storm intensity predictors.
3. Convective storm type.

Convection triggers are a category containing groups of convection predictors focused on predicting the conditions for convective storms. These predictors are of significant importance in predicting the conditions of convective storms and precipitation.

Predictors of convective storm intensity is a category designed to estimate future conditions of convective storm intensity and precipitation. Storm intensity is calculated only for the predicted areas of convective precipitation according to the convection triggering mechanisms.

Predictors for determining the type of convective storm are also predictors of selected triggering mechanisms of convection and storm intensity. The aim is to determine the type of convection, according to which the probable occurrence of dangerous accompanying phenomena can be determined.

2.2 Verification of Convective Predictors

This chapter focuses on the description of the procedure for evaluating the verification of predictors of the combined convective precipitation forecast.

Verification of convection predictors is performed on the basis of standard verification criteria Skill Scores. This verification is based on the evaluation of the numbers and frequency of cases where the phenomenon was/was not predicted and actually occurred/did not occur, in all possible mutual combinations.

Verification of convection predictors is performed on the basis of standard verification criteria Skill Scores [14, 15].

A PivotTable is a four-field table that contains categories of categories (Fig. 1):

Event forecast	Event observed		
	Yes	No	Marginal total
Yes	a	b	a + b
No	c	d	c + d
Marginal total	a + c	b + d	a + b + c + d =n

Fig. 1. PivotTable (standard method) [14, 15].

Where:

- a - is the number of cases when the phenomenon was predicted and actually occurred (correct prediction of the phenomenon),
- b - is the number of cases when the phenomenon was not predicted and actually occurred (incorrect prediction of the phenomenon),
- c - is the number of cases when the phenomenon was predicted and did not actually occur (incorrect prediction of the phenomenon),
- d- is the number of cases when the phenomenon was not predicted and did not actually occur (correct prediction of the phenomenon) [14, 15].

The main verification criterion for evaluating the verification of convection predictors is the "Accuracy" criterion with the following characteristics (Table 1):

Table 1. Characteristics of the Accuracy verification criterion

Verification criterion	Code	Formula	Limitation
Accuracy	A	$\frac{a+d}{a+b+c+d}$	0 <= A <= 1 0 <= A <= 100

Verification of convection predictors consists of the following steps:

- Selection of historical storm situation
- Evaluation of Skill Score categories using the "Eye Ball" method from the outputs of radar precipitation measurement CZRAD or X-band radar MMR50 and combined estimation of radar and station measurement of the product 1 h SUM MERGE according to the occurrence of measured precipitation (amount 2 mm/hour at approx. 28 dBZ) in order to detect a positive occurrence of convective precipitation (category a, d) or a negative occurrence of precipitation (category b or c). This evaluation applies to all categories of Skill Scores (Fig. 2).

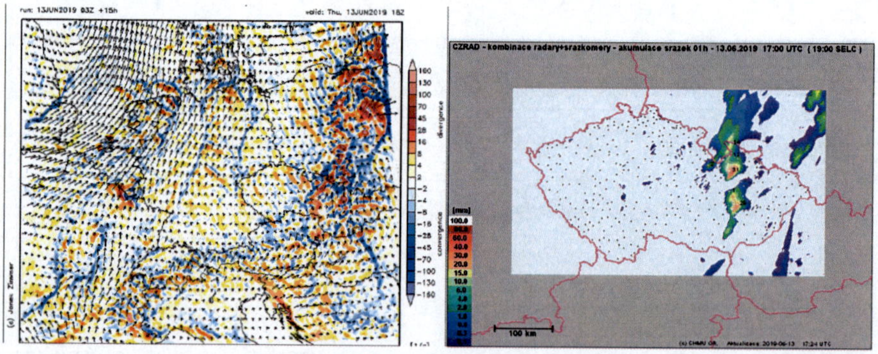

Fig. 2. Example of comparison of predictor MFDIV 0–1 km with summary of precipitation in 1 h [13].

- Verification of predictors triggering the convection mechanism.
- Verification of storm intensity predictors according to the positive occurrence of precipitation in combination with the degree of saturation.
- Verification of convection predictors for convection type
- Determining the order of success of individual convection predictors.

3 Results

Verification of convection predictors is given for three categories - Convection Triggers, Storm Intensity and Convection Types. Verification results are reported for the Accuracy verification criterion in %. The resulting values were calculated as the average Accuracy for approximately two hundred historical storm events from 2016 to 2020.

Table 2. Verification of convection predictors (triggerring mechanism of convection) after verification criteria "Accuracy“

Predictors of convection (triggering mechanism)	Triggering of convection (Accuracy - %)	Storm intensity (Accuracy - %)
Pressure MSLP	84	74
Frontogenesis 850 hPa	81	78
Relative humidity 1000–500 hPa	83	87
Direction wind shear 1000–300 hPa	88	79
Precipitable water	85	83
Moisture Flux Divergence 0–1 km	86	72
Relative vorticity 850–300 hPa	84	74
Potential orographic lifting	72	69
Relative humidity 1000 hPa	78	77
Wind velocity 1000 hPa	69	62
Temperature change of surface	71	59

Table 2 shows the average accuracy values in % for convection predictors in terms of assessing their prediction success in relation to the measured convective precipitation. The resulting values of the Accuracy criterion were evaluated for the convection triggers and storm intensity. Convection predictors that have reached the Accuracy limit above 80% are considered reliable and accurate for predicting the occurrence of convective storms and their accompanying phenomena. These predictors were included in the design of the combined convective precipitation forecast. Accuracy values above 80% were achieved in the first seven convection predictors, which are calculated mainly by the non-hydrostatic numerical model GFS and WRF ARW designed for convection modeling.

Table 3. Verification of predictor of storm intensity after verification criteria "Accuracy"

Predictors of convection (triggering mechanism)	Triggering of convection (Accuracy - %)	Storm intensity (Accuracy - %)
ML(MU)CAPE	62	82
ML(MU)CIN	55	66
LI (SI)	62	82
KI	85	85
Wetbulb temperature 850 hPa	82	82
Mixing ratio 0–1 km	75	77
Temperature 850 hPa	81	84
DLS	68	80
LLS	74	72
SREH 3 km	78	81
SREH 1 km	75	77

Table 3 presents the results of the evaluation of the Accuracy criterion for convection predictors with convection triggers and storm intensity. Compared to previous outputs, the storm intensity forecast has a different setting of convection predictors thresholds than for convection triggers. The main reason for the different settings is the characteristics of the convective precipitation forecast. Storm intensity prediction is realized through such convection predictors, which provide outputs designed for global prediction of temperature, humidity and wind conditions, usually with higher resolution without the need to predict local conditions of convection. The highest Accuracy values were achieved for traditional convection predictors such as MUCAPE, LI, KI, Wetbulb temperature 850 hPa, DLS and SREH. In predictive practice, these predictors more often provide a more accurate prediction of storm intensity than in the case of other predictors.

Table 4 shows the Accuracy values for convection predictors focused on the frontal and orographic type of convective clouds. Predictors of convection such as relative humidity in the atmosphere, Frontogenesis parameter and Relative vorticity reached the highest values, which is mainly due to their characteristics predicting atmospheric fronts. The characteristics of these predictors correspond to the characteristics of the global forecast of the occurrence of convective precipitation, which affects a large part of the territory. In contrast, predictors with local forecast properties such as K-Index, Relative vorticity at 850 hPa, Moisture Flux Divergence and Wind speed at 850 hPa achieved the most accurate results in orographic (local) convective cloud prediction.

Tables 2, 3 and 4 list the predictors in terms of triggering, intensity and type of convection that have been evaluated for their success. Based on this evaluation, the following convection predictors listed in Table 5 were selected:

The convection predictors listed in Table 5 are used in the combined prediction design, which is part of the Algorithm of statistical forecast of convective precipitation.

Table 4. Verification of predictors of type of convection after verification criteria "Accuracy"

Frontal convection	Accuracy (%)	Orographic convection	Accuracy (%)
ML(MU)CAPE+LI(SI)	70	ML(MU)CAPE+LI(SI)	84
MLCIN	75	ML(MU)CIN	66
KI	82	RH 700 hPa	87
RH 1000–500 hPa	86	KI	92
Frontogenesis 850	85	Wetbulb 850 hPa	85
RH 1000 hPa	74	Relative vorticity 850 hPa	82
RH 1000–500 hPa	84	Moisture Flux Divergence 0–1 km	84
MFDIV 0–1 km	82	DLS	86
Precipitable water	83	Wind velocity 850 hPa	87
Relative vorticity 850 hPa	80	Storm motion (Corfidi vector)	86
DLS	69		
Speed wind 300 hPa	72		
Wind shear 1000–300 hPa	79		

Table 5. Predictors of convection according to its triggering, storm intensity and type of convection

Triggering mechanisms of convection (predictors)	Storm intensity (predictors)	Type of convection (predictors)
Temperature 850 hPa	Temperature 850 hPa	Temperature 850 hPa
Frontogenesis 850 hPa	Frontogenesis 850 hPa	Frontogenesis 850 hPa
Pressure MSLP	Storm motion (Corfidi vector)	Orographic lifting
Moisture Flux Divergence 0–1 km	CAPE	Storm motion (Corfidi vector)
Relative vorticity 850–300 hPa	CIN	CAPE
Relative humidity 700 hPa	Wetbulb temperature 850 hPa	Wetbulb temperature 850 hPa
Orographic lifting	DLS	SREH 3 km
Wetbulb temperature 850 hPa	SREH 3 km	Wind velocity 850 hPa
	Wind velocity 850 hPa	Wind velocity 300 hPa
	Wind velocity 300 hPa	Precipitable water
	Precipitable water	

4 Conclusion

The aim of this article was to provide information on the methodology of verification of convection predictors and their evaluation of prediction success at real historical storm events from 2016 to 2020. The verification results were used to design combined forecast predictors in the Algorithm of Statistical Forecast of Convective Precipitation.

Convection predictors are outputs from numerical models that predict temperature, humidity and wind conditions in the atmosphere favorable for the formation and development of convective storms and their accompanying phenomena. The Accuracy limit of 80% was reached for those predictors whose characteristics best describe, for example, the conditions for starting convection, the intensity of storms and the type of convective cloud cover. Those convection predictors that reached this limit were selected to design a combined convective precipitation forecast.

In the future, the combined forecast will be improved based on verification at other historical storm events. This algorithm will be used in the more accurate prediction of convective precipitation for the purposes of crisis management of the region and municipalities with extended powers of the region. The main purpose can be understood in the use of this algorithm in terms of decision support before and during the occurrence of dangerous accompanying storm phenomena to eliminate the impact of these phenomena on society.

Acknowledgments. This work was supported by the project No. VI20192022134 - System of more accurate prediction of convective precipitation over the regional territorial unit.

References

1. Březková, L., Šálek, M., Novák, P., Kyznarová, H., Jonov, M.: New methods of flash flood forecasting in the Czech Republic. In: Hřebíček, J., Schimak, G., Denzer, R. (eds.) ISESS 2011. IAICT, vol. 359, pp. 550–557. Springer, Heidelberg (2011). https://doi.org/10.1007/978-3-642-22285-6_59
2. Šaur, D.: Support for Crisis Management of the Region in Terms of Evaluation of Flood Events. Academia Centrum TBU in Zlín, Zlín (2017). ISBN 978-80-7454-712-6
3. Rapant, P., et al.: Early warning of flash floods based on the weather radar. In: Proceedings of the 2015 16th International Carpathian Control Conference (ICCC), pp. 426–430. IEEE (2015). https://doi.org/10.1109/CarpathianCC.2015.7145117. ISBN 978-1-4799-7370-5
4. Jolivet, S., Chane-Ming, F.: WRF modelling of turbulence triggering convective thunderstorms over Singapore. In: Deville, M.O., Estivalezes, J.-L., Gleize, V., Lê, T.-H., Terracol, M., Vincent, S. (eds.) Turbulence and Interactions. NNFMMD, vol. 125, pp. 115–122. Springer, Heidelberg (2014). https://doi.org/10.1007/978-3-662-43489-5_14
5. Flora, M.L., Potvin, C.K., Wicker, L.J.: Practical predictability of supercells: exploring ensemble forecast sensitivity to initial condition spread. Mon. Weather Rev. **146**(8), 2361–2379 (2018). https://doi.org/10.1175/MWR-D-17-0374.1.ISSN0027-0644
6. Wang, Y., Belluš, M., Ehrlich, A., et al.: 27 years of regional cooperation for limited area modelling in Central Europe. Bull. Am. Meteorol. Soc. **99**(7), 1415–1432 (2018). https://doi.org/10.1175/BAMS-D-16-0321.1.ISSN0003-0007

7. Mejsnar, J., Sokol, Z., Minářová, J.: Limits of precipitation nowcasting by extrapolation of radar reflectivity for warm season in Central Europe. Atmos. Res. **213**, 288–301 (2018). https://doi.org/10.1016/j.atmosres.2018.06.005.ISSN01698095
8. Apke, J.M., Mecikalski, J.R., Bedka, K., McCaul, E.W., Homeyer, C.R., Jewett, C.P.: Relationships between deep convection updraft characteristics and satellite-based super rapid scan mesoscale atmospheric motion vector–derived flow. Mon. Weather Rev. **146**(10), 3461–3480 (2018). https://doi.org/10.1175/MWR-D-18-0119.1. ISSN 0027-0644
9. James, P.M., Reichert, B.K., Heizenreder, D.: NowCastMIX: automatic integrated warnings for severe convection on nowcasting time scales at the german weather service. Weather Forecast. **33**(5), 1413–1433 (2018). https://doi.org/10.1175/WAF-D-18-0038.1.ISSN0882-8156
10. Lakshmanan, V., Crockett, J., Sperow, K., Ba, M., Xin, L.: Tuning AutoNowcaster automatically. Weather Forecast. **27**(6), 1568–1579 (2012). https://doi.org/10.1175/WAF-D-11-00141.1
11. Beheshti, Z., Firouzi, M., Shamsuddin, S.M., Zibarzani, M., Yusop, Z.: A new rainfall forecasting model using the CAPSO algorithm and an artificial neural network. Neural Comput. Appl. **27**(8), 2551–2565 (2015). https://doi.org/10.1007/s00521-015-2024-7
12. Young, C.-C., Liu, W.-C., Chung, C.-E.: Genetic algorithm and fuzzy neural networks combined with the hydrological modeling system for forecasting watershed runoff discharge. Neural Comput. Appl. **26**(7), 1631–1643 (2015). https://doi.org/10.1007/s00521-015-1832-0
13. Modelzentralle.de, WRF ARW 4 km. http://www.modellzentrale.de/WRF4km/#0
14. Silhavy, R., Senkerik, R., Oplatkova, Z.K., Silhavy, P., Prokopova, Z. (eds.): Artificial Intelligence Perspectives in Intelligent Systems. AISC, vol. 464. Springer, Cham (2016). https://doi.org/10.1007/978-3-319-33625-1
15. Zacharov, P.: Diagnostic and Prognostic Precursors of Convection, 61 p. Faculty of Mathematics and Physics UK, KMOP, Prague (2004). https://is.cuni.cz/webapps/zzp/detail/44489/

An Empirical Study on Data Sampling Methods in Addressing Class Imbalance Problem in Software Defect Prediction

Babajide J. Odejide[1], Amos O. Bajeh[1], Abdullateef O. Balogun[1,2](✉), Zubair O. Alanamu[3], Kayode S. Adewole[1], Abimbola G. Akintola[1], Shakirat A. Salihu[1], Fatima E. Usman-Hamza[1], and Hammed A. Mojeed[1]

[1] Department of Computer Science, University of Ilorin, Ilorin PMB 1515, Nigeria
{bajehamos,balogun.ao1,adewole.ks,akintola.ag,salihu.sa, usman-hamza.fe,mojeed.ha}@unilorin.edu.ng

[2] Department of Computer and Information Sciences, Universiti Teknologi PETRONAS, 32610 Bandar Seri Iskandar, Perak, Malaysia
abdullateef_16005851@utp.edu.my

[3] Computer Services and Information Technology (COMSIT), University of Ilorin, Ilorin PMB 1515, Nigeria
alanamu.zo@unilorin.edu.ng

Abstract. With the growing rate of software systems and their applications in diverse walks of life, developing a software system that has no defects is a subject that cannot be overemphasized. Detection of software defects is one of the most prominent difficulties in the area of software engineering (SE) or software development process. Defects are usually unconscious flaws that make the software system behave unexpectedly or contrary to the specified requirements. This has made the subject of software defect prediction (SDP) a very critical one. Due to their dynamism, SDP solutions based on machine learning (ML) methods are envisaged as a viable approach. However, the latent data quality problem is a significant challenge to developing effective SDP models. The class imbalance is a classic example of the data quality problem in which there is a huge differential in the number of class (majority and minority) labels. Findings from studies have shown that data sampling methods are capable of addressing the class imbalance problem. Hence, this study conducts an empirical comparative analysis on the effect of data sampling methods in addressing the class imbalance problem inherent in SDP. Specifically, the performance of five data sampling (oversampling techniques (SMOTE, ADASYN, and ROS) and undersampling techniques (RUS and NM) methods on four software defect datasets with varying granularities are investigated. As prediction models, decision tree (DT) and random forest (RF) classifiers are deployed as well. Predictive performances of developed models were evaluated using accuracy, the area under the curve (AUC), and Matthews correlation coefficient (MCC) values. Observations from the experimental results showed that the introduction of data sampling methods in SDP processes not only addresses the class imbalance problem but also improves the prediction performances of the experimented classifiers. In addition, models based on ROS resampled datasets had superior predictive performance compared with

R. Silhavy (Ed.): CSOC 2022, LNNS 501, pp. 594–610, 2022.
https://doi.org/10.1007/978-3-031-09070-7_49

other studied data sampling-based datasets. In conclusion, it can therefore be recommended to deploy data sampling methods, particularly oversampling methods in SDP processes and other applicable machine learning tasks.

Keywords: Software defect prediction · Class imbalance · Data sampling · Machine learning

1 Introduction

The ubiquity of software systems and their magnification has made it quite inseparable from humans as it has become entrenched into our everyday life, enabling the execution of so many delicate tasks that humans may not have been able to accurately perform. This has therefore made it imperative to develop software applications that are of high quality [1, 2]. In a bid to produce a qualitative software system, several software companies have spent heavily on software quality assurance (SQA) and software testing. In fact, in 2014, the amount spent on software systems worldwide was $3.8 trillion, 23% of which was allotted to software testing and quality assurance activities [3].

Software testing is a very integral part of the software development life cycle (SDLC). SDLC is a process employed by the software industry to design, develop and test software systems. Software Quality Assurance (SQA) on the other hand, is a method used to monitor the software engineering (SE) process consisting of three concepts - quality, software quality, and SQA, to which defect prediction model is key [4]. Software engineers and development teams have employed testing methods such as unit testing, system testing, and code inspection to improve the quality and reliability of software systems. However, these conventional software testing methods are not sufficient to detect defects at an early stage of the SDLC as defects get injected into the software system at any stage of the SDLC [5–7]. From an SQA viewpoint, a defect prediction model primarily serves two manifold purposes – one is to predict modules that may likely be prone to defects in the future, the other is to know the impact of several software metrics on the defect-proneness of a module [8, 9].

A software defect is a bug in the code that causes the software system to fall short of qualitative application requirements. Although the goal of both software testing and SQA is to eliminate defects in software systems, these methods are costly, yet remain unreliable and unstable [10]. Software metrics provide a way to quantitatively measure the characteristics of the structure of a software system, as well as other mathematical measures of the software system that are sensitive to change [11]. These metrics are the key ingredients of software defect prediction (SDP). SDP is the application of machine learning (ML) techniques on software metrics to predict defective modules. It is an approach for detecting modules and/or components of a software system that are prone to defects in any way [12].

The primary aim of predicting software defects is to reduce software testing efforts by guiding the software testers through the defect classification of software systems, thereby enabling many software organizations to predict software defects in such a way as to save time, stay within the allocated budget, improve software quality, and for better planning of the resources to meet the timelines [13–15]. Although there are several SDP models used for predicting software defects, these models have been reported by several researchers to behave differently due to the class imbalance problem [5–7, 16, 17]. For instance, [6] experimented on 10 different datasets with varying imbalance ratios (IRs) using 3 resampling techniques and 9 SDP models and reported that LR and RF appear to be more stable than the other SDP models used. [17, 18] and [7] suggested that further research may be carried out using additional resampling techniques and SDP models.

Consequently, determining the appropriate data sampling method to use in a particular scenario of class imbalance will be of great benefit if carefully considered. [19] researched the relative value of data sampling methods approaches for software defects prediction. In their work, a key question was "Is balanced the best distribution?". This study builds further on the work done in [19]. This study considered datasets with a varying imbalance ratio (IR) using data sampling methods such as Random Over Sampling (ROS), Synthetic Minority Over Sampling Technique (SMOTE), Adaptive Synthetic (ADASYN), Random Under Sampling (RUS), and Near Miss (NM) sampling methods.

The remaining part of this study is organized as follows: Sect. 2 discusses the related works done in this regard. In Sect. 3, the concept of imbalanced learning is discussed, then the experimental results and analysis are presented in Sect. 4. Finally, the study is concluded in Sect. 5.

2 Related Works

This section investigates and analyses relevant developed SDP models based on ML methods and data sampling methods.

SDP is a method for detecting defects in software systems early on. It determines the characteristics of single code components to detect whether parts are prone to defect [20] or to forecast the number of defects in each component or module [21]. The latter approaches software defect prediction as a regression issue, while the former approaches it as a classification problem. This research focuses only on the classification perspective, which categorises a software module as defect-prone or non-defect-prone. A vast variety of static code features (software metrics) have been added into the realm of SDP to achieve this aim. Lines of Code-based metrics [22], McCabe and Halstead metrics [23], object-oriented metrics such as Chidamber–Kemerer [22], and Conceptual Cohesion of Classes measure [24] are examples of these metrics. [15] and [25] experimentally verified the use of static code features SDP, stating that these metrics are helpful, simple to use, and acceptable. Experts have used a variety of ways to create SDP models based on static code metrics. BN [6], DT [26], artificial neural networks (ANNs) [27], support vector machines (SVM) [28], k-nearest neighbour (KNN) [29], and logistic regression (LR) [6] are just a few of the classical classifiers that have been used directly to develop SDP models. Nonetheless, these classifiers ignore the skewness and other inherent data quality problems in software defect datasets [30]. Ignoring data quality problems has a significant negative impact on the prediction performances of SDP models. For instance, SVM and KNN tend to overlook the minority class labels as they seek to maximise the accuracy values [17]. [15] reported that software defect datasets are very susceptible to the class imbalance problem, which is an example of the data quality problem, and that strategies such as data sampling may aid in improving the classification performance of SDP models.

Class imbalance is a latent characteristic of software defect data, which consists of a small number of defective instances units and a big number of non-defective instances [31]. NASA dataset PC1 exemplifies this, with just 6.59% of cases belonging to the defective class label. Since most classifiers aim to build classifiers that maximise overall prediction accuracy, this feature has a significant influence on both model training and predictive performance. As a consequence, such models often neglect the valued minority (defective) class.

A considerable lot of study has been recommended to address the issue of class imbalance. [32] presented an overview of approaches for decreasing the detrimental impact of imbalance on prediction performance. [33] investigated whether various classifiers detect the same problems. To do this, they used NASA datasets to conduct a sensitivity analysis and evaluate the outcomes of RF, Naive Bayes (NB), RPart, and SVM. They concluded that certain flaws are more consistent in defect prediction than others and that each classifier identifies a distinct set of defects.

[34] addressed two procedures, undersampling and oversampling and then asserted that both data sampling approaches were successful. Furthermore, it was discovered that the adoption of advanced sampling procedures did not give any discernible improvement in addressing the class imbalance issue. Similarly, [35] observed that the RUS approach typically outperforms more complicated undersampling algorithms.

In addition to undersampling procedures, oversampling methods are commonly used to solve the class imbalance issue. Aside from ROS with replacement, numerous more sophisticated data sampling techniques have been created. [36] proposed a unique oversampling approach termed the SMOTE, in which fresh minority samples are generated based on feature space similarities between existing minority cases. [37] used the borderline-SMOTE to oversample minority class samples close to the borderline. Both [38] and [39] found that oversampling outperforms undersampling. It can be noted that the aforementioned research on data sampling techniques is not conducted on software defect issues, and there are few studies on evaluating data sampling algorithms for SDP. Therefore, this study attempts to investigate and comparatively evaluate the effect of data sampling methods in addressing the class imbalance problem inherent in SDP.

3 Methodology

This section introduces and explains the data sampling methods, prediction models, studied defect datasets, evaluation metrics, and the experimental approach utilised in this research work.

3.1 Data Sampling Method

In this research work, five (5) data sampling methods (SMOTE, ADASYN, ROS, RUS, and NM) are investigated. Data sampling methods are widely categorized into two types: oversampling and undersampling methods. The main concept behind the oversampling method is to balance the dataset by raising or increasing the number or frequency of the minority class instances to an equal number of classes as majority class instances [36]. On the other hand, in undersampling methods, the majority class instances are down sampled or reduced to the same number or frequency as the minority class instances [27]. ROS, SMOTE and ADASYN are types of the oversampling method while RUS and Near Miss are examples of the undersampling method.

In ROS, the samples to replicate are chosen at random. This redundancy of minority class instances often leads to overfitting [6]. In SMOTE, however, the samples are chosen using a (K-Nearest Neighbour) k-NN. A new vector is created based on the Euclidean distance between a feature vector and its closest neighbour [36]. Although ADASYN is an oversampling approach based on k-NN, it creates data adaptively depending on density distribution [40]. RUS is an undersampling strategy that eliminates examples of the majority class at random until both sets of instances are equal. Although some information may be lost in this manner, it improves computing speed and time [6]. NearMiss is another undersampling method, but instead of randomly picking samples to remove, it uses the kNN approach [41].

3.2 Prediction Models

In this research work, Decision Tree (DT) and Random Forest (RF) are deployed as prediction models. DT is a kind of non-parametric classifier whose aim is to build a model that predicts the value of a pre-determined variable using basic decision rules obtained from data variables [42]. DT is implemented as J48 in WEKA and is used to generate pruned or unpruned trees. RF is made up of a huge number of simple DTs that work together as an ensemble. Each simple tree in the RF produces a class prediction, and the class with the most votes becomes the prediction of the RF or model [43]. The choice of these classifiers is to introduce heterogeneity in the prediction models and based on their respective usage and performance in existing SDP research. Table 1 presents a brief description of the selected models and their parameter settings as used in this research work.

Table 1. Description of prediction models with parameter settings

Prediction models	Descriptions	Parameter settings
DT	It is a tree-structured classifier, where internal nodes represent the features of a dataset, branches represent the decision rules, and each leaf node represents the outcome	ConfidenceFactor = 0.25; MinObj = 2
RF	An ensemble model of random trees	bagSizePercent = 100; numIteraions = 100; seed = 1; calcOutOfBag = False; breakTiesRandomly = False

3.3 Software Defect Datasets

The software defect datasets used in this study were obtained from two distinct repositories (NASA and PROMISE). In this research work, the [44] version of the NASA corpus was employed. The NASA datasets provide software characteristics derived from static code measurements based on code size and complexity [45–47]. Datasets from the PROMISE repository are based on object-oriented metrics gathered at the class level and defects information detected in software modules. The PROMISE datasets were retrieved from Apache software and generated in the JAVA programming language [48–50]. Table 2 presents descriptions of the studied datasets with their respective imbalance ratio.

Table 2. Description of studied defect datasets

Datasets	Number of instances	Number of defective instances	Number of defective instances	Imbalance ratio (IR)
KC3	194	36	158	4
PC3	1053	130	923	7
ANT	292	32	260	8
REDKATOR	176	27	149	5

3.4 Performance Assessment Metrics

According to existing research, the choice and selection of performance assessment criteria are critical in SDP [51, 52]. This is due to the imbalanced nature of the datasets used for training and testing the SDP models. Relying just on accuracy values may be misleading. For example, in [53], an experiment was carried out to assess the biasness of measures such as Accuracy, F-Measure, and Area Under a ROC Curve (AUC). In their analysis, they determined that Matthews Correlation Coefficient (MCC) is a more credible measure since it encompasses the complete confusion metrics, as compared to others that omit the True Negative (TN). In this research work, the prediction capabilities of the resulting SDP models were evaluated using Accuracy, AUC, and MCC values. These selected assessment measures have been frequently employed and demonstrated to be trustworthy [31, 46–48, 54].

$$\text{Accuracy} = \frac{TP + TN}{TP + TN + FP + FN} \tag{1}$$

$$\text{AUC} = \frac{1 + TPr - FPr}{2} \tag{2}$$

$$\text{MCC} = \frac{TP * TN - FP * FN}{\sqrt{(TP + FP) * (TP + FN) * (TN + FP) * (TN + FN)}} \tag{3}$$

3.5 Experimental Approach

This section covers the experimental approach used in this research work as seen in Fig. 1. The approach is intended to empirically examine and validate the effect of data sampling methods on the prediction performances of SDP models. Specifically, the original defect datasets and the newly generated datasets based on studied data sampling methods (that is, ROS, SMOTE, ADASYN, RUS, and NM) are deployed on DT and RF classifiers. That is, each of the studied data sampling methods is used to alleviate the class imbalance problem in the defect datasets by balancing the number of the majority and minority class variables respectively, hence new datasets. The balancing of the original datasets is based on findings reported in existing studies. The SDP models are developed and evaluated using the K-fold (k = 10) Cross-Validation (CV) technique. The 10-fold CV was selected for its ability to build prediction models while minimizing the effect of the class imbalance issue. Besides, the K-fold CV technique allows each instance to be used iteratively for both training and testing [55–58]. The selected classifiers (DT and RF) were chosen based on their use and performance in previous research [46, 48], and Table 1 shows the parameter values of these classifiers as utilized in this research work. Thereafter, the prediction performances of ensuing models are then analyzed based on the selected evaluation metrics (accuracy, AUC and MCC values). Also, the prediction performances of the developed model are compared against each other to ascertain the effect of data sampling methods on the prediction models (DT, and RF) and the efficacy of studied data sampling methods. Python-Scikit ML library was used for implementing the data sampling methods while the WEKA ML platform was used for implementing the prediction models. These two ML resources are prominently used in SDP and ML tasks.

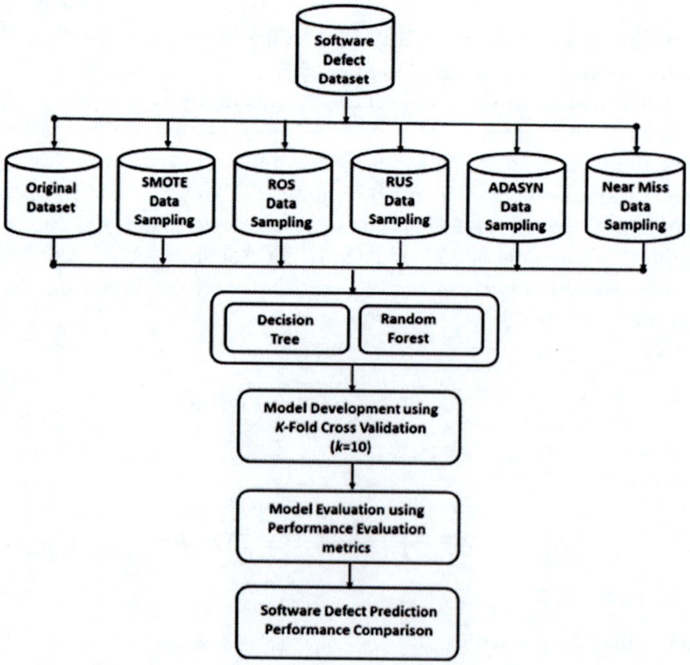

Fig. 1. Experimental framework

4 Results and Discussion

This section presents and discusses the findings obtained after assessing the different generated SDP models. It is critical to demonstrate the influence of the data sampling method on the efficacy of SDP models. Furthermore, the performance of the investigated data sampling methods is a key feature of this work. As a consequence, the results will be presented to represent these effects for experimented classifiers and software defect datasets.

Table 3 presents the prediction accuracy values of experimented prediction models on original and data sampling generated NASA defect datasets.

Table 3. Accuracy values of SDP models on original and data sampling-based NASA datasets

		DT	RF
KC3	SMOTE	81.27	89.84
	ADASYN	84.71	87.9
	ROS	87.03	93.67
	RUS	59.72	56.94
	NM	86.11	80.56
	No sampling	79.38	81.44
PC3	SMOTE	87.37	92.03
	ADASYN	88.61	93.13
	ROS	93.01	96.53
	RUS	69.62	71.15
	NM	81.15	81.92
	No sampling	84.71	87.46

As shown in Table 3, it can be observed that the prediction accuracy of experimented prediction models (DT and RF) developed with the data sampling-based datasets are superior to models developed with the original NASA datasets in most cases except in the case of RUS-balanced NASA datasets. In particular, on KC3 dataset, DT and RF models trained with SMOTE (DT: 81.27%; RF: 89.84%), ADASYN (DT: 84.71%; RF: 87.90%) and ROS (DT: 87.03%; RF: 93.67%)-balanced datasets recorded superior prediction accuracy values over when original No Sampling (DT: 79.38%; RF: 81.44%) KC3 dataset. In the case of the NM-balanced KC3 dataset, DT and RF had 86.11% and 80.56% prediction accuracy respectively. However, models based on the RUS-balanced KC3 dataset had the lowest prediction accuracy when compared to other experimented methods. Also, on PC3 datasets, DT and RF models developed with SMOTE (DT: 87.37%; RF: 92.03%), ADASYN (DT: 88.61%; RF: 93.13%) and ROS (DT: 93.01%; RF: 95.63%)-balanced datasets recorded superior prediction accuracy values over when original No Sampling (DT: 84.71%; RF: 87.46%) KC3 dataset. NM and RUS-balanced PC3 datasets generated a lower prediction when compared with other experimented models on PC3. It can be observed for both KC3 and PC3 datasets, SMOTE, ADASYN and ROS improved the prediction accuracy values of the experimented DT and RF models. That is, the experimented oversampling methods had a superior positive effect on the prediction accuracy values of DT and RF than the undersampling (RUS and NM) and when no sampling method (original dataset) is deployed. It should be noted that using only prediction accuracy values as performance metrics may not show the full effect of the data sampling methods. Consequently, Table 4 and Table 5 presents the AUC and MCC values of DT and RF models based on the studied data sampling methods.

Table 4. AUC values of SDP models on original and data sampling-based NASA datasets

		DT	RF
KC3	SMOTE	*0.814*	*0.948*
	ADASYN	*0.847*	*0.960*
	ROS	*0.893*	*0.997*
	RUS	*0.595*	*0.628*
	NM	*0.91*	*0.864*
	No sampling	*0.653*	*0.736*
PC3	SMOTE	*0.896*	*0.977*
	ADASYN	*0.902*	*0.986*
	ROS	*0.932*	*0.996*
	RUS	*0.713*	*0.786*
	NM	*0.774*	*0.893*
	No sampling	*0.591*	*0.832*

From Table 4, DT and RF models developed using the data sampling-based datasets had better AUC values than when original NASA datasets are used. Models based on the studied datasets recorded significant increments in their respective AUC values except in the case of the RUS-balanced KC3 dataset. Specifically, DT models developed with SMOTE, ADASYN, ROS and NM-balanced KC3 datasets had +25.66%, +27.71%, + 36.75%, and +39.36% increment in AUC values when compared with the DT model developed with the original KC3 dataset. A similar occurrence was observed with RF models developed with SMOTE (+28.80%), ADASYN (+30.43%), ROS (+35.46%) and NM (+17.39%)-balanced KC3 datasets and original KC3 dataset. In the case of the RUS-balanced KC3 dataset, DT and RF had poor AUC values that are lower than to other experimented models. In addition, on the PC3 dataset, DT and RF models trained with the data sampling based PC3 dataset had superior AUC values than when the original PC3 dataset is deployed with no exception. DT models developed with SMOTE, ADASYN, ROS, NM, and RUS-balanced PC3 datasets had +51.61%, +52.62%, +57.70%, + 30.96% and +20.64% increment in AUC values while RF models on developed with SMOTE (+17.43%), ADASYN (+18.51%), ROS (+19.71%) and NM (+7.33%). It can be observed that the superiority of DT and RF models developed with data sampling-based NASA datasets based on AUC values are per findings when prediction accuracy values were considered. However, the disparity in the prediction accuracy values may be due to the overfitting of models which the prediction accuracy values do not reflect. These findings further support the notion of not using only prediction accuracy values as a performance indicator.

Furthermore, as recommended by [53], the performance of the developed models was analyzed using the MCC value. Table 5 presents the MCC values of the developed DT and RF models using both data sampling-based and original NASA datasets.

According to Table 5, DT and RF models created utilizing data sampling-based datasets showed higher MCC values than original NASA datasets. Except for the RUS-balanced KC3 dataset, models based on the analyzed datasets showed a considerable increase in their respective MCC values. When compared to DT models produced with the original KC3 dataset, DT models developed using SMOTE (0.626), ADASYN (0.697), ROS (0.753), and NM (0.767)-balanced KC3 datasets showed more than +100% increase in MCC values. A similar phenomenon was seen with RF models constructed using SMOTE (0.798), ADASYN (0.758), ROS (0.881), and NM (0.620)-balanced KC3 datasets and the original KC3 (DT: 0.257; RF: 0.189) dataset. In the case of the RUS-balanced KC3 dataset, DT and RF showed lower MCC values than other tested models. Furthermore, using the PC dataset, DT and RF models trained with the data sampling based PC3 dataset consistently outperformed models trained with the original PC3 dataset. MCC values increased by +100% for DT and RF models developed with SMOTE, ADASYN, ROS, and NM-balanced PC3 datasets, respectively.

Table 5. MCC values of SDP models on original and data sampling-based NASA datasets

		DT	RF
KC3	SMOTE	*0.626*	*0.798*
	ADASYN	*0.694*	*0.758*
	ROS	*0.753*	*0.881*
	RUS	*0.196*	*0.139*
	NM	*0.727*	*0.62*
	No sampling	*0.257*	*0.189*
PC3	SMOTE	*0.747*	*0.842*
	ADASYN	*0.772*	*0.864*
	ROS	*0.866*	*0.932*
	RUS	*0.394*	*0.428*
	NM	*0.623*	*0.647*
	No sampling	*0.221*	*0.186*

For generalizability, further experiments were conducted using software defect datasets from the PROMISE repository. Table 6, Table 7 and Table 8 showed the prediction performances of DT and RF models using data sampling-based PROMISE datasets (ANT and REDKATOR) based on prediction accuracy, AUC and MCC values.

Table 6. Accuracy values of SDP models on original and data sampling-based PROMISE datasets

		DT	RF
ANT	SMOTE	*88.25*	*94.99*
	ADASYN	*89.81*	*94.42*
	ROS	*94.42*	*98.46*
	RUS	*73.44*	*75*
	NM	*79.69*	*84.38*
	No sampling	*87.67*	*89.73*
REDKATOR	SMOTE	*89.9*	*91.92*
	ADASYN	*89.26*	*91.61*
	ROS	*91.28*	*96.64*
	RUS	*72.22*	*70.37*
	NM	*74.07*	*77.78*
	No sampling	*88.64*	*90.34*

Table 6 shows that, except for prediction models (DT and RF) trained on the NM and RUS-balanced PROMISE datasets, the prediction accuracy of the prediction models produced using data sampling-based datasets is better than models generated with the original PROMISE datasets in most circumstances. On the ANT dataset, DT and RF models trained with SMOTE (DT: 88.25%; RF: 94.99%), ADASYN (DT: 89.81%; RF: 94.42%), and ROS (DT: 94.42%; RF: 98.46%)-balanced datasets outperformed those trained with the original (DT: 87.67%; RF: 89.73%). In the instance of the NM and RUS-balanced ANT dataset, DT and RF had lower prediction accuracy values ((79.69%, 84.38%) and (73.44%,75%) respectively). On the REDKATOR dataset, DT and RF models constructed with SMOTE (DT: 89.90%; RF: 91.92%), ADASYN (DT: 89.26%; RF: 91.61%), and ROS (DT: 91.28%; RF: 96.64%)-balanced datasets outperformed the original No Sampling (DT: 88.64%; RF: 90.34%) When compared to REDKATOR generated models, the NM and RUS-balanced REDKATOR datasets produced lower prediction accuracy values. SMOTE, ADASYN, and ROS increased the prediction accuracy values of the developed DT and RF models for both the ANT and REDKATOR datasets. This observation conforms with the findings observed with NASA (KC3 and PC3) datasets. That is, the studied oversampling methods (SMOTE, ADASYN and ROS) had a greater positive influence on DT and RF prediction accuracy than undersampling (RUS and NM) and when no sampling method (original dataset) was used.

According to Table 7, DT and RF models produced utilising data sampling-based datasets had higher AUC values than models generated using original PROMISE

Table 7. AUC values of SDP models on original and data sampling-based PROMISE datasets

		DT	RF
ANT	SMOTE	*0.89*	*0.981*
	ADASYN	*0.908*	*0.988*
	ROS	*0.947*	*1*
	RUS	*0.759*	*0.794*
	NM	*0.853*	*0.876*
	No sampling	*0.611*	*0.776*
REDKATOR	SMOTE	*0.919*	*0.964*
	ADASYN	*0.913*	*0.962*
	ROS	*0.919*	*0.991*
	RUS	*0.712*	*0.8*
	NM	*0.704*	*0.796*
	No sampling	*0.639*	*0.772*

datasets. DT and RF models based on the data sampling datasets saw a considerable increment in their respective AUC values. When compared to the original ANT dataset, DT models created using SMOTE, ADASYN, ROS, NM, and RUS-balanced datasets showed +45.66%, +48.6%, +55%, +39.61% and +24.22% increment in AUC values. A similar phenomenon was seen with RF models constructed using SMOTE (+26.41%), ADASYN (+27.32%), ROS (+28.86%), NM (+12.89%), and RUS (+2.32%)-balanced ANT datasets and the original ANT dataset. Furthermore, using the REDKATOR dataset, DT and RF models trained with the data sampling based REDKATOR dataset achieved greater AUC values than when the original REDKATOR dataset was used without exception. The AUC values for DT models developed with SMOTE, ADASYN, ROS, NM, and RUS-balanced REDKATOR datasets increased by +43.82%, +42.88%, +43.82%, +10.17%, and +11.42%, respectively, while RF models developed with SMOTE (+24.87%), ADASYN (+24.61%), ROS (+28.37%), NM (+3.11%) and RUS (+3.63%).

Table 8. MCC values of SDP models on original and data sampling-based PROMISE datasets

		DT	RF
ANT	SMOTE	*0.765*	*0.9*
	ADASYN	*0.797*	*0.889*
	ROS	*0.893*	*0.97*

(*continued*)

Table 8. (*continued*)

		DT	RF
	RUS	*0.475*	*0.501*
	NM	*0.594*	*0.7*
	No Sampling	*0.252*	*0.366*
REDKATOR	SMOTE	*0.8*	*0.838*
	ADASYN	*0.786*	*0.832*
	ROS	*0.832*	*0.934*
	RUS	*0.452*	*0.407*
	NM	*0.494*	*0.556*
	No Sampling	*0.501*	*0.585*

As per Table 8, DT and RF models constructed with data sampling-based datasets have higher MCC values than original ANT and REDKATOR datasets. Except for the NM and RUS-balanced REDKATOR dataset, models based on the examined datasets exhibited a significant rise in their respective MCC values. When compared to DT models created with the original ANT dataset, DT models created with SMOTE (0.765), ADASYN (0.797), ROS (0.893), NM (0.594), and RUS (0.475)-balanced ANT datasets exhibited a more than +100% increase in MCC values. A similar result was seen with RF models built using SMOTE (0.900), ADASYN (0.889), ROS (0.970), NM (0.700), and RUS (0.501)-balanced ANT datasets and the original ANT (DT: 0.252; RF: 0.366) dataset. In the RUS-balanced KC3 dataset, DT and RF had lower MCC values than the other examined models. Furthermore, when employing the REDKATOR dataset, DT and RF models trained using the data sampling based REDKATOR dataset consistently outperformed models trained with the original REDKATOR dataset. MCC values rose by +50% for DT and RF models generated using SMOTE, ADASYN, and ROS-balanced REDKATOR datasets, respectively.

From the preceding experimental results analyses, it is evident that studied data sampling methods can alleviate the class imbalance problem in SDP datasets and improved the prediction performances of experimented DT and RF models. Also, in general, it was observed that the studied data oversampling methods (SMOTE, ADASYN, and ROS) had better performance than the data undersampling methods (NM and RUS). Amongst the investigated oversampling methods, there are no significant differences in their respective performance as this varies across the experimented datasets and selected prediction models. However, it is worth noting that the RUS method had the least effect on the prediction models and in some cases performed poorly than when original datasets are deployed. This observation can be attributed to the random removal of relevant data that could be vital for the SDP process Although, the NM method is also a data undersampling method, NM removes instances based on their nearest neighbor characteristics.

5 Conclusion and Future Works

Understanding the ideas of SDP and the class imbalance problem as stated in research work is essential for effective SDP model building. In this work, data sampling methods are used on software defect datasets to relieve the latent class imbalance problem by balancing the amount of minority and majority class instances present, resulting in new defect datasets with no class imbalance problem. Three data oversampling methods (SMOTE, ADASYN, and ROS) and two data undersampling methods (RUS and NM) are utilised on defect datasets from NASA and PROMISE repositories, while DT and RF classifiers are employed on the original and newly developed software defect datasets. In summary, the experimental results demonstrated that the examined data sampling methods can overcome the class imbalance problem in SDP datasets. Furthermore, in most instances, the data sampling methods enhanced the prediction performances of the experimented prediction models. Concerning the effectiveness of the data sampling methods, the analysed oversampling approaches had a higher (positive) influence on the prediction models than their undersampling counterparts. It is consequently advised that data sampling operations, especially oversampling methods, be carried out during SDP activities. Deploying data sampling strategies may alleviate the underlying class imbalance problem and assure successful SDP models. In future work, the impact of data sampling methods on prediction performances of homogeneous and heterogeneous ensemble methods in SDP will be extensively investigated. This is based on the report that ensemble approaches outperform single classifiers.

References

1. Alazzawi, A.K., et al.: HABCSm: a hamming based t-way strategy based on hybrid artificial bee colony for variable strength test sets generation. Int. J. Comput. Commun. Control **16** (2021). https://doi.org/10.15837/ijccc.2021.5.4308
2. Ameen, A.O., et al.: Application of shuffled frog-leaping algorithm for optimal software project scheduling and staffing. In: Saeed, F., Mohammed, F., Al-Nahari, A. (eds.) IRICT 2020. LNDECT, vol. 72, pp. 293–303. Springer, Cham (2021). https://doi.org/10.1007/978-3-030-70713-2_28
3. Huda, S., et al.: An ensemble oversampling model for class imbalance problem in software defect prediction. IEEE Access **6**, 24184–24195 (2018)
4. Balogun, A.O., et al.: A novel rank aggregation-based hybrid multifilter wrapper feature selection method in software defect prediction. Comput. Intell. Neurosci. **2021** (2021). https://doi.org/10.1155/2021/5069016
5. Alsawalqah, H., Faris, H., Aljarah, I., Alnemer, L., Alhindawi, N.: Hybrid SMOTE-ensemble approach for software defect prediction. In: Silhavy, R., Silhavy, P., Prokopova, Z., Senkerik, R., Kominkova Oplatkova, Z. (eds.) CSOC 2017. AISC, vol. 575, pp. 355–366. Springer, Cham (2017). https://doi.org/10.1007/978-3-319-57141-6_39
6. Balogun, A.O., Basri, S., Abdulkadir, S.J., Adeyemo, V.E., Imam, A.A., Bajeh, A.O.: Software defect prediction: analysis of class imbalance and performance stability. J. Eng. Sci. Technol. **14**, 3294–3308 (2019)
7. Tantithamthavorn, C., Hassan, A.E., Matsumoto, K.: The impact of class rebalancing techniques on the performance and interpretation of defect prediction models. IIEEE Trans. Softw. Eng. **46**, 1200–1219 (2020)

8. Balogun, A.O., et al.: Software defect prediction using wrapper feature selection based on dynamic re-reranking strategy. Symmetry **13**, 2166 (2021)
9. Tantithamthavorn, C., McIntosh, S., Hassan, A.E., Matsumoto, K.: The impact of automated parameter optimization on defect prediction models. IIEEE Trans. Softw. Eng. **45**, 683–711 (2019)
10. Petrić, J., Bowes, D., Hall, T., Christianson, B., Baddoo, N.: The jinx on the NASA software defect data sets. In: Proceedings of the 20th International Conference on Evaluation and Assessment in Software Engineering, pp. 1–5 (2016)
11. Ali, U., Aftab, S., Iqbal, A., Nawaz, Z., Salman Bashir, M., Anwaar Saeed, M.: Software defect prediction using variant based ensemble learning and feature selection techniques. IJMECS **12**, 29–40 (2020)
12. Balogun, A.O., et al.: An adaptive rank aggregation-based ensemble multi-filter feature selection method in software defect prediction. Entropy **23**, 1274 (2021)
13. Akintola, A.G., Balogun, A.O., Lafenwa-Balogun, F., Mojeed, H.A.: Comparative analysis of selected heterogeneous classifiers for software defects prediction using filter-based feature selection methods. FUOYEJET **3**, 134–137 (2018)
14. Alazzawi, A.K., Rais, H.M., Basri, S., Alsariera, Y.A., Balogun, A.O., Imam, A.A.: A hybrid artificial bee colony strategy for t-way test set generation with constraints support. In: Journal of Physics: Conference Series, vol. 1529, p. 042068. IOP Publishing (2020)
15. Menzies, T., Greenwald, J., Frank, A.: Data mining static code attributes to learn defect predictors. IEEE Trans. Softw. Eng. **33**, 2–13 (2007)
16. Hall, T., Beecham, S., Bowes, D., Gray, D., Counsell, S.: A systematic literature review on fault prediction performance in software engineering. IEEE Trans. Softw. Eng. **38**, 1276–1304 (2012)
17. Yu, Q., Jiang, S., Zhang, Y.: The performance stability of defect prediction models with class imbalance: an empirical study. IEICE Trans. Inf. Syst. **100**, 265–272 (2017)
18. Khan, B., et al.: Software defect prediction for healthcare big data: an empirical evaluation of machine learning techniques. J. Healthc. Eng. **2021** (2021). https://doi.org/10.1155/2021/8899263
19. Bennin, K.E., Keung, J.W., Monden, A.: On the relative value of data resampling approaches for software defect prediction. Empir. Softw. Eng. **24**(2), 602–636 (2018). https://doi.org/10.1007/s10664-018-9633-6
20. Alsaeedi, A., Khan, M.Z.: Software defect prediction using supervised machine learning and ensemble techniques: a comparative study. JSEA **12**, 85–100 (2019)
21. Laradji, I.H., Alshayeb, M., Ghouti, L.: Software defect prediction using ensemble learning on selected features. Inf. Softw. Technol. **58**, 388–402 (2015)
22. El-Sharkawy, S., Yamagishi-Eichler, N., Schmid, K.: Metrics for analyzing variability and its implementation in software product lines: a systematic literature review. Inf. Softw. Technol. **106**, 1–30 (2019)
23. Hammad, M.: Classifying defective software projects based on machine learning and complexity metrics. Int. J. Comput. Sci. Math. **13**, 401–412 (2021)
24. Tiwari, S., Rathore, S.S.: Coupling and cohesion metrics for object-oriented software: a systematic mapping study. In: Proceedings of the 11th Innovations in Software Engineering Conference, pp. 1–11 (2018)
25. Turhan, B., Mısırlı, A.T., Bener, A.: Empirical evaluation of the effects of mixed project data on learning defect predictors. Inf. Softw. Technol. **55**, 1101–1118 (2013)
26. Balogun, A.O., et al.: Data sampling-based feature selection framework for software defect prediction. In: Abawajy, J.H., Choo, K.-K., Chiroma, H. (eds.) EATI 2020. LNNS, vol. 254, pp. 39–52. Springer, Cham (2021). https://doi.org/10.1007/978-3-030-80216-5_4

27. Goyal, S.: Handling class-imbalance with KNN (Neighbourhood) under-sampling for software defect prediction. Artif. Intell. Rev. **55**, 2023–2064 (2022). https://doi.org/10.1007/s10462-021-10044-w
28. Cao, Y., Ding, Z., Xue, F., Rong, X.: An improved twin support vector machine based on multi-objective cuckoo search for software defect prediction. Int. J. Bio-Inspir. Comput. **11**, 282–291 (2018)
29. Mabayoje, M.A., Balogun, A.O., Jibril, H.A., Atoyebi, J.O., Mojeed, H.A., Adeyemo, V.E.: Parameter tuning in KNN for software defect prediction: an empirical analysis. Jurnal Teknologi dan Sistem Komputer **7**, 121–126 (2019)
30. Balogun, A., Bajeh, A., Mojeed, H., Akintola, A.: Software defect prediction: a multi-criteria decision-making approach. Niger. J. Technol. Res. **15**, 35–42 (2020)
31. Balogun, A.O., et al.: SMOTE-based homogeneous ensemble methods for software defect prediction. In: Gervasi, O., et al. (eds.) ICCSA 2020. LNCS, vol. 12254, pp. 615–631. Springer, Cham (2020). https://doi.org/10.1007/978-3-030-58817-5_45
32. Mockus, A., Weiss, D.M.: Predicting risk of software changes. Bell Labs Tech. J. **5**, 169–180 (2000)
33. Bowes, D., Hall, T., Petrić, J.: Software defect prediction: do different classifiers find the same defects? Softw. Qual. J. **26**(2), 525–552 (2017). https://doi.org/10.1007/s11219-016-9353-3
34. Japkowicz, N.: The class imbalance problem: significance and strategies. In: Proceedings of the International Conference on Artificial Intelligence, vol. 56, pp. 111–117 (2000)
35. Peng, M., et al.: Trainable undersampling for class-imbalance learning. In: Proceedings of the AAAI Conference on Artificial Intelligence, vol. 33, pp. 4707–4714 (2019)
36. Chawla, N.V., Bowyer, K.W., Hall, L.O., Kegelmeyer, W.P.: SMOTE: synthetic minority over-sampling technique. J. Artif. Intell. Res. **16**, 321–357 (2002)
37. Han, H., Wang, W.-Y., Mao, B.-H.: Borderline-SMOTE: a new over-sampling method in imbalanced data sets learning. In: Huang, D.-S., Zhang, X.-P., Huang, G.-B. (eds.) ICIC 2005. LNCS, vol. 3644, pp. 878–887. Springer, Heidelberg (2005). https://doi.org/10.1007/11538059_91
38. Xie, Z., Jiang, L., Ye, T., Li, X.: A synthetic minority oversampling method based on local densities in low-dimensional space for imbalanced learning. In: Renz, M., Shahabi, C., Zhou, X., Cheema, M.A. (eds.) DASFAA 2015. LNCS, vol. 9050, pp. 3–18. Springer, Cham (2015). https://doi.org/10.1007/978-3-319-18123-3_1
39. Batista, G.E., Prati, R.C., Monard, M.C.: A study of the behavior of several methods for balancing machine learning training data. ACM SIGKDD Explor. Newsl. **6**, 20–29 (2004)
40. Satapathy, S.K., Mishra, S., Mallick, P.K., Chae, G.-S.: ADASYN and ABC-optimized RBF convergence network for classification of electroencephalograph signal. Pers. Ubiquit. Comput., 1–17 (2021). https://doi.org/10.1007/s00779-021-01533-4
41. Kamalov, F.: Kernel density estimation based sampling for imbalanced class distribution. Inf. Sci. **512**, 1192–1201 (2020)
42. Balogun, A.O., Jimoh, R.G.: Anomaly intrusion detection using an hybrid of decision tree and K-nearest neighbor. J. Adv. Sci. Res. Appl. (JASRA) **2**, 67–74 (2015)
43. Balogun, A.O., et al.: Optimized decision forest for website phishing detection. In: Silhavy, R., Silhavy, P., Prokopova, Z. (eds.) CoMeSySo 2021. LNNS, vol. 231, pp. 568–582. Springer, Cham (2021). https://doi.org/10.1007/978-3-030-90321-3_47
44. Shepperd, M., Song, Q., Sun, Z., Mair, C.: Data quality: some comments on the nasa software defect datasets. IIEEE Trans. Softw. Eng. **39**, 1208–1215 (2013)
45. Balogun, A.O., et al.: Empirical analysis of rank aggregation-based multi-filter feature selection methods in software defect prediction. Electronics **10**, 179 (2021)
46. Ghotra, B., McIntosh, S., Hassan, A.E.: A large-scale study of the impact of feature selection techniques on defect classification models. In: 2017 IEEE/ACM 14th International Conference on Mining Software Repositories (MSR), pp. 146–157. IEEE (2017)

47. Xu, Z., Liu, J., Yang, Z., An, G., Jia, X.: The impact of feature selection on defect prediction performance: an empirical comparison. In: 2016 IEEE 27th International Symposium on Software Reliability Engineering (ISSRE), pp. 309–320. IEEE (2016)
48. Balogun, A.O., et al.: Search-based wrapper feature selection methods in software defect prediction: an empirical analysis. In: Silhavy, R. (ed.) CSOC 2020. AISC, vol. 1224, pp. 492–503. Springer, Cham (2020). https://doi.org/10.1007/978-3-030-51965-0_43
49. Kondo, M., Bezemer, C.-P., Kamei, Y., Hassan, A.E., Mizuno, O.: The impact of feature reduction techniques on defect prediction models. Empir. Softw. Eng. **24**(4), 1925–1963 (2019). https://doi.org/10.1007/s10664-018-9679-5
50. Rathore, S.S., Gupta, A.: A comparative study of feature-ranking and feature-subset selection techniques for improved fault prediction. In: Proceedings of the 7th India Software Engineering Conference, Chennai, India, pp. 1–10. ACM (2014)
51. Tantithamthavorn, C., McIntosh, S., Hassan, A.E., Matsumoto, K.: An empirical comparison of model validation techniques for defect prediction models. IEEE Trans. Softw. Eng. **43**, 1–18 (2016)
52. Tantithamthavorn, C., McIntosh, S., Hassan, A.E., Matsumoto, K.: Comments on "Researcher bias: the use of machine learning in software defect prediction." IEEE Trans. Softw. Eng. **42**, 1092–1094 (2016)
53. Song, Q., Guo, Y., Shepperd, M.: A comprehensive investigation of the role of imbalanced learning for software defect prediction. IEEE Trans. Softw. Eng. **45**, 1253–1269 (2018)
54. Alsariera, Y.A., Balogun, A.O., Adeyemo, V.E., Tarawneh, O.H., Mojeed, H.A.: Intelligent tree-based ensemble approcahes for phishing website detection. J. Eng. Sci. Technol. **17**, 0563–0582 (2022)
55. Alsariera, Y.A., Adeyemo, V.E., Balogun, A.O., Alazzawi, A.K.: AI meta-learners and extra-trees algorithm for the detection of phishing websites. IEEE Access **8**, 142532–142542 (2020)
56. Alsariera, Y.A., Elijah, A.V., Balogun, A.O.: Phishing website detection: forest by penalizing attributes algorithm and its enhanced variations. Arab. J. Sci. Eng. **45**(12), 10459–10470 (2020). https://doi.org/10.1007/s13369-020-04802-1
57. Balogun, A.O., et al.: Improving the phishing website detection using empirical analysis of Function Tree and its variants. Heliyon **7**, e07437 (2021)
58. Balogun, A.O., Akande, N.O., Usman-Hamza, F.E., Adeyemo, V.E., Mabayoje, M.A., Ameen, A.O.: Rotation forest-based logistic model tree for website phishing detection. In: Gervasi, O., et al. (eds.) ICCSA 2021. LNCS, vol. 12957, pp. 154–169. Springer, Cham (2021). https://doi.org/10.1007/978-3-030-87013-3_12

Check for updates

Comparison of Different Cloud Solutions

Stefan Pocarovsky[1], Milos Orgon[1], Martin Koppl[1], and Antonin Bohacik[2](✉)

[1] Faculty of Electrical Engineering and Information Technology, Slovak University of Technology, Bratislava, Slovakia
orgon@ktl.elf.stuba.sk

[2] Faculty of Electrical Engineering and Communication, Brno University of Technology, Brno, Czech Republic
Antonin.Bohacik@vut.cz

Abstract. Nowadays, one of the major trends in IT infrastructure is virtualization and cloud computing. We can manage computing power of servers for systems we need. Computing power is transferred to the servers, so it is possible to minimize computing power of client's machines, which is one of important advantages of cloud computing. There are many technologies and approaches for virtualization of applications. The article describes two of them, full cloud application and partial cloud application. In full cloud application the whole application and the database are in the cloud. In partial cloud application, only the database is in the cloud and the application is installed at the client. In these two structures of cloud, quantities which are characterizing speed of application in the cloud for both cases were measured.

Keywords: Cloud · Virtualization · LAN · IT · Performance

1 Introduction

The boom of virtualization and cloud solutions began in 2000–2010 and the interest in this technology constantly grows [1,2]. Not only computing power of hardware (on server side) is important, but also transmission performance of data from and to the cloud is important, too [3,4]. The transmission is performed by high-speed transmission (HDT). Today there exist a lot of devices, which are connected to cloud using different technologies with many transmission parameters. For example, optic fiber (as the dominant technology for high transmission performance), virtual infrastructure [5], device to device communication technologies [6], smart grid [7–9], wireless technologies of internet of things [10], smart metering transmission technologies [11,12], smart city transmission infrastructure [13–15], and many others. Most effective is to use combinations of these technologies. They are called heterogeneous mixed networks [16], which provide combined transmission parameters. Very important in cloud solutions is cybersecurity. We might mention application layer attacks [17], cryptography primitives and end devices [18–20], or many others. As might be seen, the performance and security are the main challenges of cloud solutions [21].

R. Silhavy (Ed.): CSOC 2022, LNNS 501, pp. 611–621, 2022.
https://doi.org/10.1007/978-3-031-09070-7_50

2 Optimal Way for Providing Software as a Service

First, we wanted to observe the speed of winform applications with two types of cloud. We investigated the parameters that can change the speed of displaying application data in both cases of the cloud. Using these results, we subsequently determined an optimal way of providing software as service (SAAS - Software as a Service).

2.1 Partial Cloud

The term "database in the cloud" means the installation of a database engine, or a database out of the local environment of the application (out of the LAN). The application is connected to the database via internet (without securing the communication). For each query to the SQL server, the application is connected to a specific database (using port 1433), where we wrote the required data or from where we obtained the required data and displayed them in the application. The principle of this connection is shown in Fig. 1.

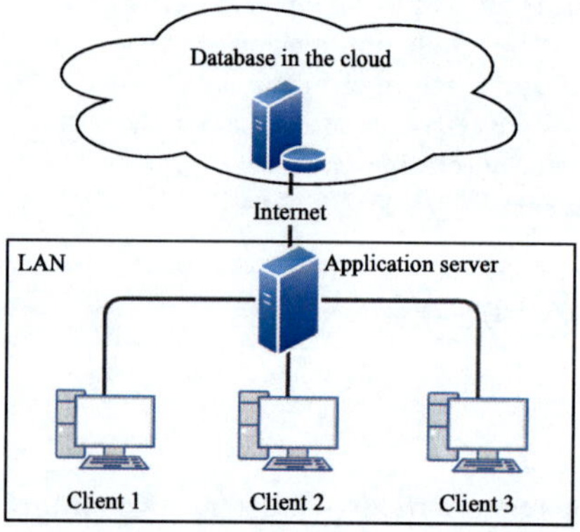

Fig. 1. The database is located in a cloud on a server outside the LAN of users. An application is installed in the LAN of users, to which users are connected and it is subsequently connected to the database.

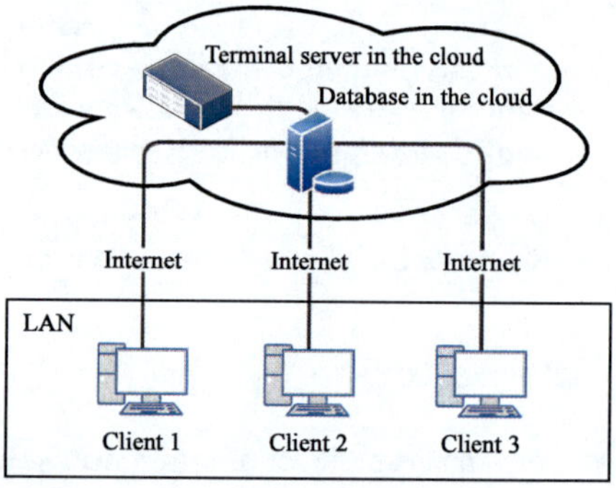

Fig. 2. Application and database will be in the cloud on the server outside the LAN. Users are connecting to the presentation layer.

2.2 Full Cloud

In this part, both database and application are together in the cloud. The application is virtualized and we connected it to the presentation layer. Currently there are not many providers of such a platform (Citrix, Microsoft, VMware) [22], so for this reason we used the "Terminal Services – RemoteApp" from Microsoft [23]. This means, that data from the database are not sent over the network. Over the network are sent only "images of display" (more precisely, changes of individual points, or vectors of the application display). This solution makes it possible to use VMware's portfolio to virtualize the presentation layer based on the PCoIP protocol [24]. Remarkably interesting results were published in article [25], where the authors compared technologies from Microsoft "Terminal services" versus VMware (with Terradici) PCoIP. Figure 2 is a graphical representation of the principle of client access to an application through the virtualized presentation layer. Application and database will be in the cloud on the server outside of the LAN. Users connect to the presentation layer.

3 Tests

For measurement we used the application PSP. This is a real application programmed in C# technology "Entity Framework". We made three tests:

- 1st test – loading 1,000 records from the database and plotting them in a "datagridview",
- 2nd test – loading a specific record and drawing it in the form,
- 3rd test – writing a new record to the database.

We were testing network data that have a direct impact on application speed. The values we measured are given below, in Table 1. Subsequently, these data were entered into tables, shown in the graph, compared with each other, and then we performed an assessment of the advantages and disadvantages of both cases.

Table 1. Network parameters affecting application speed.

Measured data	Mark	Unit
Form rendering time	tR	ms
Required bandwidth for the operation (client→server)	$B_{C\text{-}S}$	bps
Required bandwidth for the operation (server→client)	$B_{S\text{-}C}$	bps
The size of sent data (client→server)	$D_{C\text{-}S}$	B
The size of received data (server→client)	$D_{S\text{-}C}$	B
The size of sent and received data	D	B
The size of sent packets (client→server)	$P_{C\text{-}S}$	–
The size of received packets (server→client)	$P_{S\text{-}C}$	–
Total packets	P	–

3.1 1st Test

In the first test we compared the speed of reading data from database to application and subsequently rendering of datagridview in the application. The system was reading 1,000 records from the database and this data was subsequently rendered to the datagridview.

The code for reading this data was written in C# by using LINQ, but the system was changing it to T-SQL. Table 2 shows times of measurement of the 1st test for both cases of the cloud solution.

Table 2. Times of rendering the datagridview in 1st test (left 1st type of cloud, right 2nd type of cloud).

1st case		2nd case	
Mark	Value [s]	Mark	Value [s]
t_{R1}	7.599	t_{R1}	0.345
t_{R2}	5.094	t_{R2}	0.368
t_{R3}	5.357	t_{R3}	0.353
t_{R4}	5.326	t_{R4}	0.368
t_{R5}	5.264	t_{R5}	0.349
$t_{R\text{-average}}$	**5.728**	**$t_{R\text{-average}}$**	**0.357**

Table 2 shows that access in the full virtualized application is much more efficient in the terms of data display speed. In Table 3 and 4 there are data in terms of network transmission.

Table 3. Number of transmitted packets, number of transmitted bytes and the required bandwidth for transmission in 1st case.

Measurement	P_{C-S}	P_{S-C}	P	D_{C-S} [B]	D_{S-C} [B]	D [B]	B_{C-S} [bps]	B_{S-C} [bps]
1	218	355	573	17,632	486,505	504,137	26,498.1	736,008.6
2	217	355	572	18,967	486,528	505,495	26,534.0	737,006.5
3	218	355	573	17,727	490,698	508,425	26,661.2	739,568.3
4	218	355	573	17,727	490,698	508,425	26,634.2	738,016.5
5	218	355	573	17,727	490,698	508,425	26,372.1	735,819.0
Average (≈)	**218**	**355**	**573**	**17,956**	**489,025**	**506,981**	**26,54039.9**	**737,283.8**

From the data in Table 2 and Table 3, the transmission over the network is clearly more efficient in the 2nd case, when the application was virtualized and we were using "Terminal services". The values characterizing the network connection were several times lower in the 2nd case.

Table 4. Number of transmitted packets, number of transmitted bytes and the required bandwidth for transmission in 2nd case.

Measurement	P_{C-S}	P_{S-C}	P	D_{C-S} [B]	D_{S-C} [B]	D [B]	B_{C-S} [bps]	B_{S-C} [bps]
1	43	34	77	4,753	7,702	12,455	7,852.9	12,725.2
2	28	25	53	2,905	4,924	7,829	5,633.8	9,549.3
3	37	31	68	3,745	6,478	10,223	6,885.0	11,909.4
4	31	26	31	3,516	6,343	9,859	6,187.9	11,163.2
5	30	25	55	3,176	4,460	7,636	7,313.6	10,270.3
Average (≈)	**34**	**28**	**57**	**3,619**	**5,981**	**9,600**	**6,774.6**	**11,123.5**

3.2 2nd Test

The 2nd test was based on the same principle as the first test, but the system was reading only one specific record and data were displayed in the form (not in datagridview). The first tested parameters were times of rendering data from the database to the application (Table 5).

Table 5. Time of rendering of form in 2nd test in every cases.

1st case		2nd case	
Mark	Value [s]	Mark	Value [s]
t_{R1}	2.355	t_{R1}	0.588
t_{R2}	2.670	t_{R2}	0.252
t_{R3}	2.376	t_{R3}	0.246
t_{R4}	2.390	t_{R4}	0.267
t_{R5}	2.372	t_{R5}	0.531
$t_{R\text{-average}}$	**2.432**	$t_{R\text{-average}}$	**0.377**

The times t_R of rendering data are 5 times lower for the 2nd case of the cloud. In the following Table 6 we can see results of this test for network transmission.

Table 6 shows that the bandwidth is sufficient for this operation. But if the bandwidth $B \approx 122{,}684.63$ kbps is sufficient for the given procedure, it is assumed that the application is used by only one client. If the application was used by n clients and the procedure was run at the same time, the required bandwidth would increase n times.

It is clear from Table 7 that the number of sent data bytes D_{C-S} is 9 times lower as in the 1st case, but the number of received data bytes D_{S-C} is almost the same as in the 1st case. This corresponds to bandwidth $B \approx 53{,}912.628$ bps, which is necessary for the correct functionality of the procedure for one client.

Table 6. Number of transmitted packets, number of transmitted bytes and the required bandwidth for transmission in 1st case.

Measurement	P_{C-S}	P_{S-C}	P	D_{C-S} [B]	D_{S-C} [B]	D [B]	B_{C-S} [bps]	B_{S-C} [bps]
1	44	44	88	36,539	33,347	69,886	126,397.5	115,355.6
2	44	44	89	36,539	33,646	69,885	111,627.8	102,634.8
3	44	44	88	36,539	33,347	69,886	125,224.6	114,285.2
4	44	44	88	36,540	33,347	69,887	124,425.8	113,556.2
5	44	44	88	36,539	33,347	69,886	125,747.4	114,762.3
Average (≈)	**44**	**44**	**88**	**36,539**	**33,407**	**69,886**	**122,684.6**	**112,118.8**

Table 7. Number of transmitted packets, number of transmitted bytes and the required bandwidth for transmission in 2nd case.

Measurement	P_{C-S}	P_{S-C}	P	D_{C-S} [B]	D_{S-C} [B]	D [B]	B_{C-S} [bps]	B_{S-C} [bps]
1	42	37	79	4,118	28,378	32,496	10,497.1	72,337.8
2	43	39	82	4,431	27,467	31,898	10,687.0	66,246.9
3	29	28	57	3,377	14,729	18,106	10,426.3	45,476.1
4	32	30	62	3,571	19,269	22,840	8,435.8	45,519.3
5	31	30	61	3,460	15,285	18,745	9,050.8	39,983.0
Average (≈)	**35**	**33**	**68**	**3,791**	**21,026**	**24,817**	**9,819.4**	**53,912.6**

3.3 3rd Test

Another often used procedure of the database application is writing new data to a database, upgrading data in a database or their deletion. In this test we were monitoring individual parameters by writing a new record to the database. Table 8 gives required times for writing a new record to the database. The times of generating new records to the database are more effective in the 2nd case. Also, in the 3rd test speeds of writing new data to the database (Table 9 and 10) were compared and based on the measured data the 2nd case is more efficient again.

Table 8. Time of writing new record to the database in both cases.

1st case		2nd case	
Mark	Value [s]	Mark	Value [s]
t_{R1}	1.436	t_{R1}	0.076
t_{R2}	1.442	t_{R2}	0.090
t_{R3}	1.448	t_{R3}	0.084
t_{R4}	1.427	t_{R4}	0.085
t_{R5}	1.440	t_{R5}	0.081
$\mathbf{t_{R-average}}$	**1.439**	$\mathbf{t_{R-average}}$	**0.083**

Table 9. The number of transmitted packets, number of transmitted bytes and the required bandwidth for transmission in 1st case.

Measurement	P_{C-S}	P_{S-C}	P	D_{C-S} [B]	D_{S-C} [B]	D [B]	B_{C-S} [bps]	B_{S-C} [bps]
1	30	25	55	22,980	8,924	31,904	113,624.6	44,124.7
2	30	25	55	22,980	8,924	31,904	112,784.9	43,798.6
3	30	25	55	23,144	8,932	32,076	112,631.9	43,468.2
4	30	25	55	22,980	8,924	31,904	113,312.1	44,003.4
5	30	25	55	23,099	8,924	32,023	112,888.5	43,613.0
Average (≈)	**30**	**25**	**55**	**23,037**	**8,926**	**31,962**	**113,048.4**	**43,801.6**

Table 10. The number of transmitted packets, number of transmitted bytes and the required bandwidth for transmission in 2nd case

Measurement	P_{C-S}	P_{S-C}	P	D_{C-S} [B]	D_{S-C} [B]	D [B]	B_{C-S} [bps]	B_{S-C} [bps]
1	25	20	45	3,166	3,727	6,893	9,301.1	10,949.2
2	34	24	58	4,583	3,650	8,233	9,359.7	7,454.3
3	25	19	44	3,119	3,129	6,248	9,110.4	9,139.6
4	30	23	53	4,011	3,233	7,244	9,228.8	7,438.7
5	22	19	41	2,835	3,081	5,916	7,687.9	8,355.0
Average (≈)	**27**	**21**	**48**	**3,543**	**3,364**	**6,907**	**8,937.6**	**8,667.4**

4 Evaluation of Tests

In these tests we compared two types of a cloud. In the first case only a database was stored in the cloud and the computer with the application was connecting to the database via the internet. So, there was only data storage in the cloud. In the second case both database and application were stored in the cloud. The application was virtualized, and the client was connecting by "Terminal Services". The essential time parameters of the network were measured, which were listed in tables and then displayed in graphs. In Fig. 3, the graph shows the time dependences of all three tests for the first case and the second case of the cloud solution. The graph shows that in terms of speed of the application the second case is much more efficient. The displayed time of the read data is several times lower than in the first case, where only data storage is virtualized.

Figure 4 shows values, which characterize required bandwidth B [bps] for data transmission from client to server B_{C-S} [bps] and from server to client B_{S-C} [bps]. There are all three tests. It is especially important to know that in the 1st case real data are transmitted while in the 2nd case only pictures (changes of vector) are transmitted, which are encoded in the packets. It is clear from the graph that the required bandwidth for the 2nd case is much lower. The smallest difference is in the 2nd test at B_{S-C} [bps], where in the second case of the cloud solution the required bandwidth B for data transmission is about a half. However, even in this case, the bandwidth is much more efficient than

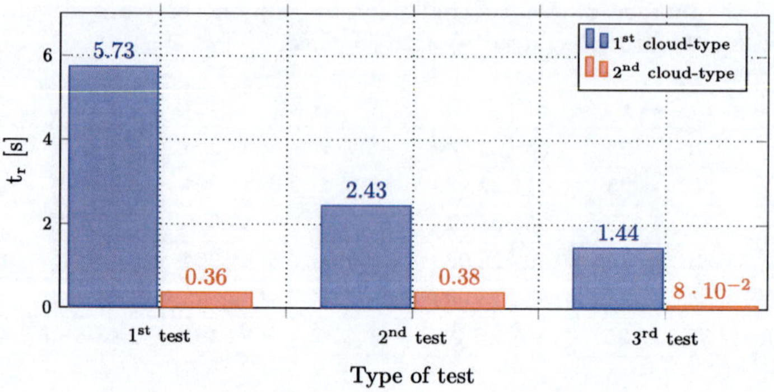

Fig. 3. Graph of the time dependence of the form rendering and writing data for both cases in all three tests.

in the 1st case. Figure 5 shows transmitted data for all three tests and for both cases. This graph also shows that it is much more efficient to use the connection to the application via "Terminal Services". The difference was especially evident in the first test for data transferred from the server to the client B_{S-C} [bps], where the difference in data transfer was up to 80 times smaller.

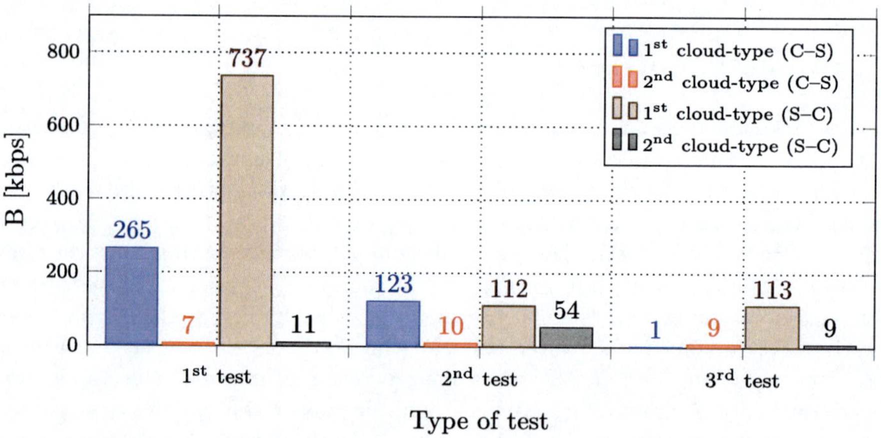

Fig. 4. Comparison of the required bandwidth for data transferred over the network for both cases in all tests.

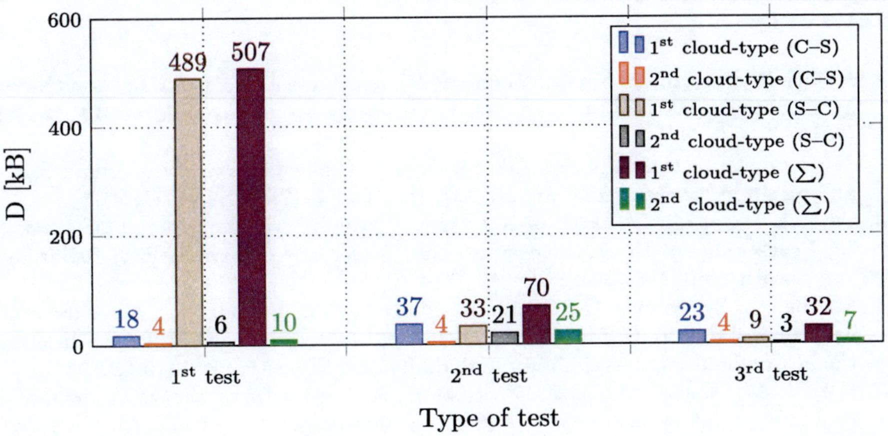

Fig. 5. Comparison of the amount of data transferred over the network for both types of cloud in all tests.

5 Conclusion

After finishing all tests we can conclude, that in terms of application speed it is much more efficient when all applications components are located in the cloud (2nd case). Connecting to this application is also faster by "Terminal Services". With using this technology all computing power is required on the server side, so the client can use a computer with smaller computing power, and it will not affect the speed of the application. The application is only rendered on the client side. The centralized management is the most important advantage of this solution. It is possible to update all applications in one place and even if the clients connect to the application from all over the world, they will always have up-to-date data and there is no problem of replication of the database. This leads to a decrease of cost of managing the IT infrastructure and it is possible to access big data with using only a small internet bandwidth. All components of the application must be installed only in the server in the cloud, so clients cannot have installed special components like 'crystal report" or ".net framework". They need to have only an application for cloud access. The aim of this research was to find more economical solutions of both cases to maximize the performance of individual devices.

Acknowledgment. The article was created with the support of the KEGA agency project "Utilization of Web-based Training and Learning Systems at the Development of New Educational Programs in the Area of Optical Wireless Technologies" – 034STU-4/2021.

References

1. Kajati, E., Papcun, P., Liu, C., Zhong, R.Y., Koziorek, J., Zolotova, I.: Cloud based cyber-physical systems: network evaluation study. Adv. Eng. Inform. **42**, 100988 (2019)
2. Jalowiczor, J., Rozhon, J., Voznak, M.: Study of the efficiency of fog computing in an optimized lorawan cloud architecture. Sensors **21**(9), 3159 (2021)
3. Fujdiak, R., et al.: On track of sigfox confidentiality with end-to-end encryption. In: Proceedings of the 13th International Conference on Availability, Reliability and Security, pp. 1–6 (2018)
4. Malina, L., Srivastava, G., Dzurenda, P., Hajny, J., Fujdiak, R.: A secure publish/subscribe protocol for internet of things, In: Proceedings of the 14th International Conference on Availability, Reliability and Security, pp. 1–10 (2019)
5. Voznak, M., Slachta, J., Rezac, F., Rozhon, J.: Exploratory analysis of virtualization tools in delay-sensitive environment. Commun. Sci. Lett. Univ. Zilina **15**(2A), 150–154 (2013)
6. Pekar, A., Mocnej, J., Seah, W.K., Zolotova, I.: Application domain-based overview of IOT network traffic characteristics. ACM Comput. Surv. (CSUR) **53**(4), 1–33 (2020)
7. Mlynek, P., Misurec, J., Koutny, M., Fujdiak, R., Jedlicka, T.: Analysis and experimental evaluation of power line transmission parameters for power line communication. Meas. Sci. Rev. **15**(2), 64 (2015)
8. Mlynek, P., Fujdiak, R., Misurec, J., Slacik, J.: Experimental measurements of noise influence on narrowband power line communication, In: 2016 8th International Congress on Ultra Modern Telecommunications and Control Systems and Workshops (ICUMT). IEEE, pp. 94–100 (2016)
9. Mlynek, P., Hasirci, Z., Misurec, J., Fujdiak, R.: Analysis of channel transfer functions in power line communication system for smart metering and home area network. Adv. Electr. Comput. Eng. **16**(4), 51–56 (2016)
10. Masek, P., Fujdiak, R., Zeman, K., Hosek, J., Muthanna, A.: Remote networking technology for iot: Cloud-based access for alljoyn-enabled devices. In: 2016 18th Conference of Open Innovations Association and Seminar on Information Security and Protection of Information Technology (FRUCT-ISPIT)
11. Mlynek, P., et al.: Performance testing and methodology for evaluation of power line communication. Elektronika ir Elektrotechnika **24**(3), 88–95 (2018)
12. Mlynek, P., Misurec, J., Kolka, Z., Slacik, J., Fujdiak, R.: Narrowband power line communication for smart metering and street lighting control. IFAC-PapersOnLine **48**(4), 215–219 (2015)
13. Fujdiak, R., Masek, P., Mlynek, P., Misurec, J., Olshannikova, E.: Using genetic algorithm for advanced municipal waste collection in smart city. In: 2016 10th International symposium on communication systems, networks and digital signal processing (CSNDSP). IEEE, pp. 1–6 (2016)
14. Fujdiak, R., Mlynek, P., Misurec, J., Slacik, J.: Simulation of intelligent public light system in smart city. In: 2017 Progress In Electromagnetics Research Symposium-Spring (PIERS). IEEE, pp. 2515–2519 (2017)
15. Fujdiak, R., Masek, P., Mlynek, P., Misurec, J., Muthanna, A.: Advanced optimization method for improving the urban traffic management. In: 2016 18th Conference of Open Innovations Association and Seminar on Information Security and Protection of Information Technology (FRUCT-ISPIT). IEEE, pp. 48–53 (2016)

16. Mlynek, P., Misurec, J., Fujdiak, R., Kolka, Z., Pospichal, L.: Heterogeneous networks for smart metering-power line and radio communication. Elektronika ir Elektrotechnika **21**(2), 85–92 (2015)
17. Krivtsova, I., et al.: Implementing a broadcast storm attack on a mission-critical wireless sensor network. In: Mamatas, L., Matta, I., Papadimitriou, P., Koucheryavy, Y. (eds.) WWIC 2016. LNCS, vol. 9674, pp. 297–308. Springer, Cham (2016). https://doi.org/10.1007/978-3-319-33936-8_23
18. Fujdiak, R., Misurec, J., Mlynek, P., Janer, L.: Cryptograph key distribution with elliptic curve diffie-hellman algorithm in low-power devices for power grids. Rev. Roumaine Sci. Tech. Sér. Méc. Appl. **61**, 84–88 (2016)
19. Raso, O., Mlynek, P., Fujdiak, R., Pospichal, L., Kubicek, P.: Implementation of elliptic curve diffie hellman in ultra-low power microcontroller. In: 2015 38th International Conference on Telecommunications and Signal Processing (TSP). IEEE, pp. 662–666 (2015)
20. Fujdiak, R., Masek, P., Hosek, J., Mlynek, P., Misurec, J.: Efficiency evaluation of different types of cryptography curves on low-power devices. In: 2015 7th International Congress on Ultra Modern Telecommunications and Control Systems and Workshops (ICUMT). IEEE, pp. 269–274 (2015)
21. Fujdiak, R., Mlynek, P., Blazek, P., Barabas, M., Mrnustik, P.: Seeking the relation between performance and security in modern systems: metrics and measures, In: 2018 41st International Conference on Telecommunications and Signal Processing (TSP). IEEE, pp. 1–5 (2018)
22. Rose, R.: Survey of system virtualization techniques (2004)
23. Anderson, C., Griffin, K.: Windows Server 2008 R2 Remote Desktop Services Resource Kit. Pearson Education (2010)
24. Perry, B., Huss, C., Fields, J., McMillan, T.: VCP VMware Certified Professional on vSphere 4 Review Guide: (Exam VCP-410). John Wiley & Sons (2010)
25. Ruest, D., Ruest, N.: Virtualization: A Beginner's Guide. McGraw Hill, McGraw Hill (2009)

Using Pagebuilder to Teach Web Development

Jiri Vojtesek(✉) and Jakub Josef Forman

Faculty of Applied Informatics, Tomas Bata University in Zlin,
Nam. T. G. Masaryka 5555, 760 01 Zlin, Czech Republic
{vojtesek,j_forman}@utb.cz
http://www.utb.cz/fai

Abstract. Web design is one of the branches of IT that is evolving very fast, like most IT fields. At the same time, it is also one of the first things students encounter when learning programming. There are two ways of creating web pages - classic using HTML, PHP, etc., and visual using editors and pagebuilders. This article aims to introduce the reader to web development and its teaching. The report summarizes the main advantages and disadvantages of traditional web development and manual programming and compares them with today's modern methods through pagebuilders in web CMS. The article concludes by answering SEO questions and the advantages and disadvantages of using pagebuilder solutions.

Keywords: Visual pagebuilders · Web design · Programming · Teaching · Shortcode

1 Introduction

The website has been with us since 1993 when everything started with the first website, info.cern.ch. Since then, much has changed in creating websites, and the current trend is to increase the development speed of websites and e-shops constantly. Since modern PHP exists along with HTML, CSS, and JavaScript, many developers are trying to simplify web development and management with many content management systems such as WordPress, Drupal, Wix, Shopify, etc. Today, each of these content management systems includes some form of what is called a "Pagebuilder" that allows you to create text content for your site and create design, logic, animations, and connect other systems. With pagebuilders, the time required to create web pages is reduced to tens of hours. In the case of simple presentations, we can talk for several minutes. [1–3]

2 How to Use Pagebuilders in Teaching at School

To create a website, you need to have visual and aesthetic sensitivity, which can be built-in children from an early age. In web content, visuals are crucial building blocks that determine the success of an article, product, or website and

R. Silhavy (Ed.): CSOC 2022, LNNS 501, pp. 622–631, 2022.
https://doi.org/10.1007/978-3-031-09070-7_51

its content as a whole. Whether a website is appropriately designed graphically and in terms of performance can also be determined from conversion rates, search engine results, and other tools for measuring the success of a website. [4]

Web development is nowadays taught in primary school, where students learn the basics of HTML, CSS, and even some pagebuilders such as Webnode. This teaching is only marginal, and they often forget that they have been through something similar. As a high school and university teacher, I often find that the answer to the question "Have you ever encountered HTML and CSS?" is "I don't know what it is." However, when students are shown a piece of HTML code, they immediately remember that they have already discussed it in primary school. It is not until high school that knowledge of these two technologies deepens and expands to include more advanced features such as PHP and JavaScript. Most students are often not fully skilled in these technologies and can quickly get lost through the inability to visualize everything before seeing the physical result of their code. In the case of using pagebuilders, graphical code can be demonstrated and should be understood by each student through quick, simple, and demonstrative actions of dragging and dropping elements. Compared to traditional teaching, where the student writes complete HTML, CSS, and PHP, there is a need to switch between code and the accurate visualization, unless the student's IDE - programming editor, has preloaded custom browser or hot reload. Imagination in creating such a design plays a significant role in students' understanding of the issues. And mastery of the material discussed. [5]

Most pagebuilders allow you to create elements from individual blocks and modify them using your HTML blocks or CSS (or even JS), and everything is displayed immediately in the browser. In addition, there is no need to manually define all CSS properties, as most elements can already eset with sliders, toggles, and other functional elements. By doing this, you can always show the manual way and the pagebuilder way, which is faster. The student can get a better idea of what's actually going on behind the code they've written, since they can see the changes from both sides and take a deeper dive into the subject.

Today's pagebuilders are trying to be more and more benevolent in building pages. Most of them already offer so many functional and editable blocks that even someone who has never seen HTML or CSS can create a page. Unfortunately, even pagebuilders are not perfect yet. Putting together sites from blocks without knowing the basics of HTML, CSS, and other technologies usually leads to deplorable page and content results on the web. So knowledge of the basics of HTML, CSS, HTML5 standards is always necessary, but practical demonstrations can be started with a minimum of these basics directly in pagebuilders.

3 History of Pagebuilders

Not so long ago, pagebuilders (Visual Page Builders) did not exist. This meant that anyone without knowledge of HTML, CSS, PHP, and JavaScript couldn't create pages because they didn't know-how. Since the days of pagebuilders, this way has changed considerably. Even reputable companies are using this way to create unique web presentations in a matter of hours using these modern tools. Back in the days when these pagebuilders did not exist, everyone had to trust in web developers, trust in their knowledge, experience and invest a lot of money in creating websites. Today, this kind of development is transforming into a web application that content management systems cannot directly manage. Previously, developers had to manually change HTML and CSS for even the slightest change to the site - a task that took much longer than anyone would like to admit. [6]

WYSIWYG editors and "shortcodes" are the first steps to creating pagebuilders in many content management systems. The WYSIWYG editor allows you to make the text content in a similar process as in MS Word or Pages. All content created in this editor is HTML code with an inline CSS structure used to style individual texts. This editor also offers the ability to switch to HTML code and edit the content manually for advanced users. [7]

Shortcodes allow predefined HTML and CSS to generate and modify using attributes automatically. However, only the developers have allowed modifications in this shortcode. One of the most well-known content management systems, WordPress, uses shortcodes in most available text fields (the text field must support shortcode resolving). Once a shortcode is inserted, many HTML, CSS, or even JS are generated based on PHP and content management system features. [8]

Shortcodes are used in most available text fields in the WordPress content management system that a web admin can fill in. The most common use can be found in the editing of "posts" (posts, pages, product, custom posts), widgets (in supported themes), and all other fields designed for this purpose. An example of the use of shortcodes in the WordPress content management system and the built-in WYSIWIG editor can be seen in Fig. 1.

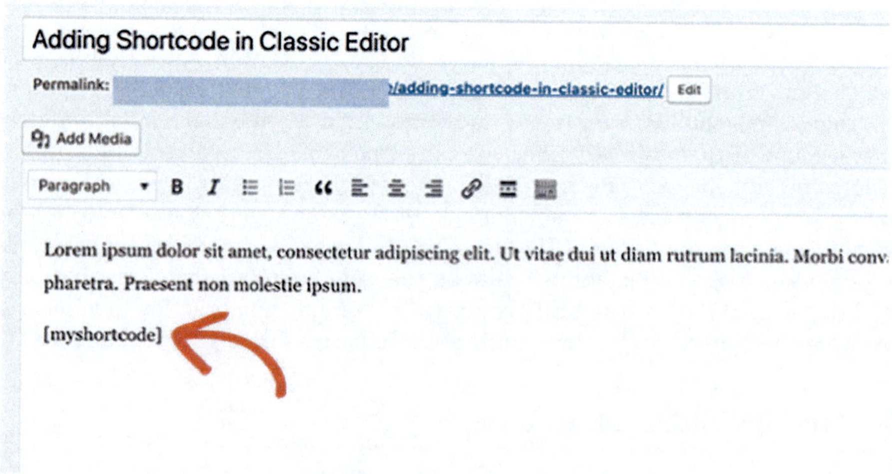

Fig. 1. Example use of shortcodes in WYSIWIG editor [8]

Another example of using the official WordPress Shortcode is the following sample:

```
[gallery ids=''47 ,86, 92, 64, 48, 75, 89, 80'' columns=''4'' size=''medium'']
```

This piece of shortcode is not translated on the administration side and is displayed as code. The post, article, and page side are already translated into their functional form. In this case, the Lightbox gallery. So, the shortcode executes a defined function in PHP that extracts the content from the inserted attributes and uses it to generate content, which can be seen in Fig. 2.

Fig. 2. Generated content using shortcodes [9]

The real revolution came when the first pagebuilders were created. A new modular block system partially replaced the original shortcodes, allowing you to create content more clearly.

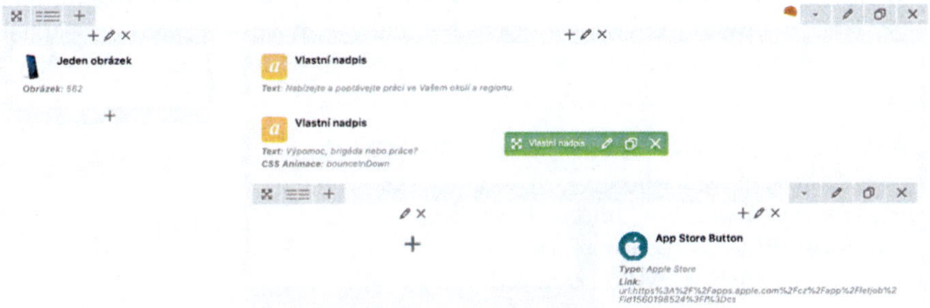

Fig. 3. Demo of pagebuilder: WPBakery

There are many pagebuilders for different content management systems, but in the most popular content management system, WordPress, the most famous are Beaver, WPBakery, Elementor, Guttenberg. The advantage is that many of these pagebuilders build on the open standards of GrapesJS or PHPageBuilder, allowing site builders to extend their functionality. Nowadays, pagebuilders also focus on design in the form of a visual page builder which leads to even more speed in web development and has taken away the need for a lot of imagination (Fig. 3). [10,11]

This allows web developers to create more sites a fraction of the time (opening the way for more customers and better funding opportunities). Moreover, beginners and budding entrepreneurs don't rely on overpriced developers. They can create content/graphics for a website using drag and drop elements or build a website from any template. There are thousands of these templates for free, and their appearance and elements can be easily modified without any knowledge of HTML, CSS, PHP, or JS. Items such as carousels (carousels), images, videos, columns, product galleries, portfolios, complex animations, and more can be easily created. [6]

Through the possibility of extending existing pagebuilders with custom code, pagebuilders can be developed with features that were missing before and that the developers missed. [12] These shortcomings can also be used in teaching web development. Not only can students learn the basic knowledge of pagebuilders, but they can also create other valuable tools for pagebuilders that will make their web development even faster.

4 SEO and Web Speed

Every pagebuilder (those designed for content management systems) already have basic SEO – "search engine optimization" or is ready to work with other popular content management system extensions for SEO. By optimizing the site, the administrator can also achieve that customers will stay longer on the site, increase brand awareness, improve conversion rate, increase profits, etc. SEO is a set of recommendations and methods that can get your business started online. So SEO is an integral part of all websites that want to breakthrough. There are a large number of techniques for proper SEO, but they all agree on what is necessary:

- Multimedia sizes and naming, alternative subtitles.
- Size of the site as a whole.
- Load speed (TTFB) - Time to First Byte
- Optimization for screen readers.
- Compliance with HTML5 standard.
- Unique content and visuals.

Most pagebuilders are already SEO ready or offer the option to link to other extensions. An example of linking dozens of popular extensions for WordPress would be the Elementor pagebuilder [13]. This pagebuilder contains one of the most numerous third-party integrations, services, and other extensions in the form of plugins or built-in functions that allow you to establish communication with a service. Connections between extensions to develop custom functionality for pagebuilders can be achieved from 2 sides. The developer modifies their extension to be compatible with Elementor or another pagebuilder, or the pagebuilder developers change the pagebuilder to link to another developer's extension. An example of an SEO extension that extends the functionality of pagebuilders is one of the most advanced and modern SEO tools, RankMath [14]. This tool extends, for example, pagebuilder Elementor with page scanning capabilities and automatic whispering of keywords, descriptions from individual blocks, or the need to fix a non-compliant site structure (Fig. 4).

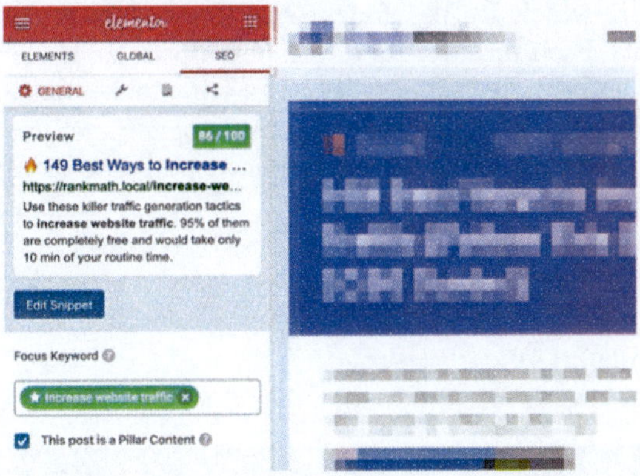

Fig. 4. Example of the pagebuilder elementor [14]

Pagebuilder speed is a pretty crucial factor in creating any web content. By the very nature of speed, it's clear that the fastest pages will be those written manually in HTML and CSS only, without any additional server-burdening calculations. Pagebuilders aim to generate valid HTML page content and speed up server responses to the maximum content management systems allow [15].

What are examples of SEO and speed optimizations in different pagebuilders:

- Dynamic HTML and CSS generation.
- Inserting CSS directly into HTML (omitting ¡link¿ tags).
- Deleting unnecessary code.
- Using browser and content management system caches.
- Minification of HTML, CSS, and JS - Minification deletes all unnecessary characters that the browser doesn't need for rendering and has to skip when reading the code. These are new lines, spaces, or comments.
- Loading dynamic content from the memory cache.
- Using browser libraries instead of server-side libraries.
- Loading external libraries that are actually used on the page.
- Responsive adaptation of graphical elements and their sizes for different resolutions.
- Local and external font loading.
- Lazy loading of images, videos and other content.
- Support for SVG, WebP, and other next-gen asset formats.

- DOM optimization - interface allowing JS access to HTML elements.
- Use of CDN - Content Delivery Network is the term for a collection of servers in different territories around the world, allowing content to be distributed as quickly as possible to people accessing from other parts of the world [16].

Each of the pagebuilders tries to approach web optimization differently. They are very similar techniques, where data is retrieved from high-speed servers, CDN, and cache memory usage. The pioneer in speedup capabilities is the pagebuilder Elementor, which is just four years of existence has been on over 6 million pages. With each update, its developers optimize the code to be smaller and faster. [17]

This speed improvement can be verified in Google PageSpeed Insights, where pages designed in Elementor by an experienced person can range from 95–100 points for desktops and 85–95 for phones and tablets. These numbers are getting closer and closer to 100 points [18]. Google PageSpeed Insights is one of the most frequently used SEO tools to determine page speed and optimize it on devices. It allows you to view error message details and essential tips for page authors to fix and optimize them (Fig. 5). [19]

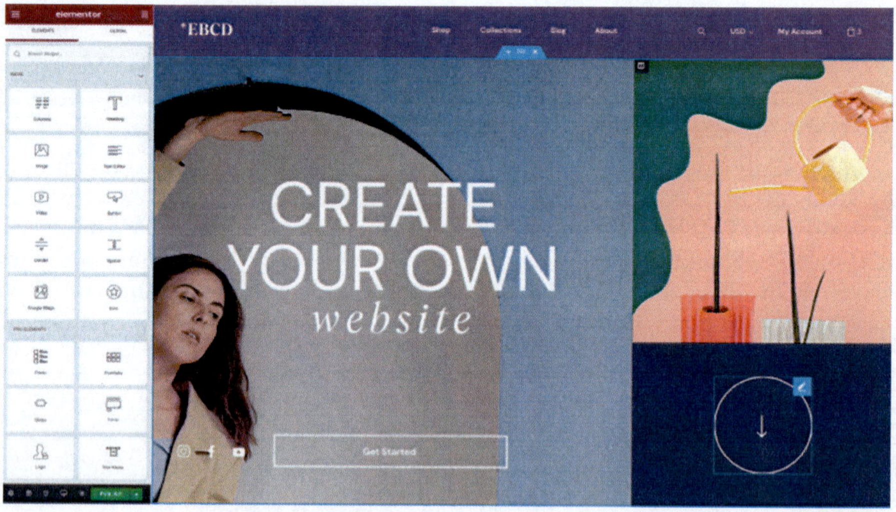

Fig. 5. Pagebuilder Elementor demo [13]

5 Claims and Concerns

With the help of modern technology, all pagebuilders have been one of the most significant advances in web content creation in the last decade.

List of the essential benefits that pagebuilders bring:

- Coding/programming is often a complex process and is constantly evolving. Pagebuilders are replacing this step. Instead of writing code, one can implement blocks to make the page according to the required criteria.
- In every pagebuilder, hundreds of predefined templates can be used and modified. So, you can create a page based on a template that is just customized according to the requirements or create a unique template from elements that can be used in each pagebuilder.
- Most pagebuilders are designed to be user-friendly and can be used by almost anyone to create content on the web. You don't need experience, a degree, or prior knowledge in the field to get started.
- Through element galleries sometimes containing over a thousand blocks, any website template can be customized to be unique and not similar to other sites on the Internet - this fact significantly affects SEO.

List of the most significant disadvantages that pagebuilders bring:

- Some pagebuilders are enormous and can slow down the web through necessary server-side calculations. Whoever builds the site has to think about the server's memory capacity, what content can be on the page and how fast it can be rendered. When limitations are not assumed of, the site often slows down, and potential customers leave the site even if it has the best content or graphics.
- All pagebuilders allow you to use up to hundreds of fonts on a page. Still, some are not optimized for specific languages or browsers and can generate dozens of errors, which in turn slow down the site's loading, prevent people from viewing the site or lead to problems with other established modules.
- Very few pagebuilders are free, and the free options are often not usable. With annual updates, prices usually range from $30 to $150 per domain. However, this limitation is much cheaper compared to the cost of custom development [20].

6 List of the Most Used Pagebuilders

There are many pagebuilders out there, and more are being added all the time. Still, this article focuses on pagebuilders for content management systems to ensure the highest level of future sustainability in a transition to other technologies [6].

Here is the list of most famous ones:

- Elementor
- WBBakery
- GeneratePress
- Beaver Builder
- Divi

- WP Page Builder
- Oxygen
- Brizy

Most of the pagebuilders mentioned above offer Free or Demo licenses to try. For testing or learning the basics this model is sufficient, but for real use it lacks a large number of functionalities that speed up web development and therefore it is necessary to choose the paid versions, which offer a large number of additional useful features.

7 Conclusion

Pagebuilders have many positive characteristics and testimonials that can be traced on the Internet. Their advanced use for creating web content and graphics makes them very convenient for anyone who wants to create some content on the web. Pagebuilders solve the problems of appearance and quick implementation of complex features and the interconnection of other systems, which are often complicated and cost a lot of financial resources. A comparison of creating websites using the traditional method and the modern method using pagebuilders demonstrates an option that will make students more interested when they open the editor.

References

1. Home of the first website. http://info.cern.ch
2. Should you use a page builder on your WordPress website? https://hollypryce.com/page-builder/
3. 4 Key Advantages of Using a WordPress Page Builder on your website. https://undsgn.com/wordpress-page-builder/
4. Average SEO Conversion Rate By Page Type. https://firstpagesage.com/seo-blog/average-seo-conversion-rate-by-page-type-fc/
5. Widgets Option in Elementor Pro and the Creation of Plugins in Wordpress CMS. https://digilib.k.utb.cz/handle/10563/46136
6. 13 Awesome WordPress Page Builders to Cut Out Coding. https://kinsta.com/blog/wordpress-page-builders/
7. What Is a WYSIWYG Editor?. https://www.howtogeek.com/752396/what-is-a-wysiwyg-editor/
8. How to Add a Shortcode in WordPress? (Beginner's Guide). https://www.wpbeginner.com/wp-tutorials/how-to-add-a-shortcode-in-wordpress/
9. Gallery Shortcode. https://codex.wordpress.org/Gallery_Shortcode
10. Free and Open Source Web Builder Framework. https://grapesjs.com
11. PHPagebuilder. https://github.com/HansSchouten/PHPagebuilder
12. The Information You Need for the Elementor You Want. https://developers.elementor.com
13. Connect Your Favorite Tools. https://elementor.com/features/integrations/
14. Swiss Army Knife of WordPress SEO. https://rankmath.com/wordpress/plugin/seo-suite/

15. Create Fast-Loading Websites With Ease. https://elementor.com/speed-performance/
16. What is a Content Delivery Network (CDN)? And When Does it Make Sense to Use One?. https://raidboxes.io/cs/blog/hosting-performance/cdn-content-delivery-network/
17. How Elementor Prioritizes and Improves Websites Performance. https://elementor.com/blog/how-elementor-prioritizes-and-improves-website-performance/
18. Elementor PageSpeed Insights - 26/100 to 92/100 in 5 Minutes. https://sagapixel.com/web-design/elementor-pagespeed-insights/
19. About PageSpeed Insights. https://developers.google.com/speed/docs/insights/v5/about
20. What Are the Pros and Cons of Using a Page Builder in 2022?. https://www.wpcity.com/pros-cons-using-page-builder/

Machine Learning Algorithm for Survival Prediction Linked to Clinical Outcome of Serous Ovarian Cancer

Varvara N. Zhurman(✉), N. G. Plekhova, and I. N. Chernenko

Pacific State Medical University, 2 Ostryakova Avenue, Vladivostok 690002, Russia
varvara2007@yandex.ru

Abstract. The aim of this study was to develop a machine learning algorithm to determine prognostic parameters that allow to categorize of patients with serous epithelial ovarian cancer according to survival. Information about 890 patients with serous ovarian cancer from the Primorsky Regional Oncological Dispensary was analyzed using statistical, regression and correlation methods. When we compiling the machine learning algorithm we used the feature selection method as Minimum Redundancy – Maximum Relevance (MRMR) and Random Forest. The Kaplan-Meier method was used to analyze patient survival. Results: High negative mean correlations between the presence of pregnancy, pathology of the gastrointestinal tract, the presence of metastases, the location of the tumor and the overall survival of patients were determined. Using the Random Forest method, out of 52 indicators, the greatest contribution to the formation of a regression model of the survival rate of the stage of the disease according to the classification of the International Federation of Gynecology and Obstetrics (FIGO) (until 17,5%), then at an unspecified degree (14,9%), stable condition after surgery (8,8%), type of cytoreduction complete (8%) and primary (6,5%) and surgery without staging (5,4%) was determined. The importance of the type of cytoreductive surgery, respectively, complete and incomplete, in the prognostic model of patient survival has been established. The machine learning algorithm proposed by us makes it possible to single out from a variety of factors accompanying the development of serous ovarian cancer those indicators that have prognostic value in determining patient survival. This machine learning algorithm may be useful to more accurately assess individual outcomes and select the best treatment options for patients with serous ovarian cancer.

Keywords: Ovarian cancer · Survival · Correlation · Random Forest · Kaplan Meier method

1 Introduction

The most common malignant neoplasms of the female genital area with a high mortality rate (up to 50.0%) are ovarian cancer. Due to the unexpressed and non-specific symptoms, it manifests itself in a variable clinical picture that is not characteristic of an oncological disease. More than two-thirds of ovarian cancer cases are diagnosed when the disease is

R. Silhavy (Ed.): CSOC 2022, LNNS 501, pp. 632–643, 2022.
https://doi.org/10.1007/978-3-031-09070-7_52

in stage III or IV, when the effectiveness of the treatments used is significantly reduced [1]. As a result, the overall survival (OS) of patients with ovarian cancer continues to be low and mortality is 4.4% of the total worldwide [2]. In Russia, morbidity and mortality, according to the International Agency for Research on Cancer (IARC) in absolute numbers will be 13898 and 8624 cases of ovarian cancer, respectively by 2030. The evaluation of the significance of numerous different markers in the primary diagnosis and the occurrence of recurrence of ovarian tumors, their contribution to the effectiveness of the therapy, as well as the prognostic value for the survival of patients is shown. For patients with ovarian cancer, according to the FIGO recommendations, risk factors include histological type and grade, residual disease status after surgery, and the level of cancer antigen 125 (CA-125) [3, 4]. One of the most commonly used indicators to predict patient survival is the FIGO stage. However, further studies are still being carried out to improve the quality of predictors in predicting the disease in patients, which is necessary for the accurate selection of treatment methods for clinicians.

Prediction methods based on a vast amount of data using intellectual analysis are effectively used in many types of health research, including in determining the effectiveness of cancer treatment [5, 6]. The use of online datasets such as machine learning UCI, SEER [7] and TCGA [8, 9] is shown. Despite the vast number of input parameters in such systems, unfortunately, they do not cover all indicators, such as the ethnicity of patients and their region of residence. For the construction of various forecasting models, Data Mining has a wide range of applications for selecting signs of minimal redundancy. It is based on the process of selecting a subset of significant features (both dependent and independent variables) and allows simplifying models, reducing training time and avoiding the influence of dimension to increase the level of interpretation [2]. Reducing the number of independent variables is designed to reduce the dimension of the model not only in order to remove all insignificant features from it to simplify the model, but also to eliminate redundant features. Building a model on all possible combinations of sets of input features with subsequent selection of the option that would have the best descriptive ability of the resulting feature Y and contain a minimum of independent variables is possible if there is a small number of contender factors for inclusion in the model. In view of this, it is necessary to build algorithms for selecting the most important factors that would require significantly less effort and, accordingly, time. From these positions, the Correlation Feature Selection (CFS) method is often used, which allows evaluating subsets of features based on the hypothesis: «Good feature subsets contain features highly correlated with the classification, yet uncorrelated to each other» [10].

The aims of present study are to identify various predictors for disease prognosis in patients with ovarian cancer. To create a model for predicting patient survival, a complex model is used that includes characteristics from different data profiles. The data of patients of the Primorsky Oncological Dispensary were collected and processed accordingly. An attempt is also made to use machine learning methods to sequentially evaluate the various effects of treatment given to patients with ovarian cancer. These treatments and the estimated time elapsed between treatments provide valuable insight into patient survival. Previously, a relationship was found between the time intervals between courses of treatment and prognosis in patients with ovarian carcinoma [11–13].

2 Materials and Methods

2.1 Data Source

The criteria for selecting patients for inclusion to the study was a morphologically verified diagnosis of epithelial ovarian cancer, who received outpatient and/or inpatient treatment or observation at the Primorsky Regional Oncology Center in the period from 2004 to 2021 (Vladivostok, Russia). The study analyzed the data of 984 patients with stage I–IV epithelial ovarian cancer. The histological type was classified according to the World Health Organization (WHO) criteria. The data was collected manually from files stored digitally in the repository of the cancer center, after obtaining the appropriate permission for use on the basis of anonymity. Ethical and research approvals were provided by the Ethics Committee of the Pacific State Medical University (Agreement No. 51/16.011.2019).

2.2 Data Collection

The parameters for the study included: date of registration and deregistration, age, date of diagnosis, censoring variable (0 - censored patient who withdrew from the study for any reason, but was alive at the time of withdrawal, 1 - uncensored patient with known outcome, death), time of appearance of distant metastases, tumor recurrence (in days), clinical stage, FIGO substage and histological type of tumor, data of molecular genetic analysis for the presence of changes in the nucleotide sequences of the BRCA gene 1 and 2, volume of surgical intervention, types of chemotherapy and radiation therapy, follow-up period after therapy (stabilization, progression, lines of therapy and their number), concomitant diagnoses, gynecological history. In total, 25 variables were analyzed in machine learning models to the base level (excluding dummy variables for missing values). Clinical data showed a high association with patient survival and were the most widely used predictors in existing studies. Unlike online datasets and other clinical studies, the present study looked at the treatments and the corresponding time intervals between each treatment approach given to each patient.

2.3 Survival Analysis

The Kaplan-Meier method was used to analyze patient survival [14]. Analysis of survival/survival (analysis of the occurrence of events) with the involvement of this method allows you to identify the pattern of occurrence of a certain outcome among representatives of the observed sample over a certain period of time. At the same time, survival can be represented as fractions of a unit or as a percentage, and the period of time before the occurrence of the event under study can be anything from seconds to decades. Kaplan-Meier survival analysis serves to evaluate time-to-event patterns in the presence of censored observations. These data characterize the process, which at the time of the last examination does not end with the onset of the studied outcome, and include indicators of patients with oncopathology who were alive at the end of the observation period. The second variant of censored cases is patients who fall out of observation (in case of moving to another city or refusal of observation, etc.). Thus, the multiplier non-parametric method (product-limit method) of Kaplan-Meier estimates the cumulative survival function at the time of occurrence of each case of outcome, which is calculated according to the formula:

$$\hat{S}(t) = \prod_{i=0}^{T} \frac{R_i - d_i}{R_i}$$

so: R_i - the number of objects surviving to a point in time t_i, excluding dropouts; d_i - number of objects for which an outcome occurred at a time t_i, $\frac{d_i}{R_i}$ - outcome probability.

The graphical representation of the Kaplan-Meier method consists in constructing a survival curve that reflects the proportion of patients in whom the expected event did not occur by the end of the observation period. Time intervals determined the frequency of control examinations of patients diagnosed with ovarian cancer. Moreover, the possibility of recalculating the proportion of the objects remaining in the study, for which the event did not occur, is reflected by a «step» down on the curve. The period of time from the initial event to the outcome under investigation is the waiting time or until the event. The database contains the dates of the last contact with the patient for censored observations or the date of the index event (even if the contact itself was later). We choose this method as the basis for the calculation, since the following criteria are met: the beginning of the observation, the moment of establishing the diagnosis; death is defined as the outcome; the period of time from the beginning of the observation to the outcome is known (censored case); the period of time from the beginning to the end of the observation is known, when the studied outcome was not registered (uncensored case); censored observations (dropped out) do not differ in survival from uncensored ones; methods for assessing survival and determining outcome are the same for objects included in the study at an early or later date; conditions that affect survival do not change over time (for example, treatments); the proportion of censored cases until the end of the observation period in the studied groups is approximately the same.

2.4 Machine-Learning Algorithms

Data analysis was carried out in three main stages (Fig. 1): data collection, preliminary processing and classification. Unfortunately, when forming a medical database, many missing and irrelevant indicators are found that cannot be used for classification. Therefore, when performing the second stage, we included a preliminary processing of clinical data in accordance with their profile, using standard imputation methods. For feature selection machine learning in order to rank and identify the most relevant features for the target class with the choice of filter type characteristics, we used the Minimum Redundancy - Maximum Relevance (MRMR) feature selection method proposed by Ding and Peng et al. [15]. Using this method allows you to find the most relevant features for the target classes, and also helps to make sure that these selected features are as different as possible from each other. Random Forest has demonstrated efficacy in clinical applications. This method allows you to predict whether observations or objects belong to a particular class of a categorical variable or the average value of a quantitative variable, depending on the corresponding values of one or more independent variables. It is based on combining a large number of weak simple predictors to get a stronger prediction.

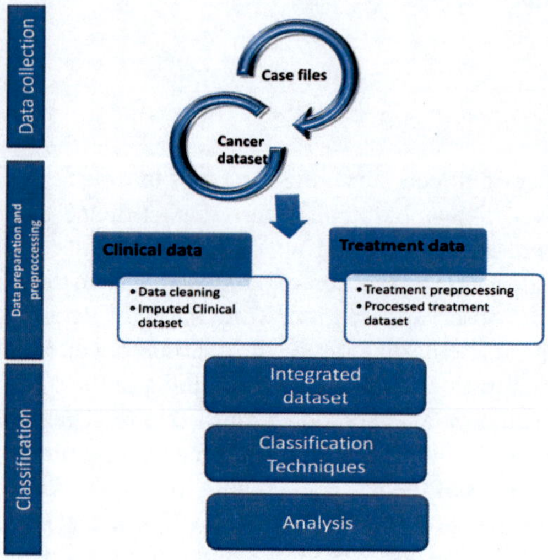

Fig. 1. Flowchart of development of the prediction model for overall survival in ovarian cancer patients.

2.5 Statistical Analysis

A descriptive characterization of the studied population indicators is presented, including the number (%) and mean value (SD) for categorical and continuous variables, respectively. The t-test or the rank-sum tests of two independent samples were used to achieve the comparison between groups. This process provided sensitivity, specificity, positive predictive value (PPV) and negative predictive value (NPV). Statistical analysis evaluating the performance of the algorithm was performed using STATA 13 MP4.

3 Result

3.1 Demographic Analysis

Demographic and clinical characteristics of patients are shown in Table 1. Medical records of 890 patients with serous ovarian cancer treated at the Primorsky Regional Oncology Center were analyzed. There were 264 cancer recurrences and 356 total deaths at a median follow-up of 47 months (range 4 to 177 months). Most patients had stage IIIc-4 FIGO (76%), high-grade FIGO (53.4%). In 139 patients (19.9%), stage IIC-4 had no residual disease after initial volume reduction surgery. In 369 patients (41.5%), no residual disease was observed after the initial volume reduction surgery.

Table 1. Patient demographics and clinical characteristics

n	Characteristics	N (%)
	Demographics	890 (100%)
1	Age at diagnosis, years	53,7 [20;70]
2	*Menarche*	890 (100%)
3	before 12 years of age	52 (5.8%)
4	12 – 15 years old	753 (84.7%)
5	after 15 years	84 (9.5%)
6	*Premenopausal*	40 (4,8%)
7	*Postmenopausal*	447 (54,0%)
8	***Pregnancy***	715 (80,4%)
	Stage	
9	I	92 (9,8%)
10	II	82 (9,2%)
11	IIIa-IIIb	40 (4,4%)
12	IIIc-IV	676 (76 %)
	Histologic subtype	
13	Serous carcinoma	
14	- Low grade	259 (29,1%)
15	- High grade	475 (53,4%)
16	***No data***	156 (17,5%)
	Ovarian involvement	
17	Unilateral	280/31,5%
18	Bilateral	336/37,7%
	Ascites	
19	Yes	29 (3%)
20	No	178 (20%)
	No data	683 (76%)
	Spread of the tumor	
21	pelvis	425 (47,6%)
22	abdominal cavity	381 (43%)
23	***No data***	84 (9,4%)
	Metastases	
24	Yes	93 (10,4%)
25	No	772 (86,8%)
26	***No data***	25 (2,8%)
	Recurrence	
27	Yes	342 (38,4%)
28	No	548 (61,6%)
	Residual disease IIIc-IV	676 (76%)
29	Nil macroscopic	341 (49%)
30	≤ 1 cm	33 (4,7%)
31	≥ 1 cm	206 (43,2%)
32	Macroscopic, size unknown	116 (13%)
	Comorbidity,	
33	presence of diseases	876 (98,4%)
34	the gastrointestinal tract	805 (91,8%)
35	diseases of the endocrine system	614 (70%)
36	obesity	506 (57,7%)
37	diseases of the cardiovascular system	
38	gynecological diseases	

(*continued*)

Table 1. (*continued*)

39	***tumor of another location***	329 (37,5%)
		150 (17,1%)
		106 (12,1%)
	Treatment IIIc-IV	676 (76%)
	Surgical	
40	Primary cytoreduction	463 (68,5%)
41	Interval cytoreduction	213 (31,5%)
42	Types of operations full	160 (23,6%)
43	Types of operations optimal residual (1 cm)	12 (1,7%)
44	Types of operations incomplete	331 (48,9%)
	Chemotherapy (IIIc-IV)	676 (76%)
45	Paclitaxel + cisplatin/ carboplatin with bevacizumab	264 (39%)
46	Paclitaxel + cisplatin/ carboplatin without bevaci-	430 (63.6%)
47	zumab	18 (2,6%)
48	Other schemes	
	Gene mutations	306 (34,4%)
	BRCA 1	
49	Yes	60 (19,6%)
50	No	246 (80,4%)
	BRCA 2	
51	Yes	6 (1,9%)
52	No	300 (98,1%)

3.2 Correlation Analyses

Cancer survival characterizes the number of sick cancers that are alive at the end of a certain time interval. This indicator measures the effectiveness of an approach to cancer treatment, monitoring of cancer control programs and stimulating decision-making in access to care. To assess survival depending on the indicators in the first stage of the study, we determined the degree of influence of various parameters on this indicator. To do this, we calculated the correlation between the survival rate of patients with ovarian cancer and 27 indicators of gynecological history and comorbidities. As shown in Fig. 2A, the highest negative mean correlations were found between pregnancy, chronic gastritis, colon pathology, urolithiasis, and overall survival. The indicators were −0.3, −0.2, −0.1 and −0.09, respectively, which show the presence of a certain constant influence of the state of the body on the course of ovarian cancer.

Other classes of diseases were weakly correlated with patient survival. In addition, we analyzed the indicators of the morphological characteristics of ovarian cancer. As shown in Supplementary Fig. 2B, the presence of metastases, the location of the tumor in the pelvis and upper abdomen has a high correlation of >0.3 with patient survival, suggesting that these indicators are important for the course of the disease. Whereas, the scores for disease stage indicators according to the FIGO classification and others were <0.1. Finally, we calculated the correlation between the type of therapy, as shown in Fig. 2C. The correlation of these indicators was very low, on average below 0.09. In conclusion, it should be noted that the presence of various comorbidities is most closely associated with patient survival.

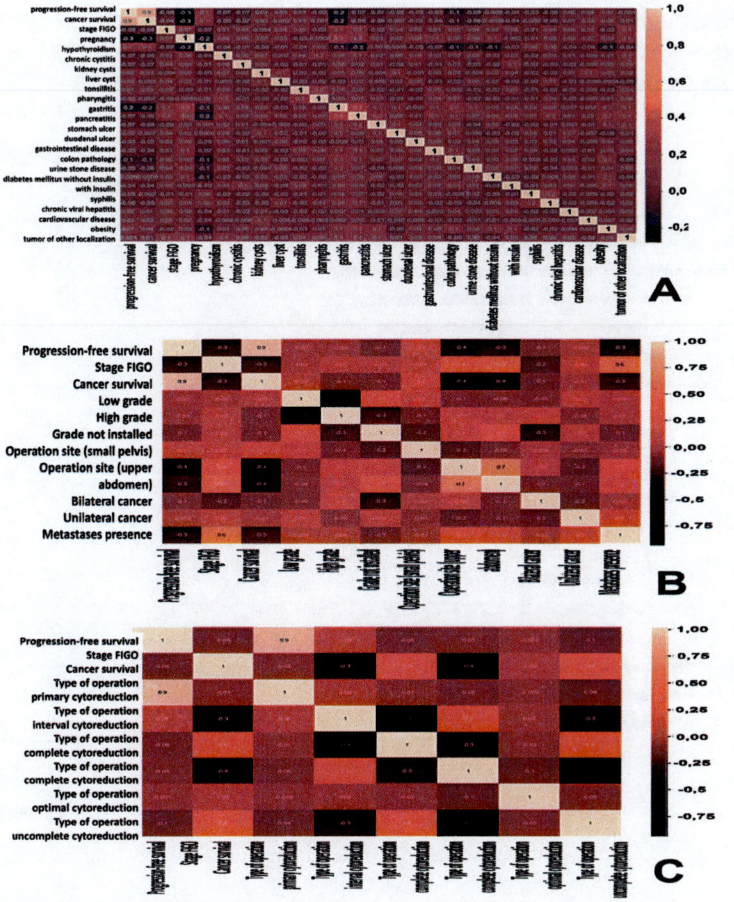

Fig. 2. Relationship between demographic, clinical parameters and survival of patients with serous ovarian cancer.

3.3 Random Forest

To perform the second step of evaluating the importance of a variable on the training dataset, we used the generated random forest training. For each element in the process of machine learning, an error was calculated on non-selected samples (out-of-bag error), the value of which was then averaged. Then, the importance of the parameter was evaluated, for which the obtained values were mixed and the out-of-bag error was calculated again. The importance of the parameter was estimated by averaging the difference in out-of-bag error rates over all trees before and after mixing the values. In this case, the normality of the value of such errors was calculated using the standard deviation. For each decision tree, evaluate the reduction in importance of each variable. The analysis result, by definition, the significance score of indicators for overall survival is shown in Fig. 3. The greatest contribution to the formation of the regression model of survival of the stage of the disease according to the FIGO classification (up to 17.5%) was determined, then with an indeterminate degree according to FIGO (14.9%), stable condition after surgery (8.8%), types of cytoreduction complete (8%) and primary (6.5%) and surgery without staging (5.4%).

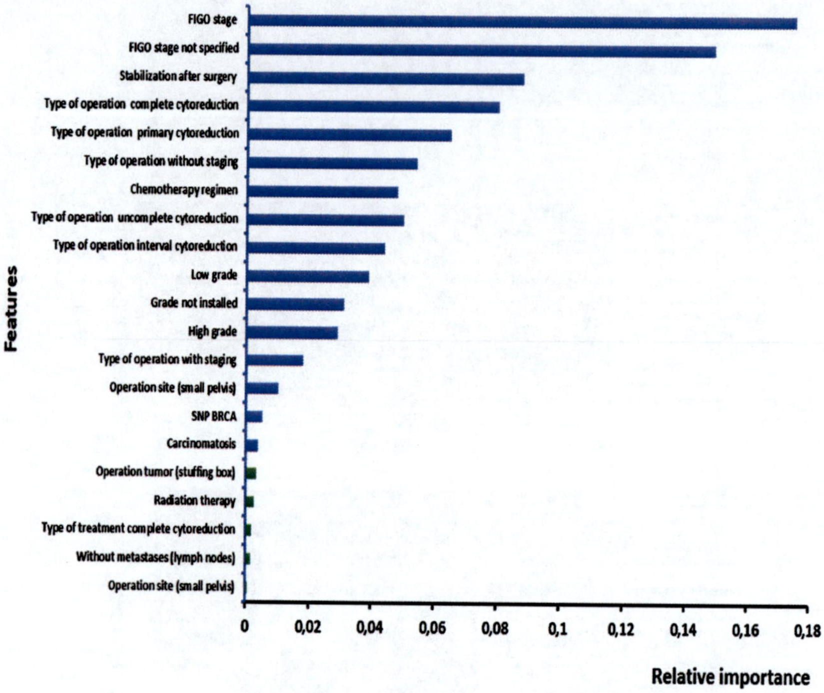

Fig. 3. Diagram of the importance of risk factors, model of the decision trees.

3.4 Prognostic Assessment of Survival

To assess the actual survival of patients with ovarian cancer, including censored data, we used the Kaplan–Meier analysis. The basis of such model is the assessment of conditional probabilities at each moment in time when an event occurs, and taking the limit from the product of these probabilities to assess survival. For calculations, we used the true number of patients in whom an event (outcome) has not yet occurred in the observed time period. As an important variable, we considered the time to the final outcome, according to which we grouped the data according to the FIGO stages diagnosed in patients, as well as the types of surgical operations. These 4 subgroups (A, B, C and D) of data were formed based on the importance of each variable in accordance with the assessment of the likelihood of survival using Kaplan-Meier analysis. As shown in Fig. 4A, OC patients are most likely to survive during the first 5 years, after which the likelihood of a favorable outcome steadily decreases. Meyer curves were clearly classified according to a subgroup, including stages A-I and B-II and separately C-IIIa-IIIb and D-IIIc-IV (Fig. 4C). As a result of log-rank tests, the differences between the curves of subgroups A, B and C, D were significant (p = 0.005), while there were no significant differences between adjacent curves by FIGO stages. We found it interesting to investigate, according to the established importance of indicators in survival, the influence of the type of cytoreductive surgery, respectively, complete and incomplete. The subgroups were distributed as follows: FIGO stages I and II - 4 subgroups (A, B, C and D, Fig. 4C) and IIIa-IIIb and IIIc-IV also (A, B, C and D, Fig. 4D). The curves of subgroups A and C (p = 0.0482) differed significantly in stages IIIa-IIIb and IIIc-IV of the disease, which indicates the reliability of the effect of cytoreduction in these patients.

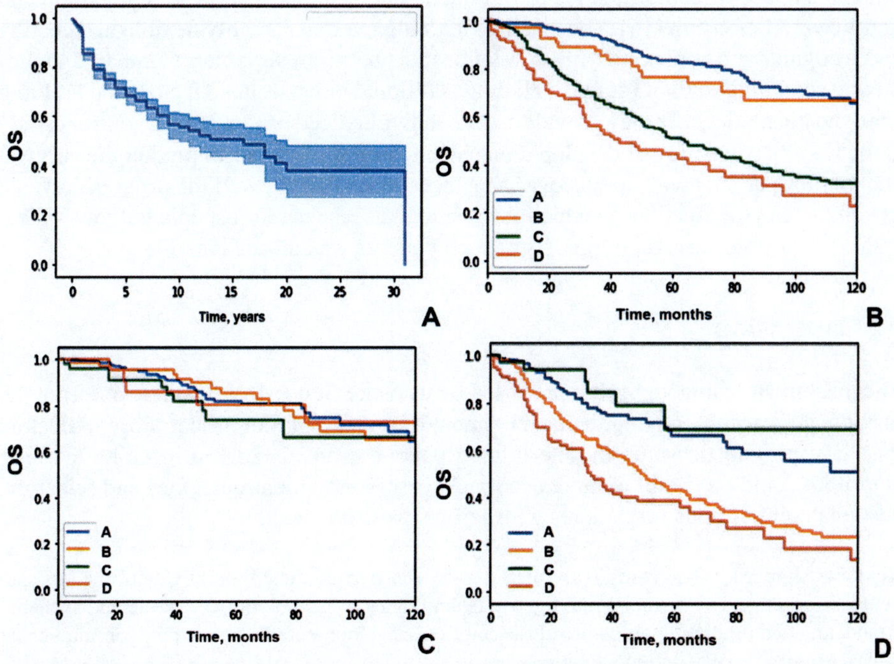

Fig. 4. Kaplan-Meier survival curves for overall survival (OS). A - general survival to the thirty-year; B – ubgroup according to FIGO stage (I, II, III, and IV) in the validation cohort based on the subgroup (A, B, C, and D); ubgroup I and II stage (C) III, and IV (D) in the validation cohort based on the subgroup complete and incomplete cytoreduction (A, B, C, and D) according to the ten-year OS probability scores predicted. FIGO, International Federation of Gynecology and Obstetrics; OS, overall survival.

4 Discussion

To analyze the effectiveness of the organization of oncological care, a population assessment of overall survival and progression-free survival of patients is carried out based on the data of the cancer registry, which has information about patients registered on its territory and has timely and complete information about their vital status. Unfortunately, these data do not fully reflect the impact of various survival parameters in ovarian cancer. In our study, we conducted a prognostic assessment of survival. To assess the actual survival of patients with ovarian cancer, including censored data, we applied the method. In this study, we applied Kaplan–Meier analysis as a machine learning classifier to predict the survival of patients with serous ovarian cancer using sequential indicators. We have demonstrated that the machine learning algorithm allows us to more accurately calculate the predictive power of indicators on the prognosis of patient survival. Moreover, staging revealed four distinct subgroups that were more clearly classified than those using only FIGO stages. An accurate prediction for an individual patient is still difficult using traditional statistical methods, since most clinical characteristics show a multivariate and non-linear relationship [16]. The machine learning method is highly flexible in detecting and recognizing complex non-linear relationships between variables. Compared to predicting according to FIGO stage, including additional tumor-related factors and building a prognostic model later may provide a more individualized prognosis. The current study is the first staging study to develop a machine learning classifier to predict the survival endpoint of patients with serous ovarian cancer. Thus, we showed the prognostic value of complete cytoreduction, in which the residual disease is microscopic without visible disease, since the survival prognosis of such patients was more favorable.

5 Conclusion

The machine learning algorithm proposed by us makes it possible to single out from the many factors accompanying the development of serous ovarian cancer those indicators that have prognostic value in determining patient survival. Such a machine learning algorithm could be useful to more accurately assess individual outcomes and select the best treatment options for patients with serous ovarian cancer.

Acknowledgments. The study was supported by the part of the Ministry Health the Russian Federation state task «Artificial intelligence technologies in phenotyping of tissue and systemic remodeling and predicting outcomes at the stages of development of chronic non-communicable diseases in people of different ethnic groups».

Declaration of Financial and Other Relationships. All authors participated in the development of the concept, the design of the study and the writing of the manuscript. The final version of the manuscript was approved by all authors.

References

1. Mizuno, M., Kajiyama, H., Shibata, K., Mizuno, K., Kawai, M., Nagasaka, T., et al.: Prognostic value of histological type in stage IV ovarian carcinoma: a retrospective analysis of 223 patients. Br. J. Cancer **112**, 1376–1383 (2015)
2. Torre, L.A., Trabert, B., DeSantis, C.E., Miller, K.D., Samimi, G., Runowicz, C.D.: Ovarian cancer statistics. CA Cancer J. Clin. **68**, 284–296 (2018)

3. Bendifallah, S., Body, G., Daraï, E., Ouldamer, L.: Diagnostic and prognostic value of tumor markers, scores (clinical and biological) algorithms, in front of an ovarian mass suspected of an epithelial ovarian cancer: Article drafted from the French Guidelines in oncology entitled "Initial management of patients with epithelial ovarian cancer" developed by FRANCOGYN, CNGOF, SFOG, GINECO-ARCAGY under the aegis of CNGOF and endorsed by INCa. Gynecologie Obstetrique Fertilite Senologie **47**(2), 134–154 (2019). https://doi.org/10.1016/j.gofs.2018.12.013
4. Paik, E.S., Lee, Y.Y., Lee, E.J., Choi, C.H., Kim, T.J., Lee, J.W., et al.: Survival analysis of revised 2013 FIGO staging classification of epithelial ovarian cancer and comparison with previous FIGO staging classification. Obstet. Gynecol. Sci. **58**, 124–134 (2015)
5. Zhordania, K.I., Kalinicheva, E.V., Moiseev, A.A.: Ovarian cancer: epidemiology, morphology and histogenesis. Oncogynecology **3**(23), 26–32 (2017)
6. Mosayebi, A., Mojaradi, B., Bonyadi Naeini, A., Khodadad Hosseini, S.H.: Modeling and comparing data mining algorithms for prediction of recurrence of breast cancer. PLoS ONE **15**(10), e0237658 (2020). https://doi.org/10.1371/journal.pone.0237658
7. Kaur, I., et al.: An integrated approach for cancer survival prediction using data mining techniques. Comput. Intell. Neurosci. **2021**, 6342226 (2021). https://doi.org/10.1155/2021/6342226
8. Surveillance, epidemiology, and end results (SEER) Program (1975). https://www.seer.cancer.gov
9. Network, T. R. The Cancer Genome Atlas Data Portal, National Institute of Health, Maryland, USA, 2010
10. Gareth, J., Witten, D., Hastie, T., Tibshirani, R.: An Introduction to Statistical Learning. Springer, Cham (2013). https://doi.org/10.1007/978-1-4614-7138-7
11. Senliol, B., Gulgezen, G., Yu, L., Cataltepe, Z.: Fast correlation based filter (FCBF) with a different search strategy. In: 23rd International Symposium on ISCIS 2008. IEEE, pp. 1–4 (2008)
12. Fagotti, A., et al.: Randomized trial of primary debulking surgery versus neoadjuvant chemotherapy for advanced epithelial ovarian cancer (SCORPION-NCT01461850). Int. J. Gynecol. Cancer Official J. Int. Gynecol. Cancer Soc. **30**(11), 1657–1664 (2020). https://doi.org/10.1136/ijgc-2020-001640
13. Kawakami, E., et al.: Application of artificial intelligence for preoperative diagnostic and prognostic prediction in Epithelial ovarian cancer based on blood biomarkers. Clin. Cancer Res. Official J. Am. Assoc. Cancer Res. **25**(10), 3006–3015 (2019). https://doi.org/10.1158/1078-0432.CCR-18-3378
14. Shen, S., et al.: Development and validation of an immune gene-set based Prognostic signature in ovarian cancer. EBioMedicine **40**, 318–326 (2019). https://doi.org/10.1016/j.ebiom.2018.12.054
15. Ding, C., Peng, H.: Minimum redundancy feature selection from microarray gene expression data. In: Computational Systems Bioinformatics CSB2003 Proceedings of the 2003 IEEE Bioinformatics Conference CSB2003, 11–14 August 2003, pp. 523–528 (2003)
16. Zhu, L., Luo, W., Su, M., Wei, H., Wei, J., Zhang, X., et al.: Comparison between artificial neural network and Cox regression model in predicting the survival rate of gastric cancer patients. Biomed. Rep. **1**, 757–760 (2013)

Maximizing Business Potential by Utilizing FinTech Innovative Technologies and Strategies

Ahmed Maghawry(✉) and Amr Elhadidi

EFinance, Cairo, Egypt
{Ahmed_Maghawry,Amr_Elhadidi}@efinance.com.eg

Abstract. The digital transformation revolution is surging, transforming almost all aspects of our lives from manually operated processes into digital ones. Nowadays, business success and expansion ability are highly influenced by the continuous adoption of technology and the ability to embrace new ones. Financial technology (FinTech) is the concept of delivering financial services through technology and innovation away from traditional methods of finance. However, an efficient FinTech firm requires not only technology, but also the understand of laws and regulations governing this domain. FinTech innovative technologies and strategies include the use of automation and algorithmic techniques to fulfill the needs of businesses. The utilization of such methods highly increases the productivity of organizations and indicate their quality. In this research, the targeted financial organization utilized a collection of methodologies that helped to extend its business domain introduced the concept of financial inclusion and increased productivity. The results include customer business enhancement with maximum benefits and minimum resources as introduced in the full-scale commodity distribution management systems. Also introduced a complete electronic payment solution with electronic settlement. In addition to payment gateway module to maximize customer payment integration. And finally, a unified billing gateway to maximize customer payment integration flexibility for customers with weak or no technology background.

Keywords: Financial technology · Digital transformation

1 Introduction

The financial services market is moving very fast with a potential that makes it very hard with conventional financial services firms to keep up with. Such firms will suffer unless they change their mind set to embrace FinTech innovative technologies and strategies. Such technologies and strategies range from the automatic stock exchange trading algorithms all the way up to the automated algorithmic Backoffice settlement processes. The utilization of such technologies helped many businesses to extend their targeted domain of customers and also to evolve their business domain [1]. Financial technology is the techniques, methodologies and innovations utilized to provide financial services competing with traditional financial methods [2, 3]. It can also be considered as an emerging

R. Silhavy (Ed.): CSOC 2022, LNNS 501, pp. 644–656, 2022.
https://doi.org/10.1007/978-3-031-09070-7_53

industry the utilizes technology to improve financial activities. There exist a variety of technologies that aims to make financial services accessible to the general public, including mobile banking apps, investment apps, cryptocurrencies and more. Financial technology firms include either startups or established firms as well as technology firms that are constantly attempting to replace or enhance the current market of traditional financial services. Many scientific studies attempted to define the term FinTech from different perspectives, and generally it was defined as an emerging financial industry that utilizes technology to enhance financial activities [4]. It is also represented in the market by the innovative applications, processes, products, or business models utilized in the financial services industry and introduced as one or more financial services provided as an end-to-end process through the Internet [5]. FinTech firms utilizes a variety of technologies to reach their goals, such technologies include the use of artificial intelligence (AI) techniques, big data methodologies, robotic process automation (RPA), and also the blockchain technology. Artificial intelligence algorithms may provide insights on customer behavior to enable financial firms to understand their customers behavior. Furthermore, Chatbots are another example of artificial intelligence inspired utility that the banking sector utilizing to support customer service [6]. The science of big data can be utilized for the prediction of customer investments and market direction to support financial firms to create new efficient strategies, analyze client interest behavior, enhance fraud detection [7]. The concept of robotic process automation (RPA) is an AI based technology that aims to automate repetitive processes that are usually done manually [8]. RPAs process financial related information more efficiently than manual processing and usually more accurately [9].

2 Background

2.1 Financial Technology (FinTech)

Financial technology (Fintech) is used to describe new tech that seeks to improve and automate the delivery and use of financial services. At its core, fintech is utilized to help companies, business owners and consumers better manage their financial operations, processes, and lives by utilizing specialized software and algorithms that are used on computers and, increasingly, smartphones. Fintech, the word, is a combination of "financial technology". When fintech emerged in the 21st Century, the term was initially applied to the technology employed at the back-end systems of established financial institutions. Since then, however, there has been a shift to more consumer-oriented services and therefore a more consumer-oriented definition. Fintech now includes different sectors and industries such as education, retail banking, fundraising and non-profit, and investment management to name a few. Fintech also includes the development and use of crypto currencies such as Bitcoin. While that segment of fintech may see the most headlines, the big money still lies in the traditional global banking industry and its multi-trillion-dollar market capitalization.

2.2 Financial Inclusion

Financial inclusion refers to efforts to make financial products and services accessible and affordable to all individuals and businesses, regardless of their personal net worth or company size. Financial inclusion strives to remove the barriers that exclude people from participating in the financial sector and using these services to improve their lives. It is also called inclusive finance. It facilitates day-to-day living, and helps families and businesses plan for everything from long-term goals to unexpected emergencies. Also, it improves the overall quality of life as people are more likely to use other financial services, such as savings, credit, and insurance, start and expand businesses, invest in education or health, manage risk, and weather financial shocks.

2.3 Electronic Payment (E-Payment)

An electronic payment is a digital transaction between two parties. E-payment types include wire and bank transfers, cards, digital wallets, mobile pay and more. This means making a payment or transaction for goods and services on an e-commerce website or electronic environment without any need to use cash or check. Electronic payment has grown increasingly over the last decades due to the growing spread of internet-based banking and shopping. As the world advances more with technology development, we can see the rise of electronic payment systems and payment processing devices. As this increase, improve, and provide ever more secure online payment transactions, the percentage of check and cash transactions will decrease.

2.4 Business Models

A business model is the doctrine that constitutes value delivery of an organization to its customers. Each organization can have one or more business models to fulfill their business needs. Generically, it describes opportunity capture as well as the creation and delivery of either tangible or intangible business value. Furthermore, Constructing and editing business models can be referred to as business model innovation and is considered a part of business strategy [10]. Electronic payment business models describe how customers can pay their service or commodity providers through electronic payment channels like point of sale (POS) machines, mobile wallets or online payment gateway. Such a model is used to describe the end-to-end process which can then be later used for study to extend the business in terms of access channels up to backend processes including settlement. Figure 1 introduces an abstracted model of automated FinTech based payment system.

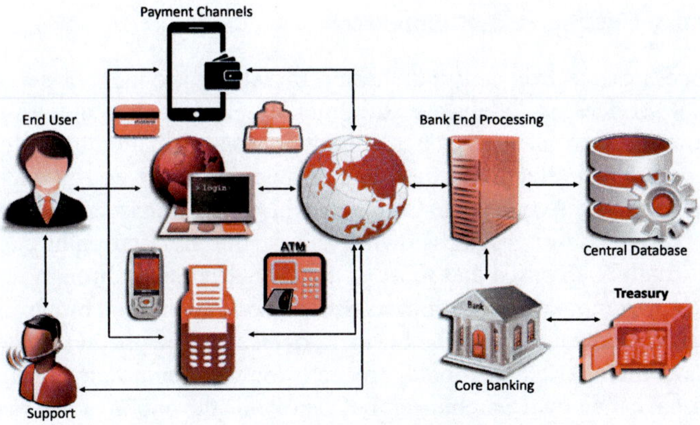

Fig. 1. End to end FinTech based payment system.

The above figure showcases end user communicating different payment channels. Each payment channel is connected securely to the internet to communicate with back-end processing server. Such servers will then communicate core banking and centralized databases for settlement. The cycle is monitored by customer support staff for maximum user experience.

2.5 Business Process Automation (BPA)

The concept of business process automation is the other face of the general concept of digital transformation. It can be described as the technology-based automation process of complicated manual business processes. BPA supports business processes to be simpler. It is considered also as a primary pillar of digital transformation. It also contributed to improving both service quality and delivery while minimizing their cost. BPA is achieved through application integration or by utilizing new software applications and also by assuring maximum utilization of labour resources. Figure 2 showcases an example of manual settlement compared to automated settlement processes.

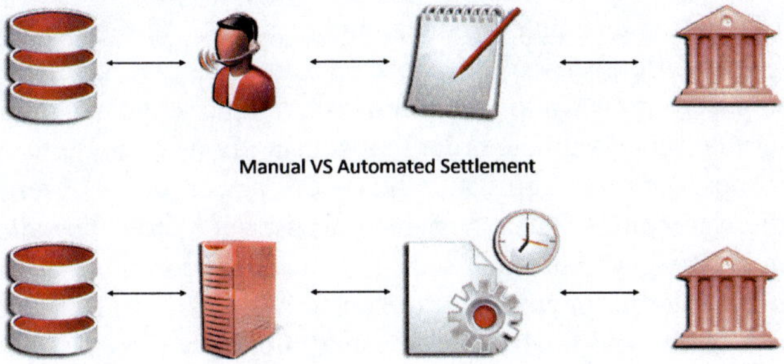

Fig. 2. Manual VS automated EOD settlement file.

In manual settlement, an employee generates a settlement file manually then transfers it to core banking for execution. On the other hand, automated settlement utilizes a workstation software that generates end of day settlement files automatically.

2.6 Electronic Commerce (E-Commerce)

It is the process of electronically purchasing or selling products or services through online special services or widely over the Internet. E-commerce utilizes technologies like smart phones, electronic money transfer, Internet-based marketing, online automatic transaction processing, electronic data interchange (EDI), inventory and warehouses management systems. A typical E-Commerce platform is based on a commodity(s) or service(s) provided by a business owner and desired by a group of customers. The platform is usually web based and most of the time accessible through more than one channels including mobile applications as well as desktop internet browsers. Such platform connects the customer with the business owner to view the available products or services, fills in their basket and finally proceeds for checkout. Many popular platforms like Amazon have their own warehouses, but also avails the option of external retailers to offer their products through the platform, in this case, the platform only acts as a broker between the product provider and the buyer. An example of a simple E-commerce based platform is shown in Fig. 3.

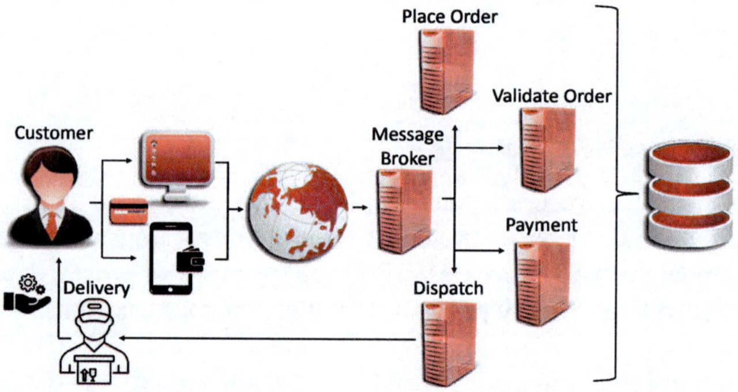

Fig. 3. Simple E-commerce platform.

The platform above consists of two channels, a web-based application that can be accessed from desktop internet browsers and a mobile application. Using the internet, a customer can view the available products and services. As the customer places an order, a message broker passes the order to a microservice that handles placed orders. Place order then posts a message to the placed orders topic in the message broker. Then another microservice called validate order that constantly listening to the placed orders topic reads this message and begins to validate the placed order in terms of product availability in the warehouses. Then the same microservice posts the valid order on the valid orders topic. The payment microservice that listens to the same topic reads the valid order and avails order payment by the customer. Once paid, the order is placed in the paid orders topic. And finally, the dispatch microservice reads the paid order of the paid orders topic and begins the dispatch process through delivery methods. All the mentioned microservices communicate with private as well as a centralized database for synchronization and reporting.

3 FinTech Enablement Prerequisites

3.1 Infrastructure

Infrastructure organizations provide the fundamental hardware, software and services to FinTech companies for their core offering to function. Electronic payment gateways, clearing systems and similar financial services require an agile and reliable infrastructure to successfully deliver services on different platforms across the world, marking the difference between an opportunity missed and an opportunity won.

3.2 Experienced Individuals

As FinTech gains momentum year-on-year, there is a huge demand for professionals with specific FinTech skills. However, when it comes to tech talent, demand definitely outstrips supply It won't be a surprise to find out that the industry that's driving innovation in the financial segment will require a lot of talent with solid hard skills. Programmers, designers, hardware engineers, and more. Fintech job skills encompass acquiring, analyzing, and utilizing the data from countless sources. So, Artificial intelligence experts are needed to help with navigating and finding the required information, allowing a more comfortable life for data scientists. However, they must know how to operate and train the AI [11].

3.3 Regulations

Maximizing business potential by utilizing disruptive technologies regardless of its form, will expose the business data to different types of processing. Afterall, the main asset targeted for such processes is the data generated by one or more financial related systems. The lazy approach is of course to prevent these forms of processing, but then, business will not benefit from FinTech and other disruptive technologies in order to secure its data. And hence, will not have a chance to expand its customer domain. Hence, a balance between regulations and processing is required to successfully secure the business data and at the same time help the business to benefits from such disruptive technologies.

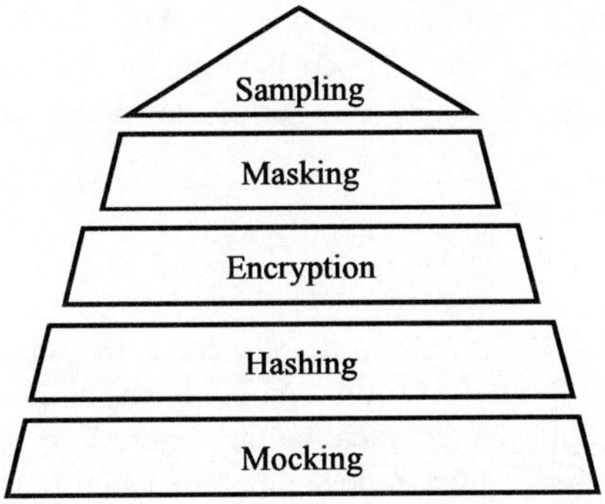

Fig. 4. Proposed secure data delivery techniques

Figure 4 describes the potential data delivery techniques for data transfer for development and testing purposes. Data is required to develop and test FinTech innovative technologies. Regulations must be set with the priority to deliver the required data in the form of mock data. If mock data will be invalid in some cases, then the possibility to deliver hashed form of the data is studied. If such form will be invalid, then encrypted data is proposed. Some technologies require real or near real data for development and testing. Such technologies will then be offered masked data, where the sensitive parts of the data like IDs or credit card numbers are masked. And finally, if all the below fails, the last resource will be to deliver an expired sample of the real data that can be used in development and testing but will be invalid for exploitation against real business value.

3.4 Allocated Capital

Capital allocation for the development of FinTech innovative technologies is a challenging process. After all, the purpose that drives a business to embrace such concepts is to expand its business domain that will be reflected in maximizing its profit. Hence, an accurate return of investment study is mandatory to elaborate the feasibility of investment in such technologies. As it is required that the money spent to develop such technologies must be at least covered buy its returned value.

4 Challenges

4.1 Technology Adoption

The FinTech industry has grown up and grown out. No longer made up of only startups, FinTech today is a host of seasoned companies that offer a broad array of financial services and operate on a global stage. Adoption of FinTech services has moved steadily upward, from 16% in 2015, to 33% in 2017, to 64% in 2019. Awareness of FinTech, even among nonadopters, is now very high. Worldwide, for example, 96% of consumers know of at least one alternative FinTech service available to help them transfer money and make payments. FinTech has caught on around the world, entering the mainstream in all markets studied. Emerging markets are leading the way: in both China and India, the adoption rate is 87%. Close behind are Russia and South Africa, both with 82% adoption. Among developed countries, the Netherlands, the UK and Ireland lead in adoption, reflecting in part the development of open banking in Europe. Consumer awareness of FinTech services is high across all categories, but particularly money transfer and payments services.

4.2 Knowledge Shortage

Despite the huge potential of FinTech, access to and retention of best-in-class talent is a growing concern within the industry on a global basis, and the greatest of these skill shortages is occurring within the technical talent sector. A recent Sparta Global survey of leading technology professionals - representing the technology, finance, media, retail, public sector, charity and legal industries – found organizations across the board are suffering from skills gaps: graduate level (38%), mid-level (55%) and management level (38%). Software development was identified by 52% of respondents as a missing skill, making it the most sought-after area of expertise. 40% recognized a gap in test automation skills, DevOps (38%), BA/PMO (19%) and manual testing (16%). Hiring and integrating new talent into an organization costs organizations time and resource. Candidate screening, conducting interviews, negotiating employment terms, and on-boarding a new hire for a role can take as long as six months – so it's important to get this right first time.

FinTech businesses must embrace collaborations with educational institutions, as a way of fostering learning and building awareness among the next generation of FinTech talent. By working hand-in-hand with educational institutions and initiatives to channel enthusiasm and learning for financial technology, FinTech will establish a platform to attract future employees who are passionate about the industry. Also, if companies have a culture that supports innovation in-house it will help to attract highly talented, curious, motivated, and innovative people. At the same time, up-skilling current workforces to develop technical skills through appropriate training programs, will equip employees with new capabilities to help drive efficiencies and improve productivity across the business.

4.3 Experienced Consultants

FinTech consultant is the evangelist of the fintech businesses. They are responsible for educating, understanding, and collaborating with the business to ensure that the solution is something that matches what is needed by the client's organization or company. Consultants have deep knowledge of technology and markets; they know how to read trends in FinTech landscape. And this reading helps them derive insights about possible solutions and identify opportunities on behalf of clients. A good consultant is a hard-working professional, with strong analysis skills and of course a perfect understanding of banking and investment regulations. A great fintech consultant is a team player and know how to negotiate with clients and fintech companies. Their expertise is then needed for the build phase, which is the connection between your existing software, habits and the new fintech company solution.

4.4 Business Integrity

The development of any disruptive technology aims ultimately to target a specific business for expansion. This requires a clear set of business requirements to be gathered and documented in a form valid for development. Hence, the business requirements phase prior the development phase is very sensitive. During the requirements gathering phase, analysists must make sure to breakdown user's requirements that are usually foggy and general, into specific business targets that will have a done definition and can be later measured and tested.

4.5 Integration Issues

During the development life cycle of FinTech innovative technologies, one or multiple integrations may be required with different parties. Such parties usually will not be using machinery from the same vendor and hence not the same technology and not the same integration interface. As a result, during the development lifecycle, many integration issues might occur that will require maximum agility from the technology development team.

4.6 Bureaucracy

It's a challenging phenomenon to almost any business and the development of FinTech innovative technologies is not an exception. As in such process, the development teams are introducing a technology that was not there before. It might be met with refusal by unexperienced users who are having learning curve issues. And also, its release might be slowed down in the process of waiting for required paperwork, the wasted time might make such technology lose its value of uniqueness with time. Hence, businesses must mitigate unneeded processes during the release of such products or services.

4.7 Information Security

The cyberspace is a vast domain that facilitated humans lives and increased its productivity. However, it can also be a domain of fraudulent activities, especially in the financial subdomain of the cyberspace (Fig. 5).

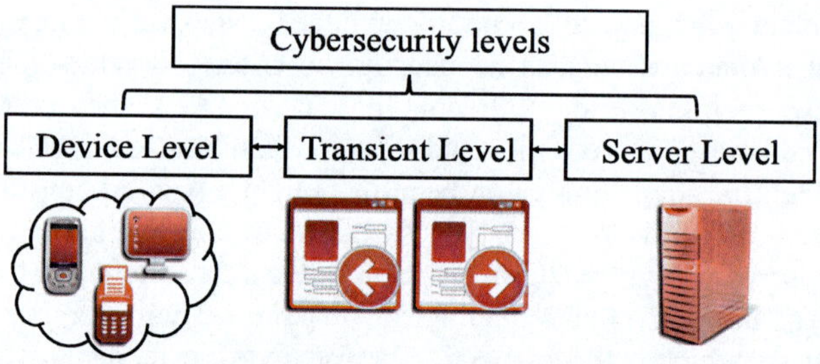

Fig. 5. Cybersecurity threat levels.

Security threats in the cyber domain can be on three levels, device, transient or server level. They can be on one or more levels at the same time, in such case they will be called complex cybersecurity threats. FinTech innovative technologies are no exception as they are hosted on the cyberspace. Technology development teams must make sure to use secure protocols like HTTPs combined with advanced encryption, hashing and masking to introduce secure FinTech innovative technologies.

5 Proposed Methodologies

Fintech companies use a variety of technologies, including artificial intelligence (AI), big data, robotic process automation (RPA), and blockchain. AI algorithms can provide insight on customer spending habits, allowing financial institutions to better understand their clients. Chatbots are another AI-driven tool that banks are starting to use to help with customer service. Big data can predict client investments and market changes in order to create new strategies and portfolios, analyze customer spending habits, improve fraud detection, and create marketing strategies. Robotic Process Automation is an artificial intelligence technology that focuses on automating specific repetitive tasks. RPA helps to process financial information such as accounts payable and receivable more efficiently than the manual process and often more accurately. Blockchain is an emerging technology in finance which has driven significant investment from many companies. The decentralized nature of blockchain can eliminate the need for a third party to execute transactions.

5.1 Electronic Settlement (E-Settlement)

Digital payments can commence from a variety of channels, including mobile wallets, online portals, or points of sale. No matter the channel, any form of digital payment will require later settlement in order for the payment to have effect on actual bank accounts. Bulk payment workstation (BPWS) is used to automate the process of notifying the settlement system about any transaction processed by any system inside the company. They collect the processed transactions from all the payment systems involved and send them to the settlement engine Also, they have failover mechanisms in place to handle any errors which occurs when sending the transactions and try again accordingly.

5.2 Payment Gateway (PGW)

The Payment Gateway system is used to collect different services' fees from consumers on behalf of various governmental and private entities which needs its services to be paid online. It's deeply integrated with all global and domestic card networks & issuers, domestic payment partners and internal settlement & reconciliation systems. By using the Payment Gateway, these entities offload the burden of integration with various payment channels and provide an easy comfortable way for the customers to pay for their services.

5.3 Unified Bill-Payment Gateway (UBG)

As part of financial technology enablement, bill payment is mandatory. Conventional bill payment includes a service provider (SP) issuing bills, on the other hand, a bill payer representing the beneficiary from such services or products. The unified bill payment gateway represents a hub connecting bill issuers to bill payers through a variety of payment channels. The UBG features a collection of micro services each representing the integration of bill issuers with a payment and collection central hub. It also provides maximum integration flexibility. And outline of the UBG solution is presented in Fig. 6.

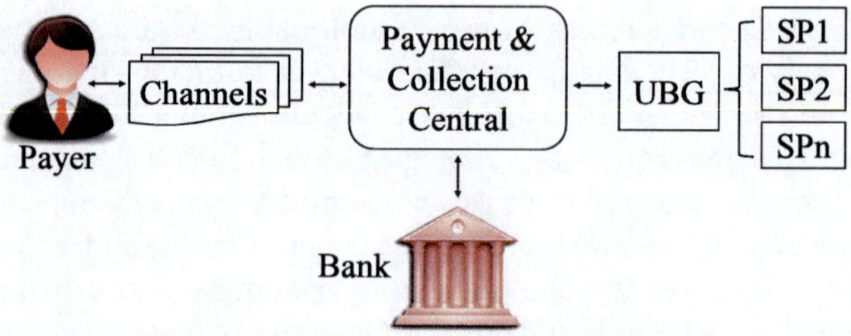

Fig. 6. Unified bill payment gateway.

5.4 Commodity Monitoring and Distribution Management Systems

The concept of digital transformation includes the digitalization of commodity distribution including business elements as well as the actual transactions. Such businesses typically consist of the commodity sources and the commodity beneficiaries and finally the shipping mechanisms. Such systems provide reliable reports that helps business owners in the decision-making process.

6 Results

FinTech have taken over the finance world by improving the overall financial services and are also very useful for retail investors to manage the stock market, small businesses, financial systems, and the insurance industry. The most thriving need is fulfilled by custom software development methods. The growth of cashless payment systems and other advanced technologies like machine learning, blockchain, automation, bank lending, or consumer lending has digitally transformed many businesses. The growth of fintech has a profound effect on many finance startups and enterprise-level businesses.

6.1 Enhanced Business Flow

At its core, fintech is all about making traditional financial processes faster and more efficient. This is achieved by eliminating the middleman and automating everything, which, in turn, leads to lower costs, happier employees, and a better customer experience.

6.2 Fraud Elimination

Better risk management is one of the crucial but lesser-appreciated roles of fintech. Fintech can help reveal these risks by giving businesses more analytics capabilities. Using automation, machine learning, and big data, apps can scour through thousands of data points and provide decision-makers with insights in seconds. Also, fintech is effective in countering money laundering and other illegal activities, which is one of the top business risks.

6.3 Increasing Productivity

Fintech improves the quality of traditional financial institutions by increasing efficiency and productivity. As banks and credit unions perceive fintech businesses as friends on this road rather than vendors, more opportunities arise. Furthermore, by delivering better and more contemporary services, firms' client retention rates are certain to rise, resulting in increased revenues. Fintech also provides customers with personalized experiences through big data and AI. That means an app can offer products and services relevant to its users based on their past purchases or even their financial standing. And personalization is a fantastic tool both for getting new customers and retaining them.

6.4 Overall Business Process Integrity

Applying FinTech innovative technologies and strategies proposes a variety of benefits. Processes automation speeds up manual tasks and promotes business capacity to industrial scale. The concept of digitalization is also connected to increased accuracy as the more digitalized and automated processes are, the more accuracy they provide. The same concepts also provide more confidence in terms of error elimination and the reduction of business cycle time.

7 Conclusion

In this research, challenges facing financial based businesses were discussed with possible best practices to overcome. Such challenges include technology adoption fears that can be overcome through campaigns and demonstrations. Knowledge shortage can also be come through digital education in addition to workshops to build experience. The lack of affordable experienced consultants can be handled by talent building of our current employees. Business integrity problems where customer business cycles suffer from defects can also be targeted for correction through workshops and demonstrations. Bureaucracy can be fought through business process improvement. Information security concerns can be handled through adopting best practices during software development and testing while assuring that software can go live only under minimum security requirements according to business needs. Furthermore, a combination of methodologies and techniques were utilized to maximized efficiency and potential of business cycles including digital payments. Such maximization helps businesses to expand to industrial levels serving massive number of customers. Manual settlement of bill payment transactions is an error prone process with a high probability of chargeback. The solid understanding of the business process in addition to automation succeeded to introduce a robust and error proof automatic settlement process. Integration problems are one big challenge that was successfully tackled by the introduction of a unified bill payment gateway that succeeded to introduce a robust yet flexible integration interface that dealt with different technologies to achieve maximum financial technology embracement. The online payment gateway also introduced an easy-to-use payer friendly interface for customer bill payment with the least possible integration effort by merchants. Monitoring and distribution management systems also introduced a full-scale solution to businesses looking for more control over their product or service distribution business cycles. Institutions succeeded by transferring their businesses from manual paperwork into such systems to generate usage data that was successfully used later to generate useful reports. Such data can also be targeted by market analysis tools and also by machine learning tools to produce further insights. Such insights can help business owners in decision making.

References

1. Razzaque, A., Hamdan, A.: Role of financial technology FinTech: a survey. In: Hassanien, A.E., Azar, A.T., Gaber, T., Oliva, D., Tolba, F.M. (eds.) AICV 2020. AISC, vol. 1153, pp. 112–117. Springer, Cham (2020). https://doi.org/10.1007/978-3-030-44289-7_11
2. Infinite Financial Intermediation, 50 Wake Forest Law Review 643 (2015)
3. Loo, V.: Rory "Making Innovation More Competitive: The Case of Fintech." UCLA Law Rev. **65**(1), 232 (2018)
4. Schueffel, P.: Taming the beast: A scientific definition of Fintech. J. Innov. Manag. **4**(4), 32–54 (2017). https://doi.org/10.24840/2183-0606_004.004_0004.ISSN2183-0606
5. Sanicola, L. "What is FinTech?". Huffington Post. Retrieved (2017)
6. Chatbot: The intelligent banking assistant. PwC. Retrieved (2021)
7. Xu, L., Gao, R., Xie, Y., Du, P.: To be or not to be? Big data business investment decision-making in the supply chain. Sustainability. **11**(8), 2298 (2019). https://doi.org/10.3390/su11082298
8. Madakam, S., Holmukhe, R.M.,: Bharati Vidyapeeth University, Pune, India; Kumar Jaiswal, Durgesh; National Institute of Industrial Engineering (NITIE), Mumbai, India. The future digital work force: Robotic Process Automation (RPA). J. Inform. Syst. Technol. Manag. **16**:1–17. https://doi.org/10.4301/S1807-1775201916001
9. Intelligent process automation: The engine at the core of the next-generation operating model | McKinsey. www.mckinsey.com. (2019)
10. Geissdoerfer, M., Savaget, P., Evans, S.: The Cambridge business model innovation process. Procedia Manuf. **8**, 262–269 (2017). https://doi.org/10.1016/j.promfg.2017.02.033.ISSN2351-9789
11. Maghawry, A., et al.: A survey on challenges facing artificial intelligence based pattern recognition for business oriented big data analytics and solutions. In: Silhavy, R. (ed.) CSOC 2021. LNNS, vol. 230, pp. 237–248. Springer, Cham (2021). https://doi.org/10.1007/978-3-030-77442-4_20

Knowledge Discovery in Language Data for the Analysis of Urban Development Project

Maria Pilgun(✉)

Russian State Social University, St. W. Pieck, 4/1, 129226 Moscow, Russian Federation
pilgunm@yandex.ru

Abstract. The article presents the analysis of the situation associated with the construction of the Biryulevskaya metro line (BML) in Moscow (Russia). The dataset included data from social networks, online media, messenger, blogs, microblogs, video hosting services, subject-matter portals, forums, shops, mapping services, TV shows dedicated to the implementation of the BML project. The data was collected between January 1, 2021 and December 23, 2021. The study involved a cross-disciplinary approach. The process of knowledge discovery and data interpretation included content analysis, sentiment analysis, neural network text analysis and analysis of lexical associations. The study of the dynamics of information activity in cyberspace involved analysis of opinions, assessments, as well as changes in the attitude of the active part of the city's population for timely response and forecasting of conflict situations.

Keywords: Language data · Knowledge discovery · Neural network technologies · Urban development project

1 Introduction

The theory, techniques and practice for extracting information from large databases is a vital scientific task. The research and practice of data mining and knowledge discovery are increasingly important in the modern scientific paradigm and enable solving a wide range of important practical problems [1–6].

In particular, for the successful implementation of urban projects, it is necessary to take into account the opinion of residents, which is never static. Decision-making based on the analysis of large amounts of real-time data makes it possible to monitor the attitude of residents and changes in the perception of the situation and to find the best solutions. Among urban projects, road construction is one of the most painful issue for residents, since, on the one hand, it is a necessary condition for the development of the urban environment, but, on the other hand, when implemented, affects the habitual way of life of some citizens, which can cause sharp discontent and protests, hinder the implementation of a project or even stop it. The article presents an analysis of the situation associated with the construction of the Biryulevskaya metro line (BML) in Moscow (Russia). The BML is a new radial line of the Moscow Metro; the construction of the first section 20.7 km long with 10 metro stations began in 2021. The first section will run south from the ZIL station

R. Silhavy (Ed.): CSOC 2022, LNNS 501, pp. 657–666, 2022.
https://doi.org/10.1007/978-3-031-09070-7_54

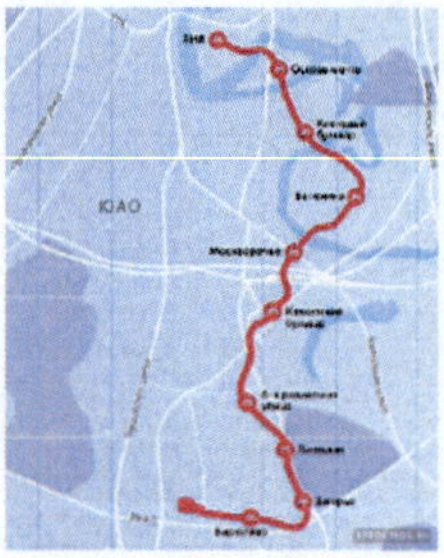

Fig. 1. BML

Table 1. Quantitative characteristics of the data

Parameter name	Quantity
Token number	49 996 462
Number of messages	15 548
Messages per day, max	452
Number of active actors	1714
Activity (posts per authors)	9,07
Number of sources	358

(Moscow Central Circle) through seven districts of Moscow: Danilovsky, Nagatinsky Zaton, Pechatniki, Moskvorechye-Saburovo, Tsaritsyno, Biryulyovo Vostochnoye and Biryulyovo Zapadnoye. The opening of the first section is scheduled for 2027.

The purpose of the study is: to analyze the dynamics of the information activity of actors in cyberspace to monitor the opinions, assessments and changes in the attitude of the active part of the population for the timely prediction of emerging and/or developing conflict situations around the BML.

1.1 Data

The dataset included data from social networks, online media, messenger, blogs, microblogs, video hosting services, subject-matter portals, forums, shops, mapping services, TV shows dedicated to the implementation of the BML project. The data was collected between January 1, 2021 and December 23, 2021. Quantitative characteristics of the data are presented in Table 1, Fig. 1.

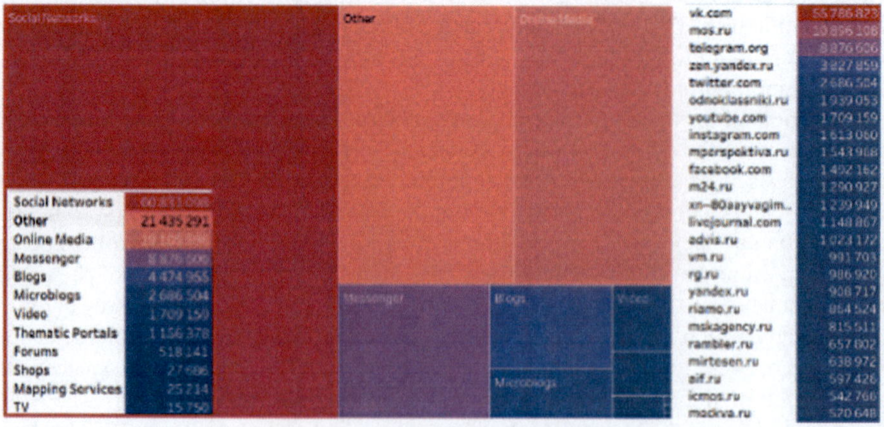

Fig. 2. Sources and digital platforms

Most of the content was generated on social media (60,831,098). The leaders among digital platforms were the social network VKontakte (56,786,823), the official website of the Mayor of Moscow mos.ru (10,896,108) and the Telegram messenger (8,876,606) (Fig. 2).

To discuss the problems associated with the implementation of the BML project, the actors preferred using social networks, as evidenced by their digital footprints (Fig. 3).

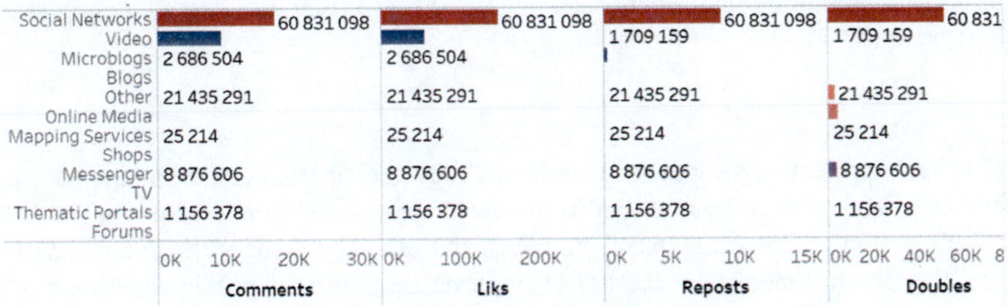

Fig. 3. Actors' digital footprints on various digital platforms

1.2 Method

The study involved a cross-disciplinary approach. For knowledge discovery and data interpretation, a technique was used that combined several methods: content analysis [7–9] using the AutoMap service (http://casos.cs.cmu.edu/), sentiment analysis using the Eureka Engine sentiment analysis module (http://eurekaengine.ru/), neural network text analysis and analysis of lexical associations [10, 11].

2 Results

2.1 General Description of the Content

The dynamics of the total number of messages demonstrates a relatively even distribution of information over the analyzed period. Three growth peaks can be observed: January 28, 2021 (452), January 22, 2021 (436), January 20, 2021 (428). The peak on January 28, 2021 (452) was caused by the announcement of the start of construction of the next metro line (Biryulevskaya). The peak on January 22, 2021 (436) is associated with information about the start of public discussions of the BML on the Active Citizen portal. The peak on January 20, 2021 (428) was generated by reports that a detailed BML design has been published, as well as the date of its construction in 2021 (Fig. 4). The dynamics of the number of unique messages also has minor peaks on July 20, 2021 (246) and July 22, 2021 (246), caused by similar information (Fig. 5).

The increase in views observed on 07/22/2021 (246) is associated with information that the construction of the Biryulyovskaya line is postponed indefinitely due to lack of funding (Fig. 6). The dynamics of the actors' activity also makes it possible to identify several growth peaks on December 22, 2021 (168), January 22, 2021 (144), and July 20,

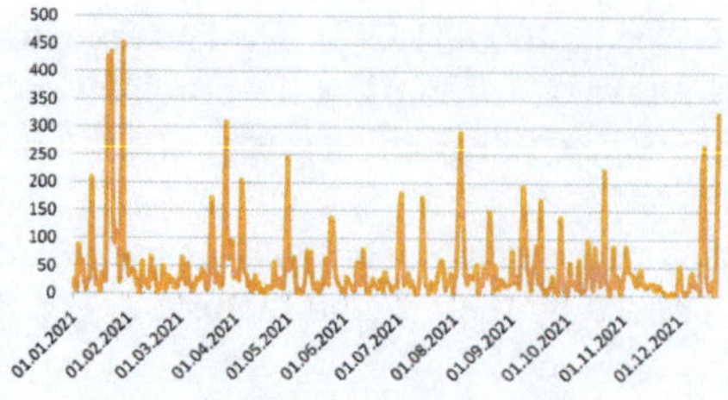

Fig. 4. Dynamics of the total number of messages

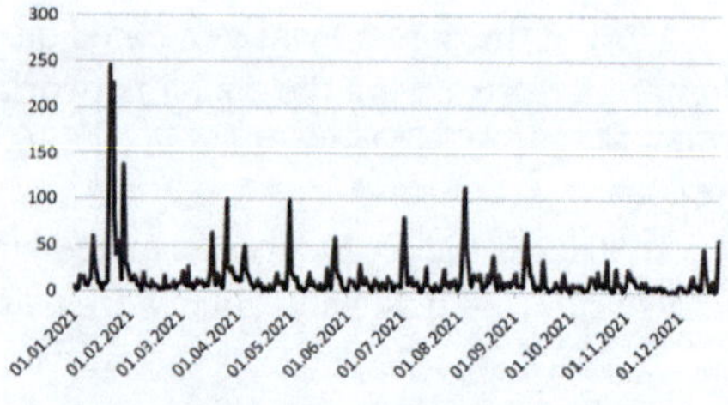

Fig. 5. Dynamics of unique messages

2021 (133). The peak of growth on December 22, 2021 (168) was caused by the statement of Moscow Mayor S. Sobyanin that in the coming years, 95% of Muscovites will live and work within walking distance from metro stations. The peak of the dynamics of the actors' activity on January 22, 2021 (144) is associated with the public discussion of the new metro line, which started on the Active Citizen portal. The peak of the dynamics of the actors' activity on July 20, 2021 (133) was caused by footages about the BML construction (Fig. 7).

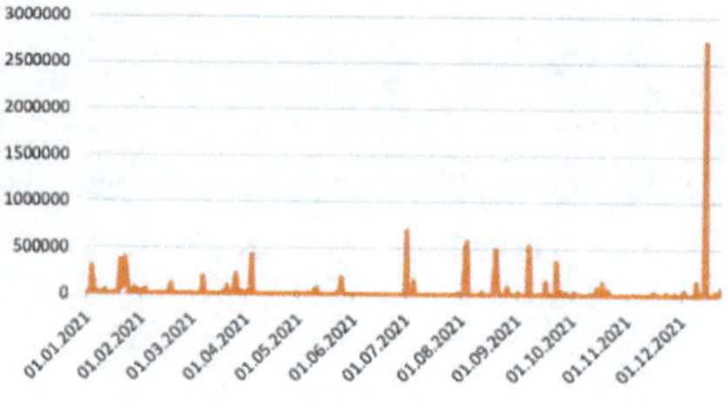

Fig. 6. Views dynamics

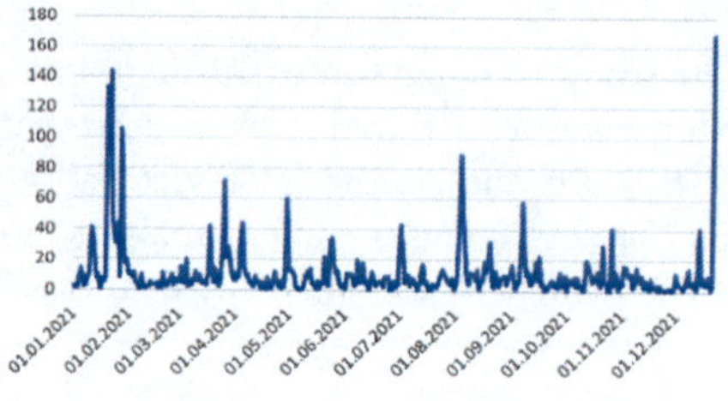

Fig. 7. Actors' activity dynamics

2.2 Analysis of Residents' Attitude

Content sentiment analysis showed a significant predominance of the neutral cluster (Fig. 8).

The study of the thematic structure, summarization and analysis of the sentiment analysis results made it possible to identify key information in three clusters:

2.2.1 Neutral Cluster

Quantitative data show the predominance of the neutral cluster, which contains information about the BML project and discussion of the BML construction progress (Fig. 9).

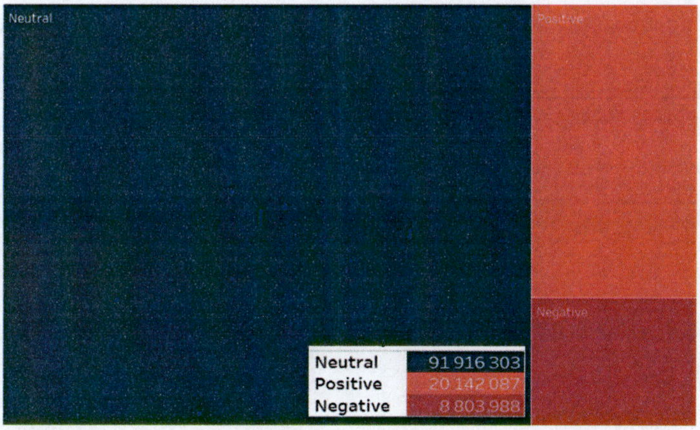

Fig. 8. Content sentiment

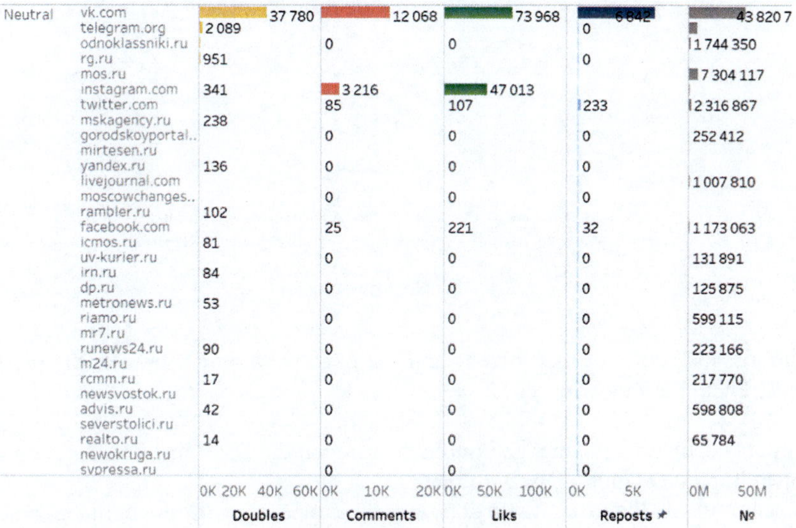

Fig. 9. Neutral cluster

Key Topics:

- Information about the BML construction plans.
- Notifications about the start of the BML construction.
- Information about the BML construction period.
- Discussion and reports on the construction progress of BML stations.
- Information that a detailed BML design has been published, as well as the date of its construction in 2021:

- Improving transport accessibility for new districts of the city after the project implementation.

2.2.2 Negative Cluster

The negative cluster includes residents' claims related to the implementation of the BML construction project (Fig. 10).

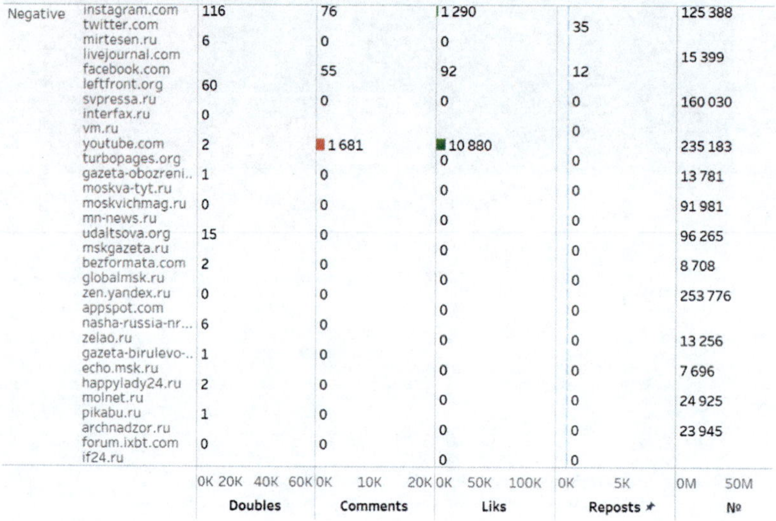

Fig. 10. Negative cluster

The dissatisfaction of the citizens and the negative attitude towards the project is associated with the following:

- Residents demand a full-fledged public hearing on the BML and coordination of the project implementation plans with them.
- Citizens express their dissatisfaction with their meeting with the builders, since the dialogue was not productive and they failed to reach accords.
- Residents believe that the BML location was chosen poorly and the planning was disadvantageous (*"Metro to nowhere"*).
- The construction period, according to residents, is unnecessarily extended.
- Negative attitude is caused by the fact that, in the citizens' opinion, the construction period has been extended indefinitely due to lack of funding.
- Residents were dissatisfied with information that the construction of the Biryulyovskaya line, which they need, has been postponed indefinitely; instead, the Moscow Region government is ready to finance the construction of a toll road that runs parallel to the Moscow Ring Road, which, according to Muscovites, is not needed.
- Changes in the performance of public vehicles, which citizens associate with the construction of the future Biryulyovskaya line and which cause serious inconvenience to residents, significantly worsen the transport situation in the district.

- Traffic and parking of trucks involved in the construction will significantly degrade living standards and mobility in the district.
- Residents of the district are concerned that the construction noise will significantly reduce the standard of living; they want administrative measures to be applied to builders during night work, as well as installation of soundproof double-glazed windows.
- Station planning provokes backlash from residents who insist on re-planning.
- The proposed merging of new metro lines draws criticism from Muscovites.
- There is dissatisfaction with the plan of crossing the river in the Saburovo district for the BML construction.
- According to residents, the BML will disrupt the logistics of freight vehicles.
- Destruction of the architectural appearance of Moscow during construction.
- Demolition of the stadium due to the construction of BML stations, lack of compensation for sports facilities in the district.
- The construction of the new metro line is associated with the construction of new residential complexes
- Irrational use of budgetary funds in the BML construction.
- Shallow laying of the future BML.
- Harm to the ecology of the city.
- Destruction of the green belt in the construction area.
- Disagreements between the Moscow and Moscow region authorities on the development of the transport system.
- Consolidation of political issues and the implementation of urban projects.

2.2.3 Positive Cluster

The content presented in the positive cluster is generated mainly by official sources or media affiliated with the Moscow authorities (Fig. 11).

Key Topics:

- Information about the start of public discussions of the BML issues on the Active Citizen portal.
- The BML construction is an important stage in the development of the Moscow metro network.
- Use of advanced technologies for the development of modern engineering and transport infrastructure of the capital.
- Development of the Moscow transport system.
- Statement by the Mayor of Moscow that in the coming years 95% of Muscovites will live and work within walking distance of metro stations.
- Design solutions for station design.
- Information about the successful progress of construction.
- Improving the transport situation in the city.
- The pace of the BML construction and the development of the metro network.
- Construction period and opening of the new line.

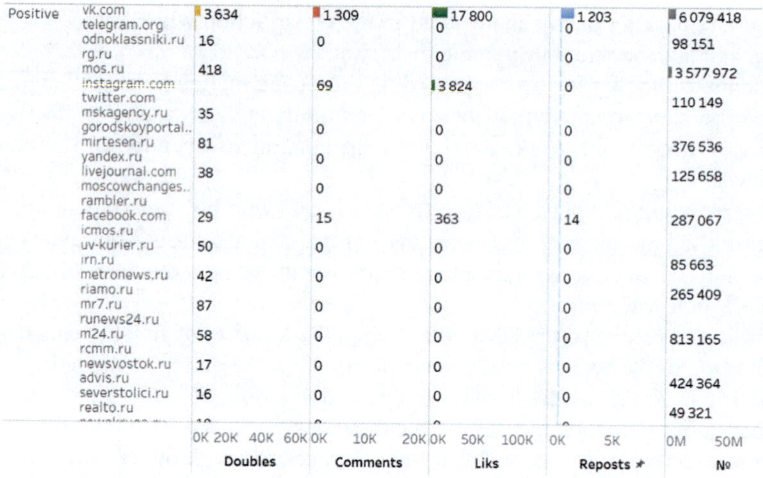

Fig. 11. Positive cluster

2.3 Social Tension Rating

The sentiment analysis of the audience coverage (Fig. 12) and digital footprints of actors (Fig. 13) made it possible to conclude that neutral sentiment prevails in discussions of the BML construction issues.

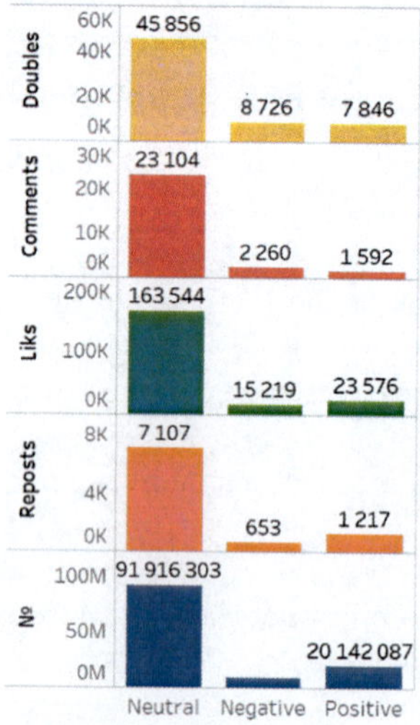

Fig. 12. Sentiment of actors' digital footprints

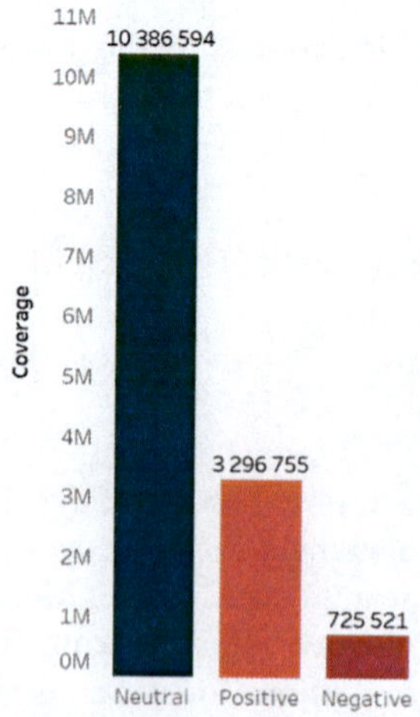

Fig. 13. Tonality of audience coverage

The result of the analysis of the consolidated database on the BML implementation showed a rather low level of social stress (6.57) and an average index of social well-being

(13.59) (Table 2). For methodology for calculating the Indices of social stress and social well-being, see [10].

Table 2. Level of social stress and well-being

Facility name	Social stress index	Social well-being index
BML	6,57	13,59

When interpreting quantitative results, social stress and well-being indices, it should be borne in mind that the consolidated database is mostly formed by neutral and positive content generated by official and affiliated sources. The reactions of users are significantly inferior in quantitative terms, but they demonstrate a different position from the official one. Residents show a very strong interest in the project implementation, as the transport situation in the district needs improvement. However, many issues of the Project implementation – planning options and the construction period, in particular – draw criticism from Muscovites.

3 Conclusion

The analysis of the data characterizing the situation related to the construction of the Biryulyovskaya metro line (BML) in Moscow showed that for the successful implementation of the project it is necessary to establish a dialogue with residents who are very interested in the most rapid BML construction, but taking into account their wishes, as well as to develop a plan for anti-crisis communications to eliminate the negative consequences of conflict interaction between citizens and builders.

It should be taken into account that the strongest negative reaction of the residents is caused by the lack of feedback from the developer and the city authorities. Residents demand a full-fledged public hearing on the BML issues and coordination of plans for the project implementation with the district residents.

The analysis also showed that negative content is generated mainly in the social network VKontakte (a small portion is also generated in Telegram and YouTube). Positive official content is also generated mainly by VKontakte (a small portion is additionally generated on the mos.ru portal and in Telegram), i.e. Youtube is little used.

When interpreting quantitative results, social stress and well-being indices, which show the predominance of neutral content and a low level of social tension, it should be borne in mind that the consolidated database is mostly formed by neutral and positive content generated by official and affiliated sources. The reactions of users are significantly inferior in quantitative terms, but they demonstrate a different position from the official one. Residents show a very strong interest in the project implementation, as the transport situation in the district needs improvement. However, many issues related to the project implementation are sharply criticized by Muscovites. The quantitative advantage of neutral official content does not reduce the conflictogenic potential of the project, which was revealed in the reactions of the district residents.

This conclusion is also confirmed by the fact that the rating of actors generating positive content is headed by official and biased sources, while the rating of actors generating negative content is headed by personal accounts of residents and activists.

For the successful implementation of the project, it is necessary to achieve compromise solutions with the district residents on issues that cause a negative attitude of the actors and are listed in Sect. 2.2.2.

Acknowledgment. The study was carried out as part of the implementation of the project "Historical City: Spaces of Patriotic Education" with funding from the Expert Institute for Social Research (EISR) to my article (Knowledge Discovery in Language Data for the Analysis of Urban Development Project).

References

1. Arco, L., Nápoles, G., Vanhoenshoven, F., Lara, A.L., Casas, G., Vanhoof, K.: Natural language techniques supporting decision modelers. Data Min. Knowl. Disc. **35**(1), 290–320 (2020). https://doi.org/10.1007/s10618-020-00718-4
2. Ribeiro, A.F.: Social media as author-audience games. Data Min. Knowl. Disc. **35**(6), 2251–2281 (2021). https://doi.org/10.1007/s10618-021-00783-3
3. Zhu, Z., Li, Y., Wang, Y., Wang, Y., Tong, H.: A deep multimodal model for bug localization. Data Min. Knowl. Disc. **35**(4), 1369–1392 (2021). https://doi.org/10.1007/s10618-021-00755-7
4. Toth, C., Helic, D., Geiger, B.C.: Synwalk: community detection via random walk modelling. Data Min. Knowl. Disc. **36**, 739–780 (2022). https://doi.org/10.1007/s10618-021-00809-w
5. Björklund, A., Henelius, A., Oikarinen, E., et al.: Robust regression via error tolerance. Data Min. Knowl. Disc. **36**, 781–810 (2022). https://doi.org/10.1007/s10618-022-00819-2
6. Rebuffel, C., Roberti, M., Soulier, L., Scoutheeten, G., Cancelliere, R., Gallinari, P.: Controlling hallucinations at word level in data-to-text generation. Data Min. Knowl. Disc. **36**(1), 318–354 (2021). https://doi.org/10.1007/s10618-021-00801-4
7. Mayring, P.: Qualitative content analysis. A Companion Qual. Res. **1**, 159–176 (2004)
8. White, M.D., Marsh, E.: Content analysis: a flexible methodology. Libr. Trends **1**, 22–45 (2006)
9. Krippendorff, K.: Content Analysis. An Introduction to Its Methodology, 3rd edn., p. 441. SAGE Publications, Inc., Los Angeles, CA, USA (2012)
10. Kharlamov, A., Pilgun, M. (eds.): Neuroinformatics and Semantic Representations: Theory and Applications. Cambridge Scholars Publishing: Newcastle upon Tyne (2020)
11. Pilgun, M., Raskhodchikov, A.N., Koreneva, A.O.: Effects of COVID-19 on Multilingual Communication. Front. Psychol. **12**, 792042 (2022). https://doi.org/10.3389/fpsyg.2021.792042

Development of an Algorithm for the Formalization of Linguistic Expert Information

Vladislav Ivanovich Danilchenko(✉) and Viktor Mikhailovich Kureichik

Federal State-Owned Autonomy Educational Establishment of Higher Vocational Education, Southern Federal University, Taganrog, Russia
vdanilchenko@sfedu.ru

Abstract. The article is devoted to the construction of specialized models for representing heterogeneous subject knowledge for information systems. The article deals with the creation of an algorithm for intelligent processing of linguistic expert information (LEI) to formalize the textual information of several collections of texts belonging to the same subject area (SbA). The algorithm is implemented on the basis of a set of analytical methods related to the class of supervised machine learning methods. The algorithm differs from the known ones in that it allows you to create a procedure for the automatic construction of hierarchical structures (dendrograms) of several collections of texts. Scientific novelty lies in the proposed analytical method for formalizing linguistic expert information, which allows you to analyze and process linguistic expert information in relation to the task of constructing a simple ontology. The fundamental difference between the proposed model is the ability to formalize the CAD subject area based on the synthesis of various already existing classifications of this subject area with the possibility of its further modification: representation by a matrix structure of hierarchical relationships of concepts, with a format of the type associated data, which will provide automation of the presentation of an ontology in an owl format.

Keywords: Learning management systems · Ontologies · The subject area of e-learning systems · A model for placing VLSI components · A model for representing heterogeneous knowledge · Formalization of LEI

1 Formalization LEI

1.1 LEI Formalization Method

To develop the LEI formalization method, tools of probabilistic-thematic modeling and hierarchical agglomerative clustering are used, the main area of application of which is the classification and clustering of text documents [1].

The paper proposes to use the analytical method for formalizing LEI. It consists in the sequential execution of the steps shown in Fig. 1.

R. Silhavy (Ed.): CSOC 2022, LNNS 501, pp. 667–674, 2022.
https://doi.org/10.1007/978-3-031-09070-7_55

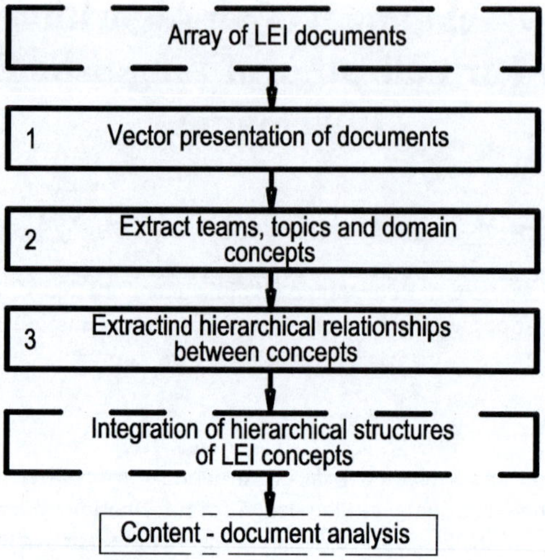

Fig. 1. The structure of the analytical method

At the first stage analytical method of LEI formalization is solved the problem of vector representation of documents with LEI. At the entrance of the analytical method of formalization of LEI from preliminary processing, an array of documents from LEI - texts, divided into terms, is received.

In the second stage analytical method of LEI formalization solves the problem of extracting terms, topics and concepts of the subject area. The initial data of the second stage is the term-document matrix obtained as a result of the solution of the first stage. The term-documentary matrix on the basis of the Dirichlet latent placement model (LDA) is used to calculate the matrix of probability distributions of terms by topics and the matrix of distribution of topics by documents. The resulting matrix of topics using a vocabulary of terms of the subject area, specified by an expert, is mapped into the space (vector) of concepts [3–5].

In the third stage analytical method of formalization LEI solves the problem of extracting hierarchical relationships between concepts. The initial data of stage 3 is a vector of concepts, which is the result of stage 2. Distances are calculated between the concepts, taking into account their probabilities, and a matrix of pairwise connections of concepts (matrix of distances) is formed. The obtained matrix is used for hierarchical cluster analysis [3–5].

1.2 Algorithm Formalization LEI

The LEI formalization algorithm forms training and control samples according to the input data - an array of LEI documents, converts LEI documents into a vector representation, identifies SbA terms, topics and concepts [5], construction of a hierarchical structure of SbA concepts [5]. Let us consider in a more similar way the transformation of LEI documents into a vector representation, the selection of terms, topics and concepts of an SbA.

At the input of the analytical method of formalization of LEI from preliminary processing, an array of LEI documents is received - a set D, consisting $N_D (N_D = \dim(D))$ of documents $d_i \in D, i = 1, N_D$. Each document $d_i \in D$ is a vector and consists of $N_{wd_i} \left(N_{wd_i} = \dim(d_i)\right)$ terms $w_n \in d_i$, $n = \overline{1, N_{wd_i}}$, that is, one can imagine $d_i = \left\{w_1, \ldots, w_n, \ldots, w_{N_{wd_i}}\right\}$ [5].

Using the "bag of terms" method [6], you can find the document are looking for d_i represent as a column vector, the length of which equal to length N_w, whose element is the coefficient u_{ij} of the term w_j in this document. A matrix D_{ij} is a numerical two-dimensional sequence containing many LEI documents. The lines are consistent with terms, columns - documents. The element of the matrix is the coefficient u_{ij}:

$$D_{ij} = \begin{bmatrix} u_{11} & \cdots & u_{1i} \cdots u_{1N_D} \\ \vdots & \ddots & \vdots\vdots\vdots \\ u_{j1} & \cdots & u_{ji} \vdots u_{jN_D} \\ \vdots & \cdots & \vdots \ddots \vdots \\ u_{N_w 1} & \cdots & u_{N_w i} \cdots u_{N_w N_D} \end{bmatrix}, \ \dim\left(D_{ij}\right) = N_w \cdot N_d, \tag{1}$$

where $j = \overline{1, N_{wd_i}}$ - row number of matrix D, ordinal number of a term; $j = \overline{1, N_d}$ - column number of the matrix D.

$$p(t|d) = \sum\nolimits_{t \in T} p(t|d) p(w|t). \tag{2}$$

The set of documents D appears to be known, and each document $d_i \in D$ - random independent sample of terms, which is created by the subset T.

Conditional distribution of terms in documents $p(w|d)$ can be found by the frequencies of the terms

$$p(w|d) = \frac{n_{wd}}{N_{wd}}, \tag{3}$$

where n_{wd} - the number of occurrences of the term in the document d; N_{wd} - the length of the document. We obtain a matrix of probability distributions of terms by topics $\Phi = \left(\varphi_{jk}\right)$:

$$\Phi_{jk} = \begin{bmatrix} p(w_1|t_1) & \cdots & p(w_1|t_k) \cdots p\left(w_1|t_{N_D}\right) \\ \vdots & \ddots & \vdots\vdots\vdots \\ p\left(w_j|t_1\right) & \cdots & p\left(w_j|t_k\right) \vdots p\left(w_j|t_{N_T}\right) \\ \vdots & \cdots & \vdots \ddots \vdots \\ p\left(w_{N_w}|t_1\right) & \cdots & p\left(w_{N_w}|t_k\right) \cdots p\left(w_{N_w}|t_{N_T}\right) \end{bmatrix}, \tag{4}$$

where $\dim(\Phi) = N_w \cdot N_T; j = \overline{1, N_w}$ - number of a row of a matrix Φ, $k = \overline{1, N_T}$ - number of a column of a matrix Φ; $p(w_j|t_k)$ - the probability of finding j term in the k section t_k; Φ^k - vector column of the section t_k.

Distribution matrix of topics by documents $O = (\theta_{ki})$ [4]:

$$O_{ki} = \begin{bmatrix} p(t_1|d_1) & \cdots & p(t_1|d_i) \cdots p(t_1|d_{N_D}) \\ \vdots & \ddots & \vdots\vdots\vdots \\ p(t_k|d_1) & \ldots & p(t_k|d_i) \vdots p(t_k|d_{N_D}) \\ \vdots & \cdots & \vdots \cdot . \vdots \\ p(t_{N_T}|d_1) & \cdots & p(t_{N_T}|d_1) \cdots p(t_{N_t}|t_{N_D}) \end{bmatrix}, \quad (5)$$

where $\dim(O) = N_T \cdot N_D$; $k = \overline{1, N_T}$ – row number of the matrix O; $i = \overline{1, N_T}$ – the number of the column of the matrix O; $p(t_k|d_i)$ – the probability of finding the k term in the i section t_k; O^i – column vector of the section t_k.

The resulting matrix of topics with the use of the SbA terms dictionary is transferred to the space (vector) of concepts.

The sequence of actions performed to formalize LEI [3–5].

Step 1. Construct training and control samples. The action allows to randomly split the array of documents LEI D into training and control samples. As a result, two arrays of LEI documents will be obtained (the training sample consists of 80–90% of documents, the rest of the LEI documents make up the control sample).

Step 2. Build the "bag of terms" model [3.4]. The action is performed for two arrays (training and control) separately. The result of the action is: a dictionary of terms [3.4].

Step 3. Calculate the frequency of occurrence of terms in documents. The result of the action is the term-document matrix.

Step 4. Train the term and topic extraction model (training sample). The output is: the number of topics in the collection; data type; concentration of topics; concentration of words; the likelihood of topics in documents; the likelihood of topics in the document; probabilities of terms in topics.

Step 5. Setting the parameters of the model. This action allows you to search for the optimal value of the parameter of the number of hidden topics, according to the specified parameters of the Dirichlet distribution [3.4]. The result of the action is a trained model, the number of topics at which the optimal value is achieved.

Step 6. Check the model for extracting terms and topics from the LEI documents, specifying the number of topics obtained in step 3. The result is a model with the minimum perplexity value.

Step 7. Build a matrix of keywords, terms in documents and topics. The action is performed by calculating the frequency of occurrence of keywords in the dictionary of terms, taking into account the inverse frequency of keywords with which this keyword occurs in the topics of the set. The result of the action is a matrix of keywords weights.

Step 8. Calculate the cosine measure of proximity between the vectors of topics and terms. The result of the action is a matrix of proximity coefficients for vectors of topics and terms.

Step 9. Rank the coefficients of proximity of vectors of topics and terms for each term. The result of the action is a matrix with the coefficients of cosine proximity of vectors of topics and terms sorted in descending order.

Step 10. Form a vector of concepts according to the rules. The result of an action is a vector of concepts.

Step 11. Calculate the distances between the vectors of themes. The function argument is the distance, which determines the choice of the metric by which the distances are calculated. The result of the action is a matrix of distances.

Step 12. Set the method for combining clusters, taking into account the function argument, which is a vector of distances between topics.

Step 13. Perform concept clustering. The function arguments are the threshold value for the selection of individual clusters [3–5].

Step 14. Build a dendrogram [3–5].

The developed algorithm and its procedure can be used both for solving particular problems of intellectual processing of textual information (extracting thematic clusters and topics from collections of texts, clustering, visualizing the content of a collection of texts, integrating hierarchical structures of concepts - for carrying out content analysis), and for constructing a simple the ontology of a collection of texts, specified in the form of a hierarchical structure of concepts in the subject area.

2 Experiment

As an example, we will give the construction of an ontology of methods for placing VLSI components using the technology described in this work.

2.1 Classification Methods for Placing VLSI

According to the classification given in the work [7–9], select a number of groups.

The first group (1) includes random, combined, exact and heuristic methods. The second group (2) methods include sequential (constructive-initial, constructive), iterative and compositional. The third group (3) includes continuous, continuous - discrete and discrete methods, which in turn are divided into classes. Note that transitions from one group to another are possible [7, 8].

Figure 2 shows the resulting hierarchy. The following ontograph (Fig. 3) [10].

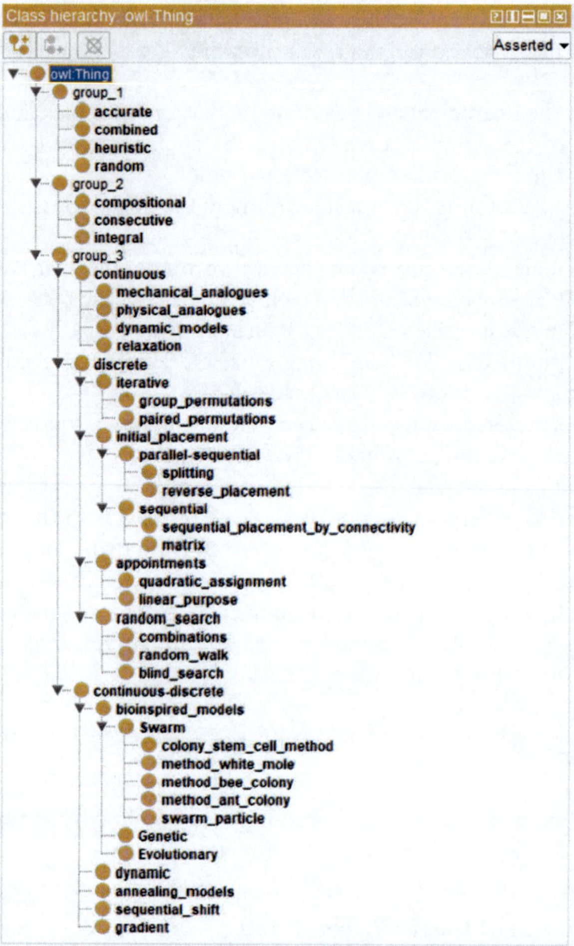

Fig. 2. Window of classes of a simple ontology, built in Protégé 4.3 according to the obtained hierarchical structure of concepts

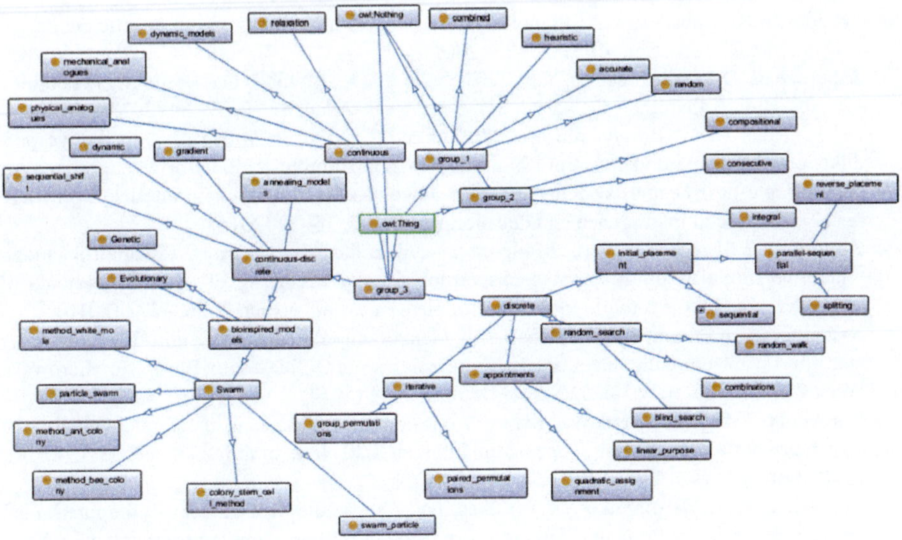

Fig. 3. Ontograph of a simple ontology "Methods for placing VLSI components"

3 Conclusion

In the course of the work, a simple ontology (taxonomy) was created for the subject area "Methods for placing VLSI components" in the form of a directed graph, the vertices of which are classes, arcs are properties representing directed relationships or connections between them. The constructed ontology is, in fact, a framework, a representation of the subject area "Methods for placing VLSI components". The ontology can be expanded by automatic integration of taxonomic structures, according to the proposed method, add new links between objects and increase the number of characteristics of each instance. The developed ontology allows you to quickly conduct information search in the knowledge base "Methods for placing VLSI components", determine the relationship between ontological concepts, and be used for content analysis.

The prospect of further development of the topic is the development of a method for reconciling the output data, represented by the matrix structure of hierarchical relationships of concepts, with the format of the linked data type, which will automate the presentation of the ontology in the owl format.

Acknowledgments. The study was performed by the grant from the Russian Science Foundation № 22-21-00316, https://rscf.ru/project/22-21-00316/ in the Southern Federal University.

References

1. Danilchenko, V.I., Kureichik, V.M.: Genetic algorithm for planning the placement of VLSI. Izvestiya SFU **2**, 75–79 (2019)

2. Lebedev, B.K., Lebedev, V.B.: Planning based on swarm intelligence and genetic evolution. Izvestia SFedU Tech. Sci. **4**(93), 25–33 (2009)
3. Zaporozhets, D.Y., Kravchenko, Y.A., Lezhebokov, A.A.: Methods of data mining in complex systems. Bull. Kabardino - Balkarian Sci. Cent. Russ. Acad. Sci. **3**, 52–54 (2013)
4. Grishchenko, M.A.: Primenenie model'no-upravlyaemogo podkhoda dlya sozdaniya produktsionnykh ekspertnykh sistem i baz znanii [Application of a model-driven approach for creating rule-based expert systems and knowledge bases]. Iskusstvennyi intellekt i prinyatie reshenii [Artificial intelligence and decision-making] **2**, 16–29 (2016)
5. Samoilova, E.M., Ignat'ev, A.A.: Modelirovanie dinamicheskoi sistemy avtomatizirovannogo tokarnogo modulya pri razrabotke ekspertnoi sistemy Modeling of dynamic systems of computer- aided lathe module in the development of expert system **2**, 268–277 (2016)
6. Kalentyev, A. A., Garays, D.V., Babak, L.I.: Genetic-algorithm-based synthesis of low-noise amplifiers with automatic selection of active elements and DC biases. In: European Microwave Week 2014, Rome, Italy, pp. 520–523, October 2014 (2014)
7. Kureychik, V.M., Safronenkova, I.B.: An ontological approach to the implementation of distributed computing technologies on the Internet 2020 News of the SFU. Tech. Sci. **4**(214), 71–82 (2020)
8. Kureychik, V.M., Safronenkova, I.B.: Creation of an ontological model of computer-aided design systems in the Protégé 4.2 environment Problems of developing promising micro- and nanoelectronic systems (MES), vol. 3, pp. 240–246 (2016)
9. Sazykin, V.G., Kudryakov, A.G.: Osobennosti podderzhki resheniya tekhnicheskikh zadach s pomoshch'yu ekspertnykh sistem [The features of support of solving technical problems with expert systems]. Put' nauki [Path of science] **8**, 21–23 (2015)
10. Danilchenko, V.I., Kureichik, V.M.: Classification and analysis of VLSI placement methods. Computer Science, Computer Engineering and Engineering Education, pp. 89–98 (2018)

Systematic Review on Online Social Media Recommender Systems

Elias Mbongeni Sibanda(✉) and Tranos Zuva

Vaal University of Technology, Vanderbijlpark, South Africa
mesibanda@gmail.com, tranosz@vut.ac.za

Abstract. The social network and the social context are two vital elements in social recommender systems. Social recommender systems are an important part of everyday life, on social media. New approaches also need to be adopted to be able to deal with the different needs of social media users. A total of twenty papers were reviewed from the years 2015–2021. There was a peak between the years 2019 and 2020 that shows that research in this area is growing. The results show that RMSE and MAE are the most used evaluation metrics. Precision and recall are the second commonly used evaluation metrics. Natural Language Processing (NLP) classifiers are becoming popular in recommender systems and are used for extracting information from textual analysis to enrich the description within a predictive algorithm. There is not enough research that focuses on online evaluation therefore it is necessary to explore online evaluations and improve results in those experiments.

Keywords: Online recommender systems · Online social network · Social network analysis

1 Introduction

Social recommender systems aim to support user preferences and help users make better decisions on social media platforms. The social network and the social context are two vital elements in social recommender systems [1]. Social recommender systems are an important part of everyday life, on social media. Users all over the world share information on social media every single minute. They post messages, news, music, and videos to draw attention from their online fans or followers. Unfortunately, those behaviors bring a plethora of information that makes users feel confused about what exactly they want [2]. Information overload is common on social media. Recommender systems are powerful tools to solve this information overload problem.

R. Silhavy (Ed.): CSOC 2022, LNNS 501, pp. 675–684, 2022.
https://doi.org/10.1007/978-3-031-09070-7_56

Social recommender systems are intended to tackle the information overload that faces social media users, thus helping users understand what they really want [3]. Numerous social media applications are stretching from Twitter, Facebook, TikTok, Reddit, etc. Their goal is to serve users with tweets or profiles that can meet users' needs. Traditional recommender systems primarily use a content-based approach and a collaborative filtering approach to serve news, books, music, and movie recommendation [4]. Social network analysis (SNA) plays a crucial role in the development of social media recommender systems.

Social media can be broken down into two main capabilities; the first capability is content generated by different users for free, while the second capability is the technological environment that is so advanced and allows these platforms to flourish. The first capability causes a problem of exponential growth because the information release is not restricted, a user can post multiple tweets on Twitter on a single day, multiple posts on Facebook, etc. [5]. Within this exponential growth, the first difficulty is that users struggle to find relevant content amongst a vast amount of alternatives. The second difficulty is faced by businesses, businesses have to provide personalized recommendations using dynamic information provided by users. According to Ai et al. [6] providing model-generated explanations in recommender systems are important to improve the experience of the user.

Businesses have a primary aim of generating profits, social media is not different from any other business model. Social media recommender systems gain popularity to increase engagements and to generate revenue for the business [7]. On social media the users play an important role, therefore constant use of the social media application is necessary. Existing recommending approaches need to be modified to deal with big data. New approaches also need to be adopted to be able to deal with the different needs of social media users. Social media users have different needs, from joining social groups or pages to using social media for serious purchase decisions and product reviews [8].

This systematic review paper focused on online social media recommender systems therefore this research paper endeavors to study the trends and identify the gaps in online social media recommender systems.

The rest of this paper is arranged as follows: Methodology, Results, Discussion, Conclusion, and Future Work.

2 Methodology

Figure 1 shows the process in which papers were retrieved, screened, and reviewed to conduct the literature review.

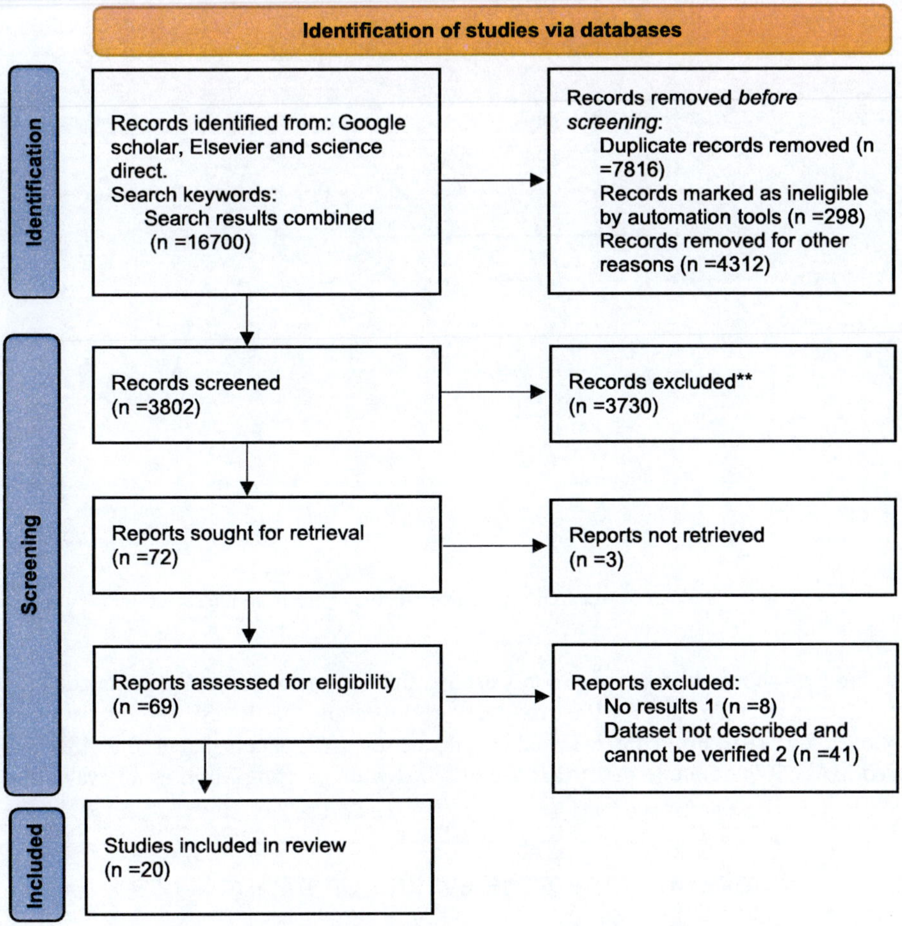

Fig. 1. PRISMA systematic review research methodology

3 Results

The 20 articles that were selected for review, were published between the years 2015 and 2021. This selection ensures that there is a wide search on the latest trends and that a proper understanding of the research area can be gained. Another reason the researchers chose this period is to ensure that the current landscape in recommender systems is adequately presented.

The results in Fig. 2 depict that from the selected papers using our search criteria only 2 papers were published in 2015, 3 papers in 2016, 2017, and 2018 respectively. There was a peak was in 2019 and 2020 where 4 papers were published that fit the research selection criteria. The graph shows a sharp decrease in 2021. The reason for having such a low number of papers that fit the criteria in 2021 could be because the study was conducted in 2021 and therefore some conferences and journals were not published.

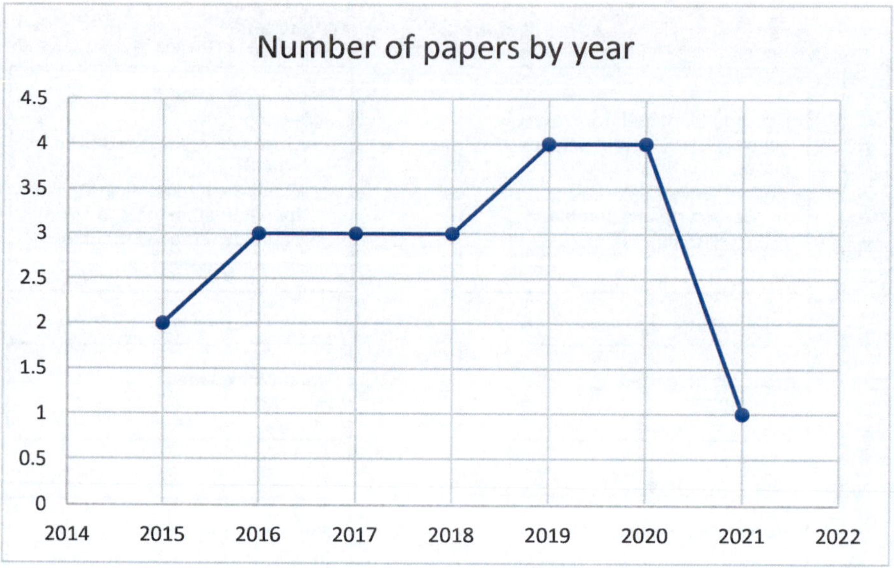

Fig. 2. Number of research papers published by year from the selection.

Figure 3 shows the most commonly used evaluation metrics from the papers that the researchers selected used for this study. The results show that RMSE is the most used, followed by MAE. From the reviewed papers, the seven (7) papers that used RMSE also used MAE, it is only one paper that used RMSE without using MAE as an evaluation

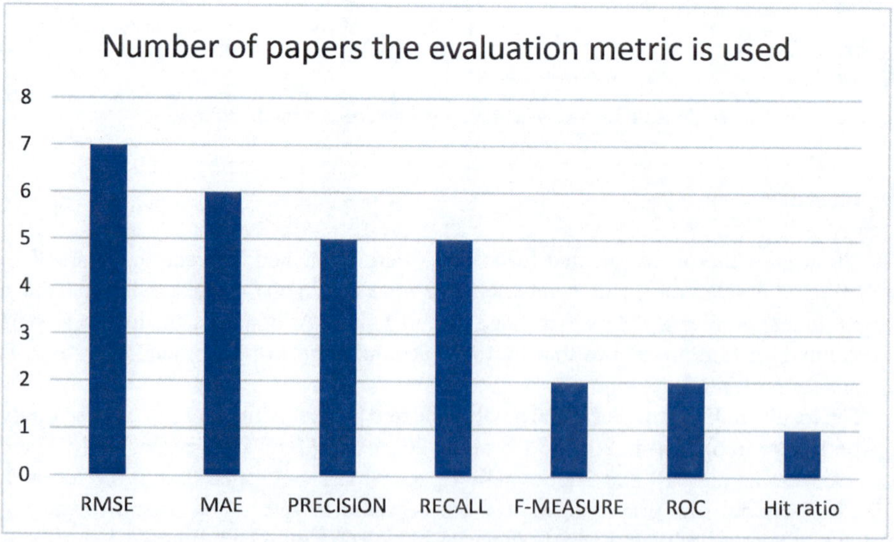

Fig. 3. Evaluation metrics used in social media recommender systems

metric. Precision and recall are the second commonly used evaluation metrics, they go together and researchers who use precision also use recall, the researcher also discovered that a precision and recall graph is popular among social recommender systems. Precision measures how relevant the results are while recall measures how many are truly relevant from the results set. F-Measure, ROC, and hit ratio have the lowest appearance from the papers reviewed in this study.

4 Discussion

In this section, related studies have been reviewed in three aspects, namely recommender systems, heterogeneous information networks, and network embedding. In the literature of recommender systems, early works mainly implement collaborative filtering (CF) methods to exploit historical interactions for a recommendation, predominantly, the matrix factorization approach as depicted in Fig. 4. Koren, et al. [9, 10] has shown its effectiveness and efficiency in many applications, which factorizes user-item rating matrix into two low-rank user-specific and item-specific matrices, and then employs the factorized matrices to make further predictions [11]. Since CF methods usually suffer from cold start problems, many works [12, 13] attempt to pull additional information to improve recommendation performance. For example, Ma et al. [14] integrate social relations into matrix factorization in recommendations.

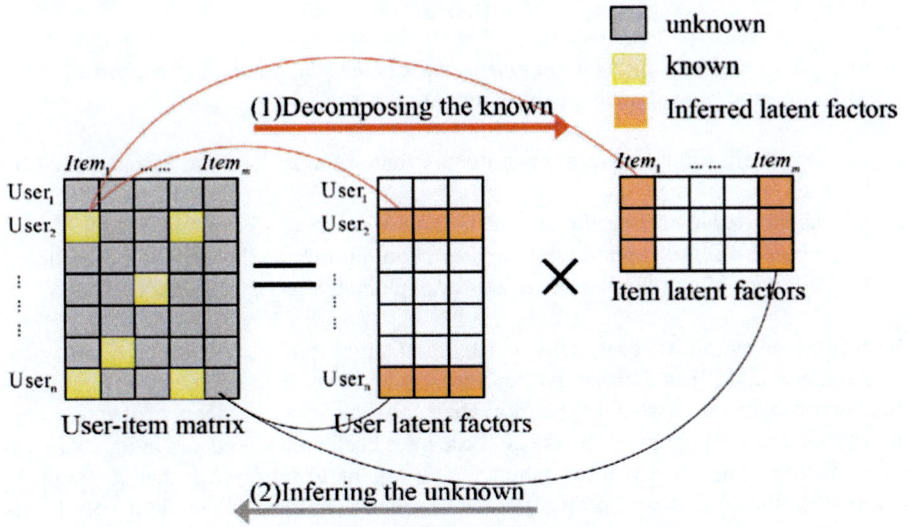

Fig. 4. Matrix factorization diagram from [15].

A personalized recommender system collects information about the preference of its users for items. With the information, it recommends items that its users may have an interest in. according to [16] in previous systems, this information could be obtained by using explicit or implicit information. As a result of using big social data such as follower-followee, friends' lists, tweets, blog posts, and tags, we can acquire more information to enhance personalized recommendations, compared with the recommendations provided by systems that do not use such data. Therefore, there has been extensive research on the use of big social data in personalized recommender systems. Big social data include various implicit data, particularly user-generated data, such as tags and profiles, because of their intrinsic nature. Seo et al. [17] proposed and described a novel recommending system for big data applications able to provide recommendations based on the interactions among users and the generated multimedia contents in one or more social media networks. The proposed system relied on a "user-centered" approach. An experimental campaign, using data coming from many social media networks, was performed and it obtained promising results.

According to Liu et al. [18], in social recommender systems, social entities and social contexts are two main apparatuses of a recommendation framework design. Based on social network analysis (SNA), social entities are used to expand social neighborhoods. Some measurements of SNA, such as a clustering coefficient, degree centrality, closeness centrality, and betweenness centrality, are calculated to analyze the network structure.

Ai et al. [6] in their study, they focused on explainable product recommendation, where the objective of the recommender system was to recommend products to users and explain why those products were recommended, they achieved their objective by first constructing a knowledgebase as a set of triplets as shown in Eq. 1

$$S = \{(eh, et, r)\} \tag{1}$$

where eh is a head entity, et is a tail entity, and r is the relationship from eh to et.

Then the goal of explainable recommendation is twofold,

1. for each user *u*, find one or a set of items *i* that are most likely to be purchased by the user
2. for each retrieved user-item pair, construct they constructed a natural language sentence based on S to explain why the user should purchase the item and considered five (5) types of entities (i.e., eh or et) for explainable recommendation:

the entities are as follows user, item, word, brand, and category.

Amato et al. [19] ın their study they proposed a novel user-centered recommendation approach for online social networks. They consider several aspects related to users' preferences; namely behavior, feedback these have been considered and integrated with items' features and context information within a general framework that can support different applications using proper particular customizations. In, modern social networking applications can take the more important advantages of their framework that can be adopted as the recommendation engine for a variety of items embedded in a social environment. In addition, the system can be easily used to provide recommendations for more than one category of items. Their experimental results revealed that the approach was promising and encouraged further research in that direction.

Ai et al. [6] proposed to learn over heterogenous Knowledge-Based Embedding (KBE) for a personalized explainable recommendation. To do so, they constructed the user-item knowledge graph to incorporate both user behaviors and item information. They further learn the KBE with the heterogenous relations collectively and leverage the user and item embeddings to generate personalized recommendations. Experimental results on real-world datasets verified the superior performance of their approach, as well as its flexibility to incorporate multiple relation types.

Javad Shayegan and Valizadeh [20] used Telegram social media application to conduct their experiments, they established a relationship between each user's trait and his words. At the same time, the personalities of the users were identified by conducting a NEO survey. The personality analysis was done by using Cosine Similarity as shown in Eq. 2 below,

In cosine-based similarity, two items are thought of as two vectors in the m dimensional user space. The similarity between them is measured by computing the cosine of the angle between these two vectors. Formally, in the $m * n$ rating matrix below the similarity between items i and j, denoted by $sim(i, j)$ is given by:

$$sim(i,j) = \cos(\bar{i},\bar{j}) = \frac{\bar{i}.\bar{j}}{\|\bar{i}\|_2 * \|\bar{j}\|_2} \tag{2}$$

where "." denotes the dot-product of the vectors and sim(i,j) is the similarity between the item i and item j.

Mughaid et al. [21] in their study explored various data mining techniques to design a social media recommendation system. Job seekers' posts and employers' job postings were collected from various fields of specialization and unsupervised learning methods were used to prepare the shortlisted database. Text mining techniques were applied on posts to structure the in-between lines information. Natural Language Processing (NLP) classifiers such as (NB, SVM, and RF) are used for extracting information from textual analysis to enrich the description within a predictive algorithm. The proposed recommendation system adapted to the cold start solution, this method proved promising results during experiments on a real data set.

Wu et al. [22] analyzed the limitations of social regularization, including three aspects: First, the social regularization only takes constraints on users with their social relation; second, much available information of users and items can be used in a heterogeneous information network (HIN), while the social regularization did not make full use of it; third, the social regularization cannot work well for users with low similarity. Therefore, we design a dual similarity regularization whose basic idea is to enforce the constraints on both similar and dissimilar objects simultaneously. Then, encouraged by the success of the low-rank matrix factorization method, we employ the similarity regularization on the low-rank matrix factorization framework and proposed the DSR method, which integrates much useful information and takes constraints on users more effectively. Experiments on three real-world data sets validated the effectiveness of DSR, especially on alleviating the cold-start problem.

Based on this study as depicted in Fig. 5 below, the researchers have discovered that between 2015 and 2020 content-based approach was prominent throughout the years in the selected papers, it had been used in a total of eight (8) papers, followed by hybrid models. Between 2016 and 2021 hybrid has been implemented seven (7) times and was the most used approach in 2018 with a total of three (3) papers. Collaborative filtering was used four (4) times between the years 2015 and 2020. İnterestingly knowledge-based approach has not been used since 2017 this could show that the approach is no longer relevant on its own, however, some hybrid models do accommodate a knowledge-based approach.

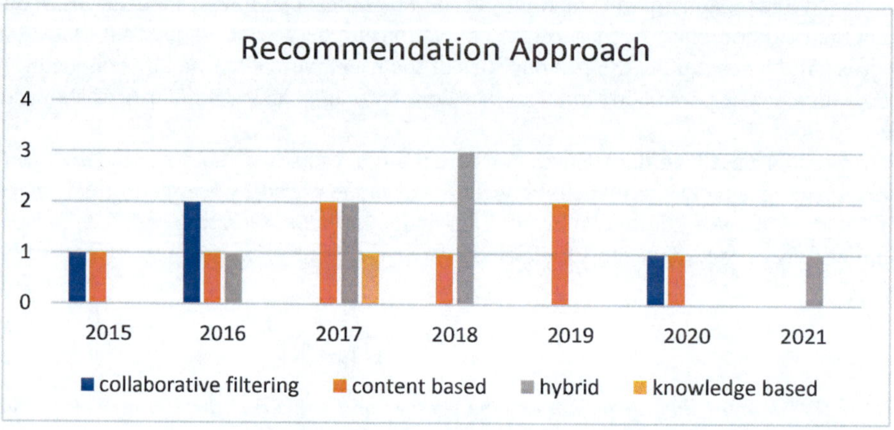

Fig. 5. Recommendation approaches

The researchers also discovered that the cold start problem is still prominent in recommender systems, however, in social media recommender systems, it is easier to overcome by using hybrids. Friendship strength in online social media recommender systems should be considered. For example, Delicious has contact information among users, then Facebook and Yelp have review information among users. These contact and review information can calculate interaction similarity of friendship strength as proposed by [17]. Different researchers use different social media sites for experiments, research that combines user profiles from different data sources to form complete user profiles from all sites would have a complete user profile.

5 Conclusion and Future Work

Most social media recommender systems research is conducted offline therefore new improved ways of conducting experiments can be adopted and practiced, although repeating the experiments may be difficult, future work should focus on online evaluation and improving results on those experiments. İn our future work we intend to explore the user-centered recommendation approach for online social networks.

This systematic literature review on social media applications shows that this research area has attracted the attention of academia. This study was carefully selected and reviewed to understand the trend of these systems. The results in this study can be summarised as follows, there is still great interest in this research area and it will grow significantly over the coming years.

References

1. Rivas, A., Chamoso, P., González-Briones, A., Casado-Vara, R., Corchado, J.M.: Hybrid job offer recommender system in a social network. Exp. Syst. **36**(4), e12416 (2019)
2. Yang, D., Huang, C., Wang, M.: A social recommender system by combining social network and sentiment similarity: a case study of healthcare. J. Inf. Sci. **43**(5), 635–648 (2017)
3. Bonhard, P., Sasse, M.A.: 'Knowing me, knowing you'—using profiles and social networking to improve recommender systems. BT Technol. J. **24**(3), 84–98 (2006)
4. Jia, R., Li, R., Gao, M.: Study on data sparsity in social network-based recommender system. Int. J. Comput. Sci. Eng. **20**(1), 15–20 (2019)
5. Alahamdi, D.H.: Recommender Systems Based on Online Social Networks-An Implicit Social Trust and Sentiment Analysis Approach. The University of Manchester (United Kingdom) (2017)
6. Ai, Q., Azizi, V., Chen, X., Zhang, Y.: Learning heterogeneous knowledge base embeddings for explainable recommendation. Algorithms **11**(9), 137 (2018)
7. Kim, N., Kim, W.: Do your social media lead you to make social deal purchases? Consumer-generated social referrals for sales via social commerce. Int. J. Inf. Manage. **39**, 38–48 (2018)
8. Ramanathan, U., Subramanian, N., Parrott, G.: Role of social media in retail network operations and marketing to enhance customer satisfaction. Int. J. Oper. Product. Manag. **37**(1), 105–123 (2017)
9. Koren, Y., Bell, R., Volinsky, C.: Matrix factorization techniques for recommender systems. Computer **42**(8), 30–37 (2009)
10. Shi, Y., Zhao, X., Wang, J., Larson, M., Hanjalic, A.: Adaptive diversification of recommendation results via latent factor portfolio. In: Proceedings of the 35th International ACM SIGIR Conference on Research and Development in Information Retrieval, pp. 175–184 (2012)
11. Koren, Y., Bell, R.: Advances in collaborative filtering. In: Ricci, F., Rokach, L., Shapira, B. (eds.) Recommender Systems Handbook, pp. 77–118. Springer, Boston (2015). https://doi.org/10.1007/978-1-4899-7637-6_3
12. Feng, W., Wang, J.: Incorporating heterogeneous information for personalized tag recommendation in social tagging systems. In: Proceedings of the 18th ACM SIGKDD International Conference on Knowledge Discovery and Data Mining, pp. 1276–1284 (2012)
13. Hong, L., Doumith, A.S., Davison, B.D.: Co-factorization machines: modeling user interests and predicting individual decisions in twitter. In: Proceedings of the Sixth ACM International Conference on Web Search and Data Mining, pp. 557–566 (2013)
14. Ma, H., Zhou, D., Liu, C., Lyu, M.R., King, I.: Recommender systems with social regularization. In: Proceedings of the Fourth ACM International Conference on Web Search and Data Mining, pp. 287–296 (2011)
15. Cai, L., Xu, J., Liu, J., Pei, T.: Integrating spatial and temporal contexts into a factorization model for POI recommendation. Int. J. Geogr. Inf. Sci. **32**(3), 524–546 (2018)
16. Yan, W., Liu, Q., Chen, R., Yi, S.: Social networks formed by follower–followee relationships on academic social networking sites: an examination of corporation users. Scientometrics **124**(3), 2083–2101 (2020). https://doi.org/10.1007/s11192-020-03553-y
17. Seo, Y.-D., Kim, Y.-G., Lee, E., Baik, D.-K.: Personalized recommender system based on friendship strength in social network services. Exp. Syst. Appl. **69**, 135–148 (2017)
18. Liu, H., et al.: A review of the smart world. Futur. Gener. Comput. Syst. **96**, 678–691 (2019)
19. Amato, F., Moscato, V., Picariello, A., Piccialli, F.: SOS: a multimedia recommender system for online social networks. Futur. Gener. Comput. Syst. **93**, 914–923 (2019)
20. Javad Shayegan, M., Valizadeh, M.: A recommender system based on the analysis of personality traits in Telegram social network. arXiv preprints arXiv:2010.00643 (2020)

21. Mughaid, A., Obeidat, I., Hawashin, B., AlZu'bi, S., Aqel, D.: A smart geo-location job recommender system based on social media posts. In: 2019 Sixth International Conference on Social Networks Analysis, Management and Security (SNAMS), pp. 505–510. IEEE (2019)
22. Wu, Q., et al.: Dual graph attention networks for deep latent representation of multifaceted social effects in recommender systems. In: The World Wide Web Conference, pp. 2091–2102 (2019)

Fundamental Skills in Support of Technical Brilliance

Anneke Harmse(✉)

Department of ICY, Vaal University of Technology, Vanderbijlpark, South Africa
anneke@vut.ac.za

Abstract. The Skills Framework for the Information Age (SFIA) is commonly used by professionals in the Information and Communication Technology (ICT) arena. SFIA is a management tool that assists educators and employers in the development of ICT skills and the search for appropriate employment. It is also considered to be a useful benchmark for consideration in the context of specific countries according to their requirements. ICT practitioners are expected to have acquired technical, together with several additional skills. Both, technical and additional (non-technical) skills were stated as important additions for an holistic ICT practitioner. The additional skills are considered essential by the South African ICT industry. A significant number of required soft skills have been expressed by SFIA as adjuncts to the implementation of listed technical skills. In order to execute technical skills, ICT practitioners must possess attributes mentioned in the technical SFIA skills descriptions. However, several additional skills are not explicitly mentioned in SFIA. These include critical thinking, curiosity, Data capturing, Entrepreneurship, Lateral thinking and IT jargon amongst others. The current study investigated the descriptions of technical skills that may or may not incorporate additional skills not listed in the SFIA. Our investigation found that these additional skills should be considered important for the holistic ICT practitioner and thus incorporated into SFIA guidelines.

Keywords: SFIA · ICT skills · Additional skills

1 Introduction

Information Communication and Technology (ICT) and related courses offered at Higher Education Institutions provide increased knowledge, skills and competencies required by the ICT industry to potential graduates. These are usually guided by the relationship between Higher Education and the industrial sector. Despite the guidance of industry regarding the offering of academic programmes the lack of soft skills are frequently reported to be lacking or that graduates are deficient in [1–4]. Harmse [5] reported on the expectation by the ICT profession for potential ICT graduates to be exposed to a combination of both technical and non-technical skills during their academic careers.

Smith [6] focused on the growing appreciation of good interpersonal skills and suggested that whilst technical skills remain critical, soft skills have become an important measure to evaluate and select potential employees. The Association for Computing

R. Silhavy (Ed.): CSOC 2022, LNNS 501, pp. 685–694, 2022.
https://doi.org/10.1007/978-3-031-09070-7_57

Machinery (ACM) refers to the "Academic Myth" whereby it is believed that an ICT qualification by itself is sufficient to attain a position in the ICT industry. Although obtaining the technical qualification is necessary, this may not necessarily be sufficient to be the only requirement for a successful ICT professional. It is generally understood that in ICT and related fields that a successful professional is expected to possess communication skills, be a strong team player, and be passionated about sucess [7]. Furthermore, the ACM is of the opinion that for appropriate student learning the curriculum should include more than only acquiring the necessary technical knowledge in order to be successful practitioners. Sabin et al. [8], have suggested that in the academic training of ICT graduates, curricula should include the development of the soft skills required from an ICT professional. A recent study by Harmse [5] reported that SFIA is useful for the identification of skills required to develop a proficient ICT graduate. The study indicated that competencies of ICT graduates in relation to the requirements of the ICT industry exceed that of only technical skills related to developing an effective IS. The study suggested that additional skills be added to the list of skills required from an ICT graduate. Consequently, the combination of technical and additional skills, that complement each other should be considered as essential skills required of an ICT professional.

Within the ICT domain, the Skills Framework for the Information Age (SFIA) is a benchmark for UCT professionals and adopted by almost 200 countries including South Africa [9]. The development of ICT skills is guided by SFIA as a management tool to assist decision making. However, it is often criticised for the lack of its focus on soft skills. Despite this lack, the SFIA is currently used as a benchmark tool by various organisations and countries [9]. A common perception is that SFIA focuses on technical skills at the expense of soft skills [10]. However, industry professionals believe that soft skills should be defined within the context of specific SFIA technical skills [10]. The study conducted by Harmse [5] included examples of the required skills including communication skills amongst others from the South African ICT industry.

The exploratory nature of this study was to gain a deeper understanding of the non-technical competencies needed by ICT students relevant to their profession. The current study aimed to compare the additional skills identified by Harmse [5] with the descriptions of SFIA technical skills. The study envisaged determining whether the identification of specific technical skills could guide the identification of additional skills required by an ICT graduate.

2 Literature Review

Due to the increasing awareness with the increased awareness of ICT and its benefits, rapid technology developments have created an expectation of change in ICT positions. New profiles such as Big Data and Cloud computing specialists have been created [11] and new employment positions are likely to be created in areas such as the development and supervision of Artificial Intelligence [12]. Although new and developing technologies create a demand for new skills sets, the redeployment of established players could result in the retrenchment of thousands of ICT practitioners. In a recent report of the Joburg Centre for Software Engineering (JCSE), Schofield [13] suggests that the global IT market could have reached an estimated US$5 trillion in 2020. McLaughlin, et al.

[11] suggested that the ICT sector realised the growth of approximately 100 000 jobs until 2015 but a deficit of more than 756 000 in the European sector jobs caused due to a lack of available talent is predicted in 2020 by CEPIS [14]. The ICT industry considers retraining and reskilling employees in the age of automation [15]. A concern reported by MacLellan [16] is the gap that threatens the supply of soft skills amongst ICT professionals where the demand is higher than for technical skills.

The pressure on ICT graduates is also greater than before. Graduates require skills with a high economic value are expected to be prepared to accept positions in industries that do not currently exist [17]. Graduates must be malleable and flexible to meet the challenges of the market and willing to continuously develop themselves [18, 19]. According to the ICT-skills and Salary Report, professionals are required "to wear many hats" [20]. Roodt and Paterson [21] suggest that in addition to the fact that the skills of ICT graduates do not meet industry requirements, the educational sector training supply institutions do not substantially contribute to raising skills among professionals skill. ManagementSA [22] has argued that the South African labour market cannot produce high-quality employees that are required by the ICT industry. Recently, Burrows (2017) reported that the ICT industry noted a deficiency of soft skills, such as emotional intelligence, cross-cultural competency and social interaction amongst ICT graduates. A further complication is that the existing educational system is not designed to prepare individuals for the workplace. Similar scenarios have been reported in countries within the European Union [23]. In this regard, Adesina et al. [24]; stated in the African Economic Outlook of 2016, that the underlying problem of skills acquisition and poor quality of the South African educational system, directly relate to the deficit of links between educational systems and employers.

There are currently several identified business and societal challenges related to ICT in South Africa. Currently, unemployment rates are at 27% [25] and a youth unemployment rate up to 55% [26]. A skills mismatch could hold back economic growth and generate significant social costs. The shortage of skilled South African ICT workers also contributed to localized knowledge gaps, such that business finds it increasingly challenging to meet the demands for ICT enabled innovation. There is a need for a continuous corporation between government, the ICT industry and the education sector [23] as the skills of South African ICT graduates do not meet expectations of the industry. This problem could be caused by the current educational system not preparing quality graduates for employability or work-readiness, along with the lack of relationships between the educational system, government and the ICT industry. Graduates will be expected to offer relevant skills to the ICT industry. Thus, ICT educational offerings should align with the expectations of this industry. Solid technical and soft skills – sometimes known as interpersonal skills - are essential for a variety of positions in ICT. These include: effective interaction with clients and suppliers, colleagues and managers, managing departments, and the ability to communicate effectively [27]. In this regard, Doyle [28] promoted the view that a combination of both hard and soft skills are evaluated during the hiring process. Furthermore, soft skills such as communication, creativity determination and teamwork were proposed as being an advantage to help a potential employee in rising above the competition.

Relevant or necessary skills can be defined by an organisation itself or by third party organisations [29]. These skills are generally demarcated in terms of a skills framework [30]. Katz [31] claimed that skill implies an ability that can be developed and not necessarily innate. Skills can therefore be demonstrated in performance, not just in potential. This has led to a prominent theory of skills that refers to three main professions, namely Conceptual, Human and Technical skills. With reference to the present study, it is necessary to understand conceptual skills such as fresh ideas and new concepts. These skills ensure the understanding of the actions or measures necessary to be followed in a particular area. Within the ICT domain, conceptual skills are explained by SFIA [32].

3 Methodology

To identify skills required by ICT graduates a qualitative framework focussing on the required skillset of ICT graduates as a phenomenon was followed. Unstructured interviews [33, 34] were utilised to collect data. Transcribed verbatim Microsoft Word - Atlas.ti™ Version 8 was used for analysing data. The theoretical approach was based on Grounded Theory with a code list, using the constant comparative method [35] to explore the skills required of ICT graduates.

The sample consisted of ICT practitioners employed in a variety of roles in industrial sectors as identified by Statistics South Africa [36]. The theoretical sampling approach was expanded and complemented with purposeful snowball sampling [37]; for the identification of ICT participants as possible interviewees. The theoretical sampling provided an avenue to saturate as many codes as possible. Purposive sampling which followed this process provided a means to identify additional ICT participants for data collection. Codes were based on skills identified by the SFIA. Where required, additional codes were created to include skills such as communications and entrepreneurship, amongst others.

Additional skills listed by Harmse [5] were compared with the descriptions of the technical SFIA skills. The latter list incorporated 31 different skills not listed in SFIA; but viewed as essential skills by South African ICT practitioners. Following on the methodology applied by Harmse [5], the current study applied unstructured interviews as the data collection method with the aim to develop a framework to determine the competencies required by ICT graduates. The population of ICT professionals were guided by two sampling techniques. Firstly, theoretical sampling was complemented by purposeful convenience sampling methods that were used to identify targeted industries based in the Gauteng province in South Africa. Secondly, acknowledging a variety of ICT professionals, purposeful snowball sampling was used to appreciate the activity of ICT professionals involved in the ICT industry. The interview transcripts were analysed by using a grounded theoretical approach with the use of a code list. The data was analysed according to the constant comparative method [35], in that data were categorised and compared to identify conceptual similarities and/or dissimilarities between the various interviews with ICT professionals. Codes were based on skills identified by the SFIA. Where required, additional codes were created to include skills required by an ICT graduate but not included in the list of SFIA skills.

4 Results

In a previous study [5] Communication skills were highlighted as important. For the skill of "Software Design", SFIA describes the creation and communication of design views to identify functional and non-functional requirements as important. Therefore, this study compared additional skills considered to be important by the South African ICT industry [5] with SFIA descriptions.

Following analysis of the interview transcripts with the use of a code list, data were categorised and compared to identify conceptual similarities and/or dissimilarities between the various interviews with ICT professionals. Whilst, in general, codes were based on skills identified by the SFIA, additional codes were created to include skills

Table 1. Additional skills within the descriptions of SFIA technical skills

Additional skill	Technical SFIA skill requiring an additional skill	
Adapt	• Innovation • Research • Project Management • Business Analysis • Requirements Definition and Management • Organisation Design and Implementation • Systems Development Management	• Systems Design • Software Design • Programming/Software Development • Competency Assessment • Learning Delivery • Product Management
Analytical Skills	• Analytics	
Attitude	• Learning Delivery	
Certification	• Learning and Development Management • Conformance Review	
Communication	• Strategic Planning • Information Governance • Information Security • Data Visualisation • Demand Management, • IT Management • Innovation • Research • Knowledge Management • Emerging Technology Monitoring • Solution Architecture • Programme Management • Business Analysis, • Business Modelling • Organisational Capability Development • Organisation Design and Implementation • Product Management	• Change Implementation Planning and Management • Systems Development Management • Software Design • Data Modelling and Design • Testing • User Experience Analysis • User Experience Design • Asset Management, • Release and Deployment • Penetration Testing, • Teaching and Subject Formation • Resourcing • Supplier Management • Relationship Management • Marketing, • Sales Support

(*continued*)

Table 1. (*continued*)

Additional skill	Technical SFIA skill requiring an additional skill	
Conflict Management	• Requirements Definition and Management • User Experience Analysis • Supplier Management	
Facilitation	• Data Visualisation • Consultancy • Demand Management • Financial Management • Innovation • Business Process Improvement • Knowledge Management • Enterprise and Business Architecture • Emerging Technology Monitoring • Project Management • Requirements Definition and Management • Organisation Design and Implementation	• User Research • User Experience Evaluation • System Software • Incident Management • Facilities Management • Learning Delivery • Performance Management • Resourcing • Professional Development • Quality Management • Relationship Management • Product Management • Software Design • Systems Development Management
Integrity	• Information Security • Information Assurance • Business Risk Management • Continuity Management • Continuity Management • Data Management • Real-Time/Embedded Systems Development	• Safety Engineering • Configuration Management • Change Management • Penetration Testing, • Storage Management • Safety Assessment • Supplier Management • Testing
IT Levels, Function and Framework	• Enterprise IT Governance • IT Management • Business Analysis • Organisation Design and Implementation • Learning Delivery • Sourcing • Supplier Management • Selling • Enterprise IT Governance • Information Governance • Information Systems Coordination • Information Security • Analytics, IT Management • Research • Business Process Improvement • Business Risk Management • Continuity Management • Marketing	• Solution Architecture • Business Analysis • Business Modelling • Organisational Capability Development • Systems Design • Software Design • Real-Time/Embedded Systems Development • Testing • Porting/Software Configuration • Configuration Management • Release and Deployment • Application Support • Network Support • Incident Management • Quality Management • Measurement • Relationship Management • Customer Service Support

(*continued*)

Table 1. *(continued)*

Additional skill	Technical SFIA skill requiring an additional skill	
Logic	• Data Management • Network Design	• Hardware Design • Database Administration
Negotiation Skills	• Performance Management • Service Level Management	• Contract Management • Selling
Patience	• Project Management	
People Relationship	• Enterprise IT Governance • Consultancy • Knowledge Management • Enterprise and Business Architecture • Continuity Management • Network Support • Testing • Information Content Authoring	• Service Level Management • Penetration Testing • Performance Management • Sourcing • Supplier Management • Relationship Management • Selling
Practical Skills	• Research • Testing	• Learning Delivery • Sales Support
Presentation Skills	• Research, • User Research	• Digital Forensics • Selling
Pro-Active	• Information Security • Information Assurance • Capacity Management	• Performance Management • Contract Management
Professional Skills	• Information Governance, • Specialist Advice • IT Management • Research • Systems Installation/Decommissioning • IT Infrastructure, • Learning and Development Management	• Learning Delivery, Teaching and Subject Formation • Performance Management • Professional Development • Safety Assessment • Sourcing
Social Skills	• User Experience Analysis	
Vendor Products	• Knowledge Management • Testing • Penetration Testing	• Resourcing, • Quality Assurance, • Sourcing • Supplier Management
Version Control	• Porting/Software Configuration • Configuration Management	• System Software

required by an ICT graduate but not included in the list of SFIA skills. The study by Harmse (2018) identified 31 additional skills to the 97 described by SFIA. Of these, 20 appeared in the SFIA descriptions. For each of the twenty additional skills, a number of SFIA skills were identified within the SFIA descriptions. Table 1 illustrates the identified additional skills essential to undertake the relevant technical SFIA skill.

5 Conclusion

As quoted above, current ICT employers consider the attainment of the skills by an ICT practitioner to be essential. This research noted the reported lack of additional by SFIA. This study aimed to determine whether the additional identified by [5] might be mentioned in the descriptions of SFIA technical skills. Thus, the study aimed to study SFIA technical skills and determine whether the required soft skills are contained in the descriptions. Based on the results, it can be deduced that soft skills are indeed used in the descriptions provided by SFIA. A significant number of required soft skills expressed by SFIA regarding the implementation of listed technical skills has been demonstrated. In order to execute technical skills, ICT practitioners must possess attributes mentioned in the technical SFIA skills descriptions. However, it should be noted that some of the additional skills identified by [5] namely Appreciation, Critical thinking, curiosity, Data capturing, Divide and conquer, Entrepreneurship, Lateral thinking, IT jargon, MS Excel Pressure and Typing are not explicitly mentioned in SFIA.

It is therefore recommended that the additional skills described herein be considered as an important attribute for performing the SFIA technical skills for the holistic ICT practitioner.

References

1. Calitz, A.P., Cullen, M.D.M., Greyling, J.H.: South African alumni perceptions of the industry ICT skills requirements. Presented at the Proceedings of the 44th Annual Conference of the Southern African Computer Lecturers' Association Johannesburg, South Africa (2015). https://www.researchgate.net/profile/Andre_Calitz/publication/281064005_South_African_Alumni_Perceptions_of_the_Industry_ICT_Skills_Requirements/links/55d3211208ae7fb244f57f7b.pdf
2. Hurrell, S.A.: Rethinking the soft skills deficit blame game: employers, skills withdrawal and the reporting of soft skills gaps. Hum. Relations **69**(3), 605–628 (2016). https://doi.org/10.1177/0018726715591636
3. Mabiza, J., Mahlalela, P., Mbohwa, C.:Reducing unemployment rate in South Africa through establishment of graduate internship programmes (GIP). In: Proceedings of the International MultiConference of Engineers and Computer Scientists, vol. 2, pp. 15–17 (2017)
4. Chetwynd, F., Aiken, F., Jefferis, H.: Reflections on the 2017 HEA STEM conference: graduate employability challenges and solutions. High. Educ. Pedag. **3**(1), 490–494 (2018)
5. Harmse, A.: Expectations of industry regarding the ICT competencies of HEI graduates. PhD Information Technology. North-West University, Vanderbijlpark (2018)
6. Smith, J.L.: Soft skills are underrated: quality professionals need strong interpersonal skills. Quality **57**(9), 18–19 (2018). https://go.galegroup.com/ps/anonymous?id=GALE%7CA557025275&sid=googleScholar&v=2.1&it=r&linkaccess=abs&issn=03609936&p=AONE&sw=w
7. Lewis, D., Mills, G.R.: The bullseye principle: mastering intention-based communication to collaborate, execute, and succeed, p. 304. Wiley, Hoboken (2018)
8. Sabin, M., et al.: Information Technology Curricula 2017: Curriculum Guidelines for Baccalaureate Degree Programs in Information Technology, p. 165. ACM, New York (2017)
9. SFIA. "About SFIA." https://www.sfia-online.org/en/framework/sfia-7/about-sfia-7. Accessed 16 Apr 2019

10. Von Konsky, B., Miller, C., Jones, A.: The skills framework for the information age: engaging stakeholders in ICT curriculum design. J. Inf. Syst. Educ. **27**(1), 37–50 (2016)
11. McLaughlin, S., et al.: E-skills: the international dimension and the impact of the globalisation. European Commission. http://www.cepis.org/media/FINAL_INTERNATIONAL_e_Skills_report_Aug_141.pdf. Accessed 26 Oct 2017
12. McLellan. IT jobs in 2020: Preparing for the next industrial revolution. https://www.zdnet.com/article/it-jobs-in-2020-preparing-for-the-next-industrial-revolution/. Accessed 8 May 2019
13. Schofield, A.: 2018 JCSE-IITPSA ICT Skills Survey. https://www.iitpsa.org.za/wp-content/uploads/2018/10/2018-JCSE-IITPSA-ICT-Skills-Survey-V1.pdf. Accessed 17 Apr 2018
14. CEPIS. Europe's widening ICT skills gap. https://www.cepis.org/index.jsp?p=3014&n=3044. Accessed 8 May 2019
15. World Economic Forum. Accelerating Workforce Reskilling for the Fourth Industrial Revolution: An agenda for Leaders to Shape the Future of Education, Gender and Work. World Economic Forum, Geneva, Switzerland. https://www.voced.edu.au/content/ngv:77198. Accessed 8 May 2019
16. MacLellan, L.: The largest skills gap managers see in IT workers isn't technical. https://qz.com/work/1558326/the-largest-skills-gap-in-it-workers-is-business-acumen/. Accessed 8 May 2019
17. Wolfe, I.: 65 Percent of today's students will be employed in jobs that don't exist yet. https://www.successperformancesolutions.com/65-percent-of-todays-students-will-be-employed-in-jobs-that-dont-exist-yet/. Accessed 24 May 2017
18. Right Management. The Flux Report: Building a resilient workforce in the face of flux. http://www.rightmanagement.co.uk/wps/wcm/connect/350a18c6-6b19-470d-adba-88c9e0394d0b/Right+Management+Flux+Report+Spread.pdf?MOD=AJPERES. Accessed 25 Oct 2017
19. Rayner, G., Papakonstantinou, T.: Employer perspectives of the current and future value of STEM graduate skills and attributes: an Australian study. J. Teach. Learn. Grad. Employab. **6**(1), 100–115 (2015)
20. Schofield, A.: 2014 JCSE ICT skills survey report. Joburg Centre for Software Engineering. http://windowsitpro.com/industry/2014-it-skills-and-salary-report. Accessed 25 Oct 2017
21. Roodt, J., Paterson, A.: ICT skills in the labour market: an occupational-level analysis focusing on computer professionals and associate professionals, 1996–2005. http://www.labour.gov.za/downloads/documents/research-documents/ICT%20skills_DoL_Report.pdf. Accessed 25 Oct 2017
22. Management SA. Understanding the skills shortage in SA? http://managementsa.co.za/understanding-the-skills-shortage-in-sa/. Accessed 9 Mar 2014
23. Kolding, M., Ahorlu, M., Robinson, C.: Post crisis: e-skills are needed to drive Europe's innovation society. http://www.microsoft.eu/linkclick.aspx?fileticket=bQQeH2nAaF8%3D&tabid=60. Accessed 26 Oct 2017
24. Adesina, A.A., Gurría, A.G., Clark, H.: African economic outlook 2016: sustainable cities and structural transformation (2016). https://www.afdb.org/fileadmin/uploads/afdb/Documents/Publications/AEO_2016_Report_Full_English.pdf. Accessed 11 Nov 2016
25. Statistics South Africa. Unemployment drops in fourth quarter of 2018. http://www.statssa.gov.za/?p=11897. Accessed 8 May 2019
26. Trading Economics. South Africa Youth Unemployment Rate. https://tradingeconomics.com/south-africa/youth-unemployment-rate. Accessed 8 May 2019
27. Schwalbe, K.: Information Technology Project Management, 8th edn., p. 672. Cengage Learning, USA (2015)
28. Doyle, A.: Top Information Technology (IT) Soft Skills. https://www.thebalancecareers.com/top-information-technology-it-soft-skills-2063781. Accessed Jan 2019

29. Ivancevich, J.: Business, Principles, Guidelines and Practices, 2nd edn, p. 242. Content Technologies Inc. (2016). Accessed 24 Oct 2017
30. Tomar, R.: Commercial Operations Management: Process and Technology to Support Commercial Activities, p. 312. Global India Publications Pvt. Limited (2009)
31. Katz, R.L.: Skills of an effective administrator. Harv. Bus. Rev. **33**(1), 33–42 (1955)
32. SFIA. Skills Framework for the Information Age. https://www.sfia-online.org/en. Accessed 30 Jan 2017
33. Lester, S.: An introduction to phenomenological research. http://www.rgs.org/nr/rdonlyres/f50603e0-41af-4b15-9c84-ba7e4de8cb4f/0/seaweedphenomenologyresearch.pdf. Accessed 25 Oct 2017
34. Marshall, C., Rossman, G.B.: Designing Qualitative Research, 6th edn, p. 352. SAGE, Thousand Oaks (2014)
35. Boeije, H.: A purposeful approach to the constant comparative method in the analysis of qualitative interviews. Qual. Quant. **36**(4), 391–409 (2002)
36. Statistics South Africa. Standard industrial classification of all economic activities (SIC). Statistics South Africa. http://www.statssa.gov.za/classifications/codelists/Web_SIC7a/SIC_7_Final_Manual_Errata.pdf. Accessed 26 Oct 2017
37. Donoso-Maluf, F.: What is the difference between theoretical sampling and snowball sampling in qualitative research? https://www.researchgate.net/post/What_is_the_difference_between_theoretical_sampling_and_snowball_sampling_in_qualitative_research. Accessed 12 Feb 2018

Consumer Readiness for Adoption of IOT-Smart Homes (CRA-IOT-SH) in South Africa Gauteng

Bryan David Julies(✉) and Tranos Zuva

Vaal University of Technology, Private Bag X021, Andries Potgieter Blvd., Vanderbijlpark 1911, South Africa

bryanjul@gmail.com, tranosz@vut.ac.za

Abstract. There are many advantages that Smart home technology can bring to the modernized household nonetheless, smart home technology is yet in early turn of events and has toiled to take up among consumers. The study's aim is to look into the readiness of IOT Smart Home technology among consumers in Gauteng South Africa. This paper looks at the relationships between determinant factors influencing the consumer's intentions. In order to achieve this relationship for the consumer's adoption the Theory Acceptance Model (TAM) was examined. The factors examined from TAM was Perceived Ease of Use (PEOU), Perceive Usefulness (PU), Relative Advantage (RA), Perceived Innovation (PI), Trust (T), Social Influence (SI), Attitude (AT) and Intention of Use (IOU). The test reflects that PU, PEOU, T, RA and PI has a significant influence on Consumer's Attitude (A) for adoption of IOT-smart home. However PEOU has an insignificant influence on Perceive Usefulness (PU) in the adoption IOT-smart home.

Keywords: Smart homes · TAM · Trust · Reliability · Internet of things · Attitude · Social influence · Innovation · Reliability

1 Introduction

A new technology in present age is called Internet of Things (IOT). IOT is an evolving technology that is expected to enhance our daily lives. As per [1] IOT as a disruptive technology where the cyber world meets the physical world. Beforehand, the Internet was a tool used to connect people. Now with IOT, the Internet is a tool that connects devices or "Things" to transmit or communicate. It includes Radio-Frequency Identification (RFID) tags, sensors, actuator and mobile phones and encompasses all technologies in SMAC (Social, Mobile, Analytics and Cloud) [1]. In other words, IOT is a network of gadgets, appliances and "electronic things" that exchange data and uses low cost Internet connected devices and sensors [1].

IOT technology has alternated the style of living in our homes. A "Smart Home" is an IOT-enabled house equipment that is more intelligent, remote controllable, and interconnected [2]. Smart Homes is also called automated homes that contains devices used to control home features such as light and heating. However, the use of technology

R. Silhavy (Ed.): CSOC 2022, LNNS 501, pp. 695–709, 2022.
https://doi.org/10.1007/978-3-031-09070-7_58

in Smart Homes has evolved that almost every electrical component can be included in the Smart Home [3].

The Smart Home technology according to [3] is presently not utilize to just remotely operate devices, but it has been developed to also track the interior environment and activities of a person inside the home.

According to [3] Smart Home communication has two types of demands.

- The urgency to connect with devices in the home.
- To transmit the home towards the rest of the cyber world.

In order to set up the technology for the smart home, one needs to comprehend the architecture of a smart home. There are two routes a consumer can follow when constructing a Smart Home. Firstly, the consumer can "Do it Yourselves" (DIY), the secondly, the consumer can make use of the services of an integrator who specialize in integration on smart homes. The integrator is part of an organisation that assists the consumer with an incorporated solution for converting your normal home into a smart home. These integration organisations make use of integral solutions like CONTOL4, SAVANT and CRESTONE HOME.

According to [4] IOT expected growth is 75.44 billion worldwide by 2025 and smart home growth is forecast to be 53.45 billion U.S. dollars by 2020.

It clear that Smart Homes has the potential of bringing advantages to homeowners but due to reluctant consumer adoption, smart home technology has been struggling to take off in the market. Some of the potential factor that contributes to the slow uptake is privacy and security, trust issues reliability and price [5].

TRUST normally comes to mind with IOT technology. Similarly as clients look to have application trust over the Internet, in the same way they anticipate dependability when the smart device correspondence by means of IOT. According to [6] IOT gives remote communication to such countless gadgets without human intercession, the danger of these weakness assaults are emanate. Privacy and security protection is the most noticeable ruining component and Smart Homes are seen as perplexing and costly, that need apparent worth and dependability cost [5].

Another factor under consideration is the discrepancy between device protocols as non-compatible communication are increasing with manufactures smart devices. This increase the user's perception to adopt to smart home technology. As per [7] the survey conducted in 2016 was to understand how ready or not U.S. manufactures was to adapt to IOT technologies. It was concluded that the majority of U.S. manufactures was not ready in 2016. Most of the U.S. manufactures back then had limited understanding or knowledge how to apply IOT to their business [7].

South Africa is a country that is faced with many challenges, like unemployment, crime, corruption, poverty/destitution, water supply, electricity supply, education, basic health care, to name a few. The major crisis the country is facing, is the issue of loadshedding.

This study's aim was to investigate whether and how consumers in Gauteng South Africa is ready for IOT-Smart Home technology. The paper looked at the relationships between determinant factors influencing the consumer's intentions. Some of the factors that were looked at was user's perceptions and consequently user's adaption decisions. This was achieved by adopting the Theory Acceptance Model (TAM) and examine the consumer readiness for IOT-smart homes within Gauteng South Africa.

The main question of the study is: What are the factors that determines the influence for consumers to adopt to IOT Smart Home Technology in Gauteng, South Africa? The secondary question is:

- What has been done in literature to identify the factors that influences readiness for adoption on IOT Smart Homes?
- What model has been proposed to measure the readiness of adoption of IOT Smart-Home- Technology within Gauteng, South Africa?
- How was the measure level applied of readiness for adoption of IOT Smart-Home-Technology within Gauteng, South Africa?

The aim and objective of the research was to:

- Study the readiness for adoption of technology in literature.
- Propose a model that would measure readiness for adoption of IOT Smart Home Technology in Gauteng, South Africa.
- Measure the effectiveness of the proposed model.
- Measure the level of readiness for adoption of IOT Smart-Home-Technology within Gauteng, South Africa.

The remaining paper parts is organized as per the following sections:

Section 2: This section discusses related work from previous literature that was reviewed on the Internet-of- Things (IOT). This section also looks at the overview of TAM.
Section 3: In this session the adopted methodology will be explained. Explaining how it was used and how it will be implemented.
Section 4: In this section we will in detail interpret the results from the questionnaire research question.
Section 5: We will present the study's conclusion and provide details to future work.

2 Related Work

The Internet-of-Things (IOT) can be defined as a network of physical objects and can be classified into three things:

- People-to-People.
- People-to-Things.
- Things-to-Things.

IOT forms part of Radio Frequency Identification (RFID) and has evolved with the increase of mobile devices, ubiquitous communication, cloud computing and data analytics [8]. Other protocols and technologies of IOT includes, Wi-Fi, Bluetooth, ZigBee and, 5G cellular, NFC and Neul. Kevin Ashton a British technology engineer was the first to use the term Internet-of-Things (IOT) in 1999 to describe how objects communicates by the use of sensors and coin Radio Frequency Identification (RFID) [9]. According to [10] the data generated by devices differ from each other and is distinguished by temperature, humidity and position (GPS). The main objective of the IOT is for things to communicate anytime, anywhere with anything. IOT uses different communication models; these consist of Device-to-Device, Device-to-Cloud, Device-to-Gateway and Back-End Data-Sharing.

The evolution of IOT technology adjusted our manner we inhabit our homes. IOT shed new light on Smart Homes, a Smart Home is a house equipped with equipment that is more intelligent, remote controllable, and interconnected [2]. Some research studies has gone about the question to asked "who are these users of smart homes, and why do they want or need them?" [11].

[5] describes a Smart Home can be remotely controlled, accessed, and monitored in order to grant assistance to its inhabitants. As per [1] the "smartness" of any system is achieved through connected devices through the IOT.

Various research studies [12–14] has tried to explain the slow up take of smart homes in the market. It became clear that Smart Home service needs to fit into the design and technology architecture of the existing home and needs to continuously evolve as the technology is use [5]. On the bases, it was found that a Smart Home owner should not grow his technological knowledge to work the smart device but rather that the device should adopt to the landlord's requirements [15].

The year 1970 saw a significance growth in system technology which contributed to the interest of numerous researchers to research the adoption of system failure that follows it [16]. As indicated by [17] by far most studies failed to convey strong evaluations that could explain system affirmation or excusal and suggested Technology Acceptance Model in 1985 [16]. Davis suggested that the client communication with any framework is a response that can be deciphered and predicted by client motivations, formed by unfamiliar impetus [16]. According to [16] client motivations are moulded through an unfamiliar impetus that abides to a system's capabilities and features. Davis dependency on going before work done by [18] proceed correctly his model and propose that Technology Acceptance Model prompts customer interest can be moved with three components:

- PEOU – Perceived-Ease-of-Use
- PU - Perceived-Usefulness
- AT – Attitude-Towards-Using-The-System

As per [16], Davis theory on the consumer's attitude against a framework is a central point that demonstrates a client desire to dismiss or utilize a system. A consumer's attitude is utilized by PEOU - perceived usefulness and PU - perceived ease of use. Literature on Information Technology (IT) shows research at the individual level [19].

Countless presumptions and models utilizes IT reception at the individual level such as Technology Acceptance Model (TAM) [16, 17, 19], Theory of Planned Behaviour (TPB) [18], TAM 2 [20] Unified Theory of Acceptance and Use of Technology (UTAUT) [20]. However, studies show that there is less investigation at an organisational level. According to [21] the whole process is an innovation from engineering developments and entrepreneurship to the adoption and implementation of those improvements by the client's context of an organization.

TAM concentrates on the tie entity of BI, adoption and technology acceptance to use [21, 22]. According to [22, 23], the fundamental variables for framework use that uses TAM is Perceived Usefulness (PU) and Perceived Ease of Use (PEOU). Perceived Usefulness (PU) can be express as "the prospective user's subjective probability that using a specific application system that will increase his or her job performance within an organisational context," and Perceived Ease of Use (PEOU) is explained as "the degree to which the prospective user expects the target system to be free of effort" [17]. According to [24], PEOU impacts PU due to its usability making TAM a famous model.

Many demonstrations by various examination concentrates that PU and PEOU factors are regularly utilized for testing and [22] stated that these two factors are referenced forty percent (40%) for individuals' objective use of technology. TAM was not only developed to foresee user adoption, but it was developed with the intention to anticipate the consumer's behaviour after they interacted with the system [25]. According to [26] the production of TAM by [17] was considering in 1989 the Theory Of Reasoned Action (TRA). As per [17] a consumer's behavioural intention to utilize a new technology is related to apparent helpfulness and usability. In the meantime, the PEOU influences PU, and PU and PEOU are influenced by unfamiliar factors.

3 Methodology

This study adopted deductive approach and quantitative strategy. After reviewing extensive literature, the research study proposed a model derived from [27] with the following eight constructs: Perceive Usefulness (PU), Perceived Ease of Use (PEOU), Relative Advantage (RA), Perceived Innovation (PI) Trust (T), Social Influence (SI), Attitude (AT) and Intention to Use as shown in Fig. 1.

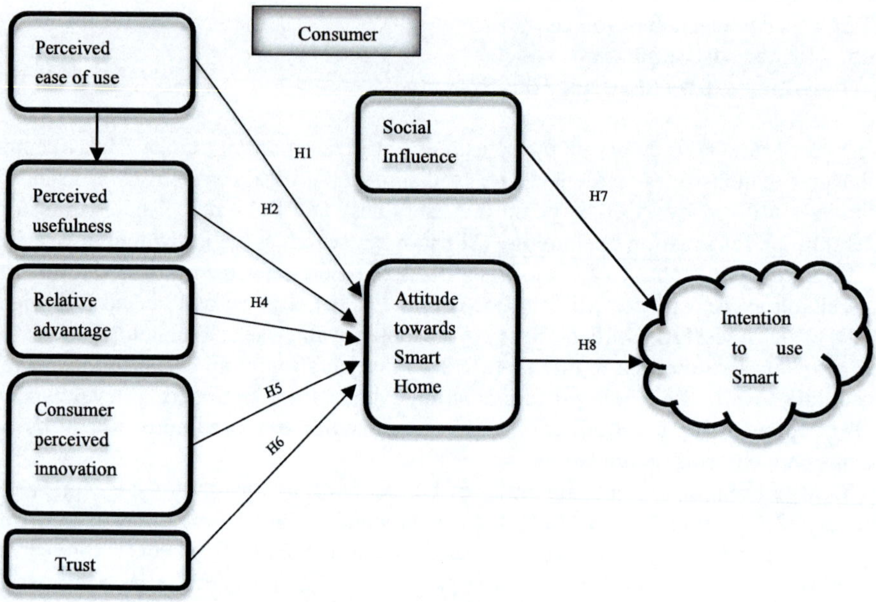

Fig. 1. The proposed consumer readiness for adoption of IOT-smart homes model (CRA-IOT-SH) derived from [27].

Perceive Usefulness (PU) and Perceived Ease of Use (PEOU) – According to [28] since its establishment, TAM had the most impact on research studies of technology usage and user acceptance. TAM is reviewed as the principal framework that is use to anticipate how a user might either accept or reject new technologies which is based on Perceive Usefulness (PU) and Perceived Ease of Use (PEOU). According to this study PEOU refer to the user's believe or how he/she perceive effort-free usage of smart home technologies. If the technology is effort-free to understand, useful and helps with daily work, the user's attitude will be more positive towards using it. PEOU is considered to have a positive impact on PU.

H1: *Perceived-ease-of-use of smart-home-technology has positive influence on attitude-toward smart-home-technology.*
H2: *Perceived-usefulness of smart-home-technology has positive influence on attitude-toward smart home technology.*
H3: *Perceived-ease-of-use of smart home technology has positive influence on Perceived-usefulness of smart home technology.*

Relative Advantage (RA) – As indicated by [29], an innovation can be depicted as a thought, practice or object that is seen new by an individual. Innovation characteristics is determine by an innovations rate of adoption. According to [29, 30] these characteristic are relative advantage, compatibility, complexity, trail ability and observability. For the purpose of this paper relative advantage "the degree to which an innovation is perceived as better than the idea is supersedes" by [31] will be use. Smart home technology is a new technology to individual and it is important to evaluate if individual's have a positive attitude towards smart home technology by evaluating the advantages between the new or existing technology. As per [32] the adoption of smart home technologies is when the user recognizes the relative advantage towards the technology.

H4: *Relative-advantage of smart-home-technology has positive influence on attitude-usefulness of smart-home- technology.*

Consumer Perceived Innovation (CPI) – CPI is defined as how the consumer perceives the innovativeness of a product. Consumer innovativeness as per [33, 34] is how quickly new products are bought by consumers, compared to other products. As per [35] CPI can be defined as how perceived innovativeness of a product influences the consumer's attitudes and adoption to the product.

H5: *Consumer-perceived-innovation of smart home technology has positive influence on attitude-usefulness of smart-home-technology.*

Trust (T) – According to [36] trust can be defined as an individual's confidence of hope in a specific service that has no security and privacy treats.

H6: *Trust of smart-home-technology has positive influence on attitude usefulness of smart-home-technology.*

Social Influence (SI) – According to [37] SI can be define as "the degree to which an individual perceives that important others believe she or he should use the new system". According to [27] "behavioural intention increases with a higher social influence".

H7: *Social-influence of smart-home-technology has positive influence on intention-to-use smart-home-technology.*

Attitude Towards Smart Home (AT) – According to [38] AT is defined as "individual's positive or negative feelings about performing the target behaviour". A general positive attitude established by consumers' perceived innovativeness, relative advantage, ease of use, and usefulness and social influence assumes a significant part in aim to embrace and utilisation smart home technologies.

H8: *Attitude towards smart-home-technology has positive influence on intention-to-use smart-home-technology.*

Intention to use Smart Home (IOU) – This paper aim is to distinguish the predecessor factors affecting an individual's goal to utilize smart home technologies. According to [39–41] intention to use is proxy for consumer acceptance as many study shows that it is an appropriate predictor of future usage.

Table 1 list the eight constructs, the number of items use per construct and sources where the items of the questionnaire were derived from.

Table 1. Measure and sources.

Constructs	Number of items	Sources
Perceive usefulness (PU)	5	[34, 35, 43]
Perceived ease of use (PEOU)	5	[35, 42, 43]
Trust (T)	7	[27, 43]
Attitude (AT)	2	[27, 43]
Intention to use (IOU)	7	[27, 42, 43]
Social influence (SI)	3	[27, 43]
Relative advantage (RA)	4	[27, 43]
Perceived innovation (PI)	4	[27, 43]

The method of data collection was performed by the use of online Google Form questionnaire. The study was only limited to individuals who lived in Gauteng Province South Africans. The survey was conducted in English. Participants consisted of adults of both genders.

The questionnaire consists of two parts

i. Consumer demographic information: Asking for age, gender, education level, etc.
ii. Consumer section: Asking multiple-choice questions about IOT smart homes on consumer level. The survey consisted of 37 questions and eight constructs from TAM. Respondents were approached to assess all items on 7-point Likert scales ranging from "Strongly agree" (1) to "Strongly disagree" (7).

3.1 Validity and Reliability

Reliability analysis testing was conducted using Cronbach's Alpha. According to [44–46], the Cronbach Alpha method is the most appropriate method to test reliability. Previous researchers like [47] state that for reliability coefficient value of 0.6 is considered as the lowest acceptable value for Cronbach's Alpha.

In this research study construct validity were utilized to give evidence how well the outcomes from the utilization of the measure fit the theories around which the test was planned. It can be said that construct validity affirmed that the instrument did tap the concept as theorized. In order to established construct validity the following can be used (1) correlational analysis (convergent and discriminant validity), (2) factor analysis, and (3) the multitrait-multimethod matrix of correlations.

According to [9], the most widely accepted forms of validity are convergent, discriminant, and nomological validity.

As per [48] item validity is a quality of estimation concerned with the degree that tests estimates what we really wish to measures. According to [49] validity refers to the degree that an instrument actually measures what it is designed or intended to measure.

Factor Analysis was performed for this study's intention. The appropriateness of FA was tested using KMO and Barlette. According to [50] to obtain acceptable sampling, the value of the Kaiser Meyer Olkin (KMO) must be larger than 0.5 and the significant value is less than 0.05.

According to [46] acceptable factor loading is 0.7 and above when sample size is 50 and more, while [51] identify factor loading of 0.6 and above as acceptable. For this study, the factor loading of 0.6 and above were used.

3.2 Regression Analysis

Linear Regression was performed to explain if there is a association between the items. Regression analysis shows an extend on the influences between items and it can also explore the effect between items [52].

The following presumptions was done preceding multiple regression analysis: (1) normality of data was checked, (2) outliers were deleted, (3) linear relationships between variables were confirmed and (4) there were no multicollinearity issues. In this study we used multiple linear regression.

For the ethical consideration consent form was sent with the online questionnaire that explained the questionnaire objective, that participation is voluntary and participants can withdraw anytime and that the all data will be treated confidentially.

4 Data Analysis

To examine the path relationships, Statistical Package for the Social Sciences software (SPSS) version 27 was used to analyse the collected data.

4.1 Demographic Information

For demographic information, out of 103 distributed questionnaires, 74 questionnaires was returned with a response rate = 50.3%.

From the participants 75.7% were male and 24.3% were female. The age group of the respondents range between 18–30 (return rate = 18.9%), 31–50 (return rate = 74.3%) and over 50 (return rate = 6,8%). The age results illustrates that the majority is in their prime of their life and has well established employment. The majority of the participant's education level was a degree with a return rate of 47.3%, followed by diploma at 27%, high school return rate = 10,8%, M Phil/PHD = 9.5% and certificate at 5,4% return rate. The minority worked in the government sector (return rate = 14,9%) and the private sector had a return rate of 85,1%. We can see here that the results is more bias towards people working in the private sector. Most of the participants was married with a response rate of 54,1%, single participants had a 45.9% return rate. Participants with a middle income had a return rate of 73%, High Income = 17,6% and return rate for lower income was 9,5%.

4.2 Validity and Reliability Analysis

The Cronbach's Alpha value was 0.879 for all the 37 items, which is acceptable [53]. All the factors was grouped according to the model. Cronbach Alpha was greater than 0.5 for all the constructs.

4.3 Regression Analysis

The regression analysis is tested on $p \leq 0.5$. The results for dependent variable Attitude (A) shows that Trust (T) does not influence Attitude (A), since Sig value (.197) is greater than 0.05. The rest of the results indicates the acceptance of the alternative hypothesis. The results for dependent variable Perceive Usefulness (PU) indicates that PEOU does not influence PU, since Sig value (.705) is greater than 0.05. The results indicates the acceptance of the alternative hypothesis is not supported. Dependent variable Intention to use (IOU) results shows that SI does not influence IOU, since Sig value (.211) is greater than 0.05. This indicates that the null hypothesis cannot be rejected. Dependent variable Intention to use (IOU) results indicates that AT has positive influence IOU, since Sig value (.037) is less than 0.05.

4.4 Hypotheses Testing Results

As seen in Table 2, the strength of the relationship between related constructs were evaluated by checking the significance of the path Beta coefficient and squared R coefficient of determination.

In Table 2 it can be seen that the third hypothesis (H3) with coefficient .048 and significance level of .705 is not accepted. PEOU has a negative influence on PU. Thus,

Table 2. Summary results of hypothesis testing:

Criterion	Beta coefficient	Sig	P-value	Results
H1: PEOU has a positive effect on Attitude towards Smart Homes	.267	.033	P-value < 0.05	Accepted
H2: PU has a positive effect on Attitude towards Smart Homes	.256	.041	P-value < 0.05	Accepted
H3: PEOU has a positive effect on Perceived usefullness	−.048	.705	P-value > 0.05	Not Accepted
H4: RA has a positive effect on Attitude towards Smart Homes	. 376	.002	P-value < 0.05	Accepted
H5: PI has a positive effect on Attitude towards Smart Homes	.358	.004	P-value < 0.05	Accepted
H6: T has a positive effect on Attitude towards Smart Homes	.264	.197	P-value < 0.05	Not Accepted
H7: SI has a positive effect on Intention to use Smart Homes	.158	.211	P-value < 0.05	Not Accepted
	.261	.037	P-value < 0.05	Accepted

the user's perception on the use of IOT smart home is seen not to be easy and useful in every day live. The results shown is consistent with previous finding within [54].

5 The Final Model with Accepted Hypothesis

See Fig. 2.

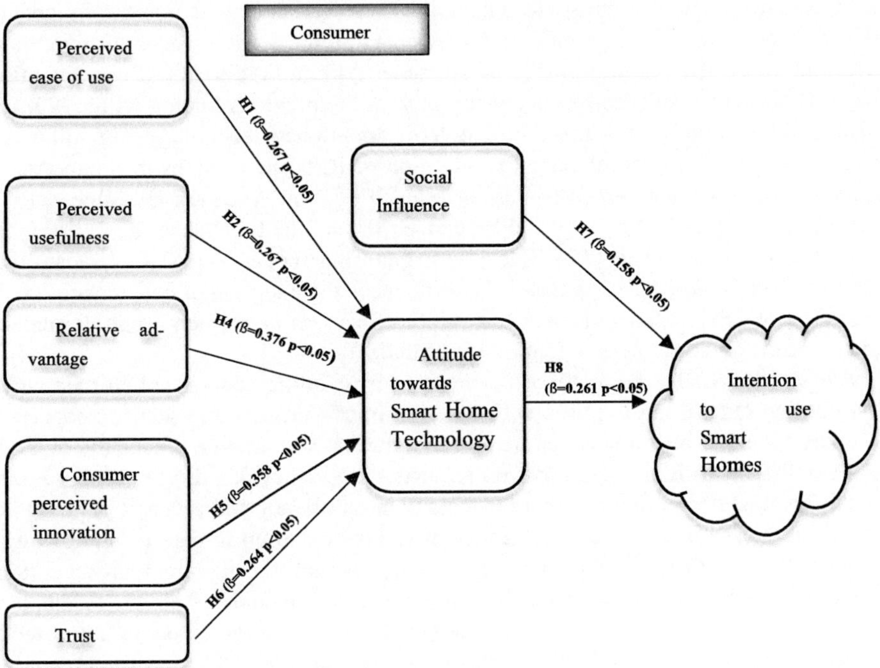

Fig. 2. The final consumer readiness model for adoption of IOT-smart homes model (CRA-IOT-SH).

6 Conclusion

The study's aim is to look into the readiness of IOT Smart Home technology among consumers in Gauteng South Africa. The paper looks at the relationships between determinant factors influencing the consumer's intentions. Factors that are looked at is the user's perceptions and consequently user's adaption decisions. In order to achieve this relationship for consumer's adoption the Theory Acceptance Model (TAM) was examined. The factors examined from TAM was Perceived Ease of Use (PEOU), Perceive Usefulness (PU), Relative Advantage (RA), Perceived Innovation (PI), Trust (T), Social Influence (SI), Attitude (AT) and Intention of Use (IOU).

Hypothesis testing was conducted through multiple regression analysis for dependent variable Attitude (A) The test reflects that PU, PEOU, T, RA and PI has a significant influence on Attitude (A) for adoption of IOT Smart Home. However PEOU has a low B coefficient towards adoption of IOT Smart Home and has an insignificant influence on Perceive Usefulness (PU) in adoption IOT-smart homes.

The findings from this study lead the researcher to draw a number of conclusions conducted to the researcher question and objectives. Hypothesis testing was conducted and it was concluded that the following factors has positive influential to the adoption of IOT-smart homes in South Africa Gauteng. These variables with positive influences are: Perceived Ease of Use (PEOU), Perceived Usefulness (PU), Relative Advantage (RA) and Perceived Innovation (PI) on Attitude (AT) and Attitude (AT) on Intention to use (IOU). The results shown is consistent with the previous finding within [27, 28, 43] study in predicting consumers' intention to adopt to new technology. The following factors was found not to influence the adoption of IOT-smart homes in South Africa Gauteng. These include Perceived Ease of Use (PEOU) on Perceived Usefulness (PU), Trust (T) on Attitude (AT) and Social Influence (SI) on Intention to use (IOU). It can be seen in the following papers [55–57] that these factors are not accepted which means that factors cannot be generalize as factor that influences the adoption of new technologies. According to [55] social influence has direct effects on consumers' attitude and an indirect effect on consumers' intention via attitude.

As per [56] firstly, the relationship between trust in the vendor and attitude when doing online purchasing is prime with the consumer's privacy and security concerns. Secondly the relationship between trust in the internet and attitude reduce the individual's confidence with high concerns on security and privacy. Lastly [56] mention that consumers trust decrease when there is a concern on privacy and security. As per [57] PEOU has great influence on the user intent to take on to up to date technology and in order for consumers to utilize new technology, the individual should have a certain degree of knowledge and skill on technology. It can be concluded from the results that consumers do not seem to have the knowledge of IOT-smart homes as this a new technology within South Africa. Based on this it can be concluded that because of the none influences of these variables there is still some open issues where there is a need to review these constructs if they really influences adoption of new technology or not.

Acknowledgment. This work is maintained by the Vaal University under the oversight of Professor Tranos Zuva in Johannesburg (Vanderbijlpark) South Africa.

References

1. Bajaj, R.K., Rao, M., Agrawal, H.: Internet of things (IOT) in the smart automotive sector?: A review. J. Comput. Eng. (2018)
2. Yang, H., Lee, W., Lee, H.: IOT smart home adoption: the importance of proper level automation. J. Sensors (2018). https://doi.org/10.1155/2018/6464036
3. Ricquebourg, V., Menga, D., Durand, D., Marhic, B., Delahoche, L., Logé, C.: The smart home concept: our immediate future. In: 2006 1st IEEE International Conference on E-Learning Industrial Electronics (ICELIE), pp. 23–28 (2006). https://doi.org/10.1109/ICELIE.2006.347206

4. Statista Research Department: Internet of Things (IOT) connected devices installed base worldwide from 2015 to 2025. Stat. Res. Dep. **44**, 2024–2025 (2016)
5. Georgiev, A., Schlögl, S.: Smart home technology: an exploration of end user perceptions. In: Innovation Lösungen für eine alternde Gesellschaft Konf. der SMARTER LIVES 18 February (2018)
6. Shaikh, E., Mohiuddin, I., Manzoor, A.: Internet of Things (IOT): security and privacy threats. In: 2nd International Conference on Computer Applications and Information Security (ICCAIS 2019), May, pp. 1–6 (2019). https://doi.org/10.1109/CAIS.2019.8769539
7. Brandt, J.: The Internet of Things has a finally arrived. In: MPI Gr., pp. 1–8 (2016)
8. Patel, K., Keyur: Internet of Things-IOT: Definition, Characteristics, Architecture, Enabling Technologies, Application & Future Challenges. Universidad Iberoamericana Ciudad México, May (2016)
9. Rose, L., Eldridge, K., Chapin, S.: The Internet of Things: an overview - understanding the issues and challenges of a more connected world. Mod. Plast. **64**(9), 54–55 (1987)
10. Tawfik, M., Almadni, A.M., Alharbi, A.A.: The risks and weakness security on the IOT a review? In: The Risks and weakness Security on the IOT, October (2018)
11. Hargreaves, T., Wilson, C.: Smart Homes and Their Users (2017)
12. Balta-Ozkan, N., Davidson, R., Bicket, M., Whitmarsh, L.: Social barriers to the adoption of smart homes. Energy Policy (2013). https://doi.org/10.1016/j.enpol.2013.08.043
13. Wilson, C., Hargreaves, T., Hauxwell-Baldwin, R.: Smart homes and their users: a systematic analysis and key challenges. Pers. Ubiquit. Comput. **19**(2), 463–476 (2014). https://doi.org/10.1007/s00779-014-0813-0
14. Premkumar, G., Roberts, M.: Adoption of new information technologies in rural small businesses. Omega (1999). https://doi.org/10.1016/S0305-0483(98)00071-1
15. Chikhaoui, B., Pigot, H.: Towards analytical evaluation of human machine interfaces developed in the context of smart homes. Interact. Comput. (2010). https://doi.org/10.1016/j.intcom.2010.08.003
16. Low, C., Chen, Y., Wu, M.: Understanding the determinants of cloud computing adoption. Ind. Manag. Data Syst. **111**(7), 1006–1023 (2011). https://doi.org/10.1108/02635571111161262
17. Zhu, K., Kraemer, K., Xu, S.: Electronic business adoption by European firms: a cross-country assessment of the facilitators and inhibitors. Eur. J. Inf. Syst. **12**(4), 251–268 (2003). https://doi.org/10.1057/palgrave.ejis.3000475
18. Teo, T.S.H., Lin, S., Hung Lai, K.: Adopters and non-adopters of e-procurement in Singapore: an empirical study. Omega **37**(5), 972–987 (2009). https://doi.org/10.1016/j.omega.2008.11.001
19. Salwani, M.I., Marthandan, G., Norzaidi, M.D., Chong, S.C.: E-commerce usage and business performance in the Malaysian tourism sector: empirical analysis. Inf. Manag. Comput. Secur. **17**(2), 66–185 (2009). https://doi.org/10.1108/09685220910964027
20. Venkatesh, V., Davis, F.D., Morris, M.G.: Dead or alive? The development, trajectory and future of technology adoption research. J. Assoc. Inf. Syst. **8**(4), 267–286 (2007). https://doi.org/10.17705/1jais.00120
21. Xu, S., Zhu, K., Gibbs, J.: Global technology, local adoption: a cross-country investigation of internet adoption by companies in the United States and China. Electron. Mark. **14**(1), 13–24 (2004). https://doi.org/10.1080/1019678042000175261
22. Au, Y.A., Zafar, H.: A Multi-Country Assessment of Mobile Payment Adoption (2008)
23. Li, Y H.: An empirical investigation on the determinants of E-procurement adoption in Chinese manufacturing enterprises. In: 2008 International Conference on Management Science and Engineering 15th Annual Conference Proceedings (ICMSE 2008), pp. 32–37. https://doi.org/10.1109/ICMSE.2008.4668890

24. Schillewaert, N., Ahearne, M.J., Frambach, R.T., Moenaert, R.K.: The adoption of information technology in the sales force. Ind. Mark. Manag. **34**(2), 323–336 (2005). https://doi.org/10.1016/j.indmarman.2004.09.013
25. Pan, M.J., Jang, W.Y.: Determinants of the adoption of enterprise resource planning within the technology-organization-environment framework: Taiwan's communications industry. J. Comput. Inf. Syst. **48**(3), 94–102 (2008). https://doi.org/10.1080/08874417.2008.11646025
26. Qin, X., Shi, Y., Lyu, K., Mo, Y.: Using a Tam-Toe model to explore factors of building information modelling (BIM) adoption in the construction industry. J. Civ. Eng. Manag. **26**(3), 463–468 (2020). https://doi.org/10.3846/jcem.2020.12176
27. Nikou, S.: Consumers' perceptions on smart home and smart living. In: 26th European Conference on Information Systems: Beyond Digitization - Facets of Socio-Technical Change (ECIS 2018) (2018)
28. Venkatesh, V.: Determinants of perceived ease of use: integrating control, intrinsic motivation, and emotion into the technology acceptance model. Inf. Syst. Res. **11**, 342–365 (2000). https://doi.org/10.1287/isre.11.4.342.11872
29. Rogers, E.M.: Diffusion of Innovations, 4th Edn. (1995)
30. E. M. Rogers, "Diffusion of Innovtions," in Diffusion of innovations, 1983
31. Rogers, E.M.: Diffusion of preventive innovations. Addict. Behav. **27**(6), 989–993 (2002)
32. Bae, Y., Chang, H.: Adoption of smart TVs: a Bayesian network approach. Ind. Manag. Data Syst. **112**, 891–910 (2012). https://doi.org/10.1108/02635571211238509
33. Midgley, D.F., Dowling, G.R.: Innovativeness: the concept and its measurement. J. Consum. Res. **4**(4), 229–242 (1978). https://doi.org/10.1086/208701
34. Roehrich, G.: Consumer innovativeness: concepts and measurements. J. Bus. Res. **57**(6), 671–677 (2004). https://doi.org/10.1016/S0148-2963(02)00311-9
35. Lowe, B., Alpert, F.: Forecasting consumer perception of innovativeness. Technovation **45** (2015). https://doi.org/10.1016/j.technovation.2015.02.001
36. Gao, S., Krogstie, J., Ther, G.: Mobile Services Acceptance Model (2008)
37. Venkatesh, V., Morris, M.G., Davis, G.B., Davis, F.D.: User acceptance of information technology: toward a unified view. MIS Q. Manag. Inf. Syst. **27**(3), 425–478 (2003). https://doi.org/10.2307/30036540
38. Ajzen, I., Fishbein, M.: Belief, Attitude, Attitude, Intention and Behavior: An Introduction to Theory of Research (1975)
39. Gao, L., Bai, X.: A unified perspective on the factors influencing consumer acceptance of internet of things technology. Asia Pacific J. Mark. Logist. **26**, 211–231 (2014). https://doi.org/10.1108/APJML-06-2013-0061
40. Lee, Y.-K., Park, J.-H., Chung, N., Blakeney, A.: A unified perspective on the factors influencing usage intention toward mobile financial services. J. Bus. Res. **65**(11), 1590–1599 (2012)
41. Venkatesh, V., Davis, F.: A theoretical extension of the technology acceptance model: four longitudinal field studies. Manage. Sci. **46**, 186–204 (2000). https://doi.org/10.1287/mnsc.46.2.186.11926
42. Carnaghan, C., Klassen, K.: Exploring the determinants of web-based E-business evolution in Canada. In: Association for Information Systems - 13th Americas Conference on Information Systems, AMCIS 2007: Reaching New Heights, pp. 3532–3543 (2007)
43. Gao, S., Krogstie, J., Siau, K.: Developing an instrument to measure the adoption of mobile services. Mob. Inf. Syst. **7**(1), 45–67 (2011). https://doi.org/10.3233/MIS-2011-0110
44. Davis, F.D.: Perceived usefulness, perceived ease of use, and user acceptance of information technology. MIS Q. Manag. Inf. Syst. (1989). https://doi.org/10.2307/249008
45. Hinton, P.R.: Statistics Explained, 3rd Edn. (2014)
46. Field, A.P.: Discovering statistics using SPSS for windows. Adv. Tech. Begin. (2000)

47. Robinson, J.P., Shaver, P.R., Wrightsman, L.S.: Criteria for scale selection and evaluation. In: Measures of Personality and Social Psychological Attitudes (1991)
48. Woollins, J.D.: Calculating, interpreting, and reporting cronbach's alpha reliability coefficient for likert-type scales. Stud. Inorg. Chem. **14**(C), 349–372 (1992). https://doi.org/10.1016/B978-0-444-88933-1.50023-4
49. Aila, F., Ombok, B.: Validating measures in business research: practical implications. Int. J. Sci. Eng. **1**(9), 11–19 (2016)
50. Ul Hadia, N., Abdullah, N., Sentosa, I.: An easy approach to exploratory factor analysis: marketing perspective. J. Educ. Soc. Res. (2016). https://doi.org/10.5901/jesr.2016.v6n1p215
51. Masrom, M.: Technology acceptance model and E-learning. In: 12th International Conference on Education, May (2007)
52. Zhao, J., Fang, S., Jin, P.: Modeling and quantifying user acceptance of personalized business modes based on TAM, trust and attitude. Sustain. **10**(2), 1–26 (2018). https://doi.org/10.3390/su10020356
53. Taber, K.S.: The use of Cronbach's alpha when developing and reporting research instruments in science education. Res. Sci. Educ. **48**(6), 1273–1296 (2016). https://doi.org/10.1007/s11165-016-9602-2
54. Gangwar, H., Date, H., Ramaswamy, R.: Understanding determinants of cloud computing adoption using an integrated TAM-TOE model. J. Enterp. Inf. Manag. (2015). https://doi.org/10.1108/JEIM-08-2013-0065
55. Malhotra, Y., Galletta, D.F.: Extending the technology acceptance model to account for social influence: theoretical bases and empirical validation. In: Proceedings of the Hawaii International Conference on System Sciences, p. 5 (1999). https://doi.org/10.1109/hicss.1999.772658
56. McCole, P., Ramsey, E., Williams, J.: Trust considerations on attitudes towards online purchasing: the moderating effect of privacy and security concerns. J. Bus. Res. **63**(9–10), 1018–1024 (2010). https://doi.org/10.1016/j.jbusres.2009.02.025
57. Agyei, J., Sun, S., Abrokwah, E., Penney, E.K., Ofori-Boafo, R.: Mobile banking adoption: examining the role of personality traits. SAGE Open **10**, 2 (2020). https://doi.org/10.1177/2158244020932918

A Model for the Adoption and Acceptance of Mobile Farming Platforms (MFPs) by Smallholder Farmers in Zimbabwe

Fine Masimba(✉) and Tranos Zuva

ICT Department, Vaal University of Technology, Vanderbijlpark 1930, South Africa
fine.masimba@gmail.com, tranosz@vut.ac.za

Abstract. The use of mobile technology has continued to rise in Zimbabwe, and farmers now can obtain agricultural information through the use of mobile technology. The potential of mobile phones in agriculture gave birth to mAgriculture, which is the use of mobile technology to provide agricultural information and services. Various technological start-up companies in Zimbabwe have developed mobile farming platforms as a means of increasing overall performance among smallholder farmers. Unfortunately, most of these mobile farming platforms have not achieved their much-anticipated impact on farmers due to low acceptance which can be attributed to the incompleteness of the available technology acceptance models. This paper contributes to the body of knowledge by proposing a more comprehensive model that extends the unified theory of acceptance and use of technology (UTAUT2) by incorporating two key factors which are attitude and cultural moderators. Structural Equation Modelling was utilized to test the hypothesized conceptual model, in a quantitative study conducted in Zimbabwe. As hypothesized, the findings of this study revealed that the constructs in the model were found to be significant determinants of behavioral intention and usage. The results also showed that cultural values have a moderating effect on user acceptance of mobile farming platforms.

Keywords: Mobile · Platforms · mAgriculture · Acceptance · Adoption · UTAUT2

1 Introduction

Mobile applications for agriculture and rural development have tremendous promise in advancing development (Qiang 2012). Mobile phones are the most essential communication tools that farmers can now use to obtain agricultural information and knowledge (Chhachar et al. 2014). According to Mittal and Mehar (2012), a 2011 survey of farmers in Indian states found that having access to information via a mobile phone allowed them to be more linked to the markets and achieve higher pricing. According to Jensen (2007), the adoption of mobile technology in Kerala reduced wastage and pricing dispersion by promoting the sharing of information amongst fishermen. Abraham (2007) observed that widespread usage of mobile phones boosted market efficiency by reducing

R. Silhavy (Ed.): CSOC 2022, LNNS 501, pp. 710–725, 2022.
https://doi.org/10.1007/978-3-031-09070-7_59

risk and uncertainty. Farmers' use of mobile phones can minimize information search costs, lower transaction costs, and allow more farmers to participate in commercial agriculture (De Silva and Ratnadiwakara 2008). According to Aker and Mbiti (2010), cell phones are a new search technique that has cut the cost of searching for farmers in Niger by nearly half. Farmers in Uganda were able to utilize mobile phones to find out the latest agricultural prices, and farmers in Tanzania were able to save time and money by using mobile phones (Adel 2005). In Niger, Aker (2008) discovered that the adoption of mobile technology reduced grain price dispersion by 10% among markets. Mobile phones are increasingly being integrated into current agricultural trade businesses, owing to the critical role they serve in facilitating information flow between farmers and buyers (Abraham 2007).

The potential of mobile phones in agriculture gave birth to mAgriculture which is the utilization of mobile phones for the provision agricultural information and services (Gichamba 2017). In Zimbabwe, various technological start-up companies have developed mobile farming platforms as a means of increasing overall performance among smallholder farmers. AgriMobi, ZIMACE, Esoko, eMkambo, Eco-Farmer and more recently e-Hurudza are examples of such platforms in Zimbabwe (Odunze and Hove 2015). These mobile farming platforms provide a variety of services, such as daily weather, rainfall pattern update, weekly best farming pricing, weekly crop statistics, marketing links and financial linkages, to name a few. Unfortunately, most of these mobile farming platforms have not achieved their much-anticipated impact on farmers. The limited acceptance or adoption of these platforms has been linked to the incompleteness of the available technology acceptance models (Alushola and Abiola 2017). Only 16% of mobile farming platforms launched in developing countries make it to widespread adoption (Qiang 2012). Most farmers, in particular, do not use these platforms for an extended length of time. Technology is worthless unless and until it is embraced and put to use (Oye et al. 2012). Reasons for discontinuing use of these mobile farming platforms must be identified and addressed, as well as factors that encourage continued use of these mobile platforms.

Although many researches in the field of technology adoption have thrown light on several key factors influencing technology acceptance, existing frameworks for successful technology adoption still need to be improved and modified. The current UTAUT 2 model is a comprehensive model that incorporates and consolidates earlier studies on technology acceptance models. Though UTAUT2 is a unified model, however recent literature suggests that there are crucial elements that are still lacking from the model, particularly attitude and culture (Alushola and Abiola 2017). UTAUT models do not account for cultural influences, which are likely to be significant in the majority of countries around the world (Alushola and Abiola 2017). The study contributes to the body of knowledge in the mobile technology adoption and acceptance area. Firstly, the study developed a conceptual framework by modifying the UTAUT 2 model that was proposed by (Venkatesh et al. 2012) by incorporating two key factors that influence user acceptance and adoption of mobile farming platforms which are attitude and cultural moderators from Hofstede thereby giving a new insight into the salient factors affecting mobile farming platforms. The study brings fresh perspectives to the subject of mobile technology adoption, which could lead to a comprehensive model for using mobile farming platforms in agriculture. Secondly, most mobile farming platforms do not achieve widespread adoption due to ineffective adoption techniques (Gichamba 2017). As a result, this study is critical since it aids in the construction of a comprehensive, workable model for the deployment of mobile farming platforms in agriculture. Finally, the findings of this study provide agricultural technology start-up enterprises and other stakeholders with a greater understanding of how mobile technologies can successfully be deployed to increase production.

This paper is organized as follows. In the next section, an overview of the theoretical background is presented, introducing mAgriculture concept, culture and technology acceptance, attitude and technology acceptance, technology acceptance models. The paper continues with the research model and hypotheses presentation, followed by data collection methodology description, results, limitations and possible future directions.

2 Literature Review

2.1 mAgriculture Concept

The third Quarter Telecommunications report by the Postal and Telecommunications Regulatory Authority of Zimbabwe (Potraz) reported an increase in Zimbabwe's mobile penetration rate from 97% in Second Quarter to 100% in Third Quarter of 2017 (Potraz Telecoms Report 2017). With a high mobile penetration rate of 100% and a strong agricultural system, mobile farming platforms have a great potential to transform the agricultural sector in Zimbabwe. According to O'Donnell (2013), mobile technologies have the potential to offer better services that can be used to have better access to, and dissemination of agricultural marketing information. The potential of mobile phones in agriculture gave birth to mAgriculture, which is the use of mobile technology to provide agricultural information and services (Gichamba 2017). Mobile faming platforms are reported to offer great benefits to the farmers such as finding new buyers, using market information to obtain higher prices, and better traceability and compliance with quality and safety standards (Engotoit et al. 2016). Market information provided to farmers through mobile farming platforms is vital to farmers as it is needed for farmers to choose what commodities to produce, what technologies to apply for production, when to produce, for whom to produce, when and what price to sell (Odunze and Hove 2015).

2.2 Unified Theory of Acceptance and Use of Technology (UTAUT)

Venkatesh et al. (2003), proposed the UTAUT model. UTAUT posits four constructs which include Performance expectancy, Effort Expectancy, Facilitating conditions and Social Influence. This model is a comprehensive model for user acceptance that resulted from eight existing models and theories namely Theory of Reasoned Action (TRA), Technology acceptance model (TAM), Motivational Model, Model of PC utilization, Social Cognitive Theory, Combined TAM-TPB, Innovation of Diffusion Theory and the Theory of Planned Behavior (TPB) (Alushola and Abiola 2017). According to Venkatesh et al. (2003), UTAUT is comprised of a range of factors that are taken into consideration when evaluating the behavioral intention to use and the level of acceptance of a particular technology.

Venkatesh et al. (2012) introduced UTAUT2 which was an extension of UTAUT. The objective of developing UTAUT 2 was to pay particular attention to the individual or consumer use context and move away from employee use context in UTAUT. UTAUT was originally developed to explain employee technology acceptance and use, but in 2012 it was extended to UTAUT2 which is in the context of individual or consumer technologies (Samaradiwakara and Gunawardena 2014). Therefore, UTAUT was expanded to UTAUT2 to suit individual context. Three more variables were included to the theory and these are hedonic motivation (HM), price value (PV), and habit (HT). UTAUT2 kept performance expectancy, effort expectancy, social influence and facilitating conditions from UTAUT, adapting them to a consumer use context.

2.3 Attitute and Technology Acceptance

Alsharif (2013) defines attitude as an individual's total emotional response to using a system. Research has been carried out to investigate the influence of attitude on behavioral intention to use technology. Venkatesh et al. (2003) argues that the effect of attitude on behavioral intention is spurious and it emerges only when performance expectancy and effort expectancy are omitted from the model but Nassuora (2012) and Jairak et al. (2009) indicate that attitude influence positively on behavioral intention with both effort expectancy and performance expectancy included. Research has also been carried out on the relationships that exist between attitude and social factors, performance expectancy and effort expectancy. Researchers report that social factors influence positively on attitude (Jairak et al. 2009; Nassuora 2012). However, Jairak et al. (2009) find that performance expectancy and effort expectancy have positive influence on attitude whereas Nassuora (2012) finds no such effects.

2.4 Culture and Technology Acceptance

Cultural variables have a substantial influence on technology acceptance and adoption (Baptista 2015). Cultural variables should be included in acceptance models, according to several researchers (Park et al. 2007), because culture influences how individuals utilize information systems (Im et al. 2011). Cultural values, according to Srite and Karahanna (2006), play a moderating role in technology acceptance. Hofstede (1980) proposed a theory of cultural differentiation, which has been widely applied in a variety of cross-cultural research, primarily through comparing cultures in different nations and even at individual level (Reisinger 2009). Hofstede's cultural framework is a complete model that has been widely utilized and referred to by a variety of studies in a variety of fields, and it is based on national culture samples (Soares 2007). Hofstede's cultural dimensions are used to expand our understanding of individual and situational aspects in mobile farming platform adoption and acceptance, as well as to shed new light on how culture affects individual behavior. According to Reisinger (2009), Hofstede's cultural theory contains five dimensions, which allow academics to examine, compare, and observe shared values across diverse national cultures. The cultural theory of Hofstede consists of five dimensions: (i) individualism/collectivism, (ii) masculinity/femininity, (iii) uncertainty avoidance, (iv) power distance, and (v) long-term orientation.

Individualism/collectivism - individualism, according to Hofstede, portrays a society in which individual relationships are loose and everyone is expected to look after himself and his immediate family solely, whereas collectivism refers to a society in which people are born into strong and cohesive groups that continue to protect them throughout their lives in exchange for unquestioned allegiance (Hofstede 2001). Masculinity/femininity - masculinity depicts a community in which social gender roles are well defined; males are expected to be tough, bold, and focused on material achievement, whereas women are expected to be humble, kind, and concerned with the quality of life in a masculine community (Podrug et al. 2017). Uncertainty avoidance refers to how a society strives to avoid uncertain situations by enforcing strong codes of conduct, believing in absolute truth, enforcing formal laws, and not accepting deviant ideas and actions (Podrug et al. 2017). Power distance is defined as the amount to which the less powerful members of a society accept inequality in power as normal, and thus respect and avoid criticizing their superiors. Long-term and short-term orientations were distinguished by Hofstede (2014), who stated that long-term orientation refers to the development of virtues that are oriented toward future benefits, such as perseverance, whereas the phrase "short-term" refers to the development of qualities relating to historical and contemporary, valuing tradition, maintaining one's appearance, and performing social obligations.

2.5 Research Model and Hypotheses

UTAUT 2 is a unified and complete model for technology acceptance and use that gives a very good and detailed model. The UTAUT model can account for 70% of the variation in usage intention, which is a significant improvement over any of the eight original models and their extensions (Olushola and Abiola 2017). This makes UTAUT the most appropriate model to adopt when studying individual technology acceptance. This is the justification why in this study the researcher chose to adopt this model in studying the adoption of mobile farming platforms by small holder farmers in Zimbabwe. Though UTAUT2 is a unified model, however recent literature suggests that there are crucial elements that are still lacking from the model, particularly attitude and culture. The theoretical underpinning for this study was based on a combination of UTAUT2 with attitude and Hofstede's cultural moderators. The major goal was to develop a more complete model based on UTAUT2 that encapsulated the essential characteristics that drive user acceptability and adoption of mobile farming platforms.

Social influence (SI) is described as the extent to which farmers believe others, particularly colleagues, friends and family members, believe they should employ mobile farming platforms (Venkatesh et al. 2003). Individuals prefer to heed the opinions of major referents when evaluating the usefulness of a technology, according to Lin (2011); thus, performance expectancy is significantly influenced by social influence. Therefore, we hypothesize that:

H1a. SI of mobile farming platforms adoption has a positive influence on PE.
H1b. SI of mobile farming platforms adoption has a positive influence on BI.

Performance expectancy is the extent to which a farmer believes that employing a mobile farming platform would aid them in their farming activities (Venkatesh et al.

2003). Farmers tend to develop a positive attitude to mobile farming platform systems if they believe that they will help them accomplish their farming activities efficiently (Jairak et al. 2009). Therefore, we hypothesize that:

H2. PE of mobile farming platforms has a positive influence on attitude (AT).

Effort expectancy is the degree of easiness connected with the use of mobile farming platforms (Venkatesh et al. 2003). If farmers find mobile farming platforms easy to use, they will develop a positive attitude toward them and they may want to use them more in sending and receiving information about their farming activities (Lin 2011).This is confirmed by Jairak et al. (2009) who highlights that effort expectancy has a positive effect on attitude. Therefore, we hypothesize that:

H3. EE of mobile farming platforms has a positive influence on attitude (AT).

Facilitating conditions refers to how people feel that technical infrastructures exist to assist them in using the technology whenever they are needed (Venkatesh et al. 2003). Nassuora (2012) argues that a person who has access to a favorable set of facilitating conditions towards a system tends to develop a positive attitude to it; therefore, facilitating conditions have a positive influence on attitude. Therefore, we hypothesize that:

H4a. FC of mobile farming platforms have a positive influence on attitude (AT).
H4b. FC of mobile farming platforms have a positive influence on behavioral intention (BI).
H4c. FC of mobile farming platforms have a positive influence on actual usage (AU).

Hedonic motivation describes the degree of enjoyment obtained from using mobile farming platforms (Venkatesh et al. 2012) and it is a significant factor in user acceptance of technology (Heijden 2004); farmers will have a higher acceptance intention if the mobile farming platform provides more enjoyment value (Zhang et al. 2008). Therefore, we hypothesize that:

H5. HM of mobile farming platforms in Zimbabwe has a positive influence on behavioral intention (BI).

Price value refers to farmers' mental trade-off between the perceived benefits of using mobile farming platform services and the monetary cost of doing so (Venkatesh et al. 2012). The price value is positive when the perceived benefits of using the mobile farming platform outweigh the related monetary cost. Therefore we hypothesize that:

H6. PV of mobile farming platforms has a positive influence on behavioral intention (BI).

Habit reflects a variety of previous experience's outcomes (Venkatesh et al. 2012) and one of the main drivers of current behavior is the frequency of previous behavior (Ajzen 2002). Therefore, we hypothesize that:

H7a. Habit (HB) of mobile farming platforms adoption has a positive influence on behavioral intention (BI).
H7b. Habit of mobile farming platforms has a positive influence on actual use (AU).

Attitude is defined as an individual's total emotional response to using a technology (Alsharif 2013). Nassuora (2012) and Jairak et al. (2009), found out that attitude has a positive influence on behavioral intention, taking into account both effort and performance expectancy. Therefore, we hypothesize that:

H8. Attitude towards using a mobile farming platform has a positive influence on behavioral intention (BI).

UTAUT2 backs up the idea that behavioral intention has a strong influence on how people use technology (Venkatesh et al. 2003), and this is consistent with all other models which revealed that individual behavior is predictable and is impacted by individual intention (Baptista 2015). Therefore, we hypothesize that:

H9. Behavioral intention (BI) of mobile farming platforms in Zimbabwe has a positive influence on actual usage (AU).

2.6 Cultural Values as Moderators

Baptista (2015) highlighted that in a study conducted by Hofstede where he gathered data in more than 50 countries on their cultural dimensions, the results indicated that most African countries were classified as strongly hierachical societies, collectivist, feminine, low uncertanity avoidance and short term orientation. Individuals in individualistic communities are more concerned with their personal interests and those of their immediate families, whereas individuals in strongly collectivistic communities are defined by the acts of the group rather than their own (Bergiel et al. 2016). People who share collectivist beliefs, according to Zakour (2004), will be interested in other people's perspectives on technology. Therefore, we hypothesize that:

H10a. Individualism/collectivism (IC) moderates effort expectancy (EE) and attitude (AT).
H10b. Individualism/collectivism (IC) moderates social influence (SI) and behavioral intention (BI).
H10c. Individualism/collectivism (IC) moderates behavioral intention (BI) and actual usage (AU).

Individuals who are masculine and feminine have opposing and distinct qualities. Masculine individuals define achievement in terms of close human relationships, consensus, valuing equality, caring for the weak, solidarity, and incentives such as free time and flexibility and quality of life, where as feminine individuals define achievement in terms of close human relationships, consensus, valuing equality, caring for the weak, solidarity, and incentives such as free time and flexibility and quality of life (Hofstede

2014). Users who are high in masculinity, according to Baptista (2015), want to maximize the value derived from using mobile farming platforms. Therefore, we hypothesize that:

H11. Masculinity/femininity (MF) moderates the relationship between price value (PV) and behavioral intention (BI).

Hofstede and Hofstede (2005) highlighted the distinction between long-term and short-term orientation, they revealed that long-term orientation fosters virtues oriented toward future rewards, such as perseverance, whereas short-term orientation fosters virtues related to the past and present, such as respect for tradition, maintaining one's face, and fulfilling social obligations. African countries have a high regard for tradition, a low proclivity to save for the future, and a drive to get things done quickly (Hofstede 2014). People with a high long-term orientation, according to Baptista (2015), demand improved performance, convenience, and speed in mobile farming platforms so that they may complete their work swiftly. Therefore, we hypothesize:

H12a. Long/short term (LT) orientation moderates performance expectancy (PE) and attitude (AT).
H12b. Long/short term (LT) moderates behavioral intention (BI) and actual use (AU).

Cultures with high degrees of uncertainty avoidance are likely to have a low rate of mobile farming acceptance because mobile farming is not well suited to uncertainty reduction compared to face-to-face encounters or dialogue with agricultural extension personnel (Straub et al. 2006). Individuals that are uncomfortable with ambiguity and uncertainty will be less likely to use technology in high uncertainty avoidance societies (Zakour 2004). Uncertainty-avoiding people are less inclined to try new things or adopt new technologies (Png and Tan 2001).Therefore, we hypothesize that:

H13: Uncertainty avoidance (UC) moderates behavioral intention (BI) and actual use (AU).

People in hierarchical cultures with high power distance values are more concerned with following the beliefs of their superiors or communities and are afraid of disagreeing with them (Hofstede 1980). Mobile farming platforms are likely to be embraced by other farmers if the hierarchy or community supports them. Therefore, we hypothesize that:

H14. Power distance (PD) moderates behavioral intention (BI) and actual use (AU).
H15. Individualism/collectivism (IC) moderates attitude (AT) and behavioral intention (BI).
H16. Masculinity/femininity (MF) moderates attitude (AT) and behavioral intention (BI).
H17. Long/short term (LT) orientation moderates attitude (AT) and behavioral intention (BI).
H18. Uncertainty avoidance (UC) moderates attitude (AT) and behavioral intention (BI).
H19. Power distance (PD) moderates attitude (AT) and behavioral intention (BI).

3 Research Methodology

The data collection was conducted in Zimbabwe, targeting the adult population that were subscribing to one or more mobile farming platforms and own one or more mobile devices such as a mobile phone, smart phone or tablet. This study used a structured research questionnaire in English to collect data from the targeted sample. The questionnaire was created and reviewed for content validity by a group of information systems academics. The questionnaire was made up of four sections, described as sections A to D. Section A included the demographic variables such as gender, age, educational level and province. Section B covered the direct determinants within the proposed conceptual framework, and Section C covered the actual usage of the mobile farming platforms while Section D covered the five cultural variables. The item and scales for the UTAUT2 constructs were adapted from Venkatesh et al. (2003) and Venkatesh et al. (2012), use behavior from Martins et al. (2014), uncertainty avoidance, power distance and individualism/collectivism from Srite and Karahanna (2006), and long/short term from Hassan et al. (2011). Each item was measured using a five-point Likert scale that ranged from 1 (strongly disagree) to 5 (strongly agree). The Martins et al. (2014) use behavior was coded from 1 (have not used) to 5 (once a month).

A survey research was employed in this study since studies of technology acceptance have traditionally been conducted using survey research (Venkatesh et al. 2003). The survey was pilot tested amongst a group of 45 small holder farmers comprising 15 farmers from each of three provinces in Zimbabwe namely Mashonaland East, Mashonaland West and Harare Metropolitan. Preliminary evidence showed that scales were reliable and valid. The pilot study was done to assess the questionnaire in terms of clarity, time and understanding. All the participants completed an initial version of the questionnaire and their assessments and comments on the questionnaire helped to refine the questionnaire. The final sample size for this study was pegged at n = 500 respondents. The collection of data was conducted between May and September 2020. The extended period of five months was attributed to the large number (500) of questionnaires that were to be distributed. The Covid-19 lockdown restrictions made it even more challenging to physically reach the people within a short period of time, hence the five months. From that initial 500 questionnaires, 411 were returned, 59 of which were unusable. After screening for missing data and duplicated responses, we retained 352 questionnaires for data analysis. More than half of the respondents were female (56.5%). The majority of the respondents were middle aged people (35–50 years) who constituted (57.1%). These are people who are believed to be economically active and therefore constitute a greater percentage of the farmers. There were a few younger (34 years or less) respondents (15.3%). More than three-fourths of the respondents (84.4%) had a diploma and below, and 4.5% had masters and above. Harare Metropolitan province had the highest number of respondents (36.4%) and Mashonaland east had the least number of respondents (30.4%), but generally all the three provinces had almost the same number of respondents.

4 Results

Overall, all means indicate that the majority of participants express generally positive responses to the constructs that are measured in this study.

4.1 Structural Model and Hypotheses Testing

A simple regression was used to test the hypotheses and the results are summarised in Fig. 1.

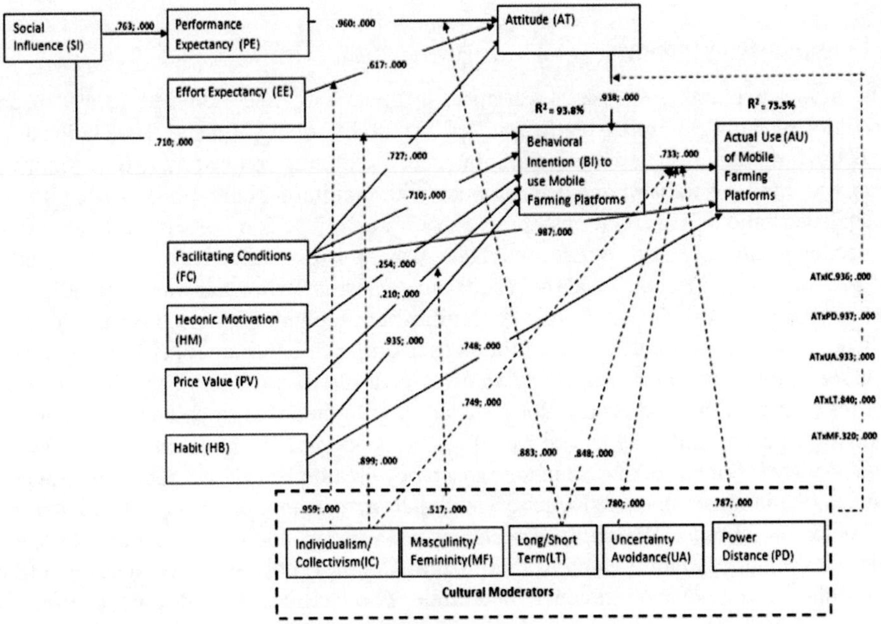

Fig. 1. Structural model results

The model explains a 93.8% of variation in behavioral intention and 73.3% in use behavior. Social influence was found to have a statistically positive influence on performance expectancy thus supporting hypothesis **H1a.** Performance expectancy, effort expectancy and facilitating conditions were found to have a statistically positive influence on attitude supporting **H2, H3** and **H4a.** Attitude was found to be have a positive influence on behavioral intention thus supporting hypothesis **H8.** Social influence, facilitating conditions, hedonic motivation, price value and habit were found to have a statistically positive influence on behavioral intention there by supporting **H1b, H4b, H5, H6, H7a.** Behavioral intention was found to have a statistically positive influence on actual use thus supporting hypothesis **H9.** There is a statistically positive influence of habit on actual use of mobile farming platforms thus supporting hypothesis **H7b.** There is also a statistically positive influence of facilitating conditions on actual use of mobile farming platforms thus supporting hypothesis **H4c.** Individualism/collectivism, masculinity/femininity, uncertainty avoidance, power distance and long/short term were found to be statistically significant, supporting their moderating effect and hypotheses **H10a, H10b, H10c, H11, H12a, H12b, H13, H14, H15, H16, H17, H18, H19**. Overall, all the nineteen hypotheses formulated were supported by the data collected.

5 Discussion

The theoretical model presented extended the unified theory of acceptance and use of technology (UTAUT2), of Venkatesh et al. (2012) by incorporating attitude and cultural moderators from Hofstede (1980), to explain the acceptance of mobile farming platforms.

5.1 Supported Findings

The model explains a 93.8% of variation in behavioral intention and 73.3% in use behavior of mobile farming platforms. The research model validated the relationships of behavioral intention namely social influence, facilitating conditions, hedonic motivation, price value and habit and use behavior with facilitating conditions and habit. The model also validated the relationships of attitude namely performance expectancy, effort expectancy and facilitating conditions. There was a positive influence of social influence on performance expectancy and the results are consistent with what was reported in earlier studies that individuals tend to take significant referents' opinions into consideration when assessing a technology's usefulness (performance expectancy) (Lin et al. 2003). The social influence relationship finding on behavioral intention is consistent with some earlier studies (Putit and Arnott 2007) which highlighted that in collectivist societies, individuals are usually influenced by other people's opinions when making decisions. The positive influence of performance expectancy on attitude is consistent with previous studies like Jairak et al. (2009) who revealed that performance expectancy has a positive influence on attitude. The effort expectancy relationship finding on attitude is consistent with previous studies, such as that of Jairak et al. (2009) who revealed that effort expectancy has a positive influence on attitude. The facilitating conditions relationship finding on attitude is consistent with previous studies such as that of Nassuora (2012) who argues that a person who has access to a favorable set of facilitating conditions towards a system tends to develop a positive attitude towards such system. The facilitating conditions relationship finding on behavioral intention is consistent with what was reported in previous studies (Miltgen et al. 2013) but contradicts others (Im et al. 2011). The hedonic motivation relationship finding on behavioral intention is consistent with what was reported in earlier studies, such as by Zhang et al. (2008) and Lee (2009). The price value relationship finding on behavioral intention is in line with earlier studies (Habbaoush et al. 2011). The habit relationship finding on behavioral intention is consistent with what has been reported in earlier studies, such as Ajzen (2002). The attitude relationship finding on behavioral intention is in line with earlier studies which reported that attitude influences positively on behavioral intention with both effort expectancy and performance expectancy included (Nassuora 2012; Jairak et al. 2009). The findings on the influence of behavioral intention on actual use are in line with earlier research (Venkatesh et al. 2003; Yu 2012).

5.2 Cultural Values as Moderators

The research model validated the influence of Hofstede's cultural moderators on different independent variables and dependent variables in the study. Individualism/collectivism had a strong and positive moderating effect on the relationships between effort expectancy and attitude, social influence and behavioral intention and between behavioral intention and actual use. The effect was found to be stronger for farmers with higher collectivism than for individualistic farmers and this is in line with earlier research findings, such as Baptista (2015). On the positive moderating effect of individualism/collectivism on effort expectancy (ease of use) and attitude as well as behavioral intention and actual use, the findings confirmed that respondents with higher collectivism encourage each other to use mobile farming platforms which are easy to use, convenient and efficient, which then leads to a higher intention to use mobile farming platforms. These findings show that the ease of use of the system or platform plays a greater role in the decision to adopt technology for collectivist users. On the relationship between social influence and behavioral intention, the findings confirmed that farmers who are high on collectivism exert a stronger social influence on behavioral intention. In a collectivist society like Zimbabwe, social influence plays a pivotal role over intention because people in collectivist societies are normally more concerned about the opinions of others (Baptista 2015). This is also in line with previous studies, including Tarhini (2013) who highlights that the opinions of significant others like peers play a greater role in the decision to adopt a technology for collectivist users. Individualism/collectivism however, had a negative moderating effect on the relationship between attitude and behavioral intention. While the negative moderating effect of individualism/collectivism on the relationship between attitude and behavioral intention was unexpected, it is however in line with other earlier studies, such as Zhou et al. (2010) who contend that collectivist cultures act as a barrier to mobile technology adoption.

Power distance moderated negatively the relationship between attitude and behavioral intention. Zimbabwe being a high power distance country, the findings confirm that individuals or farmers who are high on power distance display a significantly less favorable view towards adoption of new technology and this is consistent with previous research (Baptista 2015). Individuals high on power distance are characterized by lower rates of innovation and acceptance of technology (Tarhini 2013). On the other hand, power distance moderated positively the relationship between behavioral intention and actual use and, again, this is in line with previous research (Baptista 2015; Sriwindono and Yahya 2012). The reason for the positive moderation could be that the opinions of others, such as agritex officers and peers, play greater roles in the decision to adopt technology for users with higher power distance cultural values. It could be that farmers with higher levels of power distance were intending to use the technology to comply with their superiors and please them.

Uncertainty avoidance had a positive moderating effect on the relationship between behavioral intention and actual use. This show a low uncertainty avoidance cultural characteristic. Baptisa (2015) has indicated that people with this cultural characteristic are not averse to taking risks and have a greater acceptance of new innovations, products and services. This could be the reason why it moderated positively the relationship between behavioral intention and actual use. This is in line with previous findings, for example Chopdar et al. (2018) who highlight that low uncertainty avoidance causes low perceived risk in the context of mobile technology adoption. Long/short term orientation had a negative moderating effect on the relationship between performance expectancy and attitude as well as the relationship between attitude and behavioral intention, which is in harmony with previous research (Hassan et al. 2011). This suggests that people with short-term orientation like those in Zimbabwe, focus more on achieving quick results from a system more than the performance of that system. This then means that it is important for technology start-up companies to develop mobile farming platforms that give farmers quick results with regards to their farming activities since they have a relatively small propensity to save for the future. Masculinity moderated positively the relationship between price value and behavioral intention. This advocates that farmers high on femininity seek to have equality and solidarity with regards to prices of mobile farming platforms.

6 Limitations and Future Research

The first limitation of this study has to do with the size of the sample chosen. The sample did not include all the provinces in Zimbabwe, as the study collected data from three provinces. This limitation however does not have much effect because the data was collected from the provinces which had the highest number of mobile farming platforms users due to the fact that there is intensive farming in those provinces. The second limitation was that this research was mainly interested in studying the adoption of mobile farming platforms by individual farmers only and it did not include the opinions and views of other parties like the government, technology start-up companies and agriculture extension officers. Considering the limitations of this study, some suggestions can be made for future research. Future researches could improve on this study by expanding the size of sample from three provinces to all the provinces in the country. In this study it was not possible to use triangulation (interviews and surveys) due to time constraints. Therefore, future researchers could strengthen the findings of the questionnaire by supporting it with observation and interviews so that reliability of data can be further improved.

7 Conclusion

The study concludes that attitude positively influences behavioral intention in the adoption of technology, particularly mobile technology, and this is consistent with earlier studies, such as Nassuora (2012) and Jairak et al. (2009). On the moderating effect of Hofstede's cultural values, the study produced mixed results with regards to the moderating influences. As already disscused, some cultural values were moderating positively, others had negligible moderating effect and other cultural values were moderating negatively. These findings show that it cannot be specifically concluded that certain cultural values have a positive or negative moderating effect on the factors affecting adoption of technology. However, as a conclusion, it can be generalized that culture as a whole moderates, be it positively or negatively, in the adoption of mobile technology. The results of the study have demonstrated consistency of the variables in the model of adoption. In conclusion, it can be generalized that the variables in the model all influence positively in the adoption of mobile technology, as shown by the results. The results of this study help in understanding the factors that affect the acceptance and adoption of mobile faming platforms/for smallholder farmers in Zimbabwe.

Acknowledgement. I wish to express my sincere gratitude and appreciation to my co-author, Prof Tranos Zuva whose input helped us to produce a paper of good standard.

References

Ajzen, I.: The theory of planned behavior. Organ. Behav. Hum. Decis. Process. **50**(2), 179–211 (2002)

Alsharif, F.F.: Investigating the factors affecting on-line shopping adoption in Saudi Arabia. Ph.D. Thesis, De Montfort University (2013). https://ethos.bl.uk/OrderDetails.do?uin=uk.bl.ethos.585451

Abraham, R.: Mobile phones and economic development: evidence from the Fishing Industry in India. Inf. Technol. Int. Develop. **4**(1), 5–17 (2007)

Adel, R.: The influence of some agricultural extension television elements on diffusion of agricultural information, rural knowledge and new technologies to tenants in the Gezira Province, Gezira. J. Agricult. Sci. **3**(1), 109–114 (2005)

Aker, J.C.: Does digital divide or provide? The impact of cell phones on grain markets in Niger. In: Working Paper Number 154. Centre for Global Development, Washington, October 2008 (2008)

Aker, J.C., Mbiti, I.M.: Mobile phones and economic development in Africa. J. Econ. Perspect. **24**(3), 207–232 (2010)

Baptisa, A.M.M.: Understanding mobile banking: the unified theory of acceptance and use of technology combined with cultural moderators. Comput. Hum. Behav. **50**, 418–430 (2015). https://doi.org/10.1016/j.chb.2015.04.024

Bergiel, E., Bergiel, B., Upson, J.: Revisiting Hofstede's dimensions: examining the cultural convergence of the United States and Japan. Am. J. Manag. **12**(1), 69–79 (2016)

Chhachhar, A.R., Querestic, B., Khushk, G.M., Ahmed, S.: Impact of ICTs in agriculture development. J. Basic Appl. Sci. Res. **4**(1), 281–288 (2014)

De Silva, H., Ratnadiwakara, D.: Using ICT to Reduce Transaction Costs in Agriculture through Better Communication: A Case-study from Sri Lanka (2008). http://www.lirneasia.net

Engotoit, B, Moya, B.M., Kituyi, M.G.: A mobile based communication adoption model for agricultural market information dissemination in Uganda (2016). https://www.semanticscholar.org/paper/A-Mobile-Based-Communication-Adoption-Model-for-in-Engotoit-Moya/b88be0c5659bc3d3b77a7642d1e2246ab7889e15. Accessed 13 June 2019

Gichamba, A.: The extent of ICT Adoption by ACP Farmers: mAgriculture Adoption in Kenya. Technical centre for Agricultural and Rural Cooperation Working paper No. 15/05 (2017)

Habboush, A., Nassuora, A., Hussein, A.R.: Acceptance of mobile learning by University, Students. Am. J. Sci. Res. 119–122 (2011)

Hansen, N., Postmes, T., Vinne, N., Theil, W.: Information and communication technology and cultural change: how ICT changes self-construal and values. Soc. Psychol. **4**, 222–231 (2012)

Hassan, L.M., Shiu, E., Walsh, G.: A multi-country assessment of the long-term orientation scale. Int. Mark. Rev. **28**(1), 81–101 (2011). https://doi.org/10.1108/02651331111107116

van der Heijden, H.: User acceptance of hedonic information systems. MIS Q. **28**(4), 695–704 (2004)

Hofstede, G.: Culture's Consequences: International Differences in Work Related Values. Sage Publications, Beverly Hills (1980)

Hofstede, G.: Culture's Consequences: Comparing Values, Behaviours, Institutions and Organizations Across Nations, 2nd edn. Sage Publications, Thousand Oaks (2001)

Hofstede, G.: Cultures and Organizations. McGraw-Hill Publishing (2014)

Im, I., Hong, S., Kang, M.S.: An international comparison of technology adoption. Inf. Manag. **48**(1), 1–8 (2011). https://doi.org/10.1016/j.im.2010.09.001

Jensen, R.: The digital provide: information (technology), market performance and welfare in the South Indian fisheries sector. Quart. J. Econ. **122**(3), 879–924 (2007)

Jairak, K., Praneetpolgrang, P., Mekhabunchakij, K.: An acceptance of mobile learning for higher education students in Thailand. Spec. Issue Int. J. Comp. Internet Manag. **36**, 1–8 (2009)

Lee, M.C.: Understanding the behavioral intention to play online games: an extension of the theory of planned behaviour. Online Inf. Rev. **33**(5), 849–872 (2009)

Lin, H.-F.: An empirical investigation of mobile banking adoption: the effect of innovation attributes and knowledge-based trust. Int. J. Inf. Manage. **31**(3), 252–260 (2011)

Martins, C., Oliveira, T., Popovič, A.: Understanding the Internet banking adoption: a unified theory of acceptance and use of technology and perceived risk application. Int. J. Inf. Manage. **34**(1), 1–13 (2014). https://doi.org/10.1016/j.ijinfomgt.2013.06.002

Miltgen, C.L., Popovič, A., Oliveira, T.: Determinants of end-user acceptance of biometrics: integrating the "Big 3" of technology acceptance with privacy context. Decis. Supp. Syst. (2013)

Nassuora, A.B.: Student acceptance of mobile learning for higher education. Am. Acad. Scholar. Res. J. (2012)

Odunze and Hove. Impact of the use of mobile communication technologies. In: Proceedings of SIG GlobDev 2015 Pre-ECIS Workshop, Munster, 26 May 2015 (2015)

Oye, N., Iahad, N., Ab.Rahim, N.: The history of UTAUT model and its impact on ICT acceptance and usage by academicians. Educ. Inf. Technol. **19**(1), 251–270 (2012). https://doi.org/10.1007/s10639-012-9189-9

O'Donnell, M.: Using ICT to enhance marketing for small agricultural producers. In: USAID (2013). https://www.semanticscholar.org/paper/Using-ICT-to-enhance-marketing-for-small-producers.O%27Donnell/559c12449e8496298ae257da48049afefff7d76d. Accessed 10 Oct 2020

Olushola, T., Abiola, O.: The efficacy of technology acceptance model: a review of applicable theoretical models in information technology researches. J. Res. Bus. Manag. **2017**, 70–83 (2017). www.questjournals.org

Okazaki, S.: New perspectives on m-commerce research. J. Electron. Commer. Res. **6**(4), 160–164 (2005)

Park, J., Yang, S., Lehto, X.: Adoption of mobile technologies for Chinese consumers. J. Electron. Commer. Res. 196–206 (2007)

Podrug, N., Pavicic, J., Bratic, V.: Cross-cultural comparison of Hofstede's dimensions and decision-making style within cee context. School Econ. Bus. Univ. Sarajevco **2006**, 339–343 (2017). http://feweb.uvt.nl/center/hofstede/VSM.html

Png, I.P.L., Tan, B.C.Y.: Dimensions of national culture and corporate adoption of IT infrastructure. IEEE Trans. Eng. Manage. **48**(1), 36–45 (2001)

Qiang, C.Z., Kuek, S.C., Dymond, A., Esselaar, S.: Mobile Applications for Agriculture and Rural Development. World Bank, Washington, DC. © World Bank. License: CC BY 3.0 IGO (2012) https://openknowledge.worldbank.org/handle/10986/21892

Reisinger, Y.: Cross-cultural differences in tourist behavior. In: Handbook of Tourist Behavior, Theory & Practice. Taylor & Francis Group, Routledge, New York, London

Samaradiwakara and Gunawardena. Comparison of existing technology acceptance theories and models to suggest a well improved theory. Int. Tech. Sci. J. (2014)

Soares, F.S.: Hofstede's dimensions of culture in international marketing studies. J. Bus. Res. **60**, 277–284 (2007)

Srite, M., Karahanna, E.: The role of espoused national cultural values in technology acceptance. MIS Quart. **30**(3), 679–704 (2006)

Sriwindono, H., Yahya, S.: Towards modelling the effects of cultural dimension on ICT acceptance in Indonesia. In: International Congress on Interdisciplinary Business and Social Science 2012 (ICIBSoS 2012) (2012)
Straub, D., Loch, K., Hill, C.: Transfer of information technology to the Arab world: a test of cultural influence modeling. J. Glob. Inf. Manag. **9**(4), 6–28 (2006)
Tarhini, A.: The effects of cultural dimensions and demographic characteristics on E-learning acceptance. Ph.D. thesis, Department of Information Systems and Computing, Brunel University (2013)
Venkatesh, V., Morris, M., Davis, G., Davis, F.: User acceptance of information technology: toward a unified view. MIS Quart. **27**, 3 (2003)
Venkatesh, V., Thong, J.Y., Xu, X.: Consumer acceptance and use of information technology: extending the unified theory of acceptance and use of technology. MIS Quart. 157–178 (2012)
Zakour, A.: Cultural differences and information technology acceptance. In: 7th Annual Conference of the Southern Association for Information Systems, pp. 156–161 (2004)
Zhou, T., Lu, Y., Wang, B.: Integrating TTF and UTAUT to explain mobile banking user adoption. Comput. Hum. Behav. **26**(4), 760–767 (2010)

Trends in the Adoption and Acceptance of Technology: Challenges and Open Issues

Irene Kolo and Tranos Zuva(✉)

Vaal University of Technology, Vanderbijlpark, South Africa
irene.kolo@gmail.com, tranosz@vut.ac.za

Abstract. The year 2020 was unprecedented as the whole world went into lockdown due to the pandemic of Covid-19 resulting in loss of education. The main objective of this systematic literature review is to observe the trend within technology acceptance and adoption. Through systematic search by the use of PRISMA, the review has identified 14 relevant studies ranged between 2018 and 2021. The main finding indicates that the core variables are either exogeneous or endogenous. The findings also confirm that most variables were used over and over again with different terminologies. There were 11 core variables used prior to the pandemic and in the current pandemic situation. This study also finds gaps in current work and recommends areas for future studies.

Keywords: Adoption · Acceptance · e-Learning · Covid-19 · Variables · Technology

1 Introduction

In 2020 the whole world faced a pandemic of Covid-19 which required lockdown and a total shutdown of schools. [1] mentioned that since the beginning of 2020, when Covid-19 was first reported, many researchers have discussed the development opportunities and threats of online learning. [2] confirmed the world's re-assession of learning online to handle the challenges faced by the worldwide educational environment. The Covid-19 pandemic disrupted education in the whole world which led to changed lifestyles. The discomfort of the current pandemic is the loss of many lives lost daily in very high numbers in 2020.

Most countries around the world including South Africa, have provisionally closed educational institutions to avoid infections of Covid-19 to be viral [3]. [4] confirmed that the "indefinite closure of universities has necessitated the need to introduce alternative teaching and learning methods and tools". While [5], confirms that "schools losing long periods of learning due to disease outbreak can result to both temporal and permanent damage on the educational system."

In March 2020, South Africa faced a level 5 lockdown resulting in a total shutdown whereby all schools were temporarily closed until further notice [6].

This study aims to explore the literature review about the trends in the adoption and acceptance of technology. The main objective is to observe the trend within technology acceptance and adoption.

R. Silhavy (Ed.): CSOC 2022, LNNS 501, pp. 726–736, 2022.
https://doi.org/10.1007/978-3-031-09070-7_60

Kinds of e-learning

1.1 Simultaneous Delivery

This kind of delivery is where the [7] educators and learners connect and interact when it is convenient. This method is also called synchronous.

1.2 Asynchronous Delivery

This type of delivery is where the educator ensures material availability via videotapes or content allocation through a computer or any other means. Then the learners receive the material at their convenient time.

e-Learning is used to deliver the curriculum through the latest available technologies. The technological foundation of e-Learning is the Internet and associated communication technologies [8]. [4] confirmed that "e-learning is promising to be the future of education delivery because of disruptive innovations such as the Internet and the outbreak of pandemics like Covid-19, which emphasizes social distancing". The authors [9] confirmed that responding to the pandemic of Covid-19, e-learning was used to replace face-to-face to deliver education. While [10] confirmed that e-learning is used as a form of emergency during the pandemic.

Covid-19 exposed the digital divide as confirmed by [11] that, some institutions were ready for a smooth switch while other institutions were not even implementing e-learning. Those who used blended learning were also in a situation of learning to adopt while those learners who adopted early were busy with lessons. [12], also mentioned that due to the pandemic, most education institutions relied on internet access and the electronic devices to provide continuity of teaching and learning.

This shift to online has highlighted the stark digital divide between those who have access to electricity, internet infrastructure, data, and devices, and those that do not." Several researchers believed that the new method of learning in schools and universities became an exceptional part of home schooling and work from home [13–17]. Blended learning can therefore be summarized as using both classroom and online systems simultaneously, [5].

During a pandemic e-learning's most advantage is its ability to enable educators and learners to continue with the curriculum while in isolation. [18]. In 2020, Microsoft Teams, Google Classroom and Zoom were most used apps to continue with studies, Zoom, and Microsoft Teams to take online courses. According to [19], the virtual class was "to grow the certainty and confidence of the learners in their faculty during the Covid-19 pandemic".

The paper is organized as follows: Sect. 2 is the methodology, Sect. 3 is the results, Sect. 4 is the conclusion and Sect. 5 are the references.

2 Methodology

A systematic literature review was defined by [18] as "a process of identifying, assessing, and interpreting all available research evidence." The systematic approach for the review of the literature on the adoption and acceptance of e-learning models was adopted using

PRISMA (Preferred Reporting Items for Systematic Reviews and Meta-Analyses) to select relevant studies.

PRISMA was renamed from QUOROM statement (QUality of Reporting of Meta-analyses) The rename was to address several conceptual and practical advances in the science of systematic reviews [19].

According to PRISMA guidelines, this review was conducted on papers from 2018 to 2021, using keywords adoption, acceptance, e-Learning during covid19, pre- and post-COVID-19. All papers were screened for publication year (before or after 2017) and adopted language accepted was English only. Later then excluded papers without the results.

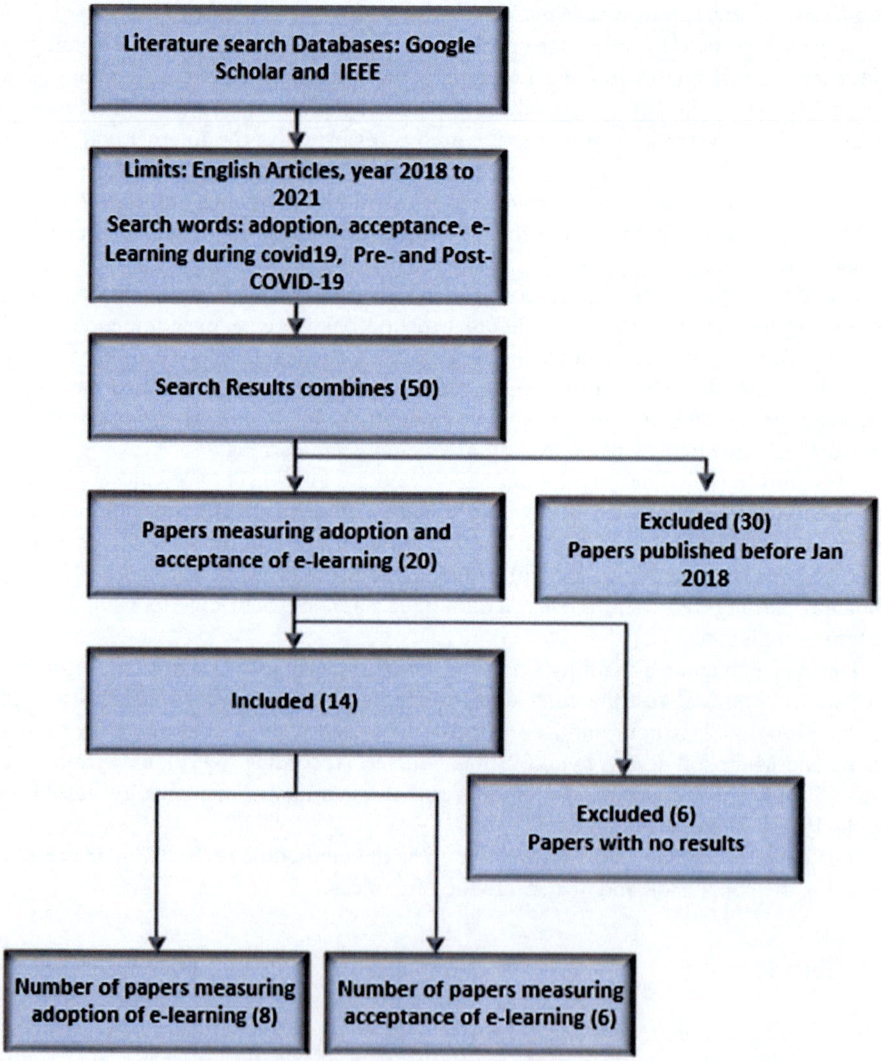

Fig. 1. Steps of the selection process

Figure 1 are the steps used to select the papers for the study. The databases used to retrieve the papers for this study are google scholar and IEEE. The papers chosen were written in English only, and published after 1 Jan 2018 to 2021. The search words used were adoption, acceptance, e-learning, covid19, pre and post covid19.

3 Results

The results were obtained from 14 different studies. The 14 selected studies gave a perspective that motivates research on adoption and acceptance of e-learning in educational background.

The results from the 14 papers are arranged as follows: number of papers measuring adoption of e-learning, number of papers measuring acceptance of e-learning, are 8 and 6 respectively as shown in Fig. 1.

The distribution of papers selected per the year from 2018 to 2021 is displayed in Fig. 2.

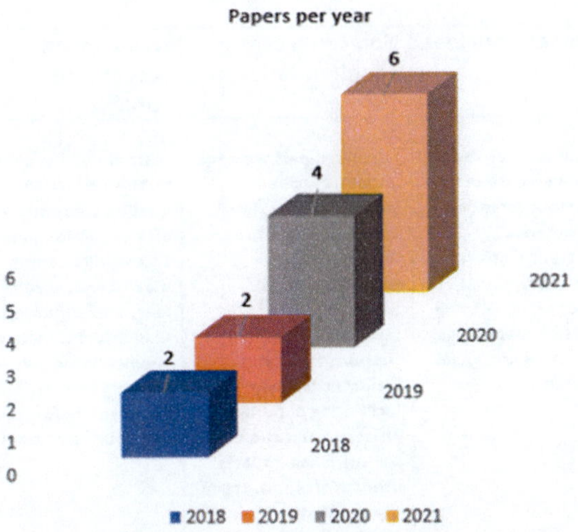

Fig. 2. Number of papers per year

The papers were selected from different years. The papers for adoption of e-learning are six papers from the year 2021, four papers from the year 2020, two papers from the year 2019 and two papers from the year 2018. The acceptance of e-learning papers in 2021 are three, two papers are in 2020 and two in 2019.

There are common and non-common variables used in acceptance and adoption of e-learning as displayed in Fig. 3.

A total of 50 variables from 14 papers were analyzed. It was identified that there are 10 common variables in acceptance and adoption papers as described in Fig. 3.

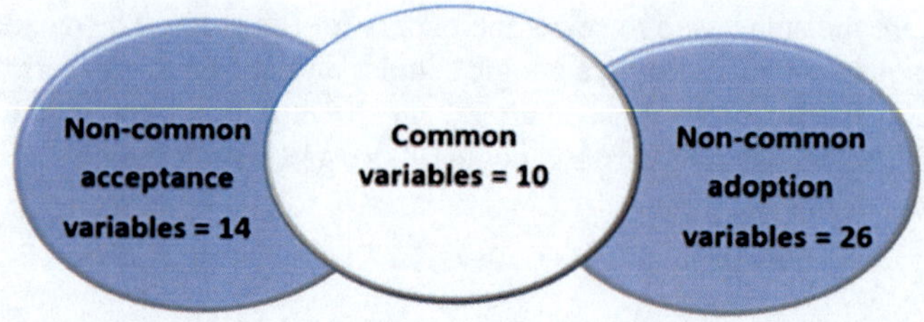

Fig. 3. Non-common variables vs common variables

While non-common acceptance variables are 14 then non-common adoption variables are 26.

Summary of common and non-common variables used in both adoption and acceptance papers are displayed in Table 1.

Table 1. Common and non-common variables summary

Common variables	Non-common adoption variables	Non-common acceptance variables
perceived usefulness, perceived ease of use, behaviour intention, performance expectancy, effort expectancy, social influence, facilitating condition, actual use, use behaviour, social isolation.	perceived performance impact, autonomy, gender, age, anxiety, experience, attitude, financial factor, technology factor, covid-19 fear, facilitating condition support, facilitating condition access, facilitating condition efficacy, utilization, performance impacts, user satisfaction, social characteristic, continuance use, confirmation, expected consequence of utilization, student satisfaction, actual social media use, task technology fit, task characteristic, technology characteristic, individual characteristic	system quality, content quality, information quality, computer efficacy, enjoyment, accessibility, computer playfulness, corona fear, peer influence, students' behaviour, voluntariness use, e-learning intention, perceived enjoyment and subjective norm

4 Discussion

From 50 variables obtained from the 14 papers of this study, there are 12 variables used with different terminologies. The actual number of variables should be 38 variables excluding the substitute variables (Tables 2 and 3).

Table 2. Variable's terminologies

Variable	Synonym variable/s
Subjective Norm	Social influence
Usage	Actual usage
Covid fear	Covid19 fear
Voluntariness of use	Voluntariness
Social characteristics	Individual characteristics
Actual social media use	Actual use/ Usage
Perceived performance impact	Performance impact/performance
E-learning intention	Intention
Student satisfaction	User satisfaction/ satisfaction
Continuance intention	Continued intention
Attitude towards use	Attitude
BI use	BI

The Core variables are the constant variables used in measuring the adoption and/or acceptance of eLearning.

Table 4 is displaying 11 core variables from 2018 to 2021. From 11 core variables, there are 7 core acceptance variables, 6 adoption variables and 2 core variables for both adoption and acceptance of e-learning PU and PEOU.

Table 3. Models and variables of the selection process paragraphs must be indented.

Components/ Factors	Wang et al. (2021)	Vladova et al. (2021)	Maphosa (2021)	Mokhtar & Karim (2021)	Qiao et al. (2021)	Pal & Patra (2021)	Raza et al. (2020)	Wan et al. (2020)	Al-Okaily et al. (2020)	Al-Rahmi et al. (2020)	Khechine & Augier (2019)	Salloum et al. (2019)	Vongjaturapat (2018)	Bere (2018)	Total
Performance expectancy			√	√	√		√	√			√				6
Effort expectancy			√	√	√		√	√			√				6
Social influence				√			√	√	√		√			√	6
Facilitating conditions				√	√		√	√			√				5
Facilitating condition support			√												1
Facilitating condition access			√												1
Facilitating condition efficacy			√												1
Social isolation					√		√								2
Corona fear/Covid19 fear					√		√								2
Perceived ease of use		√				√			√			√		√	5
Perceived usefulness	√	√				√			√			√		√	6
Perceived enjoyment/enjoyment		√										√			2
Behavioural intention		√	√	√	√		√		√	√	√	√			9
Subjective norm									√			√			2
Attitude		√				√					√	√			4
Continued intention	√							√							2
User/student satisfaction								√		√					2
Task-technology fit	√					√		√		√			√	√	6
Technology characteristics						√		√		√			√	√	5
Task characteristics								√					√	√	3
Individual characteristics						√								√	2
Social characteristics										√					1
Usage/actual use			√		√	√				√		√			5
Behaviour				√			√				√				3
Self-efficacy					√							√			2
Satisfaction	√														1
Voluntariness of use				√											1
Peer influence									√						1
Performance impact										√			√	√	3
Gender											√				1
Autonomy											√				1
Anxiety											√				1
Age											√				1
Experience											√				1
System quality												√			1
Content quality												√			1
Information quality												√			1
Accessibility												√			1
Computer playfulness												√			1
Confirmation	√														1
Performance										√			√		2
Utilisation													√		1

Table 4. Core variables from 2018 to 2021

Core variables
Actual use
Behavioural intention
Effort expectancy
Facilitating condition
Individual characteristic
PEOU
Perceived usefulness
Performance expectancy
Task characteristic
Task Technology fit
Technology characteristic

There are different types of variables used for different purposes (Fig. 4).

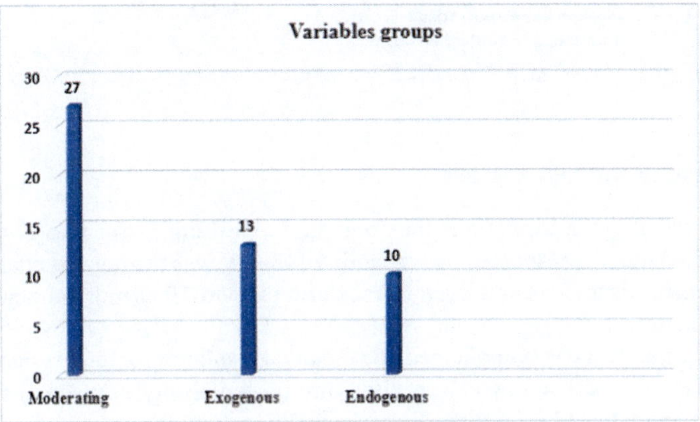

Fig. 4. Variable's classification

Above Fig. 3 shows that from the 14 models studied, 27 variables are moderating (average variables), 13 are exogenous (independent variables) and 10 are endogenous (dependent variables). This table is summarizing all the variables as per their description (Table 5).

Table 5. Variable's description

Variable/s	Description
Age, Gender, Voluntariness, Experience, Anxiety, Autonomy, facilitating condition support, facilitating condition access, Facilitating condition Efficacy, Time flexibility, Learning flexibility, System quality, Content quality, Information quality, Computer self-efficacy, Perceived enjoyment, Accessibility, Computer playfulness	Moderating variables
Task characteristic, Technology characteristic, Individual characteristic, TTF	Exogenous variables
Acceptance, Knowledge, Utilization, Use behaviour, Social isolation, Self-efficacy, Experience, Confirmation, Technology development, Function, Attitude	Endogenous variables
Perceived usefulness, perceived ease of use, behaviour intention, performance expectancy, effort expectancy, social influence, facilitating condition, task technology fit, task characteristic, technology characteristic, individual characteristic, actual use, use behaviour, social isolation, and subjective norm.	Core factors

4.1 Challenges and Open Issues

Each study has its own limitations thus opening opportunities for future work. Some studies pointed that the resources unavailability [18, 20] was the biggest challenge prior Covid-19 pandemic and became even worse during Covid-19 as most students relied on institution facilities.

Hence the pandemic was unplanned, the financial resources for remote based learning became the major challenge which resulted into unavailability of resources to support emergency remote teaching. Another general challenge was lack of non-formal training on technology-based learning with no prior experience which resulted in unfriendly learning experience [20].

Most papers had sample limitation which leads to non-generalization of the results as sampling was done in a specific country and/or a private institution. The studies were focused on education area.

Furthermore, new external variables incorporation and/or more combinations into research models might be excessive perspective which offers openings for future work.

4.2 Study Limitation

This study used only the papers with the results, published after 1 Jan 2018 to 2021 which were published strictly in English. Google scholar and IEEE databases were the only databases used due to un-subscription.

5 Conclusion and Recommendations

In this paper, acceptance and adoption of e-learning papers were reviewed. The current life of pandemic revolves around technology for smooth continuity of education. Previously e-learning was optional but due to the pandemic environment, e-learning became priority to avoid educational disruption.

Based on this study, most papers results were derived from one institution while other used convenience sampling technique. Generalization should be treated with cautious.

Another finding is that any new model that comes into adoption or acceptance derive from core variables.

The study confirmed that the resources together with the proper training on technology based are the main issues. Out of 50 variables discovered from the 14 papers, 12 variables were used with different terminologies in other papers and 11 variables are core variables. It has been discovered that the core variables are either exogenous or endogenous.

It is recommended that future research duplicate this study but with larger sample for better results.

References

1. Donitsa-Schmidt, S., Ramot, R.: Opportunities and Challenges: Teacher Education in Israel in the Covid-19 Pandemic, vol. 46, pp. 586–595 (2020)
2. UNESCO. COVID-19 Educational disruption and response (2020). https://en.unesco.org/covid19/educationresponse
3. Tria, J.Z.: The Covid-19 Pandemic Through the Lens of Education in the Philippines: The New Normal. Int. J. Pedag. Develop. Lifelong Learn. (2020)
4. Jere, N.: Investigating university academics behavioural intention in the adoption of e-learning in a time of COVID-19. South Afr. J. Inf. Manag. **22** (2020)
5. Mahaye, N.E.: The Impact of COVID-19 Pandemic on South African Education: Navigating Forward the Pedagogy of Blended Learning (2020)
6. News.gov.za. Coronavirus outbreak declared a national disaster. https://www.sanews.gov.za/south-africa/coronavirus-outbreak-declared-national-disaster
7. Obeidat, A., Obeidat, R., Al-Shalabi, M.: The Effectiveness of Adopting e-Learning during COVID-19 at Hashemite University. Int. J. Adv. Comput. Sci. Appl. **11**(12) (2020)
8. Garrison, D.R.: E-learning in the 21st century: a framework for research and practice (2011)
9. Vladova, G., Ullrich, A., Bender, B, Gronau, N.: Students' Acceptance of Technology-Mediated Teaching – How It Was Influenced During the COVID-19 Pandemic in 2020: A Study from Germany, vol. 12 (2021)
10. Wang, T., Lin, C., Su, Y.: Continuance Intention of University Students and Online Learning during the COVID-19 Pandemic: A Modified Expectation Confirmation Model Perspective (2021)

11. Alqahtani, A.Y., Rajkhan, A.A.: E-learning critical success factors during the COVID-19 pandemic: a comprehensive analysis of E-learning managerial perspectives. Educ. Sci. **10**, 216 (2020)
12. Bozkurt, A., Jung, I., Xiao, J., Vladimirschi, V., Schuwer, R., Egorov, G.: Global outlook to the interruption of education due to COVID-19 pandemic: navigating in a time of uncertainty and crisis. Asian J. Dist. Educ. **15**(1), 1–126 (2020)
13. Almaiah, M.A., Ahmad, A., Ahmad, A.: Exploring the critical challenges and factors influencing the E-learning system usage during COVID-19 pandemic. In: Education and Information Technologies, pp. 5261–5280 (2020)
14. Siriwardhana, Y., Chamitha, A., Gürkan, G., Mika, Y., Madhusanka, L.: The fight against the COVID-19 pandemic with 5G technologies. IEEE Eng. Manage. Rev. **48**, 72–84 (2020)
15. Smalley, A.: Higher Education Responses to Coronavirus (COVID-19), 17 December 2020
16. Wu, S., et al.: Pilot study of robot-assisted teleultrasound based on 5G network: a new feasible strategy for early imaging assessment during COVID-19 pandemic. IEEE Trans. Ultrason. Ferroelectr. Freq. Control **67**, 2241–2248 (2020)
17. Ye, J.: The role of health technology and informatics in a global public health emergency: practices and implications from the COVID-19 pandemic. JMIR Med. Inform. **8**, e19866 (2020)
18. Vlachopoulos, D.: COVID-19: threat or opportunity for online education? High. Learn. Res. Commun. **10**(1), 16–19 (2011)
19. Zalat, M., Hamed, M., Bolbo, S.: The experiences, challenges, and acceptance of e-learning as a tool for teaching during the COVID-19 pandemic among university medical staff
20. Khechine, H., Augier, M.: Adoption of a Social Learning Platform in Higher Education: An Extended UTAUT Model Implementation. In: Proceedings of the 52nd Hawaii International Conference on System Sciences, HICSS (2019). https://hdl.handle.net/10125/59446
21. Moher, D., Liberati, A., Tetzlaff, J., Altman, D.G.: Preferred reporting items for systematic reviews and meta-analyses: the PRISMA statement. BMJ **339**, b2535 (2009)
22. Maphosa, V.: Factors influencing student's perceptions towards E-learning adoption during COVID-19 pandemic: a developing country context. Eur. J. Interact. Multim. Educ. (2021)

The Role of Gender, Age and Cultural Differences in Online Information Disclosure and Privacy: A Systematic Review

William Ratjeana Malatji(✉), Rene VanEck, and Tranos Zuva

Department of Information and Communication Technology, Vaal University of Technology, Andries Potgieter Blvd, Vanderbijlpark 1900, South Africa
villywr@gmail.com, {Rene,Tranosz}@vut.ac.za

Abstract. Previous studies in online information disclosure and privacy have focused on gender, age and cultural differences, however, a consensus on the role of these three factors in online information disclosure and privacy was not met. Therefore, it is necessary to find out the role of gender, age and cultural differences in online information disclosure and privacy. This study aimed to systematically present the role of gender, age and cultural differences in online information disclosure and privacy. Furthermore, the methodological quality and procedures of the selected studies were assessed in this review. The procedure of the study selection and outcomes followed the PRISMA framework. The initial search resulted in 336 studies from the Scopus and ScienceDirect databases since they were freely accessible to the researchers. Later, only 12 relevant studies were included in the analysis. The results of the study revealed that female users have less, adopting behaviour that requires higher technical skills than male users, they are more vulnerable than male users in online platforms and those female users are more likely to disclose information than males. With regards to age, the results indicated that young adult users (more especially female users between 18 and 40 years) utilise online platforms regularly and have larger networks than older age group users (between 40 and 65 years). It was further indicated by the results that young adults (between 14 and 40 years) are more likely to disclose personal information on online platforms. The culture was found to be one of the influential factors towards users' information disclosure behaviour in online platforms. Collectivism relating to society and power distancing influence the strength of the association between privacy concerns and surveillance acceptance significantly, as well as the adoption of privacy protections. In assessing the methodological quality and procedures, this study used the COSMIN methodology. The results of the methodological quality evaluations proposed a necessity for a more in-depth assessment of content validity. This study provides numerous recommendations and suggestions for further research, such as an extension of the search to other domains rather than online context only, and others.

Keywords: Gender · Age cultural differences · Online information privacy

R. Silhavy (Ed.): CSOC 2022, LNNS 501, pp. 737–751, 2022.
https://doi.org/10.1007/978-3-031-09070-7_61

1 Introduction

Although previous research on "Information and Communication Technology" (ICT) mostly concentrated on the use of the internet, its impact on other related dimensions like privacy is becoming increasingly important for people and society. Privacy has various definitions across different disciplines. In the online context, privacy is defined as "the degree of privacy protection one has while connected to the internet [1]." People visit the internet for many reasons, being it, to connect with families and friends, make online purchases or any other business exchange, work, etc. Recent technological innovations, for example, social network sites, enable the collection, storage and processing of large volumes of user data, allowing a better knowledge of their behaviours and preferences [2]. However, users' privacy remains the main concern as a lot of personal information is disclosed and left behind, which can be easily accumulated, put together and utilised for manipulation or handed over in exchange for money [3, 4].

Numerous laws have been passed in recent years to safeguard individuals' rights when it comes to information privacy. For example, the "General Data Protection Regulation" (GDPR), approved by European Union in 2016 [5]. According to this law, individuals have the right to know who is accumulating or gaining access to their data, the right to control over their data, the right to give or withdraw consent on the personal data collection and processing, etc. [6]. Online discussions on security and privacy, as well as other methods to secure individual data increasingly gain attention in scientific works and well-known press reports, however, it is still unclear why people safeguard or do not protect their information online [7].

Even though information privacy is of universal concern to internet users, only a little number of studies have been conducted on how demographic differences (gender and age to be specific) have an influence on behaviours related to privacy and comprehending this could aid in identifying consumers who are more likely to have poor information privacy behaviour [8]. Starting with gender, existing studies on information privacy found that women have higher degrees of security and privacy concerns, as well as satisfactory awareness of information security [9–11]. Conversely, results of other related studies revealed that women are more vulnerable to phishing attacks, have poorer behaviour in regards to passwords and that women are less likely to adopt behaviours in regards to privacy protection [12–14]. This calls for more research to be conducted on the role of gender differences towards online information privacy and security. Individuals' perspectives have been considered as factors influencing privacy protection behaviour [14]. These incorporate response efficacy, perceived vulnerability, response cost, descriptive and subjective norm, as well as perceived vulnerability [15, 16]. Likewise, perspectives such as personal information collection concerns, subjective norms and perceived privacy risk influence privacy protection behaviour and intentions [17, 18]. The role of gender on the other hand is less obvious. Numerous studies have looked into whether it has a direct impact on privacy intentions or behaviours, while others have looked into if it has an impact on perceptions [19, 20].

With regards to age, although much study has been conducted with young adults and teenagers, less emphasis has been made on the differences in privacy management behaviour among adult age groups [21]. When it comes to the use of ICT, older adults resemble a special or unique group. They are normally more careful in accepting recent technologies and frequently less experienced. They also reveal a lack of trust in online, as well as privacy protection [22]. Teenagers are mostly perceived as vulnerable because of their immoderate use of the internet and self-disclosure more especially on social network sites [22].

There exists a growing conflict in today's society between companies' yearning for individual personal data and the significance of respecting peoples' privacy. Regardless of several studies on this subject, a general agreement on whether privacy concerns foretell privacy behaviour-the-so-called "privacy paradox" is not yet met [23]. One of the most significant concerns that people, corporations, as well as government, confronts today is online privacy and the security threats of information systems. However, meanwhile, the threats remain universal, every single society displays various values on, and a single being in every society reveals unalike attitudes and behaviours with respect to information privacy and security [24]. As a result, differences in culture can obstruct and give rise to challenges with respect to the successful development and execution of strategies, as well as interference for minimizing these concerns [24].

Cross-cultural gaps have been widely investigated in a variety of domains and differences in culture have been proven to determine a variance in phenomena relating to information privacy and security [25]. However, a survey into cross-cultural challenges is still new in this discipline and a comprehensive research topic is being untapped [24]. When it comes to information sharing and knowledge, culture is known to play an important role. Existing studies in the literature suggest that a culture to which a single being belongs ought to be considered as a significant factor in examining users' behaviour [26]. Scholars proposed models to describe and comprehend how society members share values and beliefs. Hofstede's model (2011) is the most preferred among the existing models which explain culture using six cultural dimensions; culture is described as "the collective programming of the mind that distinguishes the members of one group or category of people from others" in this model [26]. Despite the fact that every collective is made up of individuals, the members of a collective generally exhibit similar patterns across the six dimensions when seen collectively. In several fields, differences in cultural aspects are linked to disparities in privacy concerns [26].

This study aims to systematically present the role of gender, age and cultural differences in online information privacy according to the findings in the literature. Moreover, this systematic review assesses the quality of methodological procedures used in the studies of the role of gender, age and cultural differences in online information privacy. Thus, the study is specifically interested in the following research questions: 1) What is the role of gender in online information privacy according to the literature?, 2) What is the role of age in online information privacy according to the literature?, 3) What is the role of cultural differences in online information privacy according to the literature?, 4) What is the quality of methodological procedures used in online information privacy studies? This study will be arranged as follows: Background, methods, results, discussion and conclusions.

2 Materials and Methods

The PRISMA "Preferred Reporting Items for Systematic Reviews and Meta-Analysis" framework and was used in the preparation of this review [27]. Moreover, The COSMIN methodology, which was established for assessing the quality of the methodology and psychometric features of "Patient Reported outcome measures" (PROMs) in the discipline of health care, was utilised to conduct empirical analysis in this work [28]. Although this approach has mostly been utilised to evaluate the exactness of surveys measuring instrument outcomes or treatment, namely "PROMs", it can also be used in other fields of study [28].

2.1 Search Strategy

A search in this systematic review was initially conducted in 2 databases, to incorporating articles focusing on gender, age and cultural differences in online information disclosure and privacy. In line with the COSMIN methodology propositions, the eligibility criteria based on the concept of interest, population of interest, and measurement qualities were created. The following eligibility criteria were followed: a) The focus of the article's analysis must be on the individual, b) The survey scale must be solely focused on online information privacy and not include any other themes, c) Quantitative validation of one or more survey scales must be attempted in the paper. The search was conducted from 24 November to 28 November 2021. It was performed from the Scopus, as well as ScienceDirect databases. The Articles included in the search were from the beginning of 2010 to November 2021. The research utilized the following search terms: "age", "gender", "culture", "online privacy" and "information disclosure."

2.2 Study Selection

A three-stage screening approach was used to determine the eligibility of articles. The initial stage involved the removal of duplicates from the search results. The second stage involved the review of titles and abstracts. When the article's eligibility for inclusion could not be determined, it was kept for further review at the next coming screening stage. The last stage involved a full-text review of the kept articles from the first two stages. The full-texts were first evaluated to determine their eligibility for inclusion, followed by the evaluation of texts that were deemed ineligible. Then later, the list of references of the incorporated articles was examined to find other relevant articles.

2.3 Data Extraction

Only 12 articles with the following extraction characteristics were chosen in the phase of data extraction: 1) An article must be an actual paper, a conference paper, or a review paper. Published case studies, as well as reports, were all excluded, 2) All articles must be limited to English language only, 3) All extracted articles must be from 2010 to 2021 only, 4) All extracted articles must not be limited to any country.

2.4 Assessment of Measurement Properties

In line with the objectives of this study, an appropriate procedure from the COSMIN methodology for carrying out a literature search was followed and the COSMIN "Risk of Bias check list" was utilised to screen possible biases of the methodology in the development of study's scale [29]. The study concentrated on the following measurement properties: Structural validity, content validity, cross-cultural validity, internal consistency, criterion validity and reliability. The measurement properties in this study are in line with the COSMIN methodology [30].

2.5 Risk of Bias Evaluation

All studies that were included in this systematic review for analysis were assessed for risk of bias to confirm the trustworthiness of the methodological quality of each study. The assessment was achieved through the implementation of the COSM Risk of Bias checklist [29]. The "Content validity" was rated as VG- "Very Good", AD- "Adequate", DF- "Doubtful", IN-"Inadequate", and "-" which "highlights that the information was absent or not reported". This type of rating applied to other measurement constructs (i.e., Structural validity, internal consistency and cross-cultural validity, reliability, as well as criterion validity).

3 Results

3.1 Search Results

Figure 1 present the search results based on the PRISMA guidelines.

The procedure of the study selection and the outcomes using PRISMA framework is presented in a flow diagram in Fig. 1. The initial search of the two databases resulted in a total of 336 articles that are relevant, of which 104 were duplicates. After the preview of titles and abstracts, 175 articles were excluded because they were not able to meet the eligibility criteria (171) or were found as duplicates (4). Further, 15 articles were included as additional articles for full-text assessment in line with the reference list of the included articles (11), as well as those which were manually searched (4). Out of 72 articles that were fully read, 60 were excluded due to duplicates (25), articles were not concentrating on online information privacy (22), articles were not written in the English language (12), and lastly, 1 article was not accessible. As a result, the final inclusion involved only 12 articles for analysis.

3.2 Descriptive Analysis

The descriptive results in this study were divided into three sections. The first section provides the descriptive results on gender, the second section provides the descriptive results on age and the last section provides the descriptive results on cultural differences.

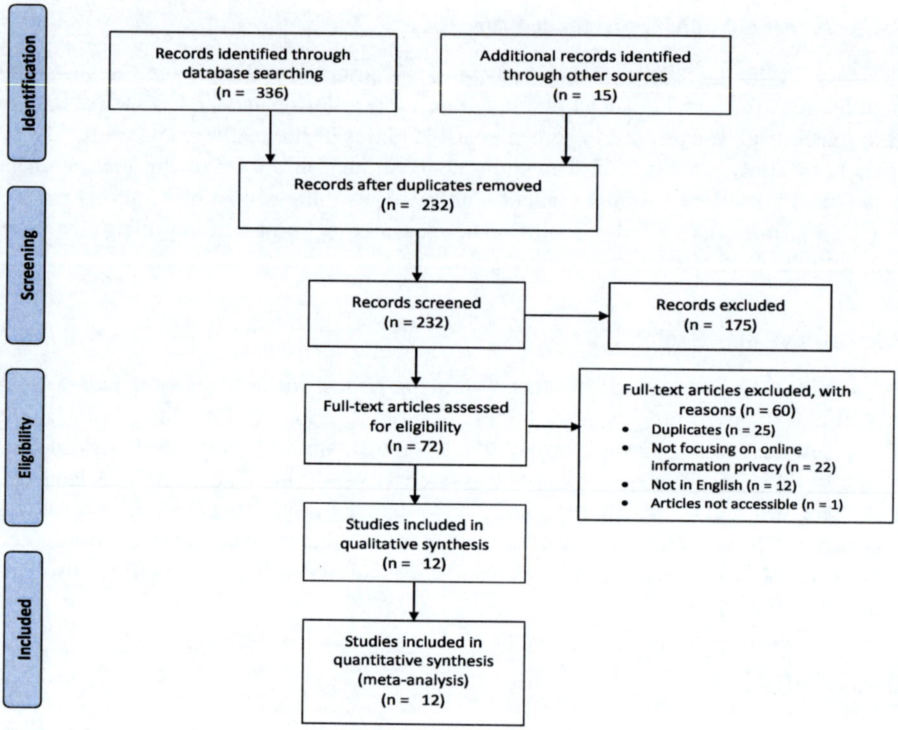

Fig. 1. Process of article selection

- Section 1: Gender

The details of the first 5 studies on the role of gender in online information privacy are provided in Table 1 (2 studies used Protection Motivation Theory (PMT), 1 study used the model of information sharing decision and the 2 remaining studies did not mention the theories or frameworks used. Only one study indicated that the pre-test (pilot study) was conducted). According to the results of this research, only 2 studies were conducted in 2018, 2 in 2020 and the last one in 2021. It was further indicated by the results that 3 studies used an anonymous online survey, 1 used Jourard Self-disclosure questionnaire and the last one used the structured online survey as a method of data collection (Number of items: 17, 6, 50, 44 and 25 respectively). In the results, 3 studies used a 7-point Likert scale ("Strongly Disagree to Strongly Agree"), 1 study utilised a 5-point Likert scale ("Little Privacy to Very Much Privacy") and the last study utilised a response scale ranging from ("0 to 2"). All studies were conducted in the English language (2 studies were conducted in the US only, 1 in both the US and Australia, 1 in Italy and the last 1 in Jordan).

Table 1. Descriptive results for gender

Article	Year	Survey type	No. Items	Response Options	Response scale	Country	Language
[14]	2021	Online survey	US: 6 Australia: 11	UAS: 1–7 Australia: 1–5	"Strongly Disagree to Strongly Agree"	US Australia	English
[31]	2018	Online survey	6	1–7	"Strongly Disagree to Strongly Agree"	US	English
[32]	2020	Jourard self-disclosure	50	–	0–2	Jordan	English
[33]	2018	Structured online survey	44	1–5	"Little Privacy to Very Much Privacy"	Italy	English
[18]	2020	Online survey	25	1–7	"Strongly Disagree to Strongly Agree"	US	English

- Section 2: Age

Table 2. Descriptive results for age

Article	Year	Survey type	No. Items	Response option	Response scale	Country	Language
[21]	2016	Online survey	18	1–5	"Strongly Disagree to Strongly Agree"	Turkey	English
[34]	2015	Online survey	22	1–5	"Never to always"	US	English
[22]	2017	Online survey	28	1–5	"I do not agree to I agree"	Germany	English

The details of 3 studies on the role of age in online information privacy are provided in Table 2 (1 study used Multicategorical Mediation Model, and the 2 remaining studies did not mention the theories or frameworks used. No study indicated if the pre-test (pilot study) was conducted or not). According to the results of this research, studies were conducted in 2016, 2015 and 2017 respectively. It was further indicated by the results that all studies used anonymous online survey (Number of items: 18, 22 and 28 respectively). In the results, 1 study used a 5-point Likert scale ("Never to Always"), 1 study utilised a 5-point Likert scale ("I do not agree to I agree") and the last study utilised a 5-point Likert scale ("Strongly Disagree to Strongly Agree"). All studies were conducted in the English language (1 study was conducted in the Turkey, 1 in the US, and the last 1 in Germany).

- Section 3: Cultural Differences

Table 3. Descriptive results for Cultural differences

Article	Year	Survey type	No. Items	Response option	Response scale	Country	Language
[35]	2017	Web-based survey questionnaire	53	1–5	"Strongly Disagree to Strongly Agree"	Australia	English
[36]	2017	Online survey	15	1–5, 1–7	"Strongly Disagree to Strongly Agree"	Germany, China, Canada, US, UK, Sweden, India, Australia	English
[26]	2020	Online survey, focus and group interviews	22	1–7	"Strongly Disagree to Strongly Agree"	US, Korea	English
[17]	2020	Online survey	18	1–5	"Strongly Disagree to Strongly Agree"	Australia	English

The details of 4 studies on the role of cultural differences in online information privacy are provided in Table 3 (3 studies used Hofstede's model of cultural dimensions and 1 study used Malhotra et al. 's model. Only one study indicated that the pre-test (pilot study) was conducted). According to the results of this research, 2 studies were conducted in 2017 and the other 2 in 2020. It was further indicated by the results that 2 studies used an anonymous online survey, 1 used web-based survey questionnaire and the last one used an online survey, focus group, as well as interviews to collect data (Number of items: 53, 15, 22 and 18 respectively). In the results, 3 studies used a 5-point Likert scale ("Strongly Disagree to Strongly Agree") and 1 study utilised a 7-point Likert scale ("Strongly Disagree to Strongly Agree"). All studies were conducted in the English language (2 studies were conducted in Australia only, 1 in both the US and Korea and the last one was conducted in 8 countries which include the US, Germany, Australia, China, UK, Sweden, India and Canada).

3.3 Risk of Bias Analysis

Table 4 present the results for the risk of bias based on the COSMIN Risk of Bias Checklist.

- Content validity

  The content validity will look into the following checkboxes; General design requirements, concept-elicitation, cognitive review, comprehensibility, as well as comprehensiveness. Starting with the "General design requirements" checkbox, this checkbox involves construct clear description, construct origin and clear measurement, target population description, context description use, as well as the sample. As indicated in Table 4, 7 studies were rated as "Very Good", 3 as "Adequate" and 2 as "Inadequate". The second checkbox is concept-elicitation which evaluates cognitive interview study quality or other pilot tests conducted. As shown in Table 4, 5 studies were rated as "Very Good", 2 as "Adequate", 3 as "Doubtful" and 2 as "Information absent or not reported." The cognitive interview checkbox assesses whether a pre-test (or pilot study) intending to assess comprehensibility, as well as comprehensiveness was performed. Rating this section inadequately indicates that the cognitive interview was not performed. The results indicate that 4 studies were rated as "Very Good", 4 as "Adequate" and the remaining 4 as "Inadequate". Studies that include either a pilot study or another type of interview were assessed to find out if researchers tested the participant's knowledge of the scale items, namely "Comprehensibility" and if they had control over the degree to which the survey scale incorporates all sides of the measured variable, namely "Comprehensiveness". The results in Table 4 indicate that with regards to "Comprehensibility", 5 studies were rated as "Very Good", 1 as "Doubtful" and the remaining 6 as "information absent or not reported." Coming to "Comprehensiveness", the results in Table 4 indicate that 3 studies were rated as "Very Good", 3 as "Adequate", 3 as "Doubtful" and the remaining 3 as "Information absent or not reported."

- Structural validity, internal consistency and cross-cultural validity

  The results in Table 4 indicate that under "Structural validity" checkbox 6 studies were rated as "Very Good", 3 as "Adequate" and the remaining 3 studies as "Inadequate". With "Internal consistency" all studies were rated as "Very Good". Lastly, with "Cross-cultural validity" 5 studies were rated as "Very Good", 5 as "Adequate" and the remaining 2 as "Information absent or not reported".

Table 4. Reviewed Methodological quality of online information privacy scales according to the COSMIN Risk of Bias checklist.

Article	General design requirements	Concept elicitation	Cognitive elicitation study	Comprehensibility	Comprehensiveness	Structural validity	Internal consistency	Cross-cultural validity	Reliability	Criterion validity
[14]	AD	DF	IN	–	–	IN	VG	AD	–	–
[31]	AD	DF	AD	DF	DF	IN	VG	–	–	DF
[32]	IN	AD	VG	VG	AD	VG	VG	AD	VG	VG
[33]	VG	–	IN	–	VG	–	VG	–	AD	VG
[18]	VG	VG	AD	VG	VG	VG	VG	AD	VG	VG
[21]	VG	DF	AD	–	DF	AD	VG	AD	–	VG
[34]	VG	–	IN	–	–	AD	VG	VG	VG	VG
[22]	VG	AD	IN	–	AD	VG	VG	VG	AD	VG
[35]	VG	VG	VG	VG	VG	VG	VG	VG	VG	VG
[36]	VG	VG	VG	VG	VG	VG	VG	VG	VG	VG
[26]	IN	VG	AD	–	DF	VG	VG	AD	AD	VG
[17]	VG	VG	VG	VG	–	AD	VG	VG	VG	VG

Abbreviations: VG- “Very Good”, AD- “Adequate”, DF- “Doubtful”, IN-“Inadequate”, “-” “highlights that the information was absent or not reported”.

- Reliability and criterion validity

  According to Table 4, the results under "Reliability" checkbox indicate 6 studies rated as "Very Good", 3 as "Adequate" and 3 as "Information absent or not reported." The results in Table 4 further indicate that under "Criterion validity" checkbox 10 studies were rated as "Very Good", 1 as "Doubtful" and the last study as "Information absent or not reported." Looking at the results in Table 4, it can then be concluded that the fourth research question of this study was answered.

4 Discussions

This review discusses the role of gender, age, as well as cultural differences in online information privacy. In this section, results will be discussed according to the four research questions of this study. To find out what role does gender play in online information privacy RQ1 (i.e., "What is the role of gender in online information privacy according to the literature?"), five of the selected articles for analysis were searched. The results of the first study revealed important variations between men and women in above 40% of the examined behaviours with respect to privacy and security, significantly, indicating lower levels for female users than males, with females adopting less behaviour that demand higher technical skills [14]. Similarly, the second study revealed that female users have much lower total levels of security behaviour than males. The results further indicated that individual perceptions as well, differ by gender in many circumstances [31]. Likewise, the third study revealed that when coming to self-disclosure, female users are more likely to disclose information than males [32]. With regards to privacy-related risks, the fourth study revealed that female users are more concerned about privacy than males [33]. In the literature, the findings of the first four studies in online information privacy are supported by the findings by [12–14]. The findings of the last study revealed that gender moderates the association between people's views regarding information disclosure and their willingness to disclose information [18]. With these findings, it can then be concluded that the first research question of this study was clearly answered. The conceptual answer for the first research question of this study is that gender plays an important role in online information privacy, with female users indicating less behaviour with regards to security and privacy, moreover, they require high technical skills than males.

Similar to the first research question, three of the selected studies analysis were searched in order to provide an answer to the second research question (RQ2) (i.e., "What is the role of age in online information privacy according to the literature?"). The findings of the first study revealed that even though the general use of the internet and online privacy literacy were comparable across age groups, young adults (between 18 and 40 years) utilised online platforms regularly and have larger networks compared to older age groups (between 40 and 60 years). They were as well more likely to disclose personal information and engage in privacy-protective behaviours. With regards to privacy attitudes, the study revealed that older age group users were more concerned regarding the privacy of others [21]. Similarly, the findings of the second study revealed that young adults can disclose their information online than older age group users [34]. Likewise, the findings of the last study revealed that older age group users are substantially more conscious of privacy concerns and safeguard their data than younger age groups [22]. The findings of these three studies are supported by the findings of [22] in the literature. It can be concluded that the second research question of this study was clearly answered by looking at the above findings. Therefore, the conceptual answer for the second research question is that age plays a significant role in online information privacy with young age group users having larger networks and easily disclosing personal information on online platforms than old age group users.

Likewise, four studies were searched from the selected studies for analysis in order to answer the third research question of this study (RQ3) (i.e., "What is the role of cultural differences in online information privacy according to the literature?"). The results of the first study revealed that familiarity and calculative beliefs of individuals have an influence on internet transactions [35]. The second study, on the other hand, revealed that cultural dimensions have a substantial influence on users' decisions with respect to privacy [36]. According to the findings of the third study [26], users' behaviour on information disclosure is positively related to an ethical culture. The findings further indicated that users' perceived ethical culture and concerns regarding privacy are the main influential factors of users' information disclosure. Looking into the findings of the last study [17], it was found that collectivism relating to society and power distancing influence the strength of the association between privacy concerns and surveillance acceptance significantly, as well as the adoption of privacy protections. These findings are in line with the findings of [26] and [24] in the literature. It can then be concluded that the third research question of this study was answered. The conceptual answer for the third research question of this study is that culture plays a significant role in online information privacy, as well as in knowledge sharing.

To provide an answer for the fourth research question of this study (RQ4) (i.e., "What is the quality of methodological procedures used in online information privacy studies?"), researchers developed Table 4 using the COSMIN Risk of Bias checklist to assess the methodological quality and procedures for the selected articles in online information disclosure and privacy.

5 Study Limitations and Future Recommendations

Like other systematic reviews, the discoveries of this study ought to be explained by the limitations of research methods and literature. In line with the systematic search, this study focused on gender, age and cultural differences in online information privacy only. This can be the first limitation of this study. It is suggested that future studies should consider other factors excluding gender, age and cultural differences to find out their role in online information privacy as they significantly contribute or play important roles in online information privacy. Moreover, the search in this study was only limited to the online context, which result in discoveries that do not generalize to other domains. Secondly, even though a comprehensive literature search was performed, bias based on publications could not be discounted completely. Thirdly, only the methodological quality of the scale procedures was assessed from the studies in this review. The evaluation and comparison of psychometric features of online information privacy was not performed. This would benefit future empirical studies in terms of decision-making in online information privacy scales for use. Lastly, the main limitation of this research involves the choice of the assessment procedure. It is well-known that the COSMIN methodology was designed to measure the properties of systematic reviews as a validated framework. However, it is mostly utilised in the health science discipline for choosing the most appropriate PROMs. While its checklists have the potential to be employed in different research domains and disciplines, certain measuring features will need to be adjusted for well-founded social science applications.

6 Conclusion

Information privacy is of universal concern to internet users. Recent ICT permits the collection, storage and processing of large volumes of user data. However, users' privacy remains the main concern as a lot of personal information is disclosed and left behind. This study systematically presented the role of gender, age and cultural differences in online information privacy and assessed the quality of methodological procedures used in online information privacy studies. The studies found that gender plays an important role in online information privacy, with female users indicating less behaviour with regards to security and privacy and requiring higher technical skills than males. Secondly, age plays a significant role in online information privacy with young age group users having larger networks and easily disclosing personal information in online platforms than old age group users. Lastly, culture plays a significant role in online information privacy, as well as in knowledge sharing. The study addressed numerous limitations which included the study's focus on online context only, the choice of the evaluation method, as well as the focus to factors such as gender, age and cultural differences, excluding other important contributing factors in online information privacy. This study can be used in future research to enhance the knowledge of the role of gender, age as well as cultural differences in online information privacy.

Author Contributions. All authors contributed to the study. The original manuscript was prepared by William Ratjeana Malatji. Tranos Zuva helped in the original draft preparation, as well as the arrangement of the paper. Rene VanEck, reviewed, provided additional ideas and edited the manuscript.

Funding. This research was not funded by any organisation or corporation.

Conflict of Interest. The authors certify that no other person or entity has a conflict of interest with them.

Informed Consent. This research does not include human participation or animals. Therefore, there was no informed consent used.

References

1. Bartol, J., Vehovar, V., Petrovčič, A.: Should we be concerned about how information privacy concerns are measured in online contexts? A systematic review of survey scale development studies. Informatics, vol. 8, no. 2: Multidisciplinary Digital Publishing Institute, p. 31 (2021)
2. Erevelles, S., Fukawa, N., Swayne, L.: Big Data consumer analytics and the transformation of marketing. J. Bus. Res. **69**(2), 897–904 (2016)

3. Sigmund, T.: Attention paid to privacy policy statements. Information, vol. 12, no. 4, p. 144, (2021). https://www.mdpi.com/2078-2489/12/4/144
4. Morey, T., Forbath, T., Schoop, A.: Customer data: designing for transparency and trust. Harv. Bus. Rev. **93**(5), 96–105 (2015)
5. Magdziarczyk, M.: Right to Be Forgotten in Light of Regulation (Eu) 2016/679 of the European Parliament and of the Council of 27 April 2016 on the Protection of Natural Persons with Regard to the Processing of Personal Data and on the Free Movement of Such Data, and Repealing Directive 95/46/Ec. In: 6th INTERNATIONAL MULTIDISCIPLINARY SCIENTIFIC CONFERENCE ON SOCIAL SCIENCES AND ART SGEM 2019, pp. 177–184 (2019)
6. G. D. P. Regulation: General data protection regulation (GDPR). Intersoft Consulting, Accessed in October, vol. 24, no. 1, (2018)
7. Sindermann, C., Schmitt, H.S., Kargl, F., Herbert, C., Montag, C.: Online privacy literacy and online privacy behavior–The role of crystallized intelligence and personality. Int. J. Human Comput. Interact. pp. 1–12, (2021)
8. Gorska, A., Korzynski, P., Mazurek, G., Pucciarelli, F.: The role of social media in scholarly collaboration: an enabler of international research team's activation? J. Glob. Inf. Technol. Manag. **23**(4), 273–291 (2020)
9. Mohamed, N., Ahmad, I.H.: Information privacy concerns, antecedents and privacy measure use in social networking sites: Evidence from Malaysia. Comput. Hum. Behav. **28**(6), 2366–2375 (2012)
10. Hoy, M.G., Milne, G.: Gender differences in privacy-related measures for young adult Facebook users. J. Interact. Advert. **10**(2), 28–45 (2010)
11. Pattinson, M. et al.: Matching training to individual learning styles improves information security awareness. Info. Comput. Secur. (2019)
12. Sheng, S., Holbrook, M., Kumaraguru, P., Cranor, L.F., Downs, J.: Who falls for phish? A demographic analysis of phishing susceptibility and effectiveness of interventions. In: Proceedings of the SIGCHI conference on human factors in computing systems, pp. 373–382 (2010)
13. Gratian, M., Bandi, S., Cukier, M., Dykstra, J., Ginther, A.: Correlating human traits and cyber security behavior intentions. Comput. Secur. **73**:345–358 (2018)
14. McGill, T., Thompson, N.: Exploring potential gender differences in information security and privacy. Inform. Comput. Secur. (2021)
15. Mwagwabi, F., McGill, T., Dixon, M.: Short-term and long-term effects of fear appeals in improving compliance with password guidelines. Commun. Assoc. Inf. Syst. **42**(1), 7 (2018)
16. Anderson, C.L., Agarwal, R.: Practicing safe computing: a multimethod empirical examination of home computer user security behavioral intentions. MIS quarterly, pp. 613–643, (2010)
17. Thompson, N., McGill, T., Bunn, A., Alexander, R.: Cultural factors and the role of privacy concerns in acceptance of government surveillance. J. Am. Soc. Inf. Sci. **71**(9), 1129–1142 (2020)
18. Lin, X., Wang, X.: Examining gender differences in people's information-sharing decisions on social networking sites. Int. J. Inf. Manage. **50**, 45–56 (2020)
19. Mamonov, S., Benbunan-Fich, R.: The impact of information security threat awareness on privacy-protective behaviors. Comput. Hum. Behav. **83**, 32–44 (2018)
20. Shah, P., Agarwal, A.: Cybersecurity behaviour of smartphone users in India: an empirical analysis. Info. Comput. Secur. (2020)
21. Kezer, M., Sevi, B., Cemalcilar, Z., Baruh, L.: Age differences in privacy attitudes, literacy and privacy management on Facebook. Cyberpsychology: J. Psychosoc. Res. Cyberspace, vol. 10, no. 1, (2016)

22. Zeissig, E.M., Lidynia, C., Vervier, L., Gadeib, A., Ziefle, M.: Online privacy perceptions of older adults. In: International Conference on Human Aspects of IT for the Aged Population, pp. 181–200. Springer (2017)
23. Fleming, P., Bayliss, A.P., Edwards, S.G., Seger, C.R.: The role of personal data value, culture and self-construal in online privacy behaviour. PLoS ONE **16**(7), e0253568 (2021)
24. Dincelli, E.: The role of national culture in shaping information security and privacy behaviors. WORLD SCIENTIFIC REFERENCE ON INNOVATION: Volume 4: Innovation in Info. Secur., pp. 47–68, (2018)
25. Khan, M.A., Law, L.S.: The role of national cultures in shaping the corporate management cultures: a three-country theoretical analysis. Organ. Cult.: IntechOpen, (2018)
26. Chai, S.: Does cultural difference matter on social media? An examination of the ethical culture and information privacy concerns. Sustainability **12**(19), 8286 (2020)
27. Moher, D., Liberati, A., Tetzlaff, J., Altman, D.G., Group, P.: Preferred reporting items for systematic reviews and meta-analyses: the PRISMA statement. PLoS Med. **6**(7), e1000097 (2009)
28. Prinsen, C.A.C., et al.: COSMIN guideline for systematic reviews of patient-reported outcome measures. Qual. Life Res. **27**(5), 1147–1157 (2018). https://doi.org/10.1007/s11136-018-1798-3
29. Mokkink, L.B., et al.: COSMIN risk of bias checklist for systematic reviews of patient-reported outcome measures. Qual. Life Res. **27**(5), 1171–1179 (2018)
30. Mokkink, L. B. et al.: COSMIN methodology for systematic reviews of patient-reported outcome measures (PROMs). User manual, vol. 78, no. 1, (2018)
31. McGill, T., Thompson, N.: Gender differences in information security perceptions and behaviour. (2018)
32. Jaradat, A.: An exploratory study of gender differences in self-disclosure: the case of college students in Jordan. N. Am. J. Psychol. **22**(3), 363–372 (2020)
33. Farinosi, M., Taipale, S.: Who can see my stuff?: Online self-disclosure and gender differences on Facebook. Observatorio (OBS*), vol. 12, (2018)
34. Shin, W., Kang, H.: Adolescents' privacy concerns and information disclosure online: the role of parents and the Internet. Comput. Hum. Behav. **54**, 114–123 (2016)
35. Heales, J., Cockcroft, S., Trieu, V.H.: The influence of privacy, trust, and national culture on internet transactions. In: International Conference on Social Computing and Social Media, pp. 159–176. Springer (2017)
36. Li, Y., Kobsa, A., Knijnenburg, B.P., Nguyen, M.H.C.: Cross-Cultural Privacy Prediction. Proc. Priv. Enhancing Technol. **2017**(2), 113–132 (2017)

Author Index

R. Silhavy (Ed.): CSOC 2022, LNNS 501, pp. 753–755, 2022.
https://doi.org/10.1007/978-3-031-09070-7